Engineering Materials

This series provides topical information on innovative, structural and functional materials and composites with applications in optical, electrical, mechanical, civil, aeronautical, medical, bio- and nano-engineering. The individual volumes are complete, comprehensive monographs covering the structure, properties, manufacturing process and applications of these materials. This multidisciplinary series is devoted to professionals, students and all those interested in the latest developments in the Materials Science field, that look for a carefully selected collection of high quality review articles on their respective field of expertise.

Indexed at Compendex (2021) and Scopus (2022)

Srikanta Moharana · Bibhuti B. Sahu ·
Arpan Kumar Nayak · Santosh K. Tiwari
Editors

Polymer Composites

Fundamentals and Applications

Springer

Editors
Srikanta Moharana
Department of Chemistry
School of Applied Sciences, Centurion
University of Technology and Management
Jatni, Bhubaneswar, Odisha, India

Bibhuti B. Sahu
Department of Basic Science
and Humanities
Nalanda Institute of Technology
Bhubaneswar, Odisha, India

Arpan Kumar Nayak
Department of Energy Engineering
Konkuk University
Seoul, Korea (Republic of)

Santosh K. Tiwari
Department of Chemistry, Centre For New
Materials & Surface Engineering
NMAMIT Nitte
Karkala, Karnataka, India

ISSN 1612-1317 ISSN 1868-1212 (electronic)
Engineering Materials
ISBN 978-981-97-2074-3 ISBN 978-981-97-2075-0 (eBook)
https://doi.org/10.1007/978-981-97-2075-0

This Springer imprint is published by the registered company Springer Nature Singapore Pte Ltd.
The registered company address is: 152 Beach Road, #21-01/04 Gateway East, Singapore 189721, Singapore

Paper in this product is recyclable.

Preface

Welcome to the world of polymer composites! Polymer composites have emerged as a revolutionary class of materials, combining the versatility of polymers with the strength and durability of reinforcing materials. The objective of this book is to offer an in-depth investigation of the fundamental concepts, progress, and practical applications of polymer composites across multiple fields. The first step in exploring the world of polymer composites is to delve into the fundamental concepts of chemistry and physics, which control the behaviour of polymer matrices and reinforcing agents. The journey moves from the fundamentals of composite materials to the most recent advancements in synthesis strategies, processing approaches, characterization techniques, and cutting-edge applications.

The book *Polymer Composites-Fundamentals and Applications* is to emphasize the processing techniques, characteristics, and numerous applications of polymer composites. This resource offers an in-depth investigation of the several categories of polymer composites, encompassing their composition, structure, and characteristics. The book also explores several processing techniques employed in the fabrication of polymer composites, including extrusion, injection moulding, and compression moulding. Moreover, it emphasizes the extensive array of uses for polymer composites, encompassing avenues such as nano-generators, biomedical, energy storage and harvesting, cosmetics, EMI shielding, and industrial, 3D printing technologies and engineering sectors. However, this book is an indispensable asset for researchers, engineers, and students who aim to comprehend and apply polymer composites in many domains. Furthermore, the book highlights the significance of comprehending the interface between the polymer matrix and the reinforcing components to enhance the performance of the composites. Also, it is carefully designed to function as a comprehensive manual for students, researchers, and professionals in order to navigate the complex field of polymer composites.

The ongoing progress in polymer composite research has resulted in the creation of diverse materials with enhanced characteristics. As a result, the use of polymer composites has greatly increased in several sectors. To stay updated on the latest developments in this field, it is crucial to have a comprehensive understanding of the current breakthroughs in polymer composite research. Because of the rapid progress in this domain, it is challenging to involve all potential applications inside a single volume. This book provides a thorough overview of the current research in polymer composites, covering a broad spectrum of topics and emphasizing important discoveries and progress. This book consists of seventeen well-structured chapters based on polymer composites. Chapter 1 "Introduction of Polymers and Polymer Composites: Basic Fundamentals" provides an overview of the essential aspects of the polymer composite sector. This chapter mainly focused on the properties and application of dielectric nanomaterials. Chapter 2 "Synthesis and Processing Techniques of Polymer Composites" deals with the synthesis of various techniques of polymer composite materials. Chapter 3 "Primary and Secondary Processing of Polymer Matrix Composites" describes the primary and secondary processing of polymer matrix composites. Chapter 4 "Characterization Techniques of Polymer Composites" accounts for various characterization of polymer composite fields. Chapter 5 "Reinforced Multiscale Polymer Composites, Properties and Applications" deals with the current overview of different properties and applications of reinforced multiscale polymer composites. Chapter 6 especially emphasizes "Glass Fiber-Reinforced Polymer Composites". Chapter 7 "Hemp Fibre-Reinforced Polylactic Acid Composites: A Sustainable Materials for Engineering and Industry" explores the use of polylactic acid (PLA) reinforced with hemp fibre (HF) is increasing rapidly among natural fibre-reinforced composites (NFRCs). Chapter 8, entitled "Polymer Composites Testing and Analysis" focuses on the several methods used to test and analyze polymer composites. Chapter 9 "3D Printing of Polymer and Polymer Matrix Composites" describes a brief overview of the 3D printing polymer composites. Chapter 10 "Natural Fibre Reinforced Composites for Industrial Applications," focuses on the use of natural fibres in composites, with a particular emphasis on its application in the industrial sector. Chapter 11 "Elastomeric-Based Composite Materials for Engineering Applications" provides a comprehensive analysis and investigation of the advancements made in different types of composite materials that are based on elastomers and their use in engineering applications. Chapter 12 "State-Of-The-Art in Textile Polymer Composites and Applications" provides an overview of the numerous textile polymer composites used in advanced applications. Chapter 13 "Polymer Composites for Use in Nano-Generators for Energy Harvesting" discusses the importance of polymer composite-based nano-generators in advancing self-sustaining electronic devices, wearable sensors, and Internet of Things (IoT) applications. Chapter 14 "Polymer Nanoparticle Composites for Modern Applications" provides an in-depth review of the utilization of polymer nanoparticles in composite systems for present-day applications. Chapter 15, titled "Polymer Composites for Electromagnetic Interference (EMI) Shielding Applications," focuses on the mechanism of EMI shielding and presents a comprehensive review of the latest research on polymer-based composite materials with properties particularly engineered for EMI shielding applications. Chapter 16 "Polymer Composites for Biomedical Applications" addresses the significant engineering and scientific challenges that must be resolved for polymer composite materials to get widespread use in the biomedical field and lastly Chapter 17 "Piezoelectric Polymer Composites for Energy

Harvesting," explores the advanced use of polymer composites as very promising materials in the field of energy harvesting.

The current book provides a clear overview of polymer science and technology, delving into many aspects of polymer composites. We are confident that after reading our book (volume) in this series, readers will have the necessary information to continue exploring the vast field of modern polymer composite research and use the knowledge gained. Moreover, the book contains comprehensive references in each chapter, which are directly linked to the latest research conducted worldwide. This book is a rich and practical resource for professionals in both academia and industry, as well as newcomers to the field of polymer composites.

The editors extend their appreciation to the pioneering individuals and contributors in the area, whose efforts have facilitated the advancement of polymer composites. We aspire for this book to act as a spark, stimulating more investigation and originality, and motivating readers to participate in the worldwide endeavour of tackling energy obstacles by utilizing the revolutionary capabilities of polymer composites.

Jatni, Bhubaneswar, India — Srikanta Moharana, Ph.D.
Karkala, India — Santosh K. Tiwari, Ph.D.
Bhubaneswar, India — Bibhuti B. Sahu, Ph.D.
Seoul, Korea (Republic of) — Arpan Kumar Nayak, Ph.D.

Contents

Editors and Contributors

About the Editors

Dr. Srikanta Moharana is presently working as an Assistant Professor, Department of Chemistry, School of Applied Sciences, Centurion University of Technology and Management, Odisha, India. He obtained his Ph.D. and M.Phil. degrees in Chemistry from the School of Chemistry, Sambalpur University, India. He received his M.Sc. degree in Chemistry from the National Institute of Technology, Rourkela. He has published more than 50 research papers in international peer-reviewed journals, 30 book chapters, and many more counting with renowned international publications such as the RSC, Elsevier, Wiley, Springer, Taylor & Francis, etc. Besides, he has edited 02 books under the banner of Elsevier and Springer Nature Publications. Moreover, he has actively participated and presented his research work at several international conferences and also, and he has been serving as a reviewer for reputed international journals. He has been awarded **Prof. G. B. Behera Best Ph.D. thesis award** under the banner of Orissa Chemical Society, Odisha, India. Currently, two (02) students are continuing Ph.D. under his guidance. He has more than five years of teaching and eight years of research experience. His research experience as well as research interests, lie in Carbon nanomaterials (graphene, carbon nanotubes) and functional ceramic-based polymer nanocomposites synthesis and Characterization for advanced energy storage applications. He is a life member of the Orissa Chemical Society.

Dr. Bibhuti B. Sahu is presently working as an Associate Professor (Physics), Department of Basic Science and Humanities, Nalanda Institute of Technology (NIT), Bhubaneswar, Biju Pattanaik University of Technology (BPUT), Odisha, India. He obtained his Ph.D. from Department of Physics, Veer Surendra Sai University of Technology (VSSUT), Burla, Odisha, India and M.Phil. degree from School of Physics, Sambalpur University, Jyoti Vihar, Burla, Odisha, India. He has published more than 24 research articles (articles and book chapters) in peer review journals. He has more than 12 years teaching and seven years research experience.

He has presented his work at various National and International conferences. His research experience and interest in ferrites, hexaferrites, dielectric and ferroelectric materials, and nanostructured materials synthesis, characterization, structure-property correlation, and application of composite materials in energy and magnetic applications.

Dr. Arpan Kumar Nayak earned his M.Sc. from Jadavpur University, India. Then, he completed his Ph.D. degree from Indian Institute of Technology Kharagpur (IIT-KGP), India. He was a post-doctoral fellow at Hanyang University, South Korea. Currently, he is working as Research Professor at Department of Energy Engineering, Konkuk University, 120Neungdong-ro, Seoul-05029, Republic of Korea. His current research mainly focuses on the synthesis of various nanostructured materials and carbon-based materials towards environment and energy applications. He has published more than 70 articles in various international journals (Google Scholar citation count: 1470, h-index: 23, i10-index-37) and edited five books in Elsevier, IOP, and IOR. He was assigned as Editor-in-chief in the International Research Journal of Multidisciplinary Technovation, Bulletin of Scientific Research and Associate Editor in Nanoscale report.

Dr. Santosh K. Tiwari received his Ph.D. (2017) from IIT Dhanbad, India in Graphene based Polymer nanocomposites and then worked with HSCL, Hanyang University, Seoul, South Korea for 1.5 years. In 2019, he joined University of Guangxi, Nanning, China as a research faculty and then moved to Poland where he got NAWA research grant from Government of Poland and worked for two years as a scientist at Faculty of Chemistry, University of Warsaw, Poland. Presently, he is working as group head, for Centre For New Materials & Surface Engineering Department Of Chemistry, NMAMIT, Nitte, Karnataka, India. He published more than 75 research articles (Citation counts more than 3000, h-index: 28) including ten books with reputed publishers, and presented his works as an invited speaker in different countries.

Contributors

Wahab Abdul Department of Civil Engineering, Ecole Centrale School of Engineering, Mahindra University, Bahadurpally, Hyderabad, India

Biswajeet Acharya School of Pharmacy and Life Sciences, Centurion University of Technology and Management, Jatni, Odisha, India

Roma Agrahari Department of Biochemical Engineering, School of Chemical Technology, Harcourt Butler Technical University, Kanpur, Uttar Pradesh, India

Akshita Department of Chemistry, Amity Institute of Applied Sciences, Amity University Uttar Pradesh, Noida, India

Soma Banerjee Department of Plastic Technology, School of Chemical Technology, Harcourt Butler Technical University, Kanpur, Uttar Pradesh, India

Amulyaratna Behera School of Pharmacy and Life Sciences, Centurion University of Technology and Management, Jatni, Odisha, India

Pravat Kumar Behera Atma Ram Sanatan Dharm College, University of Delhi, New Delhi, India

Saubhagyalaxmi Behera School of Applied Sciences, Centurion University of Technology and Management, Bhubaneswar, Odisha, India

Suchismeeta Behera State Forensic Science Laboratory, Bhubaneswar, Odisha, India

Naresh K. Chandel Department of Chemistry, RNT Govt. College Sarkaghat, Mandi, India

Monika Chaparia Amity School of Applied Physics, Amity University Rajasthan, Jaipur, India

Ravi Kant Choubey Department of Applied Physics, Amity Institute of Applied Sciences (AIAS), Amity University, Noida, Uttar Pradesh, India

Sandip P. Choudhury Amity School of Applied Physics, Amity University Rajasthan, Jaipur, India

Adyasha Dalai Department of Physics, ITER, Siksha 'O' Anusandhan Deemed to be University, Bhubaneswar, India

Hari Prasaath Durgaiahsangam Department of Civil Engineering, Ecole Centrale School of Engineering, Mahindra University, Bahadurpally, Hyderabad, India

Umesh Kumar Dwivedi Amity School of Applied Physics, Amity University Rajasthan, Jaipur, India

Ankita Subhrasmita Gadtya School of Applied Sciences, Centurion University of Technology and Management, Bhubaneswar, Odisha, India

R. Girimurugan Department of Mechanical Engineering, Nandha College of Technology, Perundurai, Tamilnadu, India

L. Girisha Department of Mechanical Engineering, PES Institute of Technology and Management, Shivamogga, Karnataka, India

Deepshikha Gupta Department of Chemistry, Amity Institute of Applied Sciences, Amity University Uttar Pradesh, Noida, India

Tejendra K. Gupta Department of Chemistry, Amity Institute of Applied Sciences, Amity University Uttar Pradesh, Noida, India

Shibanee Hota Department of Physics, ITER, Siksha 'O' Anusandhan Deemed to be University, Bhubaneswar, India

Sri Kalyana Rama Jyosyula Department of Civil Engineering, Ecole Centrale School of Engineering, Mahindra University, Bahadurpally, Hyderabad, India

Mohd Ataullah Khan Department of Civil Engineering, Ecole Centrale School of Engineering, Mahindra University, Bahadurpally, Hyderabad, India

Shilpi Kullu Department of Physics, ITER, Siksha 'O' Anusandhan Deemed to be University, Bhubaneswar, India

Sunil Kumar Department of Physics, Indira Gandhi University, Meerpur (Rewari), Haryana, India

T. Kumarasan Department of Aeronautical Engineering, J J College of Engineering and Technology, Trichy, Tamilnadu, India

Neelam Kumari Amity School of Applied Physics, Amity University Rajasthan, Jaipur, India

Rajkamal Kushwaha Department of Biochemical Engineering, School of Chemical Technology, Harcourt Butler Technical University, Kanpur, Uttar Pradesh, India

Ram Naresh Mahaling Laboratory of Polymeric and Materials Chemistry, School of Chemistry, Sambalpur University, Burla, Odisha, India

Kishore Kumar Mahato Department of Manufacturing Engineering, School of Mechanical Engineering, Vellore Institute of Technology, Vellore, India

Shivali Meena Amity School of Applied Physics, Amity University Rajasthan, Jaipur, India

Ankita Meher Department of Chemistry, School of Applied Sciences, Centurion University of Technology and Management, Bhubaneswar, Odisha, India

Debabrata Mishra Department of Physics and Astrophysics, University of Delhi, New Delhi, India

S. Mishra Department of Physics, ITER, Siksha 'O' Anusandhan Deemed to be University, Bhubaneswar, India

Smita Mohanty School for Advanced Research in Petrochemicals, Laboratory for Advanced Research in Polymeric Materials, Central Institute of Petrochemicals Engineering and Technology, Patia, Bhubaneswar, Odisha, India

Srikanta Moharana Department of Chemistry, School of Applied Sciences, Centurion University of Technology and Management, Jatni, Bhubaneswar, Odisha, India

S. Nanthakumar Department of Mechanical Engineering, PSG Institute of Technology and Applied Research, Coimbatore, Tamilnadu, India

S. K. Parida Department of Physics, ITER, Siksha 'O' Anusandhan Deemed to be University, Bhubaneswar, India

Siddharth Parthasarathy School of Applied Sciences, Centurion University of Technology and Management, Bhubaneswar, Odisha, India

Roshni Pattanayak School for Advanced Research in Petrochemicals, Laboratory for Advanced Research in Polymeric Materials, Central Institute of Petrochemicals Engineering and Technology, Patia, Bhubaneswar, Odisha, India; Utkal University, Bhubaneswar, Odisha, India

P. Pitchandi Department of Aeronautical Engineering, PSN College of Engineering and Technology, Tirunelveli, Tamilnadu, India

Sukanya Pradhan School for Advanced Research in Petrochemicals, Laboratory for Advanced Research in Polymeric Materials, Central Institute of Petrochemicals Engineering and Technology, Patia, Bhubaneswar, Odisha, India

Rinmayee Praharaj Department of Physics, Bengaluru City University, Bengaluru, Karnataka, India

Tapash R. Rautray Biomaterials and Tissue Regeneration Laboratory, Centre of Excellence in Theoretical and Mathematical Sciences, Siksha 'O' Anusandhan University, Bhubaneswar, Odisha, India

Suresh Sagadevan Nanotechnology and Catalysis Research Centre, University of Malaya, Kuala Lumpur, Malaysia

Bibhuti B. Sahu Department of Basic Science and Humanities, Nalanda Institute of Technology, Bhubaneswar, Odisha, India

S. Sakthi Department of Mechanical Engineering, Arifa Institute of Technology, Esanoor, Nagapattinam, Tamilnadu, India

R. Saravanan Department of Mechanical Engineering, J J College of Engineering and Technology, Trichy, Tamilnadu, India

Abhipsha Satpathy Department of Physics, ITER, Siksha 'O' Anusandhan Deemed to be University, Bhubaneswar, India

Mayank Tiwari Department of Physics and Astrophysics, University of Delhi, New Delhi, India

Akanksha Verma Department of Plastic Technology, School of Chemical Technology, Harcourt Butler Technical University, Kanpur, Uttar Pradesh, India

Chapter 1
Introduction of Polymers and Polymer Composites: Basic Fundamentals

S. K. Parida, Abhipsha Satpathy, Adyasha Dalai, and S. Mishra

Abstract Due to the rising demand for composite materials that integrate the advantages of both components, material science has increased in both popularity and utilization. Over the past few years, there has been an increase in interest in modifying the structural characteristics of materials on a nanometre scale. The fundamentals of polymers are deliberated about in this book chapter, with the attention being mainly focused on the properties and application of polymers and polymer composites, as well as on recent developments and the potential for further study of this class of materials and their uses. The fundamental ideas that are frequently used in polymers and polymer composites are addressed. Organic compounds having long chains are known as polymers, which are constructed from numerous nanoscopic molecules commonly known as monomers. Long continuous chains of numerous repeating monomeric units construct polymers which occasionally show cross-linking and branching between the chains. Then, a description of the physical and mechanical characteristics of polymers and polymer composites follows. The production and characterization of polymer composites are highlighted in discussions on reinforced polymer composites, and their usage is given particular significance. In this book chapter, the focus has been on discussions of the various nanofillers used in polymer composites and how they might be modified. Finally, a thorough analysis of current research and patents is provided, gradually highlighting the opportunities for the creation of a variety of polymer nanocomposite materials.

Keywords Polymer composites · Organic compounds · Monomers · Nanofillers · Nanocomposite materials

S. K. Parida (✉) · A. Satpathy · A. Dalai · S. Mishra
Department of Physics, ITER, Siksha 'O' Anusandhan Deemed to be University, Bhubaneswar 751030, India
e-mail: santoshparida@soa.ac.in

S. Moharana et al. (eds.), *Polymer Composites*, Engineering Materials,
https://doi.org/10.1007/978-981-97-2075-0_1

1.1 Introduction

The term "polymer" does indeed come from Greek roots, with "poly" meaning "many or much" and "meros" meaning "parts or units." This reflects the nature of polymers, which consist of many repeating units. Polymers mostly are high molecular weight compounds composed of macromolecules, which are made up of numerous repeating units called monomers. These monomers are bonded together by covalent bonds, forming long chains or networks. Polymers can consist of elements such as nitrogen, carbon, oxygen, and hydrogen. Polymers can have diverse structures. Some polymers have linear chains, while others can be interconnected or branched, forming 3D networks. This structural diversity permits widespread properties and applications, which enables us to experiment further with the production of polymer composites. Polymers possess several distinctive characteristics, including viscoelasticity (a combination of viscous and elastic behavior), toughness, ductility, lightweight nature, ease of processing, and the potential for unique electrical, thermal, or optical properties. These properties make polymers highly versatile materials. Polymers often exhibit glassy or semi-crystalline structures rather than fully crystalline ones. This behavior is due to the macromolecular nature of polymers and contributes to their unique properties. One can process polymers by applying pressure and heat, allowing them to be shaped into various engineered components, such as bushings, gears, and bearings. The degree of polymerization is a critical parameter in polymer science. It refers to the quantity of units that repeat or the polymer chain's monomer unit [1, 2]. These parameters are important for understanding a polymer's properties and behavior. The molecular weight of an addition polymer is the product of the Degree of Polymerization (DP) and the molecular weight of the monomer [3]. This relationship helps characterize and quantify the size of polymer molecules. Therefore, polymers are a broad class of materials with a variety of uses due to their unique structures and properties. Understanding the Degree of Polymerization and molecular weight is crucial for tailoring polymers to specific purposes and optimizing their performance. Polymers have been used by humans for a long time and continue to play a significant part in various industries and everyday life.

Biological polymers are those that are found in living organisms and play crucial roles in life processes. They are essential for the functioning of living organisms and are often associated with biological functions. Proteins are large, complex molecules made up of amino acid monomers. They serve various functions in living organisms, including enzymes, structural components, and signaling molecules. Nucleic acids, such as DNA and RNA, are polymers composed of nucleotide monomers. They carry genetic information and are responsible for the transmission of genetic traits. Carbohydrates are composed of sugar monomers and serve as a primary source of energy in living organisms. They also play structural roles in some organisms, as seen in cellulose. While not typically considered polymers, lipids are another class of biological molecules that include fats and phospholipids. They have various functions, including energy storage and cell membrane structure. Biological polymers are fundamental to life and provide the basis for numerous biological processes and structures.

Non-biological polymers, often referred to as synthetic polymers, are human-made materials. They find applications in various industries and aspects of modern life. Some examples of non-biological polymers include synthetic polymers, such as polyethylene and polypropylene, which are used extensively in the production of plastics. These materials have a wide range of applications, from packaging and consumer goods to automotive components. Elastomers are polymers with elastic properties, and examples include rubber and silicone rubber. They are used in tires, seals, gaskets, and various industrial applications. Synthetic polymers like nylon and polyester are used to make fibers for textiles and clothing. Resins, such as epoxy and polyester resins, are employed in the production of composites, adhesives, and coatings. Therefore, non-biological polymers are vital to numerous industries, including transportation, construction, communication, and clothing. They have revolutionized the way we live and have become indispensable in modern society.

In summary, the classification of polymers into biological and non-biological categories reflects the diverse roles these macromolecules play in both natural and human-made systems. Biological polymers are essential for life, while non-biological polymers have transformed our world and are integral to various aspects of modern living.

The study of polymers belongs to the field of polymer science which includes polymer physics and polymer chemistry, biophysics, and macromolecular science. Throughout the years polymer classification has been based on diverse considerations; the most abundantly used basis of classification is as follows [4]:

i. The source of origin (natural, semi-synthetic, or synthetic),
ii. The type of polymerization process used in the synthesis (addition, condensation),
iii. Type of chain and nature (linear, branched, and cross-linked),
iv. The solid-state behavior of polymer chains (homochain, heterochain),
v. Their mechanical strength and behavior,
vi. Their reaction to stress and temperature,
vii. Their molecular arrangement,
viii. The morphology of the polymeric substances, and
ix. The type of monomers involved in the polymer (homopolymer and copolymer).

The classification of polymers based on their source of origin is accurate and provides a clear distinction between natural, semi-synthetic, and synthetic polymers. Here's a comparative study on the different types of polymers, as given in Fig. 1.1, is as follows:

1. **Natural Polymers**: Natural polymers are derived from sources found in animals and plants. These polymers are formed naturally and have been used by humans for various purposes for centuries. Examples of natural polymers include

   - **Proteins**: Found in animals and plants, proteins are biological polymers composed of amino acid monomers. They have essential roles in living organisms and are used for various applications, including food production.

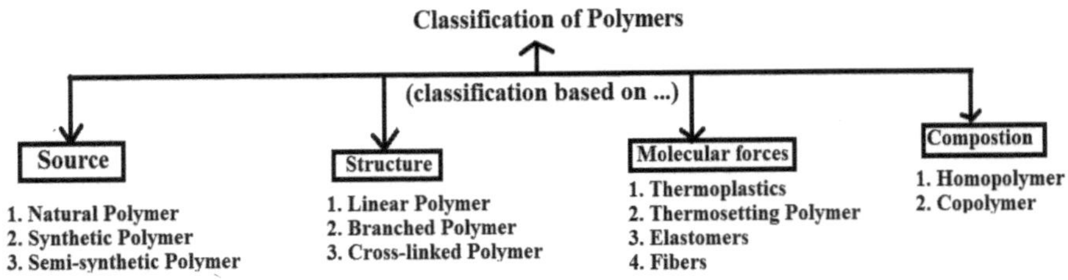

Fig. 1.1 The different types of polymers

- **Starch:** Starch is a carbohydrate polymer found in plants. It serves as an energy storage molecule and is widely used in the food industry.
- **Rubber**: Latex obtained from rubber trees are the sources of natural rubber. It exhibits unique elastic properties and is used in the production of tires, gloves, and various rubber products.
- **Cellulose**: Cellulose is a structural component in plant cell walls. It is used to produce paper and textiles.
- **Resins:** Natural resins, such as those from certain trees, can be used in coatings and adhesives.

2. **Semi-Synthetic Polymers**: Semi-synthetic polymers are produced by chemically modifying natural polymers to enhance their properties or create new materials. Examples of semi-synthetic polymers include

   - **Nitrocellulose**: Nitrocellulose is derived from cellulose and is chemically modified to increase its flammability and use in explosives and coatings.
   - **Cellulose Acetate**: This material is produced by acetylating cellulose, making it suitable for use in photography film and cigarette filters.

3. **Synthetic Polymers**: Synthetic polymers, often referred to as man-made polymers, are entirely produced through chemical synthesis. These polymers are derived from petroleum products and are created by scientists and engineers. Examples of synthetic polymers include

   - **Polyester**: Polyester is a versatile synthetic polymer used in textiles, plastic bottles, and various other applications.
   - **Nylon**: Nylon is a synthetic polymer known for its strength and durability. It is used in textiles, engineering components, and more.
   - **Teflon**: Teflon, or polytetrafluoroethylene (PTFE), is valued for its non-stick properties and chemical resistance. It is used in cookware, industrial applications, and as a lubricant.
   - **Polyethylene**: Polyethylene is a widely used synthetic polymer with various forms, including high-density polyethylene (HDPE) and low-density polyethylene (LDPE). It is used in packaging, pipes, and many other products.
   - **Plastics**: Various types of synthetic polymers are used in the production of plastics, which have a wide range of applications in everyday life.

The production of synthetic polymers from petroleum products has revolutionized industries and led to the mass production of a wide array of products. The annual production of polymers is indeed substantial and continues to grow due to their extensive use in various sectors.

The classification of polymers based on their structure is correct. Polymers are indeed categorized into three main types based on their structural characteristics: linear, branched, and cross-linked polymers.

1. **Linear Polymers**: Linear polymers consist of long and straight chains of monomer units. The polymer chains do not have significant branching or cross-linking. Examples of linear polymers include
   - **Polyvinyl Chloride (PVC)**: PVC is a common example of a linear polymer. It is used in various applications, such as pipes, cables, and vinyl flooring.
2. **Branched Polymers**: Branched polymers have linear chains, but these chains contain some branches or side chains. These branches directly affect the physical properties of the polymer. Examples of branched polymers include
   - **Low-Density Polyethylene (LDPE)**: LDPE is an example of a branched polymer. It has a more open structure with branches, making it flexible and suitable for applications like plastic bags and containers.
3. **Cross-Linked Polymers**: Cross-linked polymers are characterized by strong covalent bonds formed among various linear polymer chains. This cross-linking results in a three-dimensional network structure. Examples of cross-linked polymers include
   - **Urea–Formaldehyde Resins**: Urea–formaldehyde resins are thermosetting polymers that become rigid and infusible after cross-linking. They are used in the production of durable and heat-resistant materials.
   - **Vulcanized Rubber**: Vulcanization is a process that involves cross-linking rubber chains to increase their strength, elasticity, and resistance to abrasion. It is commonly used in tire manufacturing.

These structural differences have a noteworthy influence on the physical and mechanical properties of polymers. Linear polymers tend to be more flexible, while cross-linked polymers are typically rigid and resistant to deformation. Branched polymers fall somewhere in between, with a balance of flexibility and strength. The choice of polymer structure is critical in tailoring a material for specific applications.

Based on the structure and physical properties, organic polymers can be divided into four classes.

- **Thermoplastics**: These materials at high temperatures are flexible (e.g., polyethylene).
- **Thermosets**: These materials harden so their shape is set permanently, such as epoxy resins and some polycarbonates.
- **Elastomers**: These are stretchy like spandex and rubber, and they display elastic deformations.

- **Fibers**: These materials form threads, which are often used to make clothes such as nylon, silk, cotton, and polyester.

Based on the composition, polymers can be divided into two classes:

- **Homopolymer**: Homopolymer are a result of polymerization of single monomer (e.g., polystyrene).
- **Copolymer**: Copolymers are a result of addition of two or more types of monomers connected in the same polymer chain (e.g., ethylene–vinyl acetate) [5]. A few more of this classification basis and examples will be discussed later on in detail. This leads us to an important topic of polymer composites.

The pursuit to achieve and evolve and enlarge brand new materials with the composition of definite properties such as good fracture toughness, light weight, high substances/materials consisting of two or more chemically and physically different phases separated by a distinguishable interface is known as Composites. Nowadays, polymer composites can be tailored according to the application requirement at a low cost. The distinct available systems are combined purely to achieve a system with more useful functional or structural properties non-achievable by any of the constituents alone. Recent innovations in the polymer industry consist of fiber-reinforced polymer (FRP), glass fiber-reinforced polymers (GFRP), carbon fiber-reinforced polymer (CFRP), polymer-based nanocomposites, biomaterials, and other hybrid polymer-based composites with excellent characteristics to meet high-tech application performances. CFRP and GFRP are recognized as substituents for metal transportation networks and other applications. With the population explosion and industrialization, the global demand for polymer composites continues to increase. This leads us to our next researchable and debatable topic of polymer composites [6–8]. The addition of the proper reinforcement in the host matrix polymer creates different polymeric composites and hence enhanced the inherent qualities of the host polymers such as stiffness and strength. The conventional processing method can be used to create polymeric materials with the appropriate qualities for a given application requirement by choosing an appropriate reinforcement and matrix, with the matrix serving to hold the reinforcement firmly in place and shield it from mechanical or environmental damage [9].

Nowadays, the emerging field of supramolecular polymer composites which are formed by non-covalent bonds is an important field of research as they can be potentially utilized for material applications where traditional polymers aren't suitable. Supramolecular polymers according to the traditional definition are a polymeric assemblage of monomeric units linked by extremely directed and reversible secondary interactions that are non-covalent bonds. All these non-covalent interactions include hydrogen bonding, ionic interactions, van der Waals interactions or Coulomb interactions, metal coordination, chalcogen bonding, π–π stacking, halogen bonding, and host–guest interaction [10]. In the traditional polymer, strong covalent bonds linked with monomeric units together to deliver spectacular properties at the end. However, in melt viscous polymers, a high pressure and temperature are required for the entanglement of monomeric units to produce better physical properties. On the other hand, some polymers amalgamate good material properties with low-viscosity melts which are quite easy to handle, and these polymers are generally known as supramolecular polymers. Some supramolecular polymers showcase unique characteristics [11, 12], such as the ability to self-heal fractures. Even though the strong covalent bond never disintegrates, covalent polymers can be recycled, and hence go on to negatively affect the ecosystem as plastic wastes. These distinct characteristics and the possibility of developing responsive, self-repairing, adaptive, and environmentally friendly materials have gained supramolecular polymers a lot of attention in the field of polymer science [13, 14].

Here a few questions and confusion might arise regarding the necessity of the detailed study of the above-mentioned topics in the following sections. Polymers in general have permeated almost every aspect of our daily life. Due to its low cost of production, adaptability, and high specificity, polymers are applied in various fields. For the provided implementation, the properties of the polymer can be enhanced or tuned by amalgamation with other materials as in composites. These applications permit the preservation of other materials (coating), safeguard lives (hygiene, medical application), save the land and diminish the usage of fertilizers (synthetic fibers), save energy (thermally insulated buildings, lighter cars, and planes), protecting food and drinking water (packaging). All these lead to the thorough study of the history of polymers and polymer composites, their classification, structure, application, disadvantages, awareness programs in society for safer applications, environmental pollution, and many more such subtopics.

1.2 History of Polymers

The existence and application of polymers in the course of the evolution of human civilization is a much-debated topic. Typically, the materials that dominated human society at the time defined several eras (e.g., Stone Age, Bronze Age, and Iron Age). Numerous distinct titles of this kind have been applied to the twentieth century, such as the plastic age, oil age, and nuclear. Despite this period having no significant advancements except than the distribution of electrical power to all dwellings, it has influenced our lives more with the all-inclusive application of synthetic plastics in our automobiles, dishes, packaging, toys, clothes, construction materials, and many more. The use of polymers in human civilization has indeed played a significant role in various aspects of our lives. Here are some key points to elaborate on the information provided:

- **Polymer Application in Ancient Civilizations**: The mention of the Mayans using rubber from local trees is an important historical note. The Mayans were indeed among the first civilizations to discover and utilize polymers, in this case, natural rubber. The employment of rubber for various purposes includes making balls for

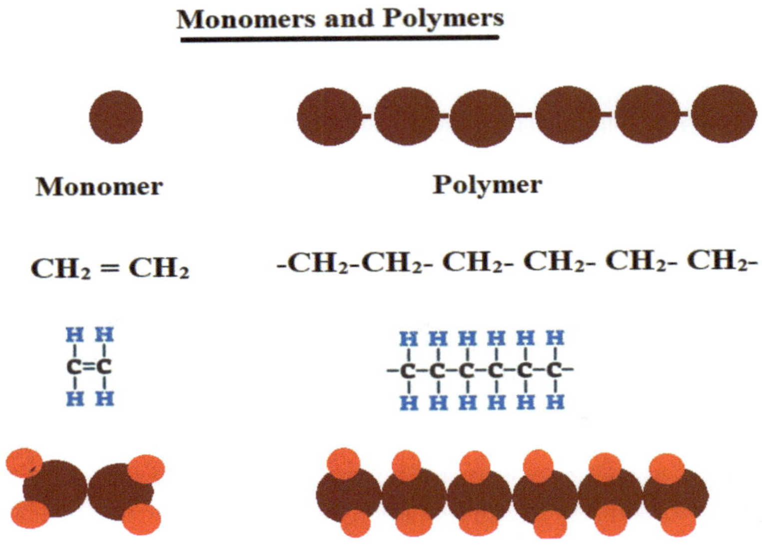

Fig. 1.2 Shows the monomers and polymers

games. The use of natural polymers like rubber predates the modern understanding and synthesis of synthetic polymers.

- **Origin of the Term "Polymer"**: The term "polymer" does have its roots in Greek. It originates from the Greek words' "poly" (meaning many) and "meros" (meaning part) [15]. This term was introduced by the Swedish chemist Jöns Jakob Berzelius in 1833, marking an important development in the understanding of these materials [16, 17]. Figure 1.2 shows the formation of the polymer using monomer as a unit building block of the polymers.
- **Isomerism and Polymerism**: The discussion about isomerism and polymerism helps clarify the distinctions between different types of chemical compounds. Isomerism is a concept where two substances have the same chemical composition but different properties due to differences in the arrangement of their constituent atoms or groups. In contrast, polymerism refers to substances with the same relative compositional formulas but different absolute compositional formulas, typically resulting from differences in the total number of atoms. This distinction is vital in understanding the diversity and complexity of polymer chemistry. With the exponential rise of organic chemistry in the 1840s and 1850s, Berzelius' original distinction became disarranged. Thus in 1888 the British chemist, Henry Armstrong, observed that [18]: Even a superficial reader of the chemical literature will soon become aware that the terms isomeric and the kindred expressions allotropic, metameric, and polymeric are by no means used in consistent senses, and he will have considerable trouble in clearly realizing their exact and relative import.

At last, the polymer was recognized as a distinct phenomenon on par with isomerism and metamerism; it was hence classified as one of the various forms of structural isomerism. Apart from all these changes in status, polymerism kept on describing a merely compositional relationship between two substances in line with Berzelius's original intent, without imposing any restrictions on the characteristics of the molecules (such as whether they are organic or inorganic), their structures, their relative sizes, or their capacity for interconversion [19]. The situation occurred as the original term polymer did not carry any undercurrent regarding the measurement; thus the German chemist, Hermann Staudinger, coined the word "macromolecule" in 1922 to describe large covalently bonded organic chain molecules consisting of more than 10^3 atoms [20]. Precisely speaking, the term polymer and macromolecule logically stand for clear-cut concepts. There exist several polymers, such as the S_3O_9 ring as a trimer of SO_3 which are not macromolecules, and several macromolecules, such as the so-called condensation and copolymers, which do not have the same composition as the polymers of the materials they are composed of. This logical distinction has long been ignored by present-day chemists, most of whom use the term polymer as a trivialized synonym for a covalently bonded organic macromolecular chain molecule [21, 22].

- **The Impact of Plastics in the Twentieth Century**: The twentieth century is often referred to as the "Plastic Age" because it witnessed a tremendous surge in the production and use of synthetic polymers, such as plastics. Plastics found applications in various industries, including automotive, packaging, construction, and consumer goods. Their versatility, durability, and relatively low cost made them an integral part of modern life.
- **Electricity and Technological Advances**: While the twentieth century is known for its plastics, it is important to note that it also saw significant technological advancements, including the widespread delivery of electrical power to homes. These developments had far-reaching effects on society, transforming how people lived and worked.

Apart from the above-mentioned events, 1839 saw the discovery of vulcanization by Charles Goodyear; it was created by heating natural rubber to 270 °F and mixing it with Sulfur. Rubber that has been vulcanized is a polymeric material that is far more resilient than natural rubber [23]. Vulcanized rubber is a polymeric substance that is much more durable than its natural counterpart. The tires on cars are where this is most commonly used. The investigation of the nature of the materials and the creation of polymer is just the beginning. Karl Nageli, a Swiss botanist, used the term molecule as the compositional unit of starches. As the term was not popularized, there was confusion as to what it meant in the science community as Avogadro and Cannizzaro had different definitions. During the Karlsruhe Conference, a meaning to the term molecule was established by the scientists after being persuaded by the research of Avogadro and Cannizzaro. This definition is the same even today. As Negali's definition did not correspond with the new usage, he changed it to micelle (an aggregate of molecules that forms a crystalline array) [24]. 1877 brought forth a fairly accurate model proposed by Kekule, which suggested that molecules are net or sponge-like masses formed by the union of many univalent molecules with polyvalent atoms acting as centers of attachment. Though he did not introduce the concept of chains of atoms, he obviously explained molecular masses. This concept was also rejected and instead micellar theory was used to explain the condition of colloids. 1882 brought forth new methods of determining molecular weight, which included Raoult's law for acquiring molecular weights by freezing, and in 1886 van't Hoff introduced a method by using osmotic pressure. Both of these methods varied in the production of results and were deemed that they don't work for colloidal solutions as the errors produced were greater than the measurement [25]. 1899 popularized the aggregate or partial valency hypothesis. Later in 1905 aggregate theory was introduced. The odd behavior of natural and synthetic molecules is attributed to the nature of molecules forming primary and secondary forces with each other and forming large structures. This was mostly advocated by Alfred Werner [26]. In 1920 Staudinger received some advice in the form of a letter which stated the following, "Dear colleague, leave the concept of large molecules well alone, organic molecules with a molecular weight above 5000 do not exist. Purify your products, such as rubber, then they will crystallize and prove to be lower molecular substances. Molecules cannot be larger than the crystallographic unit cell, so there can be no such thing as a macromolecule." Still after a lecture in Zurich in 1920, he disputed the "secondary forces of association" and proposed the concept of macromolecules for which he was highly criticized [27]. The year 1925 saw the theory of aggregation being cast aside for macromolecules as Svedberg and Adair both estimated hemoglobin weight to be around 66,000. Mass to charge is constant was shown by Tiselius in the year 1928, which proved that the partial valence theory was not likely as described by Werner [28]. Then in the year 1928, X-ray was used to determine molecular structure hence disapproving the idea that molecules had to be small and built through association. In 1929 polyester and polyamides (nylon) were synthesized sequentially by Carothers which showed wool and silk are also polyamides [29]. After going through all of these heated debates and discussions the result finally led to Staudinger receiving the novel prize for chemistry, because of his work on determining that macromolecules exist. This discovery led to further development in the field of polymers as we see today.

In summary, the use of polymers, both natural and synthetic, has had a profound impact on human civilization. The twentieth century's widespread adoption of synthetic plastics is just one example of how polymers have shaped our lives and industries. Understanding the chemistry of polymers, as demonstrated by Berzelius's contributions, has been crucial in advancing our ability to engineer materials with desirable properties for various applications [30].

1.3 Classification of Polymers

Classification of Polymers:

Due to the complex structures, varying behaviors, and broad applications of polymer, it cannot be classified under one category. So, we classified polymer based on the following categories:

A. **Classification based on the origin of accessibility:**

  i. **Natural polymer:** Natural polymers are materials that are widely used in nature and originate from plants or animals [31]. Some of the natural polymers are given as follows:
    (a) Starch: Starch is a polymer of glucose. It acts as a food reserve for plants.
    (b) Cellulose: Cellulose is also a polymer of glucose. Cellulose is made up of glucose that is produced during photosynthesis in plants.
    (c) Proteins: These are polymers of amino acids. In an animal body, these are the building blocks. These constituents are essential components of the food we eat.

  ii. **Synthetic polymers:** Synthetic polymers are polymers which are made in laboratories. These types of polymers are also known as man-made polymers [31].
    Examples: PVC, polyethylene, synthetic rubber, nylon, terylene, etc.

  iii. **Semi-synthetic polymers:** Semi-synthetic polymers originate through chemical processes from naturally existing polymers. These polymers are derived from naturally occurring polymer, cellulose. These types of polymers are also known as thermoplastic polymers [31].

B. **Classification based on structure:**

The polymers are divided into three categories linear, branched-chain, and cross-linked polymer. The schematic diagram of these polymers is given in Fig. 1.3 followed by a detailed explanation.

  i. **Linear polymers:** In linear polymer, monomeric units are linked together to form a linear chain. To create a densely packed structure, the polymeric chains are arranged on top of one another. These types of polymers have high melting temperatures, high tensile strengths, and high densities as a result of their tight packing. Some examples of linear polymers are polyethylene, polystyrene, PVC, etc. [31].

  ii. **Branched-chain polymers:** Monomers are joined to form long chains with side branches of different lengths forming branched-chain polymer. These polymers have lower tensile strength and melting point than linear polymers due to their irregular packing. Some examples of branched-chain polymers are glycogen, starch, and low-density polythene [31].

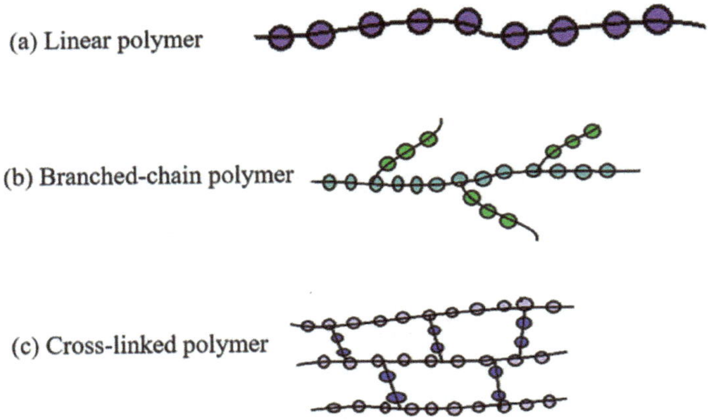

Fig. 1.3 Type of polymers: **a** Linear polymer, **b** Branched-chain polymer, and **c** Cross-linked polymer

iii. **Cross-linked polymers**: To form three-dimensional network polymer, monomers are cross-linked together which are called cross-linked polymers. Because of the network structure, these polymers are hard, brittle, and rigid. Some common examples of cross-linked polymers are melamine, formaldehyde resin, bakelite, etc. [31].

C. **Classification based on polymerization**:

i. **Addition or chain polymerization**: Addition polymerization is a process of synthesis of polymers in which direct addition of monomeric units without removal of the by-product/foreign molecules are required. These polymers are also known as chain-growth polymers. It involves just one monomer. Here monomers are unsaturated molecules [31]. Also, chain polymerization is named free-radical polymerization, cationic polymerization, or anionic polymerization.

(a) The addition polymer would be either a linear polymer or a branched-chain polymer if the monomer is ethylene. Examples: polystyrene, polyacrylonitrile, polytetrafluoroethylene, etc.

(b) The polymer is always a branched-chain polymer if the monomer is 1,3-butadiene or 2-substituted-1,3-butadiene.

Examples:

$$nCH_2\text{-}CH_2 \rightarrow \text{-}(CH_2\text{-}CH_2\text{-})_n$$

Ethene Polyethene

$$nCH_3\text{-}CH{=}CH_2 \rightarrow \left[\begin{array}{c}\text{-}CH_2\text{-}\underset{\displaystyle CH_3}{\underset{|}{CH}}\text{-}\end{array}\right]$$

Propylene Polypropylene

ii. **Condensation or step polymerization**: Condensation polymerization is a process of synthesis of polymers after the condensation of the two or more number of monomers with the removal of simple molecules like water, ammonia, alcohol, etc., during the condensation of two or more number of monomers form this type of polymer. These polymers are also called step-growth polymers. They involve more than one monomer. These monomer units must have two active functional groups [31].

D. **Classification based on molecular forces**:

i. **Elastomers**: Any stretchy substance made of long-chain molecules, or polymers, that can regain its original shape after being greatly stretched is called elastomers. In these types of polymers, the polymer chains have weak intermolecular forces holding them together. Under the weak force of interaction, by applying small stress the polymers can be stretched together and when stress is removed, they quickly return to their original position. Some examples of elastomers are natural rubber, polybutadiene, neoprene, silicone, etc. [31].

ii. **Fibers**: The intermolecular forces between the chains in these types of polymers are strong. These forces are either dipole–dipole interaction or hydrogen bonds. These polymers have high tensile strength and less elasticity because of the close packing of chains due to strong forces. These polymers thus have distinct melting points. These polymers can be woven into the fabric because they are thin, long, and thread-like. Some examples are Nylon-66, Dacron, silk, etc. [31].

iii. **Thermoplastics**: When heated these types of polymers can be softened, and hardened when cooled with little change in properties. These polymers have intermediate intermolecular forces between elastomers and fibers. Between the chains, there is no cross-linking. Due to absence of cross links in the thermoplastics polymers, the chain of monomers can move freely within the given polymers and produce the softness in the materials. They melt when heated and generate a fluid that can be shaped into any shape and then cooled to achieve that shape. Some examples are polystyrene, polythene, PVC, Teflon, etc. [31].

iv. **Thermosetting polymers**: These types of polymers undergo permanent change, and they become hard and infusible while heating. These kinds of polymers are prepared from semifluid substances which have low molecular mass. The cross-links hold the molecules in place so that heating does not allow them to move freely. So, thermosetting plastic is cross-linked and is permanently rigid [31].

Examples: formaldehyde, bakelite, melamine, etc.

E. **Classification based on monomers**:

i. **Homopolymer**: A homopolymer is formed from one kind of monomer. These polymers are simple chains. These are formed through addition polymerization.

Example: Polythene is a homopolymer of monomer ethene [31].

$$\underset{\text{Ethene}}{nH_2C{=}CH_2} \xrightarrow{\text{polymerization}} \underset{\text{Polythene}}{-(CH_2\text{-}CH_2\text{-})_n}$$

ii. **Copolymer**: Due to the copolymerization of two or more different monomers, copolymer is formed. These types of polymers are complex chains. These are formed through condensation polymerization. If a copolymer is obtained from two or three monomers, then it is called biopolymer and terpolymer, respectively [32]. Various methods such as chromatography and NMR spectroscopy are used to characterize copolymers to learn about their molecular weight, size, compositions, and characteristics [33]. Example: Nylon-66 is a polymer of two types of monomers, i.e., hexamethylenediamine and adipic acid.

Copolymers can be classified according to how the monomeric units are arranged along the chain. These are linear copolymer and branched copolymer [32]. A schematic diagram is given in Fig. 1.4 along with a detail explanation.

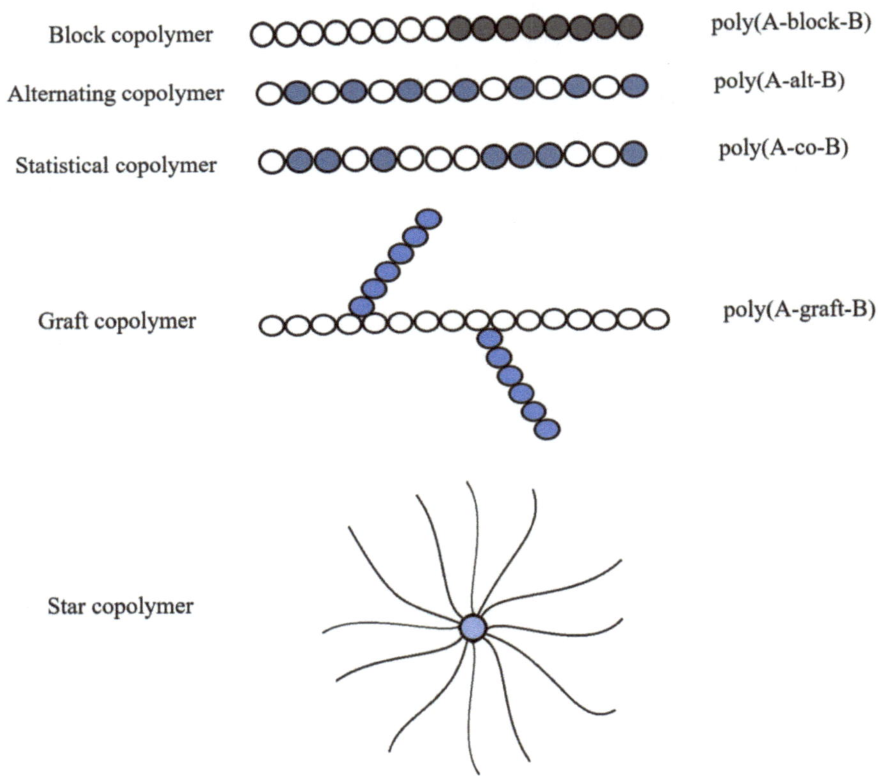

Fig. 1.4 The structures of different types of copolymers

a. **Linear copolymer**:

   i. **Block copolymer**: This type of copolymer is formed when two monomers clutch together and form blocks of repeating units. In this type of copolymer, the polymerization process occurs via sequential monomer addition. These copolymers feature lengthy chains of several monomer components [34]. Di-block copolymer has two distinct blocks and triblock copolymer has three. In a triblock copolymer, repeat units A and B might be arranged in the following order: ~ A-A-A-A-A-A-A-B-B-B-B-B-B-B-A-A-A-A-A ~ [35].
   ii. **Alternating copolymer**: The formula for an alternating copolymer, which has regular alternating A and B units is A-B-A-B-A-B-A-B-A-B- or -(-A-B-)$_n$- [34].
   iii. **Statistical copolymer**: The term "statistical copolymers" refers to copolymers in which the sequential distribution of the monomeric units complies with established statistical laws. For example, Markovian statistics of zeroth (Bernoullian), first, second, or higher order may be followed by the distribution of monomers [36]. A random copolymer is a kind of statistical copolymer in which the polymer is called genuinely random when the chance of finding a certain kind of monomer residue at a given location in the chain equals the mole fraction of that monomer residue in the chain [37].

b. **Branched copolymer**: For nonlinear copolymer, there are possibilities for different types of architectures. There are other types of branched copolymers which include comb copolymers and brush copolymers except for grafted and star polymers.

   i. **Graft copolymers**: Graft copolymers have side chains or branches with repeat units whose composition or structure is different from that of the main chain [34]. The branches are added on to perform the main chain macromolecule [35]. Generally, from one monomer the main chain is formed, and from another monomer branches are formed. Free radicals are used to polymerize styrene after dissolving polybutadiene in it [35].
   ii. **Star copolymers**: Star copolymers are the simplest type of branching polymers in polymer science, with a typical structure consisting of many linear chains connecting to a central core [38]. The polymer core might be an atom, molecule, or macromolecule and the chains of a polymer are composed of organic chains of variable lengths. Star copolymers with identical lengths and structures are regarded as homogeneous, whereas those with varying lengths and structures are called heterogeneous. Star copolymers consist of a multifunctional center that radiates at least three polymer chains [39]. These arms are chemically identical or different. Individual arms may be composed of multiple polymers, resulting in a star copolymer. The unique properties come from their chemical structure as well as the length and number of arms [39]. The unique shape and associated properties [40] of star copolymers, such as their compact structure, high arm density, efficient synthetic routes, and unique rheological properties, make them promising tools for use in drug delivery [41], other biomedical applications [42], thermoplastics [43], and nanoelectronics.

F. **Classification based on structure**:

   i. **Amorphous polymers**: The amorphous region where molecules are arranged randomly is called amorphous polymer. These types of polymers do not have uniformly packed molecules. Amorphous polymers are transparent. These polymers do not have a sharp melting point. These polymers have low shrinkage. These polymers have poor chemical resistance. These polymers have low energy and high gas permeability. Examples of amorphous polymers are atactic polystyrene, polycarbonates, polymethylmethacrylate, etc.
   ii. **Crystalline polymers**: When polymers are cooled from a molten state or concentrated from a solution, molecules are attracted to other molecules and tend to combine as tightly as possible into a solid with the least potential energy, resulting in crystalline polymer. These polymers have uniformly packed molecules. These types of polymers have sharp melting points. Crystalline polymers are opaque/translucent. They have high shrinkage. These polymers have good chemical resistance. These polymers have high energy and low gas permeability. Examples of crystalline polymers are polyethylene, polytetrafluoroethylene, polyacrylonitrile, etc.

G. **Classification based on preparative technique**:

   a. **Bulk Polymerization**

   Bulk polymerization is a polymerization technique in which a monomer unit or a mixture of monomers is polymerized in a neat, undiluted form, without the use of a solvent or diluent. This means that the monomer(s) are subjected to the polymerization reaction without any additional liquid or solvent medium. Bulk polymerization offers advantages in terms of simplicity, high purity, and ease of control, but it also has some limitations, such as the need to manage heat generation and limited options for scaling up to large production volumes. The choice of polymerization technique depends on the specific requirements of the polymer product being manufactured [44].

   Bulk polymerization is divided into two main types

   i. **Quiescent bulk polymerization**: In this kind of bulk polymerization there is no stirring. This is mostly used to produce branched and thermosetting polymers. Because the system remains quiescent, the Trommsdorff effect is prominent, resulting in longer chains and harder material. The main drawbacks of this method of polymerization include entrapped voids caused by monomer boil-off and the inability to convert all monomers [45].
   ii. **Stirred bulk polymerization**: In this kind of polymerization, the monomer is continuously stirred. Reactors with very precise designs are employed, depending on the polymer's viscosity. In certain uses, a gear pump or mild external pressure is used to remove the finished polymer melt from the reactor. In contrast to solution polymerization, it involves the use of the monomer as a solvent [45].

b. **Solution polymerization**
Solution polymerization is a technique used to polymerize monomers in a liquid solution. In this method, the monomers are dissolved in a suitable solvent, and the polymerization reaction occurs within the solution medium. Solution polymerization is a versatile and widely used process in the production of various types of polymers. Solution polymerization is valuable in the synthesis of polymers with precise properties and characteristics, making it a versatile and widely used technique in the polymer industry. The choice of solvent, monomers, initiators, and reaction conditions allows for fine-tuning the resulting polymer's structure and properties.

c. **Suspension polymerization**
Suspension polymerization is a polymerization technique that involves the formation of polymer particles within a liquid medium in the form of small droplets or solid particles suspended in a continuous phase. This method is commonly used for producing polymer particles, beads, or powders with specific properties and applications. Suspension polymerization is a versatile technique that is particularly suitable for producing polymer particles with specific properties, including controlled particle size, shape, and cross-linking. It is widely used in industries where precise control over particle characteristics is essential for the final product's performance and applications.

d. **Emulsion polymerization**
Emulsion polymerization is a polymerization technique that involves the formation of polymers within small droplets of monomers suspended in an aqueous solution, often in the presence of emulsifying agents and initiators. This method is commonly used for producing polymer latex, which consists of polymer particles dispersed in water. Emulsion polymerization is a versatile technique used in various industries to create water-based polymer products with controlled particle size and properties. The resulting polymer latex is valuable for applications that require stable dispersions, such as coatings, adhesives, and textile finishes, and it is often favored for its environmentally friendly characteristics.

H. **Classification based on end use of polymers**:

a. **Diene polymers**: A diene is a covalent molecule that has two double bonds, generally between carbon atoms. Dienes exist naturally and synthetically as a component of larger compounds and are employed in organic synthesis. In the polymer industry, conjugated dienes are commonly employed as monomers. Nutritionists are interested in polyunsaturated fats.

With respect to the relative location of the double bonds, dienes can be classified into three types.

i. **Cumulated dienes**: The double bonds in cumulated dienes share a common atom. The end outcome is referred to as an allene.

ii. **Conjugated dienes**: Conjugated dienes have one single bond between conjugated double bonds. These dienes are more steady than other dienes due to resonance.
iii. **Unconjugated dienes**: The double bonds in unconjugated dienes are divided by two or more single bonds. They often have lower stability than isomeric conjugated dienes. This is sometimes referred to as an isolated diene.

b. **Polyolefin polymers**: A polyolefin polymer is a kind of polymer with a general formula $(CH_2CHR)_n$, where R is an alkyl group. They are mostly extracted from a small set of simple olefins. Polyethylene and polypropylene are predominant in a commercial sense. Polyisobutylene and polymethyl pentene are two additional specialty polyolefins. These are all solids or oils that are colorless or white. Each polyolefin's name specifies the olefin it is formed from; for example, ethylene is the source of polyethylene, while 4-methyl-1-pentene is the source of polymethylpentene. Because of their flexibility, they constitute about 50% of plastic usage in Europe, making them the most widely used form of plastic.

   There are four types of polyolefins

   i. **LDPE (Low-density polyethylene)**: The thermoplastic known as low-density polyethylene (LDPE) is derived from the monomer ethylene. Dr. John C. Swallow and M. W. Perrin created the first grade of polyethylene in 1933 by employing a high-pressure method called free radical polymerization [46]. Till now its manufacturer uses the same process. 5.7% of LDPE is recycled in the US, according to EPA estimates [47]. LDPE has a density range of about 917–930 kg/m^3. It is not reactive at room temperature, except for strong oxidizers; some solvents cause it to swell. It can endure temperatures as high as 90 °C (194 °F) for brief periods and as low as 65 °C (149 °F) continuously. It is highly durable and flexible, and it is available in transparent and opaque versions.
   ii. **Linear low-density polyethylene (LLDPE)**: Linear low-density polyethylene (LLDPE) is a polymer with a considerable number of short branches that are usually produced through copolymerization of ethylene with longer chain olefins. Because of the absence of long-chain branching, linear low-density polyethylene varies structurally from ordinary low-density polyethylene (LDPE). LLDPE is often produced at lower temperatures and pressures by the copolymerization of ethylene with higher alpha-olefins such as butene, hexene, and octene. The LLDPE polymer created by the copolymerization process differs greatly in rheological characteristics from ordinary LDPE and has a narrower molecular weight dispersion due to its linear structure. Transition metal catalysts, specifically Ziegler or Philips kinds, are which initiate the creation of LLDPE. Reactors designed for gas phase or solution phase operation can be used for the actual polymerization process. LLDPE is more impact- and puncture-resistant and has a greater tensile strength compared to LDPE. It is very flexible and elongates under stress. It

has strong chemical resistance and may be used to manufacture thinner films. Its electrical characteristics are good. It is more difficult to process than LDPE. LLDPE has entered almost all traditional polyethylene markets; it is used for plastic bags and sheets, plastic wrap, stretch wrap, pouches, toys, covers, lids, pipes, and cable covering.

iii. **High-density polyethylene (HDPE)**: HDPE, often known as polyethylene high-density (PEHD), is a thermoplastic polymer derived from the monomer ethylene. When used in HDPE pipes, it is sometimes referred to as "alkathene" or "polythene". HDPE has a high strength-to-density ratio and is used to make plastic bottles, corrosion-resistant pipes, geomembranes, and plastic lumber. The high strength-to-density ratio of HDPE is well recognized. HDPE has a density of about 930 to 970 kg/m^3. HDPE has negligible branching, which gives it better intermolecular forces and tensile strength than LDPE, but has a density that is just slightly higher than LDPE. HDPE has a melting point of 131.8 °C. It has a temperature of crystallization of 121.9 °C. It has a latent heat of fusion of about 188.6 kJ/kg. It has a specific heat capacity of about 1331 to 2400 J/kg-k [48]. Since HDPE is extremely difficult to bond and resistant to a wide range of solvents, connections are usually created by wielding. The globally standardized testing procedures used to determine the physical characteristics of HDPE for a particular process can play a decisive role in identifying how these qualities change based on the molding process used to make a given sample. For instance, in rotational molding, the notched constant tensile load test (NCTL) is used to determine a sample's ambient stress fracture resistance. HDPE pipes are perfect for wastewater and drinking water because of their advantageous qualities. HDPE has a wide variety of applications. The uses of HDPE are 3D printer filament, arena board, ballistic plates, banners, chemical containers, ionizing radiation shields, fireworks, fuel tanks for vehicles, plastic surgery, etc. [49].

iv. **Polypropylene (PP)**: The thermoplastic polymer polypropylene, sometimes referred to as polypropene, finds extensive usage in several fields. It is made from the monomer propylene via chain-growth polymerization. Polypropylene is a kind of polyolefin that is partly crystalline and non-polar. It has comparable qualities to polyethylene, yet it is significantly tougher and more heat-resistant. It is a white, mechanically tough substance with excellent chemical resistance [50]. After polyethylene, polypropylene is the second most widely produced commodity plastic. The density of polypropylene is between 0.895 and 0.93 g/cm^3. So polypropylene is the commodity with lower density. Molding pieces with lower weight and more parts of a certain mass of plastic can be made with lower density. Unlike polyethylene, the density of crystalline and amorphous areas differs relatively little. Still, the density of polyethylene can vary greatly depending on the filler [51]. Young's modulus of polypropylene is between 1300 and 1800 N/mm^2. Depending on atactic material and crystallinity, isotactic polypropylene has a melting point of 171 °C and commercial isotactic polypropylene has a melting point of about 160 to 166 °C. Syndiotactic polypropylene with a crystallinity of 30% has a melting point of 130 °C [52]. Polypropylene is utilized in a variety of packaging formats and has replaced conventional materials such as paper, cell phones, and paper because of its low cost and versatility. Polypropylene is used to make a wide range of medical equipment [53]. It is frequently used in the production of pallets, bottles, jars, hot beverage cups, food packaging, and so on. A comparative study on polypropylene compared with polymer in the UK is given in Fig. 1.5 along with explanation.

c. **Acrylics**: Any of a series of polymers made from acrylate monomers is an acrylate polymer, also known as acrylic or polyacrylate. These polymers are known for their clarity, strength, and flexibility. As an adhesive, acrylate polymer is extensively used in cosmetics such as nail polish [54]. Acrylic elastomer refers to a form of synthetic rubber in which the principal component is acrylic acid alkyl ester. Acrylic elastomers can be characterized by two types. "Old" types include ACM (copolymer of acrylic acid ester and 2-chloroethyl vinyl ether) containing chlorine and ANM (copolymer of acrylic acid ester and acrylonitrile) without chloride. "New" types do not contain chlorine and are less prone to mold-related staining. The material is less cold-resistant, with a saturation point of 15 °C for old varieties and 28 °C

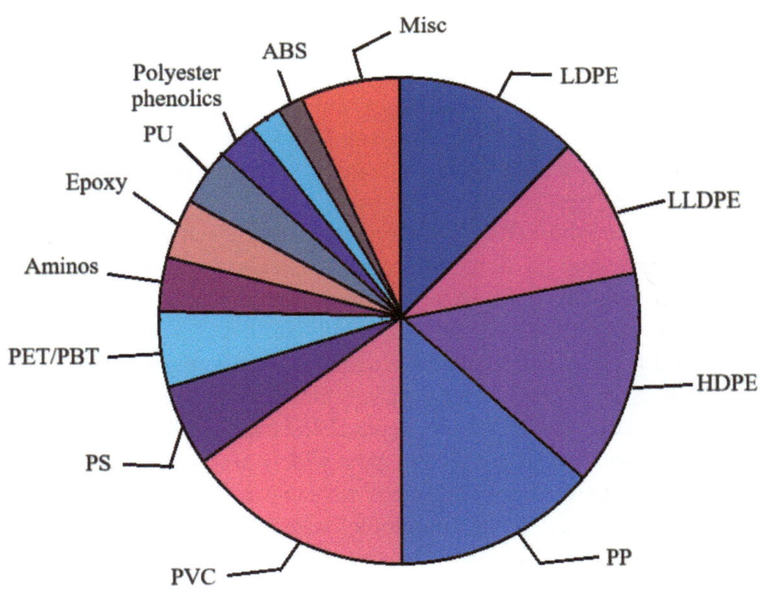

Fig. 1.5 Pie chart shows the usage of polypropylene (PP) in comparison to other polymers in the UK

to 30 °C for modern ones. Acrylates are used as artist paints, acrylic fiber, superglue, etc.

1.4 Classification Based on Tacticity

The classification of the polymer based on tacticity is discussed along with a schematic diagram of Fig. 1.6.

a. **Isotactic**: Isotactic polymers are made up of isotactic macromolecules. All substituents in isotactic macromolecules are on the same side of the macromolecular backbone. An isotactic macromolecule is made up entirely of mesodiads. Isotactic polymer polypropylene is generated using Ziegler–Natta catalysis [55]. Isotactic polymers are often semicrystalline and have a helix structure.
b. **Syndiotactic**: The substituents in syndiotactic or syntactic macromolecules move around throughout the chain. The macromolecule is made up entirely of raceme diads. Metallocene catalytic polymerization produces crystalline syndiotactic polystyrene with a melting point of 161 °C. An example of a syndiotactic polymer is gutta percha [56].
c. **Atactic**: The substituents in atactic macromolecules are distributed at random along the chain. The proportion of meso dyads ranges from 1 to 99%. Spectroscopic techniques like nuclear magnetic resonance (NMR) may be used to estimate the percentage composition of a polymer for each triad [57]. Polyvinyl chloride produced during free-radical process are usually atactic. Atactic polymers are amorphous due to their random nature. A common example is polystyrene. It is possible to obtain the syndiotactic variant of this polymer if a specific catalyst is used in its production; however, most industrial polystyrene produced is atactic.

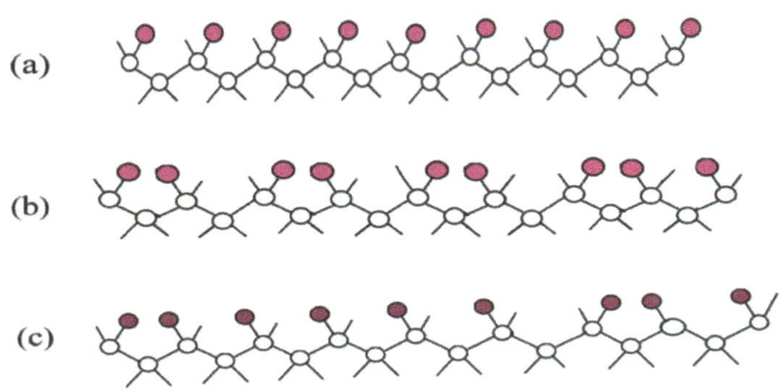

Fig. 1.6 Arrangement of macromolecules of **a** Isotactic, **b** Syndiotactic, and **c** Atactic

d. **Eutectic polymer**: Substituents may occupy any particular arrangement of places along the chain in eutectic macromolecules. Syndiotactic and isotactic polymers are examples of a wide class of eutectic polymers, which also include heterogeneous macromolecules with a series of different types of substituents: For example, the bases in nucleic acids and the side chains in proteins.

1.5 Classification Based on the Backbone of the Polymer Chain

i. **Organic polymers**: Polymer materials that contain carbon atoms in the backbone are called organic polymers. The side vacancies of backbone carbon atoms are usually those of hydrogen, oxygen, nitrogen, etc. Here most of the structures are simple. These types of polymers are environmentally friendly as they are biodegradable. Organic polymers constitute the majority of synthetic polymers. Some common examples are polysaccharides, polyesters, proteins, polynucleotides, etc.
ii. **Inorganic polymers**: In inorganic polymers, the polymer materials do not contain carbon atoms in the backbone. These types of polymers are highly branched and complex structures. These polymers are not environmentally friendly as they are not biodegradable. Some common examples are polydimethylsiloxane (silicone rubber), glass, etc.

1.6 Classification of Polymer Composite

A. **Classification based on their sources**:

   a. **Fiber-composites (Fiber-reinforced polymer)**: Fiber-reinforced polymer (FRC) is a high-performance fiber composite made feasible by a patented molecular re-engineering process that links cellulosic fiber molecules with resins in the FRC material matrix to produce a product with remarkable structural qualities. Selected physical and structural characteristics of wood are effectively cloned and incorporated into the FRC product through this feat of molecular re-engineering, together with other crucial characteristics, to provide performance values that are superior to those of modern wood. Unlike other composites, this one may be recycled up to 20 times, making it possible to reuse leftover FRC [58].
   b. **Natural-fiber composites**: Natural-fiber composites (NFCs) are materials that are composites, but at least some of the reinforcing fibers come from renewable and carbon dioxide-neutral resources like plants or wood. Composites made of natural fibers may be completely biodegradable. NFCs provide several benefits over conventional materials, including biological environmental resilience.

c. **Bio-composites**: A bio-composite is a composite material made up of a reinforcement made of natural fibers and matrix. These materials always provide biocompatibility and preserve the strengthening properties of the matrix, sometimes resembling the structure of the living components used in process. The matrix phase is composed of polymers derived from both non-renewable and renewable resources. Both non-wood and wood fibers are used in bio-composites, which all include cellulose and lignin. Additionally, fibers are the main constituents of bio-composites, which have biological origins. Examples of fibers include crop fibers (such as flax or hemp), recovered wood, wastepaper, agricultural processing by-products, and regenerated cellulose fiber (viscose/rayon) [59].

B. **Classification based on polymer matrix**:

i. **Thermoplastic composite**: Structural polymer composites gave rise to thermoplastic composites. These materials are not chemically unstable due to the utilization of a thermoplastic matrix. Thermoplastic composites soften and can be remolded without degradation when heated and when cooled, they solidify into a finished shape. In addition, thermoplastic composites could be formed utilizing methods from the metalworking and woodworking industries. Due to the matrix material's durability and strength, they have greater tolerances for recycling and damage. These types of composites can be categorized into either Glass Mat Thermoplastics (GMT) or Advanced Thermoplastic composites (ATC). Due to their high toughness, superior strength, and capacity for recycling and remolding, thermoplastic composites make a great construction material. These are employed for panels, window frames, doors, storage tanks, and lightweight buildings.

ii. **Thermoset composite**: Thermoset composites are substances whose physical and chemical characteristics remain unchanged upon initial heat exposure. They provide exceptional strength, hardness, and moldability along with outstanding chemical and heat stability. These composites are commonly used in a variety of sectors where high impact-resistant and durable performance is required. These are also employed in the automotive, aerospace, lighting, electrical, energy, and construction sectors. The recycling options for thermoset composites are limited.

iii. **Rubber composite**: Rubber is a composite material which is made from rubber and various materials such as silica. It is also a polymer. When compared to other materials, it can flex considerably more elastically under tension and regain its original shape without permanent distortion after the force is released.

It is a polymer matrix that is commonly used in industry due to its effective energy absorption capabilities [60].

1.7 Constituent of Polymers

Manufacturing processes for polymer composites involve combining polymers with reinforcing materials, such as fibers or particulates, to create a material with enhanced properties. The choice of manufacturing method depends on factors like the type of polymer, the reinforcing material, and the desired properties of the composite. Here are some common manufacturing processes for polymer composites with suitable examples.

A. **Hand Lay-Up Process**

   Description: This is a simple and low-cost method where layers of reinforcing materials and polymer resin are manually placed in a mold [61].
   Example: Fiberglass boats are often made using hand lay-up processes. Layers of fiberglass fabric are saturated with polyester or epoxy resin and laid up in a mold to form the hull [62].

B. **Spray-Up Process**

   Description: In this method, a chopper gun is used to spray a mixture of chopped fibers and resin onto a mold to build up the composite [63].
   Example: Swimming pool components like water slides are manufactured using the spray-up process.

C. **Filament Winding**

   Description: Continuous fibers, such as glass or carbon, are wound onto a rotating mandrel while being impregnated with resin to create a cylindrical composite structure [64–66].
   Example: Pressure vessels, like natural gas storage tanks, are often produced through filament winding.

D. **Pultrusion**

   Description: Pultrusion involves pulling continuous fibers through a resin bath and then through a heated die to form a constant cross-section profile [64, 67].
   Example: Window and door frames, as well as structural shapes in construction, can be made using pultrusion.

E. **Resin Transfer Molding (RTM)**

   Description: RTM is a closed-mold process where dry fibers are placed in a mold, and resin is injected under pressure [68, 69].
   Example: Automotive parts like bumper beams are often produced using RTM for their high strength-to-weight ratio.

F. **Vacuum Bagging**

   Description: In this method, layers of fiber and resin are placed in a mold, and a vacuum bag is used to remove air and consolidate the materials [70, 71].
   Example: Aircraft components, such as composite panels for interior structures, can be manufactured using vacuum bagging.

G. **Compression Molding**

Description: Compression molding involves placing pre-cut composite materials in a mold and using heat and pressure to cure them [72, 73].
Example: Composite brake pads and disks for high-performance automobiles are produced through compression molding.

H. **Injection Molding**

Description: Injection molding is typically used for thermoplastic composites, where a molten polymer is injected into a mold cavity [74–76].
Example: Many consumer products, such as laptop cases, are made from thermoplastic composites using injection molding.

I. 3D Printing/Additive Manufacturing

Description: Additive manufacturing techniques can be used to create polymer composites layer by layer, allowing for complex geometries [77–79].
Example: 3D-printed carbon-fiber-reinforced parts are used in aerospace for their lightweight and high-strength properties.

These manufacturing processes provide a wide range of options for creating polymer composites tailored to specific applications, whether it's for lightweight structures in aerospace, durable construction materials, or high-performance automotive components. The choice of method depends on factors like cost, production volume, and the required material properties.

1.8 Importance of Polymers

The existence and application of polymers in the course of the evolution of human civilization is a much-debated topic. Polymer composites have provided us with numerous advantages ranging from a reduced consumption of fuel as a result of less weight associated with excellent specific stiffness and strength, as well as being less energy-intensive for their production. Other advantages include top-notch resistance to corrosion and fatigue, good fracture toughness, high resistance to impact and abrasion, good damping capacity and high tolerance to damage, low processing and maintenance cost, flexibility in design, and many more features. Polymer and polymer composites have a wide range of advanced applications in various industries due to their versatility, durability, and lightweight properties. Here are some advanced applications of polymers and polymer composites [80].

a. **Aerospace Industry**:
Composite materials, such as carbon fiber-reinforced polymers, are used to make lightweight and strong components for aircraft and spacecraft. Polymer-based nanocomposites are used in the development of lightweight and durable materials for aerospace applications [81, 82].

b. **Automotive Industry**:

Polymer composites are used for manufacturing lightweight and fuel-efficient car parts, such as body panels and interior components. Electric vehicle battery enclosures are often made from reinforced polymer composites for their strength and light weight [83, 84].

c. **Healthcare and Medical Devices**:
Biocompatible polymers are used in medical devices, such as artificial joints, dental implants, and drug delivery systems. 3D printing of biodegradable polymers is used to create patient-specific implants and prosthetics [85, 86].

d. **Electronics and Electrical Engineering**:
Polymers are used in the encapsulation of electronic components to protect against moisture and environmental factors. Conductive polymers are used in flexible electronics and organic light-emitting diodes (OLEDs) [87].

e. **Renewable Energy**:
Polymer solar cells, also known as organic photovoltaics (OPVs), use polymer-based materials to convert sunlight into electricity. Wind turbine blades are increasingly being made from polymer composites to reduce weight and improve efficiency [88].

f. **Construction and Infrastructure**:
Polymer composites are used to strengthen and repair aging infrastructure, such as bridges and buildings. Insulated concrete forms (ICFs) made from polymer composites are used for energy-efficient building construction [89].

g. **Sports Equipment**:
Advanced polymer composites are used in the construction of sports equipment, such as tennis rackets, golf club shafts, and bicycle frames, to enhance performance and durability.

h. **Military and Defense**:
Polymer composites are used in the manufacturing of lightweight armor and ballistic protection materials. Polymers are employed in the development of stealth technology and radar-absorbing materials.

i **Environmental Protection**:
Polymer-based adsorbents and membranes are used for water purification and wastewater treatment. Biodegradable polymers are employed to reduce plastic pollution and promote sustainable packaging [90].

j. **Oil and Gas Industry**:
Polymer composites are used in offshore oil platform construction due to their resistance to corrosion and harsh environmental conditions. High-performance polymers are used for downhole tools and components.

k. **Packaging**:
Advanced polymer materials are used in the packaging industry to improve shelf life, reduce waste, and enhance product safety [91].

l. **Textiles**:
Smart textiles and fabrics made from conductive polymers are used for applications like heated clothing and wearable technology [92].

m. **Marine**:
The lightweight and good corrosion resistance have made polymer composites a useful ingredient for the construction of boats, and fabrication of components such as the bulkhead, deck, mast, and craft [93].
n. **Nanotechnology**:
Polymer nanocomposites incorporate nanoparticles to enhance material properties, enabling applications in fields like Nanoelectronics and Nano medicine [94].

These are just a few examples of the advanced applications of polymers and polymer composites across various industries. As research and technology continue to advance, new and innovative uses for these materials are constantly emerging.

1.9 Disadvantages of Polymer

There are two sides to coins. At first, we discussed the advantages of polymers; in the next phase, we will discuss the negative impact of polymers on environment, human health, etc.

A. **Environmental impact**: One of the biggest disadvantages of polymers is their impact on the environment. Polymers are non-biodegradable and take years and years to break down. Due to this, the polymers have a negative impact on air, water, and human health.

   (a) **Air pollution**: Due to the non-biodegradable nature of polymers, polymeric products like plastic bags and bottles cannot be dumped into sanitary landfills. Then these waste materials are combusted which release harmful gases like carbon dioxide, nitrogen oxides, and carbon monoxide that cause air pollution [95]. These harmful gases are toxic and cause acid rain and global warming. For instance, combustion of plastic like polyvinyl chloride releases HCl gas which converts into HCl acid with moisture availability. That HCl acid causes erosion in buildings and structures. This also affects the growth of vegetables in nearby areas [96].
   (b) **Water pollution/Effects on aquatic life**: The light plastic in marine environment degrades very slowly comparated to that on land [97]. Under environmental condition like light, wind, rain, heat, etc., there will be chemical and physical change in the degradation of plastic [97]. Under these conditions plastic turns into small wreckages but is not completely degraded [98]. According to different studies, marine plastic debris are microscopic in size, so they are called microplastics. These are present in major quantity in marine debris [99, 100]. Marine animals accidentally eat these plastics while capturing the quarry [101]. These marine animals are badly affected by eating these microplastics and plastic objects [102]. After consumption of microplastics, related polymeric products and synthetic resins by the

aquatic species, in the results 80% of the consumed aquatic animals were died due to water pollution as reported in literature.

(c) **Occupational hazards**: Workers in the polymer and plastic sectors are known to have a variety of health issues. It is predicted that workers in the polyvinylchloride business are exposed to serious health issues as its monomer causes a rare kind of liver cancer. Therefore, a limit of 1 ppm during 8 h period is assigned to vinyl chloride exposure in industry.

(d) **Health hazards**: Plasticizers that are used to harden plastic are mostly harmful to human health. For instance, phthalate ester plasticizers are removed by the blood from plastic bags used for blood storage. When a patient receives blood, it enters his bloodstream. The health risks associated with polychlorinated biphenyl are more serious. Because they are fat soluble polymers, they concentrate in aquatic creatures. When humans consume infected fish, they are vulnerable to headaches, lethargy, and joint discomfort. When the concentration of polychlorinated biphenyl reaches 12 ppm, it causes blindness.

B. **Flammability**: Polymers have a significant potential for flames to ignite and spread quickly. In some environments like buildings or aircraft, this can be dangerous.

C. **Low melting point**: Polymers can melt or deform when exposed to high temperature due to their low melting point. In some applications like the automotive industry, this can be a big problem.

D. **Limited strength**: Polymers are strong materials, but they could not be as strong as metals. This implies that applications requiring a high degree of strength or toughness might not be a good fit for them.

E. **Chemical sensitivity**: Polymers can also be chemically sensitive, causing them to break down or degrade over time. This can be a concern in applications where the polymer is exposed to chemicals, such as those used in the medical field.

1.9.1 Awareness Programs or Social Awareness

Effective education is required with the introduction of biodegradable polymers into the consumer packaging sector. To reduce the possibility for increased littering, it will be critical to educate the public that biodegradable plastics do not decompose instantly. Furthermore, unambiguous disposal routes must be found and made available to customers in order to properly manage this waste stream. For example, requiring that an item be disposed of in the consumer's green organics container, which is destined for a commercial composting operation, might cause problems in areas where households do not have access to these services [107].

Along with these existing programs various guidelines and protocols have to be followed for proper degradation of polymers. There are various methods that are studied for the polymeric materials to be degraded. For degradation of synthetic polymer, mechanical strength is reduced by different methods to solve a problem

Table 1.1 Degradation of some polymeric materials [1]

Sl. No.	Polymeric material	Degradation time
1	Disposable diapers	50–100 years
2	Tin cans	50 years
3	Painted wooden sticks	13 years
4	Plywood	1–3 years
5	Wax coated milk cartons	3 months
6	Cardboard box	2 months
7	Apple core	2 months
8	Newspaper	6 weeks
9	Paper towel	2–4 weeks
10	Plastic bottles	100 years
11	Glass bottles	1 million years
12	Plastic 6-pack holder rings	450 years
13	Aluminum cans	80–100 years
14	Nylon fabric	30–40 years
15	Plastic bags	10–20 years
16	Plastic coated paper milk cartons	5 years
17	Cigarette butts	1–12 years
18	Wool socks	1–5 years
19	Orange peels	6 months
20	Rope	3–14 months
21	Paper	2–5 months
22	Cotton rags	1–5 months

of polymer disposal. Different polymeric materials take different times to degrade. Degradation time of some polymeric materials are given below in Table 1.1.

Different types of degradation of polymers are

(a) **Thermal degradation**: This degradation process occurs by using high temperature to impairment chemical properties without the immersion of other compounds like oxygen [108]. Normally, polymers will degrade when they are heated in between 500 and 850 °C in the absence of air. It differs from thermal oxidation, which can normally occur at lower temperature [109].

(b) **Photodegradation**: This degradation process occurs due to exposure to the energetic part of sunlight, i.e., high energy radiation or UV radiation [107]. Photodegradation is frequently avoided since it ruins paintings and other artifacts. It is, however, partially responsible for biomass remineralization and is purposely employed in some disinfection systems. Polypropylene and LDPE are two common synthetic polymers that may be attacked, with tertiary carbon bonds in their chain structures. UV rays combine with these bonds to produce free radicals, which then react with oxygen in the atmosphere to produce carbonyl groups in the main chain. Products with exposed surfaces may discolor and fracture, and in severe circumstances, full product disintegration may occur.

(c) **Compostable degradation**: In this type of degradation the microorganisms present in compost increase the rate of degradation [107].

(d) **Mechanical degradation**: This mechanical process occurs because of mechanical stress strain [107].

(e) **Ultrasonic degradation**: In ultrasonic degradation, the shattering of chains occurs due to the vibration which is caused by the sound at specific frequencies [107].

(f) **Chemical degradation**: Corrosive compounds in agrochemicals, such as ozone or Sulfur, can damage the polymer chain, causing link breakage or oxidation [107].

(g) **Hydrolytic degradation**: In this process the degradation of polymer containing functional groups occurs due to the effect of water [107].

(h) **Biodegradation**: Biodegradation is described as a chemical disintegration process caused by the activity of enzymes released by living organisms (bacteria, fungus, etc.) [110]. The creation of biofilm, which is defined as a layer of deposition of microorganisms and their secreted polysaccharides, etc., on the polymer surface, is required for primary biodegradability. This is followed by the polymer being broken down into low molecular weight oligomers (presumably due to enzymes released by the microorganisms), which are then easily ingested by the bacteria. The last stage of degradation results in the creation of CO_2 and water. The microbe must be able to use the polymer as its only carbon source in order for this process to occur. Plastic deterioration in natural settings is a relatively gradual process that is affected by environmental elements such as temperature, humidity of the air and moisture in the polymer, pH, sun energy, and polymer characteristics as well as biological considerations. The most hazardous plastics polyolefins are resistant to microbial attack because of the absence of any active functional groups. Some environmental factors like physical factors (temperature and mineral structure), chemical factors (pH, salt, nutrient present, oxygen, and moisture content), and biological factors (microorganism and enzyme present) affect the rate of biodegradation [111]. Biodegradation of polymer includes the following steps which are attachment of microorganisms to the surface of the polymer, growth of microorganism utilizing the polymer as a carbon source, primary degradation of the polymer, and ultimate degradation. Most chemicals discharged into the environment are expected to be lost mostly through biodegradation. The degradation and absorption of polymers by live microbes to create degradation products is referred to as this process. Fungi, bacteria, and algae are the most significant species in biodegradation [112]. The rate of biodegradation of biodegradable polymer is dependent on the disposal environment and its conditions. The major disposal environments are composting facilities or soil burial, anaerobic digestion, wastewater treatment facilities, landfill, plastic reprocessing facilities, marine and freshwater environments, and general open environments as litter [113].

1.10 Proposal for Nanopolymer Composites as Future Trends

Composite properties refer to the characteristics and qualities exhibited by materials that are composed of two or more distinct components: a polymer matrix and reinforcing materials. These properties are a result of the interaction between the matrix and the reinforcing materials, and they can be tailored to suit specific application requirements. The key composite properties include

A. **Mechanical Properties**:

Strength: The ability of the composite to withstand loads without breaking or deforming.

Stiffness: The resistance of the composite to bending or deformation under load.

Toughness: The ability to absorb energy and deform before failure, indicating resistance to impact and fracture.

Hardness: The resistance of the composite to penetration or abrasion.

Fatigue Resistance: The ability to withstand repeated loading and unloading cycles without failure.

B. **Thermal Properties**:

Thermal Conductivity: The ability of the composite to conduct heat. This property can be tailored to be either high or low, depending on the application.

Thermal Expansion: The change in size or volume of the composite with changes in temperature.

C. **Electrical Properties**:

Electrical Conductivity: The ability to conduct electricity. Some composites can be engineered to be conductive, while others are insulating.

Dielectric Constant: A measure of a material's ability to store electrical energy in an electric field.

D. **Chemical Properties**:

Chemical Resistance: The ability of the composite to resist degradation or damage when exposed to various chemicals or corrosive environments.

Chemical Inertness: The lack of reactivity with specific chemicals.

Dimensional Stability: The ability of the composite to maintain its size and shape under different environmental conditions, such as changes in temperature and humidity.

Weight and Density: The composite's mass per unit volume, which can be significantly lower than that of traditional materials like metals, leading to weight reduction in applications.

Environmental Resistance: The resistance of the composite to environmental factors like moisture, UV radiation, and atmospheric conditions.

Aesthetic and Surface Finish: The visual and textural properties of the composite surface, which can be important in applications where appearance matters.

Cost: The cost-effectiveness of the composite compared to alternative materials, taking into account performance and longevity.

1.11 Conclusion and Future Trends

The conclusion of a topic on the "Introduction of Polymers and Polymer Composites: Basic Fundamentals" should summarize the key points discussed in the introduction and provide some insights or reflections. In conclusion, this introduction to polymers and polymer composites has shed light on the fundamentals of these versatile materials. Polymers, with their long chains of repeating units, have a wide range of applications, from everyday plastics to high-performance materials used in aerospace and health care. Polymer composites, on the other hand, combine the strengths of different materials to achieve tailored properties, making them invaluable in various industries. Understanding the basic principles of polymers, such as their molecular structure, polymerization processes, and physical properties, is essential for anyone venturing into the field of materials science and engineering. Additionally, recognizing the advantages of polymer composites in terms of strength, lightweight characteristics, and customization highlights the potential for innovation and application. As technology advances and our knowledge of polymers deepens, we can anticipate even more groundbreaking developments and applications in areas like sustainable materials, biodegradable polymers, and advanced composite materials. The fundamentals presented in this introduction lay the foundation for further exploration and research in the fascinating world of polymers and polymer composites.

Acknowledgements Authors would like to thanks to the Siksha "O" Anusandhan Deemed to be University for providing the digital library facility for collecting relevant data during the preparation of the manuscript.

Statements and Declarations
Funding: The authors declare that no funds, grants, or other support were received during the preparation of this manuscript.

Competing Interests: The authors have no relevant financial or non-financial interests to disclose.

References

1. Painter, P.C., Coleman, M.M.: Fundamentals of Polymer Science: An Introductory Text p. 1. Technomic Pub. Co., Lancaster, PA
2. Painter, P.C., Coleman, M.M.: Essentials of Polymer Science and Engineering. DES Tech Publications, Lancaster Pennsylvania (2008)
3. Staudinger, H.: Über Polymerisation [On polymerization]. Berichte der Deutschen Chemischen Gesellschaft (in German) **53**, 1073–1085 (1920)
4. Belgacem, M.N., Gandhi, A.: Monomers, Polymers and Composites from Renewable Resources. Elsevier (2011)
5. Elias, H.-G.: An Introduction to Polymer Science. VCH, Weinheim (1997)
6. Meysam, T., Nuruldiyanah, K., Zahra, D.E., Yee, L., Ashen, G., Chandima, G.: Conventional and advanced composites in aerospace industry: technologies revisited. Am. J. Aerosp. Eng. **5**, 9–15 (2018)
7. Kalirasu, S., Rajini, N., Rajesh, S., Michla, J.R.: Water absorption behaviour on natural/ synthetic fibre reinforced polymer composites. Int. J. Eng. Adv. Technol **9**, 5–9 (2019)
8. Matabola, K.P., De Vries, A.R., Moolman, F.S., Luyt, A.S.: Single polymer composites: a review. J. Mater. Sci. **8**, 445–463 (2009)
9. Sharma, V.P., Agarwal, V., Umar, S., Singh, A.K.: Polymer composites sustainability: environmental perspective, future trends and minimization of health risk. In: Proceedings of the International Conference on Environmental Science and Development IPCBEE4, Singapore, 26–28, pp. 259–261 (2011)
10. Brunsveld, L., Folmer, B.J., Maijer, E.W., Sijbesma, R.P.: Supramolecular polymers. Chem. Rev. **101**, 4071–4098 (2001)
11. Aida, T., Meijer, E.W., Stupp, S.I.: Functional supramolecular polymers. Science **335**, 813–817 (2012)
12. Aida, T., Meijer, E.W.: Supramolecular polymers- we've come full circle. Isr. J. Chem. **60**, 33–47 (2020)
13. Amabilino, D.B., Smith, D.K., Steed, J.W.: Supramolecular materials. Chem. Soc. Rev. **46**, 2404–2420 (2017)
14. Yang, L., Tang, X., Wang, Z., Zhang, X.: Supramolecular polymers: historical development, preparation, characterization, and functions. Chem. Rev. **115**, 7196–7239 (2015)
15. Berzelius, J.: Körper von gleicher Zusammen-setzung und verschiedenen Eigenschaften. Jahres-Bericht **11**, 44–48 (1832)
16. Jensen, W.B.: The origin of the term allotrope. J. Chem. Educ. **83**, 838–839 (2006)
17. Berzelius, J.: Isomerie. Unterscheidung von damit analogen Verhältnissen, Jahres-Bericht **12**, 63–67 (1833)
18. Armstrong, H.E.: Isomerism. In: Forster Morley, H., Pattison Muir, M.M. (eds.) Watt's Dictionary of Chemistry, vol. 3, pp. 79–88. London, Longmans, Green & Co. (1888)
19. Pattison Muir, M.M.: Polymerism. In: Forster Morley, H., Pattison Muir, M.M. (eds.) Watts' Dictionary of Chemistry, vol. 4, p. 297. London, Longmans, Green & Co. (1894)
20. Staudinger, H., Fritschi, J.: Über die Hydrierung des Kautschuks und über seine Konstitution. Helv. Chim. Acta **5**, 785–806 (1922)
21. Larsen, L.V.: Origin of the term oligomer. Chem. Eng. News **62**, 58 (1984)
22. Lekakou, C., Bader, M.G.: Compos. A **27**, 29–37 (1999)
23. Horgan, C.O.: The axisymmetric end problem for transversely isotropic circular cylinders. Int. J. Solids Struct. **10**, 837 (1974)
24. Moraes, V.T., Jermolovicius, L.A., Tenório, J.A., Lebrão, S.P., Lebrão, G.W.: Microwave-assisted recycling process to recover fibre from fibreglass polyester. Compos. Mater **22**, 1–8 (2020)
25. Zagho, M.M., Hussein, E.A., Elzatahry, A.A.: Recent overviews in functional polymer composites for biomedical applications. Polymers **10**, 739 (2018)
26. Pender, K., Yang, L.: Investigation of catalysed thermal recycling for glass fibre reinforced epoxy using fluidised bed process. Polym. Compos. **40**, 3510–3519 (2019)

27. Giorgini, L., Leonardi, C., Mazzocchetti, L., Zattini, G., Cavazzoni, M., Montanari, I., Tosi, C., Benelli, T.: Pyrolysis of fibreglass/polyester composites: recovery and characterization of obtained products. Fac. Mech. Eng. Trans. **44**, 405–414 (2016)
28. Turner, R.P., Kelly, C.A., Fox, R., Hopkins, B.: Re-formative polymer composites from plastic waste: novel infrastructural product application. Recycling **3**, 54–70 (2018)
29. Hugo, M.A., Scelsi, L., Hodzic, A., Jones, F.R., Joyce, D.: Development of recycled polymer composites for structural applications. Plast. Rubber Compos. **40**, 317–323 (2011)
30. Ballou, J.W., Smith, J.C.: Dynamic measurements of polymer physical properties. J. Appl. Phys. **20**, 493 (1949)
31. Salih Mustafa, N., Ali Omer, M.A., Garlnabi, M.E.M., Ismail, H.A.: Reviewing of general polymer types, properties and application in medical field. Int. J. Sci. Res. **5**, 212–221 (2016)
32. McNaught, A.D., Wilkinson, A.: Glossary of basic terms in polymer science. Pure Appl. Chem. **68**, 2287–2311 (1996)
33. Hadjichristidis, N., Pispas, S., Floudas, G.: Block Copolymers. John Wiley & Sons, Inc., Hoboken, USA (2002)
34. Cowie, J.M.G.: Valeria Arrighi, Polymers: Chemistry and Physics of Modern Materials. CRC Press Taylor and Francis Group, FL 33487–2742, 1–520 (2014)
35. Rudin, A., Choi, P.: The Elements of Polymer Science and Engineering. Academic press is an imprint of Elsevier 225 Wyman Street, Waltham, MA 02451, USA The Boulevard, Langford Lane, Kidlington, Oxford, OX5 1GB, UK, 1–561 (2013)
36. Chanda, M.: Introduction to Polymer Science and Chemistry A Problem-Solving Approach. CRC Press Taylor and Francis Group 6000 Broken Sound Parkway N W, Suite 300 Boca Raton, FL 33487–2742, 1–754 (2013)
37. Sperling, L.H.: Introduction to Physical polymer science, pp. 1–866. Canada, John Wily and Sons Inc. Hoboken Publication (2006)
38. Hadjichristidis, N., Iatrou, H., Pitsikalis, M., Driva, P., Sakellariou, G., Chatzichristidi, M.: Polymers with star-related structures: synthesis, properties, and applications. In: Polymer Science: A Comprehensive Reference, vol. 6, pp. 29–111. Elsevier, Amsterdam (2012)
39. Lapienis, G.: Star-shaped polymers having PEO arms. Prog. Polym. Sci. **34**, 852–892 (2009)
40. Chremos, A., Glynos, E., Green, P.F.: Structure and dynamical intra-molecular heterogeneity of star polymer melts above glass transition temperature. J. Chem. Phys. **142**, 1–9 (2015)
41. Zhu, W., Ling, J., Shen, Z.: Synthesis and characterization of amphiphilic star-shaped polymers with Calix [6] arene cores. Macromol. Chem. Phys. **207**, 844–849 (2006)
42. Knoll, K., Nießner, N.: Styrolux+ and styroflex+ - from transparent high impact polystyrene to new thermoplastic elastomers: syntheses, applications and blends with other styrene-based polymers. Macromol. Symp. **132**, 231–243 (2011)
43. Forman, D.C., Wieberger, F., Gröschel, A., Müller, A.H.E., Schmidt, H.-W., Ober, C.K.: Comparison of star and linear ArF resists. In: Proceedings of SPIE 7639, Advances in Resist Materials and Processing Technology XXVII, 76390P (2010)
44. Youssef, A.-R.: Solution & Bulk Polymerization **11**, 452 (2019)
45. Rodriguez, F., Cohen, F., Ober, C.K.: Lynden Archer, pp. 1–800. CRC Press, Principles of Polymer Systems (2003)
46. Malpass, D.B.: Introduction to Industrial Polyethylene: Properties, Catalysts and Processess, pp. 1–401. John Wiley and Sons (2010)
47. Birlie Genet, M., Bantie Sendekie, Z., Lemessa Jembere, A.: Investigation and optimization of waste LDPE plastic as a modifier of asphalt mix for highway asphalt: case of Ethiopian roads. Case Stud. Chem. Environ. Eng. **4**, 100150 (2021)
48. Araújo, J.R., Waldman, W.R., De Paoli, M.A.: Thermal properties of high density polyethylene composites with natural fibres: coupling agent effect. Polym. Degrad. Stab. **93**, 1770–1775 (2008)
49. Setiawan, B., Permatasari, D.C., Sumari, A.D.W., Winarno, T., Audiana, V.U., Damayanti, A.M.: HDPE plastic extruder design and control for 3D printing with plastic pellet feeding method. IOP Conf. Ser.: Mater. Sci. Eng. **1073**, 012073 (2021)

50. Gahleitner, M., Paulik, C.: Polypropylene, Ullmann's Encyclopedia of Industrial Chemistry, pp. 1–44. Wiley-VCH, Weinheim (2014)
51. Tripathi, D.: Practical Guide to Polypropylene. RAPRA Technology, Shrewsbury, 2002 1–104 (2001)
52. Maier, C., Calafut, T.: Polypropylene: The Definitive User's Guide and Data Book. William Andrew. 14 (1998)
53. Collinet, P., Belot, F., Debodinance, P., Ha Duc, E., Lucot, J.-P., Cosson, M.: Transvaginal mesh technique for pelvic organ prolapse repair: mesh exposure management and risk factors. Int. Urogynecol. J. **17**, 315–320 (2006)
54. Penzel, E.: Polyacrylates, Ullmann's Encyclopedia of Industrial Chemistry, p. 157. Wiley-VCH, Weinheim (2000)
55. Stevens, M.P.: Polymer Chemistry: An Introduction, vol. 77, pp. 234–235. Oxford Press: New York (1999)
56. Brandrup, J., Immergut, E.H., Grulke, E.A.: Polymer Handbook, vol. 46, pp. 1–2288. Wiley-Interscience, New York (1999)
57. Benjamin Brock Noble: Towards Stereocontrol in Radical Polymerization, pp. 1–251. Australian National University (2016)
58. Ladizesky, N.H., Ward, I.M.: Deformation mechanisms of low-density polyethylene with parallel lamellar morphology. J. Macromol. Sci. B **9**, 565 (1974)
59. Dunn, C.M.R., Mills, W.H., Turner, S.: Creep in thermoplastics. Br. Plastics **37**, 386 (1964)
60. Kotani, T., Sweeney, J., Ward, I.M.: The measurement of transverse mechanical properties of polymer fibres. J. Mater. Sci. **29**, 5551 (1994)
61. Baltazar-Jimenez, A., Bismarck, A., Sarikakis, K.: Green composites as panacea? Socio-economic aspects of green materials. Environ. Dev. Sustain. **8**, 445–463 (2006)
62. Oladele, I.O., Omotosho, T.F., Adediran, A.A.: Polymer-based composites: an indispensable material for present and future applications. Int. J. Polym. Sci. 1–12 (2020)
63. Avila, A.F., Paulo, C.M., Santos, D.B., Fari, C.A.: Materials Characterization **50**, 281–291 (2003)
64. Mkaddem, A., Demirci, I., Mansori, M.E.: Compos. Sci. Technol. **68**, 3123–3127 (2008)
65. Mayer, C., Wang, X., Neitzel, M.: Macro- and micro-impregnation phenomena in continuous manufacturing of fabric reinforced thermoplastic composites. Compos. A Appl. Sci. Manuf. **29**, 783–793 (1998)
66. Bednarcyk, B.A.: An inelastic micro/macro theory for hybrid smart/metal composites. Compos. B **34**, 175–197 (2003)
67. Downing, T.D., Kumar, R., Cross, V.M., Kjerengrtoen, L., Keller, J.J.J.: J. Adhes. Sci. Technol. **14**, 1801 (2000)
68. Tay, T.E., Vincent, B.C., Liu, G.: Mater. Sci. Eng. B **132**, 138–142 (2006)
69. Bergeret, A., Bozec, M.P.: Polym. Compos. **25**, 12 (2004)
70. Bledzki, A.K., Gassan, J.: Prog. Polym. Sci. **24**, 221 (1999)
71. Joseph, P.V.: Comput. Sci. Technol. **62**, 1357 (2002)
72. Geethamma, V.G., Thomas Mathew, K., Lakshminarayanan, R., Thomas, S.: Polymer **39**, 1483–1491 (1998)
73. Guo, F., Aryana, S., Han, Y., Jiao, Y.: A review of the synthesis and applications of polymer–nanoclay composites. Appl. Sci. **8**, 1696–1625 (2018)
74. Kong, I., Tshai, K.Y., Hoque, M.E.: Manufacturing of natural fibre-reinforced polymer composites by solvent casting method. In: Manufacturing of Natural Fibre Reinforced Polymer Composites, pp. 331–349. Cham, Switzerland, Springer (2015)
75. Wakeman, M.D., Cain, C.D., Rudd, C.D., Brooks, R., Long, A.C.: Compos. Sci. Technol. **59**, 1153–1167 (1999)
76. Alammar, A., Park, S., Williams, C.J., Derby, B., Szekely, G.: Oil-in-water separation with graphene-based nanocomposite membranes for produced water treatment. J. Membr. Sci. 18, 603 (2020)
77. Rane, A.V., Kanny, K., Abitha, V.K., Thomas, S.: Methods for synthesis of nanoparticles and fabrication of nanocomposites. In: Advances and Key Technologies Micro and Nano Technologies, pp. 121–139. Woodhead Publishing, Cambridge, UK (2018)

78. Shokrieh, M.M., Rafiee, R.: Comput. Mater. Sci. **50**, 437–446 (2010)
79. Huang, H., Talreja, R.: Compos. Sci. Technol. **66**, 2743–2757 (2006)
80. Arshad, M., Zubair, M., Rahman, S.S., Ullah, A.: Chapter 14—Polymers for advanced applications. In: Polymer Science and Nanotechnology Fundamentals and Applications, pp. 325–340 (2020)
81. Kesarwani, S.: Polymer composites in aviation sector. A brief review article. Int. J. Eng. Res. Technol. **6**, 518–525 (2017)
82. Rajak, D.P., Pagar, D.D., Menezes, P.L., Linul, E.: Fibre-reinforced polymer composites manufacturing. Prop. Appl. **11**, 1667–1704 (2019)
83. Okpala, C.C.: The benefits and applications of nanocomposites. Int. J. Adv. Eng. Technol. **5**, 12–18 (2014)
84. Kim, Y.S., Kim, J.K., Na, S.J., Jeon, E.S.: Experimental analysis on the application of polymer matrix composites containing Al2O3 for automotive lamp reflector. Appl. Sci. **90**, 4525–4539 (2019)
85. Agbabiaka, O.G., Oladele, I.O., Akinwekomi, A.D., Adediran, A.A., Balogun, A.O., Olasunkanmi, O.G., Olayanju, T.M.A.: Effect of calcination temperature on hydroxyapatite developed from waste poultry eggshell. Sci. Africa **8**, 1–12 (2020)
86. Gunatillake, P., Adhikari, R.: Biodegradable synthetic polymers for tissue engineering. Eur. Cells Mater. **5**, 1–16 (2003)
87. Phuong, H.A., Ayob, N.A., Blanford, C.F., Rawi, N.F.M., Szekely, G.: Nonwoven membrane supports from renewable resources: bamboo fibre reinforced poly (lactic acid) composites. ACS Sustain. Chem. Eng. **7**, 11885–11893 (2019)
88. Srinivasulu, N.V., Tejaswi, V.: Mechanical properties of polymer composite materials. Int. J. Res. Eng. Technol. **1**, 78–81 (2012)
89. Satish, G., Prasad, V.V., Ramji, K.: Manufacturing and characterization of CNT based polymer composites. Math. Models Eng. **3**, 89–97 (2017)
90. Jose, A.S., Jani, S.P., Sajin, J.B., Godwin, G., Ananthapuri, S.A.: Mechanical behaviour of coir and wood dust particulate reinforced hybrid polymer composites. Int. J. Innov. Technol. Explor. Eng. **8**, 3205–3209 (2019)
91. Gallos, A., Paës, G., Allais, F., Beaugrand, J.: Ligno cellulosic fibres: a critical review of the extrusion process for enhancement of the properties of natural fibre composites. RSC Adv. **7**, 34638–34654 (2017)
92. Cook, J.G.: Handbook of Textile Fibre: Natural Fibres, A Volume in Woodhead Publishing Series in Textiles, pp. 1–240 (1984)
93. Friedrich, K., Zhang, Z., Schlarb, A.K.: Compos. Sci. Technol. **65**, 2329–2343 (2005)
94. Nair, A.A., Prakash, S., Paul, R.C.: Synthesis and characterization of hybrid polymer composites. Int. J. Adv. Eng. Res. Sci. **4**, 126–131 (2013)
95. Alabi, O.A., Ologbonjaye, K.I., ASwosulo, O., Alalade, O.E.: Public and environmental health effects of plastic waste disposal: a review. J. Toxicol. Asses **5**, 1–13 (2019)
96. Koul, B., Yakoob, M., Shah, M.P.: Agricultural waste management strategies for environmental sustainability. Environ. Res. **206**, 112285 (2022)
97. Ho, B.T., Roberts, T.K., Lucas, S.: An overview on biodegradation of polystyrene and modified polystyrene: the microbial approach. Crit. Rev. Biotechnol. **38**, 308–320 (2018)
98. Hu, X., Thumarat, U., Zhang, X., Tang, M., Kawai, F.: Diversity of polyester-degrading bacteria in compost and molecular analysis of a thermoactive esterase from Thermobifida alba AHK119. Appl. Microbiol. Biotechnol. **87**, 771–779 (2010)
99. Jain, K., Bhunia, H., Sudhakara Reddy, M.: Degradation of polypropylene-poly L lactide blend by bacteria isolated from compost. Ann. Finance **22**, 73–90 (2018)
100. Jeon, H.J., Kim, M.N.: Functional analysis of alkane hydroxylase system derived from Pseudomonas aeruginosa E7 for low molecular weight polyethylene biodegradation. Int. Biodeterior. Biodegrad. **103**, 141–146 (2015)
101. Islam, S., Apitius, L., Jakob, F., Schwaneberg, U.: Targeting microplastic particles in the void of diluted suspensions. Environ. Int. **123**, 428–435 (2019)

102. Jeon, H.J., Kim, M.N.: Comparison of the functional characterization between alkane monooxygenases for low-molecular-weight polyethylene biodegradation. Int. Biodeterior. Biodegrad. **114**, 202–208 (2016)
103. Karimi, M., Biria, D.: The promiscuous activity of alpha-amylase in biodegradation of low-density polyethylene in a polymer-starch blend. Nat. Sci. Rep. **9**, 1–10 (2019)
104. Kawai, F., Kawase, T., Shiono, T., Urakawa, H., Sukigara, S., Tu, C.: Masaki Yamamoto, Enzymatic hydrophilization of polyester fabrics using a recombinant cutinase Cut190 and their surface characterization. J. Fiber Sci. Technol. **73**, 8–18 (2017)
105. Kawai, F., Oda, M., Tamashiro, T., Waku, T., Tanaka, N., Yamamoto, M.: A novel Ca2+-activated, thermostabilized polyesterase capable of hydrolyzing polyethylene terephthalate from Saccharomonospora viridis AHK 190. Appl. Microbiol. Biotechnol. **98**, 10053–10064 (2014)
106. Kenny, S.T., Runic, J.N., Kaminsky, W., Woods, T., Babu, R.P., O'Connor, K.E.: Development of a bioprocess to convert PET derived terephthalic acid and biodiesel derived glycerol to medium chain length polyhydroxyalkanoate. Appl. Microbiol. Biotechnol. **95**, 623–633 (2012)
107. Islam, S., Saha, S., Bakr, A., Hossain Mondal, I.: A review on biodegradations of polymers and its effect on environment. J. Compos. Biodegrad. Polym. **3**, 46–54 (2015)
108. Guaita, M., Chiantore, O., Costa, L.: Changes in degree of polymerization in the thermal degradation of polystyrene. Polym. Degrad. Stab. **12**, 315–332 (1985)
109. Peterson, J.D., Vyazovkin, S., Wight, C.A.: Kinetics of the thermal and thermo-oxidative degradation of polystyrene, polyethylene and poly(propylene). Macromol. Chem. Phys. **202**, 775–784 (2001)
110. Choi, E.-J., Park, J.-K.: Study on biodegradability of PCL/SAN blends using composting method. Polym. Degrad. Stab. **52**, 321–326 (1996)
111. Arutchelvi, J., Sudhakar, M., Arkatkar, A., Doble, M., Bhaduri, S., Uppara, P.V.: Biodegradation of polyethylene and polypropylene. Indian J. Biotechnol. **7**, 9–22 (2008)
112. Jayasekara, R., Harding, I., Bowater, I., Lonergan, G.: Biodegradability of selected range of polymers and polymer blends and standard methods for assessment of biodegradation. J. Polymer Environ. **13**, 231–251 (2005)
113. Negoro, S.: Biodegradation of nylon oligomers. Apply Microbiol. Biotechnol. **54**, 461–466 (2000)

Chapter 2
Synthesis and Processing Techniques of Polymer Composites

S. K. Parida, Shilpi Kullu, Shibanee Hota, and S. Mishra

Abstract In this chapter, composites are macroscopic-scale materials created by the fusion of two or more materials with various chemical, physical, and morphological components. These materials acquire certain brand-new or enhanced mechanical, chemical, physical, and other properties that are unrelated to those of their constituent parts. To meet some unique uses, these composite materials are created and synthesized. Biodegradable polymers have been the subject of substantial investigation by polymer scientists, and they offer a wide range of potential uses in the field of medicine. Different biopolymers have been created and used in a variety of biomedical applications nowadays. The science and technology of biopolymers are still in their infancy of development, despite the apparent explosion of biopolymers in medical science. Through diligent study and development, there are tremendous prospects for biopolymers to permeate every aspect of medical science. To prepare biopolymers, various polymerization methods and techniques are discussed in this chapter. The focus is on the fundamental characteristics, the synthesis methods, and the biomedical uses of biopolymers. Different processing procedures used for the production of devices have been addressed to produce practical biomedical devices from polymers to suit medical science's requirements. The main focus of this review is on the benefits of polymeric materials-based composite over other traditional materials and how they are used in the biomedical area. Additionally, viewpoints in this area have been highlighted, and conclusions have been drawn. The pertinent literature was gathered from a variety of sources, including books, websites, journal papers, review articles, and Google sites.

Keywords Biodegradable polymers · Biopolymers · Polymerization methods · Biomedical uses · Journal papers

S. K. Parida (✉) · S. Kullu · S. Hota · S. Mishra
Department of Physics, ITER, Siksha 'O' Anusandhan Deemed to be University, Bhubaneswar 751030, India
e-mail: santoshparida@soa.ac.in

S. Moharana et al. (eds.), *Polymer Composites*, Engineering Materials,
https://doi.org/10.1007/978-981-97-2075-0_2

2.1 Introduction

Polymers are large molecules composed of repeating structural units called monomers. These molecules can be natural or synthetic and are characterized by their long-chain structure. The process of joining monomers to form polymers is called polymerization [1]. There are two main types of polymers: (i) Natural polymers: Found in animals and plants, proteins are natural polymers made up of amino acid monomers. DNA and RNA are examples of nucleic acids, which are polymers made up of nucleotide monomers. Carbohydrates like cellulose and starch are composed of sugar monomers and form polysaccharide polymers. (ii) Synthetic polymers: Common synthetic polymers used in everyday materials, such as polyethylene, polypropylene, and polyvinyl chloride (PVC). Natural rubber (polyisoprene) can also be synthesized, and synthetic rubbers like neoprene and styrene-butadiene rubber are widely used. Polymers like nylon and polyester are used to make synthetic fibers for textiles. Polymers are often used in adhesives, such as epoxy and cyanoacrylate glues. The properties of polymers can vary widely depending on their composition, molecular weight, and structure. Some polymers are rigid and hard, while others are flexible and elastic. The versatility of polymers makes them valuable in a wide range of applications, including packaging, medical devices, electronics, and more. The study of polymers is part of a branch of chemistry and materials science known as polymer chemistry.

2.2 Role of Polymers

Interestingly, the repeating chain of monomers combined to form a very big molecular unit. The combination of oxygen, hydrogen, carbon, or nitrogen can be units. Polymers have different characteristics such as toughness and viscoelasticity, and can change into glassy and semi-crystalline structures rather than crystalline structures. It can be processed by heat and pressure and can be shaped into different structures. Polymers have lower strength and modulus when it is compared to the strength of other materials like metals and ceramics. Some common examples of polymers include plastics, rubber, cellulose, proteins, and DNA. Polymers play a crucial role in our daily lives, and their importance extends across various fields. Polymers exhibit a wide range of properties depending on their chemical structure and the choice of monomers. This versatility allows for the design of polymers with specific properties tailored for various applications. Many of the materials we encounter in everyday life, such as plastic bags, bottles, and containers, are made of polymers. The lightweight and durable nature of polymers makes them essential for packaging, reducing transportation costs and environmental impact. Polymers are used in the medical field for drug delivery systems, medical implants, and tissue engineering. Biocompatible polymers can be designed to interact with biological systems without causing harm. Synthetic polymers like polyester and nylon are widely used in the textile industry. They provide durable, lightweight, and cost-effective materials for clothing and other textile products. Polymers are used in various construction materials, including adhesives, sealants, and coatings. They contribute to the durability and performance of building materials. Polymers are employed in electronic devices and technology applications. For example, they are used in the production of insulating materials, flexible displays, and electronic components. Polymers play a significant role in the automotive sector, where they are used for manufacturing components such as bumpers, tires, interior parts, and various under-the-hood applications. Some polymers can be designed to be biodegradable, contributing to efforts to reduce environmental pollution. Additionally, recycling programs for certain polymers help mitigate the environmental impact of plastic waste. Polymers are utilized in the development of energy storage devices, such as batteries and capacitors, due to their ability to conduct ions and electrons. Continued research in polymer science and engineering leads to the development of new materials with improved properties, enabling technological advancements in various industries. In summary, the importance of polymers lies in their versatility, which allows them to be tailored for specific applications across numerous industries, contributing to advancements in technology, healthcare, sustainability, and more [2, 3]. Polymer nanocomposites are a class of materials that combine polymers (large molecules composed of repeating structural units) with nanoscale fillers. The nanoscale fillers are typically in the form of nanoparticles or nanotubes, and they are dispersed within the polymer matrix at the nanometre scale. The incorporation of nanoscale fillers into polymers can significantly enhance the mechanical, thermal, electrical, and barrier properties of the resulting composite material. Some common types of nanoscale fillers used in polymer nanocomposites include nanoparticles like clay, silica, carbon nanotubes, graphene, and other nanomaterials. The high surface area and unique properties of nanoparticles can lead to improved mechanical strength, stiffness, and toughness in the polymer matrix. Additionally, the increased surface area can enhance interactions between the filler and the polymer, resulting in improved thermal stability, flame resistance, and electrical conductivity. The processing methods for polymer nanocomposites are crucial to achieving a uniform dispersion of nanoparticles within the polymer matrix. Common techniques include melt blending, solution mixing, and in situ polymerization. Polymer nanocomposites have a wide range of applications, including automotive components, packaging materials, electronics, coatings, and biomedical devices. The ability to tailor the properties of these materials by adjusting the type, size, and concentration of nanoparticles makes them highly versatile and valuable in various industries.

2.3 Polymer Composites

Polymer-based composites also known as polymer matrix composites (PCM) main component is polymer together with fibers/fillers as the build-up. PCM has been used in a wide range of different sectors due to their unique combination of physical, chemical, and mechanical properties such as good damping capacity, corrosion and fatigue resistance, low resistance, low density, high strength, and stiffness. The characteristics of the polymer matrix have a significant bearing on their tendency to exhibit high levels of resistance to environmental degradation, such as water absorption, chemical attack, temperature, and impact forces [4]. Their examples are Polystyrene (PS), Polypropylene (PP), Polyvinyl chloride (PVC), Natural rubber, Epoxy resin, vinyls, etc. [5]. Polymers are classified into different groups such as natural, semi-synthetic, and synthetic. The property of the polymer is determined by their molecular force. They are also classified by their repeating units such as copolymers, homopolymers, and pectin. They can be classified by the type of polymerization including chain growth and step growth [6, 7].

2.3.1 Classification Polymer-Based Composites

Polymer-based composites can be classified as (i) polymer matrix and (ii) reinforcement source.

a. Classification Based on Polymer Matrix

Polymers are often as a matrix in the development of composite materials. The total composites produced today are about 75% in which the matrix is polymer because they can be processed without damaging the reinforcement at a low temperature [8]. The classification of the matrix is

- Thermoset
- Thermoplastics
- Elastomeric/Rubber

b. Classification of Reinforcement Source

- Natural Fiber
- Synthetic Fiber

Fibers can be synthetic or natural. Natural fibers are basically from plants or wood which are renewable and fully biodegradable. Natural fibers examples are silk, bamboo, jute, hemp, wheat, etc. Synthetic fibers examples are Kevlar, glass, latex, carbon, nylon, alumina, etc. The reinforcement in PCM helps the material in developing stiffness, toughness, hardness, and strength as they are the major constituents [9].

The polymer has very large uses because of its characteristics and their properties, lightweight, durability, and simplicity of processing [10]. In comparison to metals and ceramics, polymers have relatively weak thermal, mechanical, and electrical properties even though they have different modified versions of polymers such as homopolymers, blended polymers, and copolymers, which cannot eliminate metals and ceramics [11]. A novel class of composite materials called polymer nanocomposites is distributed in various organic polymer matrixes to enhance its performance qualities. The multi-phase materials are nanocomposites that must contain one component that is in the nanoscale range. Polymer nanocomposites offer a fresh and different way to the traditional filled polymers. When compared to a pristine polymer matrix, nanocomposites have remarkable capabilities and features because they contain nanoscale fillers [12, 13]. Polymer nanocomposites have many uses in every field. They are used in packing food such as processed meat, cheese, and dairy products. For carrying baby pacifiers, blood collection tubes, and drinkable water bottles polymer nanocomposites [14].

2.4 Literature Gap in Polymer Composites

There are some general insights into potential literature gaps in polymer composites. Keep in mind that the state of research is continually evolving, and new publications may have emerged since then. Here are a few areas where there might be gaps or areas for further exploration in the field of polymer composites:

Sustainable and Eco-friendly Composites: Investigating and developing polymer composites with a focus on sustainability, using bio-based or recycled materials, could be an area of interest. Research might explore the development of composites with improved environmental profiles.

Advanced Manufacturing Techniques: Exploring novel manufacturing techniques for polymer composites, such as additive manufacturing (3D printing) or advanced molding methods, could be a literature gap. These techniques could offer new possibilities for designing complex structures and improving the overall performance of composites.

Multifunctional Composites: The integration of multiple functionalities into polymer composites, such as self-healing properties, enhanced electrical conductivity, or sensing capabilities, is an evolving area. There may be opportunities to explore the development of composites with tailored multifunctional properties.

Nanocomposites: Further research into nanocomposites, which involve incorporating nanoscale reinforcements (such as nanoparticles or nanotubes) into polymers, could be an area of exploration. This may include investigating the effects of different nanofillers on mechanical, thermal, and electrical properties.

Durability and Long-Term Performance: Studying the long-term durability and performance of polymer composites under various environmental conditions could be an important aspect. This could include research on the effects of exposure to UV radiation, temperature variations, and other environmental factors on the mechanical properties of composites.

Process-Structure–Property Relationships: Understanding the relationships between processing parameters, composite structure, and resulting properties is critical. Research that delves into a more comprehensive understanding of these relationships could contribute to the optimization of polymer composite manufacturing processes.

To identify specific gaps, it is recommended to understand the role of polymer composites and find the updated literature on polymers and polymer composites. Additionally, discussions with experts in the field can provide valuable insights into emerging research directions and potential gaps in the existing literature.

2.5 Classification of Polymer Nanocomposites

A polymer composite refers to a material that consists of a polymer matrix reinforced with other materials to enhance its mechanical, thermal, electrical, or other properties. In simpler terms, it is a combination of at least two different materials, where one of them is a polymer—a large molecule made up of repeating structural units—and the other is a reinforcing material. The detail of the polymer composite is given in Fig. 2.1.

Depending on the degree of the NPs separation for polymer nanocomposite three types of morphologies are possible. Conventional composites, intercalated nanocomposites, and exfoliated nanocomposites are three possible types. The traditional composites are also called micro composites, in traditional composites we can obtain composites having separated phases normally since the polymer is incapable of being intercalated between the filler layers. A single extended polymeric chain is effectively intercalated in between the filler layers, resulting in a well-organized multilayer morphology with intercalated layers of the polymer and fillers, in intercalated-type polymer nanocomposites. Exfoliated nanocomposites are formed when the filler layers are evenly and thoroughly dispersed in a continuous polymeric matrix. The

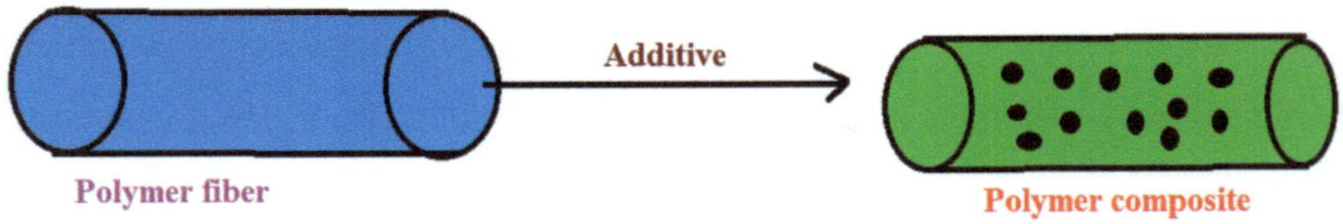

Fig. 2.1 Shows the schematic diagram of the polymer nanocomposite

nanofillers keep their crystal structure original in intercalated polymer nanocomposites. Because the nanofillers are intercalated within the polymer chains they generate a sandwich-like structure, so the distance between the sheets, interlayers, or planes is more than the original state. Because there is a lot of surface area interaction between the NPs and matrix, exfoliated nanocomposites have the maximum reinforcement of nanofillers. Since the disorder arrangement is found in the polymeric matrix the original structure of nanofillers gets lost in this type. As compared to intercalated nanocomposites, exfoliated polymer nanocomposites have significantly higher interfacial contacts between the polymeric chains and nanofillers [15, 16]. There are three types of polymer nanocomposites: zero-dimensional, one-dimensional, and two-dimensional, depending on the nanofillers used. Zero-dimensional nanomaterials include quantum dots (CdS, CdSe, Snte, ZnS, etc.) and spherical nanoparticles of metals (Au, Ag, Cu, Fe, etc.) and metal oxides (Fe2O3, CuO, ZnO, TiO_2, etc.). Carbon nanofibers, carbon nanotubes, polyaniline nanofibers, cellulose nanofibers, and nanowhiskers are a few examples of one-dimensional nanomaterials [15, 17, 18].

2.6 Synthesis Techniques

PNCs can be created via a number of methods, the most well-known of which are template synthesis (sol–gel approach), melt intercalation, in situ intercalative polymerization, and solution-driven intercalation [19]. Polymer matrix nanocomposites can be made by both mechanical and chemical methods. The uniform and homogenous dispersion of nanoparticles in the polymer matrix is one of the primary problems in the production of polymer nanocomposite. The characteristics of nanocomposites diminish due to the tendency of nanofillers to congregate and form micron-sized filler clusters that restrict the dispersion of nanoparticles in the polymer matrix. Numerous attempts have been undertaken by researchers to evenly and uniformly disseminate nanofillers in the polymer matrix through chemical reactions, challenging polymerization processes, or surface modification of filler materials [20]. The polymer itself is a key component of polymer matrix nanocomposites. Polymer matrix nanocomposites are made using a wide variety of polymers. The list of these polymers is below [21].

- Thermoplastics
- Thermosets
- Elastomers
- Natural and Biodegradable polymers

The synthesis technique/methods usually tailored for the synthesis of polymers are as described

i. **Intercalation Method**

In this method of preparation, the nanoplates such as clay materials are distributed uniformly inside the matrix of the respective polymers. In this method, necessary modifications are carried out on the surface of nanoplates to form nanocomposites. Clay is a widely chosen nanomaterials used to fill the polymer matrix to improve its various macroscopic quantities [22–24]. This involves stiffness, shrinkage, and flammability when integrated into polymer matrices. The entropy and enthalpy factors sometimes affect the outcome of such interactions [25]. These plate-like nanofillers are generally distributed uniformly during the process of intercalation which is nothing but a top-down approach. This helps to bring necessary modifying to the outer surfaces of nanoplatelets. When the chains of polymers get diffused between the spacing of layered structures results in an interacted morphology. The methods listed below are tailored to this approach to distribute the nanoplatelets evenly [26].

ii. **Chemical technique**

This approach uses in-situ polymerization, in which a monomeric solution is mixed with nanoparticles to initiate a polymerization reaction. This technique involves swelling of nanoplatelets by inserting the solution of monomers and then performing further polymerization. Further polymerization of monomer leads to interpenetrating of the platelets. In a monomer solution, the nanoplatelets swell, and a polymerization process creates a polymer between the intercalated sheets.

iii. **Mechanical technique**

In this method, the nanoplates are allowed to dissolve in the respective solvent by mechanical strategy. However, when the polymer dissolves in the respective co-solvent a solution is formed. In this technique when the polymer component gets penetrated by the nanoplates, the respective sites which are being occupied by the polymers get replaced [20, 27]. Due to direct intercalation, polymers along with nanoplatelets occur in this approach. When these two solutions are combined, the polymer chains present in the respective solution get intercalated to form layers of nanoplates and hence replace the solvent.

iv. **Melt intercalation technique**

Melt intercalation is a promising technique, it is usually a sub-type interaction where the components (polymer and nanoplates) get annealed and thereby intermix their melt solution.

Using this technique, molten polymer matrix and nanofillers (clays) are combined. This technique involves either static or shear annealing conditions for the combined product of polymer and nanofiber. This technique is very similar to the extrusion method/injection molding method entertained in industries. This technology allows the use of polymers that are not suited for in situ polymerization or solution intercalation and is compatible with modern industrial processes like extrusion and injection molding as mentioned. In order to add nanofillers to the polymer solution, a viscus solution is created by melting polymer powder or pellets at a high shear rate and high temperature. Compression molding, injection molding, or fiber production techniques can be used to construct the final shape of components. The typical composition of the melt-intercalation technique is a molten layered silicate, and neither a solvent nor a polymer matrix is necessary. Traditional techniques, including injection molding or extrusion, employ the mechanical mixing of a high-temperature thermoplastic polymer with an organophilic clay [28]. PNCs are produced in this stage by the exfoliation of polymer chains. The technique is by far the greatest option for creating thermoplastic NCs and can be a significant fix in the event that in situ polymerization fails or is caused by the wrong choice of polymer. This approach faces significant obstacles, such as its limited applicability to polyolefins, which make up the majority of utilized polymers [29–31].

v. **In-situ polymerization**

Since low-molecular-weight monomer solution can easily seep in between layers and cause swelling, during the in situ polymerization process nanofillers get swelled up. In this approach, polymerization of the monomers gets affected either by heat, UV light, or initiator. Figure 2.2 shows the process of polarization and production of polymer nanocomposite materials. This leads to the formation of exfoliated nanocomposite/intercalated nanocomposites. The resulting mixture is polymerized by thermal processing, radiation, diffusion of an initiator, or organic initiator. Exfoliated or intercalated nanocomposites are the two types that can be produced once the monomer has polymerized between the interlayers. A comparable technique is in situ template synthesis. With this technique, polymer chains are used to help create the clay layers. Aqueous solutions are used to dissolve the polymer matrix and clay layers, and gel is typically refluxed at a high temperature. Clay layer nucleation and growth occur on the polymer chains at high temperatures, trapping the polymer chains inside the clay layers. This method's only flaw is that high-temperature synthesis results in polymer decomposition. Another way to control platelet size is through in situ template synthesis, wherein the polymer matrix and platelets are refluxed in an aqueous solution at a high temperature. This allows the polymer chains to penetrate the platelets and control their size and nucleation. The drawback of this approach is that it can lead to the polymer chains breaking down at high temperatures. Using polysaccharide and polypyrene in the presence of COOH multi-walled carbon nanotubes, Creation of a nanocomposite to investigate the electrochemical characteristics of the materials [32, 33].

vi. **Direct mixing of the respective polymer and nanofillers**

Mixing of the matrix of the polymer with nanofillers directly can be furnished by the top-down process. This directly relies on the breakdown of aggregated nanofillers during mixing [34, 35]. In general, some techniques may be emphasized for mixing polymers and nanofillers and hence preparation of good nanocomposites. Among all, one is known as (i) Melt compounding, where a combination of polymers and nanofillers takes place above the glass transition temperature (without the actual involvement of any extra solvents). The alternative method, which is also known as the solvent method or solution involves combining polymers and hence the nanofillers using a solution and solvents.

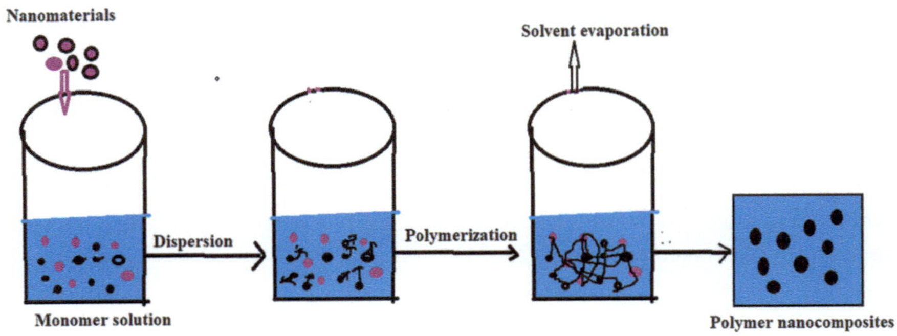

Fig. 2.2 Shows the polymerization process

vii. **Melt compounding**

In this process, nanofillers get added to the respective polymers just above the glass transition temperature. Through the employment of viscus drag, this type of approach induces shear stress (hydrodynamics force) in the polymer melt, which is then used to break up nanofiller clumps and promote homogenous and uniform nanofiller dispersion in the polymer matrix.

viii. **Solvent method**

In this process, the polymer gets dissolved within the co-solvent while the nanoparticles are disseminated in the solvent. By employing the solvent coagulation method/or by tailoring the solvent evaporation process the created nanocomposites are removed from the solvent. This method allows for better dispersion of the reinforcing materials within the polymer matrix, resulting in improved mechanical, thermal, and other properties of the composite. In comparison to the respective melt compounding technique, the shear stress lowers and helps to create matrix polymers. To break down the nanofiller aggregates, the nanofillers are first sonicated into the solvent. One of the aforementioned techniques is used to create polymer nanocomposites, which are then processed using customary manufacturing techniques as injection molding, calendaring, casting, compression molding, blow molding, rotational molding, extrusion molding, thermoforming, etc. [36].

ix. **Sol–Gel Method**

The bottom-up Sol–gel method functions in complete opposition to all previous strategies. A sol–gel relationship starts with sol and ends with gel. Solid nanoparticles are colloidally suspended in a gel in a monomer solution, which is a three-dimensional web of connections created by the phases [37]. This technique involves dispersing solid nanoparticles in a monomer solution to create a colloidal suspension of solid nanoparticles (sol), which then undergoes polymerization and hydrolysis to create a gel that connects the phases. The liquid is covered in the 3D network of polymer nanoparticles. As a nucleating agent, the polymer encourages the development of multilayer crystals. The polymer seeps between layers as the crystals grow, creating nanocomposite. In the sol–gel process, the application of an aqueous or semi-solid solution containing polymer and silicate as constituent units cause the production of clay minerals within the polymer matrix. Nucleation is the first step in processing, after which the polymer inclusion, which is encased between the layers, aids in the growth of inorganic host crystals. This approach works best for creating double-layered hydroxide-based NCs and is substantially less effective for producing layered silicates [38]. This technique involves dispersing the fillers or particles in a sol, or monomer solution, which permeates the layers of particles. It causes a network of interconnected polymers and nanoparticles to develop, which is referred to as gel. Consequently, polymer aids in the initiation and development of nanomaterials. This strategy falls within the bottom-up strategy group [20, 39]. To create a nanocomposite membrane, created a Schiff's base using chitosan and salicylaldehyde in the presence of TiO_2 nanofillers [40, 41].

2.7 Processing of the Polymer Composites

The processing of polymer composites refers to the series of steps and techniques involved in the manufacturing or fabrication of composite materials composed of polymers and reinforcing materials. Polymer composites are materials that combine a polymer matrix (such as a plastic resin) with reinforcing materials (such as fibers or particles) to create a material with improved mechanical, thermal, or other performance properties. The kind of polymer matrix, the reinforcing material, the desired final product properties, and the intended application are some of the factors that influence the processing technique selection [42].

The processing of polymer composites typically involves several key steps: (i) Matrix Selection: Choosing a suitable polymer matrix is the first step. Common polymer matrices include thermosetting resins (e.g., epoxy, polyester, phenolic) and thermoplastic resins (e.g., polyethylene, polypropylene), (ii) Reinforcement Selection: Selecting appropriate reinforcing materials, such as fibers (e.g., glass, carbon, aramid) or particles, to enhance specific properties like strength, stiffness, or toughness, (iii) Mixing/Compounding: Combining the polymer matrix and reinforcing materials through a mixing or compounding process to create a homogeneous mixture. This can be done using various methods, including melt mixing, solution mixing, or dispersion, (iv) Molding/Forming: Shaping the composite material into the desired form. Common methods include injection molding, compression molding, extrusion, pultrusion, and filament winding, depending on the application and complexity of the product, (v) Curing/Polymerization: For thermosetting polymers, initiating and completing the curing or polymerization process to solidify and harden the composite. This step often involves heat or chemical reactions, (vi) Post-Curing: Some composites may undergo a post-curing process to further enhance their properties, and (vii) Finishing/Coating: Applying surface finishes or coatings to improve aesthetics, protect against environmental factors, or enhance specific properties [43–46]. The choice of processing methods depends on the desired properties of the final product, the complexity of the part, and economic considerations. Each step in the processing of polymer composites plays a crucial role in determining the performance and characteristics of the end product. The goal is to optimize the combination of matrix and reinforcement, as well as the processing parameters, to achieve the desired balance of properties for a given application. There are various processing methods for polymer composites, each tailored to specific applications and desired properties. Here are some common processing methods with brief explanations:

(a) **Injection Molding**

Injection molding is a widely used manufacturing process for the production of polymer composite parts. It involves injecting molten polymer material into a mold cavity, allowing it to cool and solidify, and then ejecting the finished product. When it comes to polymer composites, injection molding can be a highly effective processing technique for creating complex shapes and intricate parts. Process: Molten polymer is injected into a mold cavity, where it solidifies to form the desired shape.

The detail of the injection molding of polymer composite processing technique can be explained as follows:

Choose a polymer matrix and reinforcing materials suitable for injection molding. Common polymer matrices include thermoplastics like polypropylene, polyethylene, or thermosetting resins such as epoxy. If using reinforcing materials like fibers or particles, ensure they are well-dispersed in the polymer matrix. This may involve compounding the materials in a separate process before injection molding. Design the mold to accommodate the desired part geometry. The mold typically consists of two halves, the cavity, and the core, which come together to form the final shape of the part. Set up the injection molding machine by loading the polymer composite material into the hopper. The material is then fed into a heated barrel, where it is melted to a suitable viscosity for injection. The molten polymer composite is injected into the mold cavity under high pressure. This pressure is maintained until the material fills the mold and begins to solidify. Allow the material to cool and solidify within the mold. Cooling times can vary depending on the material and part geometry. Once the part has solidified, the mold is opened, and the finished part is ejected. The cycle then repeats for the next shot. Injection molding is a high-speed process, making it suitable for large-volume production. Minimizes material waste compared to other manufacturing processes. Provides consistent and repeatable results, ensuring uniformity across production batches. Automation can be implemented, reducing the need for extensive labor. The application of this technique can be written as interior components, body panels, and various structural parts. Housings for electronic devices, toys, containers, other consumer products, and components for medical equipment. In summary, injection molding is a versatile and efficient method for producing polymer composite parts with enhanced mechanical properties, making it suitable for a wide range of applications in different industries.

(b) **Compression Molding**

Compression molding is a processing technique used for the production of polymer composite parts. This method involves the application of heat and pressure to a composite material in a mold cavity, causing the material to flow and conform to the mold's shape. Compression molding is particularly suitable for larger, more complex parts and parts that require high strength. The detail of the compression molding process for polymer composites is given below.

Choose a suitable polymer matrix and reinforcing materials for compression molding. Common polymer matrices include thermosetting resins like epoxy, and reinforcing materials may include fibers or particles. Pre-cut or pre-weigh the composite material into the desired shape and size. If using reinforcing materials, ensure they are evenly distributed within the polymer matrix. Design a mold that reflects the final shape of the part. The mold typically consists of two halves, the cavity and the core, and it may include additional features such as inserts or preforms. Place the composite material into the mold cavity. Close the mold, bringing the two halves together to enclose the composite material. Apply heat to the mold, causing the composite material to soften and flow. The heat also initiates the curing or cross-linking of the polymer matrix, depending on the type of resin used. Apply high pressure to the mold to compact the material and ensure that it conforms to the mold cavity. Maintain heat and pressure until the polymer matrix is fully cured. This step is crucial for achieving the desired mechanical properties of the composite. Allow the mold to cool, solidifying the part. Once the part has sufficiently cooled, open the mold and remove the finished part.

This technique is well-suited for producing large and complex parts. Results in parts with high strength and stiffness due to the applied pressure during molding. Generally, produces less material waste compared to some other molding techniques. Tooling costs can be lower compared to injection molding for certain applications. The application of the above-mentioned technique can be written as components such as aircraft panels and structural parts, exterior body panels, engine components, interior parts, sporting goods, helmets, and components for recreational equipment. In summary, compression molding is a versatile technique that can be employed for a wide range of polymer composite applications, especially where high strength and large part sizes are required. It is important to optimize the process parameters, including temperature, pressure, and curing time, to achieve the desired properties in the final composite part.

(c) **Resin Transfer Molding (RTM)**

Resin Transfer Molding (RTM) is a processing technique used for the production of high-quality polymer composite parts. This method is particularly well-suited for creating complex, large-scale components with intricate designs. RTM involves the infusion of liquid resin into a mold containing a pre-placed fiber reinforcement. Here, the detail of the resin transfer molding process for polymer composite is given below

Design a two-part mold to create the desired shape of the final part. The mold is typically made of metal or composite materials. Place the fiber reinforcement (usually in the form of a preform or mat) into the mold cavity. The reinforcing material may consist of glass fibers, carbon fibers, or other materials depending on the application. Close the mold to secure the fiber reinforcement in place. Liquid resin, typically a thermosetting polymer such as epoxy or polyester, is injected into the mold under low pressure. The resin flows through the fiber reinforcement, wetting out the fibers and filling the mold cavity. Allow the resin to cure or harden. This may involve the application of heat or other curing methods, depending on the resin system used. Allow the part to cool and solidify within the mold. Once the part has cured and solidified, open the mold and remove the finished composite part. It is well-suited for complex shapes and intricate designs due to the ability to precisely place reinforcing materials. Provides good control over fiber orientation and distribution, resulting in consistent mechanical properties. Operates at lower pressures compared to some other molding processes, making it suitable for lightweight structures. Tooling costs can be lower compared to processes like injection molding for certain applications.

The applications of polymer processing in this method can be mentioned as body panels, structural components, interior parts, and aircraft components, including interior and structural parts. Production of large-scale wind turbine components such as blades. In the sector of sports and recreation, bicycle frames, sporting goods, and recreational equipment are important applications. Therefore, in summary, resin transfer molding is a versatile technique that allows for the efficient production of polymer composite parts with good strength-to-weight ratios. It is suitable for applications where a combination of strength, complex geometry, and low weight is desired. As with any manufacturing process, careful optimization of parameters is crucial to achieving the desired properties in the final composite part.

(d) **Filament Winding**

Filament winding is a manufacturing technique used for the production of composite structures, particularly those with a cylindrical or tubular shape. This process involves winding continuous fibers, typically in the form of rovings or tows, onto a rotating mandrel or mold while simultaneously applying a polymer resin. Filament winding is commonly used in industries such as aerospace, automotive, and oil and gas for manufacturing components like pipes, pressure vessels, and rocket motor casings. Here's an overview of the filament winding process:

Start with the preparation of a rotating mandrel, which serves as the form around which the composite structure will be built. The mandrel is often coated with a release agent to facilitate the removal of the finished part. Use continuous fibers, such as glass, carbon, or aramid, in the form of rovings or tows. These fibers are typically impregnated with a polymer resin. Apply the polymer resin to the fibers either before winding (prepreg) or during the winding process. The resin can be applied using various methods, such as resin baths, impregnation rollers, or resin injection. The mandrel rotates, and the continuous fibers are precisely wound onto the mandrel in a specific pattern. The winding angle and tension are controlled to achieve the desired mechanical properties in the final composite structure. Build up multiple layers of fiber reinforcement to achieve the desired thickness and strength. The winding process may involve changing the winding angle and the orientation of the fibers to optimize the structural performance of the final part. Allow the resin to cure or set. Depending on the resin system, this may involve ambient curing or the application of heat. Once the composite structure has cured and solidified, remove the mandrel. The resulting tubular or cylindrical composite part retains the shape of the mandrel. Some applications may require additional post-curing to achieve optimal material properties. Finishing operations, such as trimming and surface treatments, may also be performed. Produces lightweight structures with high strength due to the continuous fiber reinforcement. Allows for precise control over fiber orientation, resulting in consistent mechanical properties. Efficient for producing cylindrical or tubular structures, such as pipes and pressure vessels. The automation process can be automated for high-volume production. Suitable for a variety of fiber types and resin systems, offering versatility in material selection. The applications of these techniques can be written as rocket motor casings, aircraft components, drive shafts, suspension components, pipes, pressure vessels, blades, and towers. Therefore, the filament winding is a versatile and efficient process for manufacturing composite structures with excellent strength-to-weight ratios. Its ability to precisely control fiber orientation makes it well-suited for applications where specific structural properties are crucial.

(e) **Pultrusion**

Pultrusion is a continuous manufacturing process used for the production of high strength, fiber-reinforced polymer composite profiles with a constant cross-section. This technique is commonly employed to manufacture structural shapes such as rods, tubes, and beams. Pultrusion offers advantages in terms of efficiency, cost-effectiveness, and the ability to produce long, continuous profiles. Here's an overview of the pultrusion process below:

The raw materials used in pultrusion typically include reinforcing fibers (such as glass, carbon, or aramid), a polymer resin matrix (commonly polyester, vinylester, or epoxy), and additives for desired properties. Continuous fibers, usually in the form of roving, are pulled from spools and fed into the pultrusion machine. The continuous fibers are pulled through a resin bath or impregnation station where they are thoroughly wetted with the polymer resin. This process ensures that the reinforcing fibers are uniformly coated with resin. The wetted fibers are then guided through a forming die or shaping die, which imparts the desired cross-sectional shape to the pultruded profile. The profile, now impregnated with resin and having the desired cross-sectional shape, passes through a heated die. The heat initiates the curing or polymerization process of the resin, causing it to harden and solidify. The cured composite material is continuously pulled through the pultrusion machine by a set of grippers or a caterpillar-like pulling mechanism. The pultruded profile is cut to the desired length, and the process can include additional cutting and shaping operations as needed. The pultruded profile may pass through a cooling section to cool and stabilize the material. Depending on the application, the pultruded profile may undergo additional surface treatments or coatings for enhanced performance. Pultruded profiles exhibit excellent strength-to-weight ratios due to the continuous fiber reinforcement. The process allows for the production of profiles with consistent and precise cross-sectional shapes. Pultrusion is a continuous process that offering high production rates and efficiency. Profiles with various shapes and sizes can be produced by adjusting the forming die. Pultruded composites are often corrosion-resistant, making them suitable for various applications in harsh environments. The application includes structural components, profiles for bridges, grating systems, insulating components, cable trays, conveyor systems, and cooling tower components. Therefore, the pultrusion is a widely used technique in industries where continuous, high strength, and corrosion-resistant composite profiles are required. Its ability to efficiently produce consistent profiles makes it suitable for a variety of structural and functional applications.

(f) **Extrusion**

Extrusion is a polymer composite processing technique that involves forcing a molten or viscous polymer composite material through a die to produce a continuous profile with a specific cross-sectional shape. This process is widely used for manufacturing various plastic and composite products, such as pipes, tubing, sheets, and profiles. Extrusion is versatile and can be adapted for the production of polymer composites by incorporating reinforcing materials or additives into the polymer matrix. Here's an overview of the extrusion process for polymer composites is given below.

The polymer composite material is prepared by blending the polymer matrix (e.g., thermoplastic resin) with reinforcing materials (e.g., fibers or fillers) and other additives to achieve desired properties. The composite material may be pelletized to facilitate feeding it into the extruder. The prepared composite material is fed into the hopper of an extruder, which consists of a rotating screw within a heated barrel. The heat and pressure generated by the screw melt and mix the composite material. The molten or viscous composite material is forced through a die that shapes the material into the desired cross-sectional profile. The die can be designed to produce various shapes, such as round, rectangular, or complex profiles. The extruded profile is rapidly cooled to solidify the material. Cooling methods may include air cooling, water cooling, or a combination of both. The extruded profile may pass through sizing tools to ensure accurate dimensions. It is then cut to the desired length. Additional post-processing steps, such as surface treatment, coating, or other finishing operations, may be performed depending on the application requirements.

Extrusion allows for the continuous production of long profiles with consistent cross-sectional shapes. The process is suitable for high-volume production with relatively high extrusion rates. Extrusion can accommodate a variety of polymer matrices, reinforcing materials, and additives to achieve specific material properties. Complex cross-sectional profiles can be extruded with precision. Minimizes material waste as it is a continuous process.

The applications include extruded profiles for window and door frames, decking, and other construction components. Extruded plastic pipes for various applications, profiles for automotive trims, seals, and other components. Extruded profiles for cable management, conduits, insulating components, furniture, and appliances. Therefore, the extrusion is a versatile and widely used processing technique in the polymer composite industry, offering efficiency and flexibility for the production of a broad range of profiles used in different applications. The method is particularly well-suited for continuous production and can be adapted to various composite formulations to meet specific performance requirements.

(g) **Hand lay-up**

Hand lay-up is a manual and traditional method for manufacturing polymer composite components. It is a relatively simple and flexible process that is commonly used for producing low to medium volumes of composite parts with various shapes and sizes. In hand lay-up, the composite materials, including a polymer resin matrix and reinforcing fibers or fabrics, are manually applied layer by layer to a mold. Here's an overview of the hand lay-up process is given below-

Begin with the preparation of a mold that defines the shape and dimensions of the final composite part. The mold can be made from materials such as metal, fiberglass, or composites.

Apply a release agent to the mold surface to facilitate the easy removal of the cured composite part. If a smooth and finished surface is desired on the outer layer of the composite part, a gel coat may be applied to the mold before the lay-up process. Start layering the composite materials onto the mold manually. This involves placing dry reinforcement materials (such as fiberglass mats or woven fabrics) and wetting them with the polymer resin. Apply the polymer resin (usually a liquid thermosetting resin like polyester, epoxy, or vinylester) onto the dry reinforcement materials to wet them thoroughly. This process can be done using brushes or rollers. Continue the layering process, adding additional layers of reinforcement materials and wetting them with resin until the desired thickness and structure are achieved. This manual build-up allows for flexibility in controlling material placement and orientation. Use tools such as rollers or brushes to eliminate air bubbles and ensure good adhesion between layers. This step is crucial for achieving a strong and void-free composite. Allow the composite to cure or harden. The curing process may involve ambient curing or the application of heat, depending on the resin system used. The process is suitable for low-volume production, and the tooling costs are relatively low compared to other methods. Allows for flexibility in changing part designs and adjusting material lay-up during production. The equipment required for hand lay-up is simple and doesn't involve complex machinery. Well-suited for the production of large and complex parts. Hand lay-up is often used for prototyping and small-scale production where flexibility is crucial. Production of boat hulls, decks, and other marine components. Manufacturing of custom components for various industries, such as automotive, aerospace, and construction. The application of this technique includes the production of architectural features, structures, manufacturing of sports equipment, and recreational products. Therefore, the hand lay-up is a versatile method, it may not be as suitable for high-volume production due to the manual labor involved and the potential for variations in quality. Other automated methods, such as resin transfer molding or pultrusion, are often chosen for larger production runs where consistency and efficiency are critical.

(h) **Autoclave Processing**

Autoclave processing is a polymer composite manufacturing technique that involves the use of an autoclave, a pressurized vessel, to cure and consolidate composite materials. This method is particularly employed for high-performance composite structures where precise control over temperature, pressure, and curing conditions is critical. Autoclave processing is often used in the aerospace and high-end automotive industries for manufacturing components with stringent quality and performance requirements. Here's an overview of the autoclave processing technique is given below.

The composite materials, including the polymer matrix (such as epoxy resin) and reinforcing materials (fibers or fabrics, e.g., carbon or glass), are prepared and laid up layer by layer to form the composite structure. The lay-up is placed into a mold that defines the final shape and dimensions of the composite part. The mold is often made of metal or composite materials. A vacuum bag is applied over the lay-up to remove air and assist in consolidating the layers. The vacuum bag helps ensure proper contact between the materials and removes voids that could weaken the final composite. The lay-up and vacuum bagged assembly are placed inside an autoclave, which is a pressurized vessel. The autoclave is sealed, and a vacuum is applied to further remove any remaining air from the composite. The autoclave is pressurized with a controlled level of pressure, and the temperature is raised to the specified curing temperature for the resin system used. The combination of heat and pressure accelerates the curing process and ensures thorough consolidation of the composite materials. The composite part is subjected to the curing temperature and pressure conditions for a predetermined period. During this time, the polymer resin undergoes curing or cross-linking, resulting in the final hardened composite structure. After the curing cycle is complete, the autoclave is gradually cooled to ambient temperature. This controlled cooling helps prevent thermal stresses in the composite part. Some high-performance applications may require additional post-curing steps to achieve optimal material properties. Autoclave processing allows for precise control over temperature and pressure, resulting in high-quality, void-free composite parts. Ensures consistent material properties and part quality, especially important for aerospace applications.

Provides excellent control over fiber alignment, leading to optimized mechanical properties. Suitable for manufacturing complex and intricate shapes with high dimensional accuracy.

The applications include the production of aircraft components, including structural parts and interior components. Manufacturing of high-performance automotive components, such as racing car parts. Therefore, the autoclave processing is a precise and reliable method for producing high-performance polymer composite parts with superior mechanical properties. While it is an effective technique, it is typically more expensive and time-consuming compared to other methods, making it suitable for applications where the highest quality and performance are paramount.

(i) **Vacuum Bagging**

Vacuum bagging is a composite manufacturing technique used to fabricate polymer composite parts, particularly those with fiber reinforcement, in a controlled environment. This process is often employed in conjunction with other molding methods like wet lay-up or prepreg lay-up. Vacuum bagging helps remove air and excess resin, ensuring a tight bond between the layers and improving the overall quality of the composite part. Here's an overview of the vacuum bagging process is given below.

Begin with the preparation of the composite materials. This typically involves layering the reinforcing materials (such as fibers or fabrics) and the polymer matrix (resin) in the desired orientation and order. This can be done using techniques like wet lay-up (applying resin manually) or prepreg lay-up (using pre-impregnated resin sheets). Place the lay-up into the mold or onto a tool surface, ensuring that it conforms to the desired shape. Cover the lay-up with a bleeder cloth and a breather fabric. The bleeder cloth absorbs excess resin, and the breather fabric provides a pathway for air to escape. Cover the entire assembly with a vacuum bagging film, creating an airtight seal around the lay-up. Seal the edges of the vacuum bag to create an airtight environment. A vacuum pump is connected to the bag, and air is evacuated, creating a vacuum. With the vacuum applied, the resin is either infused into the lay-up (in the case of resin infusion processes) or consolidated through the application of pressure and vacuum (in processes like vacuum bagging of wet lay-up or prepreg lay-up). The vacuum removes air and excess resin, ensuring the layers are well-bonded. Allow the part to cure or set. The curing process may involve the application of heat, depending on the resin system used. Some applications may require post-curing to achieve optimal material properties. Additionally, finishing operations such as trimming and surface treatments may be performed.

Provides excellent consolidation of composite layers, reducing voids and improving overall part quality. Allows for precise control of resin content in the composite, reducing excess resin that can add weight and decrease performance. Well-suited for parts with complex shapes and intricate details. Particularly cost-effective for small to medium production runs without the need for expensive autoclaves.

The application in the aeroplane components with complex shapes, such as fairings and interior parts. Again, some other applications such as blade manufacturing for wind turbines, lightweight structural components, body panels, boat hulls, decks, bicycles, sporting goods, and other recreational equipment.

Therefore, the vacuum bagging is a versatile process that contributes to the production of high-quality, lightweight polymer composite parts. It is particularly valuable in applications where precise control over resin content and consolidation of layers are critical for achieving desired material properties. These processing methods offer flexibility in terms of part geometry, production volume, and material properties. The choice of method depends on factors such as cost, complexity, required performance, and production volume [47–49].

(j) **Electrospinning technique**

Electrospinning is a versatile technique used for the synthesis of polymer composites with unique properties. This method involves the application of an electric field to a polymer solution or melt to create ultrafine fibers with diameters in the nanometer to micrometer range. These fibers can be collected to form nonwoven mats or membranes, and when combined with other materials, such as nanoparticles or other polymers, they create composite materials with enhanced properties. The

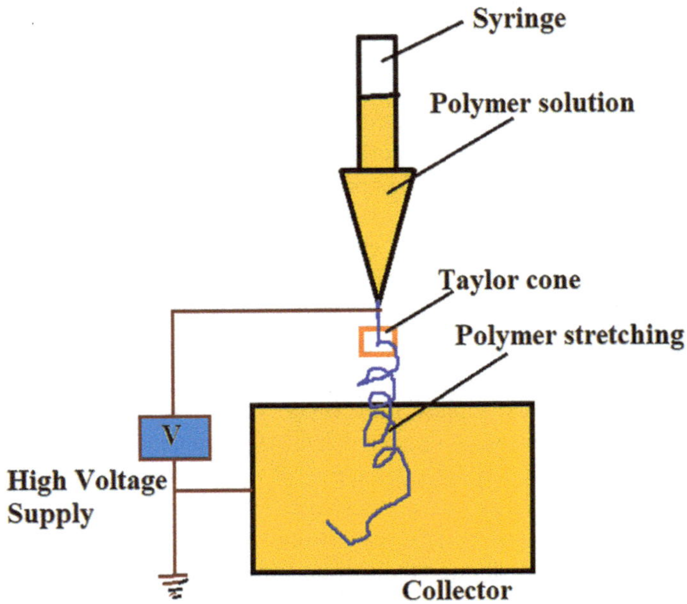

Fig. 2.3 Shows the electrospinning technique for synthesis of polymer composites

detail of the electrospinning technique for synthesizing polymer composites is given in Fig. 2.3 along with an explanation.

Polymer Solution Preparation: Select a suitable polymer that can be dissolved or melted. Common polymers include polyvinyl alcohol (PVA), polyacrylonitrile (PAN), polycaprolactone (PCL), and others. Dissolve the polymer in a solvent to create a homogeneous solution. The choice of solvent depends on the polymer and its properties.

Addition of Nanoparticles or other Polymers: Introduce nanoparticles (e.g., carbon nanotubes, graphene, silica) or other polymers into the solution to form a composite. The addition of these components enhances the mechanical, electrical, or thermal properties of the resulting material.

Electrospinning Setup: Set up an electrospinning apparatus, which typically consists of a syringe pump, a high-voltage power supply, and a grounded collector. Connect a syringe containing the polymer solution to a metallic needle or spinneret. Apply a high voltage (usually in the range of 1–30 kV) to the solution, creating an electrically charged jet at the tip of the spinneret.

Fiber Formation: The electric field overcomes the surface tension of the polymer solution, leading to the formation of a fine, continuous jet. The solvent evaporates during flight, and the charged polymer fibers are collected on a grounded or oppositely charged collector.

Collection of Nanofibers: Collect the nanofibers on a rotating drum, flat plate, or other collector surface, forming a nonwoven mat or membrane.

Post-Treatment: Optionally, perform post-treatments such as cross-linking or heat treatment to enhance the stability and properties of the electro-spun composite.

Advantages of Electrospinning for Polymer Composites

High Surface Area: Electrospun fibers have an extremely high surface area, making them suitable for various applications, such as filtration, sensing, and tissue engineering.

Tunable Properties: The process parameters can be adjusted to control the diameter, alignment, and morphology of the fibers, allowing for the customization of material properties.

Incorporation of Nanoparticles: Electrospinning facilitates the uniform dispersion of nanoparticles within the polymer matrix, leading to improved composite properties.

Versatility: The technique is compatible with a wide range of polymers and additives, offering versatility in designing composite materials with diverse functionalities.

Biomedical Applications: Electrospun polymer composites are widely used in biomedical applications, including drug delivery systems, wound healing, and tissue engineering. In conclusion, electrospinning is a powerful technique for synthesizing polymer composites with tailored properties, making it applicable in various fields, including materials science, nanotechnology, and biomedical engineering.

(k) **Solution casting**

The solution casting method is a common technique for synthesizing polymer composites. This method involves dissolving the polymer and any additional components (such as nanoparticles or other polymers) in a suitable solvent to form a homogeneous solution. The solution is then cast or spread onto a substrate, and the solvent is evaporated, leaving behind a solid composite material. The detail of the solution casting process for polymer composites is given below-

Polymer Selection: Choose a polymer that is soluble in a selected solvent. Common polymers used in solution casting include polyvinyl chloride (PVC), polyethylene (PE), polypropylene (PP), and others.

Addition of Reinforcements or Fillers: Introduce reinforcing materials such as nanoparticles (e.g., carbon nanotubes, graphene), micro-sized fillers, or other polymers to enhance the properties of the composite.

Solvent Selection: Select a solvent that can dissolve the polymer and any additional components. The choice of solvent depends on the polymer's solubility and compatibility with the other materials.

Solution Preparation: Mix the polymer and other components thoroughly in the selected solvent to form a homogeneous solution. This step is critical to achieving a uniform distribution of reinforcements in the final composite.

Casting or Coating: Pour or spread the solution onto a flat substrate or mold. The thickness of the cast film can be controlled to influence the properties of the resulting composite.

Solvent Evaporation: Allow the solvent to evaporate from the solution, leaving behind a solid film or composite. This can be done at room temperature or under controlled conditions such as elevated temperature or reduced pressure.

Post-Curing or Annealing (Optional): Depending on the polymer used, the composite may undergo post-curing or annealing processes to improve its mechanical, thermal, or chemical properties. This step is optional and depends on the specific requirements of the application.

Film or Composite Collection: Once the solvent has evaporated and the composite has solidified, the resulting film or composite can be peeled off from the substrate or used directly, depending on the application.

2.8 Advantages of Solution Casting for Polymer Composites

The solution casting method is relatively simple and cost-effective compared to other manufacturing techniques. The process allows for the uniform distribution of reinforcements throughout the polymer matrix, leading to a homogeneous composite material. Solution casting is versatile and can be used for a wide range of polymers and reinforcing materials, making it applicable in various industries. The method is easily scalable for large-scale production of composite materials. Lastly, a wide range of applications includes that solution casting is suitable for producing thin films, coatings, and bulk materials, making it applicable in diverse fields such as packaging, electronics, and biomedical devices.

In conclusion, the solution casting method is a versatile and widely used technique for synthesizing polymer composites with tailored properties. It offers simplicity, scalability, and the ability to incorporate various reinforcements to achieve desired material characteristics.

(1) **Melt blending method**

The melt blending method is a common technique for synthesizing polymer composites. This process involves mixing and blending polymer matrices with various reinforcing materials or additives in a molten state. Melt blending is particularly suitable for polymers with relatively low melting points. The detail of the melt blending process for polymer composites is discussed below.

Polymer Selection: Choose a polymer matrix that has a melting temperature suitable for the melt blending process. Common polymers include polyethylene (PE), polypropylene (PP), and polyethylene terephthalate (PET).

Addition of Reinforcements or Additives: Incorporate reinforcing materials, such as fibers (e.g., glass fibers), nanoparticles (e.g., carbon nanotubes, graphene), or other additives (e.g., plasticizers, stabilizers) into the polymer matrix. The type and amount of reinforcement depend on the desired properties of the composite.

Preparation of Feedstock: Preprocess the polymer and reinforcing materials to create a feedstock suitable for the melt blending process. This may involve pelletizing or granulating the polymer and additives to ensure uniform distribution during blending.

Melt Blending Equipment: Use appropriate melt blending equipment, such as twin-screw extruders or single-screw extruders. These machines provide controlled heat and shear forces to melt and mix the polymer matrix and reinforcements.

Loading and Mixing: Feed the polymer matrix and reinforcing materials into the extruder, where they are melted and mixed under controlled temperature and shear conditions. The shear forces help distribute the reinforcing materials uniformly in the polymer matrix.

Extrusion: The molten composite is extruded through a die to form a continuous shape, such as pellets, sheets, or profiles. The extrusion process aids in further homogenizing the composite material.

Cooling and Solidification: The extruded composite undergoes a cooling process to solidify the material and set the desired shape.

Post-Processing (Optional): Depending on the specific requirements of the application, the melt-blended composite may undergo additional processing steps such as cutting, milling, or further shaping.

Advantages of Melt Blending for Polymer Composites

Melt blending is a continuous and efficient process, making it suitable for large-scale production. The method allows for a high degree of dispersion of reinforcing materials within the polymer matrix, leading to improved mechanical properties. Melt blending is applicable to a broad range of polymers and reinforcing materials, making it versatile for different applications. The process is generally cost-effective compared to some other composite manufacturing methods. Melt blending can be easily scaled up for industrial production. By adjusting processing parameters and the type/amount of reinforcements, the properties of the resulting composite can be tailored to specific application needs. In summary, melt blending is a versatile and widely used method for synthesizing polymer composites with improved properties. It is particularly well-suited for large-scale production and allows for the customization of composite materials for various applications.

2.9 Applications

Polymer composites find a wide range of applications across various industries due to their versatility and ability to combine the desirable properties of different materials [50–59]. Here are detailed explanations of some potential applications in different sectors:

- **Aerospace**: Polymer composites, particularly carbon fiber-reinforced composites, are used in aerospace for their high strength-to-weight ratio. They contribute to lightweight structures, fuel efficiency, and improved performance. It is applied in aircraft components and spacecraft parts.
- **Automotive**: Composites help reduce the weight of vehicles, leading to fuel efficiency and lower emissions. They also offer design flexibility, corrosion resistance, and impact absorption in automotive applications. It is used in the vehicle components, body panels, and interior parts.
- **Construction**: Composites can replace traditional construction materials like steel and concrete. They offer corrosion resistance, high strength, and durability. Fiber-reinforced composites are used in the construction of bridges and other infrastructure. It is used in building materials, bridges, and pipelines.
- **Marine**: Polymer composites, especially fiberglass-reinforced composites, are widely used in marine applications due to their corrosion resistance, high strength, and low weight. They contribute to improved buoyancy and fuel efficiency. It is used in boat hulls and components.
- **Renewable Energy**: Composites are crucial in renewable energy applications. For example, wind turbine blades use fiberglass or carbon fiber composites for their lightweight yet strong properties. Composites also contribute to the durability and efficiency of solar panels. It is used in wind turbine blades and solar panels.
- **Sports and Recreation**: Polymer composites are popular in sports equipment due to their ability to provide strength, flexibility, and lightweight properties. Carbon fiber composites are frequently used in high-performance bicycles and sports gear. It is used in sporting goods, bicycles, and helmets.
- **Electronics and Electrical**: Composites offer electrical insulation properties and can be designed with specific thermal conductivity. They are used in electronic enclosures to provide protection against electromagnetic interference (EMI) and radiofrequency interference (RFI). It is used in enclosures and circuit boards.
- **Medical**: Polymer composites are used in the medical field for their biocompatibility and ability to mimic natural tissues. They are employed in prosthetics and implants, providing strength while being lightweight. It is used in the prosthetics and orthopedic implants.
- **Consumer Goods**: Composites are used in various consumer goods for their aesthetic appeal, durability, and ability to be molded into intricate shapes. Carbon fiber composites, for example, can provide a high-tech and modern appearance. It is used in the furniture, luggage, and consumer electronics.

- **Defense and Military**: Polymer composites are used in the defense sector for their high strength, impact resistance, and lightweight properties. They contribute to the development of armored vehicles, aircraft components, and protective gear for military personnel. It is used in the military vehicle components and protective gear.

In each of these applications, the specific composite materials and manufacturing processes can vary, tailored to meet the unique requirements of the intended use. The adaptability and customizable nature of polymer composites make them invaluable in addressing diverse challenges across industries.

2.10 Conclusion

In conclusion, this review has provided a comprehensive overview of the synthesis and processing techniques employed in the development of polymer composites. The synthesis methods, ranging from intercalation method to sol–gel method, have demonstrated significant advancements in enhancing the mechanical, thermal, and functional properties of these materials. While the current literature presents a robust foundation, it is essential to address the existing limitations, such as [mention limitations], to propel the field forward. Future research should focus on propose future directions, embracing interdisciplinary collaborations, and leveraging emerging technologies. By continually pushing the boundaries of knowledge in polymer composite synthesis and processing, researchers can contribute to the development of innovative materials with widespread applications in [relevant fields]. As we navigate the evolving landscape of materials science, staying attuned to emerging trends and technological breakthroughs will be crucial for shaping the next chapter in the journey of polymer composite research. Advanced composite materials are currently investigated and employed with a higher interest in several industries. Conveying the idea that composites represent the future of materials, and their unique properties make them suitable for addressing emerging challenges. Emphasizing the ongoing need for research and development to unlock the full potential of composite materials. In summary, your statement underscores the need for a concerted effort involving multiple agencies, interdisciplinary collaboration, and continuous innovation to further develop and enhance composite technology. The recognition that composites offer unique advantages over traditional materials indicates a forward-looking perspective on the role of advanced materials in addressing future challenges.

2.11 Statements and Declarations

Funding: The authors declare that no funds, grants, or other support were received during the preparation of this manuscript.

Competing Interests: The authors have no relevant financial or non-financial interests to disclose.

Acknowledgements Authors would like to thank to host Institute for providing the digital library facility for collecting relevant data.

References

1. Jeon, I.Y., Baek, J.B.: Nanocomposites derived from polymers and inorganic nanoparticles. Materials **3**, 3654–3674 (2010)
2. Oladele, I.O., Adelani, S.O., Ganiu Agbabiaka, O., Hope Adegun, M.: Applications and disposal of polymers and polymer composites: a review. Eur. J. Adv. Eng. Technol. **9**, 65–89 (2022)
3. Meysam, T., Nuruldiyanah, K., Zahra, D.E., Yee, L., Ashen, G., Chandima, G.: Conventional and advanced composites in aerospace industry: technologies revisited. Am. J. Aerosp. Eng. **5**, 9–15 (2018)
4. De, S.K., White, J.R.: Short Fiber-Polymer Composites. Woodhead, Cambridge, UK (1996)
5. Atmakuri, A., Palevicius, A., Vilkauskas, A., Janusas, G.: Review of hybrid fibre-based composites with nanoparticles-Material properties and application. Polymers **12**, 2088–2118 (2020)
6. Aldousiri, B., Shalwan, A., Chin, C.W.: A review on tribological behaviour of polymeric composites and future reinforcements. Adv. Mater. Sci. Eng. **2013**, 1–8 (2013)
7. Mustafa, N.S., Omer, M.A., Garlnabi, E.M., Ismail, H.M.: Reviewing of general polymer types, properties and application in medical field. Int. J. Sc. Res. **5**, 212–221 (2016)
8. Kmetty, A., Bárány, T., Karger-Kocsis, J.: Self-reinforced polymeric materials: a review. Prog. Polym. Sci. **35**, 1288–1310 (2010)
9. Thori, P., Sharma, P., Bhargava, M.: An approach of composite materials in industrial machinery: advantages, disadvantages and applications. Int. J. Res. Eng. Technol. **2**, 350–355 (2013)
10. Ali, N., Zhang, B., Zhang, H., Zaman, W., Li, W., Zhang, Q.: Key synthesis of magnetic Janus nanoparticles using a modified facile method. Particuology **17**, 59–65 (2014)
11. Kessler, M.R.: Polymer matrix composites: a perspective for a special issue of polymer reviews. Polym. Rev. **52**, 229–233 (2012)
12. Masoud, E.M., Liu, L., Peng, B.: Synthesis, characterization, and applications of polymer nanocomposites. J. Nanomater. **2020**, 1–2 (2020)
13. Saeed, K., Khan, I., Ahmad, Z., Khan, B.: Preparation, analyses and application of cobalt-manganese oxides/nylon 6,6 nanocomposites. Polym. Bull. **75**, 4657–4669 (2018)
14. Schmidt, G., Malwitz, M.M.: Properties of polymer–nanoparticle composites. Curr Opin Colloid Interf. Sci. **8**, 103–108 (2003)
15. Ali, N., Zhang, B., Zhang, H., Zaman, W., Li, X., Li, W., et al.: Interfacially active and magnetically responsive composite nanoparticles with raspberry like structure; synthesis and its applications for heavy crude oil/water separation. Colloids Surf. A Physicochem Eng. Asp **472**, 38–49 (2015)
16. Hanemann, T., Szabo, D.V.: Polymer–nanoparticle composites: from synthesis to modern applications. Materials **3**, 3468–3517 (2010)
17. Wang, R., Zheng, S., Zheng, Y.P.: Polymer Matrix Composites and Technology, 1st edn. Woodhead Publishing, Cambridge, UK (2011)
18. Yan, L.T., Xie, X.M.: Computational modelling and simulation of nanoparticle self-assembly in polymeric systems: structures, properties and external field effects. Prog. Polym. Sci. **38**, 369–405 (2013)

19. Rehab, A., Salahuddin, N.: Nanocomposite materials based on polyurethane intercalated into montmorillonite clay. Mater. Sci. Eng., A **399**, 368–376 (2005)
20. Tanahashi, M.: Development of fabrication methods of filler/polymer nanocomposites: with focus on simple melt-compounding-based approach without surface modification of nanofillers. Materials **3**, 1593–1619 (2010)
21. Parameswaran Pillai, J., Kurian, H. N., Yu Yingfeng, T.: Nanocomposite Materials, Taylor and Francis, USA (2016)
22. Balazs, A.C., Singh, C., Zhulina, E.: Modeling the interactions between polymers and clay surfaces through selfconsistent field theory. Macromolecules **31**, 8370–8381 (1998)
23. Balazs, A.C., Singh, C., Zhulina, E., Lyatskaya, Y.: Modeling the phase behavior of polymer/clay nanocomposites. Acc. Chem. Res. **8**, 651–657 (1999)
24. Lyatskaya, Y., Balazs, A.C.: Modeling the phase behavior of polymer- clay composites. Macromolecules **31**, 6676–6680 (1998)
25. Vaia, R.A., Giannelis, E.P.: Lattice of polymer melt intercalation in organically-modified layered silicates. Macromolecules **30**, 7990–7999 (1997)
26. Chen, T.K., Tien, Y.I., Wei, K.H.: Synthesis and characterization of novel segmented polyurethane/clay nanocomposite via poly(e-caprolactone)/clay. J. Polym. Sci.: Part A Polym. Chem. **37**, 2225–2233 (1999)
27. Alexandre, M., Dubois, P.: Polymer-layered silicate nanocomposites: preparation, properties and uses of a new class of materials. Mater. Sci. Eng. **28**, 1–63 (2000)
28. Haraguchi, K.: Synthesis and properties of soft nanocomposite materials with novel organic/inorganic network structures. Polym. J. **43**, 223–241 (2011)
29. Lee, H.S., Choi, M.Y., Anandhan, S., Baik, D.H., Seo, S.W.: Microphase structure and physical properties of polyurethane/ organoclay nanocomposites. ACS PMSE Preprints **91**, 638 (2004)
30. Roy, A., Joshi, M., Butola, B.S.: Preparation and antimicrobial assessment of zinc-montmorillonite intercalates based HDPE nanocomposites: a cost-effective and safe bioactive plastic. J. Clean Prod. **212**, 1518–1525 (2019)
31. Sinha, T., Ahmaruzzaman, M., Bhattacharjee, A.: A simple approach for the synthesis of silver nanoparticles and their application as a catalyst for the photodegradation of methyl violet 6B dye under solar irradiation. J. Environ. Chem. Eng. **2**, 2269–2279 (2014)
32. Giri, A., Bhowmick, R., Prodhan, C., Majumder, D., Bhattacharya, S.K., Ali, M.: Synthesis and characterization of biopolymer based hybrid hydrogel nanocomposite and study of their electrochemical efficacy. Int. J. Biol. Macromol. **123**, 228–238 (2018)
33. Gubin, S.P.: What is nanoparticle? Trend of development of Nanochemistry and nanotechnology. Ross Khim Zh **44**, 23–31 (2000)
34. Henrique, P., Camargo, C., Gundappa, K., Satyanarayana, W.F.: Nanocomposites: synthesis, structure, properties and new application opportunities. Mater Res **12**, 1–39 (2009)
35. Bordes, P., Pollet, E., Avérous, L.: Nano-biocomposites: biodegradable polyester/nanoclay systems. Prog. Polym. Sci. **34**, 125–155 (2009)
36. Hussain, F., Hojjati, M., Okamota, M., Gorga, R.E.: Review article: Polymer-matrix nanocomposites, processing, manufacturing, and application: an overview. J. Compos. Mater. **40**, 1511–1575 (2006)
37. Baskaran, K., Ali, M., Gingrich, K., Lyn Porter, D., Chong, S., Riley, B.J., Peak, C.W., Naleway, S.E., Zharov, I., Carlson, K.: Sol-gel derived silica: A review of polymer-tailored properties for energy and environmental applications. Microporous Mesoporous Mater. **336**, 111874 (2022)
38. Guo, X.H., Zhang, Q., Ding, X., Shen, Q., Wu, C., Zhang, L., Yang, H.: Synthesis and application of several sol–gel-derived materials via sol–gel process combining with other technologies: a review. J. Sol-Gel Sci. Technol. **79**, 328–358 (2016)
39. Thomas, M., Naikoo, G.A., Sheikh, M.U.D., Bano, M., Khan, F.: Effective photocatalytic degradation of Congo red dye using alginate/carboxymethyl cellulose/TiO2nanocomposite hydrogel under direct sunlight irradiation. J. Photochem. Photobiol. A Chem. **327**, 33–43 (2016)
40. Montaser, A.S., Wassel, A.R., Al-Shaye'a, O.N.: Synthesis, characterization and antimicrobial activity of Schiff bases from chitosan and salicylaldehyde/TiO2 nanocomposite membrane. Int. J. Biol. Macromol. **124**, 802–809 (2019)

41. Mouritz, A.P., Gibson, A.G.: Fire properties of polymer composite materials 143 (2006) Springer Netherlands, 401
42. McNally, T., Potschke, P., Halley, P., Murphy, M., Martin, D., Belled, S.E.J., Brennane, P.G., Beinf, D., Lemoineg, P., Quinng, J.P.:Polymer **46**, 8222–8232 (2005)
43. Aji, I.S., Zainudin, E.S., Abdan, K., Sapuan, S.M.: J. Compos. Mater. **47**(8), 979–990 (2013)
44. George, J., Bhagawan, S.S., Thomas, S.: Compos. Interf. **5**(3), 201–223 (1997)
45. Mark, J.E.: Ceramic reinforced polymers and polymer-modified ceramics. Polym. Eng. Sci. **36**, 2905±2920 (1996)
46. Reynaud, E., Gauthier, C., Perez, J.: Nanophases in polymers, Rev. Metall./Cah. Inf. Tech. **96**, 169–176 (1999)
47. Lerner, M., Oriakhi, C.: In: Goldstein, A. (ed.) Handbook of Nanophase Materials, pp. 1–199. Marcel Dekker, New York (1997)
48. Demir, K.D., Tasdelen, M.A., Uyar, T., Kawaguchi, A.W., Sudo, A., Endo, T., Yagci, Y.: Synthesis of polybenzoxazine/clay nanocomposites by in situ thermal ring-opening polymerization using intercalated monomer. J. Polym. Sci. Part A: Polym. Chem. **19**, 4213–4220 (2011)
49. Dizman, C., Ates, S., Uyar, T., Tasdelen, M.A., Torun, L., Yagci, Y.: Polysulfone/clay nanocomposites by in-situ photoinduced crosslinking polymerization. Macromol. Mater. Eng. **296**, 1101–1106 (2011)
50. Sarde, B., Patil, Y.D.: Recent research status on polymer composite used in concrete-an overview. Mater Today: Proc. **18**, 3780–3790 (2019)
51. Hsissou, R., Berradi, M., El Bouchti, M., El Bachiri, A., El Harfi, A.: Synthesis characterization rheological and morphological study of a new epoxy resin pentaglycidyl ether pentaphenoxy of phosphorus and their composite (PGEPPP/MDA/PN). Polym Bull. **76**, 4859–4878 (2019)
52. Oladele, I.O., Omotosho, T.F., Adediran, A.A.: Polymer-based composites: An indispensable material for present and future applications. Int. J. Polym. Sci. **56**, 1–12 (2020)
53. Neser, G.: Polymer-based composites in marine use: History and future trends. Procedia Eng. **194**, 19–24 (2016)
54. Agbabiaka, O.G., Oladele, I.O., Akinwekomi, A.D., Adediran, A.A., Balogun, A.O., Olasunkanmi, O.G., Olayanju, T.M.A.: Effect of calcination temperature on hydroxyapatite developed from waste poultry eggshell. Sci. Africa **8**, 1–12 (2020)
55. El-Aouni, N., Hsissou, R., Azzaoui, J.E., Bouchti, M.E., Elharfi, A.: Synthesis rheological and thermal studies of epoxy polymer and its composite. Chem. Data Collect **30**, 100584 (2020)
56. Yahaya, R., Sapuan, S., Jawaid, M., Leman, Z., Zainudin, E.: Mechanical performance of woven kenaf-Kevlar hybrid composites. J. Reinf. Plast. Compos. **33**, 2242–2254 (2014)
57. Valino, A.D., Dizon, J.R.C., Espera, A.H., Chen, Q., Messman, J., Advincula, R.C.: Advances in 3D printing of thermoplastic polymer composites and nanocomposites. Prog. Polym. Sci. **98**, Article 101162 (2019)
58. Hsissou, R., Bekhta, A., Dagdag, O., El Bachiri, A., Rafik, M., Elharfi, A.: Rheological properties of composite polymers and hybrid nanocomposites. Heliyon **6**, e04187 (2020)
59. Rbaa, M., Benhiba, F., Hssisou, R., Lakhrissi, Y., Lakhrissi, B., Touhami, M.E., et al.: Green synthesis of novel carbohydrate polymer chitosan oligosaccharide grafted on d-glucose derivative as bio-based corrosion inhibitor. J. Mol. Liq. **322**, Article 114541 (2021)

Chapter 3
Primary and Secondary Processing of Polymer Matrix Composites

Kishore Kumar Mahato and Srikanta Moharana

Abstract The current chapter presents the array of various processes of manufacturing related to polymer matrix composites (PMCs). PMC is defined as the combination of a polymer matrix and a reinforcement phase. Over the last decade, there has been a growing interest in PMCs, driven by their enhanced mechanical, thermal, and electrical properties, making them applicable in industries such as automotive and aerospace. Numerous studies, encompassing both experimental and theoretical approaches, have been conducted to explore the mechanical, thermal, and electrical characteristics of PMCs. Researchers have extensively evaluated the impact of primary and secondary processing on the mechanical, electrical, tribological, and thermal properties of PMCs. This chapter extensively discusses the fabrication methods, advantages, limitations, and properties associated with the primary and secondary processing of PMC.

Keywords Polymer matrix composites (PMCs) · Primary and secondary processing

3.1 Introduction and Understanding of Polymer Matrix Composite

There has been a growing recognition of the utility of polymer matrix composites (PMCs) owing to their enhanced thermal, mechanical, and electrical properties. These attributes make them suitable for diverse applications, including automotive, aerospace, oil and gas, structural areas, and various relevant fields [1, 2]. In PMCs, stronger and stiffer fibers are typically incorporated as reinforcing materials

K. K. Mahato (✉)
Department of Manufacturing Engineering, School of Mechanical Engineering, Vellore Institute of Technology, Vellore 632014, India
e-mail: kishorepce@gmail.com

S. Moharana
Department of Chemistry, School of Applied Sciences, Centurion University of Technology and Management, Jatni, Bhubaneswar, Odisha, India

S. Moharana et al. (eds.), *Polymer Composites*, Engineering Materials,
https://doi.org/10.1007/978-981-97-2075-0_3

within a polymeric matrix. This combination results in higher strength and modulus compared to the polymeric material alone, leading to the classification of such materials as fiber-reinforced polymer (FRP) composites [3, 4]. The utilization of PMCs has surpassed that of metal matrix composites in sectors like automotive, aerospace, and other structural industries due to their lower density, high modulus-to-density ratio, high strength-to-density ratio, and relatively straightforward manufacturing process. The densities of polymers in PMCs typically range from 0.9 to 1.4 g/cm^3, while reinforced fibers exhibit densities ranging from 1.4 to 2.6 g/cm^3. Consequently, the overall density of PMCs, contingent upon fiber type and volume, falls within the range of 1.2–2 g/cm^3. This lower density contributes to higher strength-to-density and modulus-to-density ratios compared to metal matrix composites, resulting in significant weight savings [5]. Generally, PMCs are composed of either a thermosetting polymer, such as epoxy and polyester, or a thermoplastic polymer like polypropylene (PP) and polyether ether ketone (PEEK). Common reinforcing fibers include glass, carbon, and Kevlar. Various fibers with different orientations, continuous or discontinuous in length, and arranged in unidirectional, bidirectional, multidirectional, or random orientations are available for commercial applications. The typical structure of a composite material involves the formation of a laminate, consisting of alternating layers of thin continuous fibers and a polymer matrix. The orientation and arrangement of fibers can vary among different layers of the laminate. The mechanical properties of PMCs, such as tensile modulus and strength, can be enhanced by varying fiber length, arrangement, orientation, stacking, and type [6]. These fibers are often combined with structures like honeycomb aluminium, wood, or plastic foam to create lightweight and stiff sandwich structures, providing reliability and flexibility in design for various structural components. PMCs exhibit exceptional features that make them suitable for diverse applications, including nonisotropic behavior that allows for tailored properties in design and specific applications. Another distinctive feature is their higher damping coefficient, surpassing that of metals. Additionally, PMCs demonstrate good anticorrosion behavior compared to their metal counterparts. Numerous theoretical and experimental studies have explored the thermal, mechanical, and electrical properties of PMCs, along with their processing methods [7–11]. This chapter offers a comprehensive overview of the introduction of PMCs and various essential processing methods employed to create structural components for diverse applications.

3.2 Processing of Polymer Matrix Composites

PMCs consist of a combination of fibers or particles within a polymer matrix, where these reinforcements are integrated to enhance various properties of the overall composite. Micron-scale particles typically contribute superior toughness and thermal stability, while nanoparticles improve mechanical and thermal characteristics. Particles with an aspect ratio resembling short fibers, whiskers, and continuous fibers are incorporated to strengthen mechanical and thermal behavior. Continuous fibers, commonly used in forms such as woven fabric, continuous random mat, unidirectional or bidirectional fabric, and stitched fiber, contribute to enhanced mechanical strength and stiffness. While the properties of the reinforcement significantly impact the composite's ultimate behavior, the inherent quality of the initial material system and the processing method employed to manufacture the bulk composite also play crucial roles in determining the final behavior. The goal in composite manufacturing is to create functional components with desired properties that synergize with the optimal characteristics of particles or fibers and the polymer, while minimizing or maximizing their respective flaws.

The manufacturing process aims to enhance the properties during the production of bulk components, a principle applicable to both continuous and short fiber composites. Short fiber composites are influenced by factors like fiber distribution and orientation density, affected by the polymer's flow direction, leading to non-homogeneous behavior in the bulk composite. Continuous PMCs face the challenge of filling voids between the polymer and fibers to establish a strong fiber/matrix interfacial bond, preventing issues like cracking and delamination under various loads.

Over the past three decades, the industry has witnessed advancements in material chemistry, composite manufacturing methods, and a shift towards more scientific processing routes, departing from traditional trial-and-error approaches. Researchers have reported significant progress in PMC advancements and processing routes, making them more versatile than metallic materials. Various manufacturing techniques have evolved to address novel matrix or fiber structures, new precursor materials, cost-effectiveness, geometrical constraints, improved mechanical, electrical, and thermal behaviors, generation of multifunctional parts, and defect reduction. The industry continues to develop novel manufacturing processes to produce PMCs with intricate structures, higher quality, cost reduction, and novel polymer and reinforcement systems. In addition to novel methods, the integration of basic science into the manufacturing process has enhanced technologies, with analytical modeling and simulations predicting physics and thermochemical reactions before prototype development. The reduction of inconsistencies across different parts is achieved through improved materials and advanced processing techniques, resulting in defect-free parts with superior properties. The future of the composite industry is envisioned to involve the integration of PMC structural design with the assembly process and manufacturing, representing a potential trend in composite technology.

3.3 Primary Processing Techniques

3.3.1 Matrix Resin Selection

The primary behavior of the matrix in a PMC is to bind or hold the fibers together, load transfer from one fiber to another, prevention from chemical attacks, moisture absorption, and ultra violet irradiation (i.e. exposure to sunlight for a longer time). Also, the tensile, compressive and shear behavior of the composite is highly influenced by the polymer phase. The changes in temperature variation (i.e. low, high, thermal spike, thermal shock, freeze–thaw) have a significant impact on the matrix properties. In the case of PMCs, the polymer phase plays a vital part in terms of strength and stiffness. The viscosity of the polymer is one of the important parameters during fabrication of PMCs. Since, it contributes to the filling of the mold with matrix, wetting of reinforcing fibers, and dismissing the entrapped air. Some other processing limits like time, temperature, pressure that controls rate of production depends on the type of polymer. In general, epoxy polymers have longer curing time as compared to polyester resin. Usually in PMCs both thermoplastic and thermosetting polymers may be used as matrix material. The time of processing and method of processing depends upon the type of polymer whether it is thermoplastic or thermosetting.

The polymers those comes under thermosetting polymers are often called as thermosets. For examples epoxy polymer in general used as thermoset polymer material for high performance material. The initial materials for several thermosetting polymers are less viscous liquid prepolymer. With is prepolymer the fiber surface usually coated and wetted before beginning of the chemical reaction and the process is called as curing. Basically during curing process the low viscous liquid polymer changes to a solid thermosetting polymer. When curing is done at higher temperature, it is called as cure temperature. The cure temperature for epoxies lies in between the range of 120–180 °C [12]. The thermosetting polymer after curing cannot be transformed to the original liquid state by either cooling or heating.

Thermoplastics or a thermoplastic polymer usually possesses very high viscosity in the liquid form. The fabrication process is much more challenging and expensive during the wetting and coating of the liquid thermoplastic with fiber surface. Generally, short fibers are used during thermoplastic polymer fabrication process. As these short fibers can be easily mixed with liquid resin using injection mold and extruders to obtain the required part at a relatively at low cost. The main advantage of thermosets is that it can be reusable and recyclable. However, recurrence of the recycling process again and again may weaken the properties.

When these thermoplastics are heated, the harder solid changes to a softer solid.

3.3.2 Reinforcement Material Selection

Fibers play a pivotal role as the primary load-bearing elements in fiber-reinforced composite materials. The mechanical and thermal properties of a composite are contingent upon several key parameters related to these fibers, including:

a. Type and properties of the fibers
b. Volume fraction of the fibers
c. Length of the fibers
d. Orientation of the fibers
e. Architecture of the fibers within the composite.

In high-performance PMCs, continuous lengths of fibers are typically utilized, constituting approximately 50–65% of the composite volume. On the other hand, random fiber PMCs employ discontinuous fibers, with a typical fiber volume fraction ranging from 20 to 40%. The most prevalent fibers in PMCs are glass, carbon, and aramid fibers. While other fibers such as boron, ceramic (e.g., silicon carbide), extended chain polyethylene, and natural fibers like jute, hemp, and sisal are available in the market, they are employed in smaller quantities. The ensuing discussion briefly outlines the general characteristics of glass, carbon, and aramid fibers.

Fibers are manufactured as slender continuous filaments with an average diameter ranging from 7 to 15 μm. In practical applications, they are supplied and utilized in the form of filament bundles, referred to as strands in the context of glass fibers and tows for carbon and aramid fibers. The number of filaments within a strand or a tow can be adjusted. In both strands and tows, the filaments maintain a parallel and uninterrupted configuration. Glass fiber strands are assembled into rovings (parallel strands) or yarns (twisted strands). The number of strands in rovings and yarns can be varied. Carbon filaments are grouped into tows of parallel filaments without twist and yarns, where some filaments are twisted. The greater the number of filaments in a strand or a tow, the higher the processing yields (in terms of production rate). However, if the strand or tow is excessively thick, the composite properties, such as strength, may be compromised since the polymer matrix might struggle to penetrate and wet out the interior filaments.

Fibers exhibit exceptionally high tensile modulus and tensile strength; nevertheless, their tensile strain at failure is notably low. Tensile stress–strain diagrams for fibers are almost linear until failure, resembling the behavior of a classic brittle material under tensile load. Tensile strength is influenced by fiber length and diameter and typically demonstrates considerable statistical variation.

Carbon fibers can have modulus values ranging from 220 to 800 GPa, with the lower end of this range being comparable to the modulus of steel. Glass and aramid fibers, while having a significantly lower modulus, exhibit tensile strength comparable to some carbon fibers. The low strain-at-failure characteristic of high-modulus carbon fibers makes them prone to breakage during processing, potentially causing interruptions in the production line. In many manufacturing processes, fiber bundles are pulled over small diameter rollers to create tension, and if the roller diameter is too small, the tensile strain in the fibers may exceed their failure strain, resulting in breakage and production delays.

Both carbon and aramid fibers exhibit a negative coefficient of thermal expansion along their longitudinal direction but a relatively high positive coefficient in the radial direction. In contrast, glass fibers have equal coefficients of thermal expansion in both longitudinal and radial directions. It's important to note that the coefficient of thermal expansion for fibers is generally much lower than that of the polymer matrix. The substantial difference in thermal contractions between fibers and the polymer matrix can lead to the development of thermal residual stresses at the fiber-matrix interface when the polymer matrix cools down from the processing temperature to room temperature.

3.4 Molding Methods

3.4.1 Bag Molding Process

The bag molding process is primarily employed for the production of laminated composite structures and components using thermoset matrix prepreg sheets or tapes as the initial material. The thermoset matrix prepreg consists of a thin sheet containing fibers coated with a partially cured (B-staged) resin. While epoxies are the principal resins used in prepregs, for high-temperature applications, and polyimides are also utilized. A typical laminate comprises multiple layers or plies arranged at various fiber orientation angles based on design requirements.

To manufacture a laminated composite part through the bag molding process, a stack of prepreg layers is placed on the mold or tool surface, covered with a vacuum bag, and then cured and consolidated under heat, vacuum, and pressure, either inside an autoclave or in a heated platen press. Autoclaves are commonly employed for vacuum bag molding of high-performance composite laminates due to their precise temperature and pressure controls.

The bag molding process is widely used in the aerospace industry, manufacturing primary and secondary structures for military and commercial aircraft, including wing flaps, ailerons, horizontal stabilizers, and radars. It also finds applications in sporting goods, motor sports, and racing yachts. The process's advantages lie in its ability to maintain precise fiber orientation, producing high-quality composite parts with relatively low void content and high reproducibility. However, drawbacks include its labor-intensive and time-consuming nature, as well as the high cost of autoclaves, particularly for large components, making it more expensive than other composite manufacturing processes like RTM.

This section begins by outlining the process of making prepreg, the starting material for the bag molding process, and subsequently details the bag molding process and its associated parameters. Although autoclaves are commonly used for curing bag-molded composites, out-of-autoclave curing is gaining acceptance, mainly for cost-reduction purposes. The chapter also covers the processing of sandwich laminates, known for their high bending stiffness-to-weight ratios.

Prepreg refers to a thin sheet (or tape) containing either unidirectional continuous fibers or woven fabric infused with a predetermined amount of partially cured thermosetting polymer matrix. Epoxy serves as the primary matrix material in prepreg sheets, with other thermosets like BMIs being utilized for high-temperature applications.

Prepreg sheets are provided by manufacturers in rolls, and their width can range from less than 25 to over 457 mm, with sheets wider than 457 mm referred to as broad goods. The typical thickness of a single ply of a cured prepreg sheet falls within the range of 0.125–0.25 mm. Commercially available prepregs generally exhibit a resin content between 30 and 45% by weight, equivalent to 38.6–54.5% by volume of resin for a T-300 carbon fiber/epoxy prepreg.

The manufacturing process for unidirectional fiber-reinforced epoxy prepregs involves pulling uniformly spaced parallel fiber rovings or tows through a liquid resin bath containing the epoxy resin–hardener mix dissolved in a solvent. Adjustments are made to the solvent content and resin bath temperature to achieve the appropriate resin viscosity for effective fiber wet-out and impregnation. Subsequently, the impregnated fibers pass through a metering device, followed by a heating/cooling chamber. In this chamber, heat is applied to eliminate most of the solvent and initiate the curing reaction, reaching the B-stage condition in the cooling chamber. At this stage, the resin is partially cured and tacky, allowing it to adhere to itself and other surfaces. The prepreg sheet is then backed up with a release film or waxed paper and wound around a take-up roll. This backup material prevents the prepreg sheet from sticking to itself during rolling and must be separated before molding. The prepreg roll is stored in a polyethylene bag in a freezer until the molding process.

Before molding, the prepreg roll is brought to room temperature, avoiding moisture condensation by allowing up to 48 h for the polyethylene bag to be opened. The shelf life of epoxy prepregs is typically 6–8 days at 23 °C, but it can be extended to 6–12 months when stored in the freezer at −18 °C.

Prepregs are specified based on fiber type, fiber architecture, resin type, fiber aerial weight, resin content, and cured ply thickness. Fiber aerial weight represents the mass of fibers per unit area in grams per square meter (g/m^2), and resin content is the weight percentage of resin in the prepreg. The recent trend is to employ a near-net resin content to minimize resin loss during molding. The cured ply thickness of unidirectional epoxy prepreg typically falls within 0.127 mm $\pm$ 6% [13]. Two critical quality characteristics of prepregs at the time of molding are tack and drapeability. Tack, or stickiness, denotes the partially cured prepreg layer's ability to adhere to the mold surface and another prepreg layer. The drapeability of prepreg refers to its ability to conform to the mold surface without fiber distortion, movement, and out-of-plane buckling.

3.4.2 Layup Techniques

A cured laminate is comprised of numerous layers, sometimes exceeding 100, each featuring fiber orientation angles ranging from $-90°$ to $+90°$. When assembling the prepreg stack in the mold, meticulous attention is required for cutting the prepreg layers to the appropriate dimensions and angular orientation, placing them on the mold surface or the preceding layer, and debulking the stack during assembly to minimize gaps between layers. To prevent errors, templates are employed to outline the shapes of the prepreg pieces to be cut, and for efficiency and waste reduction, the pieces are nested. Automated ply cutting machines and laser guidance systems are available to enhance accuracy, minimize scrap, and reduce the layup time [14].

Laying flat prepreg sheets on a flat or nearly flat surface is relatively straightforward, but challenges arise when covering compound curved surfaces or fitting into mold corners, necessitating the slitting or folding of prepreg layers. These manipulated areas may become potential weak points in the cured composite part. Extensive manipulation, including stretching, shearing, bending, and pressing, is often required to conform them to the mold contour [15, 16]. Woven fabric prepregs generally exhibit greater flexibility in such manipulations compared to unidirectional fiber prepregs, primarily due to the possibility of in-plane shear deformation known as interply shear. When working with unidirectional fiber prepregs, it is essential to ensure that fibers are not distorted or dislocated to prevent the creation of resin-rich areas or significant voids in the cured composite part.

Many defects in bag-molded laminates are associated with the ply layup and trimming operations. Precise control is necessary over the fiber orientation in each ply, the stacking sequence, and the total number of plies in the stack. As prepreg tapes are not as wide as the part itself, each layer may consist of multiple identical plies placed side by side to cover the entire mold surface. Additionally, precautions must be taken to avoid filament crossovers, and the presence of broken filaments, foreign matter, and debris should be strictly prohibited.

3.4.3 Automated Tape Laying

Hand layup entails the manual placement of each layer by hand, adhering it firmly to the preceding layer or the mold surface. This method is characterized by a slow production rate, labor intensiveness, and susceptibility to human error. Although it remains the primary approach for constructing the prepreg stack, automated tape layup machines have emerged as a more efficient alternative, capable of generating prepreg layups in significantly shorter durations and with enhanced precision. In automated tape laying machines, the prepreg roll is loaded onto the tape-laying head, mounted on a gantry that positions the tape-laying head within the mold. Both the tape-laying head and gantry feature multiple axes of movement, typically five for each, microprocessor-operated to deliver high layup flexibility and precise feedback control of translational and rotational motions. The tape head incorporates cutters, a backup paper take-up reel, a tape heating system, compaction shoes, and an optical tape flaw detector. If the prepreg lacks adequate tackiness, the tape-heating system warms it up to a temperature between 25 °C and 45 °C just before it passes through the lay-down rollers to enhance tack. Compaction shoes apply pressure on the tape during the laying process. Once a tape strip is laid, its end is cut, and the tape head moves to the next location.

Automated tape-laying machines are programmed to function based on user-defined parameters, including mold surface topology, ply boundaries, fiber orientation angle in each ply, the number of layers to be laid, and tape cutting angle. However, they are better suited for flat and moderately contoured parts, such as wing skins and ailerons, than highly contoured parts like nose cones, inlet ducts, and nozzles.

3.4.4 Autoclave Curing

Autoclave curing results in composite parts characterized by minimal void content, exceptional consolidation, and consistent resin distribution. However, the utilization of autoclaves incurs high expenses and operational costs, especially when used for small-sized parts on larger autoclaves designed for larger components. Consequently, there is a growing acceptance of out-of-autoclave curing methods, such as oven curing and heated blanket curing, within the composites industry [17, 18]. In oven curing, the vacuum bag assembly is heated in a hot air circulating oven, and vacuum application is employed to eliminate air and volatiles, with consolidation occurring at 1 bar. Due to the absence of external pressure in the oven, parts cured in ovens exhibit higher void content and lower consolidation compared to autoclave-cured parts. To achieve autoclave-quality parts with minimal void content using out-of-autoclave curing techniques, a novel generation of prepregs known as vacuum bag-only (VBO) prepregs is employed (as shown in Fig. 3.1). The dry regions serve as vacuum channels, facilitating the migration of entrapped air towards the edges and surfaces during the initial stages of curing. As the temperature rises, resin flows into these air migration channels, infiltrating and producing a void-free part.

3.4.5 Sandwich Molding

Sandwich constructions, incorporating a lightweight, low modulus core and high modulus PMC face skins, find applications in various fields. The presence of the core between the faces significantly enhances bending stiffness without adding substantial weight to the structure. Moreover, the critical load for buckling experiences a significant increase. Consequently, sandwich constructions can offer higher bending stiffness and resistance to buckling per unit weight compared to laminates made

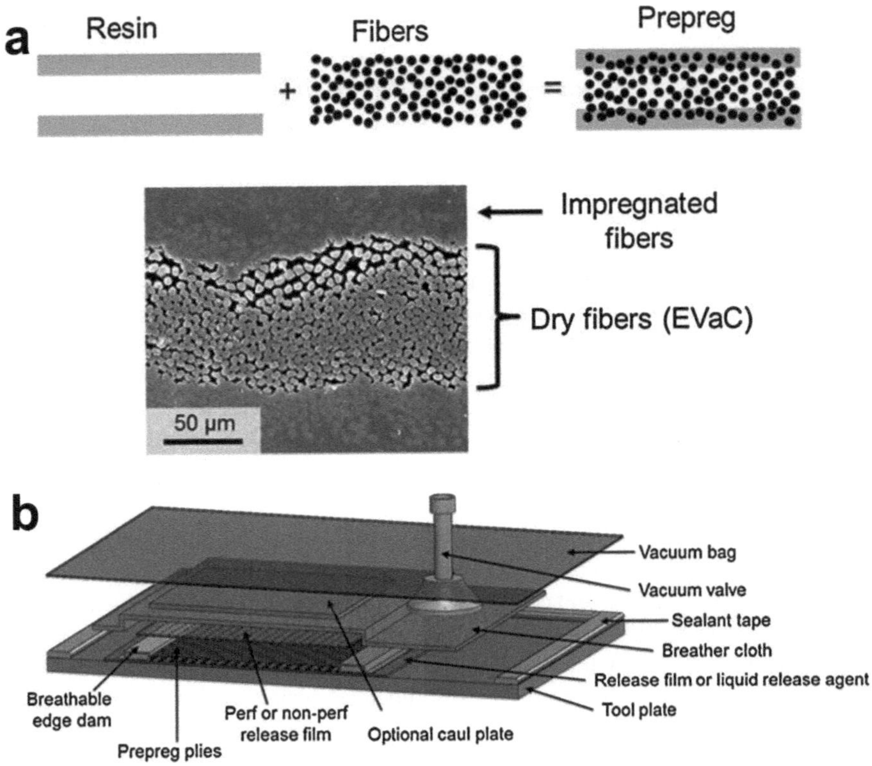

Fig. 3.1 **a** Schematic and SEM micrograph of air evacuation channels in VBO prepreg; **b** VBO layup schematic [17]. Adapted from Ref. [17] with permission from [Elsevier], Copyright [2015]

solely from the face material. Commonly utilized cores include aluminum honeycombs, aramid fiber-reinforced phenolic honeycombs (referred to as Nomex), PP honeycombs, polymeric foams (such as polyvinyl chloride and polyurethane), and balsawood. In many aerospace structures, aluminum and Nomex honeycomb cores are preferred. The strength and stiffness of these cores depend on parameters such as cell size, cell wall thickness, and properties of the cell material. It is essential for the core to possess high strength to resist transverse shear stresses and provide substantial crush resistance. Additionally, high stiffness is necessary to counter core buckling and local wrinkling under compressive stresses. Sandwich constructions can be manufactured in two ways: (1) by inserting the core between uncured prepreg layers, which will form the top and bottom face skins, followed by vacuum bag molding in an autoclave, or (2) by adhesively bonding precured face skins to the core. In the case of adhesive bonding, a film adhesive is placed between the core and face skins, undergoing curing using the standard curing procedure. During the molding of sandwich laminates, special attention is given to the ends of the core. If the ends are not sealed, moisture ingress into the laminate may occur, potentially diminishing the material properties and, consequently, the performance of the sandwich laminate.

3.4.6 Compression Molding

Compression molding is one of the oldest composite manufacturing processes and is now widely used for making PMC parts using sheet molding compounds (SMCs) and bulk molding compounds (BMCs) as the starting material. The most common forms of SMC and BMC contain randomly oriented discontinuous fibers dispersed in a highly viscous, but uncured thermosetting resin. In the compression molding process, the starting material is placed inside a preheated mold cavity of a matched die mold and then compressed or squeezed to flow sideways and outward as the top half of the mold moves downward relative to the bottom half. It requires much less preparation prior to the molding operation, and it is capable of producing parts of complex shapes at a relatively rapid rate, and therefore, it is suitable for high-volume production of composite parts. Part design features, such as non-uniform thickness, ribs, bosses, flanges, holes, cutouts, curvatures, and rounded corners, can be relatively easily incorporated during the compression molding operation. Because of this, secondary post molding operations such as drilling, machining, and joining may not be needed with compression molded parts. Very little manual labor or monitoring is needed during the compression molding operation, and therefore, the process can be automated and material handling can be done by robots. For all these reasons, compression molding has become one of the leading composite manufacturing processes in mass production industries, such as the automotive and appliance industries, where the need for fast production rates is a primary consideration in manufacturing process selection. Examples of compression-molded parts in the automotive industry include bumper beams, cross members, radiator supports, engine valve covers, hoods, and fenders. In this portion, we will discuss compression molding of SMCs and BMCs. Both materials contain randomly oriented short fibers, although in some versions of SMCs, continuous fibers are also included. The matrix material is usually a thermosetting polyester or vinyl ester resin. Epoxy is also used as the matrix material; however, since the curing time of epoxy is much longer than polyesters or vinyl esters, its application in either SMC or BMC is limited. Because of lower cost, E-glass fibers are used as the reinforcement in the SMC and BMC for automotive applications.

Carbon fibers can provide higher stiffness per unit mass, but they are much more expensive. Carbon fiber SMC is finding applications in a few aircraft components. We will start this chapter by first describing the starting materials, namely, SMCs and BMCs. Then we will discuss the compression molding process, flow and cure characteristics, and other topics related to it. More information on SMCs and compression molding process can be found in references [19–21].

3.4.7 Sheet Molding Compound

Sheet Molding Compound (SMC) is a continuous sheet of pre-prepared material, featuring fibers and mineral fillers dispersed in a highly viscous yet uncured thermosetting resin. The resin undergoes curing during compression molding within a matched die mold. In addition to these core components, SMC incorporates various other elements, including a catalyst, thickener, low-profile additive, and mold release agent. While these ingredients are utilized in relatively small quantities, they play crucial roles either during storage, before molding, or in the molding process itself. Unsaturated polyester and vinyl ester stand out as the two most commonly employed resins in SMC. The resin is blended with a reactive diluent that serves dual purposes: first, it reduces the viscosity of the unreacted resin, and second, it reacts with the resin during molding, forming cross-links between resin molecules. Styrene monomer, possessing low viscosity at room temperature, is the predominant reactive diluent used in SMC applications. Typically, the resin is provided by manufacturers in a premixed condition with 30–50% by weight of styrene monomer, resulting in a mixture viscosity ranging from 300 to 3000 mPa s.

SMCs are categorized based on the form of fibers used in them. These classifications include;

(a) SMC-R, where R signifies randomly oriented short fibers in the sheet, usually 25 mm long, but varying from 12 to 50 mm depending on the application.
(b) SMC-CR, involving continuous parallel fibers on one side of the sheet (C) and randomly oriented short fibers on the other side (R).
(c) XMC, featuring crisscrossed continuous fibers (X) in the sheet.

In both SMC-R and SMC-CR designations, the fiber content (wt.%) is specified at the end of the letter. For instance, SMC-R50 comprises 50 wt.% of randomly oriented short fibers, while SMC-C30R20 includes 30 wt.% continuous fibers and 20 wt.% randomly oriented short fibers. It's worth noting that the total fiber content in SMC-R50 and SMC-C30R20 is identical; however, due to the random fiber orientation in SMC-R50, it behaves as a planar isotropic material, exhibiting equal properties in all directions within the sheet plane. In contrast, SMC-C30R20 is an orthotropic material, and its properties are direction-dependent. For example, its strength and modulus are highest in the longitudinal (L) direction and lowest in the transverse (T) direction of the continuous fibers. Randomly oriented short fibers in SMC-C30R20 contribute to improving strength and modulus in the transverse (T) direction, though not to the extent of equaling the longitudinal values. In XMC, where crisscrossed continuous fibers are integrated with randomly oriented short fibers, the angle between the crisscrossed fibers typically falls in the range of 5°–7°. Larger angles are avoided, as they tend to diminish the longitudinal strength and modulus of the material. SMC-CR and XMC are preferred over SMC-R in applications where their superior properties in the longitudinal direction can be advantageously utilized, such as in beam or narrow plate structures.

Fillers, employed in powder or particulate form, serve various functions in SMC. They diminish volumetric shrinkage of the resin, enhance material flow during molding, and improve the surface quality of the molded part. Calcium carbonate ($CaCO_3$) is the most widely used filler in SMCs, and other fillers include kaolin clay, talc, glass spheres (both solid and hollow), and alumina trihydrate ($Al_2O_3 \cdot 3H_2O$). Most of these fillers are cost-effective compared to the resin, thus reducing the overall material cost when added to the resin. Hollow glass spheres, owing to their low bulk density, effectively reduce the weight of SMC. Alumina trihydrate serves as both filler and a fire retardant, releasing water of hydration around 220 °C, acting as an internal flame retardant and smoke suppressant. It finds application in electrical, appliance, and construction sectors where flammability and smoke generation are crucial material selection criteria.

The most prevalent fiber in SMCs is E-glass, primarily due to its cost-effectiveness. Carbon fibers are utilized in SMCs for their higher modulus, but their use is restricted to specific aerospace applications due to their high cost. The fiber content in SMC can be readily adjusted to meet application requirements. High fiber content, ranging from 50 to 70% by weight, is employed in structural applications demanding high strength and modulus. Fillers are not introduced if the fiber content exceeds 60 wt.%, as the high viscosity of SMC at such fiber content already poses challenges to material flow during molding, and adding fillers further diminishes moldability. Short fibers used in SMC-R or SMC-CR are typically 25 mm long, as longer fibers make material flow increasingly difficult during molding, despite potentially yielding higher strength in the molded composite part.

BMC, also known as dough molding compound, is a pre-formed material containing short reinforcing fibers within a thermoset matrix, specifically designed for compression molding and injection molding processes. Supplied in log form as bulk material, BMC resembles dough in appearance. It finds applications in various industrial sectors, including electrical components like fuses and switchgears, as well as in the manufacturing of pump and motor housings, appliance parts, and automotive components such as engine valve covers and timing belt pulleys. Commercial BMC formulations typically consist of 15–20 wt.% E-glass fibers embedded in a polyester resin, with fiber lengths ranging from 6 to 25 mm based on specific application requirements. Similar to SMC, the resin paste for BMC is created by blending chopped glass fiber strands and fillers in a high-intensity mixer, resulting in a highly viscous mass. This mass is extruded through a circular die to form a continuous log, which is subsequently cut into desired lengths using a pneumatic cutter and packaged in styrene-proof bags. Thickeners are not incorporated in the BMC formulation if the BMC log is molded into the end product soon after the compounding operation.

3.4.8 Compression Molding Compound

Compression molding commences with the placement of a pre-cut, assembled, and weighed quantity of the initial material, referred to as the charge, onto the lower portion of a preheated mold cavity in a matched die mold. In the case of SMC, the charge comprises a stack of rectangular sheets, die-cut from a matured SMC roll after bringing it to room temperature and removing the barrier films. The compression molding process for SMC involves the following key steps. For BMC, pre-cut sections of BMC logs are situated in the preheated mold cavity, and the subsequent steps for compression molding of BMC closely resemble those for SMC.

1. Charge preparation and placement: SMC plies are organized and stacked externally to the mold. Rectangular ply patterns are commonly utilized; however, circular, elliptic, or oblong shapes may be employed based on the part and mold design. Ply dimensions are chosen to cover 60–70% of the mold surface area. The charge is weighed before being positioned in the mold. Any weight variation in the charge, resulting from variations in SMC sheet weight per unit volume, is compensated by adding smaller cut pieces from the SMC roll. The placement location of the charge affects the flow length during molding, influencing potential defects like voids, fiber orientation patterns, and knit line formation.
2. Mold closing and compression: Following the placement of the charge in the mold's bottom half, the top half is rapidly lowered to contact the charge's top surface. Subsequently, the top half descends at a slower rate, typically 5–10 mm/min, until the pressure on the charge reaches a preset level. As the charge's temperature rises upon contact with the preheated mold surfaces, the viscosity of the SMC material decreases. With increasing pressure, the material begins to flow and fill the cavity. During this movement toward the cavity extremities, the entrapped air in the mold and charge escapes through vents and the shear edges of the mold. Achieving proper material flow is crucial to minimize air entrapment; otherwise, the molded SMC part may exhibit significant voids, blisters, and delaminations. If the mold closes too slowly, the resin in the charge may start to gel, hindering complete cavity filling and resulting in a short shot. Conversely, if the mold closes too rapidly, insufficient time for air escape may lead to entrapped air causing voids in the molded part. The flow length in the mold governs fiber orientation, thereby influencing mechanical properties and, in some instances, part distortion. Depending on the intricacy of the part, its size (which dictates the flow length), and the viscosity of the SMC material, the molding pressure can vary between 30 and 150 bar. Elevated pressure is necessary for molding components with intricate ribs and bosses. Typically, the mold temperature falls within the range of 130–160 °C for polyester and vinyl ester SMCs.
3. Curing: Once the cavity is filled, the mold remains closed for a predetermined duration to ensure thorough curing and consolidation across the part. Depending on the combination of resin, catalyst, inhibitor, part thickness, and mold temperature, the curing period may span from 1 min to several minutes.
4. Part ejection and post-molding operations: Upon achieving the desired level of cure under pressure, the mold is opened, and the part is extracted from the bottom mold using ejector pins. Subsequently, the part undergoes cooling outside the mold while the mold surfaces are cleaned of any remaining debris. A mold release agent, such as zinc stearate, is applied to the mold to prepare it for the next part. During the post-molding phase, as the part cools outside the mold, it continues to cure and undergo shrinkage. Since there are no pressure constraints on the

part during post-molding, it may display distortion or generate residual stresses due to varying cooling rates in different sections. The non-uniform temperature distribution in the molded part primarily stems from the presence of thin and thick sections. After the part cools to room temperature, there might be a need to trim its edges to eliminate any flash that may have formed during molding.

3.4.9 Liquid Composite Molding

In liquid composite molding (LCM) processes, a pre-mixed liquid thermoset resin is injected into a dry fiber preform previously placed in a closed mold. As the liquid permeates the preform, it coats the fibers, fills the inter-fiber spaces, expels trapped air, and eventually, during curing, transforms into the thermoset matrix [22, 23]. LCM processes offer several advantages over alternative composite manufacturing methods, including (1) relatively swift production cycle time, (2) low molding pressure, (3) superior surface finish, (4) close tolerance, and (5) design flexibility. The use of dry fiber preforms in LCM enhances design flexibility, allowing for various fiber architectures depending on performance requirements. The quality of liquid composite-molded parts is contingent on both preform and mold design, demanding meticulous attention for manufacturing parts with fewer defects and improved reproducibility. This chapter delineates two distinct LCM processes: resin transfer molding (RTM) and structural reaction injection molding (SRIM). While both RTM and SRIM involve the injection of liquid resin into a dry fiber preform, there are disparities between them. For instance, RTM employs epoxy, polyester, or vinyl ester resins, while SRIM utilizes polyurethane or polyurea resins. RTM, particularly with BMI resins, caters to high-temperature aerospace applications. Over the years, both processes have successfully manufactured diverse composite parts like cabinet walls, chairs, bench seats, hoppers, water tanks, bathtubs, and boat hulls, providing a cost-effective alternative to labor-intensive bag molding and capital-intensive compression molding. This has led to increased acceptance in aerospace and automotive industries.

For the aerospace sector, which predominantly employs prepregs and the bag molding process, RTM offers the following advantages:

(a) Less labor-intensive than vacuum bag molding
(b) Lower capital investment, given the cost disparity between autoclaves (used in bag molding) and molding presses (used for RTM)
(c) Reduced tooling cost, as RTM operates under lower molding pressure than bag molding
(d) Lower material cost, considering the production cost of dry fiber preforms used in RTM is lower than that of prepregs
(e) Lower material storage cost, since unlike prepregs, dry fiber preforms used in RTM don't require freezer storage, enhancing shelf life.

Examples of aerospace parts manufactured via RTM include wing structures, helicopter flaperons, various aerodynamic surfaces, and brackets. In the automotive industry, where injection molding and compression molding are prevalent for composite parts, RTM becomes a preferable option for production volumes ranging from 10,000 to 50,000 parts annually. It excels in producing composite body panels and structural components with excellent surface finish at a significantly lower manufacturing cost and a relatively high production rate. RTM, compared to compression molding, boasts low tooling costs and uncomplicated mold clamping requirements. In some instances, a ratchet clamp or a series of nuts and bolts can suffice to hold the two mold halves together. RTM's low-pressure nature allows for resin transfer molding in low-tonnage presses. Another advantage is the process's ability to incorporate ribs, metal or composite inserts, stiffeners, washers, and more.

3.4.10 Filament Winding

In the filament winding process [24], a continuous band of resin-impregnated fiber rovings or tows is tension-wound around a rotating mandrel in a predetermined fiber orientation pattern, resulting in the creation of hollow continuous FRP composite parts. The resin typically used is a thermosetting polymer, such as epoxy, vinyl ester, or polyester. Once several layers are wound to achieve the required thickness, the part undergoes curing on the mandrel, followed by the removal of the mandrel. In certain applications, like pressurized oxygen tanks, the mandrel may be retained within the filament-wound composite part to meet specific design requirements. The filament winding process accommodates the production of both axisymmetric and non-axisymmetric parts. Axisymmetric examples include golf club shafts, automotive drive shafts, rocket motor cases, cylindrical and spherical pressure vessels (e.g., oxygen tanks and compressed natural gas tanks), oil and gas pipes, and large underground gasoline storage tanks. Non axisymmetric parts, like helicopter blades and wind turbine blades, are also filament-wound, necessitating numerically controlled machines with sophisticated programming devices to control fiber motion and winding patterns around complex-shaped mandrels.

There are two primary types of filament winding processes: (1) wet winding, where the fiber rovings or tows are wetted by the liquid resin mix before winding onto the rotating mandrel, and (2) dry winding, where pre-impregnated fiber rovings or tows are wound around the rotating mandrel. Wet winding is more common and offers several advantages, including lower material costs, shorter winding times, and the ability to easily adjust resin formulation to meet design or processing requirements. Filament winding boasts several key advantages, primarily being a cost-effective semi-continuous process for manufacturing hollow composite parts. It is capable of producing cylindrical, spherical, conical, geodesic, and other curved shapes with highly repetitive and accurate fiber orientation. The resulting hollow parts can either be open (e.g., pipes) or closed at the ends (e.g., oxygen tanks with hemispherical ends).

3.4.11 Pultrusion

Pultrusion is a continuous manufacturing method designed for the production of long, straight structural composite components with a consistent cross-sectional area [25–27]. This process is capable of creating an array of solid and hollow cross sections with wall thicknesses ranging from 2 to 30 mm. Commonly pultruded products include round and rectangular tubes, solid rods, flat and corrugated sheets, as well as various profiled beams such as angles, channels, hat sections, double web sections, and wide-flange sections. These products find applications in both commercial and recreational realms, such as ladder rails, fence posts, tent poles, antenna rods, window frames, ski poles, and tool handles. Moreover, they play a crucial role in structural applications like floor gratings, loading platforms, bridge structures, and various mass transit components including overhead handrails, guardrails, and seat frames.

Typically, pultruded composites incorporate longitudinally oriented continuous fiber rovings alongside several layers of random fiber mat, including continuous filament mat (CFM) and chopped strand mat (CSM), positioned at or near the outer surfaces. The rovings impart high modulus and tensile strength in the longitudinal direction, while the random mat layers, containing randomly oriented continuous fiber strands, enhance the modulus and tensile strength in the transverse direction. The proportion of random fiber mat to continuous fiber rovings dictates the transverse to longitudinal properties ratio. In cases where superior transverse properties are essential, woven roving fabrics are employed instead of random fiber mats.

In typical pultruded products, the total fiber content falls within the range of 50–60% by weight, with a roving to mat ratio of 2.5:1 [25]. E-glass fibers are commonly used as reinforcement in the majority of pultruded products, coupled with either a polyester or a vinyl ester resin as the matrix. For applications requiring heightened strength or modulus, carbon fibers are utilized as reinforcement. Epoxy resins come into play as the matrix in scenarios demanding greater heat resistance or fatigue resistance than what polyester or vinyl ester resins can offer. However, epoxy resins require longer pultrusion dies due to their slower curing rate and do not easily release from the pultrusion die owing to their low curing shrinkage. While pultrusion with thermoplastic matrix composites is feasible, the process differs from those involving thermosetting matrix composites.

Pultrusion is a highly automated, continuous manufacturing process characterized by a rapid production rate. It demands minimal manual labor, resulting in very low material scrap. Additionally, the tooling cost is relatively economical, making pultrusion an affordable production method for manufacturing structural beam sections with a high fiber volume fraction. However, it is important to note that the dimensional accuracy and tolerances of pultruded products may not be as precise as with other processes. Furthermore, producing very thin-walled sections and intricate cross-sectional profiles is not achievable through pultrusion.

The preceding sections of this book chapter elaborate on the manufacturing processes for PMC parts. To construct an entire structure, these composite parts need to be interconnected, either with other composite parts or components made from different materials. The number of joints required depends on the structure's complexity, while the selection of the joining method is primarily influenced by the materials involved and the load transfer specifications between the interconnected parts. For instance, a large commercial aircraft incorporates millions of diverse aluminum and composite parts, necessitating an almost equal number of joints, typically comprised of rivets, pins, and other threaded fasteners. In modern automobiles, the body and chassis comprise numerous steel parts, often joined using resistance spot welds.

The fundamental purpose of a joint is to facilitate the transfer of loads from one structural member to another. This load transfer occurs through shear, tension, or a combination of both. Joints are typically the weakest points in a structure, and failures in load-bearing structures often originate at these junctures. The structural effectiveness of a joint hinges on the joining method, joint design, and overall joint quality. In the realm of PMCs, joint design assumes particular significance as these materials lack the forgiving characteristics of ductile metals, such as the ability to redistribute localized high stresses through yielding.

This section researches into the two predominant joining methods employed in PMCs: mechanical fastening and adhesive bonding. Additionally, it explores the joining of thermoplastic matrix composites through fusion bonding. The chapter concludes by addressing techniques for repairing PMCs damaged during service.

Mechanical joints are established by fastening substrates using bolts, rivets, or pins. On the other hand, bonded joints utilize an adhesive interlayer between the substrates, commonly referred to as adherends. Both types of joints are applicable for connecting composites to composites or composites to other materials, particularly metals. This chapter places emphasis on the procedural aspects of these two joining methods specifically tailored for PMC parts. In scenarios where composites are joined with metals, additional measures, such as corrosion prevention for the metal substrate, may be necessary. There are valuable [28] that offer insights into stress analysis of joints in composite materials, providing further details on joint design and performance evaluation.

3.5 Machining

The machining of polymer matrix composites requires a delicate balance between precision engineering and material science to craft components with intricate geometries. CNC machining, a fundamental aspect of composite manufacturing, utilizes computer-controlled machines for the precise shaping of complex forms. Careful consideration is given to the selection of cutting tools, with a preference for diamond-tipped or polycrystalline diamond (PCD) tools to counteract the abrasive wear induced by reinforcing fibers. The challenge lies in optimizing cutting parameters, such as feed rates and spindle speeds, to strike a balance between material removal and heat generation, thereby preventing thermal damage to the composite.

Advanced machining technologies, including abrasive waterjet cutting and laser ablation, provide alternative approaches. Waterjet cutting, employing a high-velocity stream of abrasive particles, minimizes heat buildup and tool wear, making it suitable for intricately cutting composite shapes without compromising structural integrity. Laser ablation, utilizing a focused laser beam, offers a non-contact method for precise material removal, reducing mechanical stress on the composite.

Nevertheless, machining composites presents challenges. The anisotropic nature of the material complicates the prediction of tool wear and optimization of cutting conditions. Delamination, resulting from the separation of composite layers during machining, is a concern that requires continuous attention. Ongoing research efforts focus on developing predictive models and simulation tools to better comprehend and address these challenges, ensuring efficient machining processes while preserving the mechanical properties of polymer matrix composites. With the increasing demand for high-performance composite components, continuous research and development in machining techniques aim to push the boundaries of what is achievable. Collaborations among material scientists, machinists, and manufacturers are essential for advancing cutting-edge technologies and establishing best practices in machining polymer matrix composites across various industries.

3.6 Surface Finishing

The surface finishing of polymer matrix composites is a crucial aspect of their processing, impacting not only aesthetics but also functional performance and longevity. The intricate challenge lies in achieving a smooth, flawless surface without compromising the reinforcing fibers or altering the mechanical properties of the composite. Surface finishing processes employ abrasive techniques such as sanding and abrasive blasting. However, the heterogeneous and anisotropic nature of polymer matrix composites demands a delicate balance. Increasingly, automated surface finishing technologies are utilized to ensure consistency and precision. Robotics, equipped with adaptive control systems, enable controlled material removal, reducing the risk of fiber damage and ensuring a uniform finish.

Following the finishing process, the application of protective coatings adds an additional layer of complexity. Coatings must be carefully selected based on compatibility with the composite matrix, ensuring strong adhesion and resistance to environmental factors. This protective layer enhances the composite's durability against UV radiation, moisture, and chemicals, contributing to its overall aesthetic appeal.

In recent years, innovations in surface finishing technologies have emerged to address challenges specific to polymer matrix composites. Non-contact methods, such as laser surface texturing, are explored to create controlled surface patterns, improving adhesion between the composite and coatings while minimizing damage to reinforcing fibers. Moreover, surface finishes play a crucial role in improving the fatigue resistance of composites, reducing the likelihood of stress concentration points. Techniques like shot peening, which induces compressive stresses on the surface, are investigated to enhance the overall mechanical performance of composite components. As the demand for high-performance composite materials continues to grow, ongoing research in surface finishing focuses on developing efficient, sustainable, and cost-effective methods. Collaboration among researchers, manufacturers, and surface treatment specialists is vital for advancing technologies that not only meet aesthetic expectations but also contribute to the longevity and reliability of polymer matrix composites in diverse applications.

3.6.1 Joining

Joining techniques in polymer matrix composites play a crucial role in shaping their final form and determining their structural integrity [29]. Each method presents unique challenges and considerations to ensure a reliable and robust assembly. Adhesive bonding is widely used due to its ability to evenly distribute stress across the bonded area. However, achieving a strong bond requires meticulous surface preparation, often involving abrasion and chemical treatments. Selecting the appropriate adhesive is crucial, considering compatibility with the composite matrix and specific application conditions. Advances in adhesive technology, such as toughened and high-temperature-resistant adhesives, contribute to the versatility and durability of composite structures. Mechanical fastening involves using bolts, screws, or other connectors to join composite components. Careful consideration is given to hole design, fastener material, and torque values to prevent damage to the reinforcing fibers during installation. Hybrid methods, combining adhesive bonding and mechanical fastening, leverage the benefits of both approaches. This hybridization enhances overall structural performance, providing redundancy and mitigating weaknesses inherent in each method.

The anisotropic and heterogeneous nature of polymer matrix composites requires joining techniques to adapt to variations in material properties. Tailoring joining methods to specific composite formulations, fiber orientations, and intended applications is essential. Ongoing research explores innovative joining technologies, such as automated robotic assembly and additive manufacturing, to improve efficiency, precision, and scalability in the production of composite structures. As industries increasingly rely on polymer matrix composites for lightweight and high-strength applications, ongoing developments in joining technologies aim to address challenges related to compatibility, durability, and scalability. Collaborations between material scientists, engineers, and manufacturers drive advancements in joining methods, ensuring that polymer matrix composites meet and exceed performance expectations across diverse sectors, from aerospace to automotive and beyond.

3.6.2 Adhesive Bonding

The adhesives employed for bonding structural composites include epoxies, polyurethanes, and acrylics, each offering distinct characteristics. These adhesives can be tailored to exhibit a broad spectrum of properties. Epoxies, widely utilized in the aircraft industry, are favored for their high tensile strength, temperature resistance, and fatigue durability [30]. They demonstrate effective bonding across diverse materials, encompassing composite-to-composite and composite-to-metal connections. While their impact resistance is relatively low, incorporating elastomeric tougheners can significantly enhance this aspect. Various hybrid epoxy formulations, such as epoxy-acrylic, epoxy-cyanate ester, epoxy-urethane, and epoxy-polyamide, deliver improved attributes like enhanced heat resistance or increased toughness at lower temperatures [28].

In the automotive sector, although epoxies find application, polyurethanes and acrylics are preferred due to their shorter bonding time compared to epoxies. Polyurethanes establish robust bonds with a variety of materials, offering notable impact and shock resistance, albeit with lower tensile strength and temperature resistance compared to epoxies. Acrylics are recognized for their ability to bond to surfaces with minimal or no surface preparation. While their mechanical properties rival or surpass many epoxies, acrylics exhibit higher curing shrinkage, leading to elevated residual stresses in the adhesive. Adhesives are available in liquid, paste, or film forms. The liquid and paste variants may exist as two-part systems, requiring the mixing of reactive chemicals before application on the substrates. The curing reaction, often necessitating elevated temperatures, transforms the mixture into a solid adhesive over time. Single-part systems come in two types: one where two premixed chemicals remain partially cured and nonreactive until dispensed on surfaces, and another where a single chemical relies on an external energy source such as moisture for certain polyurethanes or UV for acrylics to initiate the curing reaction. Film adhesives, also single-part premixed systems, are typically in a semisolid or solid form, supported by a thin carrier made of nylon or polyester fabric. This carrier aids in controlling bond line thickness and fortifies the adhesive joint's toughness. Film adhesives are favored for many aerospace composite joints due to their ease of handling and uniform application thickness. After applying the film adhesive between the surfaces to be joined, it undergoes curing at elevated temperatures and under pressure.

3.7 Inspecting and Testing

Essential for evaluating the internal integrity of polymer matrix composites, nondestructive testing (NDT) methods are employed. Commonly used techniques include ultrasonic testing, radiographic testing, and thermography, each presenting distinct advantages and challenges. The anisotropic and heterogeneous nature of composites necessitates specialized equipment and trained technicians to ensure accurate and reliable results. Consistent inspections conducted throughout the manufacturing process and the product's life cycle contribute to maintaining quality and performance standards. Ongoing advancements in inspection technologies, incorporating artificial intelligence and automation, are improving the efficiency and accuracy of testing processes, thereby ensuring the safety and reliability of composite materials across various applications.

3.7.1 Repair

The repair of polymer matrix composites involves the identification and resolution of various damages, ranging from delamination to fractures caused by impacts. Techniques such as patching resin injection, and scarf repairs are utilized, each requiring a nuanced approach tailored to the specific characteristics of the damage. It is essential for repair materials to match the original composite matrix, and curing conditions are meticulously monitored to ensure the restoration of structural integrity. Skilled technicians play a pivotal role in implementing effective repair solutions, taking into account the anisotropic properties of composites. Ongoing research in repair methodologies, including the development of advanced repair materials and techniques, contributes to the extended service life and sustainability of polymer matrix composite structures.

3.7.2 Assembly

The assembly of polymer matrix composites involves the identification and resolution of various damages, ranging from delamination to fractures caused by impacts. Techniques such as patching, resin injection, and scarf repairs are utilized, each requiring a nuanced approach tailored to the specific characteristics of the damage. It is essential for repair materials to match the original composite matrix, and curing conditions are meticulously monitored to ensure the restoration of structural integrity. Skilled technicians play a pivotal role in implementing effective repair solutions, taking into account the anisotropic properties of composites. Ongoing research in repair methodologies, including the development of advanced repair materials and techniques, contributes to the extended service life and sustainability of polymer matrix composite structures.

3.7.3 *Painting and Coating*

Painting and coating processes for polymer matrix composites involve selecting materials that are compatible with the composite matrix to ensure proper adhesion and durability. Surface preparation, including cleaning and priming, is critical for successful coating application. Protective coatings shield composites from environmental factors, UV radiation, and chemical exposure, contributing to enhanced longevity and aesthetics of the finished product. Coatings may also serve additional functions, such as improving fire resistance or providing anti-corrosive properties. Advances in coating technologies, including nanocomposite coatings and smart coatings, are explored to address specific performance requirements and further enhance the overall functionality of polymer matrix composites in diverse applications.

3.7.4 *Quality Control*

Ensuring the quality and reliability of polymer matrix composites involves a comprehensive approach, encompassing mechanical, thermal, and chemical testing. Standardized testing methods are employed to evaluate material properties, including tensile strength, modulus, and impact resistance. Adherence to industry standards, rigorous inspections, and continuous monitoring of manufacturing processes are imperative to maintain consistent product quality. Regular audits and quality assessments contribute to the ongoing improvement of composite materials and their applications. Advanced quality control measures, such as real-time monitoring and data analytics, are increasingly integrated into production processes to enhance efficiency and provide insights for continuous quality improvement in the dynamic field of polymer matrix composites.

3.8 Processing Techniques in Nano Polymer Matrix Composites

Traditionally, polymer matrix nanocomposites have been crafted through mechanical or chemical processes. The primary objective of employing diverse processing techniques is to achieve a homogeneous and even dispersion of nanoparticles within the polymer matrix. However, ensuring the proper dispersion of nanoparticles in the polymer matrix poses a significant challenge for these nanocomposites. Nanoparticles tend to exhibit an inclination to agglomerate, forming clusters of micron size among themselves. This tendency leads to an inadequate distribution of nanoparticles within the polymer matrix, subsequently diminishing the properties of the nanocomposites. There are various researchers have endeavored to address this issue by employing different methods, including polymerization reactions, surface modification, and chemical reactions involving filler materials [31]. The subsequent methods represent the principal approaches employed for the fabrication of polymer matrix nanocomposites.

3.8.1 Ultrasonication and Dual Mixing

This technique employs an ultrasonic apparatus to disperse various nanofillers in a polymer matrix. The ultrasonic equipment transfers high impact energy, generating lower shear energy in the ultrasonic chamber. The use of a sonicator provides the advantage of initiating the de-agglomeration of clustered nanoparticles. Qian et al. [32] successfully dispersed multi-walled carbon nanotubes (MWCNTs) in a polystyrene polymer using the high-energy sonication method. Initially, acetone is used as a solvent, and CNTs are dispersed in it, followed by sonication for a specified duration. Sonication proves instrumental in de-agglomerating the nanoparticles, given their tendency to agglomerate due to higher specific surface area. Achieving good dispersion is challenging, especially in the case of single-walled CNTs. However, the generation of high energy during sonication can potentially alter the shape and size of the nanoparticles. After a specific sonication period, the polymer and suspension are mixed until the complete evaporation of the solvent. Song et al. [33] successfully fabricated and achieved well-dispersed MWCNT/epoxy composites using this technique. The CNTs were sonicated in an ethanol solution for 2 h. Subsequently, sonication of the CNT/ethanol mixture was conducted for 1 h at 80 °C, and the suspension was vacuum-sealed for five days to ensure the complete removal of air bubbles. Researchers have advocated the mechanical stirring followed by sonication as an effective method for dispersing nanofillers in a polymer matrix [34–36]. Chatterjee A and Islam MS fabricated and characterized TiO_2/epoxy nanocomposites [37], utilizing a Vibracell ultrasonic processor for nano-TiO_2 filler dispersion in the polymer matrix. Mahato et al. [9, 38, 39] produced epoxy with nano-Al_2O_3 particles by stirring with a magnetic stirrer for 4 h, followed by sonication with a high-frequency sonicator at 60 °C, as illustrated in Fig. 3.2. The shape and size of nano-Al_2O_3 particles were examined using a field emission scanning electron microscope (FESEM).

Mahato et al. [9] have fabricated nano-Al_2O_3/epoxy mixing through magnetic stirrer for 1 h, followed by sonication with a high frequency sonicator at 60 °C presented in Fig. 3.3. The size and shape of nano-Al_2O_3 particles has been scrutinized by field emission scanning electron microscope (FESEM). Figure 3.3 depicts the dispersion characteristics of nano-Al_2O_3 particles within the epoxy matrix of GE, 0.1 wt%, and 0.5 wt% nano-Al_2O_3/GE laminated composites. In the epoxy polymer of the 0.1 wt% nano-Al_2O_3/GE composite, nano-Al_2O_3 particles exhibit a predominantly uniform distribution (Fig. 3.3a). Conversely, in the case of 0.5 wt% nano-Al_2O_3/GE composites, numerous nano-Al_2O_3 particles are observed in the form of agglomerates (Fig. 3.3b). The toughness of a material can be enhanced

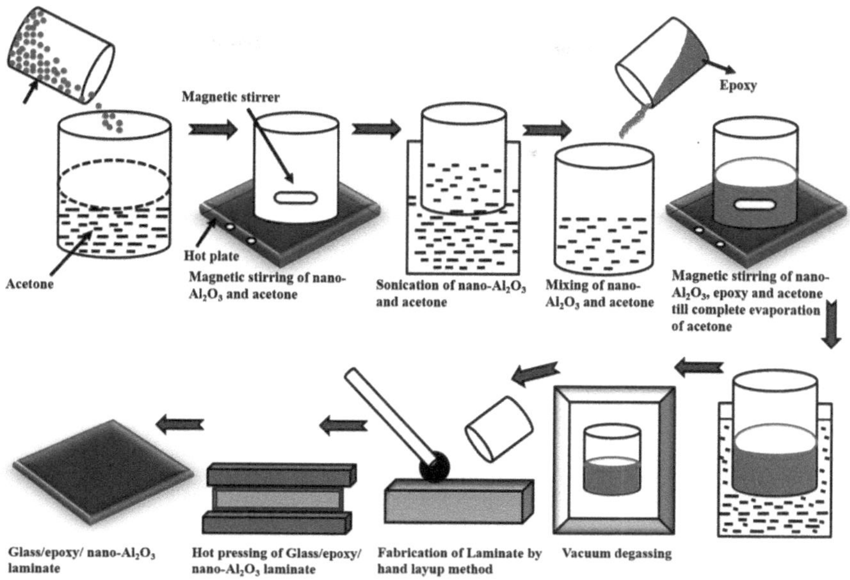

Fig. 3.2 A schematic representation of the order of processing stages involved in the fabrication of nano-Al_2O_3 in epoxy/GE composite [9]. Adapted from Ref. [9] with permission from [Elsevier], Copyright [2019]

through mechanisms such as crack bridging [40], evident in the fracture surface of the 0.1 wt% nano-Al_2O_3/GE composite (Fig. 3.3c). It is expected that the toughness of the 0.1 wt% nano-Al_2O_3/GE composite is superior compared to other composites where nanoparticles are present as agglomerates. None of the composites, except for the 0.1 wt% nano-particles enhanced composite, demonstrated the crack bridging phenomenon. In the case of FRP composites, the epoxy matrix and fiber/matrix interface govern the maximum transverse mechanical properties. Following morphological examination of the modified epoxy matrix in each nano-Al_2O_3/GE composite, the fiber/matrix interface was analyzed for control GE and 0.1 wt% nano-Al_2O_3/GE. In the GE composite (Fig. 3.3d), the bare fiber surface, devoid of any polymeric phase signatures, indicates fiber/matrix debonding, likely due to poor interfacial bonding. In contrast, the 0.1 wt% nano-Al_2O_3/GE composite (Fig. 3.3e) displays a well-established fiber/matrix interfacial bond, with a visible layer of the epoxy matrix on the fiber surface. Additionally, morphologies like mirror, mist, and hackle are observed on the fiber surface [41].

3.8.2 Sol–Gel Process

This synthesis method proves highly practical for producing nanocomposites compared to other approaches. The sol–gel method involves two key steps: sol

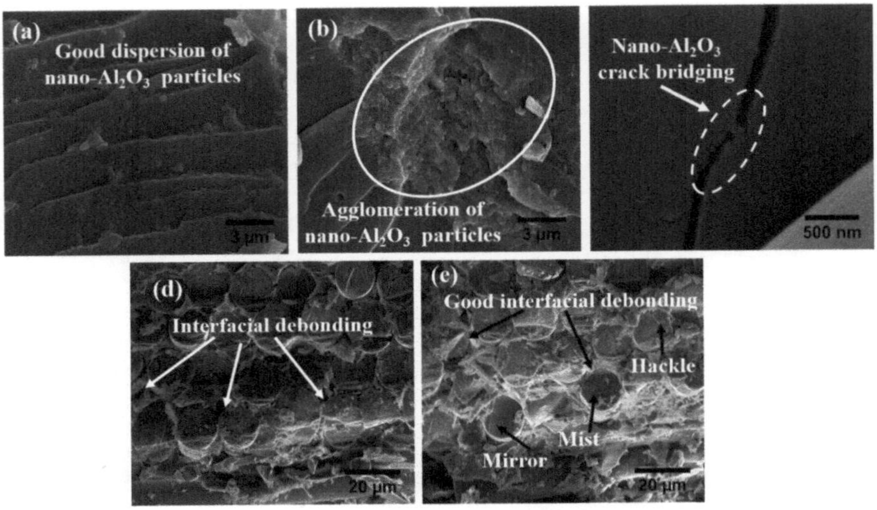

Fig. 3.3 Dispersion of **a** 0.1%, **b** 0.5% nano-Al_2O_3 in GE composite, and **c** crack bridging by nano-Al_2O_3 in 0.1% nano-Al_2O_3 composite after room temperature testing **d** GE and **e** 0.1% nano-Al_2O_3/GE composite [9]. Adapted from Ref. [9] with permission from [Elsevier], Copyright [2019]

and gel. In general, the colloidal suspension of solid nanoparticles in a monomer is referred to as the sol, while the three-dimensional connection formed between phases is termed the gel [42]. Typically, solid nanoparticles disperse throughout the monomer, creating a colloidal suspension of nanoparticles, i.e., sol. Simultaneously, through the polymerization reaction, interconnected regions between phases develop, called the gel. The three-dimensional polymer nanoparticles network spreads throughout the liquid phase, with the polymer matrix acting as the nucleation media, fostering the growth of layered crystals. This crystal development leads to the formation of nanocomposites. Numerous trials have been conducted to synthesize various polymer/filler nanocomposites, where inorganic particle phases with dimensions ranging from angstroms to nanometers are uniformly and homogeneously distributed. This includes inorganic/organic molecular hybrid materials, synthesized through reactions of metal alkoxides [42–45], as exemplified by the polycondensation and hydrolysis reactions of tetraalkoxy silane in Eq. (3.1).

$$Si(OR)_4 + 4\,H_2O \rightarrow [Si(OH)x(OR)4 - x + xROH] \rightarrow SiO_2 + 4\,ROH + 2\,H_2O \tag{3.1}$$

Here, Eq. (3.1) represents one of the typical sol–gel reactions used for synthesizing silica/polymer molecular hybrid materials. The current progress in sol–gel technology has led to widespread consideration of molecular hybrid materials and nanocomposites by numerous researchers and organizations. This technology enables the dissolution of inorganic fillers with dimensions smaller than the molecular chain length of the polymer matrix. However, in the organic area of hybrid materials in the sol–gel method, different polymers are constrained to those with hydrogen bond acceptor groups capable of forming hydrogen bonds with hydroxyl groups on the inorganic filler surface, such as alcohol- and water-soluble polymers. Moreover, the conditions of the sol–gel reactions have a significant impact on the structure of the formed inorganic network. At the molecular level, controlling the arrangement and size of the inorganic field in hybrid materials poses challenges. Recognizing this, a synthesis process known as the Site Selective Molecular Hybrid Method has been established, involving the use of silane-modified polymers with oligomers of alkoxysilane selectively introduced into suitable places in each polymer [46].

This technique allows for the use of a polymer matrix with no interactions with metal alkoxides, previously unsuitable for the conventional sol–gel process (as shown in Fig. 3.4). The ultimate advantage of this process lies in its potential to regulate the microstructure in hybrid materials, catering to specific applications. Researchers like Goda et al. have employed this technology to enhance the properties of polyurethane [46, 47]. Typically, the elastic properties of rubber in segmented urethane block copolymers are influenced by a two-phase microstructure where hard parts separate from the soft parts to form distinct areas. Through the molecular hybrid technique, a hybrid material with dispersed hybrid areas of the hard part interpenetrating with silica in the soft-zone constant phase can be synthesized. This hybrid material often exhibits excellent thermal and mechanical behaviors, as the hybrid regions are tougher and more heat-resistant than the original hard part, while maintaining the flexibility of the soft part [46].

3.8.3 Direct Mixing of Polymer and Nanofillers

The direct blending of nanofillers with a polymer matrix is a suitable approach for fabricating nanocomposites, primarily due to its effectiveness in breaking down agglomerated nanofillers during the mixing process [31]. This method encompasses two techniques for the mixing of nanofillers and polymer. The first technique involves the mixing of polymer without the presence of any solvent with nanofillers above the glass transition temperature of the polymer, commonly referred to as the melt compounding technique [31, 48, 49]. The second method of mixing includes blending nanofillers and polymer in the presence of a solvent, typically known as solution mixing or the solvent method [31, 48–50].

3.9 Applications and Future Trends

PMCs have gradually supplanted their metallic counterparts, either entirely or partially, over the past few decades, and the replacement and repair of metallic materials with PMC are well underway. Unfortunately, instances of unexpected failure

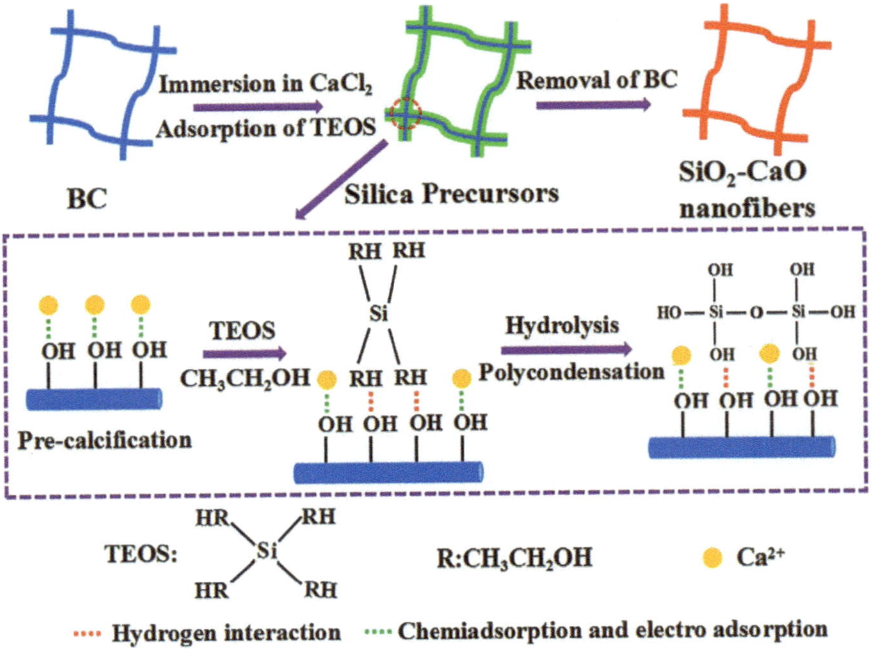

Fig. 3.4 Schematic assisted sol–gel process [47]. Adapted from Ref. [47] with permission from [Elsevier], Copyright [2017]

of PMC materials before reaching the predicted service life of components or parts are exceedingly rare. It is reasonably believed that the environment poses the most significant concern, triggering nucleation zones for early PMC component failure. Various polymer processing methods have been developed to create PMCs that meet the design requirements of different applications. However, there is a limit towards the processing of PMCs. To advance to the next level, the commercialization of PMC fabrication is imperative. The appropriate process for polymer matrix needs to be increased to unlock the full potential of PMCs. However, the environmental impact of PMCs must be carefully considered to ensure better human health. In contemporary times, various nano-fillers have been introduced into the epoxy polymer matrix to enhance the damage tolerance of composites. The impact of nano-fillers on all potential failure modes must be assessed through micro characterization techniques such as FESEM, TEM, FTIR, Raman Spectroscopy, etc. Investigating different loading modes, water uptake kinetics theory, mechanics, and the influence of nano-fillers on diffusivity at varying temperatures during conditioning and testing are crucial parameters for optimizing nano-filled glass fiber-reinforced polymer (GFRP) composites to achieve their highest potential. Micro characterization ensures the reliability, sustainability, and predictability of composites under diverse service conditions. Extensive research in various directions and dimensions of these new-generation composites may meet the emerging demands of today's technological evolution.

3.10 Conclusions

PMCs, being non-crystalline and brittle, may not benefit from the strengthening mechanisms applicable to metals due to different failure mechanisms. In PMCs, crack initiation can occur either in the matrix or at the matrix/fiber interface. If the interface bond strength is sufficient to withstand the applied load, crack initiation is more likely in the matrix than in the fiber. The crack then propagates through the matrix and fiber-matrix interface, depending on the adhesive bond between the matrix and fiber, ultimately leading to the gradual propagation of microcracks in the composites. The various manufacturing processes (primary and secondary) for the polymer matrix, such as Molding Methods, Bag Molding Process, Layup techniques, Automated Tape Laying, Autoclave Curing, Sandwich Molding, Compression Molding, Sheet Molding Compound, Bulk Molding Compound, Compression Molding Compound, Liquid Composite Molding, Filament Winding, Pultrusion, sol–gel processes, mechanical mixing, and ultrasonication has been discussed in the current book chapter. Achieving the proper processing method in the PMCs remains a challenge for both researchers and academia in the commercialization.

References

1. Ray, B.C., Rathore, D.: Durability and integrity studies of environmentally conditioned interfaces in fibrous polymeric composites: critical concepts and comments. Adv. Coll. Interf. Sci. **209**, 68–83 (2014). https://doi.org/10.1016/j.cis.2013.12.014
2. Ray, B.C., Rathore, D.: A review on mechanical behavior of FRP composites at different loading speeds. Crit. Rev. Solid State Mater. Sci. **40**, 119–135 (2015). https://doi.org/10.1080/10408436.2014.940443
3. Mahato, K.K., Dutta, K., Ray, B.C.: 6—Emerging advancement of fiber-reinforced polymer composites in structural applications. In: Samui, P., Kim, D., Iyer, N.R., Chaudhary, S., (eds.) New Materials in Civil Engineering, Butterworth-Heinemann, pp. 221–71 (2020). https://doi.org/10.1016/B978-0-12-818961-0.00006-5
4. Hollaway, L.C.: A review of the present and future utilisation of FRP composites in the civil infrastructure with reference to their important in-service properties. Constr. Build. Mater. **24**, 2419–2445 (2010). https://doi.org/10.1016/j.conbuildmat.2010.04.062
5. Mallick, P.K.: Processing of Polymer Matrix Composites. CRC Press (2017)
6. Mahato, K.K., Biswal, M., Rathore, D.K., Prusty, R.K., Dutta, K., Ray, B.C.: Effect of loading rate on tensile properties and failure behavior of glass fibre/epoxy composite. IOP Conf. Ser: Mater. Sci. Eng. **115**, 012017 (2016). https://doi.org/10.1088/1757-899X/115/1/012017
7. Hawileh, R.A., Abu-Obeidah, A., Abdalla, J.A., Al-Tamimi, A.: Temperature effect on the mechanical properties of carbon, glass and carbon–glass FRP laminates. Constr. Build. Mater. **75**, 342–348 (2015). https://doi.org/10.1016/j.conbuildmat.2014.11.020
8. Sethi, S., Ray, B.C.: An assessment of mechanical behavior and fractography study of glass/epoxy composites at different temperatures and loading speeds. Mater. Des. **64**, 160–165 (2014). https://doi.org/10.1016/j.matdes.2014.07.017
9. Mahato, K.K., Dutta, K., Chandra, R.B.: Assessment of mechanical, thermal and morphological behavior of nano-Al2O3 embedded glass fiber/epoxy composites at in-situ elevated temperatures. Compos. B Eng. **166**, 688–700 (2019). https://doi.org/10.1016/j.compositesb.2019.03.009

10. Kumar Mahato, K., Kumar Rathore, D., Dutta, K., Kumar Prusty, R., Chandra, R.B.: Effect of severely thermal shocked nano-Al2O3 filled glass fiber reinforced polymeric composites: an assessment on tensile, thermal and morphological behaviour. Mater. Today: Proc. **33**, 5521–5525 (2020). https://doi.org/10.1016/j.matpr.2020.03.334
11. Nayak, R.K., Mahato, K.K., Routara, B.C., Ray, B.C.: Evaluation of mechanical properties of Al2O3 and TiO2 nano filled enhanced glass fiber reinforced polymer composites. J. Appl. Polym. Sci. **133** (2016). https://doi.org/10.1002/app.44274
12. Kumar, D.S., Shukla, M.J., Mahato, K.K., Rathore, D.K., Prusty, R.K., Ray, B.C.: Effect of post-curing on thermal and mechanical behavior of GFRP composites. IOP Conf. Ser: Mater. Sci. Eng. **75**, 012012 (2015). https://doi.org/10.1088/1757-899X/75/1/012012
13. Dillon, G., Mallon, P., Monaghan, M.: The autoclave processing of composites. In: Gutowski, T.G. (ed.) Advanced Composites Manufacturing, pp. 207–258. John Wiley & Sons, New York (1997)
14. Lekakou MGB Constantina: Processing for Laminated Structures. CRC Press, Composites Engineering Handbook (1997)
15. Bloom, L.D., Wang, J., Potter, K.D.: Damage progression and defect sensitivity: an experimental study of representative wrinkles in tension. Compos. B Eng. **45**, 449–458 (2013). https://doi.org/10.1016/j.compositesb.2012.05.021
16. Elkington, M., Bloom, D., Ward, C., Chatzimichali, A., Potter, K.: Hand layup: understanding the manual process. Adv. Manuf.: Polymer Compos. Sci. **1**, 138–151 (2015). https://doi.org/10.1080/20550340.2015.1114801
17. Centea, T., Grunenfelder, L.K., Nutt, S.R.: A review of out-of-autoclave prepregs—Material properties, process phenomena, and manufacturing considerations. Compos. A Appl. Sci. Manuf. **70**, 132–154 (2015). https://doi.org/10.1016/j.compositesa.2014.09.029
18. Ciriscioli, P.R., Wang, Q., Springer, G.S.: Autoclave curing—Comparisons of model and test results. J. Compos. Mater. **26**, 90–102 (1992). https://doi.org/10.1177/002199839202600106
19. Mallick, P.K.: Compression molding. In: Mallick, P.K., Newman, S. (eds.) Composite materials technology—Processes and properties, pp. 67–102. Hanser Publishers, Munich, Vienna, New York. n.d. (1990)
20. Meyer, R.W.: Handbook of Polyester Molding Compounds and Molding Technology. Springer US, Boston, MA (1987). https://doi.org/10.1007/978-1-4613-1961-0
21. Kia, H.G.: In: Kia, H.G. (ed.) Sheet Molding Compounds: Science and Technology, pp. 257–257. Hanser, Munich (1993)
22. Liquid Moulding Technologies—1st Edition. n.d. https://shop.elsevier.com/books/liquid-moulding-technologies/rudd/978-1-85573-242-1. Accessed 24 Jan 2024
23. Fong, L., Advani, S.G.: Resin transfer molding. In: Peters, S.T. (ed.) Handbook of Composites, pp. 433–455. Springer US, Boston, MA (1998). https://doi.org/10.1007/978-1-4615-6389-1_21
24. Peters, S.T., Tarnopol'skii, Y.M.: Filament winding. In: Mallick, P.K. (ed.) Composites Engineering Handbook, pp. 515–548. Marcel Dekker, New York (1997)
25. Sumerak, J.E.: The pultrusion process for continuous automated manufacture of engineered composite profiles. In: Mallick, P.K. (ed.) Composites Engineering Handbook, pp. 549–577. Marcel Dekker, New York (1997)
26. Starr, T.F. (ed.): Pultrusion for Engineers. Woodhead, Cambridge, UK (2000)
27. Wilson, B.A.: Pultrusion. In: Peters, S.T. (ed.) Handbook of Composites, pp. 488–524. Springer, New York (1998)
28. Dillard, D.A. (ed.) Front matter. In: Advances in Structural Adhesive Bonding, pp. i–iii. Woodhead Publishing (2010). https://doi.org/10.1533/9781845698058.frontmatter
29. Ray, BNVSGGK, Kumar Mahato, K., Prusty, R.K., Chandra, B.: Challenges of Adhesively Bonded Joints and Their Advantages Over Mechanical Fastening. Failure of Fibre-Reinforced Polymer Composites. CRC Press (2021)
30. Budhe, S., Banea, M.D., de Barros, S., da Silva, L.F.M.: An updated review of adhesively bonded joints in composite materials. Int. J. Adhes. Adhes. **72**, 30–42 (2017). https://doi.org/10.1016/j.ijadhadh.2016.10.010

31. Tanahashi, M., Tanahashi, M.: Development of fabrication methods of filler/polymer nanocomposites: with focus on simple melt-compounding-based approach without surface modification of nanofillers. Materials **3**, 1593–1619 (2010). https://doi.org/10.3390/ma3031593
32. Qian, D., Dickey, E.C., Andrews, R., Rantell, T.: Load transfer and deformation mechanisms in carbon nanotube-polystyrene composites. Appl. Phys. Lett. **76**, 2868–2870 (2000). https://doi.org/10.1063/1.126500
33. Song, Y.S., Youn, J.R.: Influence of dispersion states of carbon nanotubes on physical properties of epoxy nanocomposites. Carbon **43**, 1378–1385 (2005). https://doi.org/10.1016/j.carbon.2005.01.007
34. Chang, F.-K., Lessard, L.B.: Damage tolerance of laminated composites containing an open hole and subjected to compressive loadings: part I—Analysis. J. Compos. Mater. **25**, 2–43 (1991). https://doi.org/10.1177/002199839102500101
35. Barbezat, M., Brunner, A.J., Necola, A., Rees, M., Gasser, Ph., Terrasi, G.: Fracture behavior of GFRP laminates with nanocomposite epoxy resin matrix. J. Compos. Mater. **43**, 959–976 (2009).https://doi.org/10.1177/0021998308100799
36. Soundararajah, Q.Y., Karunaratne, B.S.B., Rajapakse, R.M.G.: Mechanical properties of Poly(vinyl alcohol) montmorillonite nanocomposites. J. Compos. Mater. **44**, 303–311 (2010). https://doi.org/10.1177/0021998309347040
37. Chatterjee, A., Islam, M.S.: Fabrication and characterization of TiO2–epoxy nanocomposite. Mater. Sci. Eng., A **487**, 574–585 (2008). https://doi.org/10.1016/j.msea.2007.11.052
38. Nayak, R.K., Ray, B.C.: Influence of seawater absorption on retention of mechanical properties of nano-TiO2 embedded glass fiber reinforced epoxy polymer matrix composites. Archiv. Civ. Mech. Eng. **18**, 1597–1607 (2018). https://doi.org/10.1016/j.acme.2018.07.002
39. Nayak, R.K., Mahato, K.K., Ray, B.C.: Water absorption behavior, mechanical and thermal properties of nano TiO2 enhanced glass fiber reinforced polymer composites. Compos. A Appl. Sci. Manuf. **90**, 736–747 (2016). https://doi.org/10.1016/j.compositesa.2016.09.003
40. Chen, C.-H., Jian, J.-Y., Yen, F.-S.: Preparation and characterization of epoxy/γ-aluminum oxide nanocomposites. Compos. A Appl. Sci. Manuf. **40**, 463–468 (2009). https://doi.org/10.1016/j.compositesa.2009.01.010
41. Greenhalgh, E.: Failure Analysis and Fractography of Polymer Composites. Woodhead Publishing (2009)
42. Schadler, L.S.: Polymer-Based and Polymer-Filled Nanocomposites. Nanocomposite Science and Technology. Wiley-Blackwell, pp. 77–153 (2004). https://doi.org/10.1002/3527602127.ch2
43. Popall, M., Meyer, H., Schmidt, H., Schulz, J.: Inorganic-organic composites (Ormocers) as structured layers for microelectronics. MRS Proc. **180**, 995 (1990). https://doi.org/10.1557/PROC-180-995n.d
44. Claude, C., Garetz, B., Okamoto, Y., Tripathy, S.: The preparation and characterization of organically modified silicates that exhibit nonlinear optical properties. Mater. Lett. **14**, 336–342 (1992). https://doi.org/10.1016/0167-577X(92)90049-P
45. Morikawa, A., Iyoku, Y., Kakimoto, M., Imai, Y.: Preparation of a new class of polyimide-silica hybrid films by sol-gel process. Polym. J. **24**, 107 (1992)
46. Goda, H., Frank, C.W.: Fluorescence studies of the hybrid composite of segmented-polyurethane and silica. Chem. Mater. **13**, 2783–2787 (2001). https://doi.org/10.1021/cm000711w
47. Luo, H., Li, W., Ao, H., Li, G., Tu, J., Xiong, G., et al.: Preparation, structural characterization, and in vitro cell studies of three-dimensional SiO2–CaO binary glass scaffolds built ofultra-small nanofibers. Mater. Sci. Eng. C **76**, 94–101 (2017). https://doi.org/10.1016/j.msec.2017.02.134
48. Hashimoto, M., Takadama, H., Mizuno, M., Kokubo, T.: Enhancement of mechanical strength of TiO2/high-density polyethylene composites for bone repair with silane-coupling treatment. Mater. Res. Bull. **41**, 515–524 (2006). https://doi.org/10.1016/j.materresbull.2005.09.014

49. Kalaitzidou, K., Fukushima, H., Drzal, L.T.: A new compounding method for exfoliated graphite–polypropylene nanocomposites with enhanced flexural properties and lower percolation threshold. Compos. Sci. Technol. **67**, 2045–2051 (2007). https://doi.org/10.1016/j.compscitech.2006.11.014
50. Chae, D.W., Kim, B.C.: Characterization on polystyrene/zinc oxide nanocomposites prepared from solution mixing. Polym. Adv. Technol. **16**, 846–850 (2005). https://doi.org/10.1002/pat.673

Chapter 4
Characterization Techniques of Polymer Composites

Monika Chaparia, Shivali Meena, Neelam Kumari, Sandip P. Choudhury, Ravi Kant Choubey, and Umesh Kumar Dwivedi

Abstract Polymer composites significantly improve the behavior and capabilities of materials compared to conventional materials. However, in order to comprehend the behavior of polymer composites their property profile must be thoroughly understood. This profile comprises aspects such as chemistry, morphology, functionality, particle size, etc. However, applying a variety of analytical tools is significant to have a better insight into its characterization. This chapter includes the characterization techniques of polymer composites to unravel properties like mechanical performance, tribological behavior, microstructural behavior, structural behavior, physical properties, electrical behavior, rheological properties, thermal behavior, and optical properties.

Keywords Polymer composites · Characterization · Techniques · Behavior

4.1 Introduction

Polymers are widely used in advanced applications due to their versatility and ease of molding and adaptable physical and chemical properties. However, a single polymer cannot satisfy the requirements of advanced application which lead to the idea of polymer composites. Composites consist of ceramic, metal, fiber, semiconductors, and other polymers as reinforcement and as matrix. Since, many materials are present in composition, a variety of factors can influence the characteristics of polymer composites, such as type, shape, size, and fraction of reinforcing materials [1]. Polymer composites offer remarkable characteristics like good corrosion and wear resistance, good abrasion resistance, fatigue resistance economic efficiency, light

M. Chaparia · S. Meena · N. Kumari · S. P. Choudhury · U. K. Dwivedi (✉)
Amity School of Applied Physics, Amity University Rajasthan, Jaipur 303002, India
e-mail: umeshkudwivedi@gmail.com

R. K. Choubey
Department of Applied Physics, Amity Institute of Applied Sciences (AIAS), Amity University, Noida Campus, Sector-125, Noida, Uttar Pradesh 201313, India

S. Moharana et al. (eds.), *Polymer Composites*, Engineering Materials,
https://doi.org/10.1007/978-981-97-2075-0_4

weight, high strength, high stiffness, and faster molding [2–4]. Polymer composites such as reinforced plastics, fiber-reinforced polymer are of huge importance in various fields like household, construction, automotive industry, aerospace, marine, energy, military, sports, communications, electronics, civil, medical, and automobile [5]. Polymer composites are also used extensively in biomedical applications as hard (bones) and soft (skin) tissues [6].

Characterization is a comprehensive and universal technique used to investigate the structure, characteristics, and behavior of a composite, and it plays a vital role in the process of creating new composites. Through physical, chemical, mechanical, tribological, rheological, thermal, and optical characterization researchers can better understand polymer composite materials and its application in industries, automotive, aerospace, defense, and medical field [7, 8].

For Physical and mechanical testing can be done in accordance with a wide range of standards, such as ASTM (American society for Testing and Materials), ISO (International Standard Organization), BS (British Standards), DIN (Deutsches institute fürNormung - the German Institute of Standatization), or IS (Indian Standard) utilizing techniques unique to the industry.

- Physical characterization includes density, hardness, and scratch resistance.
- Electrical characterization includes conductivity, dielectric characteristics.
- Mechanical characterization include tensile, shear, flexural and compressive characteristics.
- Tribological characterization includes wear, friction, lubrication.
- Rheological characterization includes capillary, rotational, Index/Melt Flow Rate.
- Thermal characterization includes DSC (differential scanning calorimetry), DTA (differential thermal analysis), and TGA (thermogravimetric analysis).
- Optical characterization includes UV/Visible spectrometer.

4.1.1 Characterization Methods of Polymer Composites

4.1.1.1 Mechanical Characterizations

The numerous polymer composites are applicable in various areas like construction; aerospace, automobile, sporting goods, etc. are greatly influenced by their mechanical properties. These characteristics determine how the composite material will react to loads and external forces as well as other contributing elements such as toughness, resistance to fatigue, and strength.

There are several standard tests to examine the mechanical behavior of composites like tensile testing, flexural testing, impact testing, and hardness testing according to various loading circumstances (Fig. 4.1).

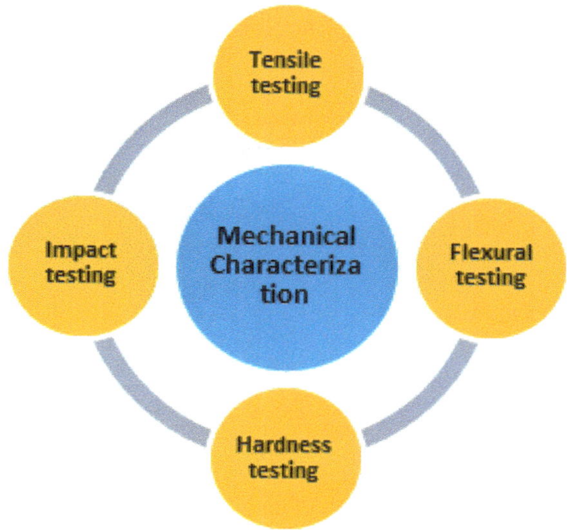

Fig. 4.1 Mechanical characterization testing techniques

4.1.1.2 Tensile Testing (Tension Testing)

In material science and engineering, tensile testing is a basic technique that is frequently used in the characterization of materials for a wide range of applications. It determines how much force is needed to break a composite or plastic specimen as well as how much the sample expands or contracts before breaking.

Tensile testing is done using flat sandwich or basic tension in compliance with standards like ASTM C 297, EN 2561, EN 2597 ISO 527-4, ISO 527-5, and ASTM D 638 [9]. With the help of these tests, stress–strain diagrams can be obtained and further tensile modulus, tensile strength, and percent elongation can be determined. Due to composites having a brittle and anisotropic nature, alignment is very critical. The anisotropic nature means its properties depend on the direction of applied force. When a composite material is tested in the direction parallel to the orientation of fibers, the tensile strength is very high and much lower if tested in any different direction. Basically, high tensile stress-based structures can be utilized in aerospace industries where maximum tensile strength with excellent axial loading is required. Generally, ASTMD638 (refer to Fig. 4.2) is used for making composite materials but ASTMD3039 is required when the polymer has more strength.

Fundamentally, stress is just about a restoring force generated per unit area of the material. It is represented by the Greek letter σ and is measured in Pascal or N/m^2 ($1\ N/m^2 = 1$ Pa). From the given experimental setup, the ratio of the amount of load applied to the specimen (F) to its original cross-sectional area (A) in the axial direction yields stress which is measured before the experiment is conducted. Mathematical expression can be written as

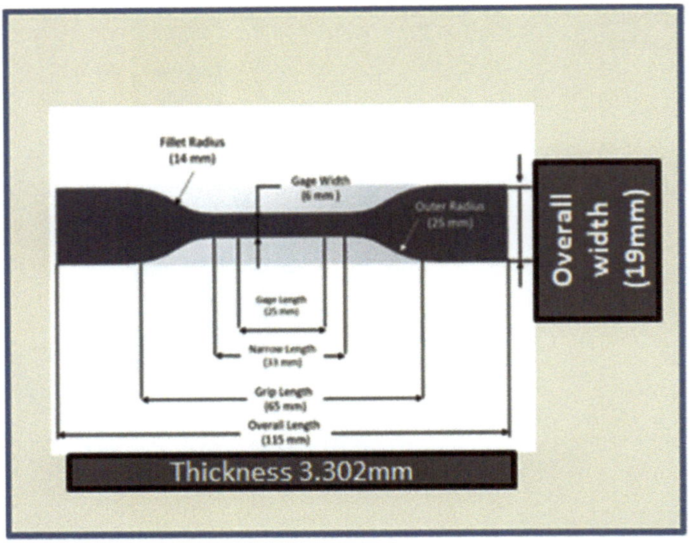

Fig. 4.2 ASTMD638 test specimen [10]

$$\sigma = \frac{F}{A} \tag{4.1}$$

In mechanics, strain determines the deformation produced due to stress applied on a material. It is a unit less quantity and can be expressed as

$$\text{Strain}(\varepsilon) = \frac{\delta \text{L}}{L} = \frac{\textit{Change in Length}}{\textit{Original Length}} \tag{4.2}$$

A stress–strain diagram illustrates the connection between the applied load on a material and the resulting deformation due to that load which is shown in Fig. 4.3 and Table 4.1.

4.1.1.3 Flexural Testing

Flexural testing is done to evaluate the ability of a material to resist or bend. To obtain a semi-qualitative understanding of the fiber-matrix interfacial strength of a composite, a flexural test is used. Generally flexural test is employed in brittle materials (both rigid and semi-rigid materials) like composites, ceramics, and plastic materials. This test procedure uses a three or four-point bend testing system. The material's property to bend before permanently deforming is indicated by its flex modulus. The following standard explains how to measure flexural properties; ASTM D 790, ASTM D 6272, ASTM D7264, ISO 178. Numerous specimen shapes are suitable for conducting this test, but the prevalent specimen size according to ISO standards is 1 cm × 0.4 cm ×

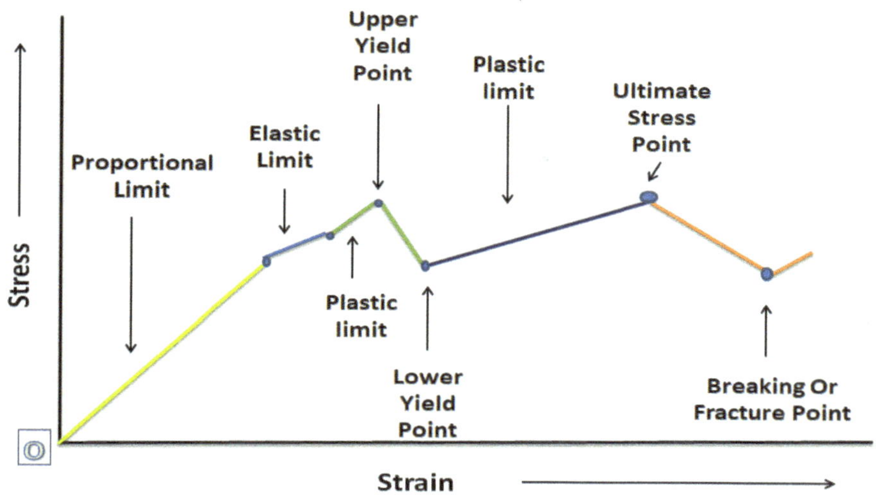

Fig. 4.3 Stress and strain curve showing significant points

Table 4.1 Showing tensile properties with an explanation

Tensile property	Justification
Proportional limit	In this region, stress is directly proportional to strain. Within this limit Hooke's law is obeyed
Elastic limit	The highest stress that a substance can endure before permanent deforming comes under elastic limit
Upper yield point	The ductile material shows plastic qualities beyond the elastic limit. At this point the material needed maximum stress to start plastic deformation
Lower yield point	At a lower yield point, the minimum load demanded for the material to behave as plastic
Ultimate stress point	The utmost stress a material can endure before breaking is known as the ultimate stress point
Breaking or fracture point	The point at which lastly material breaks. This point is also known as the breaking strength of a material

8 cm, rectangular specimens of dimension 16 cm × 2 cm × 0.8 cm × 1 cm × 0.1 cm for ASTM D790. For ASTM standards, the recommended specimen size is 0.32 cm × 1.27 cm × 12.5 cm [9] (Fig. 4.4).

For conducting the test on a universal testing machine specimen should be placed horizontally on the lower supports of the test fixture, then applying a gradually increasing load on the center of the specimen continuously as long as the specimen breaks or shows considerable deformation. All parameters like maximum deflection and speed of loading are defined separately for different standards (ISO, ASTM). Finally, data can be analyzed by calculating flexural stress and strain (at yield &

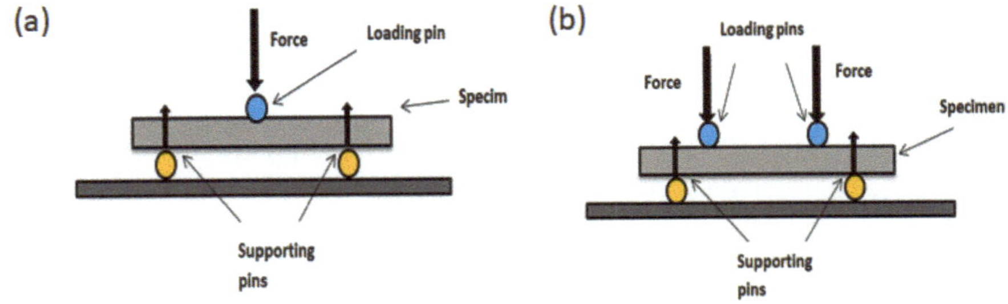

Fig. 4.4 Experimental setup of flexural test **a** 3-point flexure test, **b** 4-point flexure test

break), stress–strain curve, and flexural modulus [11]. Akhil et al. conducted experiments to determine an optimum span length for testing the flexural strength of samples of prepreg materials of glass and carbon and observed the behavior of flexural strength and flexural modulus as per the ASTM standards [12]. Saba et al. studied the comparative effects of various CNFs reinforced epoxy matrices (0.5%, 0.75%, and 1%) on different mechanical properties (flexural, tensile, and impact). The study shows a remarkable improvement for 0.75% CNFs/epoxy as compared to other epoxy nanocomposites due to uniform and homogeneous dispersion in the epoxy matrix [13] (Fig. 4.5).

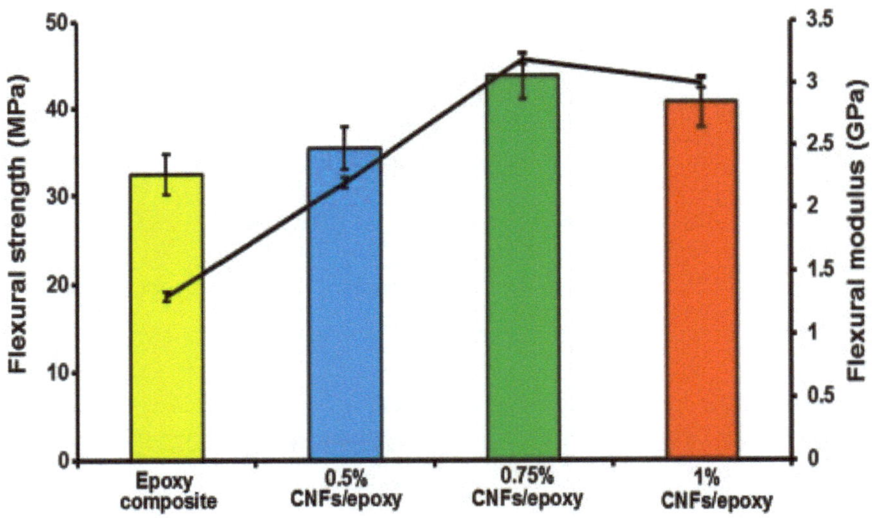

Fig. 4.5 Effect of CNFs filler loadings on flexural strength and modulus of epoxy composites

4.1.1.4 Impact Testing

The purpose of the impact test is to ascertain the reaction of a known material (i.e. ceramics, polymers, and composites) when subjected to a suddenly imposed stress. The key objective of this test is to assess the toughness, notch sensitivity, brittleness, impact strength, and as well as notch sensitivity of the engineering materials depending on the characteristics and test performed. The amount of energy that a material can withstand before it undergoes breaking or rupturing can be determined by impact strength or toughness. Most often, it takes a large amount of energy for tough materials to break while brittle materials exhibit low toughness due to their limited ability to undergo deformation. The main test configurations for measuring material characteristics are Charpy and Izod specimen configurations. These tests differ in the sample orientation which results in the changed direction of stress on the specimen and their basic procedure involves releasing a known load from a specified height to collide with the specimen in the test fixture. Usually, various notch configurations, namely U-notch, V-notch, and keyhole notch are used as the test specimens for the impact test (as shown in Fig. 4.6) [14, 15].

The dimensions of the conventional Charpy test specimen are 5.5 cm × 1 cm × 1 cm [9]. ISO 179 is the standard procedure for testing Charpy impact strength [16]. The Charpy V-notch test measures how much energy a material will absorb during fracture which indicates the material's notch toughness [17]. The experimental setup of this test involved a pendulum with a known load released from a specified height to strike a notched specimen which is placed in a horizontal position. By analyzing the difference in the hammer's height before and following the fracture occurrence, we can precisely determine the amount of energy absorbed by the material. The notch (with consistent dimensions) and size of the specimen, both factors, play crucial roles in this test as they greatly influence the conclusions drawn from the test. Izod test measures materials generally at low temperatures especially plastics. The apparatus consists of a swinging pendulum and strikes the specimen when it is held in a

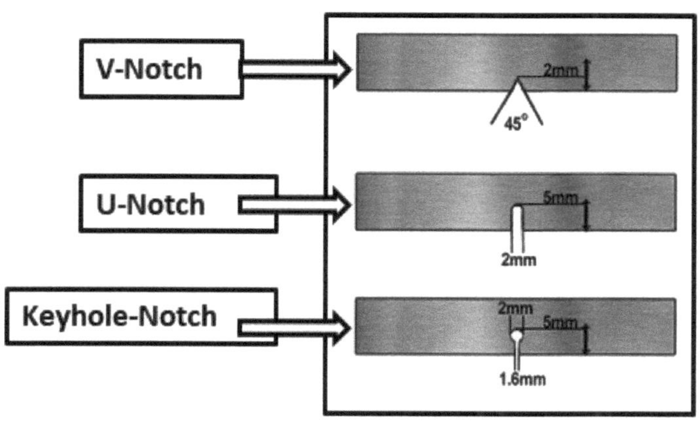

Fig. 4.6 Shows different notches in the impact test [15]

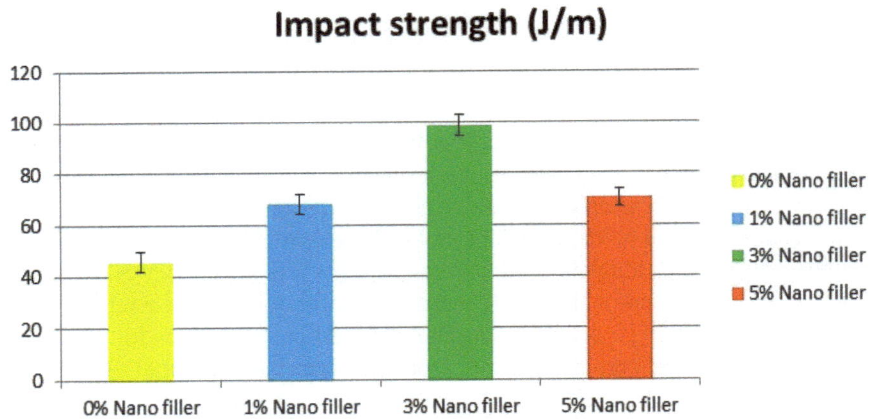

Fig. 4.7 Showing the impact strength of adding nano-fillers [18]

vertical position, besides it follows similar criteria as discussed in the Charpy notch test. This test holds particular significance in circumstances where materials must withstand sudden stresses including the development of various consumer goods, and structural gear for safety purposes. ISO 180 and ASTM D256 are the standard tests used for the Izod impact test [16]. Figure 4.7 displays the impact strength of epoxy nanocomposites on adding OPEFB nano-fillers (at 0%, 1%, 3%, 5%). In this investigation, Saba et al. concluded that 3% nano-OPEFB/epoxy nanocomposites showed improved impact strength in comparison to other nano-fillers [18].

Epoxy composites are greatly explored by the researchers due to their excellent mechanical properties. Mia et al. prepared HrGO hybrids by combining halloysite nanotubes (HNTs) and reduced graphene oxide (RGO) and incorporated them into epoxy resin which results in a notable enhancement of mechanical properties. In comparison to neat epoxy, the results revealed that 0.2 weight % of HrGO enhanced impact, flexural, and tensile strength by 1048.8%, 29.5%, and 489.9% respectively [19].

4.1.1.5 Hardness

Hardness is not a fundamental feature but rather hardness is the measurement of the material's resistance to localized plastic deformation (a small depression). Also, the value of hardness can be determined by the interaction of tensile strength, yield force, and modulus of elasticity. There are various ways to describe the hardness of a material including (1) resistance to indentation (2) resistance to abrasion or scratching (3) rebound effectiveness [20]. The most popular technique for quick and regular hardness measurements is the macro hardness test (Rockwell, Brinell, and Vickers). The range of indenting forces for the determination of macro hardness is between 50 and 30,000 N. Generally, the Rockwell hardness test measures the hardness of plastic. The ASTM D75 standards are commonly followed for examining the hardness of polymer-based composite materials [21].

Hardness measurement can be measured by the following techniques;

(1) **Macro hardness**—This is a technique for obtaining mechanical property data for the bulk material from a small example. The range of indenting forces for the determination of macro hardness is between 50 and 30,000 N. The procedures for macro indentation testing are those that include extensive area deformation, heavy loads, and extremely deep penetrations.

- **The Macro Rockwell hardness test**

When you apply load to the test object, the macro-Rockwell hardness test measures (As shown in Fig. 4.8) the indentation's depth. Hardness is inversely proportional to the depth of penetration. This test applies to all types of material from plastics to hardest steels. Therefore, various overlapping scales are used. The R, L, M, E, and K are frequently used to represent hardness numbers and hardness numbers have no units. The hardness of a material depends on the hardness number, the higher the hardness number of scales, the harder will be the material. In this test, indenters that are either ball-shaped or diamond-shaped are to be used. This test measures how much deeper an indenter can be pushed by a major load than it can be pushed by a minor (preliminary) load that was applied earlier.

- **The Macro Brinell hardness Test**

Simply Brinell hardness test is also an indentation test and it was initially developed by the Swedish physician Dr. Johan August Brinell. Forging, castings, hot-rolled and cold-rolled bars, and low- and high- alloy steels can all be tested using the Brinell method. ASTM E10-15a and ISO 6506-1 are the essential references for

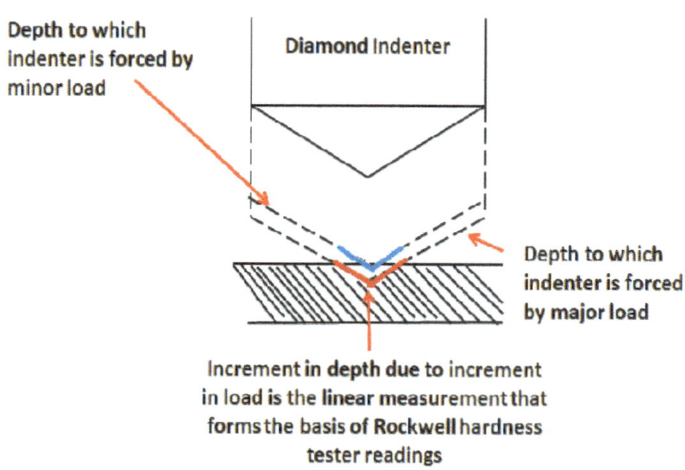

Fig. 4.8 Schematic of indentation process of Rockwell hardness test by using diamond indenter

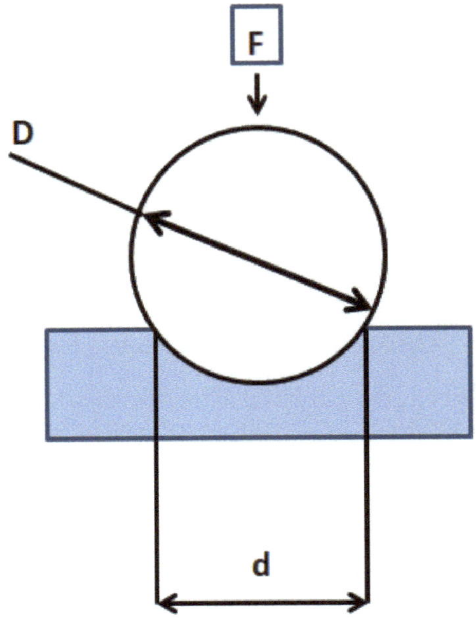

Fig. 4.9 Brinell hardness test

calculating the Brinell hardness of metallic materials and also provide instructions for consistence and precise measurement of hardness [22, 23]. In this method (As shown in Fig. 4.9), the surface of the material being examined is penetrated with a hardened steel ball, commonly composed of tungsten carbide with a specified diameter (10 mm), under a given load (L).

The Brinell hardness number (HBN) can be determined by the ratio of the test load (L) (in kilograms) to the spherical surface area of the indentation (in square millimeters) as shown below [24].

$$\mathrm{HBN} = \frac{\textit{Test Load in Kg}}{\textit{Surface Area of Indentation}}$$

$$\mathrm{HBN} = \frac{2L}{\pi D(D - \sqrt{D^2 - d^2})}\,\mathrm{kg/mm^2} \tag{4.3}$$

where,

L—Test load (in kg).

D—diameter of the steel ball (in mm).

d—diameter of the indentation or impression (in mm).

- **The Vicker's Hardness test (macro and micro)**

At Vickers Ltd., Robert L. Smith and George Sandland invented the Vickers hardness test (Fig. 4.10) in 1921. It determines the size of the impression left by a diamond indenter to determine the object's hardness. Compared to the Rockwell hardness test, the Vickers test uses a lower test load or force. In this test, a square-based diamond indenter is used [25]. It can measure the hardness of a material at both micro and macro scales. The diamond pyramid hardness (DPH) or Vickers pyramid number (HV) is the hardness measurement unit provided by the test. Vickers hardness test measures the hardness of a wide range of materials including ceramics, metals and some kinds of polymers. It follows the principle that a square-based diamond indenter with a facing angle of 136 degrees is impressed on the material surface with a pre-decided load (between 0.01 kg and 2 kg for micro hardness and 1 kg and 100 kg for macro hardness test) for a specified dwell time, usually few seconds. During this time, the indenter created a square-shaped indentation into the material and a calibrated microscope is used to measure the size of the indentation which is typically no larger than 0.5 mm. The Vickers hardness test methods are outlined in various standards like ASTM E384 (covering micro force ranges from 10 to 1 kg), and ASTM E92 (covering macro force ranges from 1 to 100 kg). In addition, ISO 6507 standards comprise of both micro and macro ranges [26].

The ratio, $\frac{force(in\,kgf)}{surface\ area\ of\ the\ indentation(in\,mm^2)}$ gives the HV as follows [28].

$$\mathrm{HV} = \frac{2F\sin\frac{\theta}{2}}{S^2}$$

$$\mathrm{HV} = \frac{2F\sin\frac{136^\circ}{2}}{S^2}$$

$$\mathrm{HV} = 1.854\frac{F}{S^2} \tag{4.4}$$

where,

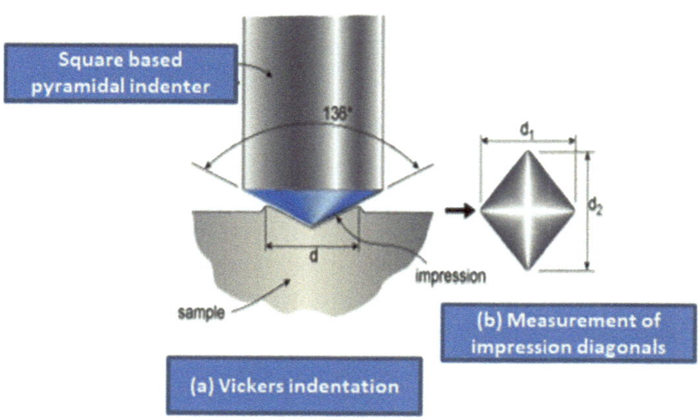

Fig. 4.10 Schematic of Vicker's hardness test [27]

F is the applied force (in kilograms-force).

S is the average length of the diagonals (in millimeters).

θ is the angle between the opposite faces of diamond.

Because HV offers a continuous scale of hardness it is well accepted in research work.

HV can be represented as xxxVHNyy, for e.g. 340VHN20 or xxxVHNyy/zz if force duration varies from 10 to 15 s.

Where,

340 represents the hardness number.

HV represents the Vickers hardness scale.

20 represents load in Kgf.

30 represents force duration if it varies from 10 to 15 s [29].

4.1.2 Tribological Characterizations

The Tribology term deals with the study and analysis of friction, wear, and lubrication between two rotating elements. By understanding what friction, wear, and lubrication one can understand the term 'tribology'. There are a lot of examples of friction and wear like automobile, skating, writing in which friction between two bodies is necessary for starting and stopping of objects. Whereas, larger amount of friction may lead to component failure and is also responsible for wear of engine parts and power loss. For reducing wear due to friction, abrasion, adhesion, and erosion lubrication techniques are used. No one can stop friction and wear completely but it can be reduced to a mini level [30]. We will discuss these concepts clearly as follows.

4.1.2.1 Wear

Wear is defined as the damage, sluggish removal of the material or its distortion from the solid surface due to colliding with the other solid surface (sliding or rolling), eventually resulting in the loss of functionality or breakdown of the material. Generally, gradual material loss caused due to relative motion between two contacting surfaces is termed as 'wear'. Tribology is the study of wear and associated processes. Both mechanical and chemical factors (such as erosion and corrosion) can be a cause of wear. Wear can be triggered by solids (metallic or non-metallic) or solid particles, by the action of fluids, and also by liquid droplets entrained in gases that are flowing. A common example for better understanding wear is worn out of the soles of the shoes by the action of friction; this reduces the life of the object.

Fig. 4.11 Adhesive wear

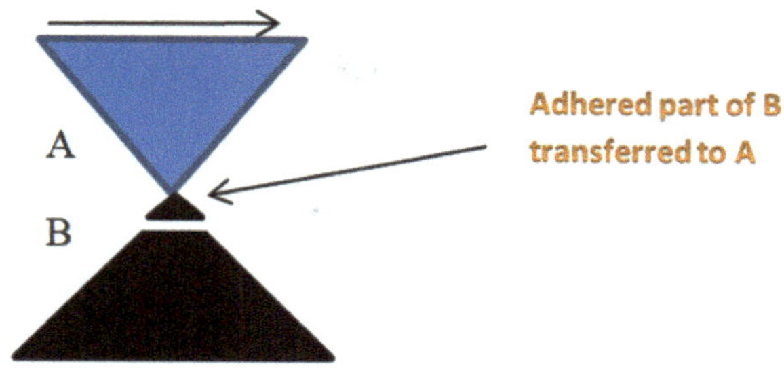

There are some broad categories in which wear can be classified depending on the tribosystem.

(1) **Mechanical Wear**—When two solid surfaces come into contact due to relative motion, removal of the material takes place from either of the surfaces that can result in mechanical wear (mainly in machine parts). It is further divided into several more parts
(2) **Adhesive wear**—Adhesive wear happens when the transfer of the material takes place due to the adhering state acquired by the two bodies when they are in sliding contact. It is because of the strong atomic bonding forces present at that particular point (as shown in Fig. 4.11). It creates depressions and indentations on the surface that can be in the form of pits, voids, or cavities [31]. Two contact surfaces may experience galling, scuffing and scoring as a result of intense adhesive wear. This type of wear leads to removal or attachment of the material from either surfaces and can be reduced by decreasing the roughness of the surface, by enhancing hardness of the surfaces or by avoiding direct metal–metal contact [32].

For adhesive wear Archard has given an equation for calculating adhesive wear;

$$V_w = \mathrm{K}\frac{Wd}{H} \tag{4.5}$$

where, V_w is the wear volume (mm^3)

K is the sliding constant.

W is the force (in Newton).

D is the sliding distance.

H is the hardness of the soft material (kg/mm^2).

- Galling—It happens when adhesive wear becomes acute i.e. forming of macroscopic protuberance.
- Scuffing—It is a moderate form of adhesive wear i.e. surface roughness or scratches.
- **Abrasive wear**—Deformation caused due to the movement of itinerary sharp and solid asperities across a smooth surface which leads to the removal of some part of

contact surfaces or loss of the material place due to the presence of macroscopic protuberances which are forced against solid surface resulting in abrasive wear. For example because of the interaction of the surface of the car tyre (smooth surface) and road (hard surface), the tyre sole is worn out with the time. Chand et al. examined the abrasive wear behavior of polymer composites reinforced with sisal fibers and published the results they discovered [33]. Abrasive wear can be understood by two modes i.e. two-body abrasive wear or three-body abrasive wear [34]. Sandpapers and grinding wheels, glass papers are commonly used as coated abrasives (paper or cloth sheets on which abrasive material like grit is attached to one face). Nowadays these abrasive materials are replaced by the materials like silicon carbide paper or aluminum oxide paper for removing the material from surfaces (for painting, removing old paint, furnishing of wood) while sandpapers are manufactured in a variety of grit sizes. According to ISO 9352 and ASTM D 4060, the Taber Abrasion Test can be used for evaluating abrasive wear as reduction of mass.

For the single abrasive wear, the volume of the wear V_w can be explained by;

$$V_w = \alpha\beta\frac{Wd}{H}\mathrm{K}\frac{Wd}{H} \tag{4.6}$$

where, α is the asperity shape factor (typically 0.1)

β gives the degrees of wear by an asperity (typically 0.1 to 1.0).

W is the force (in Newton).

K is the sliding constant.

d is the sliding distance.

H is the hardness of the softer material (kg/mm^2).

V_w gives the wear volume (mm^3).

(1) **Erosive Wear**

Erosive wear is a form of mechanical wear, sometimes known as Erosion Wear is a degradation of the surface material due to direct impingement of the liquid or solid particles on the material's surface with a very high velocity. A noble illustration of erosive wear occurs when raindrops of different velocities impact the Earth surface, leading to erosion [34]. So erosion can be (1) solid erosion (2) liquid erosion (3) liquid impact erosion (4) slurry erosion and (5) cavitation.

- **Solid erosion**

In this type of erosion (Fig. 4.12), the solid materials' surface is deformed due to the repeated impact of high-speed solid particles (carried by air) which gradually erodes the materials' surface. A number of factors affect the rate of erosive wear mainly the material properties of the particle (shape, impingement angle, and collision velocity).

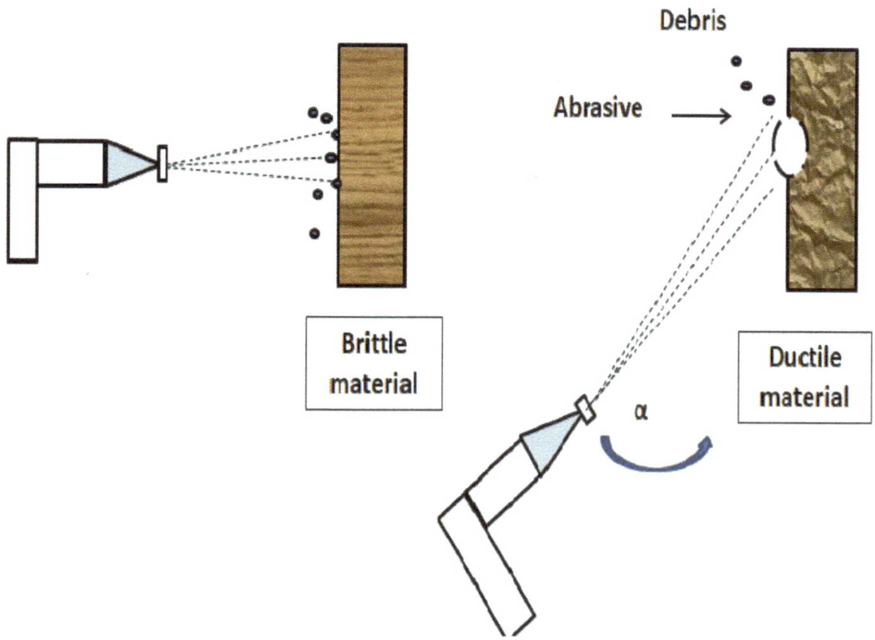

Fig. 4.12 Solid erosion

The maximum wear rate for ductile materials occurs at approx. 30° angle of impingement (α) in which the material is separated out by cutting action due to the impact of abrasive particles, while for non-ductile materials it is determined by the angle of impingement (α) being normal to the surface (90°) and due to splintering action, the surface material is removed [35]. The sharp bend or curves tend to erode more than mild bends. An Air jet of oxide particles bombarded on the surface (Abrasive particles behave like little secateurs) and amount of damage depends on the abrasive particle mass (m), impingement angle (α), impingement velocity and sharpness factor of the particle (F) [25].

$$\text{Erosion}(\text{degree of damage}) = \text{Km}v^2\text{Af} \tag{4.7}$$

where, k is the constant.

- **Liquid erosion**

In material and mechanical engineering, liquid erosion refers to mechanical damage of the solid materials' surface due to the repeated impact of high-speed liquid particles which gradually erode the materials' surface. This type of erosion mainly occurs in pipes of metals (copper, steel) in various industries including wind energy, aerospace, mainly in gas and steam turbines, nuclear power plants, power generation plants where liquid flow is involved having passive oxide films for resistance to corrosion.

The protective layer of oxide is removed by the impact of liquid. Perforation of elbows of pipe occurred due to repeated impingement of the liquid. However, this type of mechanical damage or erosion occurs above some threshold velocity [36].

- **Liquid Impact Erosion**

In this type of erosion, high velocity liquid droplet impinges on the surface of the solid. Aircraft rain erosion is the standard example, as the speed of raindrops is very high in comparison to the speed of aircraft (which is approximately 400 m/s), all the forward-facing surfaces like nose, windshields, leading edges of the spar, engine inlets are eroded or damaged.

- **Slurry Erosion**

For slurry handling equipment (i.e. valves, pumps, and pipelines) used in usually mineral processing, mining industries, systems related to oil and gas production, and in hydraulic machineries, slurry erosion is an issue of major concern due to its considerable adverse effects such as loss or removal of material [37–39]. Due to the transmission of slurries carrying abrasive particles in hundreds of kilometers long pipes (i.e. in coal pipelines), slurry erosion occurs [40].

The subsequent factors responsible are;

Particle characteristics (mass, shape factor, hardness, density).

Flowing slurry characteristics (temperature, viscosity, concentration, turbulence, corrosivity).

Impingement condition (impact angle, impact level, impact velocity or energy).

Time and Temperature.

Target surface properties (toughness, hardness, corrosion resistance, microstructures).

- **Cavitation**

Cavitation is a specialized form of liquid impact erosion in which tiny gas bubbles (or voids) are formed as a result of external influences in a liquid. These vapor cavities or bubbles (i.e. where the pressure is low) collapse when pressure increases subsequently, produce intense shockwaves and water jets that can degrade the material. In hydraulic pumps, propellers, control valves, submerged planes, pump impellers and also in spillway in dams, cavitation erosion is often experienced.

4.1.2.2 Friction

When the two surfaces are in contact then a tangential force resists the relative motion of two bodies. If the surfaces are smooth than there will be no resistive force hence, the rough surface will have more friction. Temperature, nature of the material and roughness are the major factors that affect friction. Friction can be static (rest), kinetic (dynamic), or rolling. Frictional forces can have both aspects good or bad as we are unable to walk without friction. According to Amontons equation the frictional force is directly proportional to the nominal load

$$F = \mu W \tag{4.8}$$

μ is the proportionality constant which is known as coefficient of friction.

4.1.2.3 Lubrication

Lubrication is a process in which we separate two contacting surfaces by a layer of lubrication at the interface to make them smooth and to minimize the effect of wear due to friction, abrasion, adhesion and erosion, so that metal–metal contact can be avoided. In the process of lubrication, lubricants (oil and grease) are applied to the surface for making a separation between two layers so that the substance ready to shear while adhering to the surface. Lubrication is mainly used to increase effectiveness, lifespan, and efficiency of machineries. Lubricants also serve as sealants and coolants. Therefore, the important factor is the viscosity of lubricants for the formation of film. Lubricants (Fig. 4.13) can be classified as:

Liquid lubricants—The majority of lubricants are liquid and are generally based on one type of base oil but to meet the special performance requirements mixture of base oil is also used like natural oil, water, lanolin, vegetable oil, mineral oil etc. According to their origin into two groups lubricants may be classified as:

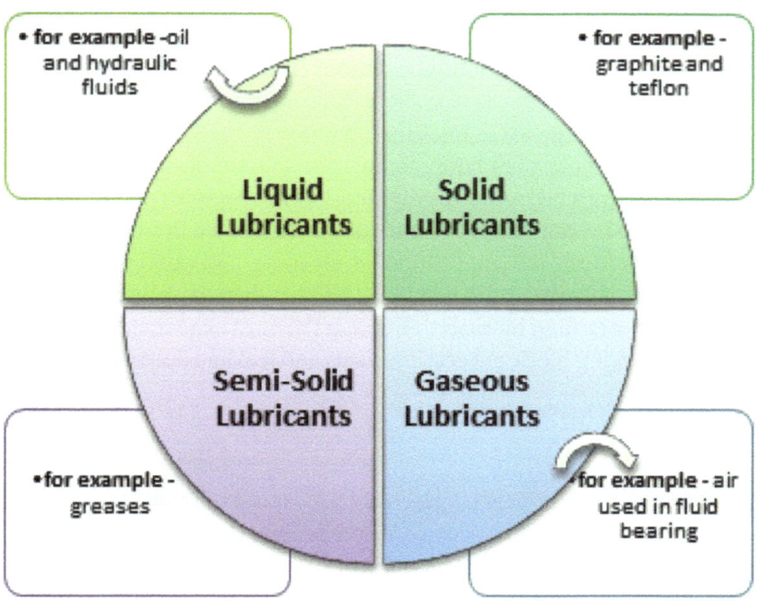

Fig. 4.13 Various kinds of lubricants

- Mineral oils: Those lubricating oils which are refined for reduction of some molecular reactions from naturally occurring crude oil are termed as mineral oils.
- Synthetic oils: Synthetics oils are man-made lubricants which are typically created by artificial chemical compounds. Mainly polyalphaolefins (PAOs) is commonly used as synthetic oil. Synthetic oils produce lower friction, less evaporation, heat resistance which led to better lifespan, nonetheless these oils are expensive.

Oil is the main ingredient for formation of grease, as greases contains 75–80% of oils.

- **Solid lubricants**

Liquid lubricants or say greases cannot be useful for lubricating bearings operated at extremely high temperatures, in vacuum, or under extremely powerful radiation. So dry lubricants or solid lubricants are used for better advantages and applied in the form of powders or thin films for reducing friction between two moving surfaces such as boron nitride, Teflon, molybdenum disulfide, graphite, tungsten disulfide. Due to their lamellar structure, oxidation prevention, thin film formation these materials possess excellent additive properties. Even some materials stay lubricious at extremely high temperature. Solid lubricants are extensively employed in various industries including aviation, marine and coastal applications, automotive etc.

- **Semi-solid lubricants**

Greases: oil is the main component in grease which is responsible for lubrication. Grease is preferred because of multitude reasons including ease of confinement, low friction and substantial better service life at reasonable cost. Three key ingredients are combined to create greases: thickener, additives, base oil (refer to Fig. 4.14). Thickeners can be simple soap base, complex soap base or non-soap base such as calcium soaps, lithium, clay, sodium, polyurea, lithium complex, aluminium complex that are mixed with some minerals or synthetic oils (as shown in Table 4.2). Generally, grease does not perform the application of cleaning and cooling like other fluid lubricants because of its semisolid nature [41]. Other than these, greases carry out all the responsibilities of a fluid lubricant. Grease is the most expedient and cost effective lubricant for preventing block out contaminants and leakage due to its water resistive nature.

- **Gaseous lubricants**

In comparison to liquid lubricants, gaseous lubricants exhibited considerably superior compressibility and lower viscosity (also viscosity increases with the increase of temperature). Molecular composition of the gaseous lubricant remains unchanged at extreme temperature range (low or high) whereas liquid lubricants are unstabble over a wide range of temperature. The liquid lubricant changes its molecular composition at high temperature and is converted into a gaseous form. Because of their chemical stability there is no chance that the lubricant would contaminate the ball bearings, which is crucial for the tools used in many industrial sectors particularly in food,

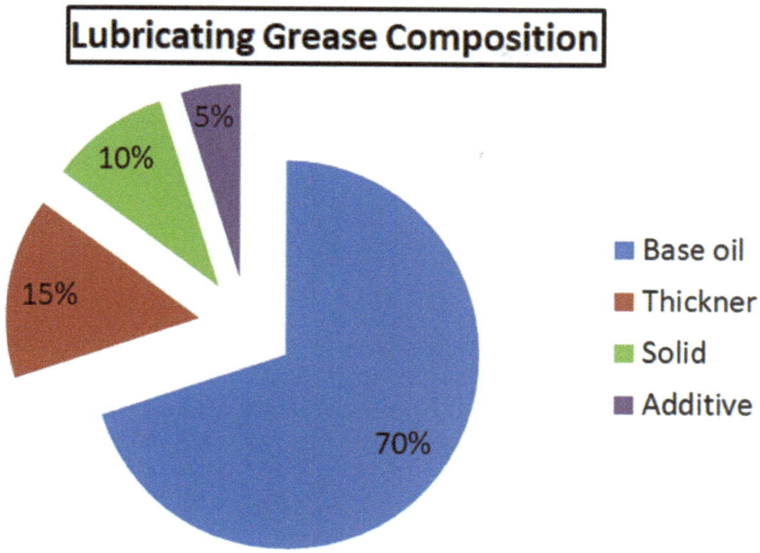

Fig. 4.14 Lubricating grease composition

Table 4.2 Different types of thickners

Simple soap			Complex soap	
Lithium	Calcium	Aluminium	Lithium	Calcium
Barium	Sodium	Strrontium	Aluminium	Barium
Mixed soap			Titanium	
Non-Soap Clay, Alumina, Polyurea, PTFE, Silica gel				

electronic and medical industries, and in many aerospace applications (such as aerostatic and aerodynamic bearings). Gases such as argon nitrogen, oxygen, hydrogen, carbon dioxide, and acetylene are most widely preferred gaseous lubricants [42]. However, air is the most viably employed gas lubricant which is preferably utilized in fluid bearing.

4.1.3 Microstructural Characterizations

A vital aspect of material science is microstructural characterization, which entails the microscopic interpretation and as well as analysis of the interior structure of a material. Understanding the mysteries concealed in the microscopic realm of materials are the goal of this area of research, known as microstructural characterization. The morphological features can be revealed by appropriate preparation of samples. The dimensions of the particles or grains, shape, and phase boundaries all are defined by the microstructural morphologies. The interaction of incident radiation with specimen is resulting in variation of image contrast which reveals the information about morphological properties. Microscopy such as scanning electron microscopy and transmission electron microscopy is one of the several methods used in microstructural characterization [43].

4.1.3.1 SEM

Scanning Electron Microscope, often referred as SEM. Instead of light microscope, SEM is more preferred as it uses high energetic focused beam of electrons scanned across the specimen for generating various signals that contains all information about the specimen's surface topology depending on the depth of the specimen resulting in the images at a much greater resolution.

The interaction of atoms and electrons in the specimen generates a number of signals that encompasses information regarding composition of the specimen, bonding, shape and size of the particles, various defects, texture, and surface topography. An external monitor is used to detect the image by correlating the electron intensity and beam position of the incident ray. Higher resolution (better than 1 nm) can be achieved by using SEM.

SEM can directly analyze a variety of solid materials, however when dealing with biological materials and insulating samples, specialized image techniques (i.e. low vacuum or low KV imaging) are often employed to mitigate surface charging and beam induced damage. For reducing the charging on insulating samples, a fine layer of conducting material like platinum or gold is sputtered.

Principle

Due to the interaction of atoms and electrons in the sample, various signals including Secondary electrons (for topographical image), Backscattered and Diffracted back scattered electrons (determines surface composition or crystal structure) and characteristic X-rays (photons), and also certain materials emit visible light via a phenomenon called cathodoluminescence (CL) are ejected from the sample which are collected by the detectors to convert them into signals for creating the final image, which is shown in Fig. 4.15.

Sample Preparation

There is a requirement of special sample preparation due to the use of electron beam and low vacuum conditions. One of the most critical precaution is to dehydrate the samples because water vaporizes in a vacuum. For obtaining a well suited sample suitable for SEM analysis firstly, ensure that it is appropriately shaped and sized according to your specimen stage. To facilitate standard SEM imaging, specimen should possess, electrical conductivity particularly on the surface, and should be grounded to prevent the buildup of electrostatic charge. All conductive specimens like metals need little preparation.

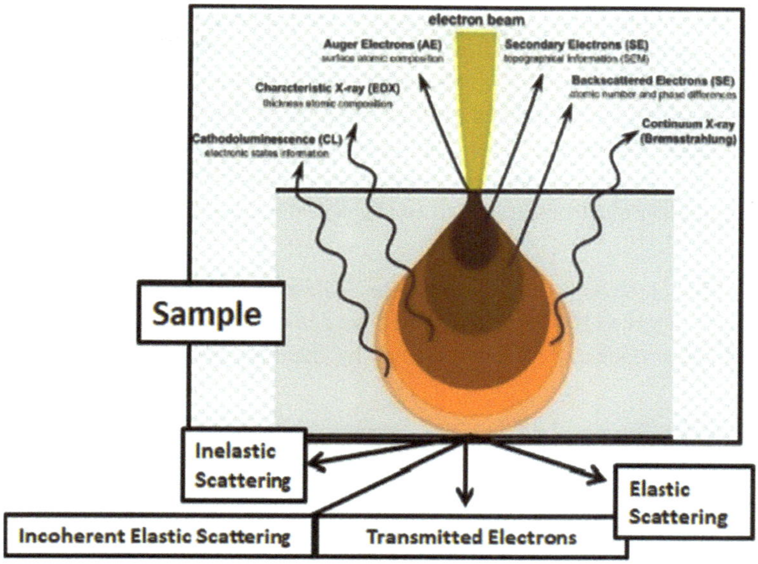

Fig. 4.15 Various levels of electron penetration [44]

Non-conductive materials are needed to be made conductive for minimize charging and to improve conductivity by generally coating with a thin layer of electrically conductive materials like platinum, gold, graphite, tungsten and chromium by the process of low vacuum sputtering. For this process a device called 'sputter coater' is used in which argon gas is ionized when exposed to an electric field in a vacuum chamber. Negatively charged gold foil is attracted by positively charged argon ions, as after collision releasing gold atoms and hence these gold atoms adhere to the surface resulting in a fine conductive layer of gold.

Instrumentation and Working

Basically SEM contains several key components including electron source (filament) for generating focused electron beam, electron lenses (objective lenses, condenser lenses), detectors for all type of signals (i.e. electron detectors, X-ray detectors), specimen chamber and a data display monitor or software.

Generally, SEM analysis process can be accomplished using the procedure outlined by Goldstein et al. [46]. Some additional clarity for SEM analysis is provided by referring to Fig. 4.16 for further understanding. At the top of the microscope a beam of electrons is produced by thermionic emission from an electron gun (tungsten filament is used due to its highest melting point and lowest vapor pressure) and subsequently accelerated through the column at a designated acceleration voltage ranging from 1 to 30 keV. To produce a fine beam of electrons (for reducing diameter) or for converging the beam it is pass through the condenser lenses and apertures and finally through an objective lens which is used to focus the beam of electrons onto the sample. A vacuum environment must be maintained throughout the entire column. Depending upon the electron source used, voltage with which the beam is accelerated and the lens system decides the diameter of the beam which ranges from less than 1–20 nm. Scan coils which are positioned above the objective lens control the position of the electron beam on the specimen and enable the electron beam to scan in a raster- like pattern across the specimen's surface in the X–Y plane.

Thus, this scanned beam interacts with the specimen producing a variety of signals such as secondary electrons, reflected and back scattered electrons, and characteristic X-rays.

- Secondary electrons resulting from the inelastic scattering are the low energy electrons (approx. 10–50 eV) by which we can obtain topographical information nearby the surface and SEs are detected by Everhart Thornley detector.

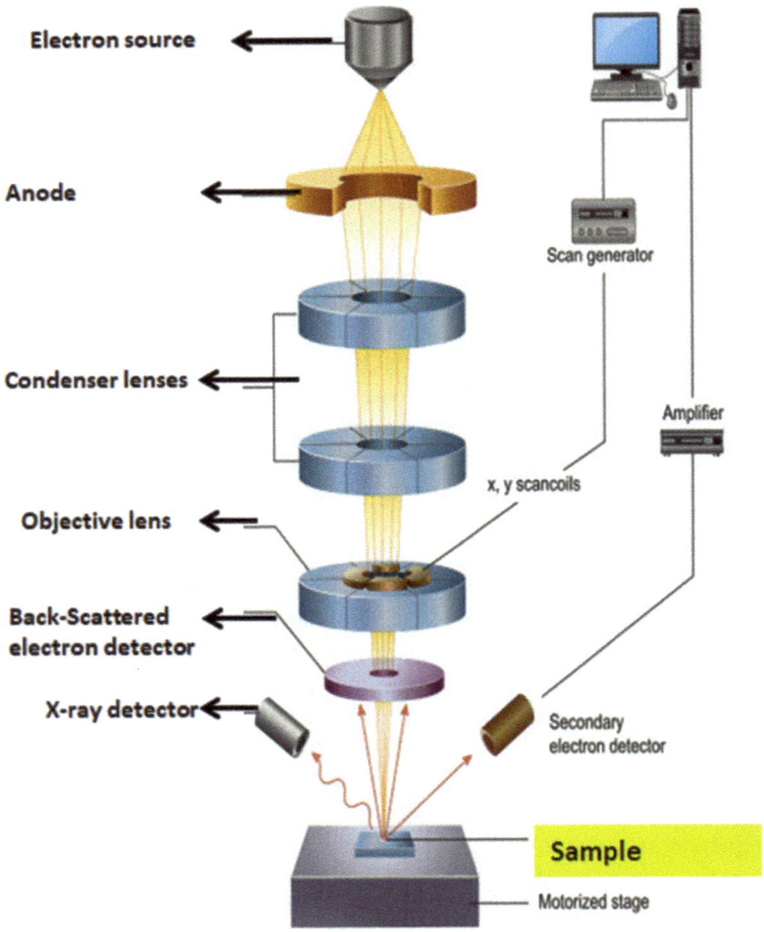

Fig. 4.16 SEM (Scanning Electron Microscope) [45]

- Backscattered electrons results from the elastic scattering at high angles (greater than 90 °) from the deeper surface and are redirected backward towards the source. Higher atomic number elements generate more backscattered electrons, hence a brighter image is created, where it provides information regarding compositional difference of a specimen.

In a study conducted by Hajsman et al. different kinds of SEM imagining were tested at pressure 30 Pa (refer to Fig. 4.17) and reported that the utilization of backscattered electrons (as shown in Fig. 4.17 is the most appropriate method for both identifying the difference in the chemical composition and also for observing morphology [47].

These signals can be detected by suitable detectors and can be seen as an image on a computer monitor. As the scanned area becomes progressively smaller, the corresponding magnification will be higher in SEM often reaching up to a million times (approximately 1,000,000 × in some modern models) [48]. Chand et al. investigated the influence of a coupling agent on abrasive wear performance of polypropylene composites reinforced with chopped jute fibers and also showed their results with SEM analysis. Fractography showed in Fig. 4.18a illustrates two notable aspects: first, the presence of twist in the fibers resulting from the melt mixing process and second, the generation of internal defects. The Fig. 4.18b provides a closeup view of a MA-g-PP treated jute fiber within a PP composite and it distinctly illustrates the improved compatibility between the surface of the coated fiber and matrix. Definitely, this treatment led to an increased elongation in the CT composite. They discovered that the utilization of the coupling agent enhances wear resistances in contrasting to not employing a coupling agent [33].

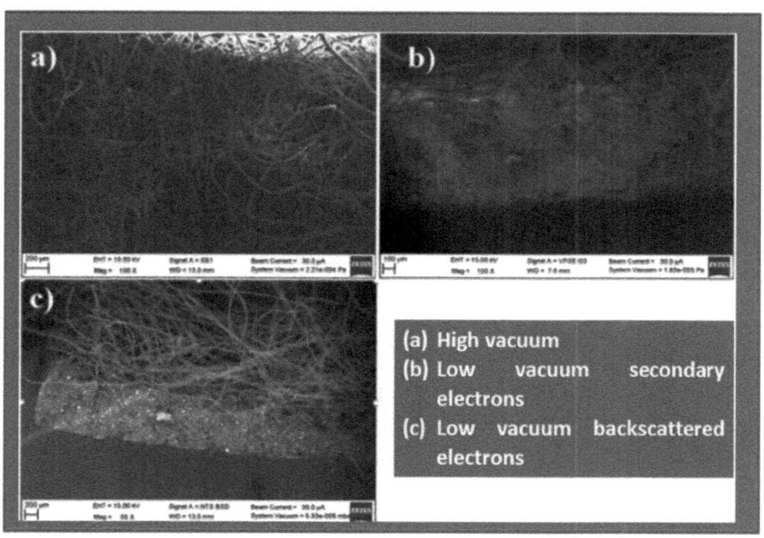

Fig. 4.17 Image of SEM, **a** high vacuum, **b** low vacuum secondary electrons, **c** low vacuum backscattered electrons [47]

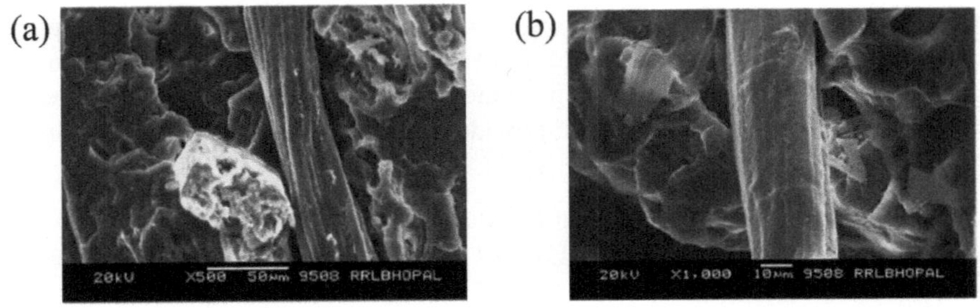

Fig. 4.18 **a** SEM showing magnified view of smooth surface of MA-g-PP solution-treated jute fiber-PP composite (CT) [33] and **b** SEM showing magnified view of fractured surface of MA-g-PP melt-mixed jute fiber-reinforced PP composites (MT) [33]

As discussed before about resolution of SEM, in contrast, TEM provides significantly greater spatial resolution ($\sim$ 0.1 nm), because of its distinctive imaging mechanism.

4.1.3.2 TEM

By zooming into the atomic scale, scientists can examine the basic components of the materials. In transmission electron microscopy, a high energetic focused beam of electron (accelerated voltage range from 60 to 300 kV) is transmitted through the ultra-thin specimen by whom we can analyze matter at the nanoscale at an extremely high resolution. Typically, the sample consists of, usually less than 100 nm in thickness; or it can be a suspension on a grid. The sample is placed in the middle of the column. After the interaction of the sample with electrons, the resulting image can be magnified on a device like as fluorescent screen or by a CCD camera (charged coupled device) on a computer screen. TEM gives us a 2D image while SEM provides a 3D image (as shown in Fig. 4.19).

Magnification of TEM can reach a maximum of 50 million and a resolution of up to 0.5 angstroms.

The basic difference between SEM and TEM is that SEM uses scattered or reflected electrons whereas TEM uses transmitted electrons. TEM provides invaluable information about its inner structure (crystal structure) and morphology at a higher resolution. TEM, due to its ability to give microstructure of the materials (polymers, metals etc.) at nanoscale, is employed in research field (such as nanotechnology, semiconductor) and many other applications like cancer research, geology, virology, biology as well as in environmental science.

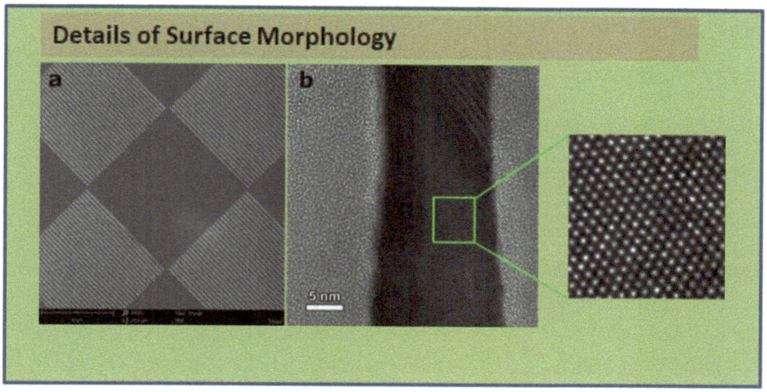

Fig. 4.19 **a** SEM image provides details on the surface's morphology, **b** TEM image illustrates the internal structure of the specimen [49]

4.1.4 Structural Characterizations

4.1.4.1 XRD

In material science, X-ray diffraction is an analytical technique of crystallography which is used to understand about arrangement of atoms in solids (i.e. atomic and molecular structure) in which constructive interference occurs between X-rays and the crystalline structure results in X-ray diffraction. The intensities and scattering angles of these diffracted beams can be determined by which we can identify the crystal structure. In 1912, Laue claimed that the x-ray wavelength used is comparable with the distances between atoms of solids, so x-rays are chosen for the technique. The diffraction pattern can be examined by using Bragg's law. When we are mentioning that x-rays are scattered by atoms, we actually mean scattered by the electrons within atoms. The intensity of x-rays is graphed against the scattered angle; by analyzing the position, angle and intensities of peaks of the diffraction patterns we can identify the structure of the material [50]. We interpreted that the crystal size is inversely proportional to the width of the x-ray diffraction meaning narrow peaks indicate larger crystals, whereas broad peaks suggest smaller crystals.

Joseph John Thomson discovered that when unpolarized x-rays impinges on itinerary electrons, then the scattered intensity at a distance d from the origin for one particle of mass m and charge e is given by

$$I = \frac{nI_o e^4}{2(mc)^2 c^2 d^2}(1 + cos^2\alpha) \quad (4.9)$$

where,

I—intensity of x-rays at some point M,

e—Charge on electrons,

n—Number of electrons.

d—Distance between origin and point M.

I_o—Intensity if x-rays incident on a free electron at origin.

m—Electron mass.

α—Angle between incident and scattered beam.

Generally, among different types of X-ray diffractometers (Fig. 4.20) that are available for crystal structure analysis, the most preferable diffractometer used for nanomaterial analysis is the Debye- Scherrer Diffractometer (Powder Diffractometer). It has the ability to provide adequate information regarding the crystal structure of the nanoparticles, polycrystalline materials and thin films. Basically, a monochromatic source of X-rays, powdered sample and detector are used for the whole setup of the diffractometer. The sample undergoes rotation within a monochromatic x-ray beam and the detector scans across the complete 2θ range. Usually a Scintillation counter, proportional counter, semiconductor counter, and gas proportional counters the signals are detected depending on the type of the sample required to be analyzed [50, 51].

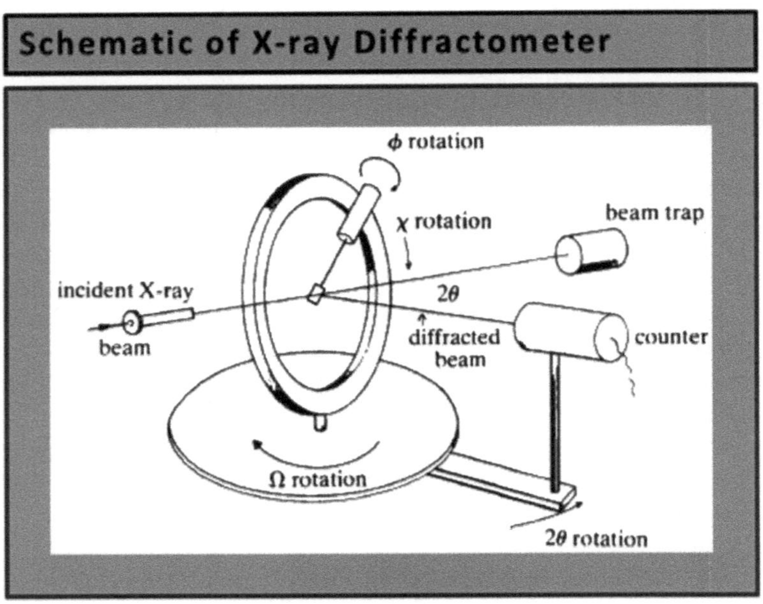

Fig. 4.20 Schematic of X-ray Diffractometer [51]

4.1.5 Physical Characterization

Physical characterization of the polymer composite entails the quantitative or qualitative measurements such as density, surface roughness, and hardness. Using this conclusive technique, size, weight, roughness, porousness, toughness, softness, strength and other characteristics can be determined. Physical characterization is typically done after fabrication of the polymer composite or material.

4.1.5.1 Density

Density is seen as a crucial factor because polymer composites are thought to be lightweight and attain higher strength to weight ratio than metals and ceramics. The density of the polymer composite is especially dependent on specific densities of reinforcement and matrix, the void content and their respective volume fractions. The measure of bulk density is weight to volume ratio that can be written as:

$$\text{Density}(\rho) = \text{m}/\text{v} \tag{4.10}$$

where "m" is the mass and "v" is the volume, and both of them can be measured through the weighing scale and measurement equipment. Another way to find out the density of the material is the Archimedes principle. According to Archimedes' principle, when an object is submerged in a liquid, its apparent weight reduces by a quantity equivalent to the weight of the liquid that is being displaced. As a result, density can be calculated using Eq. (4.11), where the mass of the substance/material in immersion liquid and air.

$$\rho_A = \frac{m_A}{[m_A - m_L]}\rho_L \tag{4.11}$$

where "m_A" is the mass of substance in air, "m_L" is the mass of substance in immersion liquid, and "ρ_L" is the density of the immersion liquid. Although it can be used to determine the bulk density of the polymer composite, it is not preferred for the exact density because the polymer composite is a two or more phase material that contains different constituents with differing densities and volume fractions. Thus, the equation for the two phase polymer composite can be expressed as following Eqs. (4.12) and (4.13) of each of the parts:

$$\rho_c = \rho_f v_f + \rho_m v_m \tag{4.12}$$

$$\rho_c = \frac{1}{\frac{m_f}{\rho_f} - \frac{m_m}{\rho_m}} \tag{4.13}$$

4.1.5.2 Particle Size

- Particle size is very important in processing, handling, and performance of polymer raw materials and final products. The size and shape of the raw components indicate the flow behavior in the feeding system in a manufacturing operation as an element of raw material handling. The kind of feeder needed to effectively transport raw materials from preservation to the processing machinery and the produced material again to the station is determined by the size and shape of the particles. For proper handling, various feeder types are needed for materials including liquid colorants, powders, cylindrical or spherical resin pellets, and glass fibres. The processing features of polymers are influenced by the particle size, so that bigger pellets might need more energy or melting time, whereas very small pellets can melt swiftly. The overall performance of parts is also affected by particle size. For instance, smaller particle sizes are desirable for faster absorption of drug in healthcare while bigger ones take more time to completely dissolve in the solvent. Numerous studies have shown that nano-fillers can be used in much smaller ratios than micro-fillers to achieve similar reinforcing effects. However, nanoparticles need special handling, processing, and safety procedures as well as precise apparatus for appropriate dispersion in the polymer matrix. The standard test technique for determining the particle size of plastic substances by employing a mechanical sieving apparatus is described in ASTM D1921. According to the standard, using an antistatic agent is necessary for substances that are likely to produce static charge in order to minimize coarse bias. Schlappa et al. [52] used emulsion polymerization to develop high-solid-content (12–55% w/w) polystyrene and polyvinyl acetate diffusion of polymeric particles with an average particle diameter of 50–500 nm. Both approaches were investigated using common methods for determining particle size, such as static light scattering (SLS) and dynamic light scattering (DLS), and their results were then compared with inline PSD (particle size distribution) observations made using PDW (photon density wave) spectroscopy on undiluted samples. The study demonstrated that there is a good agreement between analysis techniques when comparing particle size observations of undiluted polystyrene dispersions. Vinayagamoorthy et al. [53]investigated the influence of particle size on the physical and mechanical characterizations of limestone (0.2, 0.4, 0.6, 0.8, 1 mm size particles) and jute-reinforced polypropylene composite, and found best behavior for composites with particle size of 0.6 mm and 0.8 mm.

4.1.6 Electrical Characterization

One of the significant areas of interest for polymer composites is electrical applications. For monitoring large scale composite structures electrical inspection techniques (such as electrical impedance or electrical resistance methods) are most suitable techniques as they are fast, low-cost real-time techniques. Electrical characterization of a composite material refers to the material's behavior in an electric condition. Whenever a field is applied, free electron/charge carrier movement takes part and due to this there is material response in terms of electrical behavior. Different materials exhibit different electrical behavior i.e. some composite material exhibits better electrical characteristics than the others. Polymer composites with different electrical behaviors are very important in selecting materials for various defer motives.

4.1.6.1 Electrical Conductivity

Conductivity of any material exhibits the ability of that material to pass an electric current or signals through itself. It is basically the movement of free ions/charges/ electrons that move freely. In case of metals and semiconductors, electrons play the crucial role as charge carriers, while for ionic compounds ions acts as charge carriers. Conductivity is affected by temperature as it is directly proportional to temperature. The reason behind this dependency is that when temperature is increased then the kinetic energy of free charges also increases resulting in fast movement of free charges which enhances the conductivity and acts inverse in decreasing temperature. There exist very few conductivity measurement techniques that are basically categorized into two techniques, one is contact method which contains two types of techniques i.e. 2-point probe and 4-point probe, and the other is the non-contact method i.e. inductive conductivity method. Mironov et al. [54] prepared the graphite/polypropylene and graphite/polyvinylidenefluride (PVDF) composite, and studied the electrical conductivity. The comparative electrical conductivity is measured by the four-electrode (FE) and four-point probe (FPP) method for PVDF composite bar samples with varying graphite/carbon filler concentration. Here, the FE method provides better conductivity for the composite compared to the FPP method because of surface roughness and structural non-uniformity of the conductive polymer composite. Prior researcher has stated that FPP approach is not advised for any polycrystalline material because "localized impurity segregation at grain boundaries in polycrystalline material may result in large resistivity variations". Gong et al. [55] fabricated carbon nanotube (CNT)- epoxy based polymer composites and characterized the effect of CNT agglomeration on electrical conductivity by using the four-point probe method. Here, CNT was chemically treated with a strong acid to improve the dispersibility and then mixed with epoxy to make a CNT-polymer mixture which was degassed in a vacuum oven before being injected in molds. The electrical conductivity measurement was taken by four-point probe technique (ASTM F1529-97) for CNT-epoxy based polymer composite samples where conductivity exhibited rapid increased values from 10^{-14} S/m to 10^{-5} S/m with increasing CNT concentration. Cheng et al. [56] reviewed the studies on ECT (eddy current testing) of CFRP (carbon fiber-reinforced plastic) composites which comprises several unidirectional carbon fiber/epoxy plies that have been piled together. Mook et al. [57] employed a HF (high frequency) eddy-current sensor to identify small faults such as delamination, impact damage, matrix rich layers, and fiber orientation. The ECT system for CFRP detection typically consists of a function generator, lock-in amplifier, power amplifier, a 2D stage, and PC to collect data and control stage. In order to successfully extract the minor signal change caused by the weak conductivity of CFRP, the sensitivity and gain of the lock-in amplifier should be strong enough.

(1) **Contact Method**

2-point probe and 4-point probe are two well-known and mostly used contact techniques. In the 2-point probe system, a known current or voltage is applied to the material while two electrodes are attached to the opposite side of a material, typically a rectangular prism or rod. In the 4-probe technique, the two inner probes assess the voltage drop those resulting from applying a current between the two outer probes (As shown in Fig. 4.21). This method has benefit over the 2-point probe that it can get rid of the contact resistance present in the 2-point probe technique. It should be emphasized that, in contrast to four electrodes, the four-point probe approach is created for materials with excellent uniformity, such as homogenous silicon wafers or thin films that are deposited [58].

(2) **Inductive Conductivity Method**

In inductive conductivity measurement (Fig. 4.22), an electrode less conductivity meter or toroidal conductivity meter is used. Inductive conductivity meters have

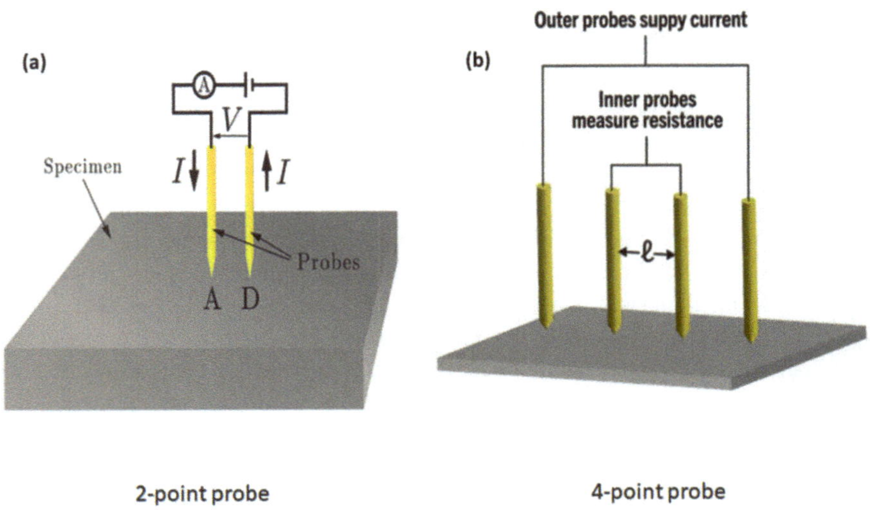

Fig. 4.21 Schematic diagram of **a** 2-point probe and **b** 4-point probe system

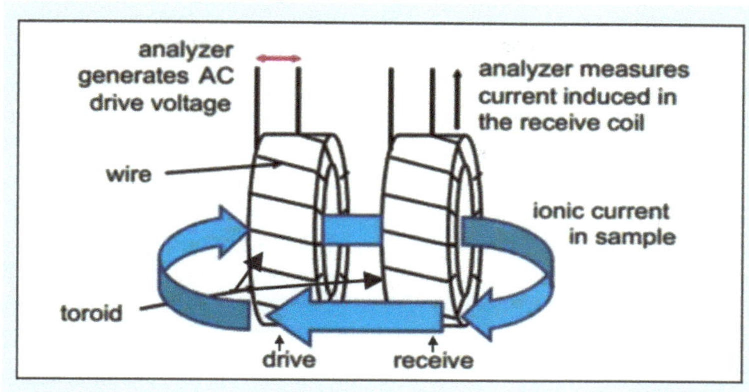

Fig. 4.22 Schematic diagram of inductive conductivity meter [60]

a corrosion-resistant casing in which 2 metal coils are present. One coil known as the drive coil generates the alternating voltage and the remaining coil acts as the receiver, and the material which will be measured acts as the core. Drive coil generated voltage induces a voltage in the receiving coil which results in the flow of current in proportion to the material's conductivity. The induced current and voltage in the receiving coil is correlated to conductivity. The material with conductivity greater than 15 micro siemen/cm works best with inductive conductivity in meters. One of the most popular non-destructive investigation methods that doesn't require any physical contact between the test sample and sensor is eddy current testing (ECT), which may be used to inspect electrically conductive materials at very high speed [59, 60].

4.1.6.2 Dielectric Measurement

Polymer composite dielectric systems are developed in response to the demand for high energy density pulse power energy storage devices [61]. These systems combine the processability and breakdown field strength of polymers with their high dielectric constants to create a composite material. Dielectric characteristics comprise complex permittivity (ε_r) and complex permeability (μ_r). And, a complex permittivity and complex permeability contain a two constituents' part. Here, we will discuss about the complex permittivity that possesses the imaginary part known as the dissipation factor and the real part known as the dielectric constant. Dielectric constant is a measure of amount of energy stored in material when an external field is applied. There are many methods to measure complex dielectric characteristics and each method attains its own constraints such as materials, specific frequency, applications etc. However, many measurement instruments are being used to measure dielectric parameters based on the application. The instrument that offers accurate calculation of the electrical parameters for the unknown material in the required frequency range might be taken into consideration. Designing of the sample holder plays a very crucial part in measuring accurate dielectric characteristics. Dielectric properties in low, medium and high frequency ranges are reviewed by Field et al. [62]. For the very low frequency range, electrode polarization phenomena will be in consideration. For higher frequency ranges, coaxial sample holders are modeled. For the microwave frequency range, transmission line, free space, and resonant cavity techniques are being used. Suresh et al. [63] investigated the electrical permittivity of P(VDF-TrFE)-CuO nanocomposites using an coaxial probe technique in the microwave frequency range which provides non-destructive and broadband observations. Where, the increase in the nanofiller concentration up to 10 wt% in the composite results in a reduced value of permittivity. Willert-Porada et al. [64] tested three different composites in the laminate structure glass-polyurethane, polypropylene-glass and woven carbon-fiber with the polymer using three different methods which are coaxial probe, SPDR and cavity perturbation. By investigating permittivity, they found that cavity perturbation provided the less value of permittivity as compared with SPDR and coaxial probe. For frequency in MHz range, Meena et al. [65, 66] prepared $CoFe_2O_4$/MWCNTs/epoxy and $BaFe_2O_4$/MWCNTs/epoxy composites with varying ferrite concentration and investigated dielectric properties using the LCR meter in the frequency range from 40 Hz to 4 MHz. The dielectric measurement techniques (Fig. 4.23) are as follows:

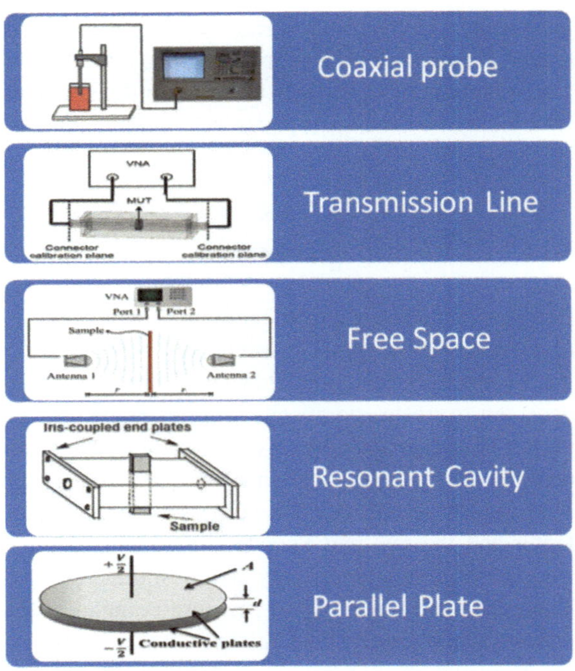

Fig. 4.23 Types of dielectric measurement techniques

(1) **Coaxial Probe Method**: This is a non-destructive testing approach also known as "open ended coaxial probe method". A conventional coaxial probe measurement set-up comprises of an impedance or network analyzer, coaxial probe and specific software. Reflection coefficient is measured through the vector network analyzer (VNA), where, the electromagnetic field (EM field) encounters an impedance mismatch between the polymer composite sample and probe resulting in reflection along coaxial line where the EM field is propagating. And, these reflected signals (reflection coefficient) are measured and converted into permittivity values at different frequencies. This method is most commonly used due to its minimal sample handling, broad frequency range, and both in vivo and ex vivo measurements. Temperature controlled environment can be accessible while measuring the parameter. This method is limited to provide reflection measurement only.

(2) **Transmission Line Method**: In this method, a specimen is placed in a coaxial line or waveguide section, and the transmission line is attached to VNA through two ports that provide the scattering parameters which are S11 (reflected signals) and S21 (transmitted signals). To determine the variation of the scattering parameter this method uses the electromagnetic wave reflected and transmitted through the samples [67]. The complex permittivity (dielectric characteristics) is then evaluated by converting these scattering parameters. There are mostly two conversion methods—Nicolson-Ross-Weir (NRW) and NIST interactive conversion methods. To obtain the direct calculation of complex permittivity from S-parameters, the NRW method is used. The transmission line method covers the large frequency range at low temperature.

(3) **Free Space Method**: This method covers the measurements on flat and large materials under high temperature in a wide band frequency range. In this set-up, two antennas are connected to VNA while facing each other and a sample holder is placed between both the antennas. The measurement of S-parameters will be taken once without the testing material on the sample holder and once with the testing material on the sample holder. The impact of the sample holder can be eliminated with the use of the de-embedding function of the network analyzer, leaving only scattering parameters for transmission and reflection coefficient of the mounted test material to be computed. Ample sample thickness is able to avoid multiple reflections between the sample and antenna. Time domain gating can be used to eliminate the energy diffraction caused by the edge of the antenna and also the multiple reflections. To ascertain the dielectric characteristics, a program can be used to post-process the measured transmission and reflection coefficient.

(4) **Resonant Method**: This method obtains permittivity with more accuracy than the others while having the limitation on frequency and loss properties of the materials. Resonant methods have so many types such as reentrant cavity, cavity resonant, split cylinder resonator etc. This section contains a general technique used in the cavity resonator. There are two resonant methods that are mostly used. All permittivity measurements and measurements of medium to high loss materials can be done using perturbation methods, and a measurement on low VNA

MUT is a low loss measuring technique. However, the perturbation method is well suited when employing TM cavity design. To ascertain the dielectric properties, it is possible to monitor the quality factor and resonance frequency of resonance characteristics dependent on MUT in a cavity. For dielectric characteristics, the first measurement can be done by measuring the quality factor and resonant frequency of an empty cavity. And, the second measurement is taken when MUT is inserted into the cavity. The frequency, Q-factor and volume can then be used to compute the material's permittivity. This kind of measurement does not require calibration of the network analyzer. Any field modifications while the sample is inside the cavity will cause the observed resonant frequency or Q-factor to shift. This measurement needs to be done using a network analyzer that has a high frequency resolution. With the use of the resonator technique, the device being studied either becomes the resonator itself or a component of a resonant structure. Resonator techniques are typically employed to measure capacitance and the dielectric constant in the high frequency range that involves injection of current or voltage into an L-C resonant circuit. A conventional capacitor-containing oscillatory circuit tuned to resonance, and the appropriate capacitance value is then calculated. The circuit then tuned to resonance once more by adding a capacitor with the dielectric in parallel with the conventional capacitor.

(5) **Parallel Plate Method**: In this method, dielectric properties of the measured dielectric material depend on electrical and physical parameters, frequency of interest and the level of accuracy. In the parallel plate method, the substance being measured is placed between the conducting parallel plates of a capacitor, which also serves as a sample holder. This approach needs an impedance analyzer or LCR meter to measure capacitor impedance. The real and imaginary components of the dielectric constant are then calculated using the measured capacitance and dissipation factor (dielectric loss). With a frequency range of up to 1.8 GHz, this approach has high accuracy with minimal sample preparation and set-up requirements. For thin flat samples this techniques works best. The knowledge of length, width, and thickness of the material and plates as well as the capacitance and dissipation factor of the measured material are helpful in calculating the dielectric constant of a material. The dielectric constant for the measured material can be calculated using the following Eq. (4.14):

$$\varepsilon_r = \frac{tC_p}{A\varepsilon_0} \tag{4.14}$$

where, t is thickness of the sample (m), C_p stands for parallel capacitance of the sample (F), A stands for the area of the electrode's surface (m^2) and ε_0 stands for permittivity in vacuum or free space (8.854×10^{-12} F/m).

4.1.6.3 Dielectric Strength and Breakdown

The electrical strength of an insulating material is referred as its dielectric strength. An applied field (maximum voltage) strong enough to break down the insulating properties of an insulator that allows charge flow through the material (breakdown) is measurement of dielectric strength expressed as Volts per unit thickness [68]. Material with high dielectric strength is proven to be better electrical insulator. To measure the dielectric strength, there exist three basic standard testing methods: the short-time methods, the slow rate-of-rise method, and the step-by-step method. All the methods use the same set-up to testing, where the test specimen is placed between two well aligned electrodes in a sample holder. After placing the sample carefully, the sample holder is mounted in an air/insulating oil bath. Some researchers investigated the dielectric breakdown strength of the polymer composite [69, 70]. Song et al. [69] prepared $BaTiO_3$ incorporated PVDF (polar) or epoxy (non-polar) composites modified by Dopamine where $BaTiO_3$ was used as nanofibers and nanoparticles correspondingly with the polymer matrix and Dong et al. [70] prepared the polyimide composite with inclusion of polydopamine coated graphitic carbon nitride. They investigated the dielectric breakdown strength of the polymer composite by using Dielectric Wihstand Voltage Test. Stoyanov et al. [71] investigated the breakdown strength of carbon black/rubber composite using a (Trek-model) high voltage amplifier.

(1) Short-time Method: It is a conventionally used test method for quality control tests. In this method, the voltage across the two electrodes is gradually increased at a constant rate (500 V/sec) until a breakdown occurs. It is a longer time taking test.
(2) Slow rate-of-rise method: In this method, the electrodes of test are subjected to a voltage that starts at 50% of breakdown voltage and lasts until breakdown happens.
(3) Step-by-step: In this method, the test electrodes are exposed to the appropriate initial voltage over the specified time until breakdown takes place.

The dielectric breakdown test (Fig. 4.24) is standardized by JIS C2110, ASTM D149, EC 60243-1, etc.

4.1.6.4 Impedance Measurement

Impedance spectroscopy is used to measure impedance of any composite that consists of applied alternating voltage and wide frequency range. The voltage response of the material is recorded and after that the resultant impedance can be used to define the electrical characteristics of the polymer and its composites.

Impedance spectroscopy is a versatile and effective method for analyzing the relationship between grain boundaries, gain, and electrode–electrolyte interface. This approach explains the electrical conduction phenomena in the presence of an external applied field (AC input signals). All impedance response exhibits a low frequency

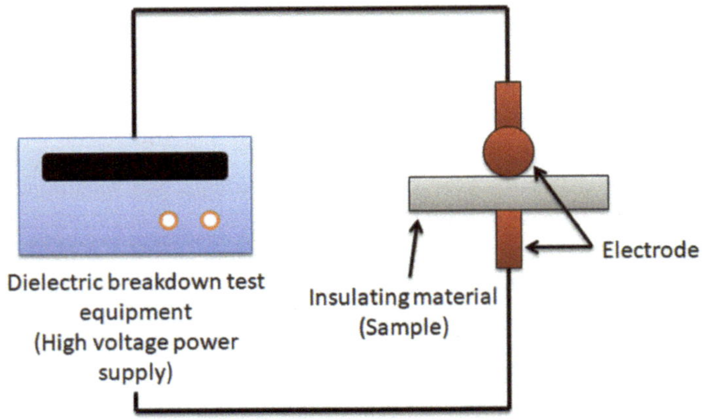

Fig. 4.24 Schematic diagram of dielectric breakdown test

spike with a high frequency semicircle where the low frequency spike represents the generation of a double layer capacitance at interface as a result of low frequency ion migration, and the high frequency semicircle represents the parallel arrangement of a resistor and a capacitor. The intercepts on the x-axis of the impedance plot are used to calculate resistance. The impedance has the same SI unit (ohm) as resistance but differs from resistance in a way that it has both phase and magnitude whereas resistance possesses only magnitude.

There are various impedance measurement techniques based on different factors such as measurement range and accuracy, operating frequency and ease in measurement. Bridge method is one of methods used in standard laboratory that covers DC to 300 MHz frequency range. This method is of low cost with high accuracy in the wide frequency range based on different types of bridge but it needs to be manually balanced. Resonant method is used in high Q device that covers 10–70 MHz frequency range but it needs to be tuned to resonance that provides low impedance measurement accuracy. I-V method is suitable for probe type tests for grounded device measurement in the 10–100 MHz frequency range that limits its operating voltage based on the transformer used in the probe. RF I-V method is used for RF component measurements, which cover 1–3 GHz with high accuracy and provide wide impedance range at high frequency. Network analysis method is used in RF component measurements at any frequency above 300 kHz with high frequency range, and provides good accuracy when characteristic impedance is close to unknown impedance. Auto balancing bridge method is applicable for generic component and grounded measurement, and provides high accuracy over a wide impedance range for 20 Hz to 110 MHz frequency, as shown in Fig. 4.25.

The study of impedance allows us to distinguish the real and imaginary part of the complex impedance which gives the structure property relationship of [72, 73]. By changing the applied frequency in an EIS measurement, a significant number of information is attained. The Nyquits and Bode plots are two graphs that are mostly used

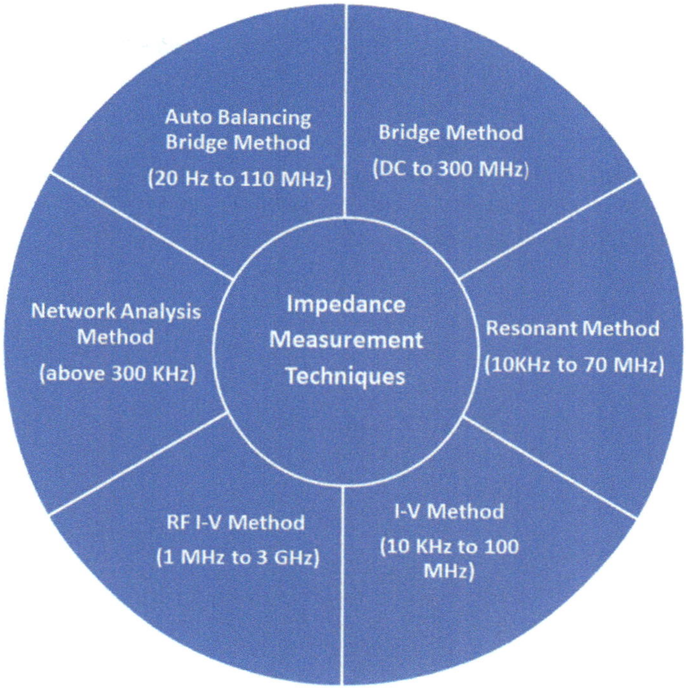

Fig. 4.25 Various impedance measurement techniques

to adequately depict this complex dataset [74]. Markevich et al. [75] presented the electric impedance spectra of CNTs-reinforced polyethylene polymer composite in the frequency range from 0.1 kHz to 0.12 GHz using the concept of a double electrical layer between the electrolyte and electrically conducting composite by EIS measurement technique. Helseth et al. [76] investigated the impedance spectra of MWCNTs-PDMS (polydimethylsiloxane) composites and found that experimental results are well fit by an electrical RC-circuit with a single relaxation time, and force-dependent capacitances and resistances. Ilgaz et al. [77] presented the impedance properties of SWCNT/VE (single walled carbon nanotube/vinylester) glass fiber-reinforced polymer composites in the frequency range from 10^{-2} to 10^{7} Hz at temperature from 300 to 420 K. They obtained Nyquist and Bode plots of impedance and interpreted the performance by Cole–Cole technique.

Complex Impedance:

$$Z^*(\omega) = (Z' - jZ'') \quad (4.15)$$

Here, $Z' = |Z|Cos\theta$ and $Z'' = |Z|Sin\theta$, where θ is phase.

$$|Z(\omega)| = \sqrt{Re(Z(\omega))^2 + Im(Z(\omega))^2} \quad (4.16)$$

(1) **Nyquist Plot-Equivalent Circuit Method**

The presented equivalent electrical circuit here is Randles circuit (Fig. 4.26), which involves R_u, charge transfer resistance (R_{ct}) and impedance Z_w called the Warburg impedance.The plot between the real component Z′ (resistive terms) and the imaginary component Z″ (inductive and capacitive terms) serves as the foundation for depiction of impedance data which is evaluated by Nyquist plot (Cole–Cole plot), which uses the formalism given in Eqs. (4.17) and (4.18) [78, 79].

$$Z' = R_u + \frac{R_{ct} + \sigma\omega^{-\frac{1}{2}}}{\left(\sigma\omega^{\frac{1}{2}}C_{dl} + 1\right)^2 + \omega^2 {C_{dl}}^2 \left(R_{ct} + \sigma\omega^{-\frac{1}{2}}\right)^2} \tag{4.17}$$

$$-Z'' = \frac{\omega C_{dl}\left(R_{ct} + \sigma\omega^{-\frac{1}{2}}\right)^2 + \sigma^2 C_{dl} + \sigma\omega^{-\frac{1}{2}}}{\left(\sigma\omega^{\frac{1}{2}}C_{dl} + 1\right)^2 + \omega^2 {C_{dl}}^2 \left(R_{ct} + \sigma\omega^{-\frac{1}{2}}\right)^2} \tag{4.18}$$

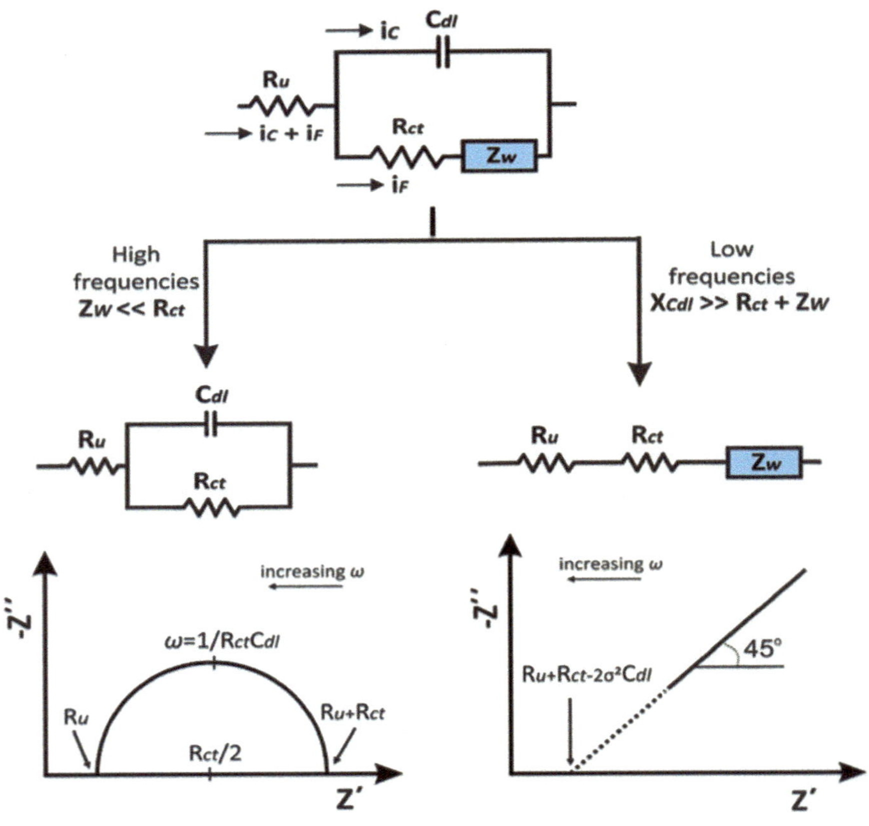

Fig. 4.26 Nyquist plot of equivalent Randles circuit

Each point in the Cole–Cole plot represents impedance corresponding to the magnitude and direction of the vector at a specific frequency. Nyquits plot exhibits a semicircular arc with an offset from the point of origin that corresponds to resistance, R_u [80]. Where, R_u is the resistance of the solution, C_{dl} is the capacitance of the electrical double layer, R_{ct} is the charge transfer resistance, and Z_w is the Warburg impedance which expresses mass transfer.

The reactance is very high ($Xc_{dl} \gg R_{ct} + Z_w$) at low frequency ($\omega \to 0$), $Z'' = \frac{-1}{\omega C_{dl}}$ is very high therefore the whole current will flow through R_u, R_{ct} and Z_w. In this frequency region mass transfer phenomena is dominated in the electrochemical process.

Thus the Eqs. (4.19) and (4.20) will be

$$Z' = R_u + R_{ct} + \sigma\omega^{-\frac{1}{2}} \tag{4.19}$$

$$Z'' = -\sigma\omega^{-\frac{1}{2}} - 2\sigma^2 C_{dl} \tag{4.20}$$

By solving both equations,

$$Z'' = Z' - R_u - R_{ct} + 2\sigma^2 C_{dl} \tag{4.21}$$

For $-Z'' = 0$, a straight line can be seen in low frequency range with slope 1 ($\theta = 45°$), which intercepts the real axis (x-axis) at

$$Z' = R_u + R_{ct} - 2\sigma^2 C_{dl} \tag{4.22}$$

When $R_{ct} \gg Z_w$ is taken into account at high frequency ($\omega \to \infty$), $Z'' = \frac{-1}{\omega C_{dl}}$ is very low and thus R_u, R_{ct} and C_{dl} all act to impede current flow. In this frequency region charge transfer phenomena is dominated in the electrochemical process.

By solving the equations under considerable condition, we obtain:

$$\left(Z' - R_u - \frac{R_{ct}}{2}\right)^2 + (Z'')^2 = \left(\frac{R_{ct}}{2}\right)^2 \tag{4.23}$$

Therefore, varying Z' with respect to Z'' results in a semicircle plot with radius of $\frac{R_{ct}}{2}$ and its centre at $Z' = R_u + \frac{R_{ct}}{2}$ and $-Z'' = 0$.

Thus, the Randles circuit exhibits resistive behavior at both low and high frequencies, and the values of R_u and $R_u + R_{ct}$ can be discovered as the semicircle's real axis intercepts.

(2) **Broad Band Dielectric Response Method**

Broad band dielectric response method known as "bode plot" is another method used to measure impedance, here, impedance is plotted between magnitude to impedance and frequency. Normally, Impedance magnitude and frequency are plotted on log

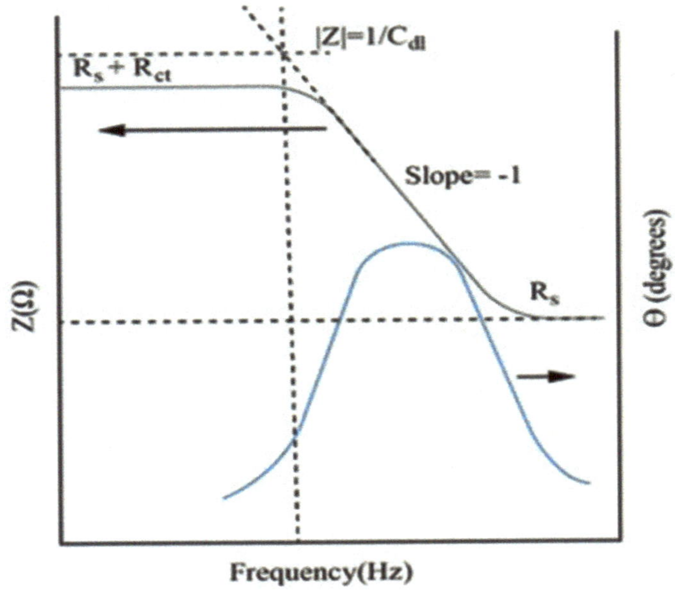

Fig. 4.27 Bode plot for Randles circuit

scale (log $|Z|$ vs log ω), while phase angle is plotted linearly. Bode plot for equivalent Randles circuit is shown in Fig. 4.27.

4.1.7 Rheological Characterization

The term "Rheology" is derived from the greek word "Rheologia" i.e. "Rheo" means flow and "logia" means study of. Under the effect of stress, a material's body experiences flow and deformation that is known as rheology and the type of deformation is influenced by the conditions of material's body [78, 81]. Elastic deformation occurs in an ideal solid (known as Hooke solid), which means that if the force is removed the solid will revert to its initial state. In this instance, the required energy for deformation will be recovered once the stress is ended. Rheology has evolved to cover a wide range of topics, with substantial advancements achieved in fields like suspensions, coating and polymer rheology. According to particular needs, rheological additives can vary the viscosity, flow and deformation characteristics. Rheology also contributes in providing information of polymer extrusion and how the process of deformation in impacted by temperature. Viscometers and rheometers have advanced in simulating industrial and handling situations to assess the performance of materials as a function of stress level and temperature. The polymers are usually melted or softened into rubbery or liquid form to get in appropriate shape and then cured (solidified) to sustain those shapes. The flow behavior of polymer melts is able to process the polymeric material into any form therefore melt flow behavior of polymeric materials plays a crucial role in understanding polymer processability. Several works [81–86] have been reported on rheological characterization of polymer composites. Lim et al. [87] investigated the nonlinear behavior of polymer composites in the presence of high magnitude oscillating shear flux. In order to investigate the polycaprolactone/MWCNTs nanocomposites in (COGA) flux, they employed a variety of analytical techniques. According to zero deformation nonlinearity and deformation magnitude, they estimated the nonlinear component through the rheology of the Fourier transformation. By increasing MWCNTs content, they were able to compare nonlinear and linear viscoelastic characteristics. Lewandowski et al. [88] studied the rheological behavior of wood-polymer composite (WPC) with polypropylene using capillary rheometer with shear rate of 10 s^{-1} to 912 s^{-1}. They found that with increasing wood flour amount, the viscosity and the flow rate of the substance comprising of the WPC matrix, and the loss of pressure of composites corresponding to entrance effects are reduced by a significant amount.

4.1.7.1 Viscosity

The term viscosity reflects the resistance to flow which is considered a critical flow characteristic of the polymeric material. Various viscosities of different polymer composites or materials are advantageous in many applications such as putty and sealants are higher in viscosity and paints are lower in viscosity (As shown in Fig. 4.28). In polymeric composites processing, a different viscosity of the material is required as low and high viscosity have their own need while working with different polymers' processing requirement. Low viscosity is required in injection molding for filling the thin walls and voids, while high viscosity is needed in blow molding and extrusion. The methods to measure viscosity are as follows.

(1) **Capillary Viscometer**

This method measures the viscosity of Newtonian fluids with more accuracy. Through this capillary, time taken for a fluid to flow can be measured. The measurement of viscosity using an appropriate capillary viscosity is done at a temperature of 20 ± 0.1 °C, unless otherwise specified. In this technique, bulb (A) is filled with enough of liquid being tested to fill the viscometer through tube (L), bringing the liquid to 20 °C beforehand unless otherwise specified but made sure that the liquid level in bulb (B) is below the exit to the ventilation tube (M). Unless otherwise specified, viscometer is placed in the water bath of temperature 20 ± 0.1 °C, kept upright, and allowed to stand for at least 30 min to allow the temperature to stabilize. The liquid level in the tube (N) should be raised to a level that is about 8 mm over the mark (E) before tube (M) is closed. By closed tube (N) and open tube (M) the liquid level can be maintained. To the nearest one-fifth of a second, time required is measured for the liquid level to fall from mark (E) to (F), with the open tube (N).

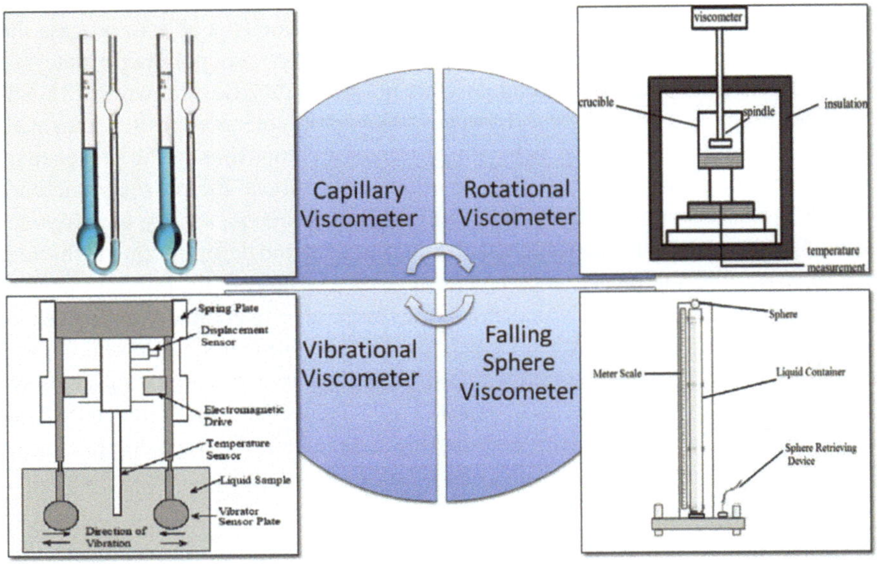

Fig. 4.28 Various methods to measure viscosity

(2) **Rotational Viscometer**

The idea behind the technique is to measure the torque or force experiences while it rotates in a liquid at a constant angular speed. Shear-independent viscosity is used to measure the viscosity of Newtonian liquids, while shear-dependent or apparent viscosity is used to measure the viscosity of non-Newtonian liquids. There are two categories of rotating viscosities that are absolute and relative viscometer. The flow within the measuring geometry is clearly defined by absolute viscometers while the flow in the measuring geometry is not defined in relative viscometers.

(3) **Falling Sphere Viscometer**

This viscometer works on "Höppler" principle and uses falling sphere to measure the dynamic viscosity of transparent Newtonian fluids. The fluid to be examined is contained inside a cylindrical tube that is inclined and contains sphere rolls and slides. The time it takes for the sphere to fall a certain distance throughout the fluid by the measurement tube directly determines the viscosity which is expressed in mPa s. In order to assess how long the sphere requires to drop back, the tube with it can then be inverted upside-down.

(4) **Vibrational Viscometer**

Vibrational viscometer generates waves inside the liquid in order to function. The Hydramotion technique uses shear waves since viscosity is a shear measure that may only be accurately in shear conditions. In vibrational viscometer, a constant displacement magnitude is applied to the vibration of an oscillator that is placed in a liquid. The power required for the vibration can be observed, and this can be used to determine the viscosity of the liquid.

4.1.7.2 Melt Flow Index

In this technique, a thermoplastic polymer's ease of melting is measured. It is described, when the polymer (in gms) flows in ten minutes across a die with a particular length and diameter under pressure from a specific weight at a specific temperature, both ISO 1133 and ASTM D1238 describe the procedure. Low molecular weight is correlated with a high melt flow rate, which is an indirect measurement of molecular weight. Melt flow rate also serves as a measure for how well melt of material flows under pressure. To measure melt flow rate, Melt Flow Index test (Fig. 4.29) is conducted in which the material is loaded in the melt flow index tester at a temperature higher than its melting point. This tester contains of an inner die and outer zone, and both are electrically heated. After loading the sample within the inner die it takes some time to melt entirely. The screw punch then begins to descend at a consistent speed while the melt flow indexer is turned on, pushing the melting polymer out of the opening at the bottom of the die. The time it takes to flow out the 10 gm of polymer across the die is used to compute the melt rate. Several material features, including processing behavior, stability, long term viscoelastic behavior throughout end usage, etc., can be predicted using the melt flow rate. As a result, it is a crucial quality control factor for many plastic and polymer products. Therefore, the melt flow index test is carried out to determine whether the polymer material is suitable to utilize.

4.1.7.3 Shear Stress

In order to evaluate the shear stress, the rheometer works on "shear mode". This rheological system consists of a coaxial clamp and a cylindrical heating furnace that can be sealed. The coaxial clamp (parallel plates clamp) is connected to a mechanical testing arrangement to manage the shear rate throughout the test. To take the measurements, the furnace is preheated to 180 °C (experimental temperature) and then the furnace is turned on to fix the parallel iron plates with the sample to the fixture; and finally, the furnace was turned off and the temperature is maintained for 15 min so that the sample melts completely.

4.1.8 Thermal Characterization

Thermal analysis evaluates the kinetic and thermophysical characteristics of the material with respect to temperature i.e. it investigates the reaction of a substance when being cooled or heated. The properties of a material can be estimated with time

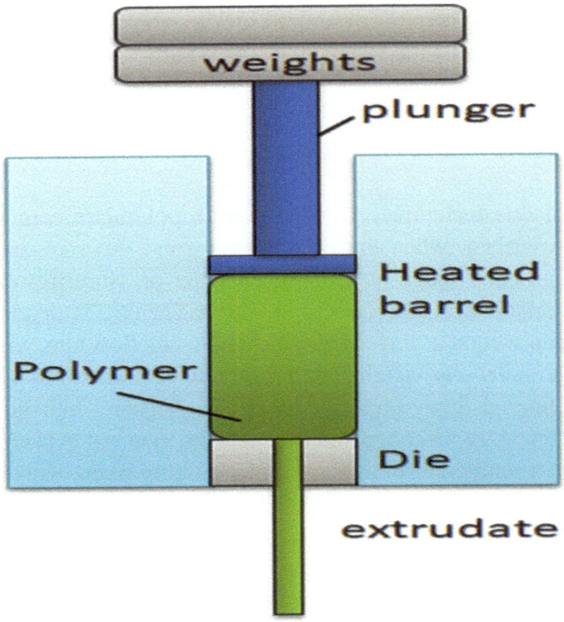

Fig. 4.29 Schematic diagram of melt flow index testing

or temperature. The most frequently used temperature dependent analysis techniques (Fig. 4.30) are differential scanning calorimetry (DSC), thermogravimetric analysis (TGA), and differential thermal analysis (DTA) [89–91].

TGA measures the sample mass in a controlled environment furnace as a function of temperature or time. DSC observes that when temperature varies chemical reactions and phase transitions produce heat effect. Finally, by observing the sample's behavior towards temperature, DTA can detect and investigate the composition of a material. Dweck et al. [92] developed and investigated the thermal characteristics of polymeric composites. The TG/DTA characterization were performed under temperature from room temperature at 800 °C. In this technique, samples from the composite's top, bottom, centre, and lateral side were obtained to examine the uniformity of polymer distribution. The polymer concentration of the composite can often be easily calculated from the overall area that comprises the corresponding DTG peaks because of its burning in oxidized environment analysis. Thamer et al. [93] reported thermal characterization techniques for epoxy/hexagonal boron nitride composites such as TGA, DTA and DSC, and the results exhibited enhanced thermal stability of nanocomposites. They used Coast-Redfern technique to analyze the deformation at 10 °C/min constant heating rate in air by obtaining TGA and DTG curves.

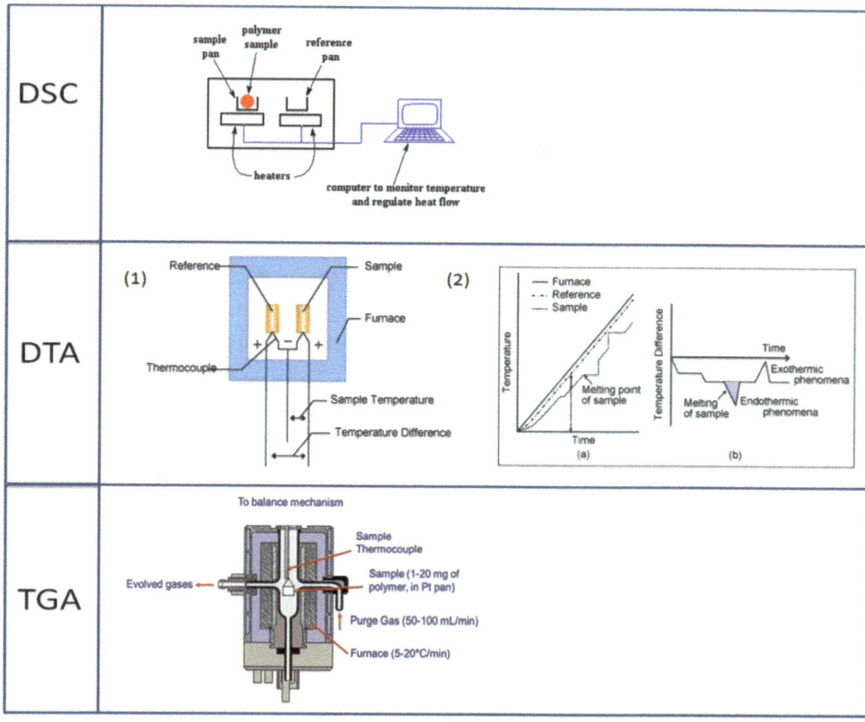

Fig. 4.30 Various thermal characterization techniques

4.1.8.1 DSC

It measures the heat flow rate (mW = mJ/sec) difference between the inert reference and material sample (between reference pan and sample pan) with varying temperature and time. Absorbed heat by the sample is a response of the endothermic process while released heat by the sample is the response of the exothermic process. Heat flows through the sample material can be the response of heat capacity (heating), Glass Transition (T_g), melting, evaporation or other endothermic processes. And, released heat out of the sample can be the response of heat capacity (cooling), crystallization, curing, oxidation, or other exothermic processes. In this technique, temperature for the sample material and reference both are kept same throughout the experiment. According to DSC analysis designed temperature program, the temperature of the sample holder should rise linearly with time. As for the reference sample, it should have a precise heat capacity over the temperature range that will be scanned. The reference sample material must also maintain its stability, high purity and lack of significant change during the temperature scan. Indium, bismuth, lead, and tin are typically in use as reference standards, although additional standards such as polyethylene and fatty acid are suggested for the study of polymers and organic molecules, respectively. There are two mostly used DSCs that are Heat-flux DSC involving heat flux difference between the reference and sample material and power differential DSC that involves the difference of power supplied to the reference and sample material.

4.1.8.2 DTA

Differential thermal analysis (DTA) is an alternate of DSC that is quite similar to DSC. Herein, the difference in temperature is the result of phase change and other processes as reference and the sample material is heated equally. In this technique, the reference and sample material are arranged uniformly in a furnace. Here, the sample and reference temperatures are varied in the furnace which is controlled by a temperature programme. In order to measure the difference in temperature between the reference and sample, a differential thermocouple is arranged as shown in Fig. 4.30(1). Additionally, the thermocouple on the sample side is used to measure the temperature of the sample material. Reference materials are substance that does not change in the measuring temperature range (often-alumina). As the furnace starts heating up, a slight delay is found in starting of heating of the reference and sample material due to their individual thermal (heat) capacity and finally they both heat up gradually in accordance with the temperature of the furnace. After the heating starts, ΔT varies until a static condition is attained, and then it changes until it achieves stability that is consistent with the difference in thermal capacity between the reference and sample material. The baseline is the signal that is in a stationary state. For instance, as the temperature increases and melting takes place in the sample material, then rise in temperature stops and ΔT rises as shown in Fig. 4.30(2a). When melting stops, ΔT reaches the peak and the temperature curve quickly returns to the baseline. This allows us to determine the reaction temperature and sample transition temperature from DTA signal (ΔT signal) as shown in Fig. 4.30(2b).

4.1.8.3 TGA

TGA evaluates weight change of a sample material as a function of temperature. It is a potent and reliable method for investigating stability and variation in the weight of a material. This change in weight of a sample though heating (at constant rate) can be attributed to a particular reaction of a material towards a thermal stress. This technique is ideal for in-depth investigation of breakdown temperatures and confirming that a material functions properly within a specific temperature range. In this process, sample and the weighing mechanism are exposed to a carrier gas. Through gas–liquid or gas–solid reactions, carrier gas can interact with sample material while also protecting the internals from oxidation or corrosion above 500 °C. Reducing environment can be helpful in preventing the oxidation of a specific material or can discover the gas–solid phase reduction processes. Exploring metal organic framework or porous materials that are catalytic is made easier with this method.

4.2 Optical Characterization

According to Shirakawa et al. [94], Macdiarmid [95], and Heeger [96], conducting polymer polyacetylene can be employed as a dopant to the material that influences conductivity of the doped material. Since then polymers are the widely investigated materials for optoelectronic applications. The electronic and optoelectronic materials are in great attention due to their implementation of different optical functions, including modulation, switching, generation, and optical waveguide structure detection generated on the substrate. Polymers are appropriate material for optoelectronics and photonic applications. Optical characteristics are one of the most crucial characteristics for investigating the electronic properties of polymer composites [97–99] that are shown in Fig. 4.31. These methods depend on either the nonlinear optical characteristics of the polymer composite or the movement of electrons from HOMO (highest occupied molecular orbital) to the LOMO (lowest unoccupied molecular orbital). Ananda et al. [100] performed UV–visible spectroscopy on PVA/Gly:Na_2SO_4 composite film to validate the characteristics absorption with regard to sodium sulphate content for 300 to 800 nm wavelength. From the UV/Vis absorption spectra, they observed higher transmission percentage for PVA/Gly which reduced with sodium sulphate content. The decrease in transmission value indicated variation in optical band gap attributed to structural change in the polymer. Larosa et al. [101] prepared polycarbonate (PC)/MWCNTs composite and performed UV–Vis spectroscopy by double beam configuration in the 300–700 nm wavelength range. They observed that the prepared composite exhibits absorption band at around 289 nm. In this section, optical measurement techniques to characterize the polymer composites are discussed.

4.2.1 UV–VIS Absorption

A method for measuring electromagnetic wave absorption as a function of wavelength and frequency is called absorption spectroscopy. The variations in the absorption spectra can be used to analyze the interaction between radiation and electrons that is caused by the absorption process. There are several literatures [102–104] on apparatus and technical elements of absorption spectroscopy that conclude that the sample holder, monochromator, light source, and optical detector are four fundamental parts. In general, a single or double beam is used by absorption spectrophotometer from which a single beam is compatible and easy to use, and the double beam comprises two beams that split into two optical pathways. In the double beam system, the first splited beam travels through the reference, while the second splited beam travels though the sample. The double beam provides more precise, stable and accurate data over the single beam system.

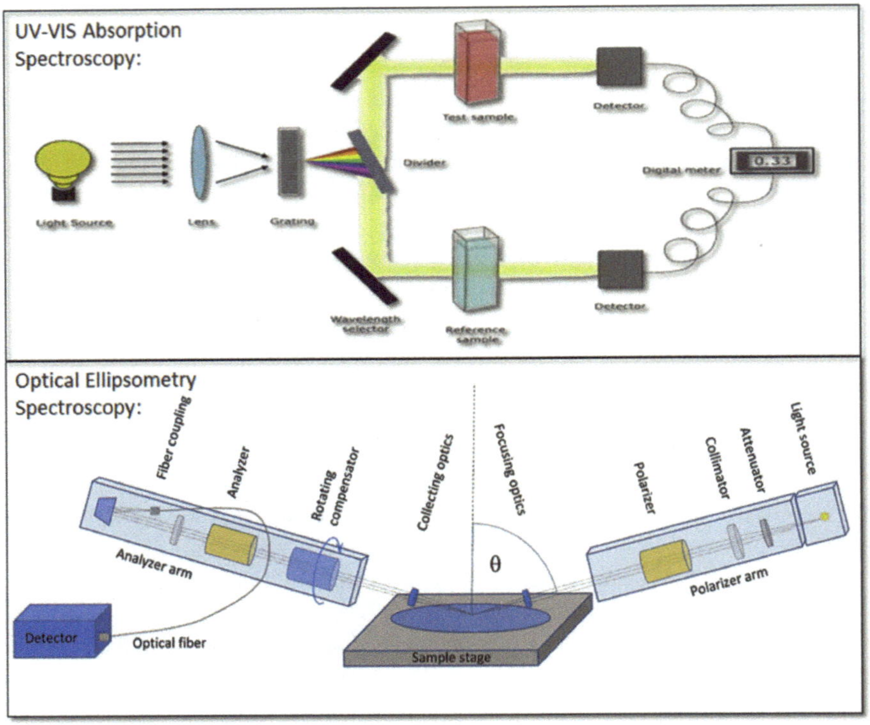

Fig. 4.31 Schematic diagrams of optical spectroscopy

In the UV–VIS absorption process, the intensity of incident light reduces when passing through the medium. The absorbed intensity by the sample can be evaluated by Lambert–Beer laws as shown in Eq. (4.24).

$$I = I_0 e^{-\alpha t} \tag{4.24}$$

where I_0 represents intensity of the incident light, I is the transmitted light, α is the absorbance coefficient that is the measurement of how strongly the light is absorbed by the sample at some particular wavelength, and t is the sample thickness. This equation represents the exponential decay of light intensity with respect to thickness of the sample.

4.2.2 Optical Reflectometry and Ellipsometry

Optical reflectometry is contactless and non-destructive and that involves interaction between sample and reflected light. Reflectometry is inexpensive and simple because the system employs non-polarized light to serve as light source i.e. tungsten bulb. While, ellipsometry involves the impact of polarization of light to investigate the optical characteristics therefore the instrument must handle delicately to observe the polarization state variation. Ellipsometer consists of some essential parts that are optical source, analyzer, polarizer, detector and phase inducer, where UV-based deuterium light serves as the light source. There are numerous different types of polarizers, including calcite, wave plate, linear, and wire grid polarizers that change the particular polarization state into an elliptical, plane, right, or left circular polarization. Phase inducers such as wave plate, phase modulators, and compensator serve to attain specified elliptic polarization. A polarized-type analyzer investigates the light which reflects from the sample. Optical reflectometer and ellipsometer measurements are used to attain thickness of the film sample and optical constant.

4.3 Conclusion

Characterization of composite materials is an essential step in the formation of new products. The characterization of polymer composite materials confirms that advanced composite materials fulfill the application performance standards through the industrial perspective. By physical and chemical characterization, quality control can be ensured to expand the understanding of materials and products. In conclusion, various techniques can be utilized for characterizations of polymer-based composites which includes mechanical, tribological, microstructural, structural, physical, electrical, rheological, thermal and optical properties.

References

1. Mohan Kumar, A., Rajasekar, R., Manoj Kumar, P., Parameshwaran, R., Karthick, A., Muhibbullah, M.: Comparative analysis of drilling behaviour of synthetic and natural fiber-based composites. Adv. Mater. Sci. Eng. **2021**, 1–13 (2021)
2. Saheb, D.N., Jog, J.P.: Natural fiber polymer composites: a review. Adv. Polymer Technol.: J. Polymer Process. Inst. **18**(4), 351–363 (1999)
3. Dhand, V., Mittal, G., Rhee, K.Y., Park, S.J., Hui, D.: A short review on basalt fiber reinforced polymer composites. Compos. B Eng. **73**, 166–180 (2015)
4. Sharma, H., Kumar, A., Rana, S., Sahoo, N.G., Jamil, M., Kumar, R., Sharma, S., Li, C., Kumar, A., Eldin, S.M., Abbas, M.: Critical review on advancements on the fiber-reinforced composites: Role of fiber/matrix modification on the performance of the fibrous composites. J. Mater. Res. Technol. (2023)
5. Sun, B., Xue, C., Shang, W., An, M., Zhao, H., Zhang, H.: The performance characterization of carbon fiber–reinforced plastic for space applications. J. Reinf. Plast. Compos. **42**(15–16), 844–853 (2023)
6. Abubakre, O.K., Medupin, R.O., Akintunde, I.B., Jimoh, O.T., Abdulkareem, A.S., Muriana, R.A., James, J.A., Ukoba, K.O., Jen, T.C., Yoro, K.O.: Carbon nanotube-reinforced polymer nanocomposites for sustainable biomedical applications: a review. J. Sci.: Adv. Mater. Dev. 100557 (2023)

7. Khan, S.B., Chen, S., Liang, J., Xiao, C., Sun, X.: Influence of nanoparticle size on the mechanical and tribological characteristics of TiO2 reinforced epoxy composites. J. Market. Res. **26**, 6001–6015 (2023)
8. Takkalkar, P., Tobin, M.J., Vongsvivut, J., Mukherjee, T., Nizamuddin, S., Griffin, G., Kao, N.: Structural, thermal, rheological and optical properties of poly (lactic acid) films prepared through solvent casting and melt processing techniques. J. Taiwan Inst. Chem. Eng. **104**, 293–300 (2019)
9. Saba, N., Jawaid, M., Sultan, M.T.H.: An overview of mechanical and physical testing of composite materials. In: Mechanical and Physical Testing of Biocomposites, Fibre-Reinforced Composites and Hybrid Composites, pp. 1–12 (2019)
10. Miller, A., Brown, C., Warner, G.: Guidance on the use of existing ASTM polymer testing standards for ABS parts fabricated using FFF. Smart Sustain. Manuf. Syst. **3**(1) (2019)
11. http://www.intertek.com/polymers/testing/flexural-properties/
12. Mehndiratta, A., Bandyopadhyaya, S., Kumar, V., Kumar, D.: Experimental investigation of span length for flexural test of fiber reinforced polymer composite laminates. J. Market. Res. **7**(1), 89–95 (2018)
13. Saba, N., Mohammad, F., Pervaiz, M., Jawaid, M., Alothman, O.Y., Sain, M.: Mechanical, morphological and structural properties of cellulose nanofibers reinforced epoxy composites. Int. J. Biol. Macromol. **97**, 190–200 (2017)
14. https://www.wmtr.com/en.impact-testing.html
15. https://www.impacttestingmachine.com/new/types-of-impact-testing.html
16. McKeen, L.W.: The Effect of Sterilization on Plastics and Elastomers. William Andrew 2018
17. Yahaya, R., Sapuan, S.M., Jawaid, M., Leman, Z., Zainudin, E.S.: Effect of fibre orientations on the mechanical properties of kenaf–aramid hybrid composites for spall-liner application. Defence Technol. **12**(1), 52–58 (2016)
18. Saba, N., Tahir, P.M., Abdan, K., Ibrahim, N.A.: Fabrication of epoxy nanocomposites from oil palm nano filler: mechanical and morphological properties. BioResources **11**(3), 7721–7736 (2016)
19. Mi, X., Zhong, L., Wei, F., Zeng, L., Zhang, J., Zhang, D., Xu, T.: Fabrication of halloysite nanotubes/reduced graphene oxide hybrids for epoxy composites with improved thermal and mechanical properties. Polym. Test. **76**, 473–480 (2019)
20. Kumar, T.V., Chandrasekaran, M.: Charactersation analysis of hardness, impact, water absorbtion nano carbon fiber reinforced polymer matrix composites [NCFRPC]
21. Das, P.P., Chaudhary, V.: Mechanical characterization techniques for composite materials. In Composite Materials, pp. 155–161. CRC Press (2021)
22. American Society for Testing and Materials, 2015. ASTM E10-15a, Standard Test Method for BrinellHardness of Metallic Materials
23. STANDARD, B. and ISO, B.: Metallic materials—Brinell hardness test (2006)
24. Broitman, E.: Indentation hardness measurements at macro-, micro-, and nanoscale: a critical overview. Tribol. Lett. **65**(1), 23 (2017)
25. Jindal, U.C.: Material Science and Metallurgy. Pearson Education India (2012)
26. https://www.scribd.com/document/295258890/6-Vickers-Hardness-Test
27. https://www.hardnessgauge.com/testing-types/vickers-hardness-testing/
28. Sugimoto, T., Kawaguchi, T.: Development of an automatic Vickers hardness testing system using image processing technology. IEEE Trans. Ind. Electron. **44**(5), 696–702 (1997)
29. https://sm-nitk.vlabs.ac.in/exp/vickers-hardness-test/theory.html
30. Chand, N.: An Introduction to Tribology of FRP Materials. Allied Publishers (2000)
31. Buckley, D.H.: Surface Effects in Adhesion, Friction, Wear, and Lubrication, vol. 5. Elsevier (1981)
32. Burwell, J.T., Strang, C.D.: On the empirical law of adhesive wear. J. Appl. Phys. **23**(1), 18–28 (1952)
33. Chand, N., Dwivedi, U.K.: Effect of coupling agent on abrasive wear behaviour of chopped jute fibre-reinforced polypropylene composites. Wear **261**(10), 1057–1063 (2006)

34. Devaraju, A.: A critical review on different types of wear of materials. Int. J. Mech. Eng. Technol **6**(11), 77–83 (2015)
35. More, S.R., Bhatt, D.V., Menghani, J.V.: Recent research status on erosion wear–an overview. Mater. Today: Proc. **4**(2), 257–266 (2017)
36. Marzbali, M., Yeganehdoust, F., Ibrahim, M.E., Tarasi, F., Jadidi, M.: Liquid-solid impact mechanism, liquid impingement erosion, and erosion-resistant surface engineering: a review. Coatings **13**(3), 577 (2023)
37. Yu, B., Li, D.Y., Grondin, A.: Effects of the dissolved oxygen and slurry velocity on erosion–corrosion of carbon steel in aqueous slurries with carbon dioxide and silica sand. Wear **302**(1–2), 1609–1614 (2013)
38. Castberg, T.S., Johnsen, R., Berget, J.: Erosion of hardmetals: dependence of WC grain size and distribution, and binder composition. Wear **300**(1–2), 1–7 (2013)
39. Arora, H.S., Grewal, H.S., Singh, H., Mukherjee, S.: Zirconium based bulk metallic glass—Better resistance to slurry erosion compared to hydroturbine steel. Wear **307**(1–2), 28–34 (2013)
40. Kumar, N., Mishra, S.: Slurry erosion: an overview. Mater. Today: Proc. **25**, 659–663 (2020)
41. Cann, P., Lubrecht, A.A.: An analysis of the mechanisms of grease lubrication in rolling element bearings. Lubr. Sci. **11**(3), 227–245 (1999)
42. https://www.tribonet.org/wiki/lubricants-formulation/#:~:text=Some%20examples%20of%20gases%20lubricants,also%20examples%20of%20gaseous%20lubricants
43. Brandon, D., Kaplan, W.D.: Microstructural Characterization of Materials. John Wiley & Sons (2013)
44. https://www.thermofisher.com/blog/materials/sem-signal-types-electrons-and-the-information-they-provide/
45. https://www.scimed.co.uk/education/sem-scanning-electron-microscopy/
46. Goldstein, J.I., Newbury, D.E., Michael, J.R., Ritchie, N.W., Scott, J.H.J., Joy, D.C.: Scanning Electron Microscopy and X-ray Microanalysis. Springer (2017)
47. Hajšman, J., Jeníček, Š.: Microscopy of polymer composites (2018)
48. Mohammed, A., Abdullah, A.: Scanning electron microscopy (SEM): A review. In: Proceedings of the 2018 International Conference on Hydraulics and Pneumatics—HERVEX, November. BăileGovora, Romania, vol. 2018, pp. 7–9 (2018)
49. https://www.thermofisher.com/in/en/home/materials-science/learning-center/applications/sem-tem-difference.html
50. Kulkarni, S.K., Kulkarni, S.K.: Nanotechnology: Principles and Practices. Springer (2015)
51. https://serc.carleton.edu/details/images/8400.html
52. Schlappa, S., Bressel, L., Reich, O., Münzberg, M.: Advanced particle size analysis in high-solid-content polymer dispersions using photon density wave spectroscopy. Polymers **15**(15), 3181 (2023)
53. Vinayagamoorthy, R.: Effect of particle sizes on the mechanical behaviour of limestone-reinforced hybrid plastics. Polym. Polym. Compos. **28**(6), 410–420 (2020)
54. Mironov, V.S., Kim, J.K., Park, M., Lim, S., Cho, W.K.: Comparison of electrical conductivity data obtained by four-electrode and four-point probe methods for graphite-based polymer composites. Polym. Testing **26**(4), 547–555 (2007)
55. Gong, S., Zhu, Z.H., Li, J., Meguid, S.A.: Modeling and characterization of carbon nanotube agglomeration effect on electrical conductivity of carbon nanotube polymer composites. J. Appl. Phys. **116**(19) (2014)
56. Cheng, J., Qiu, J., Xu, X., Ji, H., Takagi, T., Uchimoto, T.: Research advances in eddy current testing for maintenance of carbon fiber reinforced plastic composites. Int. J. Appl. Electromagnet Mech **51**(3), 261–284 (2016)
57. Mook, G., Lange, R., Koeser, O.: Non-destructive characterisation of carbon-fibre-reinforced plastics by means of eddy-currents. Compos. Sci. Technol. **61**(6), 865–873 (2001)
58. Runyan, W.R.: Semiconductor Measurements and Instrumentation McGraw-Hill Co. (1975). Schroder, D.K., Semiconductor Material and Device Characterisation. Wiley, New York (1990)

59. García-Martín, J., Gómez-Gil, J., Vázquez-Sánchez, E.: Non-destructive techniques based on eddy current testing. Sensors **11**(3), 2525–2565 (2011)
60. https://www.rjpbcs.com/pdf/2017_8(2)/[225].pdf
61. Barber, P., Balasubramanian, S., Anguchamy, Y., Gong, S., Wibowo, A., Gao, H., Ploehn, H.J., ZurLoye, H.C.: Polymer composite and nanocomposite dielectric materials for pulse power energy storage. Materials **2**(4), 1697–1733 (2009)
62. Field, R.F.: Sec 4.1, Lumped circuits, A: Permittivity, Ch. II., Dielectric measuring techniques. Dielectric Materials and Applications (1954)
63. Suresh, G., Thirmal, C., Mohan, P.N., Prasad, D.N., Raju, K.J., Vishwam, T.: Dielectric characterization of P (VDF-TrFE)-CuO nanocomposite solutions in the microwave frequency range using open-ended coaxial probe technique. Mater. Today: Proc. (2023)
64. Willert-Porada, M.A., Park, H.S., Rosin, A.: June. New method for temperature dependent dielectric property investigation in polymer composite materials. In: 2013 IEEE MTT-S International Microwave Symposium Digest (MTT), pp. 1–4. IEEE (2013)
65. Meena, S., Kumari, N., Mitra, S., Singhal, R., Verma, S., Choubey, R.K., Dwivedi, U.K.: Dielectric study of CoFe2O4/MWCNT/epoxy nanocomposites: effect of temperature, concentration and irradiation. Polym. Compos. (2023)
66. Meena, S., Kumari, N., Gahlaut, V., Shekhar, C., Mitra, S., Kumar Dwivedi, U.: Ferroelectric, dielectric, and EMI attenuation characteristics of BaFe2O4/MWCNTs/epoxy nanocomposites. In: Composite Interfaces, pp.1–19 (2023)
67. Lin, X., Seet, B.-C.: Dielectric characterisation at millimetre waves with hybrid microstrip-line method. IEEE Trans. Instrum. Meas. **66**(11), 3100–3102 (2017)
68. Karlsson, M.: Investigation of the dielectric breakdown strength of polymer nanocomposites (2014)
69. Song, Y., Shen, Y., Liu, H., Lin, Y., Li, M., Nan, C.W.: Improving the dielectric constants and breakdown strength of polymer composites: effects of the shape of the BaTiO3 nanoinclusions, surface modification and polymer matrix. J. Mater. Chem. **22**(32), 16491–16498 (2012)
70. Dong, Y., Wang, Z., Huo, S., Lin, J., He, S.: Improved dielectric breakdown strength of polyimide by incorporating polydopamine-coated graphitic carbon nitride. Polymers **14**(3), 385 (2022)
71. Stoyanov, H., Mc Carthy, D., Kollosche, M., Kofod, G.: Dielectric properties and electric breakdown strength of a subpercolative composite of carbon black in thermoplastic copolymer. Appl. Phys. Lett. **94**(23) (2009)
72. Song, C.H., Kim, M., Lee, S.M., Choi, H.W., Yang, Y.S.: J. Korean Phys. Soc. **56**, 462 (2010). https://doi.org/10.3938/jkps.56.462
73. Barik, S.K., Choudhary, R.N.P., Singh, A.K.: Adv. Mat. Lett. **2**(6), 419 (2011). https://doi.org/10.5185/amlett.2011.2228
74. Vadhva, P., Hu, J., Johnson, M.J., Stocker, R., Braglia, M., Brett, D.J., Rettie, A.J.: Electrochemical impedance spectroscopy for all-solid-state batteries: theory, methods and future outlook. ChemElectroChem **8**(11), 1930–1947 (2021)
75. Markevich, I.Y.A., Drokin, N.A., Selyutin, G.E.: Impedance spectroscopy study of a polymer composite with carbon nanotubes in contact with an electrolyte. Tech. Phys. **64**, 1324–1329 (2019)
76. Helseth, L.E.: Electrical impedance spectroscopy of multiwall carbon nanotube–PDMS composites under compression. Mater. Res. Express **5**(10), 105002 (2018)
77. Ilgaz, A.: Impedance analysis of single walled carbon nanotube/vinylester polymer composites. Kuwait J. Sci. **50**(3B) (2023)
78. Ribeiro, D.V., Souza, C.A.C., Abrantes, J.C.C.: Use of Electrochemical Impedance Spectroscopy (EIS) to monitoring the corrosion of reinforced concrete. Revista IBRACON de Estruturas e Materiais **8**, 529–546 (2015)
79. Lazanas, A.C., Prodromidis, M.I.: Electrochemical Impedance Spectroscopy–A Tutorial. ACS Measurement Science Au (2023)
80. Wolynec, S.: Técnicaseletroquímicasemcorrosão. EDUSP, São Paulo, 166 p (2003)

81. Goodwin, J.W., Hughes, R.W., Kwaambwa, H.M., Reynolds, P.A.: The phase separation behaviour and the viscoelastic properties of particles with non-adsorbing polymers: Part I—Experimental study. Colloids Surf. A: Physicochem. Eng. Asp. **161**(2), 339–359 (2000)
82. Ganguly, S., Das, N.C.: Rheological properties of polymer–carbon composites. In: Carbon-Containing Polymer Composites, pp. 271–294 (2019)
83. Münstedt, H.: Rheological measurements and structural analysis of polymeric materials. Polymers **13**(7), 1123 (2021)
84. Han, M.S., Lee, Y.K., Lee, H.S., Yun, C.H., Kim, W.N.: Electrical, morphological and rheological properties of carbon nanotube composites with polyethylene and poly (phenylene sulfide) by melt mixing. Chem. Eng. Sci. **64**(22), 4649–4656 (2009)
85. Pan, Y.: Electrical, morphological and rheological properties of carbon black filled polymer blends (Doctoral dissertation, Friedrich-Alexander-Universität Erlangen-Nürnberg (FAU)) (2017)
86. Yokohara, T., Nobukawa, S., Yamaguchi, M.: Rheological properties of polymer composites with flexible fine fibers. J. Rheol. **55**(6), 1205–1218 (2011)
87. Droval, G., Feller, J.F., Salagnac, P., Glouannec, P.: Rheological properties of conductive polymer composite (CPC) filled with double percolated network of carbon nanoparticles and boron nitride powder. e-Polymers **9**(1), 023 (2009)
88. Lim, H.T., Ahn, K.H., Hong, J.S., Hyun, K.: Nonlinear viscoelasticity of polymer nanocomposites under large amplitude oscillatory shear flow. J. Rheol. **57**(3), 767–789 (2013)
89. Lewandowski, K., Piszczek, K., Skórczewska, K., Mirowski, J., Zajchowski, S., Wilczewski, S.: Rheological properties of wood polymer composites at high shear rates–Evaluation of additional pressure losses as a result of inlet effects. Compos. Part A: Appl. Sci. Manuf. **154**, 106804 (2022)
90. Muthukumar, C., Krishnasamy, S., Thiagamani, S.M.K., Nagarajan, R., Siengchin, S.: Thermal characterization of the natural fiber-based hybrid composites: an overview. In: Natural Fiber-Reinforced Composites: Thermal Properties and Applications, pp.1–15 (2022)
91. Venkatakoteswararao, G., Vinayagamoorthy, R.: Review on thermal analysis of polymer matrix composites. In IOP Conference Series: Materials Science and Engineering, October, vol. 954, No. 1, p. 012037. IOP Publishing (2020)
92. Bommara, B., Devaiah, M., Reddy, P.L., Gandhi, M.R.: Thermal characterization of fiber reinforced polymer composites and hybrid composites. Int. J. Mech. Eng. Technol. **10**(3) (2019)
93. Dweck, J., Andrade, B., Monteiro, E., Fischer, R.: Thermal characterization of polymeric plaster composites. J. Therm. Anal. Calorim. **67**(2), 321–326 (2002)
94. Thamer, A.A., Yusr, H.A., Jubier, N.J.: TGA, DSC, DTG properties of epoxy polymer nanocomposites by adding hexagonal boron nitride nanoparticles. J. Eng. Appl. Sci. **14**(2), 567–574 (2019)
95. Shirakawa, H.: Synthesis of electrically conducting organic polymers: halogen derivatives of polyacetylene,(CH) x. Chem. Commun. **57**, 343 (1986)
96. MacDiarmid, A.G.: "Synthetic metals": a novel role for organic polymers. Curr. Appl. Phys. **1**(4–5), 269–279 (2001)
97. Heeger, A.J.: Semiconducting and metallic polymers: the fourth generation of polymeric materials. J. Phys. Chem. B **105**(36), 8475–8491 (2001)
98. Cardona, M., Merlin, R.: Light Scattering in Solids IX, pp. 1–14. Springer, Berlin Heidelberg (2007)
99. Träger, F. (ed.) Springer Handbook of Lasers and Optics, vol. 2. New York, NY, USA, Springer (2012)
100. Bower, D.I.: Infrared dichroism, polarized fluorescence and Raman spectroscopy. In: Structure and Properties of Oriented Polymers, pp. 181–233. Springer, Netherlands, Dordrecht (1997)
101. Ananda, H.T., Urs, G.T., Somashekar, R.: Preparation and characterization of conductive PVA/Gly: Na2 SO4 polymer composites. Polym. Bull. **73**, 1151–1165 (2016)
102. Larosa, C., Patra, N., Salerno, M., Mikac, L., Meri, R.M., Ivanda, M.: Preparation and characterization of polycarbonate/multiwalled carbon nanotube nanocomposites. Beilstein J. Nanotechnol. **8**(1), 2026–2031 (2017)

103. Christian, G.D., Dasgupta, P.K., Schug, K.A.: Analytical Chemistry. John Wiley & Sons (2013)
104. Perkampus, H.H.: UV-VIS Spectroscopy and its Applications. Springer Science & Business Media (2013)

Chapter 5
Reinforced Multiscale Polymer Composites, Properties and Applications

Roma Agrahari, Rajkamal Kushwaha, Akanksha Verma, and Soma Banerjee

Abstract Multiscale composites represent an innovative class of material that integrates both nanoscale as well as macroscale components, capturing the keen interest of researchers and diverse industries. These materials hold promise for application across a wide spectrum of fields, ranging from the automotive industry, aerospace industry, construction, energy, defense sector, environmental applications, etc. These composites exhibit remarkable attributes which encompass exceptional mechanical, electrical, and optical characteristics. Moreover, the nanomaterial constituents boast extraordinarily high aspect ratios, while the fibers offer uniformity, flexibility, and stability. To enhance the desired performance, it is imperative to gain a deep understanding of the selection of suitable reinforcements and precise processing techniques. Overall, this chapter focuses on the reinforcement materials and their impact on the various properties, such as mechanical, electrical, thermal, etc. The article also discusses the classification of composites, techniques, and challenges in the processing along with their utilization in various application areas.

Keywords Polymer composites · Multiscale reinforcement · Sensing · Nanoscale reinforcement · Nanofillers

---

R. Agrahari · R. Kushwaha
Department of Biochemical Engineering, School of Chemical Technology, Harcourt Butler Technical University, Kanpur, Uttar Pradesh 208002, India

A. Verma · S. Banerjee (✉)
Department of Plastic Technology, School of Chemical Technology, Harcourt Butler Technical University, Kanpur, Uttar Pradesh 208002, India
e-mail: sbanerjeepl@hbtu.ac.in

S. Moharana et al. (eds.), *Polymer Composites*, Engineering Materials,
https://doi.org/10.1007/978-981-97-2075-0_5

5.1 Introduction

5.1.1 Background and Significance of Composite Materials

In the development of the human race, material plays an important role. During that time, mankind used materials to determine the ages of history. The highest stage was found to be the Stone Age, and then we had the Bronze Age, Iron Age, and Steel Age all over again. Consequently, material progress has an important bearing on the evolution of human beings. To deal with the needs of industry and domestic use, many researchers have made efforts to improve material properties and introduce advanced materials of potential importance. In light of their intrinsic advantages, composites have been widely applied for a number of application areas. In the old days, dwellings were built with mud or clay mixed with straw to create a joint material. Mud or clay has been used as a matrix material to make the shape and structure while straw, which provides the necessary strength for a building's entire construction, acts as reinforcement fiber. Many research activities are being conducted into composite materials as new materials with all the qualities required by industry and technological developments need to be developed. It is necessary that materials have the following characteristics in terms of strength, flexural modulus, low density, low cost, ease of manufacture, environment safety and security for handling, etc. In order to meet those requirements, composite materials can be custom-designed and selected according to their respective constituents.

5.2 Constituents of Composite Material

Composite materials are basically engineered materials which are formed by physically combining two or more materials which are chemically non-identical to achieve an acquired property. In general, alloying is the process in which materials with different chemical structures can be combined on a microscopic scale, however, macroscopically the resulting material is homogeneous [1]. Due to their inherent advantages, like low weight, higher strength-to-weight ratio, corrosion resistance, impact resistance, and so on, composite materials are commonly applied in most industrial sectors. A wide range of industries, e.g., cars, defense, packaging, aircraft, sports, furniture, food processing, housing, structures, power generation, and electronics are extensively using composite materials [2, 3]. In general, composite materials are composed of a matrix material that is in a dispersed phase, which provides shape and structural stability and helps in absorbing vibrations.

In general, composite materials are composed of a matrix, reinforcement, and maybe filler components [4]. The reinforcement will be distributed and provide the composite's structural stiffness. Additionally, the reinforcement receives the external load transfer solely through the matrix phase. The filler and reinforcing materials ought to be able to be joined by the matrix material. The matrix material and other constituent materials, such as fibers and fillers, must have very good adherence. The matrix phase is in charge of providing defense against harmful environmental impacts for the fiber and fillers. The overall weight of the composite component is determined by the density of the matrix material, and the matrix materials must have little expansion during the curing process. The characteristic of the matrix material has a significant impact on the composite's thermal stability as well.

Composites made of polymer matrix materials often weigh little and are quite strong. For medium-load applications like car dashboards, seat coverings, and door panels, polymer matrix composites are utilized [5]. They are also used in home applications like packaging. However, they have greater tribological qualities, better chemical resistance, etc. Some of the most popular matrix materials include phenol-formaldehyde, polyester, and epoxy resin.

Thermosets and thermoplastic-reinforced composites are additional categories for polymer matrix composites [6]. The structural and thermal stability of thermosets are superior to that of thermoplastics. Polymer matrix composites are simple to make and don't need specialized tools or labor from highly trained personnel. Due to their superior quality and high tensile strength, reinforcements are what give composite materials the necessary strength. The mechanical qualities of the components in bulk materials are typically lowered by porosity, whereas the flaws in the fiber are reduced; as a result, the fibers have higher tensile strengths than bulk materials produced of the same material [7].

The composite material gains stiffness and toughness from the reinforcing fibers [8]. The matrix transfers the externally applied load to the fibers, and the fibers are responsible for managing the applied stress. For most cases, carbon, glass, and aramid fibers are used as reinforcement in fiber reinforced composites due to good tensile strength and simplicity of fabrication. These fibers give the composites great performance and durability. For aircraft applications that call for a high strength-to-low weight ratio, carbon fibre composites are frequently used. Glass fibers are mostly employed in boats, military equipment, and other applications that call for exceptional strength and impact resistance[9]. Aramid fiber, often known as Kevlar fiber, is primarily utilized in armored tanks, bulletproof vests, and defense applications [10].

In addition to the matrix material and reinforcing fiber, filler materials are mixed into the composite to modify its mechanical, thermal, tribological, impact, and electrical properties. Fillers, which can be particles or flakes, do not considerably increase the tensile strength of the composite material. Filler materials are used to reduce the amount of matrix material, which lowers the overall cost of the composites [11]. Micro rubber, talc powder, calcium carbonate, egg shell powder, nano silica, wood particles, carbon black, and many others are commonly used as filling material inside the polymer matrix [12]. Their aspect ratio is closer to unity due to the uniform dimension across all directions; thus, they cannot effectively contribute in stress transfer.

5.3 Composites and Its Classes

A composite is composed of a matrix and reinforcement. The matrix is the main component of the composite that consumes the most material, and reinforcements are inserted in the space between the matrices [4]. Metals, ceramics, and polymers are the most commonly utilized matrices [13]. Polymer composites were developed and employed following the advent of polymers in the early 1900s. Glues produced from plants and animals were employed as binders in the past instead of resins. Glass was incorporated as fibers in plastic in 1935, which increased the strength of the plastic significantly. The plastics utilized in the majority of polymer composites are polypropylene, polyvinyl chloride, polyamide, polystyrene, and polycarbonate [14]. The strength of a polymer composite is determined by the fiber type, plastic type, plastic-to-fiber ratio, and plastic-to-fiber orientation in the composite. Additionally, the geometry of the fiber affects the properties of composite materials [15]. Polymer composites are appealing to industrialists, material scientists, and researchers due to their uses in disciplines such as construction, electronics, and electrical industries, vehicle sectors, marine, aviation, and medical applications [16]. Figure 5.1 depicts the classification of polymer composites with respect to matrix and reinforcement categories.

5.3.1 Particle Reinforced Composite Materials

The particles and matrix materials might be a mix of metallic and non-metallic components [17]. In general, particles improve composite stiffness to some extent

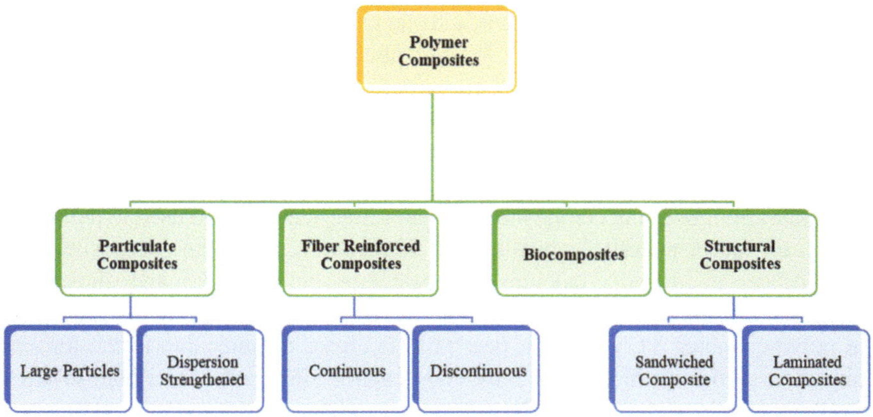

Fig. 5.1 Classification of Polymer composites

while also modifying thermal and electrical conductivities, increasing wear and abrasion resistance, machinability, performance at high temperatures, surface hardness, and decreasing friction and shrinkage [18]. The matrix serves several functions, including supporting, protecting, and transferring stress among the particles, as well as bonding the particulates so that they perform in unison [19]. Concrete, solid rocket propellants, carbides, and other particle reinforced composites are often employed. Large-particle and dispersion-strengthened composites are the two subcategories of particle-reinforcement as shown in Fig. 5.1. In large-particle reinforced composites, the particle-matrix interactions are described using continuum mechanics. These reinforcing particles tend to inhibit matrix phase movement in the region of each particle [20]. The matrix, in essence, passes some of the applied stress to the particles, which bear a portion of the load. The degree of reinforcement in mechanical behavior is dependent on strong bonding at the interface. Particles in dispersion-strengthened composites are significantly smaller, with sizes ranging between 10 and 100 nm. Particle-matrix interactions that result in strengthening take place at the atomic or molecular level. Particulate materials are commonly employed in the reinforcement of elastomers and plastics. Many current rubbers would be severely restricted if carbon black were not included [21]. Carbon black is composed of extremely small and often spherical carbon particles formed by the combustion of natural gas or oil in an environment with a restricted air supply. The addition of carbon black to elastomer enhances the overall tensile strength, toughness, tear, and abrasion resistance in vulcanized rubber. Carbon black may be found in automotive tires in concentrations ranging from 15 to 30% by volume [22]. Carbon black particles must be extremely tiny, with diameters ranging from 20 to 50 nm, to provide considerable reinforcement; also, the particles must be evenly scattered throughout the rubber matrix.

5.3.2 *Fiber Reinforced Polymer Composites*

Fibrous composites are formed by embedment and proper binding of fibers in a continuous matrix. Fiber is defined as a material having an elongated shape with an aspect ratio of 10:1. Because of its precise structure, a fiber is intrinsically stiffer and stronger than the identical material in bulk form. Glass, carbon, boron, kevlar, and graphite are examples of commercially accessible synthetic fibers. Glass fibers are made by extruding very thin strands of silica glass, and they are strong, lightweight, and durable fibers. Glass fibers are favored over metals because of how simple, inexpensive, and strong they are to produce. Glass fibers come in a variety of commercially available forms, including woven fabrics, yarns, and textiles. Glass fibers offer cost-effective adaptability and simple processing. Glass fibers are also very strong and have good fire resistance. Glass fibers have a low moisture absorption rate and don't alter when exposed to high or low temperatures. They are a fantastic choice for marine applications. Glass fibers cannot catch fire. Glass fibers are very sensitive to hot phosphoric acids and strong alkaline media even though they are unaffected by the majority of chemicals. Due to air trapped inside the glass, glass fibers exhibit superior thermal conductivity with low coefficient of thermal expansion. Glass fibers have good insulation properties and low dielectric constant, they can be utilized in printed circuit boards.

The widespread use of glass fiber reinforced polymer (GFRP) composites is a result of environmental stability, high quality, and affordability. The enormous demand for lightweight innovative materials in aerospace and automotive components points to the value of these materials across a wide range of industries. Additionally, GFRPs have excellent mechanical characteristics, including exceptional impact resistance, tensile strength that is superior to steel with lowered brittleness compared to carbon fiber, and good dimensional stability. Fishing rods, boat hulls, heat insulation in refrigerators, clothing, and sports items are typical examples of products that use reinforced glass fibers. Glass fibers are heavier than other fibers, which restricts their use as reinforcement. One of the most sought-after types of carbon is carbon fiber, which has exceptional mechanical, thermal, and electrical qualities. Carbon fibers can be short or continuous, single or braided, and they must contain at least 92 weight percent carbon. Carbon fibers can be entirely crystalline, partly crystalline, or amorphous. The graphitic structure of crystalline carbon fiber is characterized by sp2 carbon atoms arranged in a two-dimensional honeycomb lattice. These honeycomb structures include two parallel layers that are connected by van der Waals forces and are easily able to glide over one another. As a result, graphite has anisotropic properties and distinct in-plane and out-of-plane characteristics. Carbon fibers exhibit a good elastic modulus perpendicular to the plane. The multipurpose qualities of carbon fibers open the provision to fabricate lightweight carbon fiber reinforced composites (CFRPs) having potential in a variety of applications including vehicles, sporting goods, and defense equipment.

Asbestos can be replaced with aramid fibers because they were the first organic fibers to be manufactured with such great strength and stiffness. In addition to having superior mechanical qualities to steel or metal wires of the same weight, aramid fibers provide exceptional heat resistance. As a result, these fibers are used to create composites with enhanced mechanical and flame resistance capabilities making them a material for bulletproof jackets, protective apparel, aircraft, and rope for offshore oil rigs. Low density (lighter than glass fibers) along with lower creep rate remains an attractive feature for aramid fibers. Fiber reinforced polymer composites are further classified based on their fiber length as depicted in Fig. 5.2.

5.3.3 Natural and Synthetic Fiber Reinforced Biocomposites

A biocomposite is a material composed of natural fiber reinforcement and a matrix (resin) [23]. While retaining the tensile strength of the matrix, this type of material sensitively replicates the structure of live materials used in the process while always guaranteeing biocompatibility [24]. The matrix phase is made up of polymers derived from both natural and synthetic resources. Biofibers are the primary elements of

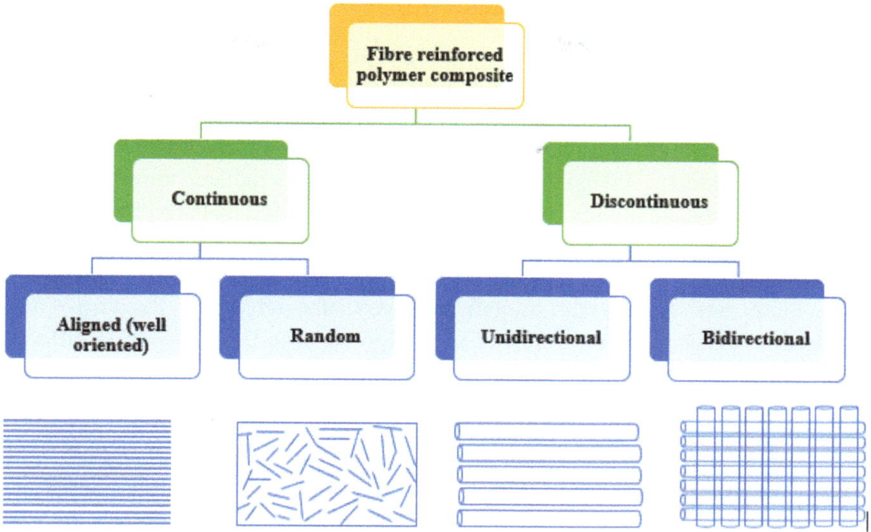

Fig. 5.2 Classification of Fiber reinforced polymer composite (FRPC)

biocomposites derived from various biological sources such as recycled wood, waste paper, agricultural processing byproducts, or regenerated cellulose fiber derived from crops such as cotton, flax, or hemp.

Natural fibers (NFs) are commonly available and easy to get. They have several outstanding material properties, including biodegradability, low cost per unit volume, with valid strength, stiffness [24]. Composites made using NF reinforcements are characterized by lightweight, cost-effectiveness, recyclability, and reduced toxicity [25]. Natural fibers differ in content; nonetheless, depending on the kind, they have equivalent structures. The inclusion of both long and short natural fibers in thermoset matrices resulted in high-performance applications [26]. Sisal fiber (SF)-based composites are widely utilized for upholstery in furniture and vehicle interiors due to their superior tribological properties [27, 28].

Hemp composites displayed a huge increment in specific flexural strength than GF-reinforced composites with a polypropylene matrix [29]. The flexural and tensile strengths of maleic anhydride grafted PP/hemp composite evidenced an increment of about 37 and 68% [30]. The tensile and flexural strengths of polylactic acid (PLA) thermoplastic composites reinforced with kenaf fiber have been reported to be 223 and 254 MPa, respectively. By removing absorbed water from the fibers prior to lamination, the flexural and tensile characteristics of kenaf fiber laminates were also enhanced [31]. Palm fibers (PFs) displayed excellent fiber-matrix interfacial interaction. Furthermore, low-density polyethylene (LDPE) containing palm fibers showed a higher Young's modulus than homopolymers [32]. At 3% fiber content, abaca fiber (AF) reinforcement provides good wear resistance with a wear rate of 2.864 × 107 cm3/Nm. Furthermore, when the abaca fiber concentration increased, the density decreased [33].

Synthetic fibers are man-made materials manufactured through chemical synthesis and classified as organic or inorganic depending on their composition [34, 35]. Glass fiber (GF) is the most common synthetic fiber having high strength and durability, thermal stability with good resistance to impact, chemicals, friction, and wear [36]. Using typical machining processes, however, machining glass fiber-reinforced polymers (GFRPs) is slow, complex, and has a reduced tool life [37, 38]. Another disadvantage of GFs is that they must be discarded once their useful life is over.

Graphene fibers are novel high-performance carbonaceous fibers that outperform carbon fibers in terms of tensile strength and electrical conductivity. Because of their increased properties, graphene fibers have the potential to be employed in different sectors, cables and wires, supercapacitors, micromotors, actuators, etc. [39].

5.3.4 Structural Composites

A structural composite is often composed of both homogeneous and composite materials. Sandwich panels and laminar composites are common structural composites. Laminar composites are fabricated with two-dimensional sheets or panels with a chosen high-strength direction and aligned fiber reinforced polymer composites [40]. Fabrics, such as cotton, paper, etc. are also utilized to make lamination. A relatively complex laminated construction is the contemporary ski. Sandwich panels are a form of structural composite that is intended to be lightweight beams or panels with moderate stiffnesses and strengths.

A sandwich panel is composed of two exterior sheets, or faces, separated by and bonded to a thicker core. The outer sheets are often formed of a stiff and strong material, such as aluminum alloys, fiber reinforced polymers, titanium, steel, or plywood, and must be thick enough to withstand stresses. Typically, the core material is light in weight and possesses low modulus of elasticity. The most popular core materials are rigid polymeric foams, wood, honeycombs, etc. The core serves several structural functions. To begin, it provides continuous support for the faces. It also has to be strong enough to withstand transverse shear forces and thick enough to provide high shear stiffness (to keep the panel from buckling). Sandwich panels are utilized in building roofs, floors, and walls, as well as aerospace and aviation (for wings, fuselage, tail-plane skins, etc.).

5.4 Multiscale Polymer Composites

A variety of sectors has shown interest in multiscale composites advanced composites reinforced at the macroscale and nanoscale both. Because of their exceptional qualities such as mechanical, electrical, and optical capabilities; exceptionally high aspect ratios of the constituent nanomaterials; and homogeneity, flexibility and stability, etc.

Multiscale composites have potential in every field. It is essential to comprehend the choice of suitable reinforcements, processing, and application of these cutting-edge materials in order to maximize the performance of these types of composites. Multiscale composites are made of nanofillers; graphene, carbon nanotube, nanoclay, metal nanoparticles, etc. combined with reinforcing fibers such as carbon, natural fibers, basalt, glass, etc. Due to multiple features emerging from the combination of both macro and nano-reinforcements, the multifunctional composites are in high demand. Additionally, the host resin and nanomaterials interact to enhance interactions and alter texture and color. Nanotechnology presents a number of unique obstacles to the creation and broader use of composites reinforced with nanomaterials.

One of the key elements of a composite that shields fillers or fibers from abrasion and adverse weather is the matrix. Additionally, it keeps the fillers and fibers together and gives the composite strength by absorbing energy through deformation under stress. The matrix transfers a load to the reinforced material when applied to composite materials. In multiscale composites, fibers are essential to defining the final performance of the material that is synthesized. It is often recognized that as a material's size reduces, so do its flaws. A fiber's greater surface area, aspect ratio, and anisotropy over the bulk material contribute to the material's superior performance.

The enhanced fiber surface area and improved interfacial contacts found in multiscale composites facilitate the interaction of the fibers with the polymer matrix. The composite exhibits exceptional strength due to the enhanced interlocking between the polymer and fiber, which progresses the stress transmission process between the fiber and matrix.

Multiscale composites have a wide range of potential uses. Modern lightweight materials are required for transportation applications (such as the automotive, aerospace, and defense industries) since component weight directly affects fuel efficiency. Advanced multiscale composites are taking the place of injection molded and heavy steel car parts in the automotive industry. Multiscale composites are very beneficial for giving structures multifunctionality, such as in-situ damage detection, signature reduction, EMI shielding, blast and impact resistance, and flame retardance. Traditional fiber reinforcements, such as glass and carbon fibers, are combined with nanoscale fillers e.g., carbon nanotubes to create multiscale composites. The idea is to use standard reinforcements for in-plane load carrying while using nanoscale reinforcement to improve through-thickness performance and other resin-dominated features of the resulting composites. Since the discovery of carbon nanotubes in 1991, there has been a great deal of interest in learning more about the unique materials they are made of, their mechanical and physical characteristics, and possible technological uses. Carbon nanotubes are known for their remarkable qualities, which include great strength and stiffness, the capacity to withstand massive elastic strain, and excellent thermal and electric conductivity. These characteristics, along with their low density and high respect ratio, imply that carbon nanotubes could be a promising reinforcing phase for nanocomposites.

5.5 Characteristics of Multiscale Polymer Composites

Researchers are constantly striving to develop advanced technologies to make human life more convenient. The conventional materials used in the automotive, aviation, space, oil, and gas industries are being replaced by polymer materials due to their lightness, ease of use, and low cost. Polymeric materials possessed exceptional erosion solidness and great mechanical properties, but the drawbacks associated with these materials like low thermal stability, second-rate chemical and natural steadiness, and low conductivity limited their applications in various sectors. To overcome these limitations, the polymer composites have been reinforced with matrices that resulted in desired performance. However, some researchers have always preferred composite materials, in addition to weight and high strength, these materials also possess excellent combinations of properties; electrical or thermal conductivity, electromagnetic shielding, barrier performance, voltage self-sensing, etc., and destruction and self-healing. To improve the performance of reinforced polymer more than one filler is needed. In aircraft and automobiles, weight is very important and narrows the width of the matrix reinforcement. Therefore, the best way would be to put fillers into the mix, thus bringing out more features without affecting the existing features. Extensive research has been conducted to improve the electrical, thermal, and electrical properties of electronic components through the use of additive materials.

These composite electrical products contain a combination of fiber materials with additional properties (electrical conductivity, sensitivity, thermal resistance) associated with special nanomaterials. Figure 5.3 shows the different features offered by reinforced multiscale polymer composites.

5.5.1 Strength Requirements

The composite material has depicted many significant improvements in mechanical properties such as wear resistance, tensile strength and modulus, flexural strength and modulus. It was reported that these polymer composites have enhanced interfacial interactions and effective interactions can provide significant durability to the composites by increasing the stress gradient between the fibers and matrix [41–43]. The number of voids obtained during processing was also reduced by the reinforced multiscale fillers leading to a compact product with high strength and hence improving the interfacial interactions.

A significant difference in the impact strength of epoxy/glass fiber with maleic anhydride grafted PE fibers and silanized nanoclay was reported by Shelly et al. The analysis with FE-SEM micrographs showed an improved interfacial interaction among various constituents of Epoxy-based glass fiber nanocomposites resulting in superior mechanical performance [44]. Dericiler et al. demonstrated enhancement in tensile properties, and flexural properties as compared to the neat PA66

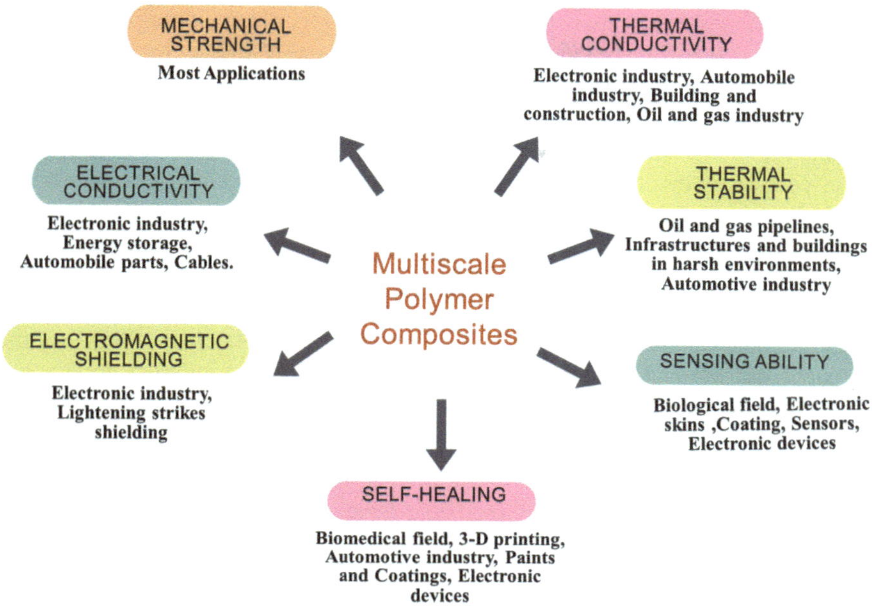

Fig. 5.3 Properties exhibited by reinforced multiscale polymer composites

by successfully integrating waste tire-driven graphene nanoplatelets and recycled carbon fibers [45]. Nimbagal et al. studied CNF aggregates with carbon nanofibers (CNF) and micron short carbon fibers (SCF), epoxy (80%)—polylactic acid (PLA) (20%) composites. The synergistic effect of CNF and SCF in the epoxy matrix was investigated by mechanical and structural analysis of the nanocomposites. Compared to CNF-reinforced nanocomposites, composites reinforced with more filler have higher tensile, flexural, impact strength, and fracture toughness. The incorporation of CNF and SCF into the epoxy-PLA matrix increased the tensile strength, flexural strength, and impact strength by 17.18–25.72%, 39.24–44.07%, and 39.87–97.05%, respectively [46]. Joseph et al. developed an epoxy-based fiber reinforced polymer composite containing multi-walled carbon nanotubes. Epoxy-based fiber reinforced polymer coating with or without multi-walled carbon nanotubes exhibits high recycling behavior in the concrete environment. Compared to unconstrained samples, restraints showed an increase in axial strain from 0.6 to 1.4% when restrained with epoxy-based fiber reinforced polymers and when multi-walled carbon nanotubes constrained with epoxy-based fiber reinforced polymer were used, axial strain support increased to 1.6% [47]. Kim et al. produced unidirectional CNT reinforced polymer composites with improved axial and transverse strength and I-mode strength through a multiscale hierarchical structure. At the macro scale, polymer-prefiltered carbon nanotube fibers are deployed in a "brick and mortar" fashion to create an additional structural interconnection in the composite. The specific axial tensile strength of unidirectional CNT is 0.85GPa/(g/cm^3), which is about 120% of the original CNT material. The measured transverse tensile strength and type I bone strength at constant stress are 81.3 MPa/(g/cm^3) and 0.670 kJ/m^2, respectively, indicating a good effect of using model engines [48]. Fereiduni et al. demonstrated the effect of nanoscale/submicron TiB on the strength of titanium matrix composites. Microstructural characterization of the TMC surface showed the formation of large columnar proto-β grains containing in-situ formed nanoscale/submicron TiB needle grains uniformly dispersed in a martensite matrix. The nanohardness and compressive yield strength of fabricated TMC are 25% and 8% higher, respectively. It has been reported that the mechanical properties of TMC are improved by the addition of nanoscale/submicron TiB reinforcement [49]. Azimpour-Shishevan et al. demonstrated the effect of multi-walled carbon nanotubes (MWCNTs) and graphene nanoplatelets (GNPs) on the electrical and thermal properties of carbon fiber reinforced polymer composites (CFRP). Temperature test results such as storage modulus, shrinkage, glass transition temperature, and weight percentage show that MWCNT and GNP nanoparticles improved the properties and properties of CFRP by increasing elastic modulus, tensile strength, and flexural modulus, critical buckling load, and natural frequencies [50].

Zakaria et al. compared the performance of single-layer graphene (SLG) and single-walled carbon nanotube (SWCNT)/ epoxy for mechanical and thermal characteristics. The tensile strength, fracture toughness, and thermal conductivity of SLG-modified epoxy evidenced increments of 7%, 14%, and 35%, respectively, while the SWCNT/epoxy reported increments of 10%, 21%, and 26%, respectively [51]. Du et al. fabricated graphene oxide-coated short glass fiber reinforced polyethersulfone-based composites. The tensile and bending strength of GO-coated SGF/PES composites were investigated under the influence of GO coating content. It was found that the GO layer on the surface of SGF effectively improved the tensile and flexural strength due to the improvement of the connection between SGF and PES through the GO layer [52]. Various studies have also reported improved interlayer shear strength (ILSS) by incorporating nanophases into fiber composites [53, 54]. Similar growth has been reported for TiO2 NPs on Aramid (AF) [55]. Experimental results showed that the development of TiO_2 NP interfacial phase increased the interfacial strength (IFSS) by 40–67%. Srivastava et al. reported the interlaminar fracture toughness of woven carbon fiber/epoxy composites in the presence of different nanofillers [43]. Composite materials containing graphene nanosheets were reported to increase Type I and Type II durability by 60.4% and 52.5%, respectively, compared to the original composite.

5.5.2 *Electrical Conductivity and Electromagnetic Shielding Effectiveness*

Besides the great mechanical properties, the reinforced multiscale polymer composites also possess electrical properties including electrical conductivity and electromagnetic interference shielding. The improved interfacial strength of these composites due to their fillers makes a good continuous network in the composite that leads towards their advanced electrical properties. Cao et al. prepared multilayer MWCNT/CF/PEEK composites and studied the effect of MWCNT incorporation on the mechanical properties, interlayer adhesion, and EMI properties of MWCNT/CF/PEEK based composites. A 76% increase in EMI shielding performance has been reported for interlayer CNT content as low as 2.5 g/m^2 (equivalent to 0.53 wt%) [56]. In another study, the electrical, thermal, and electrical properties of amine and unreduced graphene oxide/epoxy carbon fiber reinforced polymers were investigated. The high intrinsic conductivity of graphene and other materials increases electrical energy. rGO-CFRP (plasma-treated reduced graphene oxide with carbon fiber reinforced polymer with amine functional groups) outperformed pure carbon fiber reinforced polymer (CFRP) by +16% (+52%). Additionally, the apparent interlaminar shear strength and transverse conductivity increased by 12% for rGO-CFRP and 52% for frGO-CFRP [57]. In a similar study, Abushammala et al. reported the use of magnetic materials (WIF) to develop glass fiber reinforced epoxy resin (GFRE) composites. It has been reported that GFRE samples with a WIF content of 15% and above are conductive materials and their conductivity increases as the WIF content increases. The conductivity of GFRE samples containing 15% WIF is approximately 2.9 kS/m, and with the addition of 50% WIF, this value increases to 35 kS/m. It has been reported that these GFRE materials have the potential to be used in structural maintenance work in the construction industry [58].

Kang et al. developed separated network morphology (SNM)-carbon nanotube (CNT)/polyamide 6 (PA6) multiscale hybrid composites using glass foam (GB). This study analyzed the electronic properties by varying the content of carbon nanotubes and GB. The results of the four probes revealed that the conductivity improves with GB and CNT content. At 30% GB, conductivity improved 18.8 times [59]. Ramkumar and Sabarinathan used hybrid nanofiller of alumina and multi-walled carbon nanotube (MWCNT) in different weights (0.1, 0.2, 0.3, 0.4, and 0.5) to strengthen electrical, electronic, and thermal properties. It was reported that the presence of voids in the sample gradually decreased as the nanofiller content increased to 0.3% by weight. Compared to low-weight hybrid nanofillers, electrical and thermal conductivities were increased by 85% and 64%, respectively [60]. Zhang et al. successfully improved the adhesion of carbon fiber (CF) and polyetheretherketone (PEEK) substrates by adding a modified nanofiller at the coating scale. It has been reported that nanofillers increase electrical conductivity by 766% and 323% in 0° and 90° directions, respectively, by generating an electric current in the center of the fibers [61]. Modification of epoxy composites with ultrathin carbon fiber reinforced carbon nanotubes was archived by Zhang et al. The effect of carbon nanotube reinforced epoxy resin on ultrathin CFRP composites was discussed. Studies have shown that these composites exhibit excellent thermal conductivity, and electrical and electromagnetic interference resistance performance [62]. Bao et al. demonstrated the excellent strength and conductivity of metallic glass composites comprised of multi-scale configurations [63]. The composite material has a good combination of ultimate compressive strength and electrical conductivity. Bhanuprakash et al. studied the mechanical and electrical properties of graphene oxide-coated carbon fiber/epoxy hybrid composites [64]. The modified laminates showed a significant improvement in ILSS performance; GO-coated CF-based composite showed an improvement of 47%. Thermally reduced GO-coated and reduced exfoliated graphene oxide (rEGO) filler-coated CF composites show a strong increase in conductivity of 127 and 90% than uncoated composites. The increase in the conductivity value is due to the formation of electrical lines between the coated thermally reduced graphene oxide and the rEGO sheet from the non-conductive matrix of the CF fabric process.

5.5.3 *Thermal Stability and Thermal Conductivity*

Many studies have shown that composite materials have excellent thermal properties due to the excellent thermal stability of many fibers and the relationship between filler and the resulting nanofiller matrix, enhancing stability during heat treatment [57]. Ackerman et al. studied the electrical, thermal, and electrical properties of amine and unreduced graphene oxide/epoxy carbon fiber reinforced polymers [57]. After the evaluation of thermal characteristics, it was noted that the specific heat capacity was increased with higher temperatures by the inclusion of plasma-treated reduced graphene oxide. But, the thermal diffusivity for plasma-treated reduced graphene oxide with amine-functionalities carbon fiber reinforced polymer remained constant for all evaluated temperatures. It was also reported that the thermal conductivity was dependent on both the thermal diffusivity and specific heat capacity. Another study reported on alumina- multi-wall carbon nanotube hybrid nanofiller reinforced epoxy composites [60]. Alumina and multi-walled carbon nanotubes (MWCNT) hybrid nanofillers were utilized to react in pure epoxy resin at different weight ratios (0.1, 0.2, 0.3, 0.4, and 0.5). It was reported that with the addition of alumina and multi-walled carbon nanotubes, the adhesion and interface between hybrid nanofillers and epoxy resin were greatly increased, thus improving the mechanical properties. Zhang et al. investigated the electrical, thermal, and electrical properties of chitosan-doped carbon nanotubes dispersed in carbon fiber composites [65]. To improve the properties of CFRP composites, chitosan (CS) and carbon nanotubes (CNT) were used to form a multilayer structure on the CF surface. Compared to the original CFRP composite, tensile, interlayer shear and compressive strength of CS-doped CNT CFRP composite had been improved remarkably. The thermal dissipation of the as-spread CFRP composites had been increased by 92.59%, indicating good thermal conductivity. Wang et al. analyzed fiber reinforced materials for powder bed fusion process [66]. The thermal stability of the powder mixture and the actual loading of CF were measured using TGA. The DTG curve reveals that the maximum decomposition temperature of carbon nanotube-coated powder is lower than that of non-carbon nanotube powder, thus improving the thermal conductivity of the composite without affecting its thermal stability. Wu et al. studied the thermal insulation process of SiO_2 aerogels supported by glass fibers and SiC nanowhiskers [67]. The thermal conductivity of the sample was measured using a hot plate and infrared transmission. Silicon carbide nanowhiskers were found to block infrared radiation, thus reducing heat transfer and providing thermal insulation. It was found that the thermal conductivity of the airgel composite with a SiC nanowhisker volume fraction of 4% was only 0.040 W/(m·K). Ju et al. proposed ablative thermal protection using multiscale, elastic, low-density carbon fiber/silica oxycarbonate-phenolic interpenetrating airgel nanocomposites [68]. It was found that the multi-composite structure achieved high strength (4.57–5.83 MPa), good elasticity, and thermal insulation (0.068 W/(m·K)), in addition to its porous nanostructure and beautiful properties. This nanocomposite can withstand a 1000 °C flame with a butane gun for 30 min without damaging the structure, and in 1.5 MW oxyacetylene flame ablation, the linear ablation rate increases by 0.0282 mm/1 due to the increase of silicon carbon oxide content. These properties indicate that SiCF/PR nanocomposites can be used for thermal protection of re-entry aircraft and hypersonic vehicles. Carbon fiber reinforced polyimide-based composites have been grafted with carbon nanotubes to improve thermal conductivity and mechanical properties [69]. It was found that grafting carbon nanotubes onto the carbon fiber surface (CNT@CF) evidenced remarkable improvement in the surface roughness of carbon fiber, enhancing the interfacial compatibility between carbon fiber@CF and PI matrix. The PI/CNT@CF20 composite loaded with 20 wt% CNT@CF showed an improvement of 1134% compared to the good matrix. Rafiee et al. studied thermal properties of two glass fiber/epoxy composites containing graphene nanosheets, graphene oxide, and reduced graphene oxide [70]. Thermal characterization reveals that the addition of GNP, GO, rGO, and MWCNT rises the thermal conductivity. As a result, it turned out that, for the same volume fraction of nanoparticles, the thermal conductivity increase of graphene nanosheet-modified composites was more obvious than that of other nanoparticles. The distribution and interaction of nanoparticles inside the epoxy matrix play a crucial role in effecting the thermal conductivity of the nanocomposites. Su et al. investigated carbon nanotube/carbon fiber/polyetheretherketone multiscale composite materials. The addition of carbon nanotubes enhances the electrical and thermal conductivity of composite materials.

5.5.4 *Self-Healing*

In addition to the above-mentioned properties, the reinforced multiscale polymer composites also exhibit some unconventional properties. One of these unconventional properties is self-healing, where the object detects damage and takes some action to repair the damage. Self-healing can be of two types: endogenous and exogenous. In endogenous repair, the material repairs covalent or non-covalent bonds to repair the damage, while in exogenous repair, other phases or structures are added to the material. Neisiany et al. [71] produced agent-loaded core-shell nanofibers on carbon fabric. The curing agent consists of two components, epoxy resin and an amine-based curing agent embedded in styrene acrylonitrile (SAN) nanofibers through a coaxial electrospinning process. Testing and characterization studies have shown that adding nanofibers on top of the carbon layer imparts self-healing ability to carbon/epoxy composites, allowing them to self-heal for up to three consecutive cycles at room temperature. Wu et al. reported a new hybrid multiscale carbon fiber reinforced epoxy-based composite with self-healing core-shell nanofibers attached at the interfacial region [72]. Using electrospinning technology, liquid dicyclopentadiene (DCPD) was used as the curing agent and wrapped with polyacrylonitrile (PAN) to produce core-shell DCPD/PAN nanofibers to prepare ultrathin self-healing fibers. Experimental results reveal that the bending stiffness of this new self-healing material can be restored by the self-healing nanofiber sandwich. Gao et al. investigated the self-healing behavior of carbon nanotube-modified poly (ethylene-methacrylic acid) based sandwich membrane [73]. In the first stage of the treatment, the self-healing rate of the carbon nanotube-modified poly(ethylene-methacrylic acid) sandwich film was 115.3%, and in the second stage of the treatment, the self-healing rate was 129.4%.

5.5.5 Sensing Ability

Another important feature of these reinforced multiscale polymer composites is their sensing capabilities. There are many reports that nanofibers with strong sensing ability can be decorated with many nanomaterials. Qu et al. produced carbon nanotube/epoxy/glass fiber composites [74]. The electrical and mechanical properties of these composites indicated their suitability as materials for turbine blades. Hostettler and Huber studied the electrical properties and self-assembly abilities of multiphase graphene-based composites. Multiscale multiphase graphene-based composites demonstrated strong strain-sensing capabilities at low applied voltages, proving that they can be reliable and economical for self-sensing and damage applications [75]. Liang et al. created the glucose sensor by coating silk fibers with graphene and then electrodeposition of spiked platinum nanospheres [76]. Additionally, an enzyme was immobilized onto the resulting film and used as the liquid biosensor electrode; this showed a very high sensitivity to glucose (150.8 $\mu A\ mM^{-1}\ cm^{-2}$) with an ultra-low detection limit. Calestani et al. established a smart sensor by interoperating carbon fiber with ZnO nanostructures. The electrical properties of carbon fiber along with the good work properties of ZnO lead to the fabrication of piezoelectric sensors and chemically resistant gas sensors at the intersection of two strong carbon fibers [77].

5.6 Fabrication Challenges of Multiscale Polymer Composites

The fabrication of a multiscale polymer composite involves incorporating multiple types of reinforcing materials with different length scales into a polymer matrix to enhance mechanical, thermal, electrical, or other properties. Several fabrication techniques are used for multiscale polymer composites depending on the material types, properties of interest, and intended applications. Some common fabrication techniques include solution mixing and casting, extrusion, injection molding, compression molding, 3D printing, in-situ polymerization, layer-by-layer assembly, electrospinning, calendaring, etc. The details of the above-mentioned fabrication techniques are given in Table 5.1.

Each of these fabrication techniques has its own advantages and limitations, and the choice depends on factors such as the specific materials, desired properties, and intended application of the multiscale polymer composite. A significant obstacle to the broad adoption of composites is the development of an efficient mass-production method; however, their fabrication presents several challenges due to the complexity of integrating multiple scales of reinforcement. Conventional manufacturing methods like manual hand layup are characterized by slow production rates, labor-intensive processes, and potential risks to both worker's health and the environment. Even techniques such as filament winding are limited in their applicability to relatively straightforward geometries. Achieving a uniform dispersion of nanoscale fillers, such as nanoparticles or nanotubes, within the polymer matrix is crucial for enhancing the composite's properties. Agglomeration of these nanoparticles can occur, compromising uniformity and performance, often necessitating specialized mixing techniques and surfactants. Establishing a robust interface between the matrix and fillers at all scales is paramount for effective load transfer, and incompatibility between the filler and polymer matrix can hinder adhesion, ultimately diminishing mechanical properties. Additionally, alignment of reinforcements in the desired direction, especially in composites with multiple scales of reinforcement, is a complex task. Proper control of processing parameters, including temperature, pressure, and shear forces, is essential during fabrication to prevent thermal degradation and maintain filler dispersion. Overcoming these challenges while considering cost, quality control, environmental, and safety concerns is a multidisciplinary endeavor that involves materials scientists, engineers, and manufacturing experts. Nonetheless, as advances in fabrication techniques and material design continue, multiscale polymer composites hold immense promise for applications in aerospace, automotive, electronics, and beyond.

Table 5.1 Details of various fabrication techniques used to fabricate multiscale polymer composite

S. no.	Techniques	Steps Involved	Features
1	Solution mixing	1. Polymer is dissolved in a suitable solvent 2. Reinforcing material is dispersed into the polymer solution. The mixture is casted into a mold or onto a substrate 3. The solvent is evaporated, leaving behind the multiscale composite	This method is suitable for incorporating nanoscale reinforcements into the polymer matrix
2	Extrusion and Injection molding	1. The polymer pellet or resin is combined with reinforcing materials in a hopper 2. The mixture is heated to a molten state 3. The molten composite is extruded or injected into a mold with the desired shape 4. Composite gets solidified on cooling	Effective for producing multiscale composites with microscale reinforcements
3	Compression molding	1. Polymer and reinforcing materials (microscale and mesoscale) are mixed to create a homogenous blend 2. The mixture is placed into a mold cavity 3. Heat and pressure are applied to consolidate the composite	Compression molding is suitable for multiscale composites with a higher volume fraction of reinforcements
4	3D printing	1. A feedstock is prepared consisting of a polymer matrix material loaded with the desired reinforcements (microscale, mesoscale and sometimes nanoscale) 2. A 3D printer is used to deposit and layer the composite material according to the 3D model's specifications 3. Heat or UV light (depending on the printing technology) is used to cure or solidify the composite	Composite 3D printing allows precise control over reinforcement placement and is versatile for creating complex geometries
5	In-situ polymerization	1. Monomers or prepolymers are combined with reinforcing materials in a reactor 2. Polymerization takes place, allowing the polymer matrix to form in the presence of the reinforcements 3. The reaction conditions are altered to achieve the desired composite properties	In-situ polymerization enables the incorporation of nanoscale reinforcements directly into the polymer matrix as it forms

(continued)

Table 5.1 (continued)

S. no.	Techniques	Steps Involved	Features
6	Layer-by-layer assembly	1. The deposition of polymer layers with reinforcing materials (nanoscale or mesoscale) are alternated in a controlled manner, typically using techniques like spin coating or dip coating 2. The process is repeated to build up the multiscale composite layer-by-layer	Layer-by-layer assembly is useful for creating thin films or coatings with precise control over the placement of reinforcements
7	Electrospinning	1. Polymer and nanoscale reinforcements are dissolved in a solvent 2. Electrospray the solution through a charged nozzle to create nanofibers 3. Nanofibers are collected on a substrate, creating a multiscale composite mat or membrane	Electrospinning is suitable for producing multiscale composites with nanoscale reinforcements in fibrous form
8	Calendaring	1. The polymer matrix and reinforcing materials (usually microscale) are combined to create a homogenous mixture 2. The mixture is passed through a set of rollers under pressure and heat to flatten and consolidate the composite	Calendaring is useful for producing thin, flat multiscale composites

5.7 Diverse Application Areas of Multiscale Polymer Nanocomposites

Multiscale polymer composites are materials that incorporate different types of reinforcing agents at multiple length scales to enhance their mechanical, thermal, electrical, and other properties. These composites have a wide range of applications across various industries due to their versatility and superior performance. Some diverse application areas of multiscale polymer composites as shown in Fig. 5.4, include the aerospace industry, automotive industry, marine industry, defense and military, construction and infrastructure, sports and recreation, electronics and electrical, medical devices, energy sector, consumer goods, packaging industry, environmental application and so on.

Transportation industries, including the automotive, aeronautics, and defense sectors, require advanced lightweight materials due to the direct impact of component weight on fuel consumption. Within the automobile sector, advanced multiscale composites are substituting heavy-weight steel and parts manufactured by injection molding. Utilization of multiscale composites for automotive components results in lighter, safer, fuel-efficient, and low-emission products. Multiscale composites can also enhance crash safety by improving the energy absorption characteristics

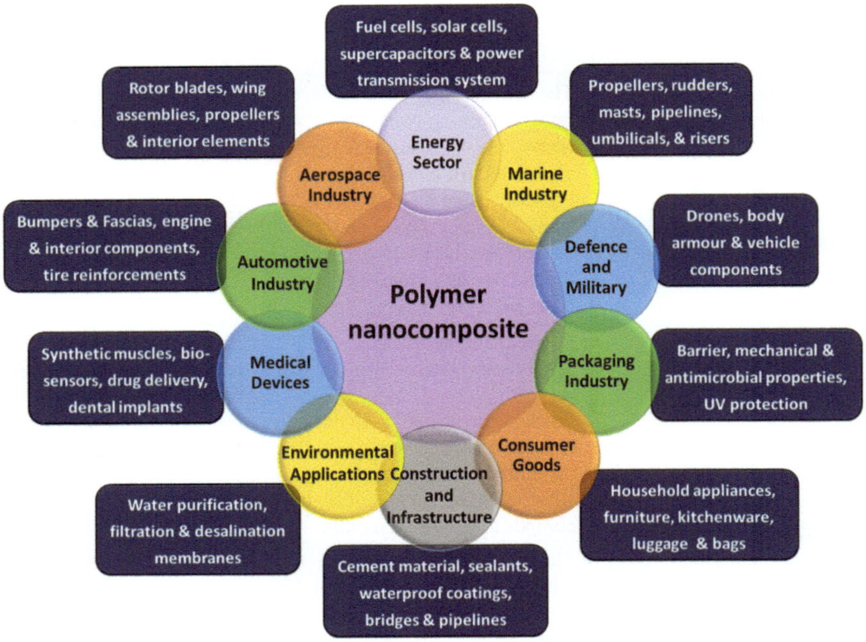

Fig. 5.4 Key applications of polymer nanocomposites in various fields

of vehicle components. Polymer nanocomposites are also used as reinforcements in tire manufacturing. They can enhance the tire's performance in terms of grip, wear resistance, and rolling resistance, leading to better fuel efficiency and safety. Components reinforced with carbon fiber offer exceptional stiffness and strength, coupled with notable stability. Furthermore, the strength can be customized through the incorporation of nanofillers. Carbon fiber-based automobiles enjoy popularity for their desirable properties and aesthetics. Nevertheless, their widespread adoption remains restricted primarily to racing cars because of the considerable production costs involved.

Like automobiles, aircraft demand lightweight structures with exceptional strength to minimize fuel consumption. Approximately half of an aircraft's components are constructed from composites resulting in 20–50% reduction in the weight in comparison to the original materials they replace. Advanced composite technologies predominantly rely on carbon fiber due to its exceptional strength-to-weight ratio. Nonetheless, the persistent challenge lies in the elevated cost associated with carbon fiber. Furthermore, aramid fiber reinforced composites find application in the design of wing components, offering both high impact resistance and excellent stiffness, which serves to safeguard the engine pylon carrying fuel. Additionally, businesses are embracing eco-friendly composites to enable the recyclability of aircraft once they reach the end of their operational life. The advanced composites also find application in fighter jets, space shuttles gliders, and hot air balloons. The typical components produced using multiscale composites include rotor blades, wing assemblies, blades, propellers, as well as seats and interior elements. In aerospace engineering, dimensional stability stands as another crucial prerequisite for composite materials. The composite layers expand or contract in response to temperature fluctuations which can result in the development of cracks within the structure of the composite. An ideal composite would possess a Coefficient of Thermal Expansion (CTE) equal to zero. Certain nanomaterials possess negative CTE, and offer the potential to create composites with minimal thermal expansion or contraction by effectively offsetting the matrix material's positive CTE fulfilling the required purpose. Also, the expansion of microflaws in multiscale composites can be effectively curtailed or considerably slowed down, thanks to the crack-bridging influence of carbon nanotubes (CNTs). This mechanism helps prevent catastrophic failures and abrupt accidents. It becomes possible to achieve self-repair of any cracks that may develop by incorporating healing agents into multiscale composites.

Composites are also utilized in marine applications, where corrosion significantly influences material longevity. Multiscale polymer composites serve a dual purpose in these settings, offering protection against corrosion while delivering ease of manufacturing and lightweight properties. Glass fiber composites are favored for their cost-effectiveness and exceptional resistance to corrosion. Similarly, composites based on aramid fibers enhance shock absorption characteristics, making them a material of interest for constructing high-performance powerboats.

Multiscale composites are employed within the sporting goods industry to meet the demand for sports material that is lightweight, strong, and cost-effective. The fibers provide the necessary flexibility and strength, while reinforced nanophases facilitate stress transfer by crack-bridging effect between the fiber and matrix. Even in small quantities, nanofillers can enhance material properties, making these materials cost-efficient. Prominent examples of sports equipment crafted from multiscale composites encompass tennis rackets, hockey sticks, skis golf clubs, and bicycle frames with enhanced performance and durability.

Similarly, within military applications, the appeal of multiscale composites lies in their lightweight properties, safety, durability, and versatility. These composites find utility in various military applications such as components of body armor, drones, and vehicles, including hoods, window hoods, window frames, wheelhouse assemblies, doors, fenders, ballistic panels, engine covers, etc. in the field of electronics. Polymer composites with enhanced thermal conductivity are used in PCBs to dissipate heat, increasing the efficiency and lifespan of electronic devices. Along with this, multiscale composites can also provide electrical insulation for high-voltage applications.

Another significant domain of multiscale composite applications involves energy management. In this context, components constructed from multiscale composites are expected to demonstrate effective energy management capabilities, enabling the storage, harvesting, and distribution of energy. Multiscale composites are sought after in various applications such as marine turbines, fuel cells, solar cells, supercapacitors, and power transmission systems due to their multifunctional attributes. Moreover, multiscale composites find utilization in diverse medical contexts, including synthetic muscles, bio-sensors, drug delivery systems, dental implants, and in the creation of artificial organs. Manufacturing of light weight and biocompatible orthopedic implants such as bone plates and joint replacements also involves polymer nanocomposites. Dental composites with improved strength and aesthetics are used in restorative dentistry. In addition to these applications, multiscale composites can also be employed in the construction of bridges, tanks, pipelines, and systems for water purification, filtration, and desalination membranes. Polymer composites can be used to reinforce concrete structures, improving their durability and resistance to corrosion. In the packaging industry, polymer composites create lightweight and durable packaging materials with improved mechanical strength, thermal stability, and barrier properties for food preservation. By incorporating nanoscale fillers, such as clay nanoparticles or graphene oxide, into polymer matrices, packing materials can achieve superior gas barrier properties. This is crucial for extending the shelf life of food products by preventing the ingress of oxygen, moisture, and other gases. These are just a few examples, and the application of multiscale polymer composite continues to expand as researchers develop new materials and manufacturing techniques, opening up new possibilities across various industries.

5.8 Conclusions

In conclusion, reinforced multiscale polymer composites represent a cutting-edge frontier in materials science and engineering, offering a blend of nanoscale and macroscale reinforcements that redefine material capabilities. Their performance is intricately influenced by a multitude of factors, including the choice of matrix, the selection of fillers, the interactions between fillers and matrix, shape and size of the fillers. Notably, when dealing with nano-reinforcements, a common challenge arises due to their propensity to agglomerate. To address this issue and unlock the full potential of composites with nano-reinforcements, various processing methods are employed. Each of these methods possesses its own set of advantages and disadvantages, ultimately shaping the physical, mechanical, thermal, and electrical properties of the resulting composites. Furthermore, these multi-dimensional fillers work in synergy to bestow these composites with exceptional properties, ranging from superior mechanical strength and electrical conductivity to optical excellence, while also introducing unconventional properties like sensing capabilities and self-healing abilities, making them a versatile solution with wide-ranging applications. Such composite materials, enriched by multiple fillers, are often called multifunctional materials. From aerospace and automotive industries to electronics and medical devices, these composites hold the potential to revolutionize diverse fields, presenting a world of possibilities for innovation and technological advancement. As research and development in this area continue to advance, the future for multiscale polymer composites appears promising, poised to meet the growing demand for high-performance materials in the modern world.

References

1. Mercier, J.P., Zambelli, G., Kurz, W.: Introduction to Materials Science. Elsevier (2012)
2. Bhatia, A., Sehgal, A.K.: Additive manufacturing materials, methods and applications: a review. Mater. Today Proc. **81**, 1060–1067 (2023). https://doi.org/10.1016/j.matpr.2021.04.379
3. Potluri, R., Chaitanya Krishna, N.: Potential and applications of green composites in industrial space. Mater. Today Proc. **22**, 2041–2048 (2020). https://doi.org/10.1016/j.matpr.2020.03.218
4. Gay, D.: Composite Materials: Design and Applications, 4th edn. CRC Press (2022)
5. Mohammed, L., Ansari, M.N.M., Pua, G., Jawaid, M., Islam, M.S.: A review on natural fiber reinforced polymer composite and its applications. Int. J. Polym. Sci. **2015**, 1–15 (2015). https://doi.org/10.1155/2015/243947
6. Sonnenfeld, C., Mendil-Jakani, H., Agogué, R., Nunez, P., Beauchêne, P.: Thermoplastic/thermoset multilayer composites: a way to improve the impact damage tolerance of thermosetting resin matrix composites. Compos. Struct. **171**, 298–305 (2017). https://doi.org/10.1016/j.compstruct.2017.03.044
7. Al-Maharma, A.Y., Patil, S.P., Markert, B.: Effects of porosity on the mechanical properties of additively manufactured components: a critical review. Mater. Res. Express **7**, 122001 (2020). https://doi.org/10.1088/2053-1591/abcc5d
8. Balakrishnan, P., John, M.J., Pothen, L., Sreekala, M.S., Thomas, S.: Natural fibre and polymer matrix composites and their applications in aerospace engineering. In: Advanced Composite Materials for Aerospace Engineering, pp. 365–383. Elsevier (2016)
9. Wang, B., Gao, H.: Fibre Reinforced Polymer Composites, pp 15–43 (2021)
10. Nurazzi, N.M., Asyraf, M.R.M., Khalina, A., Abdullah, N., Aisyah, H.A., Rafiqah, S.A., Sabaruddin, F.A., Kamarudin, S.H., Norrrahim, M.N.F., Ilyas, R.A., Sapuan, S.M.: A review on natural fiber reinforced polymer composite for bullet proof and ballistic applications. Polymers (Basel) **13**, 646 (2021). https://doi.org/10.3390/polym13040646
11. Fan, J., Njuguna, J.: An introduction to lightweight composite materials and their use in transport structures. In: Lightweight Composite Structures in Transport, pp. 3–34. Elsevier (2016)
12. Fan, Y., Fowler, G.D., Zhao, M.: The past, present and future of carbon black as a rubber reinforcing filler—a review. J. Clean. Prod. **247**, 119115 (2020). https://doi.org/10.1016/j.jclepro.2019.119115
13. Chawla, K.K.: Composite Materials: Science and Engineering, 3rd ed. Springer Science & Business Media (2012)
14. Thakur, S., Verma, A., Sharma, B., Chaudhary, J., Tamulevicius, S., Thakur, V.K.: Recent developments in recycling of polystyrene based plastics. Curr. Opin. Green Sustain. Chem. **13**, 32–38 (2018). https://doi.org/10.1016/j.cogsc.2018.03.011
15. Nambiar, S., Yeow, J.T.W.: Polymer-composite materials for radiation protection. ACS Appl. Mater. Interfaces **4**, 5717–5726 (2012). https://doi.org/10.1021/am300783d
16. Hsissou, R., Seghiri, R., Benzekri, Z., Hilali, M., Rafik, M., Elharfi, A.: Polymer composite materials: a comprehensive review. Compos. Struct. **262**, 113640 (2021). https://doi.org/10.1016/j.compstruct.2021.113640
17. Rohatgi, P.K.: Metal matrix composites. Def. Sci. J. **43**, 323–349 (1993)
18. Yadav, R.: Fabrication, characterization, and optimization selection of ceramic particulate reinforced dental restorative composite materials. Polym. Polym. Compos. **30**, 096739112110627 (2022). https://doi.org/10.1177/09673911211062755
19. Buragohain, M.K.: Composite Structures: Design, Mechanics, Analysis, Manufacturing, and Testing. CRC Press (2017)
20. Subhajit, K., Mitra, D., Das, M.: Toughened Composites: Micro and Macro Systems. CRC Press (2022)
21. Pirityi, D.Z., Bárány, T., Pölöskei, K.: Hybrid reinforcement of styrene-butadiene rubber nanocomposites with carbon black, silica, and graphene. J. Appl. Polym. Sci. **139** (2022). https://doi.org/10.1002/app.52766

22. Kudapa, V.K.: Investigation of carbon black nanoparticles on effectiveness of cementation in oil and gas well. Mater. Today Proc. (2023). https://doi.org/10.1016/j.matpr.2023.07.111
23. Panthapulakkal, S., Raghunanan, L., Sain, M., Birat, K.C., Tjong, J.: Natural fiber and hybrid fiber thermoplastic composites. In: Green Composites, pp. 39–72. Elsevier (2017)
24. Nair, A.B., Joseph, R.: Eco-friendly bio-composites using natural rubber (NR) matrices and natural fiber reinforcements. In: Chemistry, Manufacture and Applications of Natural Rubber, pp. 249–283. Elsevier (2014)
25. Omrani, E., Menezes, P.L., Rohatgi, P.K.: State of the art on tribological behavior of polymer matrix composites reinforced with natural fibers in the green materials world. Eng. Sci. Technol. Int. J. **19**, 717–736 (2016). https://doi.org/10.1016/j.jestch.2015.10.007
26. Ouarhim, W., Zari, N., Bouhfid, R., Qaiss, A. el kacem: Mechanical performance of natural fibers–based thermosetting composites. In: Mechanical and Physical Testing of Biocomposites, Fibre-Reinforced Composites and Hybrid Composites, pp. 43–60. Elsevier (2019)
27. Senthilkumar, K., Saba, N., Rajini, N., Chandrasekar, M., Jawaid, M., Siengchin, S., Alotman, O.Y.: Mechanical properties evaluation of sisal fibre reinforced polymer composites: a review. Constr. Build. Mater. **174**, 713–729 (2018). https://doi.org/10.1016/j.conbuildmat.2018.04.143
28. Chand, N., Dwivedi, U.K.: Sliding wear and friction characteristics of sisal fibre reinforced polyester composites: effect of silane coupling agent and applied load. Polym. Compos. **29**, 280–284 (2008). https://doi.org/10.1002/pc.20368
29. Shahzad, A.: Hemp fiber and its composites—a review. J. Compos. Mater. **46**, 973–986 (2012). https://doi.org/10.1177/0021998311413623
30. Sullins, T., Pillay, S., Komus, A., Ning, H.: Hemp fiber reinforced polypropylene composites: the effects of material treatments. Compos. B Eng. **114**, 15–22 (2017). https://doi.org/10.1016/j.compositesb.2017.02.001
31. Ochi, S.: Mechanical properties of Kenaf fibers and Kenaf/PLA composites. Mech. Mater. **40**, 446–452 (2008). https://doi.org/10.1016/j.mechmat.2007.10.006
32. Chollakup, R., Smitthipong, W., Kongtud, W., Tantatherdtam, R.: Polyethylene green composites reinforced with cellulose fibers (coir and palm fibers): effect of fiber surface treatment and fiber content. J. Adhes. Sci. Technol. **27**, 1290–1300 (2013). https://doi.org/10.1080/01694243.2012.694275
33. Liu, Y., Ma, Y., Yu, J., Zhuang, J., Wu, S., Tong, J.: Development and characterization of alkali treated abaca fiber reinforced friction composites. Compos. Interfaces **26**, 67–82 (2019). https://doi.org/10.1080/09276440.2018.1472456
34. Sathishkumar, T., Naveen, J., Satheeshkumar, S.: Hybrid fiber reinforced polymer composites—a review. J. Reinf. Plast. Compos. **33**, 454–471 (2014). https://doi.org/10.1177/0731684413516393
35. Begum, K., Islam, M.: Natural fiber as a substitute to synthetic fiber in polymer composites: a review. Res. J. Eng. Sci. **2**, 46–53 (2013)
36. Sathishkumar, T., Satheeshkumar, S., Naveen, J.: Glass fiber-reinforced polymer composites—a review. J. Reinf. Plast. Compos. **33**, 1258–1275 (2014). https://doi.org/10.1177/0731684414530790
37. Komanduri, R.: Machining of fiber-reinforced composites. Mach. Sci. Technol. **1**, 113–152 (1997). https://doi.org/10.1080/10940349708945641
38. Gaugel, S., Sripathy, P., Haeger, A., Meinhard, D., Bernthaler, T., Lissek, F., Kaufeld, M., Knoblauch, V., Schneider, G.: A comparative study on tool wear and laminate damage in drilling of carbon-fiber reinforced polymers (CFRP). Compos. Struct. **155**, 173–183 (2016). https://doi.org/10.1016/j.compstruct.2016.08.004
39. Xu, Z., Gao, C.: Graphene fiber: a new trend in carbon fibers. Mater. Today **18**, 480–492 (2015). https://doi.org/10.1016/j.mattod.2015.06.009
40. Liu, Q., Paavola, J.: Lightweight design of composite laminated structures with frequency constraint. Compos. Struct. **156**, 356–360 (2016). https://doi.org/10.1016/j.compstruct.2015.08.116
41. Shekar, K.C., Prasad, B.A., Prasad, N.E.: Interlaminar shear strength of multi-walled carbon nanotube and carbon fiber reinforced, epoxy—matrix hybrid composite. Proc. Mater. Sci. **6**, 1336–1343 (2014). https://doi.org/10.1016/j.mspro.2014.07.112

42. Boroujeni, A.Y., Tehrani, M., Nelson, A.J., Al-Haik, M.: Hybrid carbon nanotube–carbon fiber composites with improved in-plane mechanical properties. Compos. B Eng. **66**, 475–483 (2014). https://doi.org/10.1016/j.compositesb.2014.06.010
43. Srivastava, V.K., Gries, T., Veit, D., Quadflieg, T., Mohr, B., Kolloch, M.: Effect of nanomaterial on mode I and mode II interlaminar fracture toughness of woven carbon fabric reinforced polymer composites. Eng. Fract. Mech. **180**, 73–86 (2017). https://doi.org/10.1016/j.engfracmech.2017.05.030
44. Shelly, D., Nanda, T., Mehta, R.: Synergistic effect of compatibilized nanoclay/polyethylene fibers on the impact strength of epoxy-glass fiber nanocomposites. Polym. Compos. **44**, 6528–6541 (2023). https://doi.org/10.1002/pc.27577
45. Dericiler, K., Zafar, S., Sas, H.S., Saner Okan, B.: The effect of reinforcement aspect ratio on the performance of <scp>PA66</scp> composites with recycled carbon fiber and upcycled graphene nanoplatelets: An interface characterization from process to modeling. J. Appl. Polym. Sci. **140** (2023). https://doi.org/10.1002/app.54352
46. Nimbagal, V., Banapurmath, N.R., Umarfarooq, M.A., Revankar, S., Sajjan, A.M., Soudagar, M.E.M., Shahapurkar, K., Alamir, M.A., Alarifi, I.M., Elfasakhany, A.: Mechanical and fracture properties of carbon nano fibers/short carbon fiber epoxy composites. Polym. Compos. **44**, 3977–3989 (2023). https://doi.org/10.1002/pc.27371
47. Joseph, L., Madhavan, M.K., Jayanarayanan, K., Pegoretti, A.: Evaluation of hybrid fiber multiscale polymer composites for structural confinement under cyclic axial compressive loading. J. Compos. Sci. **7**, 152 (2023). https://doi.org/10.3390/jcs7040152
48. Kim, J.-W., Gardner, J.M., Sauti, G., Wincheski, R.A., Jensen, B.D., Wise, K.E., Siochi, E.J.: Multi-scale hierarchical carbon nanotube fiber reinforced composites towards enhancement of axial/transverse strength and fracture toughness. Compos. Part A Appl. Sci. Manuf. **167**, 107449 (2023). https://doi.org/10.1016/j.compositesa.2023.107449
49. Fereiduni, E., Ghasemi, A., Elbestawi, M.: Microstructural characterization and mechanical properties of nano-scale/sub-micron TiB-reinforced titanium matrix composites fabricated by laser powder bed fusion. J. Alloys Compd. **896**, 163054 (2022). https://doi.org/10.1016/j.jallcom.2021.163054
50. Azimpour-Shishevan, F., Akbulut, H., Mohtadi-Bonab, M.A.: Synergetic effects of carbon nanotube and graphene addition on thermo-mechanical properties and vibrational behavior of twill carbon fiber reinforced polymer composites. Polym. Test. **90**, 106745 (2020). https://doi.org/10.1016/j.polymertesting.2020.106745
51. Zakaria, M.R., Abdul Kudus, M.H., Md Akil, H., Thirmizir, M.Z.M., Abdul Malik, M.F.I., Othman, M.B.H., Ullah, F., Javed, F.: Comparative study of single-layer graphene and single-walled carbon nanotube-filled epoxy nanocomposites based on mechanical and thermal properties. Polym. Compos. **40** (2019). https://doi.org/10.1002/pc.25173
52. Du, S.-S., Li, F., Xiao, H.-M., Li, Y.-Q., Hu, N., Fu, S.-Y.: Tensile and flexural properties of graphene oxide coated-short glass fiber reinforced polyethersulfone composites. Compos. B Eng. **99**, 407–415 (2016). https://doi.org/10.1016/j.compositesb.2016.06.023
53. Tüzemen, M.Ç., Salamcı, E., Avcı, A.: Enhancing mechanical properties of bolted carbon/epoxy nanocomposites with carbon nanotube, nanoclay, and hybrid loading. Compos. B Eng. **128**, 146–154 (2017). https://doi.org/10.1016/j.compositesb.2017.07.001
54. Udupa, G., Rao, S.S., Gangadharan, K.V.: Functionally graded composite materials: an overview. Proc. Mater. Sci. **5**, 1291–1299 (2014). https://doi.org/10.1016/j.mspro.2014.07.442
55. Wang, B., Duan, Y., Zhang, J.: Titanium dioxide nanoparticles-coated aramid fiber showing enhanced interfacial strength and UV resistance properties. Mater. Des. **103**, 330–338 (2016). https://doi.org/10.1016/j.matdes.2016.04.085
56. Cao, S., Tao, Y., Li, H., Ren, M., Sun, J.: Multiscale hybrid CNT and CF reinforced PEEK composites with enhanced EMI properties. Nanocomposites **8**, 184–193 (2022). https://doi.org/10.1080/20550324.2022.2100683
57. Ackermann, A.C., Demleitner, M., Guhathakurta, J., Carosella, S., Ruckdäschel, H., Simon, S., Fox, B.L., Middendorf, P.: Mechanical, thermal, and electrical properties of amine- and non-functionalized reduced graphene oxide/epoxy carbon fiber-reinforced polymers. Polym. Compos. **44**, 4937–4954 (2023). https://doi.org/10.1002/pc.27461

58. Abushammala, H., Mao, J.: Waste iron filings to improve the mechanical and electrical properties of glass fiber-reinforced epoxy (GFRE) composites. J. Compos. Sci. **7**, 90 (2023). https://doi.org/10.3390/jcs7030090
59. Kang, G.-H., Kim, M., Park, Y.-B.: Improvement of electrical conductivity in glass bubble-carbon nanotube/polyamide 6 hybrid scale composite through novel mechanical forming and segregated network morphology. Polym. Test. **126**, 108138 (2023). https://doi.org/10.1016/j.polymertesting.2023.108138
60. Ramkumar, R., Chellamuthu, S.: Enhancing the mechanical, thermal and electrical properties of alumina-MWCNT hybrid nanofiller reinforced epoxy composites. Materiali in Tehnologije **57** (2023). https://doi.org/10.17222/mit.2022.684
61. Zhang, C., Duan, Y., Xiao, H., Wang, B., Xin, Z., Liu, G., Wang, F., Cui, W.: Preparation of MWCNTs/CF/PEEK multi-scale composites with good mechanical and electrical conductivity by a two-step process of AFP and out-of-autoclave tempering. Compos. Part C Open Access **9**, 100321 (2022). https://doi.org/10.1016/j.jcomc.2022.100321
62. Zhang, C., Ling, Y., Zhang, X., Liang, M., Zou, H.: Ultra-thin carbon fiber reinforced carbon nanotubes modified epoxy composites with superior mechanical and electrical properties for the aerospace field. Compos. Part A Appl. Sci. Manuf. **163**, 107197 (2022). https://doi.org/10.1016/j.compositesa.2022.107197
63. Bao, W.-Z., Chen, J., Li, J.-Z., Yu, B.-H., Liu, C.-Y., Jiang, P., Liu, Z.-J., Hu, K.-T., Louzguine-Luzgin, D.V., Xie, G.-Q.: Outstanding strength and conductivity of metallic glass composites with multiscale configuration. Rare Met. **42**, 3099–3113 (2023). https://doi.org/10.1007/s12598-023-02308-x
64. Bhanuprakash, L., Parasuram, S., Varghese, S.: Experimental investigation on graphene oxides coated carbon fibre/epoxy hybrid composites: Mechanical and electrical properties. Compos. Sci. Technol. **179**, 134–144 (2019). https://doi.org/10.1016/j.compscitech.2019.04.034
65. Zhang, C., Zhang, X., Ling, Y., Liang, M., Zou, H.: Chitosan-doped carbon nanotubes encapsulating spread carbon fiber composites with superior mechanical, thermal, and electrical properties. Compos. Sci. Technol. **230**, 109755 (2022). https://doi.org/10.1016/j.compscitech.2022.109755
66. Wang, Z., Zhou, M., Xiao, H., Yuan, S.: Development and evaluation of multiscale fiber-reinforced composite powders for powder-bed fusion process. Chin. J. Mech. Eng. Addit. Manuf. Front. **2**, 100079 (2023). https://doi.org/10.1016/j.cjmeam.2023.100079
67. Wu, Q., Yang, L., Chen, Z., Yang, M., Liu, T., Li, M., Mukhopadhyaya, P.: SiO2 aerogel multiscale reinforced by glass fibers and SiC nanowhiskers for thermal insulation. J. Porous Mater. **30**, 1587–1596 (2023). https://doi.org/10.1007/s10934-023-01432-4
68. Jin, X., Liu, C., Huang, H., Pan, R., Wu, C., Yan, X., Wang, H., Pan, Y., Hong, C., Zhang, X.: Multiscale, elastic, and low-density carbon fibre/siliconoxycarbide-phenolic interpenetrating aerogel nanocomposite for ablative thermal protection. Compos. B Eng. **245**, 110212 (2022). https://doi.org/10.1016/j.compositesb.2022.110212
69. Wu, Z., Dong, J., Teng, C., Li, X., Zhao, X., Qin, X., Ji, C., Zhang, Q.: Polyimide-based composites reinforced by carbon nanotube-grafted carbon fiber for improved thermal conductivity and mechanical property. Compos. Commun. **39**, 101543 (2023). https://doi.org/10.1016/j.coco.2023.101543
70. Rafiee, M., Nitzsche, F., Laliberte, J., Hind, S., Robitaille, F., Labrosse, M.R.: Thermal properties of doubly reinforced fiberglass/epoxy composites with graphene nanoplatelets, graphene oxide and reduced-graphene oxide. Compos. B Eng. **164**, 1–9 (2019). https://doi.org/10.1016/j.compositesb.2018.11.051
71. Neisiany, R.E., Lee, J.K.Y., Khorasani, S.N., Ramakrishna, S.: Towards the development of self-healing carbon/epoxy composites with improved potential provided by efficient encapsulation of healing agents in core-shell nanofibers. Polym. Test. **62**, 79–87 (2017). https://doi.org/10.1016/j.polymertesting.2017.06.016
72. Wu, X.-F., Rahman, A., Zhou, Z., Pelot, D.D., Sinha-Ray, S., Chen, B., Payne, S., Yarin, A.L.: Electrospinning core-shell nanofibers for interfacial toughening and self-healing of carbon-fiber/epoxy composites. J. Appl. Polym. Sci. **129**, 1383–1393 (2013). https://doi.org/10.1002/app.38838

73. Gao, Y., Liu, L., Wu, Z., Zhong, Z.: Toughening and self-healing fiber-reinforced polymer composites using carbon nanotube modified poly (ethylene-co-methacrylic acid) sandwich membrane. Compos. Part A Appl. Sci. Manuf. **124**, 105510 (2019). https://doi.org/10.1016/j.compositesa.2019.105510
74. Qu, M., Cai, J., Li, X., Wu, J., Chen, H., Zheng, Z., Nilsson, F., Liu, S., Gao, Q.: Mechanical and electrical properties of carbon nanotube/epoxy/glass-fiber composites intended for <scp>nondestructive</scp> testing. Polym. Adv. Technol. **34**, 2554–2563 (2023). https://doi.org/10.1002/pat.6071
75. Hostettler, N., Hubert, P.: Electrical characterization and sensing capabilities of self-assembly multi-scale multi-phase graphene-based composites. Carbon N Y **208**, 131–139 (2023). https://doi.org/10.1016/j.carbon.2023.03.005
76. Liang, B., Fang, L., Hu, Y., Yang, G., Zhu, Q., Ye, X.: Fabrication and application of flexible graphene silk composite film electrodes decorated with spiky Pt nanospheres. Nanoscale **6**, 4264–4274 (2014). https://doi.org/10.1039/C3NR06057H
77. Calestani, D., Villani, M., Culiolo, M., Delmonte, D., Coppedè, N., Zappettini, A.: Smart composites materials: a new idea to add gas-sensing properties to commercial carbon-fibers by functionalization with ZnO nanowires. Sens. Actuators B Chem. **245**, 166–170 (2017). https://doi.org/10.1016/j.snb.2017.01.109

Chapter 6
Glass Fiber-Reinforced Polymer Composites

Akshita, Tejendra K. Gupta, Deepshikha Gupta, Naresh K. Chandel, Sunil Kumar, and Ravi Kant Choubey

Abstract There has been an increasing fascination in studying lightweight polymer composites in recent years. Consequently, the use of glass fibers and carbon fibers has been widespread in the production of high-performance polymer composites owing to their customizable attributes and unique features, such as exceptional strength, flexibility, and stiffness. A range of methods, including extrusion, manual layup, wet layup, filament winding, and vacuum-assisted resin transfer molding (VARTM), were employed to produce glass fiber-reinforced polymer (GFRP) composites. These composites have found extensive usage in many industries. The inception of continuous glass fiber production occurred in the 1930s, mainly for applications involving high temperatures and electricity. Currently, these fibers play a crucial role in electronics, aviation, and automotive industries, showcasing exceptional characteristics such as robustness, adaptability, rigidity, and immunity to chemical harm. These materials are offered in many formats, such as rovings, chopped strands, yarns, textiles, and mats, each designed for specialized uses in polymer composites. This chapter presents an overview of the several techniques used to create GFRP composites, as well as their tribological, mechanical, vibrational, water absorption, and thermal properties. These qualities demonstrate the continuous investigation and improvement of these materials. The chapter concludes by addressing the significant uses and problems associated with these composite materials.

Keywords Glass fiber · Composites · Mechanical properties · Thermal properties

Akshita · T. K. Gupta (✉) · D. Gupta
Department of Chemistry, Amity Institute of Applied Sciences, Amity University Uttar Pradesh, Noida 201313, India
e-mail: tkgupta@amity.edu

N. K. Chandel
Department of Chemistry, RNT Govt. College Sarkaghat, Mandi 175024, India

S. Kumar
Department of Physics, Indira Gandhi University, Meerpur (Rewari), Haryana 122502, India

R. K. Choubey
Department of Applied Physics, Amity Institute of Applied Sciences (AIAS), Amity University, Noida Campus, Sector-125, Noida, Uttar Pradesh 201313, India

S. Moharana et al. (eds.), *Polymer Composites*, Engineering Materials,
https://doi.org/10.1007/978-981-97-2075-0_6

6.1 Introduction

The combined effect of many constituents in a composite material imparts it with attributes that cannot be achieved by any individual substance in isolation. Young's modulus, fiber strength, matrix strength, and stability all affect how optical fiber-reinforced polymer (GFRP) composites behave mechanically. The necessary characteristics and functional capabilities of GFRP composites can be achieved by intelligently arranging and mixing glass fibers [1]. This enables the composites to have qualities comparable to steel, including better rigidity than aluminum and lower relative density than steel. Glass fiber alignment increases the stiffness and strength of the composites, and it also has a significant effect on their mechanical properties. Furthermore, the composition of the glass fibers, such as the inclusion of SiO_2 in the form of continuous or staple fibers, might influence the characteristics of the GFRP composites. Each variety of glass fiber, such as S-glass with more silicon and aluminum content than E-glass, has its own set of qualities for certain applications [2]. Moreover, the matrix material, responsible for both binding the fibers and transferring loads and stresses, significantly influences the material's maximum working temperature. GFRP composites are extensively employed in numerous industrial applications owing to their favorable characteristics [3]. The selection of the matrix material, which binds the fiber reinforcement and facilitates load transmission between the fibers, is based on the glass transition temperature. The enhanced characteristics and reduced expenses of composites are attributed to the incorporation of fillers, which traditionally hindered the widespread usage of polymers due to their high cost [4]. Among many other industrial applications for GFRP composites, laminated GFRP composite materials are used in the plumbing and marine industries due to their exceptional environmental resistance, high specific strength, and superior damage tolerance for impact loading [3]. The aeronautical industry makes extensive use of polymeric composites because of their low weight and excellent fatigue resistance. Nevertheless, it is essential to study the storage modulus and damping factors, which are examples of dynamic stability, in order to determine how GFRP composites behave at different temperatures, as these mechanical properties are temperature-sensitive for all polymeric composites [1–4]. The mass manufacture of glass fiber commenced in the late eighteenth century, attributed to the alleged invention of the material by René Ferchault de Reaumur. The popularity of glass fiber composite did not arise until 1935, when fiberglass was converted into yarn. The aviation industry was the first to recognize the potential of this technology, and it has subsequently discovered several profitable applications [5–8]. Glass fibers are a highly adaptable substance employed in a wide range of industrial and household uses [9]. Glass fibers have many industrial and home uses due to their adaptability and abundance. Glass fibers, which may be found in a wide range of commercially accessible chemical compositions, are mostly composed of silica [10]. Glass fibers are needed to make many different products, including printed circuit boards, structural composites, and other specialty items. Fiberglass, sometimes referred to as glass fiber, is a highly sought-after fiber in the reinforced polymer industry because of its remarkable flexibility and ability to be formed into sheets. Fiberglass is produced by the process of liquefying glass and subsequently passing the molten substance through minuscule openings. Because of their remarkable specific strength, stiffness, and resistance to damage from impact loads, glass fiber-reinforced composites are used in the plumbing and marine industries [11]. For applications like insulation and soundproofing, they can be crushed and combined into thin sheets. GFRC, or glass fiber reinforced concrete, is utilized in a wide range of applications including automobiles, aircraft, pressure tanks, military gear, and everyday objects. Fiberglass, a kind of glass fiber, provides more adaptability at a reduced expense when compared to carbon fiber. Furthermore, it exhibits both malleability and durability, surpassing the performance of other metals. The numerous practical applications and beneficial properties of GFRC have resulted in its swift adoption in the business sector [12]. Because of their adaptability and affordability, glass fiber-reinforced composites (GFRC) have captured a large portion of the market. Fillers have improved the characteristics of GFRP composites while decreasing their preparation and total cost. Many different industries make use of GFRC, including the aerospace, chemical, construction, marine/shipping, car, consumer products, and pipe sectors. Because of its high specific strength, stiffness, impact loading and damage tolerance, and excellent corrosion resistance, laminated GFRC materials find widespread usage.

6.2 Fundamentals of Glass Fiber-Reinforced Composites

6.2.1 Definition and Composition

Glass fiber-reinforced matrix composites are composites that use glass fibers for the purpose of reinforcement. A matrix substance is employed to bind the fibers together. Composites get their form, surface polish, environmental tolerance, and durability through the use of metal, ceramic, or polymer matrix components [13]. To join the glass fibers to the matrix, chemical bonding agents called silanes are used. This is carried out following the extrusion of silica fibers into very fine filaments [14]. The concept of fiber filler reinforcement is based on the principle that the total strength of the combined composite material is achieved by transmission of stress from the matrix to the fibers, which then carry the majority of the structural load [13, 14]. The excellent properties of glass fibers—such as their great strength, pliability, rigidity, stability, and inertness—make them useful for manufacturing structural composites, printed circuit boards, and a variety of other specialized products. Fibers made of glass are among the most adaptable industrial materials currently known, and their supply is practically endless [13, 15]. The final characteristics of glass fiber-reinforced matrix composites are influenced by several parameters, such as the kind, quantity, and content of the resin systems and reinforcements, as well as the complicated chemical activity that occurs during processing [14]. The composite

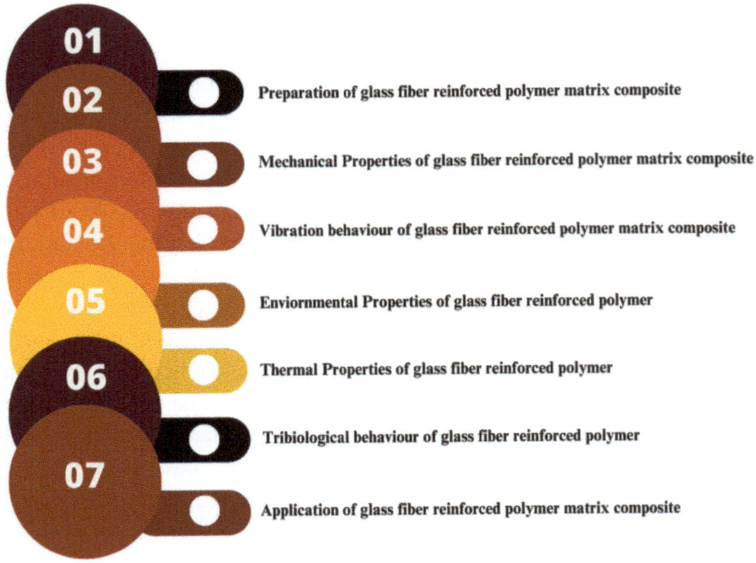

Fig. 6.1 Flow diagram of glass fiber-reinforced polymer composites: preparation method, properties, and applications

material's characteristics are very susceptible to the additives used in its production. Although there are many benefits to using glass fiber-reinforced matrix composites, there are also some drawbacks, such as difficulties in processing, recycling, and durability. Glass fiber-reinforced matrix composites have some constraints that are being actively researched and developed to improve their mechanical qualities, processing efficiency, and sustainability [13, 14, 16]. Some of the methods used to manufacture Glass Fiber Reinforced Polymer (GFRP) composites include pultrusion, vacuum-assisted resin transfer molding, hand-layup, and extrusion. Because of this, the next sections go over the items shown in Fig. 6.1, such as preparation techniques, mechanical and thermal properties, environmental effects, tribological and vibrational behavior, and applications.

6.2.2 Types of Glass Fibers

Different types of glass fibers have different constituents that result in unique properties. Glass fibers are a versatile material that can be used in various applications. With the exception of a few raw ingredients, all forms of glass fibers share the same essential component. Every variety of glass fiber has a different set of properties because each type has a unique amount of each raw element. The main raw materials used to make glass fibers are silica, soda ash, and limestone. Other ingredients include feldspar, borax, magnesite, calcined alumina, and kaolin clay [1]. Glass is formed in

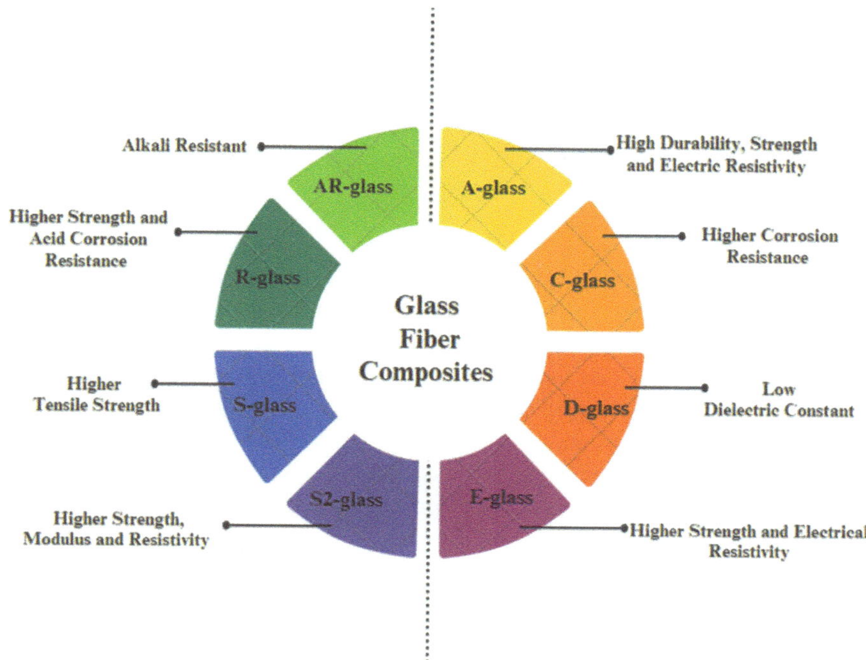

Fig. 6.2 Different types of glass fibers and their physical characteristics

part by silica sand, and as Fig. 6.2 illustrates, the various forms of glass fibers have different compositions of basic materials, giving them distinct characteristics.

Silica sand is essential to the process of glass formation because it helps to shape glass [1–3]. Soda ash and limestone are also essential components, with soda ash acting as a fluxing agent that reduces the furnace temperature necessary to melt the silica, thus lowering the energy required to produce glass. Additional components can be added to enhance various properties, such as borax, which increases the glass's ability to resist chemical attack [1].

6.2.2.1 A-Type

About 90% of all glass produced is alkali glass, sometimes referred to as soda-lime glass or A-type glass. It is the most widely available type of glass fiber. It is extensively utilized in the production of baking utensils, glass jars, windowpanes, and bottles for food and drink. Baking utensils are made of glass that has been tempered using alkali glass. This kind of glass is comparatively less expensive, very workable, chemically stable, and fairly hard. It is reusable multiple times [4]. The primary components that go into making A-glass are soda (Na_2CO_3), lime, silica (Si_2O_3), dolomite, and alumina (Al_2O_3). Sodium chloride (NaCl) and sodium sulfate (Na_2SO_4) are regarded as the fining agents, and they're also utilized to make soda-lime glass. The composition of various types of glass fibers varies, resulting in unique properties. Silica sand is the primary ingredient in glass formation, while soda ash and limestone reduce the melting point. Other constituents, such as borax, can enhance specific properties, such as chemical resistance [4, 5].

6.2.2.2 C-Type

One kind of glass fiber that offers remarkable defense against chemical attack is compound glass, sometimes referred to as C-glass. It has a sizable amount of calcium borosilicate in it, which provides structural stability in aggressive conditions. As surface tissue for pipes and tanks that hold water and synthetic compounds, this kind of glass is frequently used in external coatings on covers. Strong, workable, chemically stable, and reasonably priced, C-glass is well-known for these qualities. It is recyclable several times. The constituents of different types of glass fibers vary, giving them distinct properties. Borax, for example, can strengthen resistance to chemical attack. [6].

6.2.2.3 D-Type

Its low dielectric constant is attributed to the presence of boron trioxide (B_2O_3) in D-glass, also referred to as borosilicate glass. It is appropriate for use in optical cable applications because of this feature. Because of its extremely low coefficient of thermal expansion, D-glass fiber-reinforced composite is appropriate for use in cookware and electrical appliances. [7].

6.2.2.4 E-Type

Due to its superior properties, electrical glass, or e-glass, is a type of alkali glass that is widely used in marine, aerospace, and other industrial applications. The constituent elements that go into making E-type glass fiber composite are silica (SiO_2), alumina (Al_2O_3), magnesium oxide (MgO), sodium oxide (Na_2O), boron trioxide (B_2O_3), potassium oxide (K_2O), and calcium oxide (CaO). Better strength, increased stiffness, heat resistance, low density, fire resistance, chemical endurance, relative inertness to moisture, improved electrical isolation, and the capacity to maintain structural integrity under a variety of conditions are among the qualities of e-glass that make it well-known. Different kinds of glass fibers have different compositions, which give them distinct qualities [2].

6.2.2.5 R-, S-, and T-Type

They are more rigid and have a higher modulus than E-type GFRP. They are also stronger in acidity and have better wetting properties. These properties are obtained by decreasing the fiber filament radius. The design of this specific variety of glass fiber takes aerospace and defense applications into consideration [2].

6.2.2.6 S2-Type

This kind of glass has higher silica content than other types of glass fiber. Consequently, it provides enhanced properties such as higher compressive strength, elevated thermal endurance, enhanced weight performance, and advanced impact resistance. Above all, with an 85% increase in tensile strength, S2-type GFRPs are more affordable than traditional fiberglass. It provides improved hardness and impact capabilities along with increased damage tolerance and composite longevity [2].

6.2.2.7 M-Type

Due to the beryllium content, this particular variety of glass fiber is more flexible.

6.2.2.8 Z-Type

It is reasonably transparent and has a use in a number of industries, especially 3D printing and the building industry. This type of glass fiber has improved mechanical strength, resistance to UV rays, salt, alkali, acid, wear, temperature, and scratches. It is arguably the strongest and most stable variety of glass fiber [2]. Table 6.1 displays the percentages of the constituents of the previously mentioned glass fibers [13]. Table 6.2 lists the mechanical and physical characteristics of glass fiber. Poisson's ratio, Density, Tensile strength, Young's modulus of elasticity, refractive index and coefficient of thermal expansion are a few of these [13].

6.3 Classification Based on Form

The utilization of glass fibers to strengthen polyester, phenolic resins, and epoxy is a common and established procedure in various industrial environments. They possess chemical resistance and exhibit exceptional strength, flexibility, and stiffness. A multitude of configurations exist for glass fibers, and the amalgamation of individual fibers is referred to as a continuous strand. The roving form consists of a group of similar strands, whereas chopped fibers are a group of strands that are around 5–50 mm in length. Fiberglass, including tow, chopped strand, veil, and nonwoven

Table 6.1 Constituent percentage for different types of glass fiber

	SiO_2	Al_2O_3	TiO_2	B_2O_3	CaO	MgO	Na_2O	K_2O	Fe_2O_3
Basalt	52	17.2	1	–	8.6	5.2	5	1	5
EGR-glass	61	13	–	–	22	3	–	0.5	–
R-glass	60	24	–	–	9	6	0.5	0.1	–
D-glass	74		–	22.5	–	–	1.5	2	–
A-glass	67.5	3.5	–	1.5	6.5	4.5	13.5	3	–
S-glass	65	25	–	–	–	10	–	–	–
C-glass	64.6	4.1	–	5	13.4	3.3	9.6	0.5	–
E-glass	55	14	0.2	7	22	1	0.5	0.3	-

Table 6.2 Exploring the mechanical and physical characteristics of glass fiber

Fiber	Density (gcm^{-3})	Tensile strength (GPa)	Young's modulus (GPa)	Elongation (%)	Coefficient of thermal expansion (10^{-7}/°C)	Poisson's ratio	Refractive index
E-glass	2.58	3.445	72.3	4.8	54	0.2	1.558
C-glass	2.52	3.310	68.9	4.8	63	–	1.533
S2-glass	2.46	4.890	86.9	5.7	16	0.22	1.521
A-glass	2.44	3.310	68.9	4.8	73		1.538
D-glass	2.11–2.14	2.415	51.7	4.6	25	–	1.465
R-glass	2.54	4.135	85.5	4.8	33	–	1.546
ECR-glass	2.72	3.445	80.3	4.8	59	–	1.579
AR-glass	2.70	3.241	73.1	4.4	65	–	1.562

mats, encompasses a wide range of commercially accessible varieties [22]. Polymeric resins like epoxy and unsaturated polyester rely heavily on glass fibers, a very adaptable material class. As a replacement for heavy metal components, they are the most popular and oldest kind of reinforcement in many industrial settings, including the automobile and construction sectors [11]. In many different industrial applications, glass fibers are utilized as reinforcement for polyester, epoxy, and phenolic resins. Roving is a collection of threads that are quite similar to one another, whereas continuous strands are composed of groups of individual fibers. Whether in the form of roving or tow, glass fiber may be used to its fullest potential. When acquiring fiberglass, it comes in reels that can be wound into filament winders or unrolled and cut as needed. For fiberglass to keep its mechanical qualities, the fibers need to be kept in tension. Veil mats are made up of thin layers of randomly looped continuous strand fibers woven throughout a roll of fabric [8]. Glass fibers that had been chopped and arranged in an arbitrary pattern are assembled into chopped strand mats using a binder. Glass fibers that are continuously oriented at random and bound together with a binder make up nonwoven mats. A bunch of interconnected, twisted continuous filaments is called a tow [8, 9]. Chopped strand mats consist of randomly arranged 3–5-inch-long fibers. This variety of glass fiber is also the most widely used because it is less expensive. Yet considering the fiber is shorter in length than others, it is not as strong [10].

6.4 Preparation of GFRP Matrix Composites

6.4.1 Silicon Rubber Mold

Using a silicone rubber mold, Aramide [11] and associates created a woven-mat glass fiber-reinforced unsaturated polyester composite. To make it easier to remove the composite, the mold's surface was cleaned, dried, and then coated with a hard wax-releasing agent. First, the mold surface was coated with unsaturated polyester resin that contained curing additives. Next, the glass fiber was fully wetted onto the resin. After saturating the fiber with resin using a steel roller, a concluding coat of resin was applied. Subsequently, the cured laminated composite was extracted from the mold after undergoing a sufficient period of solidification [25]. The edges of the cured composite plate were filed to the appropriate size using a hand file. Composites with varying fiber contents (between 5 and 30%) were made.

6.4.2 Hand Lay-Up Method Followed by Compression Molding

The woven roving mat E-GF/modified and unmodified polyester matrix composites were prepared by Erden et al. [12] using a manual lay-up technique. The laminates were then compressed at 120 bars of pressure for 120 min at room temperature, with a 37% fiber volume fraction (Vf) and a 3.5 mm thickness. In Fig. 6.3, the process schematic is mentioned.

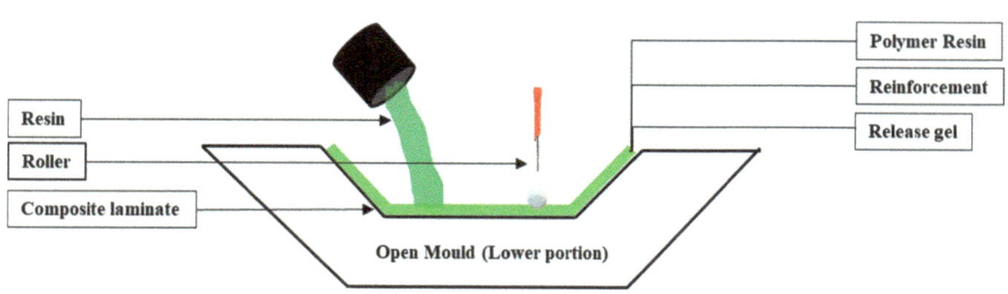

Fig. 6.3 Composite manufacturing using a hand lay-up method

6.4.3 Hot Press Technique

Atas et al. produced woven-mat glass fiber-reinforced epoxy composites by using the hot press technique. A 305 mm × 305 mm non-orthogonal fabric was prepared by shearing an orthogonal fabric with different weaving angles between the warp and fill to create the non-orthogonal fabric. The weight ratio of the hardener to epoxy resin was 10:3. The composite panels were cured at 60 °C for 2 hours and 93 °C for 4 hours during the curing process. Laminated composites were subjected to a pressure of 0.35 MPa [13]. Aktas et al. (2019) used the hot press method to create unidirectional glass fiber (GF)-reinforced epoxy laminates. Two stacking sequences were used to produce laminates with a 65% weight fraction: [0°/90°/0°/90°] and [0°/90°/+45°/+45°]. After combining CY225 epoxy resin and HY225 hardener, the mixture was heated to 120 °C for 2 hours under the hot press method, maintaining a constant pressure of 15 MPa. The manufactured composite plate had a thickness of 3 mm [14].

6.4.4 Molding and Mixing

Using fly ash as a filler, Gupta et al. produced discontinuous E-glass fiber-reinforced epoxy composites. For composite preparation, the 10 mm-diameter glass fiber was cut into 2.54 cm lengths. The ratio of the hardener to epoxy resin was 100:10. In the process of mixing, the resin and hardener, calcium carbonate fillers were added to the flyash filler at concentrations of 2.5% and 5% by volume. The spherical shape and small size of the flyash particles helped to facilitate their mixing and wetting of the matrix and fiber. The dimensions of the mold were 154 mm × 78 mm × 12 mm. The fiber and matrix were placed into the mold and allowed to cure for a full day at room temperature for a total of 24 hours [15]. A 30 cm square E-glass fabric lamina was impregnated with epoxy resin by Kajorncheappunngam et al. and then cured between two 0.32 cm thick Teflon-coated layers. After using a roller to remove extra resin, the composite was given 3 days to cure. The cured composites were then post-cured in an oven for 3 hours at 60 °C. The resultant composite had a 47% weight fraction and a thickness of 1.5 mm [16].

6.4.5 Compression Molding

Chopped stand mat E-glass fiber-reinforced modified epoxy composites with different fiber volumes (10–60%) were made by Hameed et al. using compression molding. The fiber mats were sized and heated to 150 °C in an air oven to eliminate moisture. To create a 3 mm thick composite, pre-weighed fiber mats and epoxy resin were combined with hardener. After being compressed in the mold, the laminates

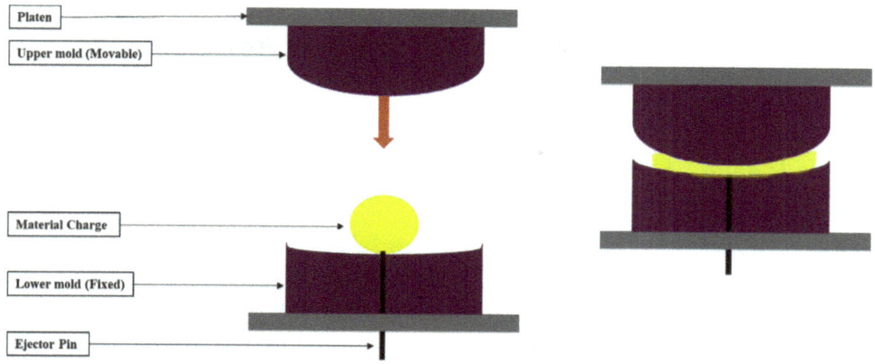

Fig. 6.4 Compression molding of polymer composites

were heated to 180 °C for three hours. The cured composite was post-cured at 200 °C for 2 hours, and then it was gradually cooled to room temperature to produce the final composites [17]. The schematic of the process of compression molding in polymers is described in Fig. 6.4.

6.4.6 Manual Assembly Technique Followed by a Hydraulic Press

To create woven mat glass fiber-reinforced epoxy composites, Suresha et al. employed the hand lay-up method. 100:12 weight ratio was used to combine the hardener and epoxy resin. The fibers and resin were combined, and a 0.5 MPa hydraulic press was used to press the mixture into a sample that was 3 mm thick. For a day, the specimen was allowed to cure at room temperature. Following demolding, post-curing was carried out in an electrical oven for two hours at 120 °C. The prepared laminate measured 250 × 250 × 3 mm [18].

6.4.7 Dry Hand Lay-Up Method

Glass fiber-reinforced vinyl ester composites were made by Suresha et al. using the dry hand lay-up technique and woven into fabric. The diameter of each fiber was 8–12 mm. The resin mixture was supplemented with methyl ethylketone peroxide (MEKP) as a catalyst, cobalt naphthenate as an accelerator, and N-dimethyl aniline as a promoter. The weight ratio of cobalt naphthenate, vinyl ester resin, and MEKP was 1:0.015:0.015. Two different fillers were mixed with the resin: 25 mm-sized silicon carbide and 50 mm-sized graphite. The resin mixture was applied to the textiles by dry hand lay-up molding after the woven mats were stacked one on top of the other. The complete assembly was compressed at a fixed pressure of 0.5 MPa using a hydraulic press. The ready slabs measured 250 mm by 250 mm by 3 mm [19].

6.4.8 H-type Press

Mohan et al. prepared oil cake-filled woven fabric glass fiber-reinforced epoxy composites with individual fiber diameters of 18 mm. Using a magnetic stirrer, the epoxy resin and hardener were combined at a weight ratio of 100:38. After applying the resin mixture to the prepared mat with a roller and brush, the laminate was cured for 24 h at 0.0965 MPa in an h-type press. Post-curing was carried out at 100 °C for 3 hours. The laminated preparation measured 300 × 300 × 2.6 mm [20].

6.5 Mechanical Properties of Glass Fiber-Reinforced Composites

The behavior and material characteristics of unidirectional glass fiber-reinforced composites at low and static temperatures were investigated by Torabizadeh et al. [21]. They employed G11 woven glass composite laminates, which exhibit material properties that change with temperature when subjected to mechanical loading at low temperatures. The matrix material was epoxy resin, and the reinforcing material was unidirectional glass fibers. Unidirectional laminates were created using the hand lay-up method, cured for seven days at room temperature, and then cut. Compressive, Tensile and shear tests were performed on these unidirectional laminates at both room temperature and low temperature. The study discovered that as the temperature dropped, the laminates' tensile strength and Young's modulus increased until they broke. At cryogenic temperatures, brittleness was also found to increase rapidly. The relationship between shear stress and strain was discovered to be nonlinear, and the catastrophic failure that transpired in each case was ultimately caused by the fiber-matrix interface. Compressive loading at low temperatures, bonding imperfections brought on by unidirectional laminate curing, and this behavior were linked to the polymeric matrix's plastic deformation. The in-plane shear stress results showed very nonlinear behavior at all temperatures.

The impact behavior of sandwich structures composed of cross-piled E-glass fiber-reinforced polymer matrix composites with a polyurethane foam core was investigated by V.K. Srivastava et al. and colleagues [22]. Sandwich structures consist of a soft, thick core that is adhesively bonded to two thin, stiff faces. Principal loads are primarily carried by the faces, while strain energy is absorbed by the core and shear loads are transmitted to the faces. The results show that tensile failure in the faces, compressive and shear failures in the core, and delamination between the face and core are the most frequent failure modes. It was found that during the sandwich structures' service life, these damages decreased their residual strength and stiffness. To address these issues, the study proposed to reinforce thickness through stitching. Weight drop impact energy also showed a lower dynamic fracture toughness in comparison to Izod and Charpy impact energies. Additionally, the results demonstrated that, in comparison to Izod and weight drop impact energy, the dynamic fracture toughness was more prominent under Charpy impact energy. The characteristics of glass fiber-reinforced polymer composites (GFRP) produced with different processing technologies have been investigated by Sathishkumar et al. [23]. The mechanical, thermal, vibrational, and water absorption characteristics of several GFRP composites produced by hand-lay, hydraulic press, hot press, compression molding, and matrix mixing and molding were examined in this study. The highest impact strength was found at 0% Vf of fiber, while the maximum flexural and tensile strengths were found at 25% Vf of fiber. The study found that when the Vf of glass fiber was increased to 25%, the elastic strain of the composite increased to 0.26 and then decreased with further increases in Vf. The composite modulus of elasticity also increased with fiberglass Vf. These characteristics were compared to those of other composites and steel as a reference using tensile and shear stresses. The study discovered that the natural frequencies of glass fiber-reinforced composites increased with the concentration of oligomeric siloxane. Adekomaya et al. [24] looked into how impact strength, tensile orientation, and fiber loading affected composite polymeric materials. The study found that fiber loading has an impact on the mechanical, corrosion-resistant, and strength characteristics of polymer composites. Strength and stiffness are enhanced by the fiber volume percentage in GFRP composites. The thermal conductivity of the composite material increases by 10% with an increase in fiber volume. The volume of the fiber increases with impact strength regardless of orientation. The direction effect is demonstrated by the rise in impact energy.

The fatigue behavior of unidirectional glass fiber-reinforced polyester (GFRP) composites under combined torsional and bending loads at room temperature was examined by El-Assal et al. [25]. To establish the GFRP composite failure model using different failure theories, the study included a series of pure torsional fatigue tests. The results show that, in comparison to pure bending fatigue strength, unidirectional GFRP composites have low torsional fatigue strength. Moreover, the number of stress cycles increased along with the fiber volume. The breakdown curves from the study are very helpful to designers, and the experimental results matched the predictions well at various cycles.

In order to evaluate mechanical damage in glass fiber-reinforced polymer (GFRP) composite laminates subjected to static tensile loading, Harizi et al. [26] employed passive infrared thermography. The lay-up method was utilized to prepare the cross-ply GFRP laminates used in the study, which were made from unidirectional E-type glass fibers. The researchers recorded infrared rays and the behavior of composites at high temperatures using a CEDIP JADE III (MW) IR camera. The scientists produced heat maps with information on damage characterization for each experiment. These maps can be used to track the state of damage in mechanical parts that are subjected to different static loading amplitudes.

Pandu et al. [27] used FEA software to conduct experimental research on glass fiber-reinforced composite materials under mechanical loading. The primary objective of the research was to design and fabricate a front axle for a two-wheeler using glass fiber and epoxy resin materials. Using a traditional steel axle, the study's findings were compared under various mechanical loading scenarios. The investigation's composite wheel axle was made via the pultrusion process. Using FEA software, the wheel axle's results from tensile, bending, impact, and fatigue tests were analyzed. According to the results, there was a notable 64% weight reduction when composite materials were substituted for traditional mild steel shafts. While the tensile strength was lower, the impact and bending strengths of the composite axle were higher than those of the traditional steel axle.

A study on the impact behavior of glass fiber-reinforced polymer composites and its relationship to thickness and orientation of fiber was carried out by Sikarwar et al. [28]. The study employed a conical bullet to impact test cross-ply GFRP laminates composed of woven glass fabric and epoxy resin. The laminates' ballistic limit and energy absorption capacity were demonstrated by the results. We examined and calculated the energy absorbed as well as the ballistic limits. The impact of thickness on the glass/epoxy laminates was also examined in this study. The projectile was fired for varying fiber thicknesses and directions at a constant velocity of 170 m/s. When the modulus of elasticity at higher strain rates was taken into consideration, the experimental analysis and mathematical analysis showed better agreement with impact loading for the glass/epoxy composites. The damage area shrank as projectile speed increased for velocities above the ballistic range. The modulus and failure strain of the dynamic young increased along with the energy absorption capacity and ballistic limit of the laminates.

Aiello and Focacci's [29] study is concerned with the high transverse thermal expansion coefficients (TECs) of non-metallic materials, particularly concrete. They stress that a rise in temperature may result in splitting cracks in the concrete due to tensile stress. To solve this problem, the researchers suggested an analytical model to calculate the temperature increase values at the beginning of the cracking phenomenon. The study highlights how crucial it is to comprehend and lessen the effects of TECs on the mechanical integrity of non-metallic materials, especially when it comes to temperature-induced tensile stress and cracking.

The incidence of fatigue damage in composite materials was examined by Bhat et al. [30], with an emphasis on the different failure mechanisms that lead to damage. The experiments showed that it was possible to characterize and minimize three distinct fault mechanisms that contribute to damage at a particular moment in time. In the context of fatigue and damage tolerance, this study highlights how critical it is to comprehend and mitigate the effects of these mechanisms on the mechanical integrity of composite materials. The study conducted by Briccoli Bati et al. [31] concentrated on the reinforcement of masonry arches through the use of two different composite materials. One of the damaged arch models used in the experimental analysis showed higher collapse load, stiffness, and ductility. The study highlighted that composites can be used to connect structural elements even when there is seismic loading. But the study also made clear that distinguishing between the mechanical properties of the polymeric matrix and the masonry is crucial to the usefulness of composites in this situation. The study adds to the continuing investigation of novel strengthening techniques for masonry constructions by offering insightful information about the advantages and disadvantages of employing composite materials to reinforce masonry arches.

El-Habak et al.'s research [32] indicates that the structural properties of composite materials are influenced by their geometric makeup. The impact characteristics of woven fabric composites have garnered significant attention recently, primarily due to their potential utility in understanding the results of impact compression load high-strain experiments. The instrumental Izod and Charpy experiments were performed on composite material. Wahid Ferdous, Allan Manalo, and colleagues [33] investigated the effects of stress level and stress concentration on GFRP composite fatigue life at a stress ratio of 0.1, using GFRP composite laminates and a static tensile test setup. An MTS machine is used at different loads, amplitudes, and frequencies to determine the specimen's ultimate tensile strength, stiffness, and fatigue tests. Vahid Movahedi-Rad et al. [34] looked into fatigue models and cycle time prediction techniques for fiber-reinforced polymer composites. In this study, fatigue models were split into three groups. These fatigue life models employ stress versus number of cycles (SAN) curves or Goodman life cycle diagrams to introduce a state of fatigue failure rather than taking into account the real mechanisms of deterioration. Table 6.3 shows the mechanical properties of glass fiber-reinforced polymeric (GFRP) composites.

6.6 Thermal Properties of GFRP Matrix Composites

The thermal properties of E-glass-reinforced soy-based polyurethane composites were investigated by Husic et al. [39] using soybean oil and petrochemical polyol Jeffol, two different types of polyurethane. According to the study, petrochemical polyol Jeffol-based polyurethane was less thermally stable than soy-based polyurethane. The thermal behavior of modified epoxy composites reinforced with chopped strand E-glass, with varying fiber volumes (Vf) ranging from 10 to 60%, was studied by Hameed et al. [17] Thermogravimetric analysis (TGA) was carried out at temperatures ranging from 30 °C to 900 °C in a nitrogen atmosphere. The results showed that the 60% Vf composite had better thermal stability; its degradation temperature was shifted from 357 °C to 390 °C.

The thermal behavior of chopped strand mat E-GF-reinforced unsaturated polyester composites with different numbers of glass mat layers and types of glass fibers was investigated by Budai et al. [50] using TGA and heat distortion temperature (HDT) analysis. At different temperatures, the tests were carried out in purging atmospheres of nitrogen and oxygen. The study discovered that the composite's thermo-oxidative breakdown was postponed when the amount of glass fiber increased. Lopez et al. [51] investigated the thermal behavior of an E-GF waste polyester composite

Table 6.3 Mechanical properties of glass fiber-reinforced polymeric (GFRP) composites

Types of glass fiber	Resin	Current Agent	V_f	Testing Standard	Tensile strength (MPa)	Tensile modulus (MPa)	Elongation at break (%)	Flexural strength (MPa)	Flexural modulus (MPa)	Impact strength	Interlaminar shear strength (MPa)	Ref
Woven mat	Polyester	MEKP/ Cobalt	0.25	ASTM D412 (T)	1.601	80.5	20.0	–	–	41.850 (J)	–	[11]
Woven mat	Polyester with (3% oligomeric siloxane)	Naphthalene	0.37	ASTM D-3039 (T), ASTM D-790 (F), ASTM D-2344 (S)	395.8	18,000	3.9	399.4	18,800	–	44.7	[35]
Woven mat	Polyester	–	0.33	ASTM D 638–97 (T)	249	6240	–	–	–	–	–	[23]
Woven mat	Polyester	–	-	2810 E6 (T)	189.0	–	–	–	–	–	–	[23]
Chopped strand	Polyamide66 (PA66)/ poly-phenylene sulfide (PPS) blend	–	0.30	GB/T I 6, 42I-I 996 (T), GB/T I 6,419-I 996 (F), GB/T I 6,420-I 996 (I)	–	124	–	159	–	98.2 (KJ/ m^2)	–	[36]
Woven mat (0°/ 90°)	Isophthalic/ neopentyl glycol polyester	–	0.42	PS 25C-0I I8 (T)	200	–	–	–	–	10 (J)	–	[37]
Woven mat (non-symmetric)	Polyester	-	-	362 F (BS, 1997) (T)	220	7000	0.055	–	–	–	–	[38]

(continued)

Table 6.3 (continued)

Types of glass fiber	Resin	Current Agent	V_f	Testing Standard	Tensile strength (MPa)	Tensile modulus (MPa)	Elongation at break (%)	Flexural strength (MPa)	Flexural modulus (MPa)	Impact strength	Interlaminar shear strength (MPa)	Ref
Woven	Polyurethanes	–	0.49	ASTM D3039 (T), ASTM D790M (F), ASTM D2344 (S)	278	18,654	–	444	27,075	–	27	[39]
Chopped strand mat	Polyester	–	0.60	ASTM D638 (T)	250	325	0.022	–	–	–	–	[40]
Woven mat	Polyester (acid resistant resin)	–	–	ASTM D 2344 (S)	–	–	–	–	–	–	30	[41]
Chopped strand mat	Polyester	–	0.015	ASTM E 399 (T)	–	3000	–	16.5	–	–	–	[42]
Chopped strand+vertical roving	Polyester	–	–	ASTM D 3039 (T), ASTM D 5379 (I)	103.4719	–	–	–	–	37.926 (J)	–	[43]
Virgin fiber	Polyester	–	–	ASTM D256 (T), ASTM D2240 (I)	64.4	7200	1.8	–	–	645.1 (J/m)	–	[44]

(continued)

Table 6.3 (continued)

Types of glass fiber	Resin	Current Agent	V_f	Testing Standard	Tensile strength (MPa)	Tensile modulus (MPa)	Elongation at break (%)	Flexural strength (MPa)	Flexural modulus (MPa)	Impact strength	Interlaminar shear strength (MPa)	Ref
Glass	Polyester (3 wt% Na-MMT)	–	0.40	ASTM-D638 (T), ASTM-D790 (F), ASTM-D256 (I)	130.03	–	–	206.15	–	153.50 (KJ/m^2)	–	[45]
Chopped strand	Epoxy (5.1 V_f Flyash)	Hardener	3.98	ASTM standard	–	–	–	–	–	0.0176 (j/mm^2)	–	[15]
Woven (biaxial stitch)	Epoxy	–	0.57	ASTM D 2355 (S)	–	–	–	–	–	–	18.2	[46]
Randomly oriented	Epoxy (10 wt% SiC)	–	0.5	ASTM D 3039.76 (T), ASTM D 256 (I)	179.4	6700	–	297.82	–	1.840 (J)	18.99	[47]
Woven	Epoxy (0.5 wt% MWCNTs)	–	0.73	ASTM D 2344 (S)	–	–	–	–	–	–	41.46	[48]
Unidirectional	Epoxy	–	0.55	ASTM D3039 (T)	784.98	–	0/032	–	–	–	–	[21]
Woven	Epoxy (6 wt% joc)	–	0.60	ASTM D 3039 (T)	311	18,610	3.8	–	–	–	–	[20]
Woven + (35 wt% short borosilicate)	Epoxy	–	–	–	355	43,700	1.65	–	–	–	–	[49]

without filler. The TGA/DTG curve showed that the mass loss changed from 1.8 wt% to 4.4 wt% and the degradation temperature changed from 209.8 °C to 448.7 °C.

6.7 Thermo-Mechanical Properties

The degradation of stiffness in GFRP composites under combined thermo-mechanical loading conditions was studied by Shi, Shengbo, and colleagues [52]. The stiffness degradation of FRP composites under such circumstances was predicted by the study using a mesomechanical model. The material's stiffness properties are significantly reduced and it becomes more porous and permeable as the pyrolysis reaction advances, the researchers found. High-temperature compression research was conducted, and the proposed model was validated by comparing numerical results with particular experimental results. Additionally, a three-zone annular face was installed in a universal testing apparatus to assess how GFRP composites behaved at high temperatures in relation to temperature. Important details regarding the stiffness degradation of GFRP composites under combined mechanical and thermal loading conditions are provided by the study's findings. Glass fiber-reinforced polymer (GFRP) composites' fire resistance and multifunctionality have been studied by Zhou, Aixi, and colleagues [53]. The integration of structural loading, interior building climate control, and fire resistance in GFRP building elements was the main focus of the study. Delaying the thermally-induced structural failure of GFRP members during a fire was found to be an effective use of internal liquid cooling. The temperature distributions during fire exposure were precisely predicted by the researchers' 2D and 3D thermo-mechanical models, which also matched experimental data from structural modeling of liquid-cooled specimens. The study's findings have significant implications for the development of multifunctional, fire-resistant GFRP composite building components. Publications in several scientific journals highlighted the research for its contributions to the domains of composite materials and structural engineering.

Ye et al. conducted research on the failure of fiber-reinforced composites under thermo-mechanical processing [54]. Determining how mechanical properties and variations in ambient temperature impact composite laminate failure was the main objective of the research. The study considered three different temperature conditions and found that a larger temperature variation in the ambient generally causes stress degradation and an increase in failure intensity for both uniaxially and biaxially charged laminates. Under both uniaxial and biaxial tensile loadings, the highest stresses in the composites were often found near the interface, regardless of whether temperature differences were taken into consideration. The study's findings provide significant insight into the behavior of fiber-reinforced composites during thermo-mechanical processing, particularly in relation to temperature fluctuations and stress distribution. David Hui, Piyush K. Dutta, and associates studied the tension and compression behavior of fiber-reinforced polymer (FRP) composites under elevated temperatures and sustained load [55]. A number of tension and compression tests were conducted on 0.25 mm thick FRP plates as part of the study. In order to calculate the time to failure at three distinct temperatures—25, 50%, and 80%—the study set out to ascertain the ultimate tensile strength, elastic modulus, Poisson's ratio, and strain at failure at room temperature and under sustained loads ranging from 60 to 80% of the ultimate load. This study's conclusions offer insightful information about the mechanical behavior of FRP composites at high temperatures and sustained loads, which enhances the materials' usefulness.

6.8 Vibration Characteristics of GFRP Matrix Composites

A matrix modification technique was employed by Erden et al. [12] to investigate the vibrational characteristics of glass/polyester composites. To unsaturated polyester, they added oligomeric siloxane in concentrations ranging from 1 to 3 weight percent. The mechanical and interlaminar properties of modified matrix composites reinforced with woven roving glass fabric were compared to those of untreated glass/polyester composites using tensile, flexure, and short-beam shear tests. The study discovered that as siloxane concentration rose, so did the composites' natural frequencies. This study adds to our understanding of the behavior and possible uses of glass/polyester composites by shedding light on how matrix modification affects these composites' vibrational properties. The damping behavior of woven fabric GFRP composites in saline water treatment was studied by Sridhar et al. [23] using the logarithmic decrement method. The fiber volumes studied were 20, 25, 30, 35, and 40, and the time periods were also varied. As the percentage of fiber volume increased, so did the untreated specimen's damping and stiffness, as indicated by the damping factor value. The specimen with 40% fiber volume saw the greatest decrease in damping values. Shape memory alloy (SMA) and piezoelectric (PZT) actuators were employed by Yuvaraja et al. [56] to investigate the vibration characteristics of a flexible glass fiber-reinforced polymer (GFRP) composite beam. The study used two cases: one was a GFRP beam with SMAs attached externally, and the other was a GFRP beam with surface-bonded PZT patches. An ANSYS vibration suppression analysis was conducted in addition to a mathematical model of the smart beam's behavior. The experiment's goals were to evaluate the effectiveness of the recommended actuators and the flexible beam's ability to control vibration. The study found that as the concentration of siloxane increased, the natural frequencies of the composites also increased. The study's conclusions highlight the potential uses of GFRP composites in infrastructure design and civil engineering by deepening our understanding of their corrosion resistance and fatigue behavior.

Bledzki et al. [57] examined the elastic constants of a unidirectional E-glass-reinforced epoxy matrix composite by vibrating plates with two diverse fiber-surface treatments. In order to encourage fiber/matrix adhesion, the first was treated with an epoxy dispersion containing aminosilane, and in order to inhibit fiber/matrix adhesion, the second was sized with polyethylene. For polyethylene composites, the elastic properties of the epoxy dispersion were poor, but good for aminosilane composites.

Vibration analysis was used by Mishra et al. [23] to investigate unidirectional glass fiber-reinforced resol/vinyl acetate-2-ethylhexyl acrylate (VAC-EHA) composites with varying fiber volume fractions. The focus of the study was on the effects of fiber/matrix interactions on the stiffness and damping properties of the composites. It was found that the VAC-EHA copolymer content dropped after being blended with resol, despite the fact that increasing the glass fiber content in the composite plate increased the damping properties. Additionally, glass fibers were incorporated into the matrix to enhance its tensile, stiffness, and damping properties.

Polyethylene fiber composites were subjected to vibration analysis and damping tests at temperatures between 10 and 60 degrees Celsius by Colakoglu et al. [58] In the study, a finite element program was used to compute the frequency numerically and an experimental damping monitoring method was used to measure the frequency response. The damping properties were found in terms of the damping factor using the half-power bandwidth technique. The experimental findings showed that the natural frequency and elastic modulus decreased with increasing temperature. The vibration damping of glued laminated beams reinforced with various epoxy matrix composite layups reinforced with graphene fibers was investigated by Naghipour et al. [59]. A range of methods, such as the half band power method, moving block method, logarithmic decrement method, and Hilbert transform method, were used in the study to assess the vibration damping of the composites. The experimental results demonstrated that adding GRP reinforcement to the bottom surface of cantilever beams significantly improved their stiffness and strength characteristics. The study provides crucial insight into how vibration is attenuated by composite materials and how beam mechanical properties are impacted by reinforcement.

6.9 Environmental Behavior of GFRP Matrix Composites

The water absorption behavior of polyester composites reinforced with fiberglass wastes that included different amounts of fiber waste—20, 30, and 40%, for example—was examined by Araujo et al. [60]. We used the test specimen that was immersed in distilled water for up to 600 h at different intervals to plot the time versus water absorption curve. The polyester/fiberglass wastes (40%) composite showed the lowest water absorption, with the composite's water absorption declining as the fiber content increased.

The effects of weathering on the mechanical properties of thermoset plastic composites reinforced with graphene fibers were investigated by Abdullah et al. [61]. Numerous weathering factors, including humidity, temperature, UV radiation, and pollution, led to the degradation of mechanical properties.

Botelho et al. investigated the behavior of woven mat glass fiber-reinforced poly ether-imide thermoplastic matrix composites in environmental conditions [62]. As part of the study, the composites were tested in seawater at various temperatures and

90% relative humidity for 60 days. The behavior of moisture absorption was found to be primarily influenced by temperature and relative humidity, with weight gain initially increasing linearly with respect to time. The maximum moisture absorption of 0.18% was observed after 25 days. By providing valuable insights into the behavior of glass fiber-reinforced thermoplastic matrix composites in the environment, the study advances our understanding of the potential applications and performance of these materials. The environmental deterioration of GFRP composites at various temperatures, including 45 °C and 55 °C, was examined by Chhibber et al. [63] After 1 and 2 months, the tests were carried out in regular water and sodium hydroxide (NaOH) baths. The weight gain percentage rose as bath time and temperature increased; the weight gain in the NaOH bath was greater than that of the water bath. The study sheds important light on how temperature and bath conditions affect the performance and durability of GFRP composites and sheds insight on how these materials deteriorate in the environment.

The behavior of E-GF-reinforced isophthalic polyester composites was investigated by Renaud et al. [64] in a range of conditions, including deionized water, strong acids, cement extract, saltwater, and tap water. Various types of glass fibers, including boro-silicate and boron-free ones, were used in the composites. After 50 years of testing at 60 °C, it was discovered that the boron-free E-GF composite increased the resistance of moisture absorption in all environmental conditions. The study provides insightful information about the behavior of these materials in the environment, which broadens our understanding of the potential applications and performance characteristics of glass fiber-reinforced composites.

Kajorncheappunngam et al. [16] investigated the effect of aging environment on the degradation of woven fabric E-glass-reinforced epoxy using four different liquid media: distilled water, saturated salt solution, 5-molar NaOH solution, and 1-molar hydrochloric acid solution. The study found that soaking in water had fewer negative effects than soaking in acid or alkali, and that at lower immersion levels, the mechanical properties were unaffected.

The environmental effects of randomly oriented E-GF-reinforced polyester composites in freezing conditions, Ganga water, brine, acid solution, and kerosene oil were investigated by Agarwal et al. [65]. The tests were run for a variety of durations, including 64, 128, and 256 h. Tensile strength decreased as a percentage after each time interval; the largest percentage reduction was observed in NaOH solution, while the lowest percentage reduction was found in freezing conditions.

Ellyin et al.'s [66] examined the moisture absorption process of fiber epoxy composite tubes reinforced with E-glass. The tubes were immersed in distilled water at two different temperatures (20 °C and 50 °C) for a period of 4 months. The weight increased by 0.23% at 20 °C and 0.29% at 50 °C, based on the time versus moisture absorption curve.

Abbasi et al. [67] investigated the environmental behavior of GFRP composites with different combinations, such as GF/isophthalic polyester, GF/vinyl ester, and GF/urethane-modified vinyl ester. The tests were conducted at different temperatures ranging from 20 °C to 120 °C for 30, 120, and 240 days under standard water and alkaline conditions. It was found that the mechanical properties of the composites, such as strength and modulus, declined in alkaline environments at higher temperatures.

Visco et al. looked into the mechanical characteristics of polyester composites reinforced with e-glass both before and after they were submerged in seawater [68]. The study used two different types of polyester resins—orthodontic and isophthalic—as well as two different types of laminates for the preparation of composites. The experimental findings showed that shear modulus, flexural strength, and flexural modulus all decreased with immersion time. Isophthalic resin bonded with GFs more effectively than orthophthalic resin and resisted absorption of seawater.

6.9.1 Tribological Behavior of GFRP Matrix Composites

In a study conducted at room temperature, El-Tayeb et al. [69] looked at the worn surfaces of unsaturated polyester composites reinforced with chopped glass fiber (GF) at various sliding velocities and loads. The chopped GF orientations in the composites were parallel/anti-parallel (P/AP). At lower loads and higher speeds, sliding in the P-orientation demonstrated a lower friction coefficient than the AP-orientation, according to the experimental results. On the other hand, sliding in the AP-orientation showed a lower friction coefficient at higher loads, speeds, and distances than the P-orientation. Furthermore, the AP-orientation demonstrated a lower mass loss (16%) in comparison to the P-orientation.

El-Tayeb et al. [70] investigated the multipass two-body abrasive wear behavior of chopped glass fiber (CGRP) composites with varying sliding velocities (0.157 and 0.314 m/s) while applying normal loads ranging from 5 to 25 N. The tests were conducted with sliding against waterproof SiC abrasive paper in dry contact conditions. The wear rate dropped in tandem with the rotational speed and load increases. By strengthening its resistance to abrasion, the AP orientation of the CGRP composite decreased wear rate when compared to other orientations. The scanning electron microscopy (SEM) results indicated that the AP orientation did not exhibit any fiber damage.

Quintelier et al. [71] examined the wear and friction behavior of GF-reinforced polyester composites under dry conditions, loading ranges of 60–300 N, and constant speed of 10 mm/s. Scanning electron microscopy (SEM) observations revealed that the parallel orientation had less friction than the transverse orientation. A range of thermoset resins, such as polyester, vinyl ester, and epoxy resin, were employed by Mathew et al. [72] to investigate the tribological properties of E-GF-reinforced polyester composites with various directly oriented warp knit fibers, such as biaxial, biaxial non-woven, tri-axial, and quad-axial fabric. Biaxial non-woven-reinforced vinyl ester composite outperformed other combinations in terms of tribological properties, according to the study. Kishore et al. looked at the effects of sliding velocity and load on the friction and wear behavior of epoxy composites reinforced with plain weave bi-directional E-glass fabric and different fillers, such as oxide and rubber particles [73]. The study found that the oxide particle-filled composite outperformed rubber particles in terms of wear resistance at low load levels. On the other hand, under higher load conditions, the rubber particles performed better in terms of wear resistance than the oxide particles. When chopped strand mat GF-reinforced unsaturated polyester composites slid against smooth stainless steel in dry contact, Yousif et al. [23] investigated the friction and interface temperature behavior of the materials under different loads and sliding velocities. After measuring parallel and anti-parallel chopped GF orientations at room temperature, the researchers found that the anti-parallel orientation had higher interface temperatures (29–50 °C) and friction coefficients (0.5–0.6) than the parallel orientation. Chand et al. [74] investigated the three-body abrasive wear behavior of short E-GF-reinforced polyester composites with and without filler at varying sliding speeds, abrasive particle sizes, and applied loads. The study's findings demonstrated that a higher glass fiber content reduced wear loss because it reduced the composite's volume loss by raising the weight percentage of fiber in the composite.

The impact of fillers on the wear and friction behavior of GF-reinforced woven mat epoxy composites under dry sliding conditions at various loads and sliding velocities was examined by Suresha et al. [18]. In the study, two different inorganic fillers were used: graphite (5 weight percent) and SiC particles (5 weight percent). Scanning electron microscopy (SEM) observations revealed that graphite-filled composites had lower coefficients of friction than unfilled SiC-filled composites. SiC-filled composites demonstrated enhanced resistance to wear as well. This study provides insightful information about the effects of fillers on the wear and friction characteristics of the material, which advances our understanding of the potential applications and performance characteristics of glass fabric-reinforced epoxy composites. Sampathkumaran et al. [75] examined the wear behavior of GF-reinforced epoxy composite under dry sliding conditions by varying the applied load (20–60 N), velocity (2–4 m/s), and sliding distance (0.5–6 km). The experimental results showed that while the debris rate increased with increasing distance traveled, weight loss increased with increasing load and velocity. In order to investigate the wear and friction behavior of chopped strand mat GF polyester composites with varying loads (30–100 N) under wet contact conditions, Yousif et al. [10] used two different test methods: pin-on-disc and block-on-ring. The presence of water content in the study was found to increase the roughness value in both the parallel and anti-parallel fiber orientations. Additionally, the anti-parallel orientation demonstrated higher wear and frictional resistance than the parallel orientation. Suresha et al. [76] looked at the effect of silicon carbide (SiC) fillers on the wear and friction behavior of epoxy composites reinforced with woven mat (E-GF). The study discovered that the composite with five weight percent SiC particles filled it had a higher coefficient of friction and wear resistance at sliding distances between 2000 and 4000 m compared to the composite without SiC fillers. Pihtili et al. [77] study of the wear behavior of E-CGRP composite used sliding distances ranging from 235.5 mm to 1884 mm. The plain polyester began to lose weight more when the sliding distance was greater than 942 mm, according to the study. The results showed that compared to plain polyester, the polyester matrix composite reinforced with glass fibers was more wear-resistant.

The sliding wear behavior of Jatropha oil cake-filled woven fabric E-GF-reinforced epoxy composites was investigated by Mohan et al. [20] at different loads (10 and 20 N). According to the study, wear loss increased with sliding distance; wear loss was noted at a sliding distance of 2000 m under a 10 N applied load. The glass epoxy composite filled with Jatropha oil cake showed a high coefficient of friction and good wear resistance at different sliding distances. Suresha et al. looked at the three-body abrasive wear behavior of woven E-GF-reinforced vinyl ester composites with particulate-filled fillers like SiC and graphite [19]. In the study, tests were conducted at two distinct loads (22 and 32 N) and sliding distances ranging from 270 to 1080 m. The experimental results demonstrated that filled composites with silicon and graphite had higher abrasion resistance and lower specific wear rates (1.95 10 $\times$ 11 $m^3/(Nm)$) than unfilled composites. Patnaik et al. looked at the wear behavior of randomly oriented epoxy composites reinforced with E-GF that also included particulate fillers like SiC, Al_2O_3, and pine bark dust [47]. The composite configurations used in the study included alumina, pine bark dust, and 50 weight percent GF/40 weight percent epoxy/10 weight percent SiC. In the experimental tests, loads of 50 and 75 N were applied, and sliding distances ranged from 200 to 600 m. The results demonstrated that the GF(50 wt%)/Epoxy(40 wt%)/Pine bark dust (10 wt%) composite had superior wear resistance (0.000881–0.001365 $mm^3/(N\text{-}m)$) for all sliding distances.

Chauhan et al. examined the wear and friction characteristics of bi-directional woven fabric S-GF-reinforced vinyl ester composites with 65 weight percent fiber and the resin mixed with different co-monomers, such as butyl and methyl acrylate [78]. Three different composite compositions were made: GF + (vinyl ester + styrene), GF + (vinyl ester + methyl acrylate), and GF + (vinyl ester + butyl acrylate). The tests were conducted under dry sliding conditions with varying sliding velocities (1, 2, 3, and 4 m/s) and loads (10, 20, 30, and 40 N). The GF + (vinyl ester + butyl acrylate) composite exhibited a higher specific wear rate and a lower coefficient of friction at lower sliding speeds.

6.10 Other Properties

6.10.1 Aluminum in Glass Fibers

Kumar, Anil, Arshad Mehmood, and associates [79] conducted an experimental analysis to look into the mechanical characteristics of reinforced composites composed of aluminum, glass fiber, and nylon. As part of the research, the mechanical properties of nylon, glass fiber reinforced plastic (GFRP), and their composites with aluminum were calculated using a computer-controlled universal testing machine at room temperature. The study found that the aluminum-GFRP composites had the highest tensile strength but the lowest ductility. Furthermore, in terms of ductility and tensile strength, the composite GFRP performed better than pure GFRP composites. By providing valuable insights into the mechanical behavior of these reinforced composites, the study advances our understanding of their potential applications and performance characteristics.

6.10.2 Kevlar Fiber Hybridization in Composites

The tensile and damping properties of hybrid laminates made of Kevlar composite were examined by Bulut, Mehmet, and associates [80]. To illustrate the impact of hybridization, the study combined various fiber volume fractions to create hybrid Kevlar-epoxy resin composites. A battery of vibration tests was used to evaluate the composites' dynamic properties, and an instrumented tensile testing system was used to examine their mechanical characteristics. The research specimens' S-type glass fibers were found to be protected by Kevlar in the study, and following the vibration tests, the frequency response curves were recorded and tabulated. The findings showed that there appeared to be a higher frequency of naturally occurring KFRP composites than GFRP composites. The study's insightful observations help us understand the behavior of Kevlar composite hybrid laminates and their potential uses and performance characteristics.

6.10.3 Damping Characteristics

Roy Choudhury and associates conducted the Mridusmita A et al. [81] study on the mechanical properties of green composites, specifically bamboo fiber-reinforced polylactic acid (PLA) composites. These composites were created in the research using hot compression molding and the film stacking technique. The primary objective of the study was to analyze the compressive and tensile failure load in the single-lap bolted and adhesive joints of these green composites. The combination of bamboo fibers' natural and synthetic qualities with PLA, a polymer made from natural resources like starch and maize, showed the results' biodegradable nature. Additionally, the research looked at the drilling behavior of unidirectional bamboo fiber-reinforced green composites to shed light on their machinability and potential applications. The mechanical characteristics and possible uses of green composites, particularly those reinforced with PLA and bamboo fibers, are better understood thanks to the study's findings.

6.10.4 Corrosion in FRP Composites

Yunyi Zou and Arthur Huckelbridge conducted a study [82] on the corrosion resistance of glass fiber-reinforced polymer (GFRP) composites in bridge slabs. Plots of hysteresis were examined, experiments were conducted on different FRP bars coated with a single steel bar, and the energy loss per period was computed. The study indicates that GFRP exhibits a robust resistance to corrosion when exposed to steel frames, rendering it a viable solution for mitigating corrosion in bridge slabs. By examining the hysteresis plots and energy loss per period (Table 4.1) and by observing the experiments on different FRP bars covered with a single steel bar, the research also demonstrated GFRP charging under constant cyclic fatigue amplitude. The findings expand our knowledge of GFRP composites' fatigue behavior and corrosion resistance while showcasing their potential.

6.10.5 Applications of Glass Fiber-Reinforced Composites

GFRC have found various applications in the automotive industry, aerospace applications, construction and infrastructure, sporting goods, and other industrial applications.

6.10.6 Automobile Industry

In the automotive industry, glass fiber-reinforced composites are used in body panels, bumper systems, instrument panels, leaf springs, and drive shafts. Glass fiber-reinforced thermoplastics and plastic polymers have shown great potential in the automotive industry due to their weight reduction potential and advantageous properties. The automotive industry has seen a rise in the use of glass fiber-reinforced composites as a result of consumer demand for lightweight materials that can lower emissions and increase fuel efficiency. Furthermore, strict government laws pertaining to emission standards and vehicle safety are increasing the demand for glass fiber-reinforced composites in the automotive sector [83].

6.10.7 Aerospace Application

Aerospace applications include the use of glass fiber-reinforced composites for internal parts, engine components, and aircraft structures [84].

6.10.8 Construction and Infrastructure

In construction and infrastructure, GFRC are utilized making of bridge decks, reinforcement bars, and pipes [85].

6.10.9 Sporting Goods

In sporting goods, GFRC are used in golf club shafts, tennis rackets, and bicycle frames [86]. In other industrial applications, GFRC are used in electrical and electronic components, marine applications [87], and wind turbine blades [88].

6.11 Other Industrial Applications

6.11.1 Medical Applications

In the medical field, glass fiber-reinforced composites are essential components of many medical devices and applications, including biologically resistant or reactive composites that provide protection against hazards posed by excessive moisture, which promotes the growth of bacteria and fungi [89].

6.11.2 Applications in the Food Industry

In the food industry, GFRC are useful in the manufacturing of food processing equipment, where they offer high strength and durability, making them suitable for applications that require hygienic and corrosion-resistant materials [90].

6.11.3 Advancements in Glass Fiber Technology

Various advancements in glass fiber technology have been made in recent years, including the use of nanotechnology in glass fiber composites, hybrid composites, and smart materials integration.

6.11.4 Nanotechnology in Glass Fiber Composites

The incorporation of nanotechnology, such as the use of nanoclay in hybrid composites, has led to the development of materials with elevated erosion resistance. The performance and durability of glass fiber composites could be enhanced by this innovation in a number of applications [91].

6.11.5 Hybrid Composites

The benefits of various materials, such as glass and natural fibers, are combined in hybrid composites to produce composites with better qualities. These hybrid composites offer higher strength, flexibility, and environmental benefits compared to traditional glass fiber composites. The use of hybrid composites in various applications has been extensively researched and documented in the literature, highlighting their potential to revolutionize the field of composite materials [92].

6.11.6 Smart Materials Integration

The integration of sensor technology into composite materials, such as the development of smart composites, possesses the capacity to enhance performance and safety of various structures and components. By incorporating sensors into the materials, it is possible to monitor the health and status of the composites, leading to more efficient maintenance and repair strategies. This breakthrough could enhance the functionality and robustness of glass fiber composites in a range of industries, including infrastructure, sporting goods, automotive, and aerospace [93].

6.12 Challenges and Limitations

6.12.1 Processing Challenges

The manufacturing of glass fiber matrix composites involves complex processing techniques, such as resin impregnation, curing, and molding. The high processing temperatures required for these composites can lead to increased energy consumption and production costs. Furthermore, the abrasive nature of glass fibers can result in significant tool wear and maintenance challenges during processing, impacting the overall manufacturing efficiency and cost-effectiveness [5].

6.12.2 Recycling Issues

Glass fiber matrix composites pose challenges in terms of recycling due to the difficulty in separating the glass fibers from the matrix material. The presence of contaminants, such as resins and coatings, further complicates the recycling process. As a result, the recycling of glass fiber composites is limited, leading to concerns regarding environmental sustainability and end-of-life disposal [94].

6.12.3 Durability Concerns

While glass fiber composites offer high strength and stiffness, they are susceptible to impact damage and moisture absorption, which can compromise their long-term durability. Additionally, the brittleness of glass fibers can lead to reduced impact resistance and fatigue performance, particularly in dynamic loading conditions. These durability concerns are significant, especially in applications where the composites are exposed to harsh environmental conditions and mechanical stresses [95].

6.13 Future Trends and Prospects

6.13.1 Emerging Technologies

The development of new and innovative technologies, such as advanced fiber materials, manufacturing processes, and recycling techniques, is expected to revolutionize the glass fiber matrix composites industry. These emerging technologies aim to address the challenges of processing, recycling, and durability associated with glass fiber matrix composites, ultimately driving innovation and growth in this field.

6.13.2 Research and Development Directions

Ongoing research and development efforts focus on improving the mechanical properties, processing efficiency, and sustainability of glass fiber matrix composites. These efforts include the exploration of chemical modifications, novel fiber materials, and advanced manufacturing techniques to enhance the performance and durability of these composites. Additionally, researchers are investigating the integration of emerging technologies, such as artificial intelligence and the Internet of Things, to optimize the design and manufacturing of glass fiber matrix composites for various applications.

6.13.3 *Sustainability in Glass Fiber Composites*

As the global demand for sustainable and eco-friendly materials increases, there is a growing emphasis on incorporating sustainability principles in the production and use of glass fiber matrix composites. This includes the use of bio-based and recycled materials, as well as environmentally friendly manufacturing processes and recycling strategies. Reducing the environmental effect of glass fiber matrix composites at every stage of their life cycle—from the procurement of raw materials to the final applications—is the aim.

6.14 Conclusion

GFRC are a great option for a variety of applications in the automotive, aerospace, construction, and sports industries because of their exceptional strength, stiffness, and thermal resistance. The continuous advancements in glass fiber technology, including the development of innovative manufacturing processes, chemical modifications, and the integration of smart materials, are driving the evolution of these composites. Furthermore, the incorporation of sustainability principles, such as the use of bio-based and recycled materials, is enhancing the environmental profile of glass fiber composites. While challenges related to processing, recycling, and durability exist, ongoing research and development efforts are focused on addressing these limitations. Overall, the future prospects of glass fiber-reinforced matrix composites are promising, with emerging technologies and a strong emphasis on sustainability positioning these materials for continued growth and innovation in the coming years.

Acknowledgements Authors are thankful to Director, Amity Institute of Applied Sciences (AIAS), and Head, Department of Chemistry for providing the facilities. One of the authors, Tejendra K Gupta is thankful to the Council of Science and Technology, Lucknow, Uttar Pradesh, India for the financial support (Vide No. CST/CHEM/D-1431).

References

1. Foruzanmehr, M., Elkoun, S., Fam, A., Robert, M.: Degradation characteristics of new bio-resin based-fiber-reinforced polymers for external rehabilitation of structures. J. Compos. Mater. **50**(9), 1227–1239 (2016). https://doi.org/10.1177/0021998315590262
2. Introduction to Glass Fiber-Based Composites and Structures (2022)
3. Tokyo, S.Y.: Application of Glass Fiber-Reinforced Plastics to Electrical and Electronic Apparatus in Japan
4. Rajabipour, F., Asce, M., Maraghechi, H., Fischer, G.: Investigating the Alkali-Silica Reaction of Recycled Glass Aggregates in Concrete Materials. https://doi.org/10.1061/ASCEMT.1943-5533.0000126

5. Morampudi, P., Namala, K.K., Gajjela, Y.K., Barath, M., Prudhvi, G.: Review on glass fiber reinforced polymer composites. Mater. Today Proc. 314–319. Elsevier Ltd (2020). https://doi.org/10.1016/j.matpr.2020.11.669
6. Yilmaz, G.: Structural characterization of glass-ceramics made from fly ash containing SiO2-Al2O3-Fe2O3-CaO and analysis by FT-IR-XRD-SEM methods. J. Mol. Struct. **1019**, 37–42 (2012). https://doi.org/10.1016/j.molstruc.2012.03.028
7. Spaghtz Akd, M.E., Clark, J.D.: Studies on glass. The electrical conductivity of boron trioxide-sodium borate glasses
8. Negawo, T.A., Polat, Y., Akgul, Y., Kilic, A., Jawaid, M.: Mechanical and dynamic mechanical thermal properties of ensete fiber/woven glass fiber fabric hybrid composites. Compos. Struct. **259**. https://doi.org/10.1016/j.compstruct.2020.113221
9. Chawla, K.K.: Glass Fibers. In: Reference Module in Materials Science and Materials Engineering. Elsevier (2016). https://doi.org/10.1016/B978-0-12-803581-8.02325-0
10. Yousif, B.F., El-Tayeb, N.S.M.: Wear and friction characteristics of CGRP composite under wet contact condition using two different test techniques. Wear **265**(5–6), 856–864 (2008). https://doi.org/10.1016/j.wear.2008.01.016
11. Aramide, F.O., Atanda, P.O., Olorunniwo, O.O.: Mechanical properties of a polyester fibre glass composite. Int. J. Compos. Mater. **2**(6), 147–151. https://doi.org/10.5923/j.cmaterials.20120206.06
12. erden2010
13. Atas, C., Liu, D.: Impact response of woven composites with small weaving angles. Int. J. Impact Eng **35**(2), 80–97 (2008). https://doi.org/10.1016/j.ijimpeng.2006.12.004
14. Aktaş, M., Atas, C., Içten, B.M., Karakuzu, R.: An experimental investigation of the impact response of composite laminates. Compos. Struct. **87**(4), 307–313 (2009). https://doi.org/10.1016/j.compstruct.2008.02.003
15. Gupta, N., Singh Brar, B., Woldesenbet, E.: Effect of filler addition on the compressive and impact properties of glass fibre reinforced epoxy (2001)
16. Kajorncheappunngam, S., Gupta, R.K., Gangarao, H.V.S.: Effect of Aging Environment on Degradation of Glass-Reinforced Epoxy. https://doi.org/10.1061/ASCE1090-026820026:161
17. Hameed, N., Sreekumar, P.A., Francis, B., Yang, W., Thomas, S.: Morphology, dynamic mechanical and thermal studies on poly(styrene-co-acrylonitrile) modified epoxy resin/glass fibre composites. Compos. Part A Appl. Sci. Manuf. **38**(12), 2422–2432 (2007). https://doi.org/10.1016/j.compositesa.2007.08.009
18. Suresha, B., Chandramohan, G., Prakash, J.N., Balusamy, V., Sankaranarayanasamy, K.: The role of fillers on friction and slide wear characteristics in glass-epoxy composite systems. J. Miner. Mater. Charact. Eng. **05**(01), 87–101 (2006). https://doi.org/10.4236/jmmce.2006.51006
19. Suresha, B., Chandramohan, G.: Three-body abrasive wear behaviour of particulate-filled glass-vinyl ester composites. J. Mater. Process. Technol. **200**(1–3), 306–311 (2008). https://doi.org/10.1016/j.jmatprotec.2007.09.035
20. Mohan, N., Natarajan, S., Kumareshbabu, S.P., Siddaramaiah: Investigation on sliding wear behaviour and mechanical properties of Jatropha oil cake-filled glass-epoxy composites. JAOCS J. Am. Oil Chem. Soc. **88**(1), 111–117 (2011). https://doi.org/10.1007/s11746-010-1657-z
21. Tensile, compressive and shear properties of unidirectional glassepoxy composites subjected to mechanical loading and low temperature services
22. Srivastava, V.K.: Impact behaviour of sandwich GFRP-Foam-GFRP composites. Int. J. Compos. Mater. **2**(4), 63–66 (2012). https://doi.org/10.5923/j.cmaterials.20120204.04
23. Sathishkumar, T.P., Satheeshkumar, S., Naveen, J.: Glass fiber-reinforced polymer composites—a review. J. Reinforc. Plast. Compos. **33**(13), 1258–1275. SAGE Publications Ltd. https://doi.org/10.1177/0731684414530790
24. Adekomaya, O., Adama, K.: Glass-fibre reinforced composites: the effect of fibre loading and orientation on tensile and impact strength. Niger. J. Technol. **36**(3), 782–787 (2017). https://doi.org/10.4314/njt.v36i3.17

25. El-Assal, A.M., Khashaba, U.A.: Fatigue analysis of unidirectional GFRP composites under combined bending and torsional loads. Compos. Struct. **79**(4), 599–605 (2007). https://doi.org/10.1016/j.compstruct.2006.02.026
26. Harizi, W., Chaki, S., Bourse, G., Ourak, M.: Mechanical damage assessment of glass fiber-reinforced polymer composites using passive infrared thermography. Compos. B Eng. **59**, 74–79 (2014). https://doi.org/10.1016/j.compositesb.2013.11.021
27. Pandav, M.P.A.: Experimental evaluation and analysis of Glass Fiber Reinforced Composite under mechanical loading by using FEA software. http://www.irphouse.com
28. Sikarwar, R.S., Velmurugan, R., Gupta, N.K.: Influence of fiber orientation and thickness on the response of glass/epoxy composites subjected to impact loading. Compos. B Eng. **60**, 627–636 (2014). https://doi.org/10.1016/j.compositesb.2013.12.023
29. Aiello, M.A., Focacci, F., Nanni, A.: Effects of Thermal Loads on Concrete Cover of FRP Reinforced Elements: Theoretical and Experimental Analysis
30. Bhat, M.R., Majeed, M.A., Murthy, C.R.L.: Characterization of fatigue damage in unidirectional GFRP composites through acoustic emission signal analysis (1994)
31. Briccoli Bati, S., Rovero, L., Tonietti, U.: Strengthening Masonry Arches with Composite Materials. J. Compos. Constr. **11**(1), 33–41 (2007). https://doi.org/10.1061/(asce)1090-0268(2007)11:1(33)
32. Tarfaoui, M.: Dynamic composite materials characterisation with hopkinson bars: design and development of new dynamic compression systems. J. Compos. Sci. **7**(1) (2023). https://doi.org/10.3390/jcs7010033
33. Ferdous, W., et al.: Testing and modelling the fatigue behaviour of GFRP composites—effect of stress level, stress concentration and frequency. Eng. Sci. Technol. Int. J. **23**(5), 1223–1232 (2020). https://doi.org/10.1016/j.jestch.2020.01.001
34. Movahedi-Rad, A.V., Keller, T., Vassilopoulos, A.P.: Fatigue damage in angle-ply GFRP laminates under tension-tension fatigue. Int. J. Fatigue **109**, 60–69 (2018). https://doi.org/10.1016/j.ijfatigue.2017.12.015
35. Enhancement of the Mechanical Properties of Glasspolyester Composites via Matrix Modification Glasspolyester Composite Siloxane Matrix Modification
36. Chen, Z., Liu, X., Lü, R., Li, T.: Mechanical and tribological properties of PA66/PPS blend. III. Reinforced with GF. J. Appl. Polym. Sci. **102**(1), 523–529 (2006). https://doi.org/10.1002/app.24253
37. Yuanjian, T., Isaac, D.H.: Combined impact and fatigue of glass fiber reinforced composites. Compos. B Eng. **39**(3), 505–512 (2008). https://doi.org/10.1016/j.compositesb.2007.03.005
38. Singh Chauhan, P., Kumar Sharma, A., Saraswat, A., Chauhan, P.S., Sharma, A.K., Prakash Pandit, P.: Tensile and bending analysis of composite material. Int. J. Adv. Sci. Technol. **29**(10S), 6598–6606. https://www.researchgate.net/publication/342170287
39. Husić, S., Javni, I., Petrović, Z.S.: Thermal and mechanical properties of glass reinforced soy-based pol yurethane composites. Compos. Sci. Technol. **65**(1), 19–25 (2005). https://doi.org/10.1016/j.compscitech.2004.05.020
40. Leonard, L.W.H., Wong, K.J., Low, K.O., Yousif, B.F.: Fracture behaviour of glass fibre-reinforced polyester composite. Proc. Inst. Mech. Eng. Part L J. Mater. Des. Appl. **223**(2), 83–89 (2009). https://doi.org/10.1243/14644207JMDA224
41. Putić, S., Bajčeta, B., Vitković, D., Stamenović, M., Pavićević, V.: The interlaminar strength of the glass fiber polyester composite. Chem. Ind. Chem. Eng. Q. **15**(1), 45–48 (2009). https://doi.org/10.2298/CICEQ0901045P
42. Avci, A., Arikan, H., Akdemir, A.: Fracture behavior of glass fiber reinforced polymer composite. Cem. Concr. Res. **34**(3), 429–434 (2004). https://doi.org/10.1016/j.cemconres.2003.08.027
43. Effect of orientation of glass fiber on mechanical properties of GFP composites
44. Araújo, E.M., Araújo, K.D., Pereira, O.D., Ribeiro, P.C., De Melo, T.J.A.: Fiberglass wastes/polyester resin composites: mechanical properties and water sorption. Polimeros **16**(4), 332–335 (2006). https://doi.org/10.1590/S0104-14282006000400014

45. Yakoub, N.G.: Mechanical characterization of polyester/e-glass fiber reinforced/MWCNTS nanocomposites. IIUM Eng. J. **23**(1), 329–338 (2022). https://doi.org/10.31436/IIUMEJ.V23I1.1848
46. Yang, B., Kozey, V., Adanur, S., Kumar, S.: Bending, compression, and shear behavior of woven glass ®ber±epoxy composites. www.elsevier.com/locate/compositesb
47. Patnaik, A., Satapathy, A.: Investigations on Three-Body Abrasive Wear and Mechanical Properties of Particulate Filled Glass Epoxy Composites (2010). www.fkkksa.utm.my/mpj
48. Liu, Y. et al.: Role of matrix modification on interlaminar shear strength of glass fibre/epoxy composites. Compos. Part B Eng. 95–98 (2012). https://doi.org/10.1016/j.compositesb.2011.04.037
49. Godara, A., Raabe, D.: Influence of fiber orientation on global mechanical behavior and mesoscale strain localization in a short glass-fiber-reinforced epoxy polymer composite during tensile deformation investigated using digital image correlation. Compos. Sci. Technol. **67**(11–12), 2417–2427 (2007). https://doi.org/10.1016/j.compscitech.2007.01.005
50. Budai, Z., Sulyok, Z., Vargha, V.: Glass-fibre reinforced composite materials based on unsaturated polyester resins. J. Therm. Anal. Calor. 1533–1544 (2012). https://doi.org/10.1007/s10973-011-2069-5
51. López, F.A., Martín, M.I., Alguacil, F.J., Rincón, J.M., Centeno, T.A., Romero, M.: Thermolysis of fibreglass polyester composite and reutilisation of the glass fibre residue to obtain a glass-ceramic material. J. Anal. Appl. Pyrol. **93**, 104–112 (2012). https://doi.org/10.1016/j.jaap.2011.10.003
52. Shi, S., Gu, L., Liang, J., Fang, G., Gong, C., Dai, C.: A mesomechanical model for predicting the degradation in stiffness of FRP composites subjected to combined thermal and mechanical loading. Mater. Des. **89**, 1079–1085 (2016). https://doi.org/10.1016/j.matdes.2015.10.060
53. Zhou, A., Keller, T.: Structural responses of FRP elements under combined thermal and mechanical loadings: experiments and analyses. https://www.researchgate.net/publication/37449915
54. Ye, J., et al.: Failure analysis of fiber-reinforced composites subjected to coupled thermo-mechanical loading. Compos. Struct. **235** (2020). https://doi.org/10.1016/j.compstruct.2019.111756
55. Dutta, P.K., Hui, D.: Creep rupture of a GFRP composite at elevated temperatures. www.elsevier.com/locate/compstruc
56. Yuvaraja, M., Senthilkumar, M.: Comparative study on vibration characteristics of a flexible GFRP composite beam using SMA and PZT actuators. Proc. Eng. 571–581. Elsevier Ltd. (2013). https://doi.org/10.1016/j.proeng.2013.09.132
57. Bledzki, A.K., Kessler, A., Rikards, R., Chate, A.: Determination of elastic constants of glass/epoxy unidirectional laminates by the vibration testing of plates
58. Colakoglu, M.: Damping and vibration analysis of polyethylene fiber composite under varied temperature. https://www.researchgate.net/publication/279888558
59. Naghipour, M., Taheri, F., Zou, G.P.: Evaluation of Vibration Damping of Glass-Reinforced-Polymer-Reinforced Glulam Composite Beams. https://doi.org/10.1061/ASCE0733-94452005131:71044
60. Araújo, E.M., Araújo, K.D., Pereira, O.D., Ribeiro, P.C., de Melo, T.J.A.: Fiberglass Wastes/Polyester Resin Composites: Mechanical Properties and Water Sorption (2006)
61. Abdullah, H., Al Araimi, S., Siddiqui, R.A.: The effects of weathering on mechanical properties of glass fiber reinforced plastics (GRP) materials (2000)
62. Botelho, E.C., Bravim Júnior, J.C., Costa, M.L., de Faria, M.C.M.: Environmental effects on thermal properties of PEI/Glass fiber composite materials. J. Aerosp. Technol. Manage. **5**(2), 241–254 (2013). https://doi.org/10.5028/jatm.v5i2.193
63. Chhibber, R.: Integrated wind turbine design. Environmental Degradation of Glass Fibre Reinforced Polymer Composite
64. Renaud, C.M., Greenwood, M.E.: Effect of Glass Fibres and Environments on Long-Term Durability of GFRP Composites

65. Agarwal, A., Garg, S., Rakesh, P.K., Singh, I., Mishra, B.K.: Tensile behavior of glass fiber reinforced plastics subjected to different environmental conditions (2010)
66. Ellyin, F., Maser, R.: Environmental effects on the mechanical properties of glass-fiber epoxy composite tubular specimens. Compos. Sci. Technol. **64**(12), 1863–1874 (2004). https://doi.org/10.1016/j.compscitech.2004.01.017
67. Abbasi, A., Hogg, P.J.: Temperature and environmental effects on glass fibre rebar: modulus, strength and interfacial bond strength with concrete. Compos. B Eng. **36**(5), 394–404 (2005). https://doi.org/10.1016/j.compositesb.2005.01.006
68. Visco, A.M., Calabrese, L., Cianciafara, P.: Modification of polyester resin based composites induced by seawater absorption. Compos. Part A Appl. Sci. Manuf. **39**(5), 805–814 (2008). https://doi.org/10.1016/j.compositesa.2008.01.008
69. El-Tayeb, N.S.M., Yousif, B.F., Yap, T.C.: An investigation on worn surfaces of chopped glass fibre reinforced polyester through SEM observations. Tribol. Int. **41**(5), 331–340 (2008). https://doi.org/10.1016/j.triboint.2007.07.007
70. El-Tayeb, N.S.M., Yousif, B.F.: Evaluation of glass fiber reinforced polyester composite for multi-pass abrasive wear applications. Wear **262**(9–10), 1140–1151 (2007). https://doi.org/10.1016/j.wear.2006.11.015
71. Quintelier, J., De Baets, P., Samyn, P., Van Hemelrijck, D.: On the SEM features of glass-polyester composite system subjected to dry sliding wear. Wear **261**(7–8), 703–714 (2006). https://doi.org/10.1016/j.wear.2006.01.006
72. Mathew, M.T., et al.: Tribological properties of the directionally oriented warp knit GFRP composites. Wear **263**(7–12), 930–938 (2007). https://doi.org/10.1016/j.wear.2006.12.001
73. Sampathkumaran, P., Seetharamu, S., Vynatheya, S., Murali, A., Kumar, R.: SEM observations of the effects of velocity and load on the sliding wear characteristics of glass fabric-epoxy composites with different fillers. www.elsevier.comrlocaterwear
74. Chand, N., Naik, A., Neogi, S.: Three-body abrasive wear of short glass fibre polyester composite. www.elsevier.comrlocaterwear
75. Sampathkumaran, P., Seetharamu, S., Murali, A., Kumar, R.: On the SEM features of glass-epoxy composite system subjected to dry sliding wear (2001)
76. Suresha, B., Chandramohan, G., Rao, S., Sampathkumaran, P., Seetharamu, S., Venkateswarlu, V.: Friction and slide wear characteristics of glass-epoxy and glass-epoxy filled with SiCp composites (2006)
77. Pıhtılı, H.H., Tosun, N.: Investigation of the wear behaviour of a glass-fibre-reinforced composite and plain polyester resin. www.elsevier.com/locate/compscitech
78. Chauhan, S.R., Kumar, A., Singh, I.: Study on friction and sliding wear behavior of woven S-glass fiber reinforced vinylester composites manufactured with different comonomers. J. Mater. Sci. **44**(23), 6338–6347 (2009). https://doi.org/10.1007/s10853-009-3873-1
79. Mehdi, H., Kumar, A., Mahmood, A., Saini, M.: Experimental Analysis of Mechanical Properties of Composite Material Reinforced by Aluminium-Synthetic Fibers (2014). http://www.ijmejournal.com
80. Bulut, M., Erkliğ, A., Yeter, E.: Experimental investigation on influence of Kevlar fiber hybridization on tensile and damping response of Kevlar/glass/epoxy resin composite laminates. J. Compos. Mater. **50**(14), 1875–1886 (2016). https://doi.org/10.1177/0021998315597552
81. Choudhury, M.R., Debnath, K.: Experimental analysis of tensile and compressive failure load in single-lap bolted joint of green composites. Compos. Struct. **225** (2019). https://doi.org/10.1016/j.compstruct.2019.111180
82. Zou, Y., Huckelbridge, A.: Experimental Analysis of Crack Growth in GFRP Reinforced Concrete. https://doi.org/10.1061/ASCE1084-070220071 2:2246
83. Prasad, T.S.: Characterstics of Glass Fiber Reinforced Epoxy Composites for Automotive Engine Hood
84. Thamizh Selvan, R., Vishakh Raja, P.C., Mangal, P., Mohan, N., Bhowmik, S.: Recycling technology of epoxy glass fiber and epoxy carbon fiber composites used in aerospace vehicles. J. Compos. Mater. **55**(23), 3281–3292 (2021). https://doi.org/10.1177/00219983211011532

85. Keller, T., Bai, Y., Vallée, T.: Long-Term Performance of a Glass Fiber-Reinforced Polymer Truss Bridge. https://doi.org/10.1061/ASCE1090-0268200711:199
86. Subie, A., Mouritz, A., Troynikov, O.: Sustainable design and environmental impact of materials in sports products. Leisure/Loisir **2**(3–4), 67–79. Routledge (2009). https://doi.org/10.1080/19346182.2009.9648504
87. Gellert, E.P., Turley, D.M.: Seawater immersion ageing of glass-fibre reinforced polymer laminates for marine applications (1999)
88. Rani, M., Choudhary, P., Krishnan, V., Zafar, S.: A review on recycling and reuse methods for carbon fiber/glass fiber composites waste from wind turbine blades. Compos. Part B Eng. **215**. Elsevier Ltd. (2021). https://doi.org/10.1016/j.compositesb.2021.108768
89. Lapa, A., Cresswell, M., Jackson, P., Boccaccini, A.R.: Phosphate glass fibres with therapeutic ions release capability–a review. Adv. Appl. Ceram. **119**(1), 1–14 (2020). Taylor and Francis Ltd. https://doi.org/10.1080/17436753.2018.1564413
90. Choudary, M.V., Nagaraja, A., Sai, K.O.C., Balasubramanian, M.: Characterization of laminate sandwiched with stainless steel and glass fibre. Mater. Today Proc. 847–852. Elsevier Ltd. (2020). https://doi.org/10.1016/j.matpr.2019.11.030
91. Li, M., et al.: Recent advancements of plant-based natural fiber–reinforced composites and their applications. Compos. Part B Eng. **200**. Elsevier Ltd. (2020). https://doi.org/10.1016/j.compositesb.2020.108254
92. Mochane, M.J. et al.: Recent progress on natural fiber hybrid composites for advanced applications: a review. Express Polym. Lett. **13**(2), BME-PT and GTE, 159–198 (2019). https://doi.org/10.3144/expresspolymlett.2019.15
93. Fotouhi, S., Jalalvand, M., Wisnom, M.R., Fotouhi, M.: Smart hybrid composite sensor technology to enhance the detection of low energy impact damage in composite structures. Compos. Part A Appl. Sci. Manuf. **172** (2023). https://doi.org/10.1016/j.compositesa.2023.107595
94. Maiti, S., Islam, M.R., Uddin, M.A., Afroj, S., Eichhorn, S.J., Karim, N.: Sustainable fiber-reinforced composites: a review. Adv. Sustain. Syst. **6**(11) (2022). John Wiley and Sons Inc. https://doi.org/10.1002/adsu.202200258
95. Minhaj Pirzada, M.: Recent trends and modifications in glass fibre composites-a review. Int. J. Mater. Chem. **5**(5), 117–122 (2015). https://doi.org/10.5923/j.ijmc.20150505.03

Chapter 7
Hemp Fibre-Reinforced Polylactic Acid Composites: A Sustainable Materials for Engineering and Industry

L. Girisha, R. Saravanan, T. Kumarasan, P. Pitchandi, S. Sakthi, S. Nanthakumar, and R. Girimurugan

Abstract In recent years, composites made from hemp fibre (HF) reinforced polylactic acid (PLA) have seen rapid growth between the Natural Fibre Reinforced Composites (NFRCs). The engineering and industrial sectors have taken a keen interest in HF-PLA composites as a sustainable alternative material. The outstanding qualities of HF-PLA composites, such as their low specific weight, better mechanical capabilities, and biodegradability are increasing their popularity. Creep behaviour, thermal, and mechanical properties of HF-PLA composites are briefly examined. The biodegradability and recycle-ability of HF-PLA composites made using various

L. Girisha
Department of Mechanical Engineering, PES Institute of Technology and Management, Shivamogga, Karnataka 577 204, India

R. Saravanan
Department of Mechanical Engineering, J J College of Engineering and Technology, Trichy, Tamilnadu 620 009, India

T. Kumarasan
Department of Aeronautical Engineering, J J College of Engineering and Technology, Trichy, Tamilnadu 620 009, India

P. Pitchandi
Department of Aeronautical Engineering, PSN College of Engineering and Technology, Tirunelveli, Tamilnadu 627 152, India

S. Sakthi
Department of Mechanical Engineering, Arifa Institute of Technology, Esanoor, Nagapattinam, Tamilnadu 611103, India

S. Nanthakumar
Department of Mechanical Engineering, PSG Institute of Technology and Applied Research, Coimbatore, Tamilnadu 641 062, India

R. Girimurugan (✉)
Department of Mechanical Engineering, Nandha College of Technology, Perundurai, Tamilnadu 638 052, India
e-mail: dr.r.girimurugan@gmail.com

S. Moharana et al. (eds.), *Polymer Composites*, Engineering Materials,
https://doi.org/10.1007/978-981-97-2075-0_7

manufacturing techniques have also been investigated. Potential usage in additive manufacturing and sustainable packaging is also mentioned, along with structural, dielectric, and automotive applications for HF-PLA composites in the future. Mechanical and physical characteristics of HF-PLA-produced components have also been summarized, along with the impacts of 3D printing sceneries. This work sheds information on the potential and constraints of HF-PLA composites, which is an important contribution. Reinforcing natural fibres in a polymer matrix enhanced mechanical, thermal, and tribological behaviour by increasing interfacial adhesion between the fibre and the matrix. The composites' fibre-matrix interface stress transmission performance was improved by chemical treatment of the natural fibres. To expand their usefulness to high-end sectors like the aerospace and marine industries, improvements must be made to the thermal characteristics and moisture response.

Keywords Hemp fibre · Ploylactic acid · Mechanical properties · Tribological properties · Biodegradability · Creep behaviour · Sustainability

7.1 Introduction

Plastic is used extensively because of its wonderful qualities, such as its low cost, durability, malleability, resistance to corrosion, and good thermal and electrical insulation [1]. Most plastics are made from petroleum and chemicals that are known or suspected to be harmful [2]. Plastic pollution or garbage is another concern since it can pose a risk to human health if it enters the food supply. It takes a very long time for plastics to disintegrate because they are mostly made of non-biodegradable materials [3]. In order to solve the world's waste management and pollution issues and kick off a more sustainable civilization, eco-friendly polymers that break down in nature are desperately needed [4]. The difficulty lies not only in finding suitable replacements, but also in developing and supplying materials with the required mechanical, physical, and thermal performance qualities at the lowest possible production cost [5–7]. One eco-friendly option to meet the strict requirements is to use natural fibres as reinforcement. Natural fibres are gradually replacing synthetic ones due to their numerous benefits. There are two main types of natural fibres, derived from plants or animals [8].

Natural fibres can be derived from either plants or animals; examples of the former include the aforementioned plant-based materials and the animal-derived silk, wool, etc. The several types of composites are shown in Fig. 7.1 [9]. It is possible to substitute synthetic fibres with natural fibres [10–12]. Natural fibres have better specific mechanical qualities than synthetic fibres, less cost to produce, are more user-friendly, and need only around 20–40% of the manufacturing energy [13]. Towards the goal of sustainability, the use of natural materials and cutting-edge production processes reduces building waste and optimizes energy efficiency [14, 15].

Renewable resources including starch, sugar, vegetable oils, and soy oils are used to create green composites that are then reinforced with natural fibres [16, 17]. PLA is

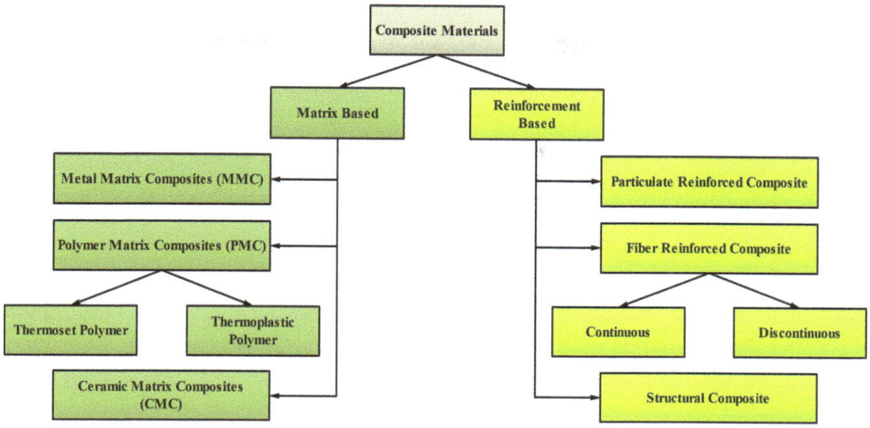

Fig. 7.1 Composites are categorized according on the matrix and reinforcing materials used [9]. Adapted with permission from Ref. [9] Copyright @ 2023, Wiley

the most common form of green composite. When compared to alternative reinforcements, including vetiver grass fibre or coconut fibre, HF was proven to be the most effective [18]. The thermal and mechanical properties of hemp fibre and eco-friendly composites based on hemp fibre and polylactic acid resin have been the subject of several investigations. [19]. The impact of cooling rate and hot-pressing heat on thermal and physico-mechanical characteristics, as well as the impact of polyethylene glycol on mechanical characteristics in hemp fibre-polylactic acid composites, are only some of the phenomena that have been studied [20]. To improve the hemp fibre-PLA composites' mechanical and thermal characteristics, a wide variety of surface modification treatments were applied. This included alkali (NaOH) and an alkali-silane coupling agent. Properties of HF-PLA composites were also investigated while being subjected to a mild NaOH solution as a pretreatment [21, 22].

In addition, a variety of HF including hemp charcoal particles in nonwoven membranes and biodegradable polymer reinforcement have been utilized to increase the latter's impact resistance [23–25]. Researchers have investigated the effects of nano-SiO_2 and MA on HF-PLA composites, and PLA-grafted-HF PLA composites. BCNW PLA bio-composites' mechanical, thermal, and morphological characteristics regenerated hemp chopstick fibre-reinforced PLA green composites' fire resistance [26]. Organo-montmorillonite-modified fibres and polylactic acid were used to assess six WPCs [27]. The effects of electron beam irradiation on the physical and chemical characteristics of hemp powder (BP) and BP-PLA composites were investigated [28]. Mechanical characteristics of composites derived solely from renewable resources were analysed [29]. Composites made from PLA were analysed to investigate how the addition of natural fibres and an artificial cellulose fibre affected their qualities. The elements and obstacles in creating and defining PLA-based green composites were discussed [30]. In the evaluation of HF-reinforced composites, focused on the production and resulting attributes of HPs with polymer matrices tailored to specific uses. Researchers discussed the several ways in which HF may be modified on a structural, chemical, and physical level, as well as the effect this interface has the final composite's performance. A recent literature [31] looked at how different factors in the production process impact the final behaviour and qualities of polylactic acid (PLA) and PLA composites reinforced with natural fibres. Natural fibres with a variety of desirable features were also highlighted as an alternative to synthetic ones, and it was established that composites made from these materials are comparable to their synthetic counterparts. Mechanically and chemically extracted hemp fibre's fabrication techniques, uses and types of composites were discussed [32]. Based on the aforementioned works, it is safe to assume that only rudimentary studies of HF-PLA composites' thermo-mechanical characteristics, sustainability factors, and potential uses have been done [33]. The primary goal of this research was to establish a standard against which subsequent studies could be measured, as well as to pave the way for their eventual industrialization and eventual commercialization. To further improve environmental sustainability, the biodegradability, recycling, and limited use of HFPLA composites are also considered. The role that 3D printing parameters have in the fabrication of HF-PLA composites is also discussed.

7.2 PLA Hemp Fibre Composites

The use of natural fibre-reinforced composites (NFRCs) in the business sector has increased significantly in recent years. One of the most popular types of NFRCs is HF-PLA composites [34, 35]. In addition to cost-effective, HF-PLA composites also have a low environmental impact. Engineers and manufacturers may benefit from HF-PLA composites because of its lightweight, durability, and high thermal and mechanical resistance [36–38].

7.3 Polylactic Acid (PLA)

The biodegradable thermoplastic polymer polylactic acid (PLA) is also known as polylactide. The manufacture, utilization, and recycling of PLA are depicted in Fig. 7.2. When compared to the manufacture of conventional polymers derived from petroleum, PLA uses less water, less energy, and produces less greenhouse gas emissions [39–41]. Compostability is only one of the many benefits of PLA, which also has a favourable melting temperature of 160–180 °C. Physical and mechanical properties of PLA as well as its biocompatibility and low immunogenicity, make it an excellent engineering material [42, 43].

The medical, tissue engineering, additive manufacturing/3D printing, sustainable packaging, textile/structural/automotive/ drug delivery/technical/surgical masks

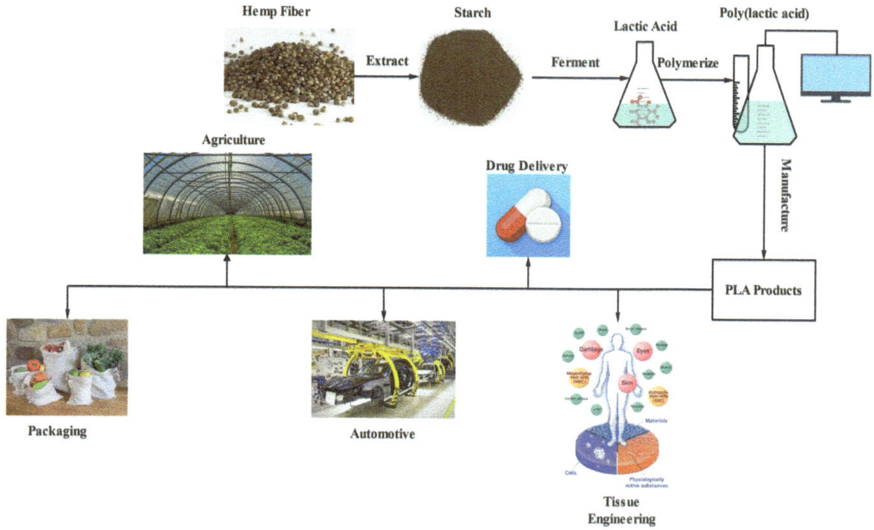

Fig. 7.2 The manufacture, recyclable nature, and potential uses of PLA [44]

industries are just a few examples of the many possible applications for PLA. Figure 7.3 displays few of the most common uses of PLA [44].

In addition, PLA produced using extrusion, compression moulding, injection moulding, and thermoforming methods and is a good polymer matrix for the reinforcements with natural fibres [45–47]. Despite its promising properties and wide range of possible uses, PLA is limited by a number of technical constraints, including as its flimsy structure, slow breakdown under natural conditions, limited cell contact, and excessive hydrophobicity [48, 49].

7.4 Merits from Hemp Fibre (HF)

Hemp is a massive grass that grows almost everywhere save Europe and Antarctica. China has the greatest hemp resource base in the world, with 2.8% of its total forest acreage devoted to hemp growing (over 4,210,000 hectometres). For comparison, steel has a tensile strength of 391 MPa, whereas hemp's tensile strength (400–481 MPa) may reach up to 1000 MPa. In contrast to conventional Western building materials like concrete, steel, and timber, hemp culm has a 20-fold greater lifespan [28, 29, 50]. In places where hemp is more cost-effective than steel, such as in this case, utilising hemp instead of steel for reinforcement in building materials might reduce the overall cost of housing projects. Hemp be used as a substitute for WPCs, but it could also increase the commercial use of hemp and lead to the development of cleaner, greener, and more sustainable goods. Hemp is considered environmentally benign since it is derived from natural, renewable resources [51].

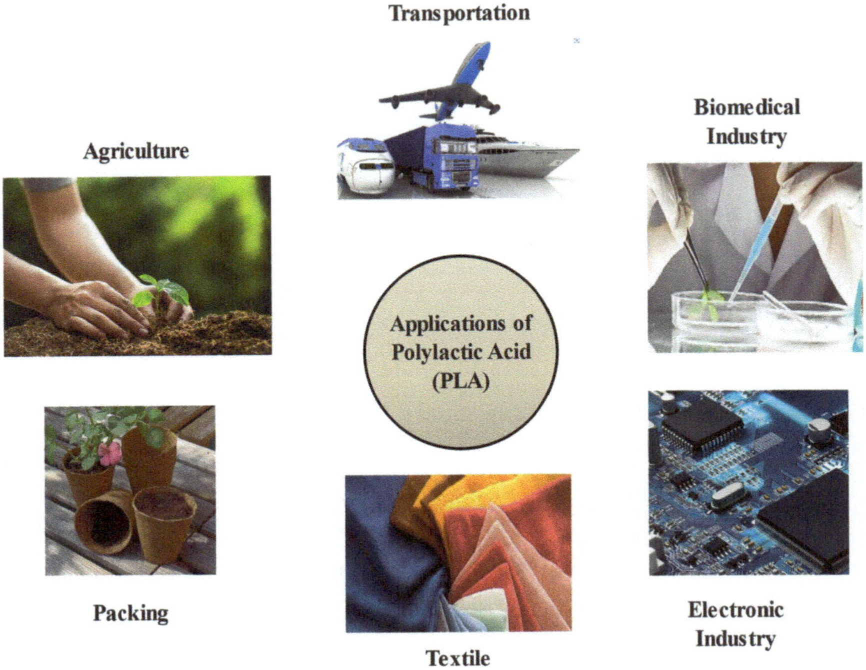

Fig. 7.3 A variety of PLA uses [44]

In addition, the carbon storage capacity of a hemp forest is promising, especially when the harvested hemp culms are transformed into useful goods. Because hemp products last so long, the trapped carbon will not be released to the environment as soon. Therefore, it plays a crucial part in carbon sequestration and has a tremendous impact on reducing the negative consequences of climate change [52]. Hemp fibres provide a barrier against the spread of bacteria. Hemp materials are hypoallergenic and do not cause skin irritation [53]. The main components of hemp are cellulose (73.83%), hemicellulose (12.49%), lignin (10.15%), aqueous extract (3.1%), and pectin (0.37%). The process of hemp fibre extraction is depicted in Fig. 7.4. HFs are beneficial since they are cheap, eco-friendly, long-lasting, recyclable, and biodegradable. However, HF has a high water absorbency, which has aided in the creation of biodegradable, inexpensive, environmentally friendly, recyclable, and sustainable materials [54]. Table 7.1 displays the carbon-savings percentages of many different bio-composites.

Due to its many useful properties, including ease of extraction, strength, thermal properties, structural malleability, and amenability to chemical modification, hemp fibre may find usage in the composites industry. However, previous research using mechanical testing indicated that HFs are capable of creating the mechanical product [55]. Composites made with it might also serve as a replacement for carbon and glass. According to the US Department of Energy's analysis, just 17% of the energy

Fig. 7.4 Extraction of hemp fibres [57]

Table 7.1 Bio-composites and the relative carbon dioxide reduction

Type of Bio-composite	Applications	Carbon Savings (in CO_2 equivalent)
Wood-plastic composite	Decking	Up to 50% reduction compared to pure plastic decking
Hemp fibre-reinforced PLA	Automotive parts	Approximately 20–30% reduction compared to traditional plastic components
Bamboo fibre-reinforced concrete	Construction materials	15–20% reduction compared to conventional concrete
Flax fibre-reinforced epoxy	Aerospace components	Around 10–15% reduction in carbon emissions compared to traditional aerospace materials
Rice husk and polypropylene	Packaging	Up to 40% reduction in carbon emissions compared to conventional plastic packaging

required to generate glass fibre is used in the production of an HF mat [56]. Hemp grows quickly, needs little water, and can be grown without the use of pesticides or herbicides, making it one of the most desirable crops. Hemp fibre has a relatively large diameter in relation to its length, and its surface is uniformly smooth and spherical. Natural fibres have several advantages over man-made alternatives like metal and

Table 7.2 Evaluation among natural fibres, glass, and metal

Property	Natural Fibres	Glass Fibres	Metals
Composition	Plant or animal sources	Silica and additives	Elements with metallic bonds
Strength	Lower	High	High
Weight	Lightweight	Lightweight	Heavier
Density	Low	Moderate	High
Flexibility	Flexible, elastic	Stiff	Ductile and malleable
Thermal conductivity	Low	Low	High
Electrical conductivity	Non-conductive	Non-conductive	Conductive (varies)
Corrosion resistance	Generally resistant	Resistant to most chemicals	Susceptible, varies
Magnetic properties	Non-magnetic	Non-magnetic	Magnetic (varies)
Environmental impact	Biodegradable, renewable	Non-biodegradable, energy-intensive	Environmental impact varies
Cost	Often cost-effective	Moderately expensive	Costs vary
Applications	Clothing, textiles, paper, some composites	Composites, insulation, electrical applications	Construction, transportation, machinery, electronics

glass. An evaluation of metal, glass fibres, and natural fibres is shown in Table 7.2 [57].

7.5 Production of HF-PLA Composites

Compression moulding, hot-press moulding, extrusion using a solution mixing, injection moulding, solution casting and twin-screw extruder are just few of the many producing methods used in the creation of HF-polylactic acid composites. Figure 7.5 depicts the methods used in the manufacture of HF-PLA composites.

7.6 Properties of Hemp Fibre-Polylactic Acid Composites

7.6.1 Thermal Properties of Hemp Fibre-Polylactic Acid Composites

The thermal behaviours of composites are also important for their practical applications [59]. Different kinds of tests are used to determine thermal properties. When PLA-g-glycidyl methacrylate (GMA) was added to HF-PLA composites, the thermal

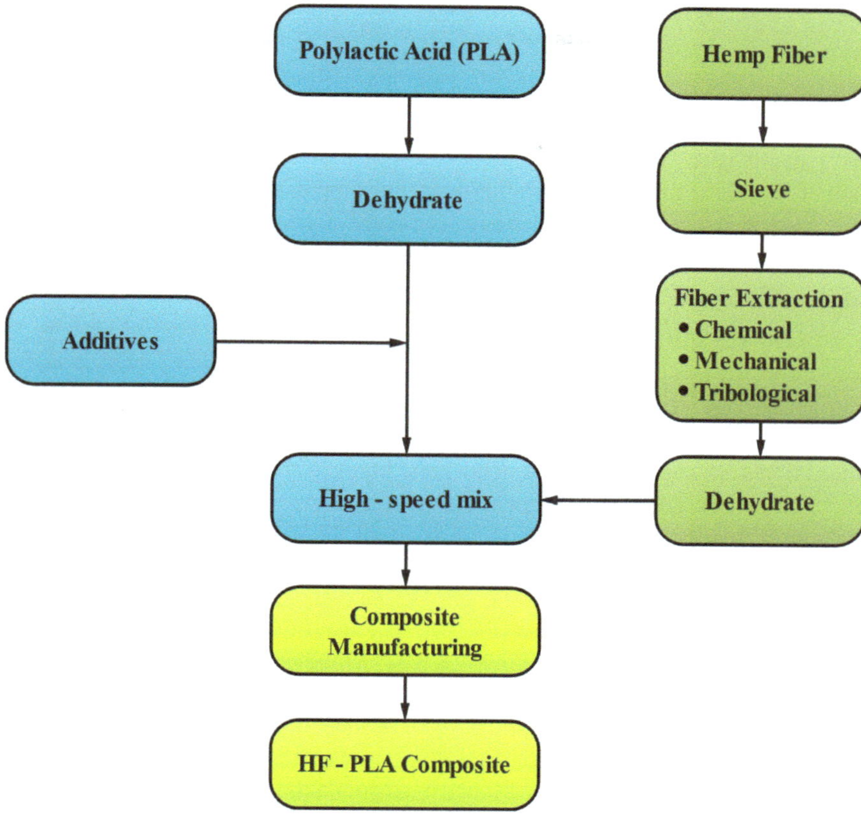

Fig. 7.5 Fabrication of hemp fibre-polylactic acid composites [44]

breakdown temperature increased. Titanate coupling agent-treated HF-PLA composites outperformed those treated with sodium hydroxide in terms of thermal stability. Since cPLA2 had a too-low Tg (27.1–27.6 °C), a thermal test was carried out to illustrate the variation in glass transition temperature between the different types of polylactic acid plasticizer, and the results showed that cPLA1 was more relevant for 3D printing [60–62]. Compared to PLA, HF-PLA composites show 16.74–28% and 6% decrease in thermal stability after being treated with NaOH [24]. Overall, the thermal stability of the HF-PLA composites was enhanced by the graft modification treatment, with crystallinity increasing by 5.14% compared to the unmodified version [63]. NaOH-treated HF acted as nucleating agents and disrupted polymer crystallization, as shown by differential scanning calorimetry (DSC) and X-ray diffraction (XRD). NaOH-treated HFs have better thermal stability, according to TGA [64, 65]. Experimental results using injection-moulded and extruded HF-reinforced composites made from starch-polypropylene (PP) showed that even a little amount of HF may significantly increase the composites' thermal stability [66]. Xc, which stands for crystallinity was calculated using the following formula (7.1)

$$X_c = \frac{\Delta H_f}{\Delta H_0} \times 100\% \tag{7.1}$$

where,

ΔHf fusion heat
ΔH0 fusion heat for 100% crystallinity of the composites.

7.6.2 *Mechanical Behaviour of Hemp Fibre-Polylactic Acid Composites*

Natural fibres have gained popularity as a reinforcing material because they are inexpensive, less weight, and have excellent mechanical qualities. Mechanical qualities of natural fibres are superior to those of artificial fibres [67]. However, impact strength, tensile properties, and flexural properties all be used to get a sense of the material's mechanical qualities. The following are some of the conditions upon which these qualities depend [68]. In the realm of composite materials utilizing natural plant fibres like hemp, flax, or jute, several key factors emerge as crucial determinants of the resulting material's properties [69, 70]. The identification of the specific plant fibre used, whether it be hemp, flax, or jute, plays a fundamental role in understanding the composite's characteristics and performance. Secondly, the alignment of these fibres within the composite is a critical factor influencing tensile strength. Improved alignment leads to stronger reinforced composites. Another pivotal aspect is the choice of matrix material, which can vary from polylactic acid (PLA) to acrylonitrile butadiene styrene (ABS) to polypropylene (PP), among others. The matrix material significantly impacts the overall properties of the composite [71–73]. Additionally, if fillers are incorporated, their specific features and concentrations are essential considerations. Fibre content within the composites, expressed as a percentage, is a key factor affecting mechanical strength. For instance, in HF-PLA composites, mechanical characteristics were optimal at 20 wt % HF. Moreover, the geometry of the fibres, whether in the form of cylinders, spheres, prisms, or platelets with a rectangular cross-section, plays a role in shaping composite properties [74–76].

Further enhancing the compatibility between hemp fibres (HF) and the PLA matrix is essential for maximizing composite performance. Techniques such as the addition of coupling agents like KH during fibre separation, extraction, and treatment processes significantly impact the mechanical characteristics of HF-PLA composites [61]. Especially, NaOH and KH treatments yield similar enhancements in tensile modulus and strength, while graft-modified HF further enhances the interfacial harmony between HF and PLA, resulting in improved tensile and flexural properties. Manufacturing techniques are also instrumental in shaping the final composite's properties. The precise order and mixing procedures of fibres, matrix materials, and additives are essential to prevent processing issues and achieve efficient and effective composite production [75]. This amalgamation of factors illustrates the intricate interplay in developing high-performance plant fibre-reinforced composites, as

Table 7.3 Mechanical properties of hemp fibre [57]

Specimen	Gauge length (mm)	Tensile modulus (GPa)	Tensile strength (MPa)	Max. strain (mm/mm)
1	25.4	54.467	941.245	1.728
2	20.1	46.726	309.371	0.662
3		60.135	777.002	1.292
4		72.438	886.711	1.224
5		59.386	606.389	1.021
6	50.8	51.925	634.057	1.221
7		61.344	647.245	1.055
8		59.391	710.371	1.196
9		54.751	722.217	1.319
10	76.2	47.76	532.093	1.114
11	70.2	77.397	883.951	1.142
12		57.238	375.539	0.656
13		64.734	302.372	0.467
Mean		59.053	640.659	1.084
Standard deviation		8.448	205.26	0.318

shown in Table 7.3, which provides mechanical performance comparisons for various composites incorporating hemp fibre.

7.6.3 Tensile Properties of HF-PLA Composites

HF-PLA composites treated with NaOH solution and KH550 had tensile strengths that were twice as high as those of HF-PLA composites containing untreated HF [76]. NaOH treatment also increased the tensile modulus of the composites. The tensile characteristics of HF-PLA composites were enhanced when HF reinforcement was woven in the weft direction. After being treated with NaOH for 3 h, the BPPLA composites reached their maximum range of 411.36 MPa, 45.36 MPa, and 7.56% elongation at break. When natural fillers were added to PLA, its tensile strength dropped by 55.6%. Mechanical strength increased from 34.6 MPa to 54 MPa when a coupling agent was employed. The ideal conditions for moulding, which provide a high tensile strength-to-weight ratio, were determined to be 4 min of moulding time, 1 MPa of pressure, and 190 °C [77]. With only 1% MAH, the BP-PLA composite's tensile strength increased to 48.1 MPa and its elongation at break increased to 7.13%. Sodium hydroxide-treated HF has tensile strength comparable to other tough natural fibres like flax and hemp.

The tensile strength of the kenaf-hemp-coir-PLA was around 20% greater than that of the hemp-coir-PLA and 78% higher than that of the kenaf-coir-PLA, despite the fact that all three composites had low Young's moduli. The HF-PLA composite material's tensile strength and tensile modulus increased by 135% and 44% after being treated with 15 phr polylactic acid-g-glycidyl methacrylate (PLA-g-GMA). When compared to an E-glass/epoxy composite, the HFPLA composite had a Young's modulus that was twice as high [78, 79]. The tensile properties of HF-polylactic acid/HF-g-LA composites increased by 30% with an HF-g-LA concentration of 2 phr. When compared to untreated HFs, HFs that were chemically treated with NaOH exhibited considerably greater tensile strength and elongation at break [80]. When maleated polypropylene (MAPP) was added to HF-PP-PLA composites at a concentration of 5%, the tensile strength raised by 13%. The resulting pressure was found to be 33.73 MPa. The modulus increased from 91 to 230% when PLA filament was reinforced by short HFs (l/d = 45), whereas the dust-like fractions increased by only 39% [81]. When applied to HF, a graft alteration increased both tensile strength and elongation at break by 30.1% and 19.3%, respectively. Tensile studies on HFPLA composites demonstrated that reinforcement with 20% recycled HF led to mechanical or adhesive failure due to poor fibre-matrix interaction. The optimal percentage of BSSFs in ternary composites including starch and PLA resulted in a tensile strength of 33.1 MPa [82]. Figure 7.6 depicts the stress-strain relationship for hemp/PLA composites under tensile testing. Table 7.4 details the mechanical properties of hemp/PLA composites.

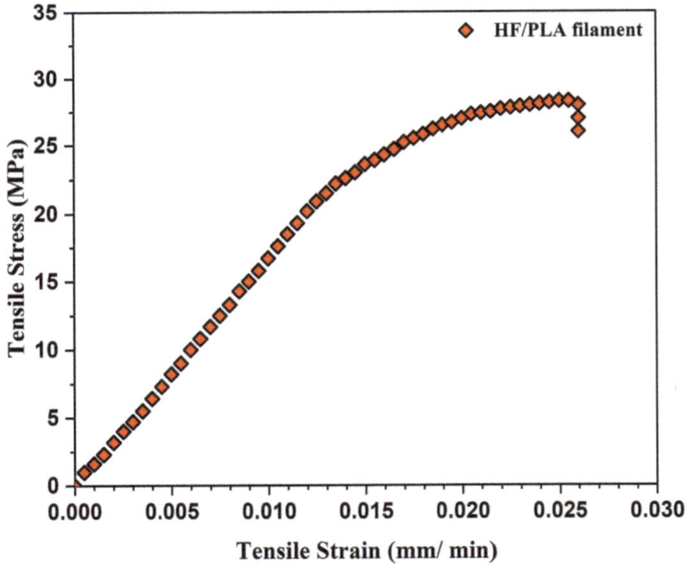

Fig. 7.6 Tensile stress vs strain curvature of hemp/PLA composites [83]. Adapted with permission from Ref. [83] Copyright @ 2020, Springer Nature

Table 7.4 Mechanical characteristics of Hemp/PLA composites [83]. Adapted with permission from Ref. [83] Copyright @ 2020, Springer Nature

Tensile modulus (MPa)	Tensile strength (MPa)
1542.53 ± 23	29 ± 4.1

A visual assessment of how high-fibre-volume-percentage composites outperform those with a lower fraction of fibres in each test is provided in Fig. 7.7. The increase in tensile modulus may be traced back to the fibres' function as load bearers, as increased the stiffness of the composite. In addition, an increase in stiffness will reduce the mobility of the matrix, that is, the PP molecular chains. The Young's modulus of natural fibres is affected by a number of variables. These include fibre orientation, micro fibril angle, cellulose concentration, crystallinity level, etc. The material's resistance to force, measured as tensile strength, was found to rise in direct proportion to the fibre content. But if the fibre volume percentage is too high, the composite will have less tensile strength. [84]

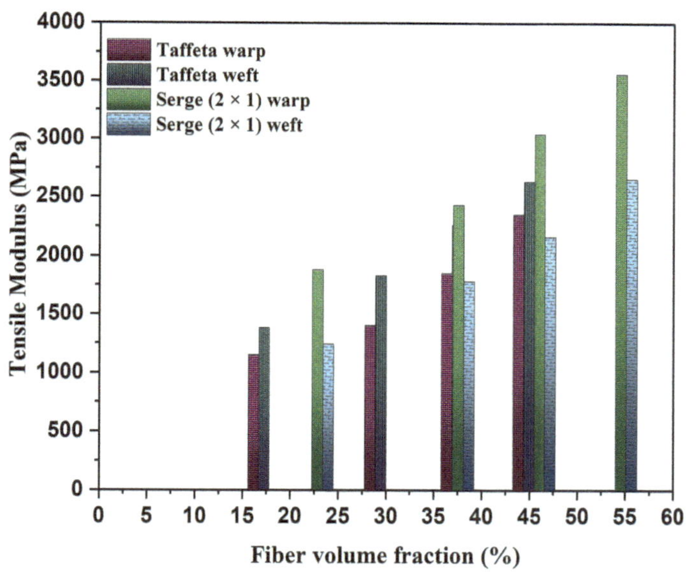

Fig. 7.7 Evaluation of tensile strength for hemp fibre and polypropylene composites with variant percentage volume of fibre [84]

7.6.4 Flexural Properties of HF-PLA Composites

Chemically treated HF-PLA composites have flexural characteristics that are very sensitive to HF concentration and treatment method. For instance, as HF concentrations rose, flexural moduli also raised. NaOH treatment enhanced the flexural moduli, and both NaOH and silane coupling agents boosted flexural strength [85]. BP-PLA composites reached their maximum flexural modulus of 4.50 GPa and flexural strength of 83.85 MPa after being treated with NaOH for 3 h. With a fibre percentage of 70% and a moulding temperature of 160 °C, biodegradable HF-PLA composites showed a significantly increased flexural strength of 273 MPa [86]. Flexural characteristics improved up to an HF level of 40%, but thereafter declined. Flexural strength raised by 3.52–48.46% and modulus by 11.46–11.59% when weft direction HF reinforcement was applied in PLA [87]. Maximum flexural modulus was 4.72 GPa and flexural strength was 72.61 MPa with 0.5% MAH, however, both values decreased with increasing MAH concentration. By increasing the natural fillers, the flexural strength decreased by 24.5–60% compared to pure PLA, but the mechanical strength increased by 70% with the adding of join agent MDI [88]. Hemp-coir-PLA (206 MPa) and kenaf-hemp-coir-PLA (199 MPa) composites had a 20% and 16% increase in flexural strength, respectively, above kenaf-coir-PLA [89]. The flexural properties of hemp fibre-polypropylene-polylactic acid composites were enhanced by 11.7% when MAPP was added at a concentration of 5%, reaching a value of 47.18 MPa. HF's flexural strength in PLA-based composites was enhanced by 19.3% after undergoing graft modification. Figure 7.8 shows the flexural stress-strain curve of hemp/PLA 3D printed structure.

Core bending begins at u = 1.5 mm in the flexural test, and skin detachment from the core begins at u = 3 mm. As the experiment continues, more and more of the structure's skin peels away, eventually leaving only the skeleton. As a result, the construction has excellent compression properties but poor flexural properties.

7.6.5 Impact Strength of Hemp Fibre-Polylactic Acid Composites

Impact strength (maximum 23.8% for HF) decreased with increasing natural fibre content in PLA-based composites but increased by 40% compared to clean PLA after being treated with a flexible epoxy. When compared to hemp grass fibre and coconut fibre, HF has been demonstrated to be the superior reinforcing material [90]. The Izod impact energy of hemp fibre-polylactic acid composites were calculated to be 6.530.23 kJ/m^2 with 50% mass of micro-fibre hemp fibre, and it was shown to decrease with further increase in micro-fibril hemp fibre content [58]. The impact characteristics of PLA were improved by 57% when HF reinforcement was woven in along the weft direction. We laminated HF-PLA with Cliosite 30B nano-clay using the film stacking process, resulting in a hybrid composite with an impact strength that

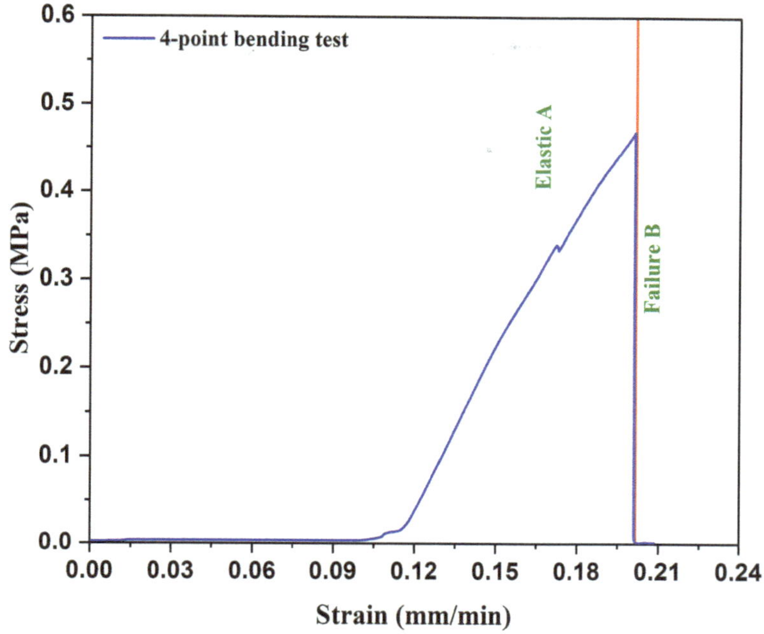

Fig. 7.8 Evaluation of flexural test between force and displacement curve of hemp/PLA 3D printed structure [83]. Adapted with permission from ref. [83] Copyright @ 2020, Springer Nature

is 26.58% more than that of virgin HF-PLA composites [91]. Compared to pure PLA, the HF-PLA composite showed a 240% improvement in impact strength along the warp direction. Adding 2 phr of HF-g-LA content to HF-PLA-HF-g-LA composites raised the impact strength to 9.56 kJ/m^2 [92]. Composites made from HF-PLA have a Charpy impact strength of 126 J/m, much greater than that of pure polymers. The impact strength of HF-PP-PLA composites was 3.15 kJ/m^2, a 23.5% improvement over the impact strength of composites without MAPP.

7.6.6 Creep Behaviour

The term "creep behaviour" is used to describe the deformation of a polymer when subjected to external forces, most notably when the temperature increases. Previous research looked on the mechanical characteristics and creep resistance of HF-rPLA composites. Using a compression moulding process, researchers showed that HF-rPLA composites with 60 wt % of fibre had the maximum creep resistance, whereas composites with more than 70 wt % of fibre had a lower creep resistance. Over the course of 30 years, the modulus of all HF-rPolylactic acid composites examined reduced by somewhere between 27 and 40%, according to the same research [93,

94]. However, the time scale creep curves and creep master curves for HF-rPLA composites at room temperature are shown in Fig. 7.9 a, b [58].

7.6.7 Sustainable Aspects

Bio-based products or composites that meet the criteria of sustainability include being recyclable, exhibiting triggered biodegradation, being environmentally acceptable, and having marketable viability [95]. The concept of sustainability is depicted in Fig. 7.10. The HF-PLA composite material used in packaging is safe for the planet and may be recycled indefinitely. Because of its outstanding structural and mechanical capabilities and environmental benefits, hemp has a lot of room to grow as a raw material for composites that may be used in the long run [52].

7.6.8 Biodegradability of HF-PLA Composites

Biodegradability is the property of a material or composite to decompose into harmless components such as water and carbon dioxide when exposed to microorganisms such as algae, bacteria, and fungus [96, 97]. The biodegradability of composites can be tested in a variety of ways, including burial in soil and compost, immersion in river or seawater, and aerobic decomposition. When compared to the soil burial test, the rate of deterioration in aerobic testing under water and compost conditions is significantly higher. Deterioration rate of composites requires measuring their weight loss. Weight loss of the composites may be calculated using the following Eq. (7.2),

$$\text{Weight loss } (\%) = \frac{W_0 - W_1}{W_0} \times 100 \tag{7.2}$$

where

W0 sample weights before biodegradation test.

W1 sample weights after the biodegradation test.

Using HFs as reinforcements in biodegradable composites has shown encouraging results [98]. Using HF as reinforcement delayed the breakdown rate of HF-PLA composites, which was discovered by scientists studying the biodegradability of HFPLA composites for eco-friendly packaging. Reinforcing PLA with HF that has been treated with NaOH allowed the creation of biodegradable composites with practical uses in industry. There are several benefits to using HF-PP-PLA composites, including the fact that they are inexpensive and biodegradable [95]. Adding a specified amount of HF to starch-PP-HF composites improves both their biodegradability and thermal stability. Natural ageing of HF-polylactic acid composites (prepared by injection moulding process) is serious, as evidenced by a 78% and 69.6% reduction in impact strength and tensile strength, respectively, after 137 days of ageing. This

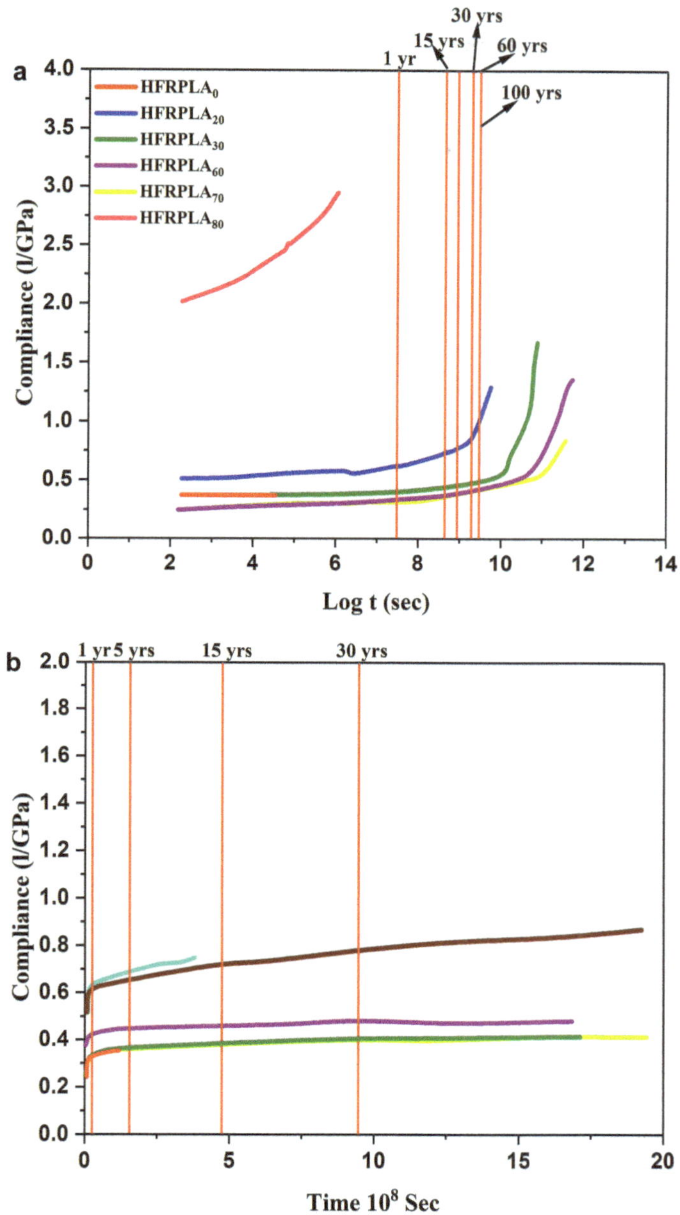

Fig. 7.9 **a** Master creep curves for a typical time period [58]. Adapted with permission from ref. [58] Copyright @ 2015, Elsevier. **b** Master creep curves at 20°C for a number of HF-rPLA composites [58] Adapted with permission from Ref. [58] Copyright @ 2015, Elsevier

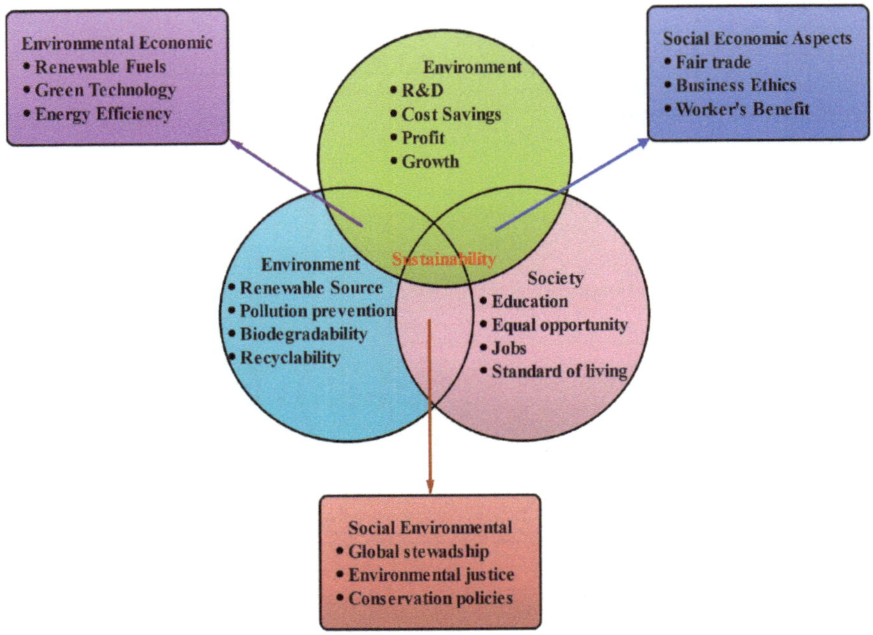

Fig. 7.10 The idea of eco-friendly composites [44]

emphasizes the importance of adopting firm practical techniques to further enhance the out-of-doors behaviours of the composites.

Composites' biodegradability is crucial since it is necessary to dispose of the product once its useful life is over or after it has been damaged. When characteristics like biodegradability and sustainability are weighed, NFCs clearly come out on top when compared to the vast majority of synthetic fibre-reinforced composites. Integrant materials, fibre fraction, interface, and environment all contribute to the rate of biodegradation. Biodegradation occurs in important degrading habitats such soil, water, landfill, and compost. These composites are fully biodegradable due to the use of plasticized (benzylated) fir sawdust and reinforcing (also biodegradable) sisal fibres in the matrix. Fibre volume content of the benzylated fir sawdust and unidirectional sisal laminate composites correlated with the amount of weight lost following enzymatic cellulase treatment [28]. The results showed that banana fiber-polypropylene, pineapple leaf-polypropylene, and hemp-polypropylene composites were all biodegradable, with maximum biodegradation levels of 5%, 10%, and 15%, respectively. Due to the high fibre volume content, degradation of hemp fibre composite was more pronounced, as the exposed fibres are more deteriorated than those completely covered in the matrix, which was unaffected in scanning electron microscopy pictures of the fibre-matrix interface [52].

Physical and thermal parameters (reduced weight and heat conductivity) of the hand-laid-up HF-reinforced polyester biodegradable composites suggest their possible application in the automotive, electrical, and construction sectors. Electrospinning was used to create a biodegradable double-layered filter membrane from natural fibres (banana, tamarind, sweet potato, and lufa silk squash), and the resulting water had a pH between 6.5 and 8.5 and a TDS concentration of 150 mg/L to 300 mg/L, meeting the standards for human consumption. When comparing the biodegradability of treated and untreated fibre composites with Aspergillus flavus (fungal enzyme), it was found that the treated composites had less biodegradability but better mechanical qualities. Low-density polyethylene and citrus tree trimmings were used to create these composites. Hardwood, softwood, kraft pulp fibres, and PLA were used in the production of totally biodegradable composites using the film stacking method (laminates) [34]. Mechanical characteristics of composites made from 50% kraft pulp fibre were found to be better. Recent research has shown that including cellulose derivatives improves the biodegradability and overall characteristics of the material. The lignocellulosic and starch-rich components added to improve the characteristics serve as filler and potential matrices, respectively, and affect the component's biodegradability to varying degrees [37].

Degradation conditions were applied to a mixture of sawdust and starch bound together in PP (a synthetic polymer matrix). Soil with moisture, climatic conditions, liquids (such water and salt solution 10%). This allowed the researchers to examine the effects of various factors, including microorganisms, hydrolysis, oxidation, chloride ions, sunlight, moisture, and humidity. The 15-day time frame is considered. The outcome demonstrates the composite's biodegradability in a variety of settings. Reduced tensile characteristics of the composites led to an increase in biodegradability, as shown by the results. The most corrosive chloride and microorganisms caused the greatest biodegradation when it was buried in the soil [24]. An alternate cellulose matrix is cellulose ester, which might be used. The research on the biodegradability of composites made from combinations of flax and ramie single fibre with cellulose acetate butyrate (CAB) and Cellulose Acetate Propionate (CAP) matrix [75]. Areca fibres soaked in a diluted NaOH solution and ground maize were the subject of another investigation, which involved subjecting the mixture to a hydraulic hot press at 120 °C and a pressure of 2 MPa. To test the component's biodegradability, a sample is submerged in compost for 60 days while maintaining the ideal pH and other environmental conditions [73]. There is a rise in weight at an earlier stage due to the presence of moisture content, as shown by the results. In addition, the interaction between the organism's enzymes and the polymer causes a gradual loss of weight over the period of a few days. This shows that biodegradation occurs over time and is ongoing. Biodegradable composite staples have been proposed for medical use, using cellulose acetate nanofiber and poly-L-lactic acid (PLLA) as potential components.

7.6.9 Tribological Characteristics

Tribological qualities are modified by wear, friction, and lubrication when two surfaces are in contact and undergo relative motion (all illustrated in Fig. 7.11). The rate of energy dissipation and material breakdown is amplified by friction and wear. Depending on the sliding direction and applied force, antiparallel jute fibre-reinforced epoxy composites were found to have the maximum wear resistance.

The mats are arranged in an anti-parallel, such that they are parallel to the force's application but perpendicular to the direction of sliding. The antiparallel oriented mat experienced a maximum volume loss of 10 mm over a sliding distance of 11 km when subjected to a force of 30 N and a sliding velocity of 3 m/s against a stainless-steel surface. A PLA-natural fibre composite containing 20% nettle, sisal, and Grewia optiva fibre weight percentage improved in wear resistance and had a reduced maximum value of friction coefficient. The enhanced hardness and porosity of the brown fibres, as well as the smaller median pore diameter, were found to provide the brown fibre-reinforced composite an edge over the white fibre-reinforced composite when it came to wear resistance. Woven glass fibre reinforcement may be replaced by polyester-reinforced seed-oil palm fibres, which have a lower specific wear rate and interface temperature.

Flax fibres incorporated in a polypropylene matrix have mechanical and frictional properties that change with sliding speed and temperature. At greater sliding speeds (10 m/s), the friction coefficient increases from 25 °C to 60 °C before decreasing again, whereas the inverse is true at lower sliding speeds (2 m/s). As the load was raised from 5 to 20 N, the volume loss and specific wear rate of a coir particulate epoxy composite treated with 5% wt NaOH decreased, and the effects were even more pronounced for samples with a smaller particle percentage (2.5%, 5%, and 7.5%). When exposed to simulated acid rain, composites made from polyvinyl chloride reinforced with sorghum straw fibre fared better than those made from wheat, rice, or maize straw fibre. The sorghum straw fibre is more superior due to its higher concentrations of crystallinity, cellulose, lignin, and interfacial strength. At modest

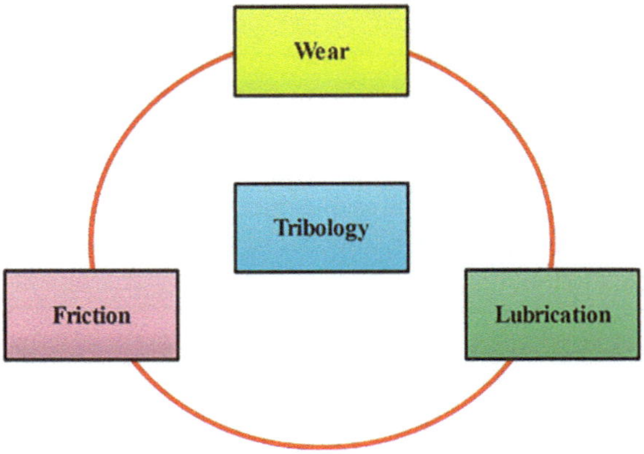

Fig. 7.11 Factors effecting on tribological properties [9]. Adapted with permission from Ref. [9] Copyright @ 2023, Wiley

sliding velocities (1.7–2.2 m/s), the antiparallel-oriented hemp fibre epoxy composite showed good wear resistance. At a sliding speed of 2.2 m/s, the orientation's influence on reducing the shear resistance experienced by the fibres in the contact/rubbing area boosted wear resistance by about 60% compared to plain epoxy. Epoxy composites produced from coccinia fibre were shown to have improved wear resistance and friction reduction after being pretreated with 5% NaOH and 3% silane (wt %). A sample with 35% fibre reduced maximum specific wear by 94.95% compared to PLA alone, demonstrating its promise as a lightweight bearing material. Composites were fabricated using an epoxy matrix reinforced with flax fibre (30%), Calotropis gigantea stem fibre (40%), and glass fibre, and wear tests were performed on these materials (10%). Under a load of 20 N, the mass loss was found to be as low as 0.0034 g, showing a linear relationship between force and the rate of mass loss [75]. The tribological properties of fusion fibre polymer composites were superior to those of coir pith polymer composites.

7.7 Recyclability of HF-PLA Composites

Previous studies have shown that HF-PLA composites may be made using compression moulding and then used via the thermoforming method for eco-friendly packaging. Granulating, extruding, and pelletizing were some of the later techniques used to recycle the composites. Once again, recyclable HF-PLA composites were used in the injection moulding of eco-friendly packaging goods. Based on their findings, the researchers determined that the resulting HF-PLA composites were durable enough to be recycled several times. The recyclable nature of HF-PP-PLA composites is another perk [79].

7.8 Applications of HF-PLA Composites

7.8.1 Packaging Applications

Plastic's remarkable attributes make it a good choice for packaging, including its low cost, portability, durability, adaptability, resistance to corrosion, and thermal and electrical insulation [58]. However, plastic use results in significant emissions of pollutants. Plastic packaging is notorious for being single-use, meaning it contributes to an endless stream of trash. The thermoplastic polymer PLA is biodegradable and has several applications, including usage as a material for plant containers and trash bags [93]. NFRCs have been included into packaging due to their sustainability, renewability, and efficiency as a plastic substitute. The impact strength and heat resistance of HF-PLA composites are much greater than those of virgin PLA. As a

Fig. 7.12 The versatile applications of the hemp plant; each part serves a specific purpose in various industries [99]

result, HF-PLA composites show significant promise for usage in environmentally friendly packaging.

HF-PP and HF-PLA composites developed for packaging purposes have been studied in the past, and their performance characteristics have been analysed. Charpy impact strength was found to be higher for both HFPP and HF-PLA than for virgin polymers. Drop weight impact testing showed a similar pattern. Adding HF improved thermal resistance, as shown by DSC, TGA, and heat deflection temperature studies. When compared to HF-PLA composites (13% water absorption), HF-PP composites (7.3% water absorption) performed better. To learn more about the potential of HF-polylactic acid composites for green wrapping applications, a prior review research explored many effective features and examples of packaging items. Potential uses for the generated bio-composites included packaging for food and other liquids. When MCC is removed from hemp, the MCCPBS- PLA composites may be used as a good packaging material. Figure 7.12 depicts the use of HF-PLA composites in the environmentally friendly packaging and building industries.

7.8.2 Structural Applications

When compared to conventional building materials like concrete, steel, and lumber, hemp culm (hemp in its natural state) is about 20 times more sustainable [96]. Significant potential exists for hemp to replace concrete and steel as a dependable construction material, especially for organically curved structures. When utilized as a building material, hemp may be easily pliable and shaped. As a result, hemp has legitimate and dependable applications. For structural uses, a biodegradable HF-PLA composite was developed. The hemp-based HF-PLA composite was made with a manual lay-up and melting process at 185 °C for 20 min to preserve the biodegradability of the material. Resin film infusion (RFI) was used to bind hemp strips that were mechanically extracted at a 63% HF concentration and a 33.5% lignin level. Similar to aerospace composites, the compressive strength of this HF-PLA material is around 187 MPa cm^3/g, or roughly 84% of the stated tensile strength. The Young's modulus of HF-modulus PLA was significantly higher than that of E-glass/epoxy, coming up at 38 GPa cm^3/g. The flexural strength was around 273 MPa cm^3/g, making it comparable to E-glass/epoxy.

7.8.3 Additive Manufacturing

Rapid prototyping methods, such as 3D printing and additive manufacturing, have seen exponential growth in popularity in recent years. Sustainability, ease of use, cheap cost, waste cleanup, and the capacity to make complicated designs without adjustments have all contributed to AM's rise in popularity [60]. Using a CAD file or a computed tomography (CT) scan as a starting point, this rapidly developing approach enables for fully automated product fabrication [64]. The use of 3D printing technology has spread to many other fields, such as transportation, architecture, construction, design, metals and alloys, digital art, electronics, medicine, sports, textiles, apparel, and even the fashion business. Figure 7.13 shows several different 3D printing schematics. Materials jetting, powder bed fusion, sheet lamination, vat photopolymerization, directed energy deposition, binder jetting, and material jetting are the seven types of 3DP listed by ASTM [4]. Stratasys, headquartered in Eden Prairie, Minnesota, pioneered FDM, a 3D printing technique, to produce functional prototypes and 3D printed objects with superior strength to weight ratios and chemical and heat resistance. FDM 3D printers are sometimes referred to as home 3D printers due to their inexpensive price and ease of design and assembly. Due to its low cost, ease of use, dependability, material availability, and little waste, FDM 3D printing has becoming more popular. Extruding a filament via a nozzle, deposition modelling allows for the printing of three-dimensional objects in layers on a printing bed (Table 7.5).

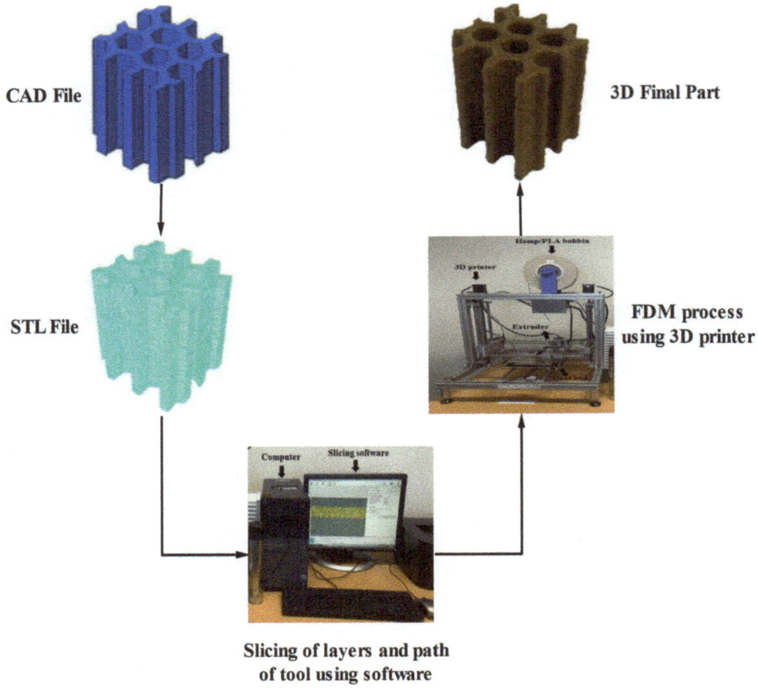

Fig. 7.13 Schematic of 3D printing [83]. Adapted with permission from Ref. [83] Copyright @ 2020, Springer Nature

Table 7.5 3D printing Factors [83]. Adapted with permission from Ref. [83] Copyright @ 2020, Springer Nature

Factors	Values
Shell thickness	0.8 mm
Layer thickness	0.2 mm
Printing speed	20 mm/s
Extruder temperature	180 °C
Top/bottom thickness	0.8 mm
Bed temperature	55 °C
Infill density	100%
Minimum nozzle diameter	0.6 mm
Extrusion multiplier	100%

7.8.4 Fabrication and Characterization of Hemp Fibre-Based 3D-Printed Honeycomb Sandwich Structure by FDM Process

In order to create the layers, the filament must be heated to just over its melting point by resistive heaters in the nozzle. Each new layer is placed over the previous one until the final component is created [11]. Most thermoplastic filaments on the market today are not biodegradable and might be harmful to human health if ingested. Natural fibres including wood, hemp, flax, kenaf, coir, and sisal are currently the subject of cutting-edge research into the production of new environmentally friendly, cost-effective filaments. A more sustainable and long-lasting option than additive manufacturing is composite materials fused together with natural fibres. Greenhouse gas emissions from producing synthetic fibres for 3D printing filaments can be reduced by switching to natural fibres [22]. Hemp's many desirable qualities—ample supply, rapid growth, high water absorption, antibacterial characteristics, cheap cost, minimal maintenance, high tensile strength, and little environmental impact make it an ideal material for a wide variety of applications. Hemp is a superior material for making FDM filaments. But PLA, a bioplastic derived from maize, decomposes in the environment. Novel composite materials based on PLA reinforced with HF have been the subject of state-of-the-art research.

Where the formula for determining the weight percent (3),
$\text{wt}\% = \frac{\text{weight dry fibers after dissolving}}{\text{filament weight}}$ (3).

In cases where Eq. (7.4) [46]was used to determine the amount of HF present in the filament material,

$$\frac{\rho_{\text{composite}} - \rho_{\text{matrix}}}{\rho_{\text{reinforcement}} - \rho_{\text{matrix}}} = \%\text{ of filling material} \tag{7.4}$$

7.8.5 Impact of 3D Printing Variables on Finished Products

It is known that the parameters used to make a 3D-printed object have a major impact on its mechanical properties. Some of the most important factors affect the mechanical qualities of 3D-printed objects. Printing smoothly using HFPLA composites requires adjusting the appropriate settings [24]. Larger nozzle diameters result in noticeably greater tensile strength; thicker layers, faster extrusion and filling velocities, and higher surface roughness all contribute to shorter build times. The surface roughness can also be affected by the printing settings. Increasing wall thickness or layer height causes surface roughness to deteriorate. The printing procedure went off without a hitch due to the 0.27 mm-sized hemp powder and the 0.20–0.40 mm thick ink layers. The raster angle orientation is directly related to the material's ability to absorb impact energy and remain flexible. From the four possible orientations (0°, 90°, 0° and 90°, and −45° and 45°), the 0° raster angle was the most effective [100].

7.8.6 Limitations of HF-PLA Composites

Natural fibres provide a number of benefits, like being inexpensive, lightweight, biodegradable, strong, insulating, causing less skin and respiratory irritation, etc. The incompatibility of natural fibres' hydrophobic surfaces with synthetic polymers is a major limitation of these materials. Fibres in polymer matrices are generally unevenly dispersed due to their hydrophilic nature. Fibre's mechanical characteristics can be altered by its atypical growth [101]. Natural fibres are less practical than synthetic ones because they wear out faster, absorb more moisture, have weaker mechanical qualities, are more flammable, are more difficult to produce, and cost more. There have been a number of attempts to address these problems by improving the interfacial interaction between natural fibres and their matrix by chemical surface modification. Authors [102] found few uses for HF-PLA composites in aircraft, while authors [103] advised against using HF-PLA composites in environments with high humidity for packing.

7.8.7 Conclusions

In this analysis, the thermal characterization, mechanical parameters, and mechanical properties of HF-PLA composites were discussed, as well as their sustainability and possible applications. Further, the 3D printing parameters and their effects on HF-PLA composite applications in AM were explored. There were also very few known restrictions on the use of HF-PLA composites. The research findings on Hemp Fibre (HF) reinforced polylactic acid (PLA) composites reveal a significant divergence in their mechanical properties and thermal characteristics, hinging on various factors such as fibre extraction, surface modification, and chemical treatments. PLA's tensile properties experience considerable enhancement when HF reinforcement is applied, particularly in the weft direction. Additionally, graft-modified HF proves instrumental in substantially improving the mechanical attributes of HF-PLA composites, although caution is necessary with regenerated HF reinforcement as it can lead to mechanical failure. Furthermore, the study underscores the impact of treatments like NaOH and PLA-glycidyl methacrylate, where the former diminishes thermal stability but the latter elevates the thermal breakdown temperature of HF-PLA composites. Notably, HF-PLA composites with 60 wt % HF exhibit exceptional creep resistance and flexural characteristics, while an excess of 70% HF fibres yields adverse effects on these properties. The integration of HF reinforcement into the PLA matrix exhibits the ability to retard deterioration, and the use of NaOH-treated hemp fibres showcases promise in achieving complete biodegradability, rendering these composites suitable for diverse industrial applications. In a complementary vein, these HF-PLA composites offer substantial potential for recycling, bolstered by their thermal stability and mechanical strength. The research underscores the positive impact of reinforcing natural fibres in a polymer matrix, enhancing mechanical, thermal, and tribological qualities by augmenting the interfacial adhesion between fibres and the matrix, particularly when subjected to chemical treatments. However, for the successful penetration of high-end sectors like aerospace and marine industries, improvements in thermal characteristics and moisture response are deemed imperative. Despite limited prior utilization, nonwoven fibre composites (NFCs) present an opportunity for enhanced characteristics and performance, paving the way for potential applications in biological, electrical, and construction fields. The study further emphasizes the pivotal role of 3D printing factors in shaping the mechanical strength and surface quality of HF-PLA products. Thus, the versatility of HF-PLA composites extends across several domains, including the automotive industry, construction, environmentally friendly packaging, dielectrics, and musical instruments, with aerospace currently being the foremost industry adopting these composites, albeit with precautions against high humidity environments. In conclusion, this research underscores the vast untapped potential for growth in HF-PLA composites research and commercialization, with a substantial positive impact on environmental sustainability and long-term viability.

References

1. Sarikaya, E., Çallioğlu, H., Demirel, H.: Production of epoxy composites reinforced by different natural fibers and their mechanical properties. Compos. Part B Eng. **167**, 461–466 (2019)
2. Neoh, K.W., Tshai, K.Y., Khiew, P.S., Chia, C.H.: Micro palm and kenaf fibers reinforced PLA composite: effect of volume fraction on tensile strength. Appl. Mech. Mater. **145**, 1–5 (2012)
3. Siva, R., Nemali, S.S., Gokul, K.: Comparison of mechanical properties and water absorption test on injection molding and extrusion-injection molding thermoplastic hemp fiber composite. Mater. Today Proc. **47**, 4382–4286 (2021)
4. Bendourou, F.E., Suresh, G., Laadila, M.A., Kumar, P., Rouissi, T., Dhillon, G.S., Zied, K., Brar, S.K., Galvez, R.: Feasibility of the use of different types of enzymatically treated cellulosic fibres for polylactic acid (PLA) recycling. Waste Manage. **121**, 237–247 (2021)
5. Yang, Y., Murakami, M., Hamada, H.: Molding method, thermal and mechanical properties of jute/PLA injection molding. J. Polym. Environ. **20**, 1124–1133 (2012)
6. Smoca, A.: FTIR spectroscopy analysis of PLA biocomposites reinforced with hemp fibers. Key Eng. Mater. **850**, 112–117 (2020)
7. Yang, X., Fan, W., Ge, S., Gao, X., Wang, S., Zhang, Y., Foong, S.Y., Liew, R.K., Lam, S.S., Xia, C.: Advanced textile technology for fabrication of ramie fiber PLA composites with enhanced mechanical properties. Ind. Crops Prod. **162**, 113312 (2021)
8. Murakami, M., Yang, Y., Hamada, H.: Mechanical properties of jute/PLA injection molded products-all natural composites. ASME Int. Mech. Eng. Congr. Expos. **54891**, 761–766 (2011)
9. Akhil, U.V., Radhika, N., Saleh, B., Aravind Krishna, S., Noble, N., Rajeshkumar, L.: A comprehensive review on plant-based natural fiber reinforced polymer composites: fabrication, properties, and applications. Polym. Compos. **44**, 2598–2633 (2023)

10. Girimurugan, R., Senniangiri, N., Krishnan, B.P., Kavitha, S., Vairavel, M.: Tensile behaviour of hybrid polymer composites–an experimental study. IOP Conf. Ser. Mater. Sci. **1059**, 012033 (2021)
11. Murugan, G., Loganathan, G.B., Sivaraman, G., Shilaja, C., Mayakannan, S.: Compressive behavior of tamarind shell powder and fine granite particles reinforced epoxy matrix based hybrid bio-composites. ECS Trans. **107**, 7111 (2022)
12. Dinesh Kumar, D., Balamurugan, A., Suresh, K.C., Suresh Kumar, R., Jayanthi, N., Ramakrishnan, T., Hasane Ahammad, S.K., Mayakannan, S., Venkatesa Prabhu, S.: Study of Microstructure and wear resistance of AA5052/B_4C nanocomposites as a function of volume fraction reinforcement to particle size ratio by ANN. J Chem. (2023)
13. Girimurugan, R., Maheskumar, P., Sahoo, G., Sivalingam, A., Mayakannan, S.: Effect of nano alumina powder and water hyacinth stem powder addition on tensile properties of polypropylene matrix hybrid composites–An experimental study. Mater. Today Proc. **60**, 2099–2104 (2022)
14. Seile, A., Spurina, E., Sinka, M.: Reducing global warming potential impact of bio-based composites based of LCA. Fibers. **10**, 79 (2022)
15. Zawawi, N.A., Ismail, A., Khalina, A., Mahdi, M.A.: Investigation on the microwave properties of kenaf and rice-husk fiber reinforced PLA composite utilizing one-port coaxial transmission line reflection method. Key Eng. Mater. **471**, 868–873 (2011)
16. Hanson, M., Yang, C.H.: Characterization of polylactic acid filament with biodegradable hemp fiber infused during additive manufacturing process. Int. Manuf. Sci. Eng. Conf. **85802**, V001T01A009 (2022)
17. Debeli, D.K., Qin, Z., Guo, J.: Study on the pre-treatment, physical and chemical properties of ramie fibers reinforced poly (lactic acid) (PLA) biocomposite. J. Nat. Fibers. **15**, 596–610 (2018)
18. Zhang, J., Khatibi, A.A., Castanet, E., Baum, T., Komeily-Nia, Z., Vroman, P., Wang, X.: Effect of natural fibre reinforcement on the sound and vibration damping properties of biocomposites compression moulded by nonwoven mats. Compos. Commun. **13**, 12–17 (2019)
19. Motaung, T.E., Linganiso, L.Z., Mohomane, S.M.: Agricultural waste fibers and biopolymer matrices used in biocomposites. Biocompos. Prop. Perform. Appl. 11–26 (2017)
20. Yaguchi, Y., Takeuchi, K., Waragai, T., Tateno, T.: Durability evaluation of an additive manufactured biodegradable composite with continuous natural fiber in various conditions reproducing usage environment. Int. J. Autom. Technol. **14**, 959–965 (2020)
21. Seile, A., Beļakova, D.: Nonwoven development by the multilayer structure. ide. Tehnologija Resursi - Environment, Technology, Resources **3**, 292–297 (2017)
22. Bogard, F., Bach, T., Abbes, B., Bliard, C., Maalouf, C., Bogard, V., Beaumont, F., Polidori, G.: A comparative review of Nettle and Ramie fiber and their use in biocomposites, particularly with a PLA matrix. J. Nat. Fibers. **19**, 8205–8229 (2022)
23. Smoca, A.: Water absorption properties of hemp fibres reinforced PLA bio-composites. Eng. Rural Develop. **18**, 1079–1083 (2019)
24. Jamadi, A.H., Razali, N., Petrů, M., Taha, M.M., Muhammad, N., Ilyas, R.A.: Effect of chemically treated kenaf fibre on mechanical and thermal properties of PLA composites prepared through fused deposition modeling (FDM). Polymers **13**, 3299 (2021)
25. López-Alba, E., Schmeer, S., Díaz, F.: Energy absorption capacity in natural fiber reinforcement composites structures. Materials **11**, 418 (2018)
26. Girimurugan, R., Pugazhenthi, R., Maheskumar, P., Suresh, T., Vairavel, M.: Impact and hardness behaviour of epoxy resin matrix composites reinforced with banana fiber/camellia sinensis particles. Mater. Today Proc. **39**, 373–377 (2021)
27. Girimurugan, R., Senniangiri, N., Adithya, K., Velliyangiri, B.: Mechanical behaviour of coconut shell powder granule reinforced epoxy resin matrix bio composites. J. Dyn. Control Syst. **10**, 533–541 (2018)
28. Azlin, M.N., Sapuan, S.M., Zainudin, E.S., Zuhri, M.Y., Ilyas, R.A.: Natural polylactic acid-based fiber composites: a review. In: Advanced processing, properties, and applications of starch and other bio-based polymers, pp. 21–34 (2020)

29. Alias, N.F., Ismail, H., Ishak, K.M.: Poly (lactic acid)/natural rubber/kenaf biocomposites production using poly (methyl methacrylate) and epoxidized natural rubber as co-compatibilizers. Iran. Polym. J. **30**, 737–749 (2021)
30. Muhandes, H., Kalácska, Á., Székely, L., Keresztes, R., Kalácska, G.: Abrasive sensitivity of engineering polymers and a bio-composite under different abrasive conditions. Materials **13**, 5239 (2020)
31. Moorthy, A.A., Girimurugan, R., Prakash, E., Madheswaran, S., Sankar, A.N., Sriman, P.: Investigations on flexural performance of epoxy matrix composites strengthened with chemically modified and unmodified banana fiber/Used Camellia Sinensis (UCS) particles. Mater. Today Proc. **45**, 8115–8119 (2021)
32. Girimurugan, R., Saravanan, K.G., Manickavasagam, P., Gurunathan, G., Vairavel, M.: Experimental studies on water absorption behaviour of treated and untreated hybrid bio-composites. IOP Conf. Ser. Mater. Sci. Eng. **1059**, 012017 (2021)
33. Girimurugan, R., Vairavel, M., Shanjai, S.D., Manikandan, S., Manikkumar, R., Manojkumar, R.: Experimental impact on mechanical characteristics of banana fiber reinforced, groundnut shell powder filled epoxy resin matrix bio composites. Int. J. Eng. Innov. Technol. **9**, 2279–2282 (2020)
34. Gurunathan, G., Paramadhayalan, P., Purushothaman, S., Girimurugan, R.: An experimental study on mechanical properties of jute fiber reinforced coconut shell particles filled epoxy resin matrix biocomposites. J. Xi'an Univ. Archit. Technol. **12**, 2773–2783 (2020)
35. Thiruvasagam, R.G., Selvaraj, M., Saravanan, S., Vairavel, M., Girimurugan, R.: Art of review-fiber reinforced polymer composites. J. Inf. Comput. Sci. **9**, 461–468 (2019)
36. Sahoo, G., Kamalakannan, R., Pradeep, G.M., Manivelmuralidaran, V., Girimurugan, R.: A process of analyzing the performance evaluation of sisal fiber in fiber reinforced composites. Mater. Today Proc. **56**, 3201–3206 (2022)
37. Girimurugan, R., Maheskumar, P., Adithya, K., Kamalakannan, K.: Water absorption behaviour of tamarind shell powder and marble dust particles reinforced hybrid bio-composites. Int. J. Mech. Eng. **6**, 1073–1078 (2021)
38. Rahaman, G.A., Girimurugan, R., Vairavel, M., Karthi, S., Karthikeyan, K., Prakash, G.T.: Effect of eggshell particles addition on mechanical properties of jute fiber rein-forced epoxy resin matrix bio-composites–an experimental study. J. Crit. Rev. **7**, 2726–2734 (2020)
39. Jaiganesh, V., Senthilnathan, K., Kumarasan, T., Saravanan, R., Nanthakumar, S., Girimurugan, R.: Influence of fibre on mechanical behavior of Ramie fibre/polystyrene hybrid composite. Mater. Proc, Today (2023)
40. Babu, B., Muruganandhan, P., Girimurugan, R., Sakthi, S., Nanthakumar, S., Vignesh, S.: Influence of powdered chick eggshell (PECS) on mechanical and wear properties of kenaf fiber (KF) reinforced composites. Mater. Today Proc. (2023)
41. Jayaraman, R., Viknesh, M., Girimurugan, R.: Experimental investigations on mechanical and water absorption properties of epoxy resin matrix treated sugarcane and tamarind shell powder reinforced bio-Composites. Mater. Today Proc. **74**, 636–641 (2023)
42. Bothiraj, T., Boopathi, K., Kalaiselvan, K., Benham, A., Mayakannan, S.: Experimental investigations on mechanical and wear behavior of waste marble dust and coconut fiber reinforced hybrid bio composites. Mater. Today Proc. **68**, 2239–2242 (2022)
43. Thangaraj, S., Pradeep, G.M., Dani, M.H., Mayakannan, S., Benham, A.: Experimental investigations on tensile and compressive properties of nano alumina and arecanut shell powder reinforced polypropylene hybrid composites **68**, 2243–2248 (2022)
44. Wang, Y., Sultana, J., Rahman, M.M., Ahmed, A., Azam, A., Mushtaq, R.T., Rehman, M.: A sustainable and biodegradable building block: review on mechanical properties of bamboo fibre reinforced PLA polymer composites and their emerging applications. Fibers Polym. **12**, 3317–3342 (2022)
45. Vijayasekaran, G., Kamal, M., Gopi, M., Nanthakumar, S., Girimurugan, R.: Enhancement of natural fiber-reinforced plastics by polyester and seaweed waste fibers. Mater. Today Proc. (2023)

46. Muruganandhan, P., Jothilakshmi, S., Vivek, R., Nanthakumar, S., Sakthi, S., Mayakannan, S., Girimurugan, R.: Investigation on silane modification and interfacial UV aging of flax fibre reinforced with polystyrene composite. Mater. Today Proc. (2023)
47. Thangavel, P., Vignesh, S., Sureshkumar, R., Gowrishankar, A., Girimurugan, R.: Study on mechanical properties of polylactic acid matrix added with fly ash and tamarind kernel powder micro fillers. NanoWorld J. **9**, S626-630 (2023)
48. Mayakannan, S., Raj, J.B., Raja, V.L., Nagaraj, M.: Effectiveness of silicon nanoparticles on the mechanical, wear, and physical characteristics of PALF/sisal fiber–based polymer hybrid nanocomposites. Biomass Convers. Biorefin. **14**, 13291–13305 (2023)
49. Smoca, A., Zelca, Z.: Optimal use of industrial hemp for PLA biocomposite and LLDPE composite reinforcement. Solid State Phenom. **320**, 131–138 (2021)
50. Nurhafizah, S.M., Anuar, H., Mel, M., Sapuan, S.M., Aimi, M.N.: Polylactic acid-based Kenaf biomass synthesized via ring opening polymerization. Biomass Bioenergy Appl. 211–231 (2014)
51. Jayaraman, R., Girimurugan, R., Suresh, V., Shilaja, C., Mayakannan, S.: Improvement on tensile properties of epoxy resin matrix sugarcane fiber and tamarind seed powder reinforced hybrid bio-composites. ECS Trans. **107**, 7265 (2022)
52. Girimurugan, R., Shilaja, C., Mayakannan, S., Rajesh, S., Aravinth, B.: Experimental investigations on flexural and compressive properties of epoxy resin matrix sugarcane fiber and tamarind seed powder reinforced bio-composites. Mater. Today Proc. **66**, 822–828 (2022)
53. Loganathan, G.B., Maheskumar, P., Jayanthi, N., Sureshkumar, R., Sakthi, S., Girimurugan, R.: Development of novel environmental proficient hybrid composites based on marble dust and poultry's eggshell. Mater. Today Proc. (2023)
54. Thangaraj, S., Vigneshwaran, M., Muthuraj, M., Shenbagaraj, R., Girimurugan, R.: Distinctive study on banana/sisal fiber hybrid composites filled with nano marble dust particles. Mater. Today Proc. (2023)
55. Girimurugan, R., Muthuraj, M., Ahammad, S.H., Vijayakumar, N., Appadurai, M.: Experimental study of mechanical properties of sisal/jute fibers hybrid sandwich composite. Mater. Today Proc. **68**, 1742–1749 (2022)
56. Karthikeyan, R., Girimurugan, R., Sahoo, G., Maheskumar, P., Ramesh, A.: Experimental investigations on tensile and flexural properties of epoxy resin matrix waste marble dust and tamarind shell particles reinforced bio-composites. Mater. Today Proc. **68**, 2215–2219 (2022)
57. Jasti, A., Biswas, S.: Characterization of elementary industrial hemp (Cannabis Sativa L.) fiber and its fabric. J. Nat. Fibers. **20**, 2158982 (2023)
58. Yang, T.C., Wu, T.L., Hung, K.C., Chen, Y.L., Wu, J.H.: Mechanical properties and extended creep behavior of bamboo fiber reinforced recycled poly (lactic acid) composites using the time–temperature superposition principle. Constr. Build. Mater. **93**, 558–563 (2015)
59. Cisse, O., Placet, V., Guicheret-Retel, V., Trivaudey, F., Boubakar, M.L.: Creep behaviour of single hemp fibres. Part I: viscoelastic properties and their scattering under constant climate. J. Mater. Sci. **50**, 1996–2006 (2015)
60. Durante, M., Formisano, A., Boccarusso, L., Langella, A., Carrino, L.: Creep behaviour of polylactic acid reinforced by woven hemp fabric. Compos. Part B Eng. **124**, 16–22 (2017)
61. Durante, M., Langella, A., Formisano, A., Boccarusso, L., Carrino, L.: Dynamic-mechanical behaviour of bio-composites. Proc. Eng. **167**, 231–236 (2016)
62. Guicheret-Retel, V., Cisse, O., Placet, V., Beaugrand, J., Pernes, M., Boubakar, M.L.: Creep behaviour of single hemp fibres. Part II: Influence of loading level, moisture content and moisture variation. J. Mater. Sci. **50**, 2061–2072 (2015)
63. Placet, V., Cisse, O., Guicheret, V., Trivaudey, F., Boubakar, L.: Viscoelastic behaviour of single hemp fibre under constant and cyclic humidigy environment-experiment and modelling. In: International Conference on Composite Materials (2015)
64. Franciszczak, P., Taraghi, I., Paszkiewicz, S., Burzyński, M., Meljon, A., Piesowicz, E.: Effect of halloysite nanotube on mechanical properties, thermal stability and morphology of polypropylene and polypropylene/short kenaf fibers hybrid biocomposites. Materials **13**, 4459 (2020)

65. Santulli, C.: Creep behaviour of plant fibre composites. In: Advanced High Strength Natural Fibre Composites in Construction, pp. 459–477 (2017)
66. Sala, B., Watjanatepin, P., Zarafshani, H., Guicheret-Retel, V., Trivaudey, F., Scarpa, F., Van Acker, K., Placet, V.: Creep behaviour of eco-friendly sandwich composite materials under hygrothermal conditions. Compos. Part B Eng. **247**, 110291 (2022)
67. Senthilkumar, K., Subramaniam, S., Ungtrakul, T., Kumar, T.S., Chandrasekar, M., Rajini, N., Siengchin, S., Parameswaranpillai, J.: Dual cantilever creep and recovery behavior of sisal/hemp fibre reinforced hybrid biocomposites: effects of layering sequence, accelerated weathering and temperature. J. Ind. Text. **51**, 2372S-2390S (2022)
68. Chiasson-Poirier, L., Lecompte, T., de Menibus, A.H.: Static and long term compression behavior of hemp shiv for floating floor application. Constr. Build. Mater. **268**, 121220 (2021)
69. Azadi, F., Jafari, S.H., Khonakdar, H.A., Arjmand, M., Wagenknecht, U., Altstädt, V.: Influence of graphene oxide on thermally induced shape memory behavior of PLA/TPU blends: correlation with morphology, creep behavior, crystallinity, and dynamic mechanical properties. Macromol. Mater. Eng. **306**, 2000576 (2021)
70. Waseem, M., Salah, B., Habib, T., Saleem, W., Abas, M., Khan, R., Ghani, U., Siddiqi, M.U.: Multi-response optimization of tensile creep behavior of PLA 3D printed parts using categorical response surface methodology. Polymers **12**, 2962 (2020)
71. Ji, X., Cai, Y., Xie, A., Liu, J., Zhu, P., Zhang, M.: Mechanical properties of hydroxyapatite/polylactic acid composites based on nanoindentation method. Acta Mater. Compos. Sin. **35**, 553–563 (2018)
72. Fischbach, M., Weinberg, K.: Effect of physical aging on the flexural creep in 3D printed thermoplastic. Adv. Struct. Mater. **194**, 115–130 (2023)
73. Todkar, S.S., Patil, S.A.: Creep and heat deflection temperature (HDT) study of Pineapple leaf fibre (PALF) and Montmorillonite (MMT) nanoclay reinforced Poly-lactic acid (PLA) based laminated hybrid biocomposite. Mater. Today: Proc. **62**, 7534–7539 (2022)
74. Jayswal, A., Liu, J., Harris, G., Mailen, R., Adanur, S.: Creep behavior of 3D printed polymer composites. Polym. Eng. Sci. **63**, 3809–3818 (2023)
75. Morreale, M., Mistretta, M.C., Fiore, V.: Creep behavior of poly (lactic acid) based biocomposites. Materials **10**, 395 (2017)
76. Amza, C.G., Zapciu, A., Baciu, F., Vasile, M.I., Nicoara, A.I.: Accelerated aging effect on mechanical properties of common 3D-printing polymers. Polymers (Basel) **13**, 4132 (2021)
77. Liang, D., Zhi, H., Sun, Z., Zhang, B., Hu, G., Liu, X., Wang, S.: Influence of titanate coupling agent on rheological properties of PLA/PHA/nanometer HA blends. Polym. Mater. Sci. Eng. **29**, 86–93 (2013)
78. Dogan, O.: Short-term creep behaviour of different polymers used in additive manufacturing under different thermal and loading conditions. Stroj. Vestnik/J. Mech. Eng. **68**, 451–460 (2022)
79. Chen, X., Chen, F., Wang, G., Ma, X., Wang, J., Deng, J.: Eco-friendly, disposable bamboo fiber dishware: static and dynamic mechanical properties and creep behavior. Ind. Crops Prod. **187**, 115305 (2022)
80. Jhao, Y.S., Ouyang, H., Yang, F., Lee, S.: Thermo-mechanical and creep behaviour of polylactic acid/thermoplastic polyurethane blends. Polymers **14**, 5276 (2022)
81. Adibeig, M.R., Vakili-Tahami, F., Saeimi-Sadigh, M.A.: Numerical and experimental investigation on creep response of 3D printed Polylactic acid (PLA) samples. Part I: The effect of building direction and unidirectional raster orientation. J. Mech. Behav. Biomed. **145**, 106025 (2023)
82. Cha, J.H., Ahn, S.J., Jeon, H.Y.: Analysis of dynamic and creep behaviors of FLA blends with blending ratio. Polym. **39**, 566–571 (2015)
83. Antony, S., Cherouat, A., Montay, G.: Fabrication and characterization of hemp fibre based 3D printed honeycomb sandwich structure by FDM process. Appl. Compos. Mater. **27**, 935–953 (2020)
84. Antony, S., Cherouat, A., Montay, G.: Effect of fibre content on the mechanical properties of hemp fibre woven fabrics/polypropylene composite laminates. Polym. Polym. Compos. **29**, S790-802 (2021)

85. Doğan, O., Kamer, M.S.: Experimental investigation of the creep behavior of test specimens manufactured with fused filament fabrication using different manufacturing parameters. J. Fac. Eng. Archit. Gazi Univ. **38**, 1839–1848 (2023)
86. Kim, H.J., Miyamoto, S., Takada, Y., Takemura, K.: Effect of surface modification on flexural properties of jute fiber green composites. ICCM Int. Conf. Compos. Mater. **6300**, 2–6 (2011)
87. Xia, S., Liu, X., Wang, J., Kan, Z., Chen, H., Fu, W., Li, Z.: Role of poly (ethylene glycol) grafted silica nanoparticle shape in toughened PLA-matrix nanocomposites. Compos. Part B Eng. **168**, 398–405 (2019)
88. Peixoto, T., Carneiro, S., Pereira, F., Santos, C., Fangueiro, R., Duarte, I., Paiva, M.C., Lopes, M.A., Guedes, R.M.: Hybrid structures for Achilles' tendon repair. Polym. Adv. Technol. **33**, 2362–2373 (2022)
89. Molina, R., Pender, G., Moro, L., Piovan, M.T.: Creep behaviour of a polymer used for 3D printing. Rev. Mater. **23** (2018)
90. Wang, P., Xiong, Z., Fei, P., Cai, J., Walayat, N., Xiong, H.: An approach for compatibilization of the starch with poly (lactic acid) and ethylene-vinyl acetate-glycidyl-methacrylate. Int. J. Biol. Macromol. **161**, 44–58 (2020)
91. Novotna, K., Zajdlova, M., Suchy, T., Hadraba, D., Lopot, F., Zaloudkova, M., Douglas, T.E., Munzarova, M., Juklickova, M., Stranska, D., Kubies, D.: Polylactide nanofibers with hydroxyapatite as growth substrates for osteoblast-like cells. J. Biomed. Mater. Res. Part A. **102**, 3918–3930 (2014)
92. Tezel, T., Kovan, V., Topal, E.S.: Effects of the printing parameters on short-term creep behaviors of three-dimensional printed polymers. J. Appl. Polym. Sci. **136**, 47564 (2019)
93. Widiastuti, I., Sbarski, I., Masood, S.H.: Creep behavior of PLA-based biodegradable plastic exposed to a hydrocarbon liquid. J. Appl. Polym. Sci. **127**, 2654–2660 (2013)
94. Esmaeilzadeh, J., Hesaraki, S., Ebrahimzadeh, M.H., Asghari, G.H., Kachooei, A.R.: Creep behavior of biodegradable triple-component nanocomposites based on PLA/PCL/bioactive glass for ACL interference screws. Arch. Bone Jt. Surg. **7**, 531 (2019)
95. Yang, T.C., Hung, K.C., Wu, T.L., Wu, T.M., Wu, J.H.: A comparison of annealing process and nucleating agent (zinc phenylphosphonate) on the crystallization, viscoelasticity, and creep behavior of compression-molded poly (lactic acid) blends. Polym. Degrad. Sta. **121**, 230–237 (2015)
96. Ye, J., Yao, T., Deng, Z., Zhang, K., Dai, S., Liu, X.: A modified creep model of polylactic acid (PLA-max) materials with different printing angles processed by fused filament fabrication. J. Appl. Polym. Sci. **138**, 50270 (2021)
97. Cláudio, R.A., Dupont, J., Baptista, R., Leite, M., Reis, L.: Behaviour evaluation of 3D printed polylactic acid under compression. J. Mater. Res. Technol. **21**, 4052–4066 (2022)
98. Georgiopoulos, P., Kontou, E., Christopoulos, A.: Short-term creep behavior of a biodegradable polymer reinforced with wood-fibers. Compos. Part B Eng. **80**, 134–144 (2015)
99. Farinon, B., Molinari, R., Costantini, L., Merendino, N.: The seed of industrial hemp (Cannabis sativa L.): nutritional quality and potential functionality for human health and nutrition. Nutrients **12**, 1935 (2020)
100. Lamsaf, H., Singh, S., Pereira, J., Poças, F.: Multifunctional properties of PBAT with hemp (cannabis sativa) micronised fibres for food packaging: cast films and coated paper. Coatings **13**, 1195 (2023)
101. Gheribi, R., Taleb, Y., Perrin, L., Segovia, C., Brosse, N., Desobry, S.: Development of chitosan green composites reinforced with hemp fibers: study of mechanical and barrier properties for packaging application. Molecules **28**, 4488 (2023)
102. Köselerli, S.Ü., Gücüş, M.O., Uslu, M.K.: Production of biodegradable composite plates from cross-linked starch and cellulosic fibers. BioResources **18**, 6772–6783 (2023)
103. Teli, G., Mahakur, V.K., Paul, R., Bhowmik, S.: Investigation of dry sliding tribological behaviour of epoxy composites filled with hemp particulates using artificial neural networks. Arab. J. Sci. En. **48**, 3989–4001 (2023)

Chapter 8
Polymer Composites Testing and Analysis

Sukanya Pradhan, Roshni Pattanayak, and Smita Mohanty

Abstract Researchers have embraced polymer composites because of their lightweight, favorable mechanical characteristics, and alternative manufacturing processes as compared to other composite designs. This chapter initially summarizes the research efforts in analyzing the polymeric composites through chromatography and spectroscopy analysis along with a few case studies. In the further section, the effect of the incorporation of fillers and other reinforcements on the degree of thermal stability of the composite materials and its analytical techniques have been discussed followed by the measurement of significant mechanical parameters and associated testing. The last section of the chapter deals with the morphological techniques (SEM and TEM) that provide information about the microstructure of the polymeric composites and X-ray diffraction for investigations of crystal structures.

Keywords Composite · Nanocomposite · Testing · Analysis · Characterization · Techniques

8.1 Introduction

Polymer testing is an important aspect of materials science and engineering that involves the evaluation of physical, chemical, thermal, mechanical, morphological, and electrical properties of polymers. Polymers are large molecules composed of repeating structural units that are widely used in a variety of industries because of their versatility, durability, and cost-effectiveness. Testing of polymeric materials ensures that they meet specific quality standards and performance requirements for their intended use. The primary purpose of polymer testing is to understand and

S. Pradhan (✉) · R. Pattanayak · S. Mohanty
School for Advanced Research in Petrochemicals, Laboratory for Advanced Research in Polymeric Materials, Central Institute of Petrochemicals Engineering and Technology, Patia, Bhubaneswar, Odisha 751024, India
e-mail: pradhan.sukanya2@gmail.com

R. Pattanayak
Utkal University, Vani Vihar, Bhubaneswar, Odisha 751004, India

S. Moharana et al. (eds.), *Polymer Composites*, Engineering Materials,
https://doi.org/10.1007/978-981-97-2075-0_8

quantify the properties of a polymer to confirm its suitability for a specific application. Testing is useful for quality control, research and development, and troubleshooting problems related to polymer materials. Polymers include many different types of materials such as plastics, rubber, and fibers, each with unique properties. Testing and analysis help characterize the behavior of polymers under different conditions and ensure that specific performance requirements are met. Polymer composites are synthetic materials that combine a polymer matrix with reinforcements such as fibers or particles. The synergistic combination of these components improves the mechanical, thermal, and structural properties of the composite.

8.2 Analytical Methods of Polymer Composites

Polymer composites are materials that combine a polymer matrix with reinforcing elements such as fibers, particles, and other additives. Since polymer composite analysis helps to understand the properties, behavior, and performance of polymeric materials, it is a crucial part of materials science and engineering. In order to adequately describe materials and their interfaces, which often contribute significantly to material properties, the analysis of composite materials often requires the use of various analytical techniques.

To learn more about a polymer and its composites, whether they are known or unknown, a combination of techniques such as chromatography, spectroscopy, microscopy, mechanical testing, thermal analysis, and biodegradability/compostability can be applied. Through polymer testing, researchers can predict how a polymer would behave under certain circumstances and environmental factors, such as extreme heat or stretching forces. Using tests that look at a polymer's composition, structure, and other characteristics, researchers can assess a polymer and its composites based on its molecular weight distribution, mechanical capabilities, morphology, flammability, biodegradability, and compostability. In order to make a product on a big scale, manufacturers use polymer composition analysis to determine whether materials have the necessary properties for the desired environment or function. For example, researchers/manufacturers may seek to determine the chemical composition of polymers in order to identify unknown substances or exploit the functionality of composite materials. In addition, compositional analysis helps predict the lifetime of the material.

This is important when developing products that need to withstand high loads or have a long lifespan. Composite analysis of polymers can also predict larger-scale degradation patterns and rates.

8.3 Chromatography

Chromatography is an analytical technique with significant analytical capacity that can be utilized in the analysis of polymer composites. Gas chromatography (GC), liquid chromatography (LC), and gel permeation chromatography (GPC), also referred to as size exclusion chromatography (SEC), are the most often utilized chromatographic techniques in polymer analysis. Each of these techniques has unique applications and advantages in the analysis of polymer composites.

- **Gel Permeation Chromatography (GPC) or Size Exclusion Chromatography (SEC)**

GPC/SEC separates polymers based on size. Large molecules cannot penetrate the pores of the stationary phase and therefore elute first, while smaller molecules enter the pores and elute later. GPC/SEC is commonly used to measure the molecular weight distribution and average molecular weight of polymer composites.

Case study: Polylactic acid (PLA) and PLA composites

In a study reported by Zhao et al. [1], a novel sulfamic acid intercalated MgAl-LDH (SA-LDH) was prepared by intercalating NH_3SO_3 into MgAl layered double hydroxides (LDH), and it was then introduced into poly (lactic acid) (PLA) resin in association with intumescent flame retardant (IFR) by melt blending to prepare a flame-retardant biodegradable PLA composite. The addition of IFR reduces the molecular weight of PLA somewhat, which is quite typical in melt processed samples (Table 8.1). Mechanical and heat stress are generated during the extrusion process, which can induce molecular chain rupture and thus decrease molecular weight. PLA is known to be particularly susceptible to hydrolysis, resulting in the scission of ester linkages and a decrease in molecular mass. As a result, the decomposed H_2O from LDH and SA-LDH can speed up the hydrolysis of PLA.

Gas Chromatography (GC)

GC separates volatile compounds based on their affinity for the stationary phase. It is suitable for the analysis of low molecular weight compounds that can be vaporized without decomposition. GC is often used for analyzing additives, low molecular weight polymers, and monomers in polymer composites.

Liquid Chromatography (LC)

LC separates compounds based on their affinity for the mobile and stationary phases. It can be further categorized into various techniques such as reverse-phase chromatography, normal-phase chromatography, and ion-exchange chromatography.

Table 8.1 GPC results of PLA and its composites (results adapted from Ref. [1])

Sample	Mw × 10^4	Mn × 10^4	PDI
PLA	16.8	11.0	1.5
PLA/IFR	15.8	10.5	1.5
PLA/IFR/LDH	10.5	6.7	1.6
PLA/IFR/SA-LDH	9.4	5.9	1.6

LC is versatile and can be used for separating a wide range of compounds, including polymers and additives in polymer composites.

High-Performance Liquid Chromatography (HPLC)

HPLC is a type of liquid chromatography that uses high pressure to improve separation and speed. It is commonly used for the analysis of polymer additives and degradation products.

HPLC can provide high-resolution separation of polymer components in composites.

Pyrolysis–Gas Chromatography (Py–GC)

Pyrolysis–Gas Chromatography (Py–GC) is a valuable analytical technique used for the characterization of polymers, as well as other complex materials. This method involves the thermal decomposition of a sample in the absence of oxygen (pyrolysis) followed by the separation and identification of the volatile pyrolysis products using gas chromatography. Py–GC involves the pyrolysis of polymers followed by gas chromatographic separation of the resulting pyrolysis products. In industrial scenarios, Py–GC is used for quality control of polymer products to ensure consistency and compliance with specifications. It provides insights into the composition and thermal stability of polymers, helping manufacturers optimize their production processes. Py–GC can be applied in the analysis of geological samples, such as shale and other organic-rich rocks [2, 3]. It provides information about the organic composition of these materials and aids in understanding their thermal history.

In the field of energy, Py–GC is employed to analyze the thermal degradation of polymers used in batteries, fuel cells, and other energy storage devices. It helps researchers understand the behavior of polymers under high-temperature conditions relevant to energy applications [4, 5].

Case Study

In a study [6], the thermal-oxidative degradation of polymers using a gas chromatograph was analyzed simultaneously to study both the kinetics of oxygen absorption and the kinetics of the release of the main gaseous products of thermal-oxidative degradation. Carbon fiber and glass fiber were used as fillers.

The compositions of composite materials were as follows:

1. PEI + 30% filler + 10% plasticizer;
2. PEI + 30% filler + 15% plasticizer;
3. PEI + 30% filler + 20% plasticizer.

Since no substantial amounts of additional thermo-oxidative by-products (methane, nitrogen oxides, etc.) were discovered in the kinetic curves, it can be assumed that only structural changes occur in the composites under these conditions.

Thin-Layer Chromatography (TLC)

TLC separates compounds based on their affinity for a thin layer of stationary phase on a flat support. TLC is a quick and cost-effective method for qualitative analysis of polymer components in composites.

When conducting chromatography analysis of polymer composites, it's crucial to choose the appropriate technique based on the properties of the polymers and the specific information needed. Additionally, sample preparation and detection methods play a significant role in the success of the analysis.

8.4 Spectroscopy

Spectral analysis is a powerful tool for characterizing polymer composites, offering insights into their chemical, structural, and physical properties. Researchers often use a combination of these techniques to gain a more thorough understanding of the complex nature of polymer composite materials.

A well-prepared sample is essential for obtaining accurate and precise results in polymer composite analysis. It demands knowledge of the particular spectroscopic technique to interpret the spectrum data. A more thorough examination of polymer composites can be achieved by combining various spectroscopic techniques. The study of how a material interacts with electromagnetic radiation at different wavelengths is called spectral analysis of polymer composites. The structure, composition, and characteristics of polymer composites can all be learned through the use of various spectroscopic techniques.

Here are some commonly used spectroscopic methods:

UV–Vis Spectroscopy

The technique known as UV–Vis spectroscopy, or ultraviolet–visible spectroscopy, measures the amount of ultraviolet and visible light that a material absorbs. UV–Vis spectroscopy, when used on polymer composites, can shed light on the material's optical characteristics and electronic transitions. When analyzing the electrical and optical characteristics of polymer composites, UV–Vis spectroscopy represents a useful technique, especially when working with conjugated polymers or applications that need transparency and color control. To acquire a thorough comprehension of the subject matter, it is frequently utilized in conjunction with additional analytical methods.

8.5 Application of UV–Vis Spectroscopy to Polymer Composites

8.5.1 Conjugated Polymer Analysis

Conjugated polymers benefit greatly from UV–Vis spectroscopy because their extended π-electron systems cause electronic transitions that occur in the visible or ultraviolet spectrum. The polymer's electrical structure is shown via the spectra.

8.5.2 *Quantification of Additives*

The concentration of colorants or additives in polymer composites can be measured using UV–Vis. The amount of a specific component can be found by measuring the absorbance at different wavelengths.

8.5.3 *Curing and Crosslinking Monitoring*

UV–Vis spectroscopy is employed to monitor curing and crosslinking processes in polymer composites. Changes in absorbance over time can indicate the progression of these chemical reactions.

8.5.4 *Assessment of Transparency and Opacity*

UV–Vis spectroscopy can be used to assess the transparency or opacity of polymer composites in the UV or visible range, depending on the application requirements.

8.5.5 *Analysis of Optical Properties*

The material's optical characteristics, such as transmittance, reflectance, and absorbance, can be determined from the UV–Vis spectra. This is important for use in optical devices, films, and coatings.

In a study [7], a polymer nanocomposite was developed by dispersing multi-walled carbon nanotubes (MWCNTs) in a polycarbonate (PC) matrix with the help of ultrasonic aid. It examined to determine how MWCNTs affected the properties and composition of the composites. The UV–Vis absorption spectra of the PC/MWCNT composite films are shown in Fig. 8.1 which has been adapted from the study. The composition with a 3-wt% loading of MWCNTs exhibits the greatest absorbance, and the absorbance band intensity increases with increasing filler loading. Strongly bundled MWCNTs do not exhibit an absorption band in the 200–1200 nm wavelength range as either their photoluminescence has been quenched or the carriers are tunneling between the nanotubes. This absorption is specific to separately dispersed MWCNTs.

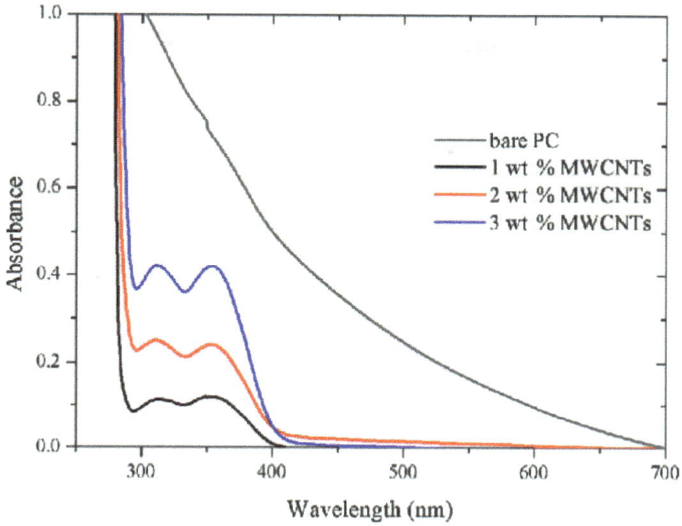

Fig. 8.1 UV–vis absorbance of composite samples with different MWCNT loading (Adapted from Ref. [7]

8.6 IR Spectroscopy

IR spectroscopy relies on the interaction of infrared radiation with molecular vibrations. When a molecule absorbs infrared light, it undergoes vibrational transitions between different energy states. Different functional groups in polymers exhibit characteristic absorption bands in the IR spectrum. These bands correspond to specific vibrations, such as stretching and bending modes of chemical bonds. Spectral peaks are assigned to specific functional groups based on known vibrational frequencies. Vibrational modes include stretching (ν) and bending (δ) modes. The intensity and shape of peaks provide additional information about the concentration of specific functional groups and the molecular environment. The qualitative and quantitative investigation of nanofillers and their nanocomposites has made considerable use of FTIR spectroscopy. FTIR spectroscopy can be used in qualitative investigations to characterize the nanofillers, which include metal oxide nanoparticles, carbonaceous nanostructure, clay, nanofiller synergy, and their polymer nanocomposites. FTIR could also be used in quantitative analyses to demonstrate how nanofillers affect the characteristics of nanocomposite films by observing variations in the FTIR spectra's band intensity.

8.6.1 Bamboo and Flax (BFBF) Mat Reinforced Epoxy Hybrid Composite

In a study reported by Shukla et al. [8], natural fibers reinforced epoxy composite was developed using the hand-lay-up technique. The systematic view of the developed hybrid composite is represented in Fig. 8.2. The IR spectral results represented in Fig. 8.3 have been used in the study as a valuable tool for both qualitative analysis and quantification of these processes. It has been shown how to extract relevant information from spectra to identify typical components of resins and hardeners. The hydroxyl (OH) stretching vibration of free and hydrogen-bound OH groups is represented by a significant peak in the FTIR spectra at 2918.56 cm^{-1}. The bending vibration of absorbed water is responsible for the absorbance peak at 1239.39 cm^{-1}. The atmospheric carbonation on the matrix surface where it reacts with carbon dioxide is indicated by the existence of bands in the range of 1180.97–1003.74 cm^{-1}. The peak at 1418 cm-1, which is ascribed to the CH_3 bending vibration of cellulose, indicates the presence of flax fibers in the composites. Due to the presence of absorbed water in the fibers, the bands at 3385 and 1653 cm^{-1} become more intense.

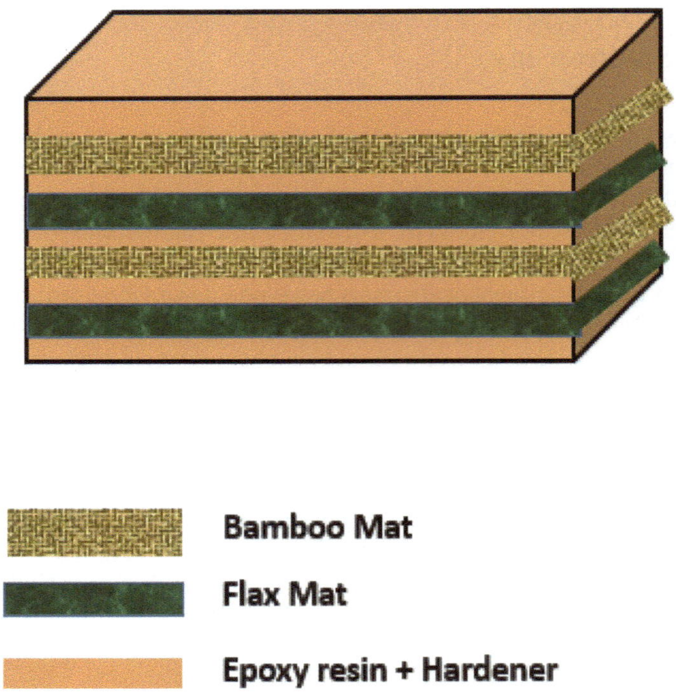

Fig. 8.2 Schematic representation of epoxy/natural fiber hybrid composite

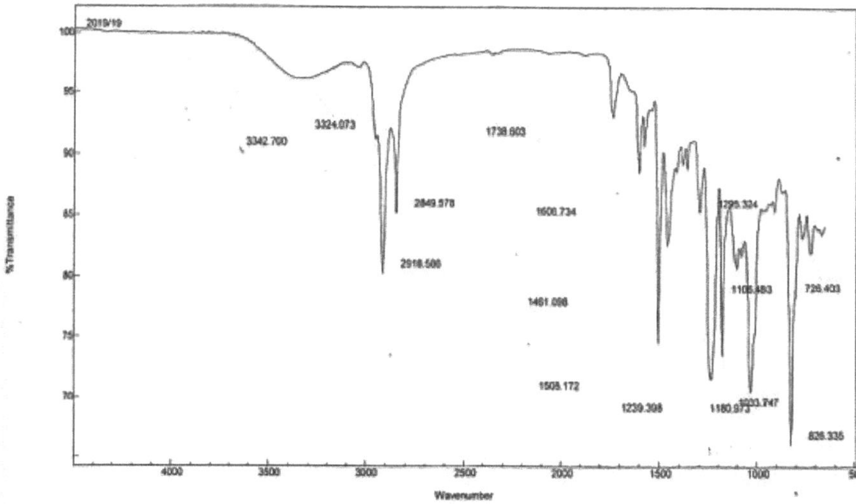

Fig. 8.3 FTIR spectrum of Natural fiber reinforced epoxy composite (Adapted from Ref. [8])

8.7 Raman Spectroscopy

Raman spectroscopy is an additional effective method for analyzing the structural characteristics and molecular vibrations of polymer composites. The inelastic scattering of monochromatic light is measured by Raman spectroscopy, as opposed to infrared (IR) spectroscopy, which detects light absorption. In order to get important insights into the chemical and structural properties of polymer composites, Raman spectroscopy is frequently employed in this field of study. It provides a thorough understanding of material properties in conjunction with other analytical methods including IR spectroscopy and microscopy.

8.7.1 Interpreting Raman Spectra

Peak Assignments

Peaks in the Raman spectrum are assigned to specific vibrational modes. The positions and intensities of these peaks provide information about molecular structure and composition.

Comparative Analysis

The identification and comprehension of the composition of polymer composites can be facilitated by contrasting their Raman spectra with reference spectra of their constituent parts.

8.8 Application of Raman Spectroscopy to Polymer Composites

I. Identification of Components

The polymer matrix, fillers, and additives are among the constituents of a polymer composite that can be recognized and distinguished using Raman spectroscopy. Distinct Raman spectra are displayed by various materials.

ii. Polymer Conformation

The orientation and conformation of the polymer chains inside the composite can be determined using Raman spectroscopy. Modifications in peak positions and intensities could be a sign of structural changes in the molecule.

iii. Crystal Structure and Crystallinity

The method is useful for researching the crystallinity and crystal structure of polymers. In the Raman spectrum, peaks associated with crystalline and amorphous regions can be identified.

iv. Stress and Strain Analysis

The distribution of stress and strain in polymer composites can be examined using Raman spectroscopy. Mechanical characteristics can be understood through variations in molecule orientation and strain-induced shifts in Raman bands.

v. Detection of Degradation

Over time, chemical alterations and degradation in polymer composites can be identified using Raman spectroscopy. Changes in the chemical composition or the existence of degradation products may be indicated by shifts in the Raman peaks.

vi. Mapping and Imaging

Chemical maps of polymer composites can be made by combining imaging methods with Raman spectroscopy. This makes it possible to analyze data spatially resolved, exposing compositional changes throughout the sample.

8.9 NMR Spectroscopy

For the investigation of polymer composites, nuclear magnetic resonance (NMR) spectroscopy is a useful method. It offers comprehensive details regarding the dynamics, composition, and chemical structure of the polymer matrix and any reinforcing elements. All kinds of NMR techniques enable semi-quantitative monitoring of changes in segmental dynamics along the polymer-filler interface and bonding inside the polymer chain. Here are some key aspects of applying NMR spectroscopy to polymer composites:

i. Polymer Matrix Analysis:

Proton NMR (^{1}H NMR): Commonly used to characterize the polymer matrix. It provides information about the different types of hydrogen environments in the polymer structure.

Carbon-13 NMR (^{13}C NMR): Useful for studying the carbon backbone of the polymer and can provide insights into the polymer chain architecture.

ii. Reinforcing Material Analysis:

If the composite includes materials like fibers, fillers, or nanoparticles, NMR can be applied to study their composition and interactions with the polymer matrix.

iii. Crosslinking and Curing:

NMR can be employed to monitor the progress of polymerization, crosslinking, or curing reactions. For example, in the case of thermosetting polymers, changes in chemical shifts and peak intensities can provide information about the degree of curing.

iv. Quantification:

NMR can be used for quantitative analysis, helping to determine the relative amounts of different components in the composite.

v. Dynamics and Morphology:

NMR relaxation times can provide information about the molecular dynamics of polymers within the composite material.

Solid-state NMR techniques can offer insights into the morphology and crystallinity of polymers in composite materials.

vi. Interaction Studies:

NMR can be applied to investigate interactions between the polymer matrix and reinforcing materials, such as intermolecular forces or chemical bonding at the interface.

vii. Quality Control:

NMR can be used as a non-destructive tool for quality control, helping to ensure the consistency and integrity of polymer composite materials during manufacturing.

viii. Variable Temperature Studies:

By varying the temperature during NMR experiments, researchers can study how temperature influences the properties and behavior of polymer composites.

ix. Multi-Nuclear NMR:

If the composite contains elements other than hydrogen and carbon, multinuclear NMR, such as fluorine-19 or phosphorus-31 NMR, can be employed to gain additional information.

x. In-situ Studies:

NMR is a real-time tool that may be used in-situ to investigate changes in the composite's characteristics and structure under various circumstances.

The precise nature of the composite material and the information sought will determine which NMR approach is best to use when using NMR spectroscopy to polymer composites. Careful sample preparation is necessary.

8.10 Mass Spectroscopy

Polymer nanocomposites can be analyzed using mass spectroscopy techniques, which can also reveal details on the dispersion and interactions of nanoparticles inside the polymer matrix. It provides a flexible range of methods for the thorough examination of polymer composites, allowing scientists to learn more about the composition, structure, and thermal behavior of these materials. [9–11]. Mass spectroscopy in polymer research involves either degrading the sample prior to analysis or vaporizing/ionizing high molecular weight molecules into the mass spectrometer. Mass spectroscopy was primarily used for degradation in polymer research until the introduction of soft ionization techniques like electrospray ionization (ESI) and matrix-assisted laser desorption/ionization (MALDI). The advancement of soft ionization technologies has made mass spectrometry (MS) an extremely potent tool for characterizing polymers. These days, the most popular ionization techniques include desorption electro spray, electrosonic spray, matrix-assisted laser desorption/ ionization (LDI and MALDI), direct laser or ESI and its derivatives, and electrosonic spray, which cause little to no fragmentation [12–14]. Applications of polymer research include structural characterizations of synthetic polymers of various chemical types, including their composition, sequence, presence and kind of additives, end groups, linear or cyclic geometry, composition, and molecular weight distributions.

8.11 Inductively Coupled Plasma Mass Spectrometry (ICP-MS)

ICP-MS is a flexible technique that is useful for materials characterization and quality control in sectors including electronics, automotive, and plastics. It can yield quantitative and qualitative data on a broad variety of elements in polymer composites. ICP-MS is effective for screening and thorough characterization of polymer composites because it enables the simultaneous determination of several elements in a single analysis. ICP-MS can perform elemental analysis as well as isotope ratio analysis, which is useful for identifying sources and figuring out the components present in a polymer composite.

Various major thermo-analytical tests such as Differential Scanning Calorimetry (DSC), Thermal gravimetric Analysis (TGA), and Dynamic Mechanical Analysis (DMA) are there to analyze the thermal properties of the developed composite samples. In this context, we have tried to describe briefly the thermal properties of composite materials using the above-mentioned techniques.

8.12 Terminology Associated with the Thermal Properties

8.12.1 *Melting Temperature (T_m)*

Melting temperature (T_m) is a type of first-order thermodynamic transition in which instantaneous change in volume occurs after exchanging the heat between the system and surroundings. At this point, the degradation of material takes place from solid to liquid form. Preferably, crystalline material shows a sharp T_m value whereas amorphous material shows a wide range value. It is a very essential factor that varies from material to material which gives a clear vision from processing to finished product. The thermal stability of the developed composite samples not only depends upon the T_m value but also on the adjacent environment atmosphere, exposure to beyond time duration, and contamination to other hazardous chemicals.

8.12.2 *Glass Transition Temperature (T_g)*

Glass transition temperature (T_g) is a type of second-order thermodynamic transition in which a change in heat capacity takes place instead of exchanging heat. It is directly connected to disorderness in the molecular structure of the sample affecting the increased motion in wiggling the chains (not segmental motion) preferably found in an amorphous region. In simple form, it is described as a point in which a rubbery state (Flexible) exists above this point whereas a glassy state (Brittle) exists below it. The mobility of the polymer chain in the polymeric composite affects the T_g value which totally depends upon various factors such as chain stiffness, intermolecular forces, pendant group, crosslinking chains, and plasticizers. The increment in the immobile chain with restricting the motion also influences the higher T_g value. The easy mobility in the polymeric chain brings softness and flexibility to the sample at lower temperatures whereas a high temperature is required to form a rubbery region if the polymeric chain cannot shift easily to molecules that are in steady mode. In a DSC curve, a step-like baseline shift indicates the T_g point. With a specified scanning rate, specific heat changes during the T_g phase, and this change in specific heat is responsible for the height of the baseline shift. Nealy no baseline shift is noticed in the crystalized polymer sample whereas the height shift is observed in a completely amorphous sample. Figure 8.4. clearly explained the T_m and T_g point present in crystalline (first-order transition) and amorphous (second-order transition) regions.

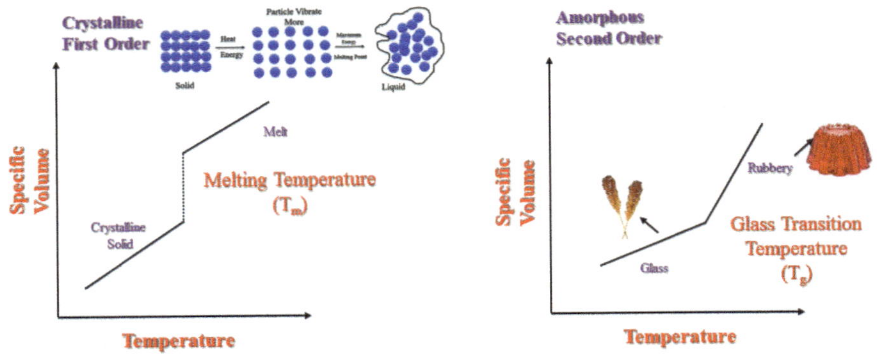

Fig. 8.4 Specific volume vs. temperature graph showing T_m and T_g point

8.13 Characterization Test Methods for Thermal Analysis

Various standardized test methods are present to evaluate the thermal analysis of the composite materials. Here, essential characterization test methods are discussed below:

8.13.1 Differential Scanning Calorimetry

Working Principle

Differential Scanning Calorimetry (DSC), ASTM D3418-08 is a thermo-analytical technique in which two crucibles (sample and reference pan) are subjected to the difference in thermal capacity at nearly the same specified temperature throughout one single run for the specific samples. The reference is measured as a function of temperature and the temperature differences in the sample crucible are observed due to the difference in the applied heat to the system. Here, a DSC curve is obtained in the controlling display by plotting the graph between heat flux (dH/dt or dQ/dt) and temperature as shown in Fig. 8.5. This represents the DSC graph plot for the PET polymeric sample. Here, the thermal capacity (C_p) is represented at the baseline step in the temperature of 77 °C which is also called as T_g temperature. The exothermic peak is observed at 136 °C which represents the crystallization temperature (T_c). The endothermic peak at 261 °C is explained as the T_m temperature of the PET sample. The shaded area in the curve explains the amount of heat absorbed or released during the phase transition in the sample.

Summarily, the study of polymeric composite materials in liquid, powder, and film of sheet form is frequently conducted using this calorimetric technique which yields data on phase transformations, melting temperature, glass transition temperature, curing temperature, and crystallization process.

Instrumentation and Sample Preparation

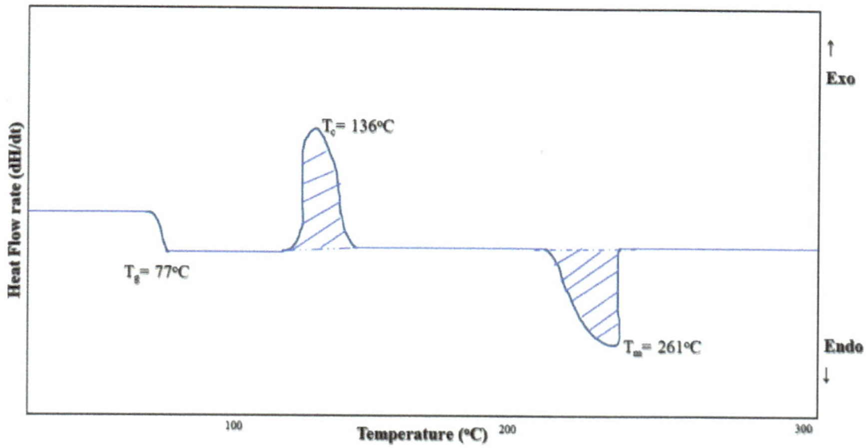

Fig. 8.5 DSC graph plot of a random PET polymer (Adapted from Ref. [15])

An overview of the DSC instrument having a sample section and monitoring section was mentioned in Fig. 8.6.

a. Instrumentation:

A DSC instrument unit includes a furnace with two pans (sample and reference), a temperature controller (monitoring the temperature), a power compensation unit (heating system and purge gas for inert atmosphere), and a controlling panel (data collection).

b. Sample Preparation:

Generally, 1–10 mg of solid samples like powder, sheets, films, etc. are tested in this. The taken sample mass is totally dependent upon the behavior of the sample composition. Specifically, composite samples are taken in larger amounts to overcome their heterogeneity property. Apart from this, liquid samples also can be tested

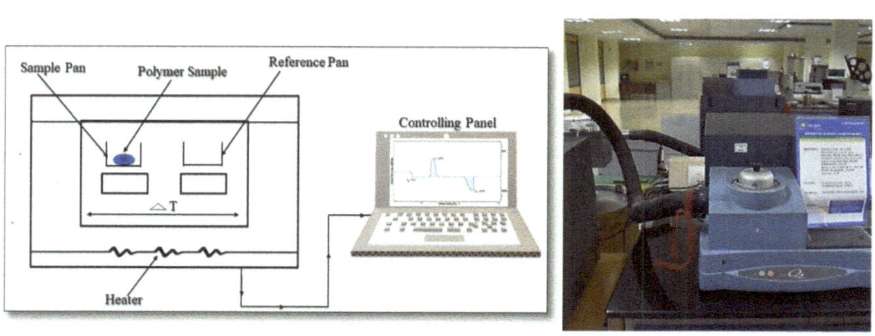

Fig. 8.6 An overview of DSC Instrument (DSC Q 20, TA INSTRUMENTS, USA)

in this test prior to the annealed crucible. Two types of crucibles are being used such as aluminum crucibles for less-temperature samples and alumina or platinum crucibles for higher-temperature samples. In general, polymeric composite materials are analyzed under the temperature of 350 °C because they cannot sustain and start degrading above this limit.

8.13.2 *Factors Associated with DSC*

a. Temperature:

From the DSC curve, one can obtain knowledge regarding the melting or crystallization temperature and glass transition temperature of the polymeric composite material. Melting temperature and crystallization temperature always exhibit endothermic and exothermic peaks, respectively, whereas a jump in baseline peak in a specified temperature range indicates the glass transition temperature point of the polymeric composite sample.

b. Scanning Rate:

Scanning rate always plays an important role in the DSC curve specifically for the transition temperature, but does not affect the endo or exothermic peak area. As polymeric composite materials show poor thermal conductivity, it affects the peak in the DSC plot. Rapid scanning creates an overlapping between two peaks and may hinder the result as compared to the actual one. Besides this, slow scanning is mostly time-consuming and polymers show low thermal transition which may affect the cleared peak value. Thus, the scanning rate should be maintained according to the thermal conductivity of the tested sample. The scanning rate of around 5–10 °C/min is preferably taken for the polymeric samples.

c. Thermal history:

Thermal history including processing temperature parameters, storage conditions, and heating and cooling rate of the specified sample affects the DSC result.

d. Purge gas:

Purge gas including He and N_2 prevents the oxidization process in the sample and also protects the detector from corrosion in case of volatile sample is used to test. Besides this, high purge gas flow also affects the baseline plot. Thus, the flow rate of 20–40 mL/min maintaining the consistency throughout the experiment is recommended.

Case Study

DSC graph can be easily interpreted from a random case study. The study by Wen et al. [16] employed DSC characterization to evaluate the thermal characteristics of the neat epoxy and fiber reinforced plastics (FRP) composites as described in Fig. 8.7. Since Glass fiber restricts the mobility of epoxy molecules, a definite increment

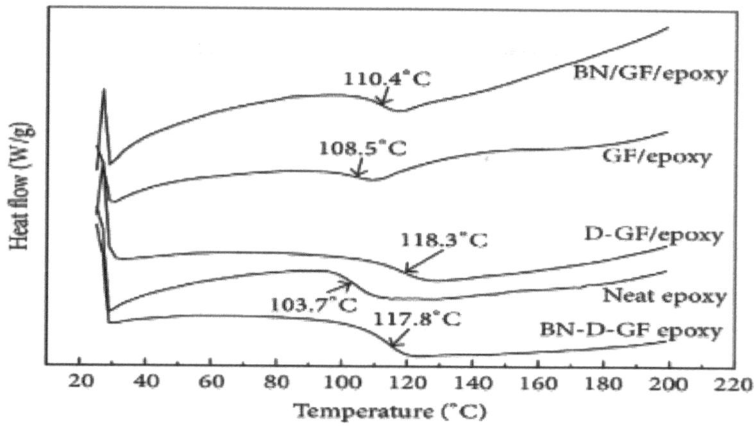

Fig. 8.7 DSC plot of neat epoxy resin and FRP composites. Adapted from Ref. [16]

in T_g value was seen in GF loaded epoxy composite. Wen et al. have reported a significant interaction between dopamine-treated composite that had a higher T_g than the neat epoxy one. Thus, we concluded that the presence of filler or any chemical modification can affect the thermal characteristics of a composite sample.

8.14 Thermal Gravimetric Analysis

Working Principle

Thermal gravimetric analysis (TGA), ASTM E 1868 is a thermal analysis method that measures the weight change of a sample with a specific temperature range in an inert atmosphere (nitrogen). This provides significant information regarding composition and weight loss according to the temperature and thermal stability of the sample. A TGA graph may contain a step or multi-step weight loss plot in which the mass loss percentage of the given samples is calculated from the height of the steps that appeared in the curve. In general, initial mass loss and its corresponding temperature are denoted as m_i and T_i respectively, whereas final mass loss and its corresponding temperature are denoted as m_f and T_f, respectively. The DTG graph is the differential graph of the TGA plot in which the maximal weight loss rate of the sample can be obtained from the T_{max} point. This is clearly visualized in Fig. 8.8, which explains a random TGA-DTG plot of FEP (fluorinated ethylene propylene) material.

Instrumentation and Sample Preparation

An overview of a typical TGA instrument is shown in Fig. 8.9.

a. Instrumentation:

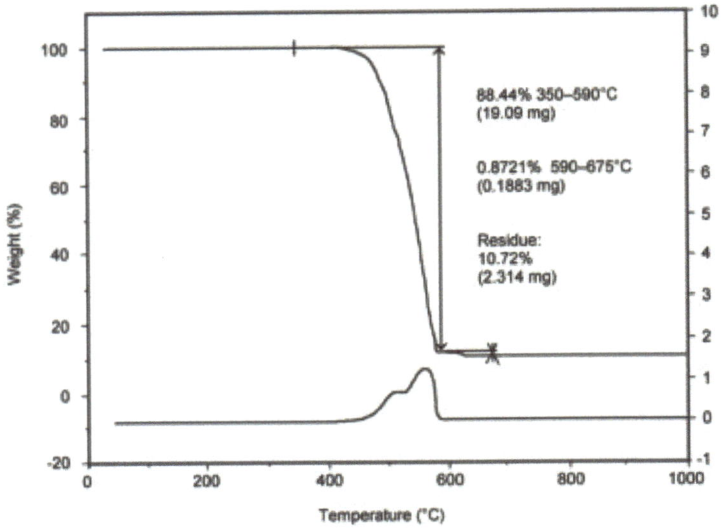

Fig. 8.8 TGA and DTG graph plot of pigmented FEP (fluorinated ethylene propylene) resins. Adapted with permission from Ref. [17]

Fig. 8.9 An overview of TGA instrument (TGA Q50, TA INSTRUMENTS, USA)

TGA instrument comprises a sample pan, temperature control panel, and data collector. The sample is positioned in the crucible, the temperature is applied through a control panel with a specific range, and the data is collected from the collector. The used crucible for the test is generally made up of platinum or alumina which have characteristics to resist the high-temperature range.

b. Sample Preparation:

Composite Powder, thin film, or sheet samples of 5–10 mg are appropriate for the test. This test can be carried out at a maximum of up to 900–1000 °C.

8.14.1 Factors Associated with TGA

1. Chemical Constituents:

A temperature region for the specific sample should be identified for the smooth running of the instrument. Before examining the sample, basic individual constituents should be priorly known for safety purposes.

2. Effect of experimental parameters:

The heating rate is the crucial parameter in the TGA test result. A high heating rate affects the test result as polymers have poor thermal conductivity properties. Thus, a moderate heating rate of 5–10 °C/min is typically used for the polymeric composite material. Generally, degradation behavior is analyzed in the N_2 atmosphere, whereas thermal oxidation is analyzed in the O_2 atmosphere. The purge gas flow rate is approximately 40 mL/min as a high flow rate creates an unstable result. As a whole, various experimental parameters such as sample mass, purge gas flow rate, and temperature control affect the TGA test results.

8.14.2 Case Study

TGA graph can be easily interpreted from a random case study. The study by Ahmed Moosa et al. [18] employed TGA characterization to evaluate the weight loss characteristics of the neat epoxy and its nanocomposites as described in Fig. 8.10. The TGA curve (left side: Fig. 8.10) showed three stages of degradation including initial decomposition, glass transition curve, and decomposition plot. At the initial stage, a very small amount of degradation is noticed due to the presence of heavy constituents in the composite material. The maximum (50%) decomposition temperature was noticed at around 395 + 5 °C which was visualized in Fig. 8.9: right side. The presence of nanofillers in the composite sample evidently affects the thermal stability at the third stage. As a whole, epoxy and modified epoxy showed apparently similar values as noticed from the graph. From this, we can conclude that filler content may or may not affect the thermal stability of the composite material, and thermal properties totally depend upon the composition of materials.

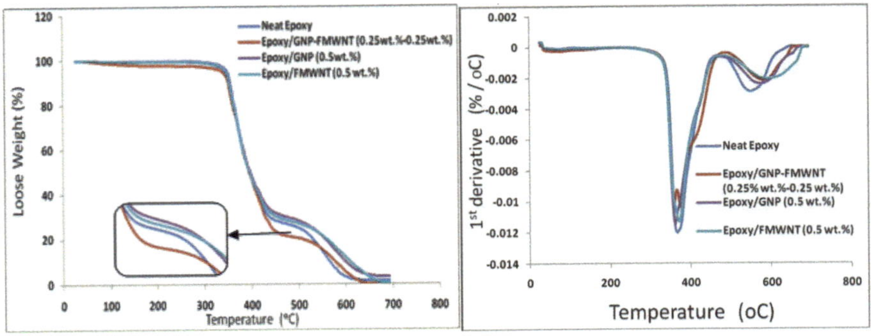

Fig. 8.10 TGA Plot of neat epoxy and its nanocomposites (Left) and DTG Plot of neat epoxy and its nanocomposites (Right) [19]

8.15 Dynamic Mechanical Analysis

- **Working Principle:**

Dynamic Mechanical Analysis (DMA) D4065, D4440, D5279 is a thermal analytical technique to measure the change in viscoelastic behavior of a sample with respect to temperature, time, or deformation frequency after a certain amount of oscillating force applied to it. The elastic modulus, loss modulus, and damping coefficient as a function of temperature, frequency, or time can be analyzed as sinusoidal or fixed rate or constant fashion mechanical deformation applied to the sample. Additionally, the quantitative analysis of the sample can be carried out by detecting the glass transition temperature and other sub-T_g transition values of the sample. From the DMA graph, the degree of phase separation in multi-component material, filler amount, and processing parameter behavior can be measured. Figure 8.11 represents the typical graph of DMA storage modulus vs. temperature in which various transitions or regions were noticed. Initially, some amount of molecular motion along with bending and stretching in the polymeric chain occurs in the initial phase. β-transition explains the toughness behavior of the sample. When the ratio between T_g and T_β is less than or equal to 3, the sample shows good impact strength. After that, the glass transition region is noticed in which chain mobility and phase transition are observed. The curve behavior describes the nature of the sample. At the last phase, mobility of a large segment was observed in which the polymer starts melting and the T_m value is obtained.

8.15.1 Instrumentation and Sample Preparation

A clear overview of the DMA instrument is shown in Fig. 8.12.

a. Instrumentation:

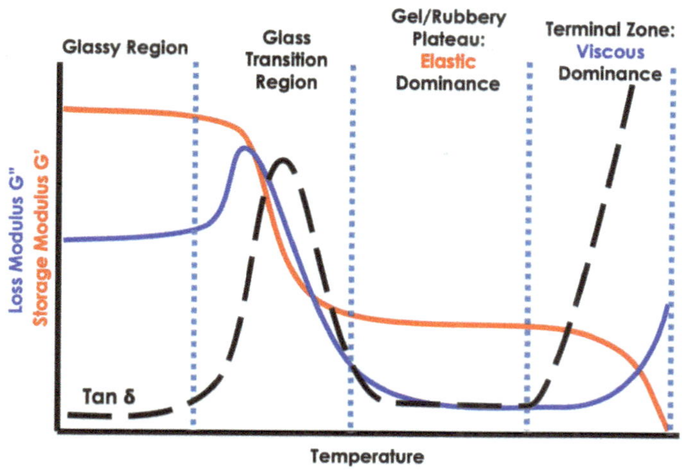

Fig. 8.11 A general DMA plot of storage modulus (red), loss modulus (blue), and tan delta (black dashed) as a function of temperature parameter [20]

Fig. 8.12 An overview of DMA Instrument (DMA Q 800, TA INSTRUMENTS, USA)

The DMA instrument consists of sample holding panels, an Oscillatory deformation device, a detector temperature controller, and an oven. The sample is clamped between the ends of two parallel arms which is adjusted mechanically by a motor attached to one arm drive of the sample panel. Here, after mechanical vibration force is applied, the sample undergoes deformation. The sample is placed in the temperature-controlled chamber in which a precise temperature is controlled and a continuous flow of purge gas is maintained. Before running the instrument, calibration is a must.

b. Sample Preparation:

Solid samples including sheets, films, fibers, rods, or rectangular shape bars can be analyzed with this instrument. Uniformity in thickness should be maintained in a thin sample to avoid the uneven aspect ratio and fluctuating results. Therefore, it should be cut in a parallel edge of 5–8 mm wide by using a parallel blade cutter or straight edge or razor blade for smooth cutting. Thinner-shaped fibers should be twisted uniformly for easy running. Thick-shaped single fiber can be easily tested. Sheet-shaped samples must be cut into bar shapes of appropriate size.

8.15.2 Factors Associated with DMA

Moisture content present in the sample affects the modulus result for which samples should be pre-dried before testing. Excess filler present in the composite material affects the storage modulus result. The rise of buckling during applied load causes serious errors in test results which should be minutely examined.

- **Case study:**

DMA graph can be easily interpreted from a random case study. The study by Wen et al. [16] employed DMA characterization to evaluate the elastic modulus and loss modulus of the neat epoxy and FRP composites as described in Fig. 8.13. In this graph (left: Fig. 8.13), it was clearly visible that the modulus of the matrix got increased after the increment in filler constituents that allowed easy stress transfer throughout the samples. The D-GF-loaded epoxy composite sample exhibited the highest E value of 376.24 MPa as compared to the neat one. Approximately more than 75% enhanced storage modulus was noticed in the coated sample at room temperature. The presence of physical crosslinking by a hydrogen bond between the coated filler and neat epoxy samples is the reason for this increased modulus value. In the right-sided graph of Fig. 8.10, the variation of loss modulus for all the samples with respect to temperature was visualized. The addition of fiber content affects the broadening of the loss modulus peak which could be ascribed to the inhibition of the relaxation process within the samples. The composite sample showed a higher loss modulus value as compared to the neat one due to the rise of increased dissipation energy during internal friction. As a whole, the behavior of the used filler is fully responsible for the storage or loss modulus value in the composite sample.

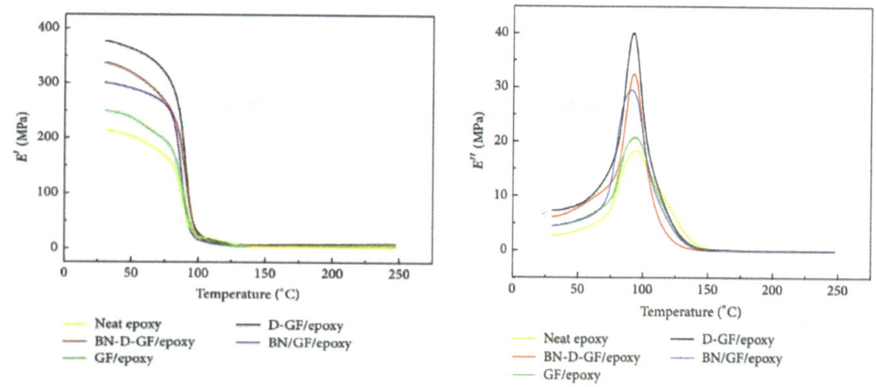

Fig. 8.13 DMA plot: Dynamic storage modulus of neat epoxy and FRP composites (left); Loss modulus of neat epoxy and FRP composites (right) [16]

8.16 Mechanical Testing

The integration of reinforcements in the matrix is primarily responsible for the remarkable mechanical properties of composites. Because of their nature, composites exhibit mechanical behavior that differs greatly from that of typical materials such as metals. As a result, for tensile, compressive, and shear tests, other standards established for directly measuring the bulk mechanical properties of composites are usually preferable. Composite materials, particularly lightweight composite structures, are increasingly being used in a wide range of products and applications, including aerospace, construction, ground transportation, and environmentally sustainable energy systems, all of which require extensive mechanical and physical testing of composite material components and structures prior to their application. Tensile (tension), flexural, impact, shear, and compression with open and closed holes are the most typical standardized mechanical tests for polymer composites, and physical testing includes water absorption, density, void content, hardness, and scratch resistance. Researchers also tested the compression, shear, and interlaminar strength of composite samples to determine the optimal parameters of the failure criterion model considering ultimate strength and failure prediction.

- **Tensile**

Tensile testing is a destructive test that provides information about the material's tensile strength, yield strength, and ductility. Tensile testing according to ASTM D3039 is used to determine the force required to break a polymer composite specimen as well as the percentage to which the specimen stretches or elongates to that breaking point. Tensile testing generates a stress–strain diagram that is used to calculate tensile modulus. Tensile testing of polymer composites is typically performed using basic tension or flat-sandwich tension standards such as ISO 527-4, ISO 527-5, ASTM D 638, ASTM D 3039, and ASTM C 297.

- **Hardness**

Hardness testing is a non-destructive test procedure that involves applying a steady force to a metal surface using a rounded or pointed object under controlled conditions to form an indentation. This is then measured to establish the material's hardness. It is a measure of the plastic deformation that a material can undergo when subjected to external stress. As a result of the improved resistance to plastic deformation, particle reinforcement increases the hardness of the material. Polymer hardness is an important property that determines a material's resistance to abrasion, scratching, and other forms of mechanical deformation. It is a key factor to consider when selecting and creating polymers for a variety of applications such as technical parts, consumer goods, and biomedical equipment.

The three most often used procedures for determining the hardness of polymers are as follows.

- **Shore hardness examination**

The Shore hardness test is a popular way to assess the hardness of elastomers, rubbers, and other soft materials. It works on the idea of measuring the depth of penetration of a standardized indenter into the material's surface under a set force. A durometer is used in the test to evaluate the resistance of the material to indentation by a spring-loaded indenter. The durometer measures hardness on a scale of 0–100, with higher readings indicating more hardness. The Shore hardness test is simple and quick to perform, and it provides an accurate assessment of the hardness of soft materials. However, it isn't appropriate for measuring the hardness of hard polymers and composites since the indentor might damage the material's surface.

Rockwell Hardness Test

The Rockwell hardness test, which can also be used to evaluate the hardness of composite materials and polymers, is a standard technique for determining the hardness of metals. It is based on the concept of determining how much material a diamond or tungsten carbide ball can penetrate under a certain amount of stress. The test employs a Rockwell hardness tester, which measures a material's resistance to indentation by a diamond or tungsten carbide ball. The Rockwell hardness test can be used to accurately and easily determine the hardness of hard polymers and composites.

Nanoindentation

Nanoindentation is an advanced technology for assessing material hardness at the nanoscale level. It works on the premise of measuring the depth of penetration of a sharp indenter into a material's surface under a defined force. The test employs a nanoindenter, which is an instrument that applies a very small force to the material (usually in the micronewton range) and measures the consequent displacement of the indenter. Nanoindentation enables highly accurate measurements of material hardness, as well as other mechanical parameters such as elastic modulus and creep resistance.

A nanoindentation test evaluates the elastic and plastic properties of materials on the nanoscale by using electromagnetic force and capacitive depth measuring. It is an adaptable method for characterizing a wide range of materials, particularly in a homogenous continuum system. Polymers loaded with functional reinforcements such as CFs and nanoparticles (i.e., CNTs, CNFs, and graphene) have garnered significant interest and extensive application in recent decades. The nanoindentation can characterize inhomogeneous composite materials formed of discrete regions with unique material properties due to the extremely small force and displacement resolutions. Pyramidal indenters such as Berkovich and Vickers are often used for the characterization of composite materials, and conical and cylindrical punches with flat-ended cylindrical geometry are also frequently used for characterizing the polymers [21].

Impact Test

An important test for determining a material's ability to tolerate shock loading or unexpected forces is the impact test for polymer composites. Impact tests on polymer composites can be performed using a variety of standard techniques, the selection of which is frequently influenced by the particular characteristics and intended use of the composite material. Two popular techniques for impact testing are as follows.

Charpy Impact Test

A standardized notched specimen of the polymer composite is positioned horizontally in a Charpy impact test, and it is struck by a pendulum, which fractures the specimen. It is measured how much energy the specimen absorbs during fracture. A V-notch or U-notch is used to prepare the specimen in order to establish a point of stress concentration. It is noted how much energy the specimen used to break. The amount of impact energy absorbed reveals something about how resilient the material is to abrupt loads.

Izod Impact Test

The Izod impact test entails striking a notched specimen with a pendulum, just like the Charpy test. The primary distinction is in the specimen's vertical placement in the Izod test. The Izod impact test yields data on the material's resistance to impact loading, just like the Charpy test does. These tests are often conducted at varying temperatures in order to evaluate the material's performance under various environmental conditions. Impact test results make it easier to comprehend a material's behavior under sudden loads and assess its suitability for applications where impact resistance is essential, such as in sports equipment, automotive components, or structural components.

Flexural

Flexural testing, also known as bending or transverse testing, is a common method for evaluating the mechanical properties of polymer composites. This test helps determine the material's flexural strength, flexural modulus, and other related characteristics. These tests are essential for understanding how a polymer composite behaves under bending loads, providing valuable information for structural design and material selection.

Compression Test

Compression testing of polymer composites is a standard method for evaluating the behavior of materials under compressive loading. This test helps determine important mechanical properties such as compressive strength, compressive modulus, and deformation properties. This section provides an overview of compression testing. Compression test results are important for understanding how polymer composites behave under compressive loading conditions. This understanding is useful when designing parts or structures that may be compressed during use.

8.17 Morphology Analysis

The relation between structure and property in multi-component polymeric materials is crucial; that is, fundamental research on the subject leads to the creation of new polymer material designs that meet the various demands of industry. Thus far, different microscopic and scattering techniques have been used to conduct morphological studies of the multi-component polymeric materials. Among the commonly used and commercially accessible microscopes are optical, transmission electron, scanning electron, and atomic force microscopes (TEM and AFM). The primary benefit of microscopy is its ability to present diverse morphologies in an intuitive real-space form. But because of its narrow field of view, microscopy can occasionally be unreliable in terms of statistical correctness when it comes to "measurements," particularly quantitative ones. Conversely, because the scattering methods' observation volume is more than that of the microscopes, they offer significantly higher statistical accuracy than microscopy.

8.17.1 Scanning Electron Microscopy

Scanning Electron Microscopy (SEM) is a powerful technique used to analyze the surface morphology and structure of materials at a high resolution. When it comes to polymer composites, SEM can provide valuable information about the distribution of fillers, the adhesion between matrix and reinforcement, and the overall microstructure. Here's a general guide on how SEM analysis of polymer composites can be conducted:

Sample Preparation

Cutting and Mounting

Cut the polymer composite sample into small, representative pieces. Mount the samples on suitable SEM stubs using conductive adhesives.

Surface Cleaning

Clean the samples to remove any contaminants or residues. This may involve solvent cleaning or other appropriate methods.

Coating (Optional)

Coating with Conductive Material:

Some polymer composites may be insulating, leading to charge build-up during SEM analysis. To address this, a thin conductive coating (e.g., gold, palladium) can be applied using a sputter coater.

Instrumental Analysis

Instrument Setup

Adjust the SEM parameters, such as accelerating voltage, beam current, and working distance, based on the characteristics of the polymer composite.

Image Acquisition

Capture secondary electron (SE) and/or backscattered electron (BSE) images.

SE images provide surface topography details, while BSE images can give information about compositional variations.

Elemental Analysis (EDS)

If equipped with an Energy-Dispersive X-ray Spectroscopy (EDS) system, perform elemental analysis to identify the distribution of different elements within the polymer composite.

Mapping

Conduct elemental mapping to visualize the spatial distribution of elements, particularly if the composite contains fillers, reinforcements, or additives.

Magnification

Use various magnifications to study both the overall morphology and finer details of the composite structure.

8.17.2 Transmission Electron Microscopy (TEM)

TEM is a type of microscopy in which an electron beam is passed through a very thin material and interacts with it along the way. Strong magnetic fields are the focus of the electron stream, and electrons are located in a spiral trajectory. Electrons transmitted through the material are magnified and focused by an objective lens to generate a picture that is displayed on an imaging screen. One crucial and fundamental topic for the usage of nanoparticles in various fields is their size and size distribution. The photos were used to calculate the size of the nanoparticles and the size homogeneity of the particles.

Investigating nanostructures is necessary for gaining information about the properties of nanostructured materials, which is necessary for understanding materials at the nanoscale. Nonetheless, the study and development of products based on nanotechnology comes with difficulties. Sophisticated characterization techniques are required to precisely manage the size of nanoparticles, grains, size distribution and homogeneity, lattice type, crystal structure, dispersion, and chemical and physical properties of phases, such as number, morphology, and structure of the phases at the nanoscale. For this reason, the most effective method for characterizing nanotechnology is transmission electron microscopy (TEM). Through many years of development, the TEM has become a very sophisticated tool with extensive use in many scientific areas. TEM is a perfect tool for characterizing nanostructures, and it has been helpful in providing high-resolution 3D information on a range of polymeric structures, such as carbon black/natural rubber composites,

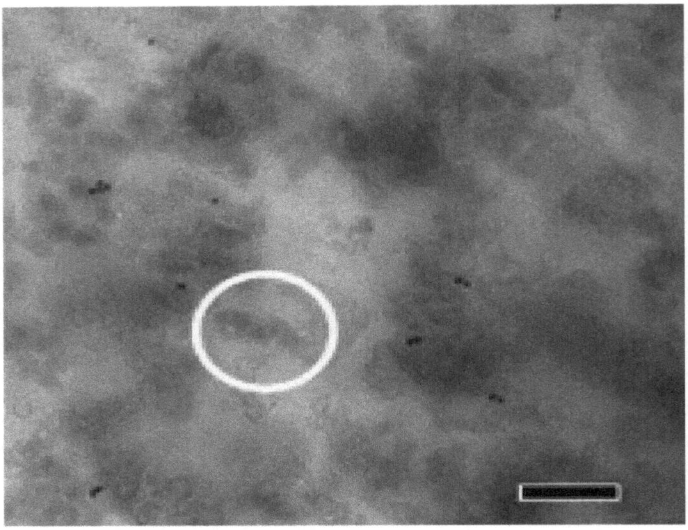

Fig. 8.14 TEM micrograph of the CB & Si/NRBR system (Adapted from Ref. [23])

clay/polymer nanocomposite materials, and block copolymer nanoscale microphase-separated structures. [22, 23]. In addition to 3D images, some of these investigations also included quantitative structural analyses.

One typical example (Fig. 8.14) is a composite of CB and Si nanoparticles in a rubbery matrix (Represented further as CB & Si/NRBR).

8.18 Atomic Force Microscopy (AFM)

AFM is a powerful and versatile microscopy technique that can be used to characterize the surface morphology and mechanical properties of polymer composites at the nanoscale. AFM offers a real 3D surface profile, in contrast to the electron microscope 2D projection or 2D image of a sample. Furthermore, samples examined by AFM do not need to undergo any unique treatments (such as metal or carbon coatings) that could harm or alter the sample permanently. While most AFM modes operate flawlessly in ambient air, electron microscopes require an expensive vacuum environment to function properly. AFM has the potential to offer better resolution than SEM. True atomic resolution in ultra-high vacuum (UHV) and, more recently, in liquid conditions has been proven to be achievable with AFM. The resolution of high-resolution AFM is similar to that of TEM and scanning tunneling microscopy. The image size is a drawback of the AFM method in contrast to the SEM. The SEM has a depth of field of millimeters and can image an area that is millimeters by millimeters. The maximum height and maximum scanning area that the AFM can image are both in the order of micrometers and roughly 150 by 150 μm, respectively.

Using AFM, the surface of clay particles scattered inside a polymer matrix has been imaged in the context of clay-containing polymer nanocomposites. Yalcin and Cakmak [24], for instance, employed AFM to study the surface morphology of the intercalated silicate layers that were distributed throughout the nanocomposite films. The effect of adding nanoclay particles on the shape and crystal growth behavior of immiscible polymer blends has also recently been studied by researchers employing AFM.

8.18.1 X-ray Diffraction (XRD)

X-ray diffraction spectroscopy (XRD) is an important technique for the structure–property correlation study of polymer nanocomposites. X-ray Diffraction (XRD) analysis is a powerful technique used to study the crystalline structure of materials. It provides information about the arrangement of atoms in a crystal lattice, crystal size, and phase identification. Atoms are arranged in a periodic pattern in crystalline materials. Fixed wavelength X-ray light interacts with the electrons, causing an elastic collision that causes the electrons to oscillate. In order to produce a coherent source of electromagnetic radiation at the same frequency and phase as the entering X-ray, the electron cloud functions as a secondary source. Different atoms' radiation will interfere with one another either constructively or destructively.

The desired set of crystal characteristics is determined by measuring, recording, and analyzing the diffraction peaks for constructive interference. While destructive interference defies Bragg's law, constructive interference does not. The development of nanocomposites, as well as the characteristics and applications of these materials employing diverse blends of polymers and nanoparticles have been extensively studied. XRD is a primary technique used to understand the impact of nanoparticles on the alteration of the polymer structure. Because of their distinct layered structure, clays and organoclays have a distinctive diffraction peak in their XRD spectra. Peak position, d-spacing, and peak intensity changes indicate the structural alterations caused by polymer intercalation or surface modification. By applying Bragg's law, the interlayer spacing may be determined based on the XRD spectra's peak position (2θ) and peak breadth at half strength. The following equation represents Bragg's law which states that a rise in d-spacing causes the associated XRD peak to expand and shift toward lower diffraction angles (2θ).

$$n\lambda = 2\mathrm{d}\sin\theta$$

where λ = wavelength;
θ = angle of incidence;
d = path difference;
n = a whole number.

8.19 Conclusion

In the present scenario, novel polymeric composites are being designed based on the structure–property relationship. These cutting-edge composite materials are made with a variety of fillers, including organic, mineral, and metallic ones, and reinforced by organic and/or inorganic fibers. The novel materials are targeted for various applications, and hence, aimed at a desired property, it is necessary to analyze the structure and properties for extensive research. Excellent mechanical performances, high heat resistance, outstanding flammability behavior, high impact resistance, superior abrasion resistance, remarkable electric insulation, and polymer-fillers compatibility are all benefits of making composite materials. Herein, in the present chapter, comprehensive analytical methods viz. spectral, mechanical, thermal, and morphological have been presented for a complete analysis of advanced polymer composites.

References

1. Zhao, C.X., Liu, Y., Wang, D.Y., Wang, D.L., Wang, Y.Z.: Synergistic effect of ammonium polyphosphate and layered double hydroxide on flame retardant properties of poly (vinyl alcohol). Polym. Degrad. Stab.. Degrad. Stab. **93**(7), 1323–1331 (2008)
2. Zang, X., Hatcher, P.G.: A Py–GC–MS and NMR spectroscopy study of organic nitrogen in Mangrove Lake sediments. Org. Geochem. **33**(3), 201–211 (2002)
3. Fuentes, M., Baigorri, R., González-Vila, F.J., González-Gaitano, G., García-Mina, J.M.: Pyrolysis–gas chromatography/mass spectrometry identification of distinctive structures providing humic character to organic materials. J. Environ. Qual. **39**(4), 1486–1497 (2010)
4. Stenzel, Y.P., Horsthemke, F., Winter, M., Nowak, S.: Chromatographic techniques in the research area of lithium ion batteries: current state-of-the-art. Separations **6**(2), 26 (2019)
5. Devi, M., Rawat, S., Sharma, S.: A comprehensive review of the pyrolysis process: From carbon nanomaterial synthesis to waste treatment. Oxford Open Mater. Sci. **1**(1), itab014 (2021)
6. Shabaev, A.S., Rzhevskaya, E.V., Khakulova, D.M., Khashirova, S.Y., Tkhakakhov, R.B.: Thermal oxidative stability, physico-mechanical and rheological properties of composite materials based on polyetherimide. Key Eng. Mater. **816**, 78–83 (2019)
7. Larosa, C., Patra, N., Salerno, M., Mikac, L., Meri, R.M., Ivanda, M.: Preparation and characterization of polycarbonate/multiwalled carbon nanotube nanocomposites. Beilstein J. Nanotechnol. J. Nanotechnol. **8**(1), 2026–2031 (2017)
8. Shukla, R., Srivastava, A.K.: FTIR analysis of bamboo and flax (BFBF) mat reinforced epoxy hybrid composite. Int. J. Sci. Res. Sci. Eng. Technol. **6**, 119–124 (2019)
9. Occhiello, E., Garbassi, F., Morra, M., Nicolais, L.: Spectroscopic characterization of interfaces in polymer composites. Compos. Sci. Technol. **36**(2), 133–151 (1989)
10. Wesdemiotis, C., Williams-Pavlantos, K.N., Keating, A.R., McGee, A.S., Bochenek, C.: Mass spectrometry of polymers: a tutorial review. Mass Spectrom. Rev. (2023)
11. Price, D., Milnes, G.J., Tayler, P.J., Scrivens, J.H., Blease, T.G.: Degradation of polymers studied by mass spectrometry. Polym. Degrad. Stab.. Degrad. Stab. **25**(2–4), 307–323 (1989)
12. Harvey, D.J.: Matrix-assisted laser desorption/ionization mass spectrometry of carbohydrates. Mass Spectrom. Rev.Spectrom. Rev. **18**(6), 349–450 (1999)
13. Banerjee, S., Mazumdar, S.: Electrospray ionization mass spectrometry: a technique to access the information beyond the molecular weight of the analyte. Int. J. Anal. Chem. (2012)
14. Miao, Z., Chen, H.: Direct analysis of liquid samples by desorption electrospray ionization-mass spectrometry (DESI-MS). J. Am. Soc. Mass Spectrom.Spectrom. **20**(1), 10–19 (2011)

15. Using Differential Scanning Calorimetry to Characterize Polymers. https://www.azom.com/article.aspx?ArticleID=15458. Accessed 14 Dec 2023
16. Wen, X., et al.: Constructing novel fiber reinforced plastic (FRP) composites through a biomimetic approach: connecting glass fiber with nanosized boron nitride by polydopamine coating. J. Nanomater. **2013** (2013). https://doi.org/10.1155/2013/470583
17. Ebnesajjad, S.: Surface and material characterization techniques. In: Surface treatment of materials for adhesion bonding, pp 43–75 (2006). https://doi.org/10.1016/B978-081551523-4.50006-7
18. Moosa, A.A., Kubba, F., Raad, M., Ramazani, A.: Mechanical and thermal properties of graphene nanoplates and functionalized carbon-nanotubes hybrid epoxy nanocomposites. Am. J. Mater. Sci **6**(5), 125–134 (2016)
19. Shen, M.Y., et al.: Mechanical properties and tensile fatigue of graphene nanoplatelets reinforced polymer nanocomposites. J. Nanomater. **2013** (2013). https://doi.org/10.1155/2013/565401
20. (DMA) Dynamic Mechanical Analysis: Tension, Torsion, Compression. https://www.rheologylab.com/services/dynamic-mechanical-analysis-dma/. Accessed 06 Dec 2023
21. Zhang, Z., Bellisario, D., Quadrini, F., Jestin, S., Ravanelli, F., Castello, M., Dong, H.: Nanoindentation of multifunctional smart composites. Polymers **14**(14), 2945 (2022)
22. Asadi Asadabad, M., Jafari Eskandari, M.: Transmission electron microscopy as best technique for characterization in nanotechnology. Synth. React. Inorg. Metal Org. Nano-Metal Chem. **45**(3), 323–326 (2015)
23. Jinnai, H., Spontak, R.J.: Transmission electron microtomography in polymer research. Polymer **50**(5), 1067–1087 (2009)
24. Yalcin, B., Cakmak, M.: Nanocomposites based on poly (vinyl chloride). In: Recent Advances in Polymer Nanocomposites: Synthesis and Characterisation, pp 137–154 (2010)

Chapter 9
3D Printing of Polymer and Polymer Matrix Composites

Wahab Abdul, Hari Prasaath Durgaiahsangam, Sri Kalyana Rama Jyosyula, and Mohd Ataullah Khan

Abstract This chapter provides information on the efficient utilization of polymer and polymeric composites with the aid of 3D printing. Polymer materials encompass a broad range of materials, including thermoplastics, photopolymers, and composites. This diversity offers a versatile palette for creating complex structures with specific mechanical, thermal, and chemical characteristics. It is easier to produce prototypes, specialized parts, and end-use components for industries with rapid design evolution and on-demand production capabilities, which meets the changing demands of the market and encourages the shift towards more environmentally friendly and productive manufacturing techniques. Furthermore, it explored the significant influence of polymers in the field of additive manufacturing, emphasizing their distinct characteristics and the diverse range of uses they provide. Polymer integration in 3D printing technologies has transformed production in a numerous sector and allowed for many novel applications and uses. The overview of polymer-based 3D printing procedures emphasizes the importance of material choice, processing settings, and design details for achieving the functional requirements. Thus, the incorporation of polymers into 3D printing has completely changed the standards of production and has an extensive number of uses in a wide range of sectors. The flexibility, adaptability, and material variety of polymers in additive manufacturing highlight their critical role in influencing the direction of production and providing opportunities for innovative ideas and continual advancements.

Keywords Polymers · Polymer composites · 3D printing · Additive manufacturing

---

W. Abdul · H. P. Durgaiahsangam · S. K. R. Jyosyula · M. A. Khan (✉)
Department of Civil Engineering, Ecole Centrale School of Engineering, Mahindra University, Bahadurpally, Hyderabad 500043, India
e-mail: mohdataullah.khan@mahindrauniversity.edu.in

S. Moharana et al. (eds.), *Polymer Composites*, Engineering Materials,
https://doi.org/10.1007/978-981-97-2075-0_9

9.1 Introduction and the Development of Polymer 3D Printing

The creation of feedstock materials, particularly polymers, is critical to the innovation of 3D printing. During the 1920 and 1940s, polymers like polylactic acid (PLA), polyamides, and epoxy, were synthesized and developed [1–3]. Therefore, it is no surprise that, according to a recent survey, polymers make up more than half of that Additive manufacturing (AM) presently produces, as illustrated in Fig. 9.1.

The inception of AM could be dated back to 50 years ago. 3D printing is also referred to as AM, as the process is associated with the addition of phases in contrast to various conventional manufacturing processes that are subtractive in nature. It is also referred to as rapid prototyping. In the 1980s and 1990s, the emergence and expansion of the 3D printing industry was seen. Stereolithography, the first 3D printing process, was commercialized in 1987. During this time, different 3D printing techniques like fused filament fabrication (FFF) and selective laser sintering (SLS) were developed [4]. Earliest versions of 3D printing techniques arose and evolved quickly in the late 1990s, owing to the advancement of computer technology. 3D printing technology was extensively investigated in fields such as medical and aerospace in early 2000. Simultaneously, several patents for older versions of 3D printing technology expired, and several organizations began to produce their own brands of 3D printers, making 3D printers available to the general public. Following the broad availability of 3D printing techniques in the 2010s, refined printing practices having improved accuracy and precision have emerged. Initially, AM approaches mostly adopted polymers to

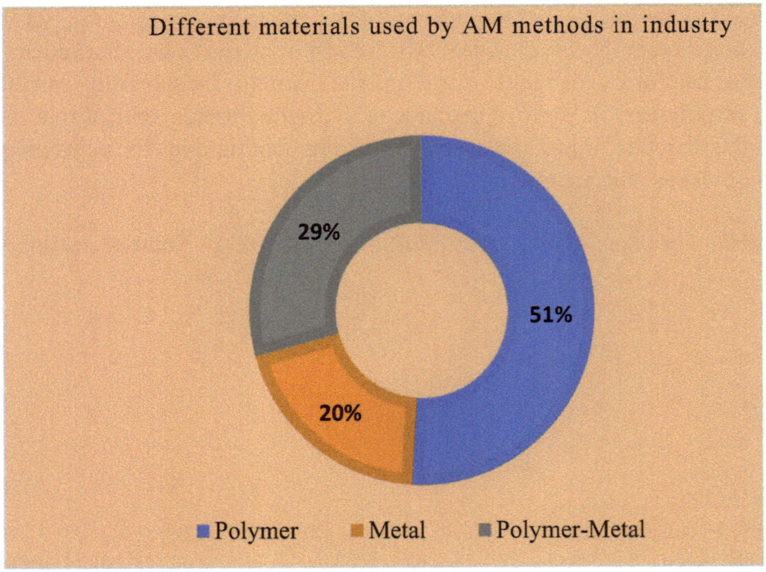

Fig. 9.1 Different materials used by AM methods in industry

make them acceptable. Currently, there is a lot of focus on improving 3D printing polymers. Therefore, AM emerged as a vital technique leading to a new industrial revolution in numerous industries due to the rapid improvement of 3D printing high-performance materials like advanced engineering plastics and smart polymers. The physical characteristics of polymers and composites of 3D printing are shown in Table 9.1. Furthermore, Fig. 9.2. Shows the typical stress–strain curve of the polymer.

Table 9.1 The physical characteristics of polymers, composites of 3D printing [1]

Principle for printing	Polymer/ composite state	Polymer materials	Benefits	Drawbacks
Extrusion	Solid filament	Thermoplastics like PLA, ABS, polycarbonates and nylon	Easy, low cost, and rapid (from 50 to 150 mm/h up to 500 mm/h)	Clogging in the nozzle and inferior in quality
Pressurized extrusion	Liquid polymer	Hydrogel, colloidal suspension & liquid polymer	Higher resolution (till 200 mm) & material flexibility	Low speed
Layer by layer adhesive	Polymer sheet	Polymer composites with bonding agents	Effortlessness handling of material (45 in 3/ h: CBAM)	Restricted mechanical property diverse finish

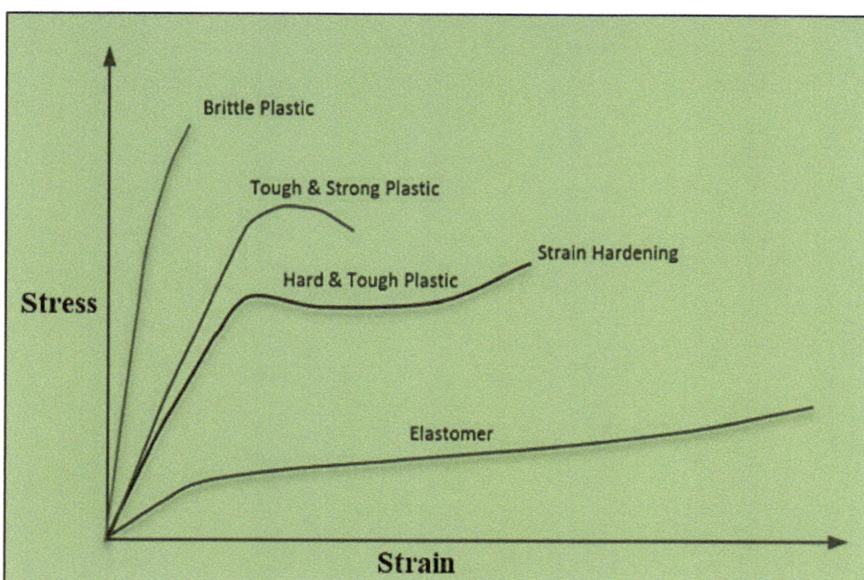

Fig. 9.2 Shows the typical stress–strain curve of polymer [5]

However, 3D printing is one of the pillars of the fourth industrial revolution. In general, the industry is reorganizing its manufacturing process with the use of robotics and digital technologies to minimize manpower and enhance competitiveness. The construction industry is also looking towards 3D printing considering its efficiency, productivity, high geometric freedom, automated operations, and accuracy which minimizes the role of dependency on unskilled workers [6]. For example, new design polymers like smart hydrogel [7–10], shape memory polymers (SMPs) [11–14], liquid crystal elastomers (LCEs) [15–18], and liquid crystal polymers (LCPs) have given industries the opportunity to make new smart devices [19], robots [20], and biomedical products [21]. 3D printing is currently transitioning from prototyping to end-use items, with more societal influence. On the contrary, a maker movement that is projecting the 3D printing technology is affordable and simple to use.

3D printing or AM creates 3D objects of the required shape by adding layers of material. Unlike traditional manufacturing techniques such as molding and milling, AM is quick and can directly generate complicated shapes from a digital model. When using traditional manufacturing procedures, creating a different component or modifying a partial design necessitates the creation of a new tool, mold, or jig [22]. In contrast, 3D printing does not require physical alterations to the machine after the object is designed using software. Furthermore, when compared to subtractive manufacturing methods, they reduce material wastage promoting sustainable practices using very low waste and renewable energy resources. The 3D printing step-by-step procedure is illustrated in Fig. 9.3.

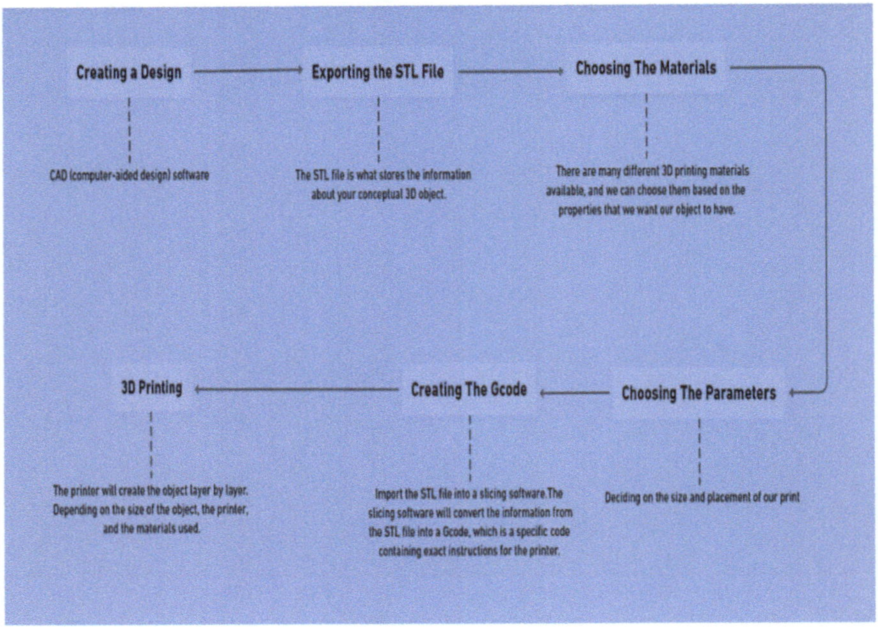

Fig. 9.3 A step-by-step procedure for 3D printing

9.2 Types of 3D Printers

Although the process of 3D printing relies on layer-by-layer deposition, however, realization of such a process for different typologies of materials demands exclusive processes. Although these techniques have reached a high level of development, they are still considered to be specialized and limited in their use. The following are some categories of printers designed for polymeric materials.

9.2.1 Extrusion Printers

Extrusion-based 3D printing, also called Fused Filament Fabrication (FFF), was developed and patented by S. Scott Crump in the late 1980s and played a pivotal role in promoting AM technology. A binder must be blended before being fed into a nozzle to extrude the surface. This combination was then solidified as it exited the nozzle, and the 3D structure was formed by moving the nozzle in the requisite direction. In comparison with other 3D printing processes, the extrusion printing process is quite slow, and to ensure the accuracy and quality of the printed product, regular cleaning of the nozzle is important, as materials may block the nozzle. Extrusion printing takes longer than other technologies like selective laser sintering (SLS) and stereolithography (SLA). It is due to the material being fed slowly through a hot nozzle which gradually builds up each part of the product. In 3D printing techniques, extrusion printing process components include fused direct light processing (DLP), deposition modeling (FDM), continuous liquid interface formation (CLIP), and stereolithography (SLA). DLP directs light to the surface of the resin, hardening it and forming the required 3D structure. Generally, a laser is used in SLA for applying liquid resin and producing a desirable 3D structure. The CLIP technique evaluates phase progression at each layer by using an oxygen permeability window [23].

9.2.2 Inkjet Printers

Inkjet printing is a popular and adaptable technique for printing digital images and text on a variety of materials, including plastic, paper, and fabric, to name a few. A brief overview of how the inkjet printing process usually works is to spray liquid ink droplets onto a surface. These printers operate based on the principle of propelling liquid ink drops on a surface. The print head's small nozzles are used to discharge these droplets. The ink printing system's components are as follows: 1. Ink cartridges: This is the component of the printer that holds liquid ink, 2. A movable gadget occupied with several colored inks, 3. The print head is a component of a printer that contains nozzles, jets that spray ink onto paper. The print head is moved front and back of the paper while printing. 4. The printing process is controlled by electronics. It is in charge of firing nozzles, monitoring ink flow, and transmitting data to the printer. Some of the author's previous works [24]. Utilized inkjet printing to demonstrate multi-material simultaneous 3D printing of soft and stiff phases which allows the development of compliance tailored interfaces in multi-layered assemblies. As multiple nozzles are jetted simultaneously, it is easy to realize the printing of various phases. Object260 Connex compact printer was used which comprises two cartridges consisting of two miscible photopolymers commercially named Verowhite and Tangoplus. The miscibility of these polymers allowed the creation of photopolymers with property values ranging between the two. Verowhite is relatively stiff with Young's modulus about 2 GPa and Tangoplus is very soft with elastic modulus about 1 MPa. Such multi-material 3D printing with various phases could provide the potential to mimic nature-inspired materially-tailored architecture [25]. The Objet260 Connex can have a print resolution as low as 30 μ m to about 42 μ m.

The inkjet printing primary benefit is low cost. When compared to other 3D printing technologies, inkjet printing is comparatively affordable. The inkjet printing permits the fabrication of complicated patterns and colorful graphics even when time or resources are restricted. As discussed, the inkjet printers can print composites.

9.2.3 Selective Laser Sintering Printers

Selective laser sintering (SLS) is used for composite selections, plastics, and metals. For every single cross-section and along the powder bed, selected particles are scanned to one thickness, and new material is deposited, continuing the procedure until the material is complete [26]. To construct consecutive layers, laser scan cross sections created from 3D digital descriptions of component geometry and selectively fuse powder particles. The primary benefit of SLS over traditional engineering methods is, that it permits items that would be difficult or impossible to build using standard waste-intensive processes to be produced at a lower cost and with more complicated geometries [27]. SLS is a 3D printing technique which involves fusing tiny bits of plastic, glass, ceramic, metal, and supplementary materials into three-dimensional objects using a laser. As no supporting system is required, this 3D printing is suitable for complicated and complex form parts. SLS is capable of producing components for a comprehensive range of industries such as aerospace, automotive, and consumer goods. SLS's products are durable and detailed which makes an excellent option for high-performance components. SLS employs a far more accurate laser to fuse powder material into predefined form layer by layer. The SLS process starts with a big idea that subsequently divides into slices by computer-aided design (CAD) program. The slices are utilized to guide a laser beam, which fuses powder particles to form each layer. When the layer is finished, the powder bed is lowered and a new layer of powder is applied on top of the previous bed. This procedure is continued until the completion of the project. After printing, the printed model is taken from the powder bed and cleaned and a final product is obtained.

9.3 Benefits of SLS (Selective Laser Sintering)

1. It is low-cost 3D printing technology that produces high-quality, complicated products with strict tolerances.
2. SLS is capable of producing components with complicated geometries and forms, such as thin walls and hollow holes.
3. Technology enables the manufacture of stronger and more durable items than previous 3D printing methods.
4. SLS is a powder-based 3D printing technology that may use a variety of materials such as metals, polymers, and composites.
5. Procedure is quick and efficient, with components completed in a matter of hours.
6. Owing to minimal post-processing, SLS is an ideal solution for creating components with complicated geometries.
7. Moreover, SLS is a good choice in terms of low-volume production and prototyping [28].

9.4 Continuous Fiber Printing

A sophisticated kind of AM known as continuous fiber 3D printing combines continuous strands of reinforcing fibers along with conventional 3D printing techniques including stereolithography (SLA) and fused filament fabrication (FFF). Usually, materials like fiberglass, Kevlar, carbon fiber, and other high-strength polymers are used to make these fibers. Fig. 9.4 shows continuous fiber 3D printing with continuous strands of reinforcing fibers of the specimen. In comparison to parts manufactured exclusively utilizing conventional 3D printing techniques using plastic or metal, the procedure inserts the continuous fibers into printed pieces, strengthening them and enhancing strength, stiffness, and longevity. Figure 9.5 illustrates Young's moduli and strength of continuous carbon fiber composites fabricated and related to composites fabricated by FDM with commercial 3D printers like FDM, SLS, and SLA. In 3D printing, continuous fibers can be added in a variety of ways.

9.4.1 Filament Winding

This technique involves wrapping continuous threads around a printed item or core material. Fibers are capable of being strategically placed to give strength in particular directions, improving the finished object's mechanical properties.

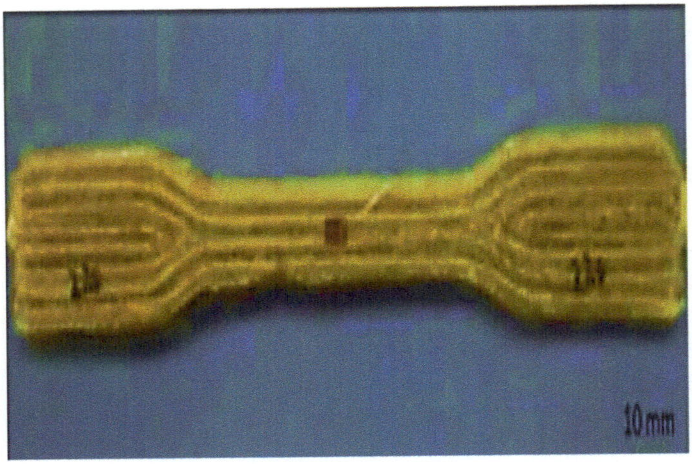

Fig. 9.4 Shows the continuous fiber 3D printing with continuous strands of reinforcing fibers [29]. Open access under a CC BY 4.0 license

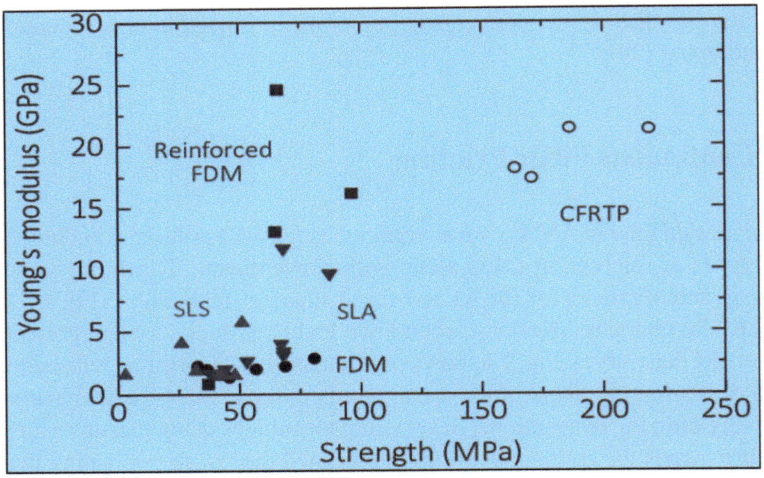

Fig. 9.5 Illustrates the Young's moduli and strength of carbon fiber composites like Continuous Fiber-reinforced thermoplastic (CFRTP) fabricated and compared with composites fabricated by FDM with commercial 3D printers like FDM, SLS, and SLA [29]. Open access under a CC BY 4.0 license

9.4.2 *Extrusion-Based Printing with Continuous Fiber Reinforcement*

This method incorporates continuous fibers right into the extrusion procedure. The 3D printer adds continuous fiber strands in predetermined orientations layer by layer while it extrudes a thermoplastic material, usually a polymer matrix such as nylon or ABS.

1. *Resin-based Printing with Fiber Placement*: During the printing process, continuous fibers can be manually or robotically inserted into the resin layers in resin-based 3D printing techniques like SLA or DLP. The mechanical qualities of the finished printed object are optimized thanks to this method's ability to precisely insert and align fibers.

9.5 Various Printers and Their Accuracy

There are different types of 3D printers, each with its own level of precision. A 3D printer's accuracy may be impacted due to a variety of variables, with the technology it employs, the quality of its components, and the calibration. Here are some examples of common 3D printers and their expected accuracy levels:

9.5.1 Fused Deposition Modeling FDM

They are commonly used and affordable as well. Although accuracy varies, most consumer-grade FDM printers can attain layer resolutions of 0.1–0.4 mm. Higher-end FDM printers and industrial versions may achieve much more precision, generally in the 0.05–0.2 mm range.

9.5.2 Stereolithography SLA

They employ a liquid resin which is cured by UV laser or other light source. Typically, these printers have exceptional precision, with layer resolutions ranging from 0.025 to 0.1 mm. SLA printers are well-known for generating prints that are both detailed and smooth.

9.5.3 Digital Light Processing DLP

These are comparable to SLA printers but employ a different resin curing technique. They can attain comparable degrees of precision, with layer resolutions generally ranging from 0.025 to 0.1 mm.

9.5.4 Multi Jet Printing (MJP) 3D Printers

To deposit and cure materials, MJP printers employ several jets. They are capable of achieving excellent precision with layer resolutions as low as 0.016 mm. MJP is frequently used to create precise and intricate pieces.

9.5.5 Binder Jetting 3D Printers

Binder jetting method selectively bonds powder materials using a liquid binder. The precision varies, but it is often in the 0.1–0.2 mm range.

9.6 Factors Affecting 3D Printers

3D printing is a versatile technique that allows you to build three-dimensional objects by layering material. The accuracy and quality of 3D printing are determined by a variety of factors that are discussed as follows:

Quality and type of 3D Printer: This has a significant impact on the quality and accuracy of 3D printed specimens. The accuracy and precision of various 3D printers, such as Stereolithography (SLA), Fused Filament Fabrication (FFF), Selective Laser Sintering (SLS), and others, vary. High-end industrial 3D printers are more precise and detailed than desktop or hobbyist devices.

Print Parameters: Print parameters are the settings and configurations that you select for any particular printing job. These characteristics may be changed using slicing software, which turns 3D models into 3D printer instructions. The following are some important print settings that impact accuracy.

Layer height: Smaller layer heights (e.g., 0.1 mm) provide more detailed and precise prints, but they significantly lengthen printing time.

Wall thickness and infill density: Thicker walls and increased infill density can enhance a print's structural integrity and accuracy. Support structures are required for overhanging elements that impact the surface smoothness, and accuracy of a print.

Print temperature: The temperature of the print nozzle and print bed might affect layer adherence and quality. Note that for the FDM printers, the print temperature would depend on the melting point of the materials being printed and could be very high for a few materials such as PEEK. Handing higher temperatures at the print head would require printing to occur in a well-planned manner. Inclusion of sacrificial towers, to avoid overheating in the chamber may be required [30]. Note that warping could be induced by overheating.

Cooling fan: Appropriate cooling reduces the warping and increases the precision of microscopic details.

Print orientation: Overhangs and the accuracy of some characteristics can be affected by the model's orientation on the build platform.

Material: The kind of material that is selected for 3D printing also has an impact on the precision of your prints. Shrinkage, warping, and thermal expansion are all features of different materials that might weaken overall accuracy. Polylactic Acid (PLA) and Polyethylene Terephthalate Glycol-Modified (PETG) are well-known for their simplicity in use and precision, but more interesting materials may necessitate careful control and skill to get correct results.

Calibration: Calibration of 3D printer on a regular basis is critical for accuracy. Calibration of the print bed level, extruder temperature, and flow rate are all included. Calibration ensures that the printer lays the material in the intended location and layer height.

Post-processing: Post-processing techniques like sanding, painting, or joining several components can improve the precision and aesthetics of the final 3D printed product.

Printing speed: The printing speed might vary based on the equipment and procedure employed. The size and complexity of the item being printed, as well as the qualities of the material combination being used, can all have an impact on printing speed [31].

9.7 Printing Challenges

It might be challenging to get the appropriate directional behavior in 3D printing. Parameters like print speed, nozzle temperature, layer adhesion, and fiber alignment need to be carefully controlled to produce composites with the intended properties.

Layer orientation: The mechanical characteristics of the finished product are affected by the layer orientation in 3D printing. For instance, a printed part's tensile strength can be stronger in a direction perpendicular to layer lines and lower in another direction.

Print settings: Directional behavior may be affected by 3D printing parameters such as print speed, infill density, and layer height. To some extent, controlling these variables will assist in mitigating the anisotropic behavior of the printed object.

Support structures: Support structures are employed in certain 3D printing procedures to aid in the manufacturing of complicated or overhanging geometries. The existence of these support structures may have an impact on the directional behavior particularly, if such components are not eliminated from the printed part.

Design considerations: Designing components for 3D printing requires careful consideration of the directional behavior. The shape and orientation of the part can be optimized by engineers and designers to match the required mechanical qualities and effectiveness.

Validation and testing: Conducting testing and analysis of 3D printed components in various orientations and situations is sometimes important to completely understand and control directional behavior. This aids in determining the ideal material selections and printing conditions for a specific use.

Adhesion of layers: Layer adhesion is a significant consideration in 3D printing since it impacts the overall strength and reliability of the completed item. The bond strength between interlayers is related to the anisotropic response of printed specimens. Lack of interlayer adhesion reduces the bond strength between the printed layers, introducing anisotropy in the printed objects. The force that binds layers of a 3D printed item together is referred to as interlayer force. The composition, printing circumstances, and design of the item being printed are all factors that might influence interlayer force. Improving interlayer force can aid in improving 3D printed item buildability and overall quality.

9.8 Materials for Polymer 3D Printing and Their Use

9.8.1 Various Polymers for 3D Printing

The materials used in 3D printing are chosen on a variety of criteria, including specific applications, desired qualities, and the type of 3D printing method employed. Here are some commonly used materials and their optimal applications which are also illustrated in Fig. 9.6.

PLA (Polylactic Acid): Because of its simplicity of use and biodegradability, PLA is one of the most popular 3D printing materials. It's ideal for prototyping demonstrations and educational projects. PLA should not be used in high-temperature or load-bearing applications.

PLA+ : PLA+ is an upgraded type of PLA that is stronger and more resistant to heat. It is appropriate for items that require more strength than normal PLA, such as mechanical components and functioning prototypes.

ABS (Acrylonitrile Butadiene Styrene): ABS is well-known for toughness and impact resistance. Frequently used for functional prototypes, automotive parts, and consumer goods. Sometimes, it is difficult to print with ABS due to its sensitivity to temperature changes, particularly at low temperatures as it has the tendency to wrap.

PETG (Polyethylene Terephthalate Glycol-Modified): PETG combines the printability of PET with the durability of materials such as ABS. It's an excellent choice for

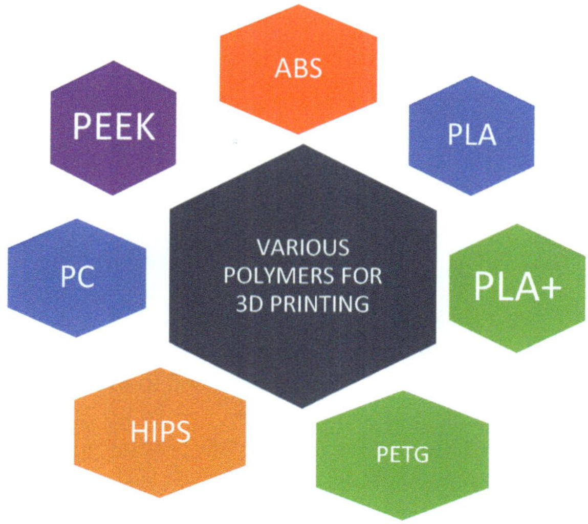

Fig. 9.6 Illustrates different types of polymers for 3D printing

printable products that must endure mechanical stress as well as some environmental variables. PETG is frequently utilized in engineering practices.

Nylon: Nylon gets attention for its strength, flexibility, and long life. It is utilized in parts that require higher mechanical strength and wear resistance, such as gears, bearings, and functional prototypes.

TPU (Thermoplastic Polyurethane): It is a flexible filament recognized for abrasion resistance and elasticity. Often used for making soft, rubbery elements such as phone covers, shoe bottoms, and seals.

HIPS *(High-Impact Polystyrene):* For dual-extrusion 3D printers, HIPS is frequently utilized as a support material. It dissolves in D-Limonene and is therefore easily removed from complicated components. It isn't often utilized in the finished product, although it does help in the printing process.

PC *(Polycarbonate):* PC is well-known for its superior temperature resistance and impact strength. It is used in automotive components and electronic enclosures which have to be functional in harsh environments, such as automotive components and electronic enclosures.

PEEK *(Polyether Ether Ketone):* PEEK is a thermoplastic engineering grade noted for its high strength, heat resistance, and chemical resistance. It is utilized in sectors such as aerospace, medical, and oil and gas for high-performance applications.

Carbon Fiber-Reinforced Filaments: These filaments blend PLA or other polymers with carbon fibers to create lighter and stronger products. They can be used in aerospace, automotive, and drone applications.

Table 9.2 Physical and mechanical properties of polymers used in 3D printing [1]

Materials	Glass transition temperature(°C)	Melting temperature (°C)	Elastic modulus (Gpa)
ABS	102–115	–	1.8–2.39
Nylon6	47–57	220	2.8–3.1
Polyether Ether Ketone	137–152	335–343	3.56
PLA	53–64	145–186	1.2–3.0

Wood and Composite Filaments: These are made up of a combination of wood and other natural fibers. They are used to make pieces that look like wood, such as ornamental objects and furniture.

It is crucial to note that the 3D printing method utilized (e.g., Fused Deposition Modeling, Stereolithography, SLS) and characteristics of individual 3D printers might impact material selection. It is essential to consider pricing, availability, and post-processing needs when deciding the choice of the right material for 3D printing project. Some of the useful properties of these materials are listed in Table 9.2.

9.9 Effect of Directional Behavior in 3D Printing

3D printing objects show anisotropic mechanical response [21] owing to poor bond strength between the printed layers. Directional dependency of a material properties, like hardness/strength/stiffness, is referred to as its anisotropic nature. It could be generalized that the 3D printed polymers as well as 3D printed polymeric composites show directional behavior. The source of anisotropy for polymeric composites is also attributed to the fiber orientation [32]. Also, note that the source of anisotropy for the printed polymers is from the lack of interlayer adhesion and layer-by-layer deposition [33]. The anisotropy contributed from lack of interlayer adhesion is expected to be present also in the polymeric composites. This phenomenon can influence many factors of both the printing procedure and the characteristics of printed specimens.

The ultimate characteristics and functionality of the produced items can be greatly impacted by the directional behavior of composites during 3D printing. In 3D printing, composites typically consist of a polymer matrix combined with particles or fibers for reinforcement. During printing, the orientation of these reinforcing components can affect several different features of the finished product. Anisotropic behavior is one of the most crucial aspects of 3D printed composite. In summary, in order to completely realize the future potential of 3D printing, it will be necessary to address its anomalous behavior. More study is required to provide practical solutions that will improve system performance and minimize the anisotropic behavior. Among the main consequences, the effects of directional behavior in composite 3D printing are the following.

9.9.1 Anisotropic Mechanical Properties

Anisotropic mechanical characteristics, in which the direction of load application affects the material's strength, stiffness, and other mechanical properties, can arise from the orientation of the components inside the composite [34]. Parts printed with filaments aligned in one way, for instance, could be weaker in other directions while being more durable along that axis.

9.9.2 Improved Strength and Stiffness

The tensile strength and stiffness of the composite can be enhanced. Applications that require a high level of structural integrity would especially benefit from it.

9.9.3 Reduced Print Time and Material Usage

It is feasible to decrease the amount of composite material required and shorten the printing time by optimizing the directional behavior. This is accomplished by carefully arranging reinforcing elements in the most critical areas.

9.9.4 Enhanced Thermal Conductivity

Thermal conductivity of composite materials can be affected by their orientation. For applications such as heat sinks, parts having fibers orientated in a certain direction which can transmit heat more efficiently in that direction.

9.9.5 Weight Reduction

By using directional behavior to minimize material use in non-critical locations like supportand surface which varies accordingly, parts may be made lighter. This is particularly important for applications in aerospace and automotive technology where weight reduction is essential.

Fig. 9.7 3D printing specimens using ABS at Mahindra University 3D printing lab

9.10 Applications of Polymer Composites

The use of polymer composites has resulted in considerable breakthroughs in recent years. This cutting-edge technology has been used in a variety of ways starting, from permitting creative architectural designs to improving the structural integrity and durability of structures. Furthermore, the uses of polymer composites in various industries are discussed in this section, emphasizing how they might completely transform the industries.

Due to the demand for sustainable and effective practices as well as technical advancements, there has been a rapid change. Using cutting-edge 3D printing technology, polymer composites have become a paradigm shift. These materials provide a wide range of applications with the adaptability of polymers. Fig. 9.7 shows the 3D printing specimens developed using ABS polymer at Mahindra University 3D printing lab.

9.11 Polymer Composites Applications in Various Industries

Aerospace: The aircraft industry makes extensive use of polymer composites due to their exceptional strength-to-weight ratio. It is employed in the production of fuselage and wing components for aeroplanes. Carbon-fiber reinforced polyamide 12 (Carbon amide), aluminium-filled polyamide 12 powder (Alumide), glass bead-filled polyamide 12 powder (PA 3200 GF), and white polyamide 12 powder with a flame-retardant additive (PA 2210 FR) (Techno-Grafica GmbH, n.d.) are the polymers that are typically used to print various aircraft components using SLS technology Polyphenyl-sulfone (PPSF/PPSU) (Stratasys Ltd., n.d.a), Stratasys Ltd., n.d.b), ABSi (Stratasys Ltd., 2022), and Nylon 12 (Stratasys Ltd., n.d.a) PEAK-Polyarylether-ketone (Quantevo-CF) (AREVO Ltd., n.d.), ULTEMTM 9085 (European Space Agency [ESA], 2013; are the polymers that are used to print various aircraft

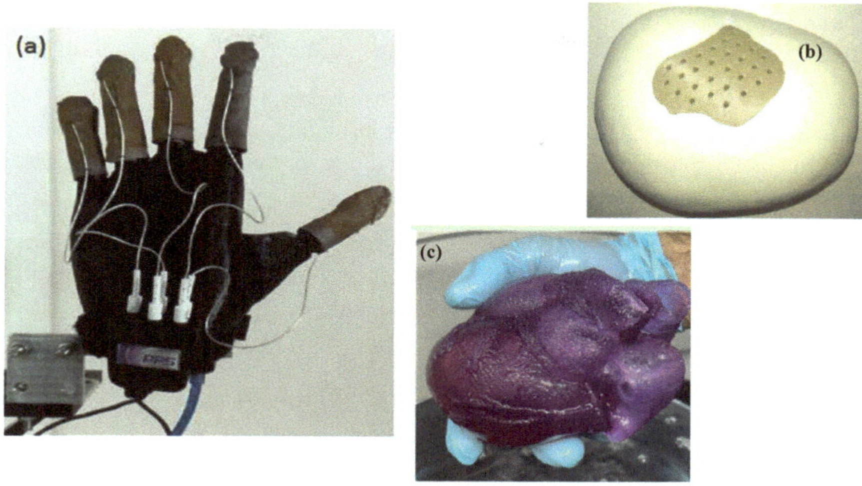

Fig. 9.8 Shows 3D printing applications in different industries **a** Prosthetics, **b** Customized implants are printed in PEEK locally for skull and **c** 3D printed heart [36–38]. Open access under a CC BY 4.0 license

components using FDM technology. Aluminium alloy AlSi10Mg with DMLS (EOS GmbH—Electro Optical Systems, 2017) and martensite-hardenable steel (EOS Maraging Steel MS1) with DMLS (EOS GmbH—Electro Optical Systems, 2017) are two examples of 3D printing technology used to manufacture various aircraft parts [35]. Figure 9.8 shows 3D printing applications in different industries (a) Prosthetics, (b) UAVs, (c) customized implants are printed in Titanium and PEEK printed locally, and d) Medical applications of 3D printing.

Automotive: Automobiles employ lightweight polymer composites to increase efficiency and save fuel. They are present on the body panels, interior parts, and suspension parts.

Sports and sports equipment: Polymers are used in making hockey sticks, skis, and helmets, to enhance performance, durability, and safety.

Marine: Because the materials are resistant to salt water, they are perfect for maritime applications. Also used in nautical applications such as sails and hulls.

Wind energy: Composite materials are commonly used in wind turbine blades because of their higher strength and lightweight, which allows for efficient energy generation.

Electronics: Due to their insulating and thermal qualities, polymer composites are utilized in the electronics industry for printed circuit boards (PCBs) and as encapsulation materials for electronic equipment.

Medical Devices: Implants and medical equipment adopt biocompatible composite materials because they are strong and suited to the human body.

Oil and Gas Industry: Although polymer combinations are strong and resistant to corrosion, they are employed in the oil and gas industry for pipelines and offshore buildings and tanks.

Construction Industry: Owing to its enhanced insulating, thermal, and mechanical properties polymer composites are used in cladding, piping, flooring, and glazing applications. They are also used as a repair material.

Patient-specific implants: 3D printed heart is used by medical practitioners to practice surgeries in order to reduce the original surgery time. Further, medical applications such as patient specific implants could aid in reducing the number of implants and optimizing the surgical process.

9.12 Summary

This chapter covers the fundamentals of polymer 3D printing, such as its history, printer types and precision, influencing variables, material selection, and the influence of directional behavior, and its numerous applications across a range of sectors. In addition, it is essential to understand the conditions, limits, and technological restrictions that impact 3D polymeric printers. Furthermore, the drawbacks of 3D printing such as problems with surface quality, printing speed, and structural integrity must be investigated. Lastly, polymers have a wide range of uses in 3D printing. They vary from producing complex designs and customizable items to manufacturing prototypes in a variety of sectors, including automotive, consumer goods, aerospace, construction, and healthcare (including medical implants).

References

1. Park, S., Shou, W., Makatura, L., Matusik, W., Fu, K.K.: 3D printing of polymer composites: Materials, processes, and applications. Matter **5**(1), 43–76 (2022)
2. Feldman, D.: Polymer history. Des. Monomers Polym., **11**(1), 1–15 (2008)
3. Dizon, J.R.C., Espera, A.H., Jr., Chen, Q., Advincula, R.C.: Mechanical characterization of 3D-printed polymers. Addit. Manuf. **20**, 44–67 (2018)
4. Wohlers, T., Gornet, T., Mostow, N., Campbell, I., Diegel, O., Kowen, J., Huff, R., Stucker, B., Fidan, I., Doukas, A., Drab, B.: History of additive manufacturing. (2016)
5. Zhang, X.: 3D printing of bio-inspired, multi-material structures to enhance stiffness and toughness. Doctoral dissertation, The University of Waikato (2021)
6. Paritala, S., Singaram, K.K., Bathina, I., Khan, M.A., Jyosyula, S.K.R.: Rheology and pumpability of mix suitable for extrusion-based concrete 3D printing–A review. Constr. Build. Mater. **402**, 132962 (2023)
7. Sydney Gladman, A., Matsumoto, E.A., Nuzzo, R.G., Mahadevan, L., Lewis, J.A.: Biomimetic 4D printing. Nat. Mater. **15**(4), 413–418 (2016)
8. Schwartz, J.J., Boydston, A.J.: Multimaterial actinic spatial control 3D and 4D printing. Nat. Commun. **10**(1), 791 (2019)

9. Raviv, D., Zhao, W., McKnelly, C., Papadopoulou, A., Kadambi, A., Shi, B., Tibbits, S.: Active printed materials for complex self-evolving deformations. Sci. Rep. **4**(1), 7422 (2014)
10. Zhang, Q., Zhang, K., Hu, G.: Smart three-dimensional lightweight structure triggered from a thin composite sheet via 3D printing technique. Sci. Rep. **6**(1), 22431 (2016)
11. Van Manen, T., Janbaz, S., Zadpoor, A.A.: Programming 2D/3D shape-shifting with hobbyist 3D printers. Mater. Horiz. **4**(6), 1064–1069 (2017)
12. Long, K.N., Scott, T.F., Qi, H.J., Bowman, C.N., Dunn, M.L.: Photomechanics of light-activated polymers. J. Mech. Phys. Solids **57**(7), 1103–1121 (2009)
13. Wei, H., Zhang, Q., Yao, Y., Liu, L., Liu, Y., Leng, J.: Direct-write fabrication of 4D active shape-changing structures based on a shape memory polymer and its nanocomposite. ACS Appl. Mater. Interfaces **9**(1), 876–883 (2017)
14. Van Oosten, C.L., Bastiaansen, C.W., Broer, D.J.: Printed artificial cilia from liquid-crystal network actuators modularly driven by light. Nat. Mater. **8**(8), 677–682 (2009)
15. Yuan, C., Roach, D.J., Dunn, C.K., Mu, Q., Kuang, X., Yakacki, C.M., Qi, H.J.: 3D printed reversible shape changing soft actuators assisted by liquid crystal elastomers. Soft Matter **13**(33), 5558–5568 (2017)
16. Kotikian, A., Truby, R.L., Boley, J.W., White, T.J., Lewis, J.A.: 3D printing of liquid crystal elastomeric actuators with spatially programed nematic order. Adv. Mater. **30**(10), 1706164 (2018)
17. Gantenbein, S., Masania, K., Woigk, W., Sesseg, J.P., Tervoort, T.A., Studart, A.R.: Three-dimensional printing of hierarchical liquid-crystal-polymer structures. Nature **561**(7722), 226–230 (2018)
18. Rim, Y.S., Bae, S.H., Chen, H., De Marco, N., Yang, Y.: Recent progress in materials and devices toward printable and flexible sensors. Adv. Mater. **28**(22), 4415–4440 (2016)
19. Mao, S., Dong, E., Jin, H., Xu, M., Low, K.H.: Locomotion and gait analysis of multi-limb soft robots driven by smart actuators. In: 2016 IEEE/RSJ International conference on intelligent robots and systems (IROS), pp 2438–2443. IEEE, (2016)
20. Gao, B., Yang, Q., Zhao, X., Jin, G., Ma, Y., Xu, F.: 4D bioprinting for biomedical applications. Trends Biotechnol. **34**(9), 746–756 (2016)
21. Khan, M.A., Kumar, S., Cantwell, W.J.: Additively manufactured cylindrical systems with stiffness-tailored interface: Modeling and experiments. Int. J. Solids Struct. **152**, 71–84 (2018)
22. Wang, X., Jiang, M., Zhou, Z., Gou, J., Hui, D.: 3D printing of polymer matrix composites: A review and prospective. Compos. B Eng. **110**, 442–458 (2017)
23. Singh, M., Haverinen, H.M., Dhagat, P., Jabbour, G.E.: Inkjet printing—process and its applications. Adv. Mater. **22**(6), 673–685 (2010)
24. Sola, A., Bellucci, D., Cannillo, V.: Functionally graded materials for orthopedic applications–an update on design and manufacturing. Biotechnol. Adv. **34**(5), 504–531 (2016)
25. Schmid, M., Amado, A., Wegener, K.: Polymer powders for selective laser sintering (SLS). In: AIP Conference proceedings, vol. 1664, no. 1. AIP Publishing, (2015)
26. Xia, Y., Zhou, P., Cheng, X., Xie, Y., Liang, C., Li, C., Xu, S.: Selective laser sintering fabrication of nano-hydroxyapatite/poly-ε-caprolactone scaffolds for bone tissue engineering applications. Int. J. Nanomedicine, 4197–4213 (2013)
27. Akhoundi, B., Behravesh, A.H., Bagheri Saed, A.: An innovative design approach in three-dimensional printing of continuous fiber–reinforced thermoplastic composites via fused deposition modeling process: in-melt simultaneous impregnation. Proc. Inst. Mech. Eng., Part B: J. Eng. Manuf. **234**(1–2), 243–259 (2020)
28. Lescot, T., Lebel-Cormier, M.A., Seniwal, B., Gros-Louis, P., Bellerive, C., Landreville, S., Fortin, M.A.: Tumor shape-specific brachytherapy implants by 3D-printing, precision radioactivity painting, and biomedical imaging. Adv. Healthcare Mater. **12**(25), 2300528 (2023)
29. Matsuzaki, R., Ueda, M., Namiki, M., Jeong, T.K., Asahara, H., Horiguchi, K., Hirano, Y.: Three-dimensional printing of continuous-fiber composites by in-nozzle impregnation. Sci. Rep. **6**(1), 23058 (2016)
30. Utela, B., Storti, D., Anderson, R., Ganter, M.: A review of process development steps for new material systems in three-dimensional printing (3DP). J. Manuf. Process. **10**(2), 96–104 (2008)

31. Somireddy, M., Singh, C.V., Czekanski, A.: Mechanical behaviour of 3D printed composite parts with short carbon fiber reinforcements. Eng. Fail. Anal. **107**, 104232 (2020)
32. Hong, S.Y., Kim, Y.C., Wang, M., Kim, H.I., Byun, D.Y., Nam, J.D., Suhr, J.: Experimental investigation of mechanical properties of UV-Curable 3D printing materials. Polymer **145**, 88–94 (2018)
33. Chen, J., Liu, X., Tian, Y., Zhu, W., Yan, C., Shi, Y., Zhou, K.: 3D-Printed anisotropic polymer materials for functional applications. Adv. Mater. **34**(5), 2102877 (2022)
34. Gadelmoula, A., Aldahash, S.A.: Tribological properties of glass bead-filled polyamide 12 composite manufactured by selective laser sintering. Polymers **15**(5), 1268 (2023)
35. Karkun, M.S., Dharmalingam, S.: 3D printing technology in aerospace industry–a review. Int. J. Aviat., Aeronaut., Aerosp. **9**(2), 4 (2022)
36. Alhaddad, A.Y., AlKhatib, S.E., Khan, R.A., Ismail, S.M., Shehadeh, A.S.S., Sadeq, A.M., Cabibihan, J.J.: Toward 3D printed prosthetic hands that can satisfy psychosocial needs: grasping force comparisons between a prosthetic hand and human hands. In: Social Robotics: 9th international conference, ICSR 2017, Tsukuba, Japan, November 22–24, 2017, Proceedings, vol. 9, pp. 304–313. Springer International Publishing
37. Honigmann, P., Sharma, N., Okolo, B., Popp, U., Msallem, B., Thieringer, F.M.: Patient-specific surgical implants made of 3D printed PEEK: material, technology, and scope of surgical application. BioMed Res. Int., (2018)
38. Mirdamadi, E., Tashman, J.W., Shiwarski, D.J., Palchesko, R.N., Feinberg, A.W.: FRESH 3D bioprinting a full-size model of the human heart. ACS Biomater. Sci. Eng. **6**(11), 6453–6459 (2020)

Chapter 10
Natural Fibre Reinforced Composites for Industrial Applications

Neelam Kumari, Shivali Meena, Monika Chaparia, Sandip P. Choudhury, Ravi Kant Choubey, and Umesh Kumar Dwivedi

Abstract Natural fibre reinforced (NFR) polymer composites set an avenue for the development of new sustainable materials. These materials led a paradigm shift from synthetic materials to natural fibre reinforced materials. Natural fibres are easily and excessively available in nature, biodegradable, cost effective, lightweight, environment friendly, and no irritation to eyes or skin and hence gained the attention of researchers as a promising candidate for the development of natural fibre reinforced composites with enhanced properties. NFR-based composites research field is continuously evolving. Various industries such as aviation, military, sports, sensors, electrical engineering, construction, and food are some major sectors to utilize NFR composites for designing different components and materials due to their distinct properties as conferred in the text. Jute, sisal, kenaf, hemp, cotton, flax, coir, etc. are commonly used fibres as reinforcement. Inherent characteristics of natural fibres such as moisture, strength, thermal stability, incompatibility with embedded matrices, etc. are major challenges while developing and applicability of NFR composites. However, these shortcomings can be minimized by chemical treatments. Furthermore, considerable research is going on the elevation of the performance of NFR composites.

Keywords Natural fibre · Reinforcement · Polymer composites · Industrial applications · Automobile · Construction

N. Kumari · S. Meena · M. Chaparia · S. P. Choudhury · U. K. Dwivedi (✉)
Amity School of Applied Physics, Amity University Rajasthan, Jaipur 303002, India
e-mail: umeshkudwivedi@gmail.com

R. K. Choubey
Department of Applied Physics, Amity Institute of Applied Sciences (AIAS), Amity University, Noida Campus, Sector–125, Noida, Uttar Pradesh 201313, India

S. Moharana et al. (eds.), *Polymer Composites*, Engineering Materials,
https://doi.org/10.1007/978-981-97-2075-0_10

10.1 Introduction

Over the last decades, natural fibres have been a great substitute for synthetic fibres due to their tremendous advantages such as lightweight, low cost, abundance in vegetation, non-toxic, not prone to corrosion, easy processability, etc. [1, 2]. Composite materials developed by reinforcement of natural plant fibres in the polymeric matrix have received significant interest from researchers, owing to increased global awareness and the quest for sustainable technologies. Natural fibre reinforced (NFR) composites are a sustainable market driver which significantly decreases the cost of materials as energy consumption during the production of natural fibres (NF) composites is only 17% as that of synthetic fibres [3]. The NFR composites have several environmental benefits which include CO_2 assimilation (reducing carbon content in the atmosphere), positive impact on cultivation, and reduced fossil-based resource exhaustion [4, 5]. Natural fibres are biodegradable and low-cost. NFs comprise lignin, cellulose, hemicelluloses, etc. The time span and strength of fibres depend on the amount of cellulose present and therefore strength and durability are directly affected by the proportion of cellulose loss. Moreover, the quality of fibres is considerably impacted by environmental conditions, harvesting, and processing parameters. The use of NFR composites significantly increases the pace of manufacturing of reliable, lightweight products, which are recyclable as well as environment friendly and recover energy through incineration. The use of NFs must be modernized for comprehending the performance and other related attributes of NFR composites. Generally, the NFR composites comprise different constituents based on the formation of complete and partial green composites such as biopolymers/synthetic fibres, NF/synthetic fibres, and NF/synthetic polymers offer good performance and have positive environmental impacts [6]. The orientation of fibres in woven, non-woven, and mat stacking composites offers better mechanical behaviour.

As suggested by several reports [7, 8] use of NFR composites shows immense growth in the market for the development of several industrial components most notably in the military, automobiles, and construction sectors. Commercial applications of NFR composites include door and roof panels, helmets, antiballistic vests, and outdoor furniture. NFR composites have found extensive applicability in designing low-cost houses, civil structures, insulating walls, portable lightweight tents, etc. [9]. Plant-derived fibres such as hemp, jute, pineapple, flax, and kenaf exhibited good mechanical strength, while being considerably lighter as compared to synthetic fibres such as aramid, Kevlar, and glass fibres.

10.2 Type of Natural Fibres and Their Processing

Natural fibres are classified on the basis of their sources such as plants, minerals, and animals. These sources are enormously available in nature. A wide range of fibres are obtained from a plant's stem (flax, jute, ramie, hemp, etc.), fruit fibres (coir fibres),

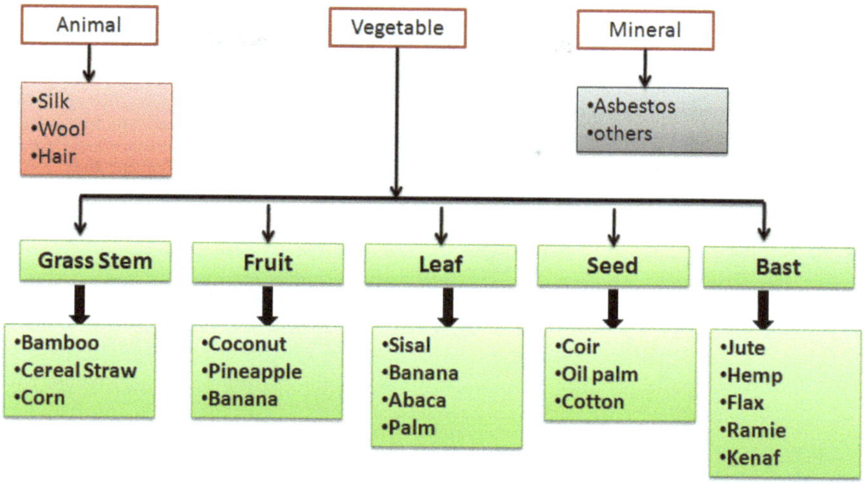

Fig. 10.1 Classification of natural fibres obtained from different sources

leaf (palm, sisal, etc.), stalks (corn, bamboo, etc.), and seeds (cotton, etc.) as shown in Fig. 10.1. Natural fibres like hemp, jute, flax, etc. are normally used reinforcement due to their high mechanical stability, and the longest length. Based on the habitation of plants, different fibres are obtained in different parts of the world such as flax and hemp are common in Europe, Jute, and cotton are mainly produced in Bangladesh and India. Natural fibres are harvested once a year. Afterwards, NF undergoes processing which involves washing, drying, and heckling as shown in Fig. 10.2. Heckling is a process of mechanical carding of short and long fibres to separate from fibre bundles where contamination and debris are removed [10]. The loose short fibres are directly woven into textiles, mats, ropes, and insulating materials. Long and short fibres are used for industrial applications while economic viability is a concern. To achieve desirable properties of composites, the NF should be continuously densely aligned in the same direction at which load is applied.

10.3 Properties of Natural Fibres

Natural fibres have several advantages such as low density, high tensile strength, etc. as depicted in Fig. 10.3. NF exhibited almost half the density of synthetic fibres and similar mechanical tenacity. While considering the density of NF account, the absolute modulus of NFR composites is similar to synthetic fibres which makes them an appropriate candidate for industrial applications. It can be estimated that replacing glass fibres with NFs subsequently decreases the 40% weight of composite in the automobile sector. Reduced weight of NFR composites considerably decreases the fuel loss [11, 12]. The value of mechanical properties such as tensile strength,

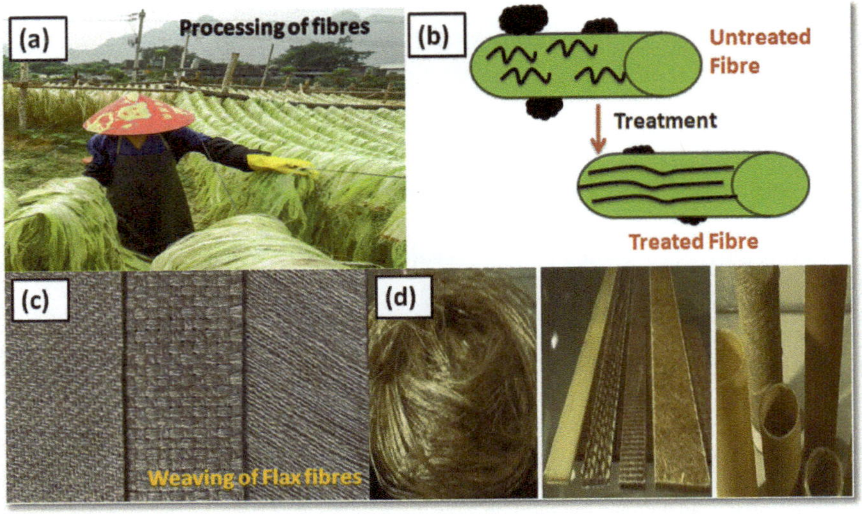

Fig. 10.2 Represents processing of natural fibres **a** Drying natural fibres in a bask **b** Treatment of natural fibres to remove impurities and **c** Different weaving of flax fibres **d** Applications of fibre

stiffness, Young's modulus, and impact resistance are measured according to non-standard parameters however proper compliance to measurement standards can bring better clarification and increase the use of NFR composites in structural applications where high mechanical tenacity is required [13]. Usually, NFs are lignocellulosic as they consist of 71–43% of cellulose microfibrils integrated into a matrix of hemicelluloses, wax, and lignin which are associated with hydrogen bonds [14]. The presence of cellulose content determines the strength of NF with a low microfibrillar angle. The more will be the content of cellulose, the higher the strength of the fibres. In addition, NFs are non-toxic, recyclable, biodegradable, easy to handle, and non-abrasive which helps in reducing the wear of manufacturing equipment. The properties of NFs are summarized in Table 10.1.

There are several shortcomings of NF such as being prone to absorb moisture, poor compatibility with polymer matrix, and poor interfacing with polymers. There is always a probability of the presence of waxy compounds on NF's surface which leads to poor matrix fibre bonding. In order to overcome these shortcomings, treatment of fibres with acetylene, ethanol and several organic agents reduces the moisture uptake and enhances the interfacing with polymers which subsequently offers tempting prospects that NFR composites possess high consistency and durability to composites. The surface modification and treatment of NFs do not deteriorate the fibre strength and stiffness but enhance the overall properties. In the open air, the presence of various microorganisms, temperature, and ultraviolet (UV) light leads to the degradation of natural fibres due to the breakdown of inherent compositions such as lignin, and cellulose. A study was conducted to investigate the effect of water showers, environmental dampness, and UV light on the mechanical properties of

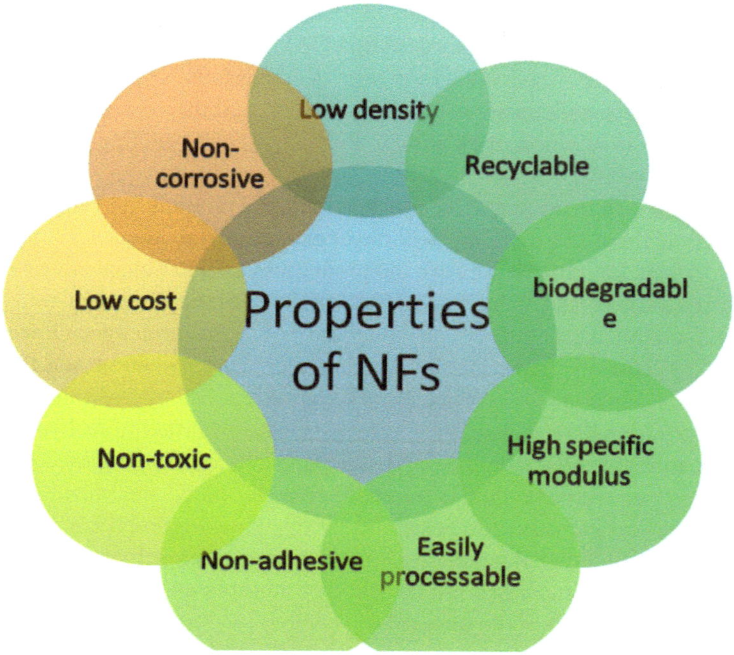

Fig. 10.3 Properties of natural fibres

Table 10.1 Mechanical strength, density, and young modulus of NFs [13]. *Adapted with permission from ref* [13] *Copyright (2016) (Elsevier)*

Fibre	Moisture (wt%)	Density (g/cm^3)	Mechanical strength (MPa)	Young's modulus (GPa)
Kenaf	10–12	1.4	200–930	15–54
Coir	8	1.1–1.4	95–220	2–6
Flax	8–12	1.5	340–2000	27–104
Jute	1.0–1.9	1.5	320–800	8–78
Bamboo	8.9	0.6–1.1	150–800	11–32
Sisal	10–22	1.3–1.5	360–700	9–38
Pineapple	12	1.5	413–1625	1.5
Cotton	7.9–9	1.6	288–800	5–12.6
Hemp	1–4	1.5	260–900	23–90
Banana	1.5–9	1.3	449–500	12
Ramie	7.5–18	1–1.5	400–1000	24.5–128

kenaf reinforced composites. The results indicated lower mechanical properties due to the deprivation of the hemicelluloses matrix [15]. The overexposure of composites by UV radiation was ascribed to the formation of cracks in NFs. Similarly, the elastic properties of jute phenolic resin were reduced to one-half due to excessive exposure to UV radiation. In addition, dark stigma, fibrillation, and swelling were observed after UV exposure [16].

Different kinds of treatment of NFs such as additive, physical, and chemical treatment overcome these shortcomings without depreciating the properties. In addition, several works are ongoing to improve the properties of NFs. Enhancement in properties of NFs carried out by the mean of irradiation, corona discharge, and plasma discharge. Functional properties of NFs increase by plasma discharge at low temperatures that consequently cause etching, implantation, crystallization, and formation of free radicals. Moisture absorbance of NFs can be reduced by hydrothermal treatment which additionally enhances the thermal insulation. Furthermore, hydrothermal treatments cause lower degradation of NFs as compared to untreated ones. Chemical treatment such as mercerization is an alkali (NaOH) based treatment that modifies the surface and increases the possibility of dye uptaking in NFs-based textiles. Polymer and fibre interfacial bonding become stronger as a result of alkali treatment [17]. Alkali treatment increases the presence of amorphous cellulose and inactive the active sites as a result of the removal of water.

$$\text{Fibre} - \text{OH} + \text{NaOH} \rightarrow \text{Fibre} - \text{ONa} + \text{Water} \tag{10.1}$$

Another important chemical treatment is acetylation where acetyl anhydride successfully replaces the hydroxyl group of NFs. Acetylation attributed to the rendering of more hydrophobic surfaces thus reduces the risk of biological attacks and enhances fibre compatibility with polymer matrix.

The polymer matrix is used to develop high-performance composites while considering synthetic and natural fibres. Polymer-based composites exhibited high mechanical performance with no chances of corrosion. Impregnation of NFs in a well-suited polymer matrix can increase the interface bonding of polymer-NFs. The degree of adhesion between natural fibres and polymer determines the efficiency of load distribution and stress transfer at interfaces. Development stages of natural fibre reinforced composites have a certain impact on the environment which affects the product popularization of NFR composites. Each stage has different parameters that should be considered by designing products therefore contribution of each stage to the environment is different hence to address the best possible method is necessary. However, there are several techniques to manufacture NFR composites [18]. Hand layup technique simplest method to produce complex NFR composites due to the low cost of equipment and low production time. Conversely, the product formed by this method is compromised in quality i.e. one dimension of the product is smooth. Due to many layers in composites being non-uniform which consequently deteriorated the properties. Another more preferred used is the resin transfer moulding method (RTM) in which resin and reinforcing fibres are filled in two tight moulds and compressed with high force. The RTM products are better in quality as compared to other methods.

Some other methods such as contact method, spray method, open mould method, and pultrusion are considered depending on the quality of the product required. Products obtained by pultrusion are of high quality with higher reinforcement of fibres (nearly 80%) which leads to excellent mechanical properties [19].

10.4 Industrial Applications of Natural Fibre Reinforced Composites

10.4.1 NFR Composites in Automotive Industries

- **In automobile industries**

With gaining insight into lightweight weight and environment friendly automotive components, the use of natural fibres is extensively increased. Not all naturally occurring fibres are suitable for automotive applications, but only a few exhibit desirable properties. The use of NFR polymer composites provides momentum to research in NFR polymer composites. The advantage of NFR composites over metal-embedded composites is to produce recyclable and reusable lightweight automobile components. Polyester-based natural reinforced fibres are predominantly used by German-based automotive industries [20]. The NFR composites are primarily used in door panels, seat covering, and other insulating parts [21]. Figure 10.4 shows the use of NFR composite in various parts of automobiles. Banana-reinforced polymer composites fabricated and used in car panels named 'MANACA' exhibited excellent fibre-polymer bonding hence there was no proof of delaminating and de-bonding [12]. The mechanical strength of the car's structural components increased by using kenaf–glass fibre reinforced epoxy composites in bumper beams [22]. The glass fibre-replaced jute fibres are utilized in the front bonnets of vehicles. These NFR-based bonnets are environmentally, economically, and technologically better. Additionally, Jute fibres are capable of establishing excellent structural components [23]. Pineapple mixed with cassava flour reinforced in polylactic acid is used in interior components of cars. The pineapple-reinforced composites efficiently reduce the odour emission in the interior of cars as compared to synthetic polymers [24]. Mechanical strength of jute, abaca, and glass fibre composites are reported in literature where abaca composites exhibited excellent tensile strength, impact, and flexural strength. The jute and abaca composites can be a good replacement for components of automotive applications such as mudguards and engine covers [25]. Sisal fibres reinforced in formaldehyde matrix exhibited superior flexural strength and provided good impact resistance. The different loading of sisal fibres is considered an important criterion for the fabrication of mechanically stable composites through the compression moulding method which can be utilized in practical applications [26]. Furthermore, kenaf and sisal fibres loaded epoxy composites are suitable for lightweight applications which reduce more than 40% weight of car components hence lessening fuel consumption [27]. The exterior and interior part of automobiles is developed by kenaf fibre reinforced composites. Firstly, fibres are treated with polypropylene, unsaturated polyester, and epoxy resin to minimize the moisture content. The resin transfer moulding (RTM) method is used to fabricate automotive components manufacturing. About 40% fibre loading (unidirectional) indicates a higher mechanical strength of 55 MPa which is good for the replacement of other synthetic fibre-based composites [28]. Apart from the mechanical durability of NFR composites, a lot of literature has reported on the acoustical and thermal properties of composites in interior components, heat shields, ducts, engine covers, etc. due to porosity. Therefore, these composites exhibited great sound-absorbing ability and thermal properties. These composites can be a great substitute for synthetic fibre-based composites. According to Prabhu et al. [29] combination of tea waste and glass fibres impacts the thermal, acoustical, and mechanical characteristics of sisal fibre-based epoxy composites. As per results, this combination improves the characteristics of NFR composites which makes them suitable for brake pedals, door panels, soundproofing materials in speaker compartments, bumper, and interior components, etc. Palm fibres and sisal fibre composites significantly possess acoustic characteristics [29]. Roof and door panels, and wall and floor lamination are prepared by hybrid sisal fibre reinforced composites.

Leading automobile industries are using hybrid sisal-based composites as given in Table 10.2. Sisal hybrid material composed of flax and cotton fibres is incorporated in the epoxy matrix for the manufacturing of rear panels and door panels of Mercedes Benz as shown in Fig. 10.4 [30]. A composition of 70% sisal fibres with 30% acrylic polymer is used by BMW for the manufacturing of door panels [31–33]. NFR door panels lower the risk of passengers during accidents by considerably reducing the weight of panels by 20%. Demand for components made of natural fibre composites increased in European countries, such demand will considerably increase in the future due to the recyclability of NFR composites and their goodness for the protection of the environment.

Fig. 10.4 Use of natural fibre reinforced composites in automobile components [30]. *Open access under a CC BY 4.0 license. Copyright@2015 Layth Mohammed* et al.

Table 10.2 Companies using NFR composites in automobile parts

Components	Materials	Automobile
Door panels	Sisal-flax composites	Mercedes Benz (E-class)
Rear panel	Sisal-cotton with epoxy	Mercedes Benz (C-class)
Interior panel (door)	Sisal-flax	Volkswagen
Interior door panel	Sisal-jute with flax	BMW

10.4.2 In Aviation

Nowadays, NFR composites are majorly used in the aviation sector to determine the future scope of these composites. However, a lot of research deficit is there to form appropriate NFR-based components. The use of NFR brings triumph to lightweight, low consumption, and reduced cost, corrosion resistance aviation components. In general, NFR composites significantly enhanced physical as well as mechanical properties. Stacked and sandwich panels are generally employed in flooring, gallery walls, compartments, and ceilings of aircraft. The sandwich panels offer excellent flame retardation, and heat resistance, and significantly reduce weight which results in less fuel consumption. The airliner using natural fibre-based components reduces the weight of 200–300 kg of panels used in aircraft cabins. The overall weight equates to 2000–7000 tonnes reduced emission of CO_2 during the lifetime of aircraft. Placement of a heavy system in low earth orbit nearly costs €28,000/kg and the cost of fuel is 50% of the operational cost. However, the tremendous use of NFR composites reduces the cost of the product to 50% by reducing structural weight to 50% which substantially reduces fuel consumption to 20% (Boeing 787) as compared to aluminium composites. Generally, flax textiles in phenolic resin are used for the development of different interior parts of aircraft by Airbus Company. The flame-retardant test was performed before the employment of flax fabrics [34] the results exhibited excellent flame retardation which complies with Federal Aviation Administration (FAA).

10.4.3 In Railways

The NFR composites are utterly applicable in Indian railways for the development of components like, floor/floor panels, modular toilets, partitions, seat covers, handles, etc. [35]. To achieve fast speed, weight reduction, less inertia, lower track wear, and low power consumption, the use of natural fibres in railway industries is essential. The NFR composites have paved the way for the development of lightweight (about 40% weight reduction), cost-effective (reduce fuel consumption), and sustainable technology which is applicable in almost all industries. Nowadays, NFR composites are being used by automotive industries for the fabrication of different components listed in Table 10.3

Table 10.3 Different natural fibres are used in different automobile parts

Automotive components	Fibres used
Roof and door panel	Jute fibres
Rear desk and wall, door panels	Kenaffibres
Roof panel	Oil palm fibres
Door panel cover, armrest	Hempfibres
Seat, instrument panels, Engine cover	Flax fibres
Insulation	Ramiefibres
Interior panel, door panel, roofing sheets	Sisal fibres
Floor and body panels	Abacafibres
Floor and seat mats, Mattresses	Coirfibres
Roof panels	Rice huskfibres
Luggage compartment	Coconut fibres
Soundproofing, Insulating material	Cotton fibres
Hard armrest, seat back	Wood fibres
Wrapping paper	Banana and pineapple fibres
Seat covers	Wool fibres

10.4.4 Military

Since ancient times, the intrusion of foreign troops, and insurgencies evolved the scenario of arms conflicts which constantly enhanced the exploitation of weapons and armour. The armour used in the beginning was heavy because of increased protection capacity. Lighter composites are in demand time by keeping the mobility of soldiers as a concern. Currently, natural fibres are used by armed forces for the development of lightweight and flexible materials which offer protection against bullet strikes, and fragments of explosives. Researchers are prompted to develop cost-effective ballistic protective devices that can be readily available for military troops and vulnerable people who are susceptible to get targeted by terrorists. Therefore, the production of vegetable-based fibrous material with less cost and low weight is abundantly increased. The use of NFR composites is of great concern to the military in uniform and other textiles, personal protection devices, field housing construction (fireproof camouflage tents), weapons and hard armour, transport vehicles and packaging, etc. [36]. Military includes protective clothing and systems which incorporate textiles within. Use of natural fibres such as hemp and flax in clothing applications was prevalent before the Second World War. During that time, troops of the Russian army wore red-coloured uniforms made of hemp fibres. After World War, all the textiles were replaced by synthetic woven fibres such as nylon, polyester, etc. Nowadays, textiles are integrated with smart devices and intelligent fibres for military and civilian safety in hard zones [37]. In 2001, a project was started by the Défense Advanced Research Projects Agency (DARPA) for the development of electronics-integrated

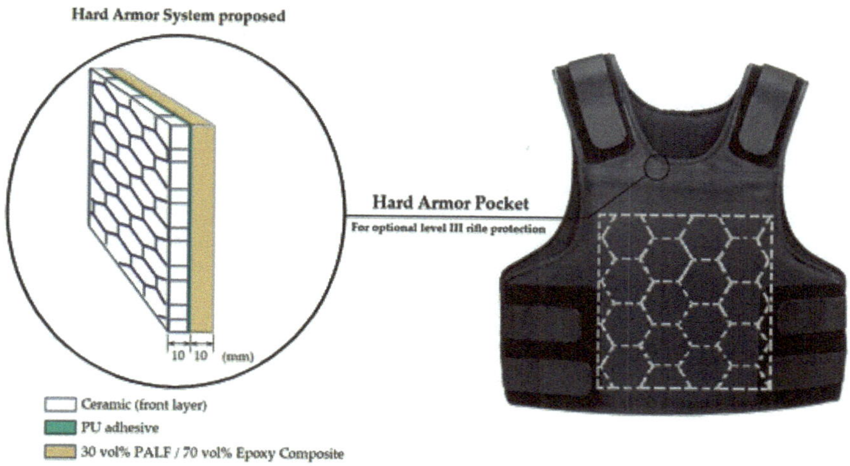

Fig. 10.5 Use of pineapple leaf fibres in hard armour vests [44]. *Open access under a CC BY 4.0 license. Copyright@2020*

textiles. Special conductive cellulosic fibres are integrated into electronic textiles for identifying special pharmaceuticals on a soldier's body.

Low-density composites with high strength gained a lot of attention for the development of armour vests as shown in Fig. 10.5 [38]. Presently, synthetic fibres such as aramid fibres, zylon fibres, and Dyneema fibres are utilized in the development of bulletproof vests [39]. Benzait et al. have suggested the use of natural fibres in military applications as they possess excellent characteristics which can serve as an excellent substitute for aramid fibres [40]. Most researchers explored epoxy-based composites but were not found preferable due to the high stiffness and low ductility of epoxy. Rohen et al. have investigated the ballistic resistance property of sisal-reinforced epoxy composites. They found that NFR composites show comparable results to aramid fibrous composites with the same thickness [41].

Excellent ballistic performance of NFR composite dissipates high impact energy and depends on the ability of fibres to capture ballistic fragments as a contribution of lignocellulosic fibres. The presence of higher lignin and low cellulose absorbs high impact energy with randomly oriented dense weaving of coir fibres [42]. Pineapple leaf fibre (PALF)/ceramic-based composites are extensively investigated as one of the strongest naturally occurring fibre with a stress of 1.6 GPa which substantially enhances the mechanical strength of composites. Ballistic tests for these composites are carried out at Brazilian Army Assessment Centre. Surprisingly, ballistic performance of PALF/ceramic mat was comparable with stringer synthetic aramid/kevlar fibres. PALF-based composites not only displaced similar ballistic performance but also the same mechanism of fragment capturing. The mechanism responsible for fracture capturing is due to Van der Waals forces, and short static surface charge generated after the interaction of the targeted projectile with the surface of composites [43]. Besides, good strength, the cost of natural fibres is 70 times reduced as of dyneema fibres. The protection level of ballistic and bulletproof armours can be enhanced by double layers of PALF/epoxy composites. These hard armours can work as a guard against 7.6 mm calibre rifles without any requirement of alloys [44]. Commercially available fibres are quite expensive and the development of personal protective devices is a fairly complex process. Therefore, these fibres are substituted by fibres obtained from natural sources. Integration of natural fibres with synthetic fibres is prevalent due to protection of armed forces against high impact of bullets i.e. natural fibres are promising candidates for impact resistance in personal protective wears and are being benefited by abundant quantity in nature. During the selection of natural fibres for high impact resistance-based applications, some of the characteristics of the natural fibres should be considered. These characteristics are greatly dependent on growth conditions, chemical composition, and structural strength of natural fibres.

The natural fibres have relatively low mechanical strength as that of synthetic fibres therefore modification and integration of natural fibres with resin provides comparative strength to armour devices. Moreover, interfacial adhesion between fibres and polymer resin plays a vital role in enhancing the mechanical response of NFR composites. Improvement of mechanical properties of NFR composites is carried out by coating the surface of NF with polymer resin before polishing through graphene oxide for high impact resistance against ballistics [45]. Liu et al. developed a model to investigate the mechanism of transfer of energy during collisions of bullets and high energy projectiles. They suggested that high impact energy dissipation capacity and material integrity can be enhanced by adjusting viscosity and shear modulus of the matrix. In ductile materials, energy absorption leads to plastic deformation whereas dissipation of energy substantially occurs through fracture mechanism in ceramics. Hence, some small cracks are responsible for fragmentation and expansion of cracks in the material. In first stage, the expansion of cracks initiates the deformation of matrix until complete transfer of load to the length of fibres as shown in Fig. 10.6. In second stage, cracks propagate with slower pace than before as fibres are under strain which causes the delamination of fibres and matrix. At last, catastrophic penetration spread rapidly due to breakage of fibres and failure of matrix [46]. Inappropriate match between fibres and matrix is responsible for the propagation of cracks which further causes delamination of matrix and fibres. Any type of bubble casting and holes leads to the generation of microcracks when tested with high kinetic energy projectiles. The propagation of cracks can be resisted by dense arrangement of fibres. Haro et al. investigated the dissipation of high impact energy by the arrangement of fibres and hybridization of material during ballistic collision. Initially impact energy of weapons was compared with impact resistance ability of composite targets. The outcomes of crack propagation in composite targets measured at different stages are accomplished by the National Institute of Justice (NIJ) requirement for level II, IIA, and IIIA for protective armours [47]. Hajer et al. studied the mechanical properties of flax fibres with and without natural viscous-elastic layer. A different stacking sequence was followed to perform uni-axial stack loading test. The results for different mechanisms categorized the obtained cracks in various categories according to with and without visco-elastic stack [48]. Natural fibres are hydrophilic and therefore absorb moisture, have low thermal stability, and poor integration with commercially available fibres. Some fibres are thermally stable such as kenaf and coir fibres. Modification of kenaf and aramid fibres with polymer resin enhances the mechanical properties which appropriate weaving patterns. Enhanced mechanical performance of kenaf hybrid fibres makes them applicable for military vehicle spall liners. Yahaya et al. studied the kenaf-Kevlar (semi-aramid) hybrid laminates in woven, mat, and unidirectional patterns for spall liners which provide protection against small fragments. They found that woven pattern of kenaf hybrid laminates exhibited high flexural, tensile strength of 100.3 MPa which is 52.07% higher as that of mat laminates [49]. Interest in natural fibres is prompted for exploitation of personal protective equipments such as helmets, hard hats, and shin guards for the protection of civilians and military troops [50, 51].

Hemp and flax fibres reinforced polypropylene was tested for ballistic tests in Royal Military Academy in Brussels. The composites are stacked with 26 layers of fabrics made of hemp and flax polypropylene laminates. The results indicated plain flax did not meet the standard criteria. Flax with steel integration irregular shape decreased the possibility of penetration of armour material. NFR composites have the advantage of repairing damages in composites. In emergency, a soldier can patch up the damaged area of composites by following the simple process of drying plant materials; extracting the fibres and binding fibres with proper binders. During World War Second, phenolic resin impregnated unidirectional flax fibre composites

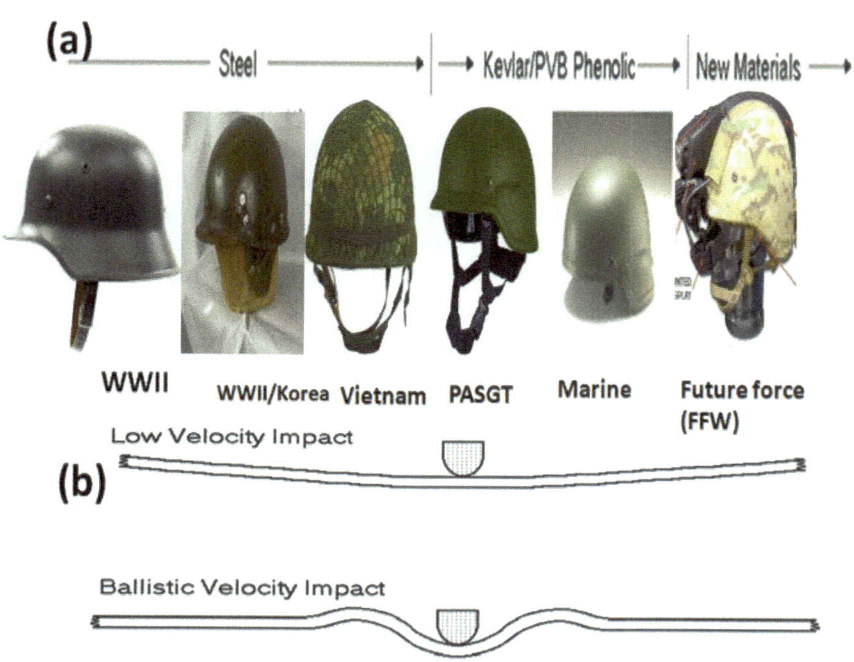

Fig. 10.6 **a** Depicts the history of Army helmet design and materials. **b** Shows ballistic impact tests

named as Gordon-Aerolite were used as fuselage in aircrafts and in helmets [50, 51]. In military transport vehicles, NF based composites are used in panels, passenger walkways, doors, bulkheads, nose cones, etc. Unmanned remote controlled vehicles can be produced by NFR composites having reduced cost, low weight and integrated with several electronic devices which is a current demand of time.

According to the prompted use of NFR composites in several applications, these composites can also be used in military applications in large scale. Although, several materials are driven by natural fibres are also being used in military. NFR composites would be suitable materials for the development of weapon components such as knife handles, gun barrels, rifle bodies, and stocks. Depending on weight requirement of weapons bast and wood fibres are suitable materials however great challenge would be to ensure weather tolerance of these materials such as tolerance to moisture, and temperature.

Marsyahyo et al. developed high impact resistance bulletproof panels by using naturally occurring ramie fibres integrated into epoxy resin. Before the reinforcement of fibres in polymer resin, fibres were treated with several chemical solvents, such as ethanol, acetone, methyl ethyl ketone, and silane used as coupling agents. The panels were tested for NIJ standards II, IIA, and IV. The panels were able to oppose the penetration of high energy bullets in stage II. However, high energy projectile for stage IV was not resisted by ramie fibres-based composite [51, 52].

Currently, lightweight and low-cost bridges are a requirement by the military for crossing short-span gaps (nearly 4 m) with supporting military loads of 27,000 kg carried by military trucks. The bridges must be of low thickness as it can be used in other applications such as roadway matting and ramping systems for loading weight in aircraft and ships. For instance, the short track bridge (4 m) is located in Germany, which is a commercially available mobile bridge for multitasking that also bears stringent environmental conditions and supports a military load of 27,000 kg at a time. Robinson et al. studied the possibility of exploiting fibre reinforced short-span lightweight bridge/treadways for the military. The bridges were field tested in the U.S. army with a load of 27,000 kg and also tested for extreme weather conditions. The proof testing revealed the bridge as having adequate mechanical strength which results in no sign of stiffness and permanent damage. These lightweight bridges are viable candidates for employment in emergency response where a short-span gap is encountered [53].

10.4.5 Sports Industries

Traditionally, metals and alloys-based sports equipments were prevalent although, these equipments were quite heavy which restricted the movement of equipment to a limit. Lightweight equipments are required preferably for proper movement and better athlete's performance. Currently, research on lightweight sports equipments designed by natural fibres reinforcement is continuously evolving with improved mechanical functionalities i.e. NFR composites have an unmatched advantage in this regard [54]. Use of NRF composites protects environment by advocating green sports and implementing sustainability. Before the production of sports equipments, strength, elasticity and other mechanical properties should be considered. NFR composites have good tenacity, specific modulus, and stiffness making it good candidates to replace the synthetic fibres. NFR composites-based sports equipments are easily repairable as compared to other fibre-based sports equipment. Traditional sports equipments were hard to repair, and have excessive maintenance costs. Moreover, formation of NFR sports equipment is quite easy. Furthermore, performance to price ratio was the problem during metal materials. NFR composites have good wholesomeness and a good performance to price ratio which increases product popularization and revenue.

According to the American Chemold Company, carbon-reinforced fibres utilized in table tennis exhibited poor damping absorbance properties. Currently, Flax is commonly employed as fibre for the development of sports equipment in industries. To enhance the properties of natural fibre-based composites flax hybrid composites are used for producing sports equipment. For instance, tennis rackets, fishing rods, and bicycle frames are commercially manufactured with flax and carbon reinforcement appropriate proportions as flax fibres provide excellent damping properties compared to synthetic fibres [55]. Better damping properties provide sportsmen good firmness while striking the ball and increase the contact time between tennis bat and ball, it increases the probability of striking the ball with a high speed by the collective force of sportsmen.

Due to the excellent impact resistance of NFR composites, these composites are being successfully used in sports industries, particularly for the development of components for outdoor sports. Generally, flax fibres with integration in carbon fibres are successfully employed in the development of snowboards, canoes, surfboards, bats, ice hockey, and bike frames due to excellent vibrating damping behaviour. Carbon/flax fibres-based composites are prone to high impact resistance. Companies like Magine, and Flaxland manufactured snowboards and canoes using flax fabrics. Yusup et al. prepared alkaline-treated oil palm hollow bunch fibres in the epoxy matrix for potential applicability in hockey sticks. The mechanical properties of alkaline-treated oil palm fibres show promising results for prepared hockey sticks [56]. Some special equipment is required for sport climbing including climbing shoes with body and waist gripers, helmets, ropes, carabiners, etc. with high strength and durability to reach the highest summit. Generally, cliffs and figure eight are made from metal and alloys and are prone to corrosion. However, NFR composites are chosen in place of metal and alloys as NFR composites are lightweight, and corrosion resistance. Furthermore, manufacturing processes and parameters also affect the properties of final products.

10.5 Construction

In ancient times, straw and clay composites were used by Egyptians for the construction of walls of houses. In 1973, jute reinforced fibre composites were employed for the construction of primary schools and were considered the first use of NF by a developing country [57]. Phenolic resin coated bagasse fibres are used by Jamaicans and Ghanaians in the construction of houses by roofing sheets and panels. Currently, Polymer reinforced natural fibres have won the attention of several industries. Indian government supported the production and development of jute fibre composites as a substitute to wooden composites. The use of NFR composites is surging in popularity among construction industries to produce low cost, environment friendly building materials. Natural Fibres have voids/space and naturally consist of cellulose, lignin pectin etc. which enhances the thermal insulating properties. In addition, the production of building materials leads to environmental pollution whereas NFR composites cause 0.1% of pollution to the environment. Internationally, 95% of NFR composites are being utilized in market [58]. The NFR polymeric composites significantly substituted the traditional cement, steel, and concrete like materials. Different natural fibres obtained from different plants are explored by users which include, cereal straw, kenaf, bamboo, rice, husk, corn stalk, etc. for different structural applications listed in Table 10.4 [59]. The characteristics of NFR in the building industry include; high stiffness, lightweight, low cost, high strength to weight ratio, and biodegradability. Material like asbestos fibres is hazardous to the air atmosphere or requires controlled parameters to work on, but natural fibres are neither hazardous nor required any controlled environmental parameters. Development of green building is a suitable mindful place to work and live. These composites are remarkably used in construction industries for manufacturing of structural components such as roofs and bridges, and non-structural components which include windows, panels, door frames exterior parts, etc. Besides good mechanical stability, natural fibres have some disadvantages like degradation of polymer at 200–250 °C temperature, hydrophilic nature which resulted in inappropriate adhesion, and moisture absorbing nature. To get excellent environmental performance, natural fibres are treated with proper resins which decrease the hydrophilic characteristic and derive hydrophobic fibres by blocking the pores. As natural fibres are prone to moisture, a fatigue test was conducted on a 50 years old bridge surprisingly; NF composites revealed almost negligible moisture content owing to a special coating of epoxy applied on fibres which prevents moisture penetration. Resin coating is applied as a protective layer against waterproofing and moisture absorption.

10.5.1 Structural Beams and Panels

Composites are widely employed in the development of bridges in Netherlands because these bridges require low maintenance. On the other hand, Concrete bridges

Table. 10.4 Use of different fibres in construction [59]. *Open access under a CC BY 4.0 license. Copyright@2021. Al-Azad* et al.

Fibres	Uses in construction
Oil palm	Window and door frames, ceiling, insulating panels, fencing, etc
Wood	Fencing, doors, window frames, and boards etc
Rice husk	Bricks, railing, panels, window frames, etc
Flax	Window frames, door frames, laptop cases, railing
Sisal	Beams, panels, roofing boards, floor laminates,
Stalk	Building panel, drain construction, bricks
Kenaf	Insulating material, boards, panels
Cotton	Wardrobes, cupboards, cabinets,
Jute	Roofing, flooring, panels, door and window frames
Coir	Structural beams, panels, door shutters, tanks, roofing sheets, fencing, window frames, cabinets, walls, wardrobes, bricks

are low cost alternatives however high weight of concrete based composites makes it inapt in many locations. Therefore, NFR composites are widely tested for structural applications such as beams, bridges, panels, and columns production as shown in Fig. 10.7 [60]. One of the fundamental components of bridge, and building is a beam due to its load bearing capacity. Before construction, a beam is tested for bending capacity which yields three-to-four-fold bending. Beams are comprised of wood, cellulose, reinforced concrete, fibres, iron, and steel. Viable ways to develop a beam with great load bearing capacity with low cost and low pollution are demonstrated in several literatures [61]. Sandwich beam and I-shaped beam production is the best idea for construction where thin top and bottom layers are embedded with a thick core material. Acrylated epoxidized soybean oil (AESO) as skins and cellulosic fibres foam core is investigated in which AESO skins exhibited great mechanical strength as of core [62]. In some cases, Flax fibres are considered for structural applications over glass fibres as a cost concern however possess low stiffness as of glass fibres. To achieve the desired stiffness, the thickness of bridge is solely modified. A Bridge in Eindhoven University of Technology has been made by composition of hemp and flax fibres in polylactic acid (PLA) [63]. Poletanovic et al. reported the incorporation of hemp fibres in a geopolymer reduces the density by 5% due to high porous fibre interfaces which are ascribed to entrapping of air [64]. Hybridisation of sisal and jute fibres in geopolymer such as in fly ash enhances the ductility and mechanical strength of composites which can further be applicable in the construction of structural beams [65].

The NF reinforcement benefited the construction industries with their low density, high corrosion resistance, and low cost. Generally, NFR composites can be divided into four divisions on the basic of size and alignment of fibres; such as long fibres, short fibres, nano cellulosic fibres, and consolidate fibres composites. Hemp, flax,

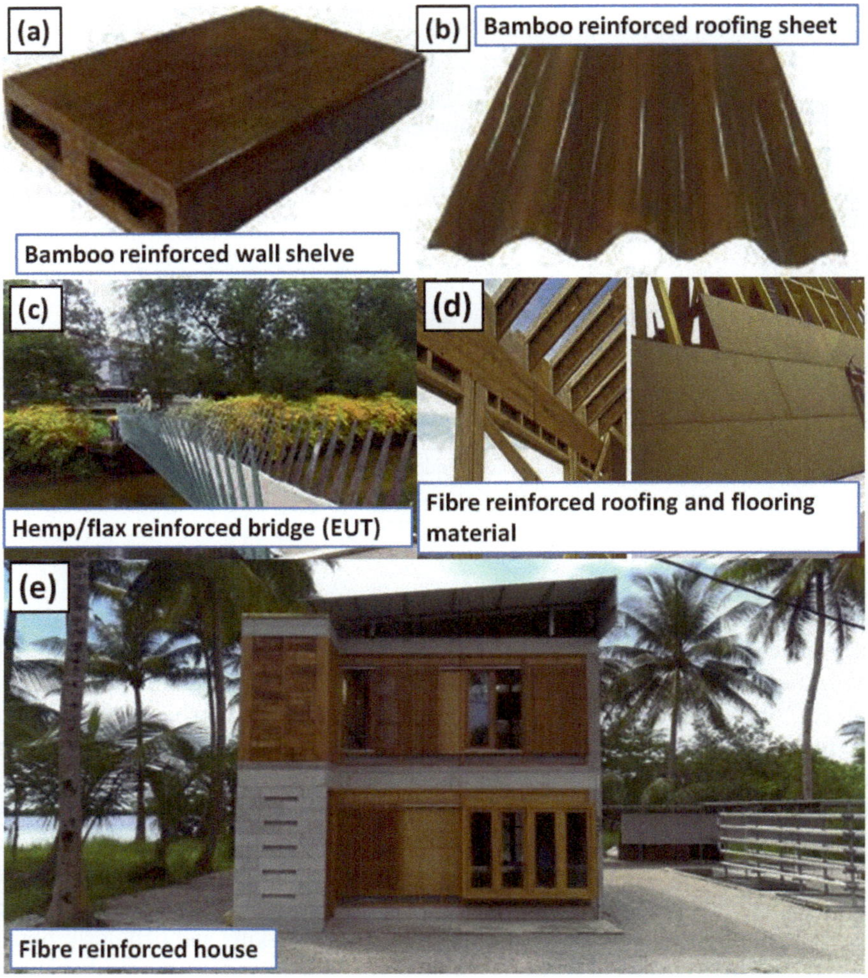

Fig. 10.7 Use of natural fibre reinforced composites in structural components

sisal, and jute fibres are the long fibres which are used as reinforcement in polymer matrix. Long NFR composites depict excellent mechanical characteristics like flexural or tensile which make them recognized as potential fibres in structural beams, load bearing components, roof panels, multipurpose panels, etc. [66]. Wood based composites are used in panels such as plywood panels in the form of blocks and stressed skin roofing panels for enhancement of the efficiency of building [67]. Woven mat sisal fibres embedded in cashew nut shell liquid with AESO as core material in the form of sandwich structure are used to make roof panels. Corrugated roof panels are also possible to develop other than flat panels in either the form of sandwich or I-shape structure. Sandwich structure fibre-based composites have been developed and tested by LOC Composites Pty. Ltd for the production of balcony walls, roofs, floors, etc. [68, 69].

Jute reinforced composites are utilized for the construction of pipes in structural reconstruction whereas synthetic fibres are not suitable for it because synthetic fibres are viable to corrosion, high weight, low strength, etc. Further, some research on the use of jute/polypropylene composites are reported for the manufacturing of structural beams and panels [70]. Moreover, jute, hemp, and flax polyester composites can be used for manufacturing of load bearing composites such as cellular beams and panels with enhanced efficiency which also depend arrangement of cellular materials. A NFR-based construction structure can be developed by several international and Indian companies. Permali Wallace Pvt.Ltd, an Indian company is specialized in the development of NFR composites for the development of houses, buildings, automobiles, railways components, etc. [71]. Bamboo fibre reinforced bethink is constructed by CSIR-AMPRI, Bhopal in which bamboo composites are employed for structural components of housing sectors. The use of bamboo composites in housing sectors is recognized as a good earthquake resistance construction material [72]. In addition, some interlocking blocks can also be manufactured for mortar free structures. These blocks may smooth the progress of energy dissipation during earthquakes and ground excitations.

10.5.2 Building Non-Structural Components

NFR composites are recognized as a great substitute for wall building. The NFR built composites have excellent shear resistance. NFR panels reduce the time of erection of wall and prove to be economical due to great weight to strength ratio, which is not the case with brickwork. NFR fibres are widely used in the insulation of walls due to compatible properties such as flexural strength, thermal insulation, acoustic insulation, etc. Li et al. stated that tensile properties of natural fibres such as flax fibres have comparable properties to that of glass fibres [73]. Hasan et al. suggested that flexural strength of coir reinforced cementitious is 10 times greater as compared to steel reinforced composites due to coir fibre bridging effects [74]. Flexural strength of wooden sheets with plastic composites is investigated for hollow cross sections [75]. NFR composites are great replacement for cementitious paste used in walls to form mortar free walls. The uses of NFR composites are helpful in making sustainable construction industries. These natural fibres include flax, coir, sisal, pineapple leaf, kenaf, bamboo, abaca, hemp, banana, sugarcane, and cotton fibres. All these fibres are vegetable fibres and naturally occurring in many countries. The use of NFR increases the properties of composites and costs very little when compared. NFR composites are used in tropical earthquake regions as low cost construction materials. Natural fibres reinforced composites are gaining insight in non-structural applications such as window frames and floor lamination, etc. [76, 77] Different fibres reduce the loss of building material and enhance the energy efficiency which promotes the idea of sustainability and provide the solution for infrastructure need. Natural fibres are much cheaper as compared to other materials required for construction. Coir fibre reinforced composite boards are used in applications like bricks, tiles, hollow blocks plywood, etc. which are recognized as a material for exterior walls, roofing, frames, etc. [78]. More importantly, coir fibres in any matrix possess high durability and excellent tensile strength. Coir fibre consists of 46% lignin which results in resistance to weather conditions such as wet and dry and prevents rotting. Coir/rubber/jute boards can be used for making wardrobes, surface lamination, ceiling, furniture, etc. Das et al. investigated the use of jute reinforced polymer composites in structural applications such as floor, wall partitions, window frames, ceiling, etc. [79]. Li et al. used wood fibre reinforced composites in the manufacturing of window frames, floor lamination, ceiling, etc. [80]. Sisal fibres reinforced in polymer matrix are very suitable for the development of roofing and ceiling sheets. Cement and gypsum blocks can be developed by using rice husk. Whereas, rice straws are useful for the production of cement bonded boards, medium density boards, particle boards, etc. Cotton stalks can be used in wall, roofing, furniture, door shutters, and in wall plastering.

10.6 Packaging

Currently, non-biodegradable polymers are used in a majority of packaging materials such as food containers, etc. due to their indestructible nature. However, environmental perseverance of these materials resulted in excessive litter, pollution, and waste dumping problems [81]. With growing environmental problems have triggered a spark in researchers to produce environmentally compatible products by keeping the preservation of natural resources and environmental safety a concern. NFR composites are readily manufactured products with desired properties. However, use of biodegradable resin in packaging material enhances the cost of overall products. Whereas, incorporation of natural fibres in biodegradable resin substantially reduces the cost. In addition, NF reinforcement relatively enhances the strength and thermal deformation of material without loss of degradability. Some polymers derived from plants such as polylactic acid (PLA) and starch are commercially available and being used as packaging material for food and drinks which significantly preserve the food items. Bast fibres (flax, hemp, etc.) composites are mostly employed in packaging materials because of their lightweight and adequate natural availability. For instance, mobile companies are using flax fibre-based packaging materials for packing their smart, high-end mobile phones. In coming times, it would be possible to develop plant-based fibres for the production of packaging material for food and drinks which subsequently may replace the poly bags, plastics, and other paper based packaging materials. However, some of the plastic beer bottle's interior surface is clad with glass layer. In a study oil palm bunch fibres derived cellulose and thermoplastic starch hybrid composites are used in packaging materials with enhanced mechanical properties [82]. Varghese et al. prepared strawberries packaging material for preservation of fruit's freshness from bark fibres. At higher bark loading in polymer matrix, the packing material also exhibited antibacterial properties due to the presence of phenolic compound increasing the evolution of certain microbial organisms in lignin. This packaging is an eco-friendly one which protects the life of perishable fruits [83].

10.7 Consumer Goods

For the development of cost-effective consumer goods, short natural fibres are considered promising candidates due to easy moulding of natural fibres in any shape. Due to high demand of environment friendly goods, NFR composites are readily adopted for the development of products with high performance and good life span. NFR composites are suitable elements to utilize in furniture applications like desks, windows, household furniture, etc. as they resemble the wood and require minimal maintenance [84]. Wood products are prone to being captured by termites whereas NFR composite based products are not prone to it. Hence, NFR composites-based products exhibited a better life span as compared to wooden products. Parameter such as moisture absorbance is significantly affecting the mechanical tenacity of products. Increase in moisture rate considerably deteriorates the tensile and flexural strength of products. Wooden products are apt to moisture uptaking whereas NFR are better in this regard. Fruit based fibres are suitable substitutes for wood in plywood and boards. Better acoustical properties of NFR composites devoted the plant and fruit-based fibres a good candidate to employ in furniture segment with better properties. Household goods like tables, kitchen cabinets, decks, electrical goods, mobile phones, planters, and chairs are produced by using jute reinforced polymer for development of lightweight, easy movable, cost effective, and finished products as depicted in Fig. 10.8. Back support plate of chair is feasibly produced by using bagasse/polypropylene granules due to excellent support and elastic properties [85].

NFR composites seldom serve as catalysts for products for maintaining sustainability and low consumption of natural resources. A lightweight cabinet was developed by Sheffield Hallam University Art and design Research Centre and Net composites using flax/Polypropylene woven textile panels. A variety of furniture can be created by modular arrangement of panels [86]. In a study, Hemp and epoxy resin are used to produce chairs with complex geometry and good mechanical behaviour [87]. Electrical goods such as electrical switchboards, coolers, air conditioner bodies, and fens are nowadays made up of NFR composites with better finishing, and different colours. The colours of electrical goods can be obtained be polishing or coating NFR composites.

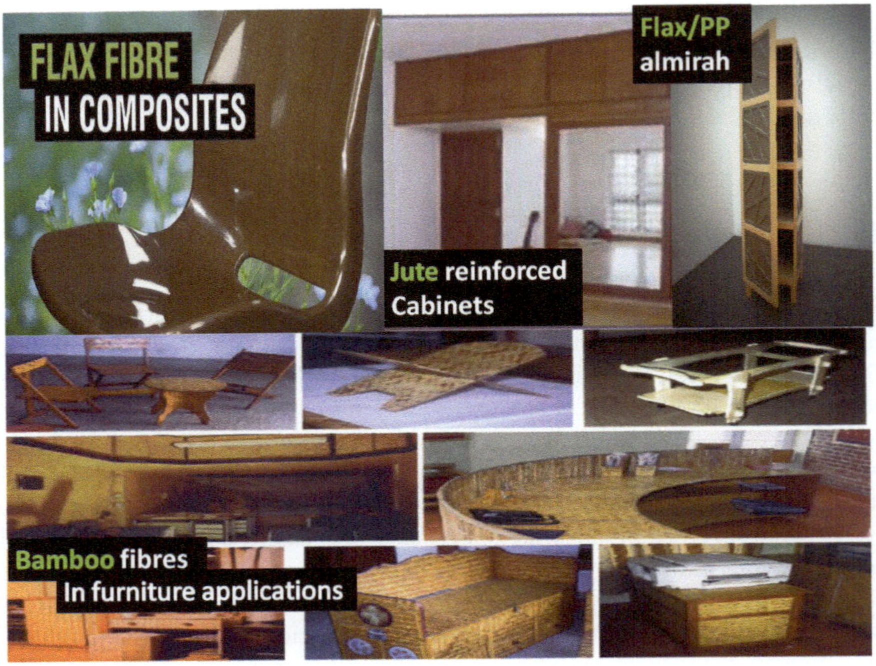

Fig. 10.8 Use of natural fibre reinforced composites in household applications

10.8 Conclusion

A wide range of natural fibres exists in nature such as plants-based fibres, animal-based fibres, and mineral-based fibres. Due to utmost mechanical properties of flax, hemp, and jute, these fibres are widely employed in industrial applications for the development of various components. Besides excellent specific strength of NFs, they also exhibited excellent stiffness to weight ratio, strength to weight ratio, sustainability and complete biodegradability which made them successfully replace commercial fibres. NFs consume negligible energy during production and are sustainable mediums due to absorption of CO_2 i.e. NFs have considerable positive impact on the environment. NFs also have the potential to apply in load bearing applications. Fibre quality is considerably affected by several parameters such as growing environment, and processing parameters. The type of NFs used determines the actual cost of the final product and performance of composites.

For better strength of NFR composites, impurity and waxy substance must be low. Long length fibres offer excellent mechanical properties. For full impregnation of fibres in polymer resin, fibres must have low bends and twists. Selection of NFs for NFR composites greatly depends on the required properties for a particular application. NFR composites exhibited tremendous applications in industries such as applicability in sports sector, automobile sector, marine structure, food industries, and construction sector. However, some shortcomings occurred including matrix-polymer adhesion, moisture uptaking, and incompatibility with the matrix. Tayota Car Company makes interior components using Kenaf/PLA mats. Most notably, short length fibres in different polymer matrices are used to produce outdoor furniture with excellent water resistance property, automobile parts with less weight, structural applications, and decking with excellent mechanical properties. Fibre treatment is considered as an appropriate option for enhancing the compatibility of fibres with matrix. Short fibres are compounded for small screws injected into automotive components. Woven flax/epoxy prepregs exhibited higher mechanical strength due to alignment of NFs. While considering environment safety, commercial resins have a greater negative environmental impact as compared to NFs. Therefore, development of bioresin is also an important aspect to be considered to facilitate the manufacturing of 100% natural industrial products.

References

1. Mahir, F.I., Keya, K.N., Sarker, B., Nahiun, K.M., Khan, R.A.: A brief review on natural fiber used as a replacement of synthetic fiber in polymer composites. Mater. Eng. Res. **1**(2), 86–97 (2019)
2. Sanjay, M.R., Madhu, P., Jawaid, M., Senthamaraikannan, P., Senthil, S., Pradeep, S.: Characterization and properties of natural fiber polymer composites: A comprehensive review. J. Clean. Prod. **172**, 566–581 (2018)
3. Shalwan, A., Yousif, B.F.: In state of art: mechanical and tribological behaviour of polymeric composites based on natural fibres. Mater. Des. **48**, 14–24 (2013)
4. Gejo, G., Kuruvilla, J., Boudenne, A., Sabu, T.: Recent advances in green composites. Key Eng. Mater. **425**, 107–166 (2010)
5. Pandey, J.K., Ahn, S.H., Lee, C.S., Mohanty, A.K., Misra, M.: Recent advances in the application of natural fiber based composites. Macromol. Mater. Eng. **295**(11), 975–989 (2010)
6. Khalid, M.Y., Al Rashid, A., Arif, Z.U., Ahmed, W., Arshad, H., Zaidi, A.A.: Natural fiber reinforced composites: Sustainable materials for emerging applications. Results Eng. **11**, 100263 (2021)
7. Asim, M., Abdan, K., Jawaid, M., Nasir, M., Dashtizadeh, Z., Ishak, M.R., Hoque, M.E.: A review on pineapple leaves fibre and its composites. Int. J. Polym. Sci., (2015)
8. Petchwattana, N., Covavisaruch, S.: Mechanical and morphological properties of wood plastic biocomposites prepared from toughened poly (lactic acid) and rubber wood sawdust (Hevea brasiliensis). J. Bionic Eng. **11**(4), 630–637 (2014)
9. Satyanarayana, K.G., Arizaga, G.G., Wypych, F.: Biodegradable composites based on lignocellulosic fibers—An overview. Prog. Polym. Sci. **34**(9), 982–1021 (2009)
10. Ghosh, T.: Handbook on jute (No. 51). Food & Agriculture Org., (1983)
11. Kumar, R., Ul Haq, M.I., Raina, A., Anand, A.: Industrial applications of natural fibre-reinforced polymer composites–challenges and opportunities. Int. J. Sustain. Eng. **12**(3), 212–220 (2019)
12. Al-Qureshi, H.A.: The use of banana fibre reinforced composites for the development of a truck body. In: Second international wood and natural fibre composites symposium, Kassel/Germany, pp. 1–8. (1999)
13. Symington, M.C., Banks, W.M., West, O.D., Pethrick, R.A.: Tensile testing of cellulose based natural fibers for structural composite applications. J. Compos. Mater. **43**(9), 1083–1108 (2009)
14. Zafeiropoulos, N.E., Williams, D.R., Baillie, C.A., Matthews, F.L.: Engineering and characterisation of the interface in flax fibre/polypropylene composite materials. Part I. Development

and investigation of surface treatments. Compos. Part A: Appl. Sci. Manuf., **33**(8), 1083–1093 (2002)
15. Abdullah, M.Z., Dan-mallam, Y.,Yusoff, P.S.M.M.: Effect of environmental degradation on mechanical properties of kenaf/polyethylene terephthalate fiber reinforced polyoxymethylene hybrid composite. Adv. Mater. Sci. & Engineering., (2013)
16. Azwa, Z.N., Yousif, B.F., Manalo, A.C., Karunasena, W.: A review on the degradability of polymeric composites based on natural fibres. Mater. Des. **47**, 424–442 (2013)
17. Adekomaya, O., Majozi, T.: Sustainability of surface treatment of natural fibre in composite formation: challenges of environment-friendly option. Int. J. Adv. Manuf. Technol. **105**, 3183–3195 (2019)
18. Dwivedi, U.K., Kumari, N.: Foundation of composites. In: Advances in biomedical polymers and composites, pp. 31–60. Elsevier, (2023)
19. Kozlowski, R., Wladyka-Przybylak, M.: Uses of natural fiber reinforced plastics. In: Natural fibers, plastics and composites, pp. 249–274. Springer, US, Boston, MA (2004)
20. Karus, M., Kaup, M.: Natural fibres in the European automotive industry. Journal of Industrial Hemp **7**(1), 119–131 (2002)
21. Schuh, T.G.: Renewable materials for automotive applications. Daimler-Chrysler AG, Stuttgart (1999)
22. Davoodi, M.M., Sapuan, S.M., Ahmad, D., Ali, A., Khalina, A., Jonoobi, M.: Mechanical properties of hybrid kenaf/glass reinforced epoxy composite for passenger car bumper beam. Mater. Des. **31**(10), 4927–4932 (2010)
23. Alves, C., Silva, A.J., Reis, L.G., Freitas, M., Rodrigues, L.B., Alves, D.E.: Ecodesign of automotive components making use of natural jute fiber composites. J. Clean. Prod. **18**(4), 313–327 (2010)
24. Kim, K.W., Lee, B.H., Kim, S., Kim, H.J., Yun, J.H., Yoo, S.E., Sohn, J.R.: Reduction of VOC emission from natural flours filled biodegradable bio-composites for automobile interior. J. Hazard. Mater. **187**(1–3), 37–43 (2011)
25. Ramnath, B.V., Manickavasagam, V.M., Elanchezhian, C., Krishna, C.V., Karthik, S., Saravanan, K.: Determination of mechanical properties of intra-layer abaca–jute–glass fiber reinforced composite. Mater. Des. **60**, 643–652 (2014)
26. Zhong, J.B., Lv, J., Wei, C.: Mechanical properties of sisal fibre reinforced urea formaldehyde resin composites. Express Polym Lett **1**(10), 681–687 (2007)
27. Ramasubbu, R., Madasamy, S.: Fabrication of automobile component using hybrid natural fiber reinforced polymer composite. J. Nat. Fibers **19**(2), 736–746 (2022)
28. Uzoma, A.E., Nwaeche, C.F., Al-Amin, M., Muniru, O.S., Olatunji, O., Nzeh, S.O.: Development of interior and exterior automotive plastics parts using kenaf fiber reinforced polymer composite. Eng **4**(2), 1698–1710 (2023)
29. Prabhu, L., Krishnaraj, V., Gokulkumar, S., Sathish, S., Ramesh, M.: Mechanical, chemical and acoustical behavior of sisal–tea waste–glass fiber reinforced epoxy-based hybrid polymer composites. Materials Today: Proceedings **16**, 653–660 (2019)
30. Mohammed, L., Ansari, M.N., Pua, G., Jawaid, M., Islam, M.S.: A review on natural fiber reinforced polymer composite and its applications. Int. J. Polym. Sci., (2015)
31. Dhandapani, N., Megalingam, A.: Mechanical and sound absorption behavior of sisal and palm fiber reinforced hybrid composites. J. Nat. Fibers **19**(12), 4530–4543 (2022)
32. Beg, M.D.H., Pickering, K.L.: Mechanical performance of Kraft fibre reinforced polypropylene composites: Influence of fibre length, fibre beating and hygrothermal ageing. Compos. A Appl. Sci. Manuf. **39**(11), 1748–1755 (2008)
33. Kalia, S., Dufresne, A., Cherian, B.M., Kaith, B.S., Avérous, L., Njuguna, J., Nassiopoulos, E.: Cellulose-based bio-and nanocomposites: a review. Int. J. Polym. Sci., (2011)
34. Balakrishnan, P., John, M.J., Pothen, L., Sreekala, M.S., & Thomas, S.: Natural fibre and polymer matrix composites and their applications in aerospace engineering. In: Advanced composite materials for aerospace engineering, pp. 365–383. Woodhead Publishing, (2016)
35. Saxena, M.: Sisal: potential for employment generation and rural development. In: Proceedings 3rd international conference rural india-achieving millennium development goals and grassroots development. (2005)

36. Hazell, P.J.: Armour: materials, theory, and design. CRC Press (2022)
37. Wretfors, C., Svennerstedt, B.: Bio fibre technology used for military applications: an overview. Sverigeslantbruksuniv, (2006)
38. Abrate, S.: Impact on composite structures. Ed. Revised, pp. 215–220. Cambridge University Press, Cambridge, UK (2005)
39. Morye, S.S., Hine, P.J., Duckett, R.A., Carr, D.J., Ward, I.M.: Modelling of the energy absorption by polymer composites upon ballistic impact. Compos. Sci. Technol. **60**(14), 2631–2642 (2000)
40. Benzait, Z., Trabzon, L.: A review of recent research on materials used in polymer–matrix composites for body armor application. J. Compos. Mater. **52**(23), 3241–3263 (2018)
41. Rohen, L.A., Margem, F.M., Monteiro, S.N., Vieira, C.M.F., Madeira de Araujo, B., Lima, E.S.: Ballistic efficiency of an individual epoxy composite reinforced with sisal fibers in multilayered armor. Mater. Res. **18**, 55–62 (2015)
42. Naveen, J., Jawaid, M., Zainudin, E.S., Sultan, M.T., Yahaya, R.: Evaluation of ballistic performance of hybrid Kevlar®/Cocos nucifera sheath reinforced epoxy composites. J. Text. Inst., (2018)
43. Monteiro, S.N., Drelich, J.W., Lopera, H.A.C., Nascimento, L.F.C., da Luz, F.S., da Silva, L.C., dos Santos, J.L., da Costa Garcia Filho, F., de Assis, F.S., Lima, É.P., Pereira, A.C.: Natural fibers reinforced polymer composites applied in ballistic multilayered armor for personal protection—an overview. In: Green materials engineering: an EPD symposium in honor of sergio monteiro, pp. 33–47. Springer International Publishing, (2019)
44. Luz, F.S.D., Garcia Filho, F.D.C., Oliveira, M.S., Nascimento, L.F.C., Monteiro, S.N.: Composites with natural fibers and conventional materials applied in a hard armor: A comparison. Polymers **12**(9), 1920 (2020)
45. Costa, U.O., Nascimento, L.F.C., Garcia, J.M., Monteiro, S.N., Luz, F.S.D., Pinheiro, W.A., Garcia Filho, F.D.C.: Effect of graphene oxide coating on natural fiber composite for multilayered ballistic armor. Polymers **11**(8), 1356 (2019)
46. Liu, J., Zhu, W., Yu, Z., Wei, X.: Dynamic shear-lag model for understanding the role of matrix in energy dissipation in fiber-reinforced composites. Acta Biomater. **74**, 270–279 (2018)
47. Haro, E.E., Szpunar, J.A., Odeshi, A.G.: Ballistic impact response of laminated hybrid materials made of 5086–H32 aluminum alloy, epoxy and Kevlar® fabrics impregnated with shear thickening fluid. Compos. A Appl. Sci. Manuf. **87**, 54–65 (2016)
48. Hajer, D., Jean-Luc, R., Charfeddin, M., Mohamed, T., Mohamed, H.: Experimental analysis of the linear and nonlinear vibration behavior of flax fibre reinforced composites with an interleaved natural viscoelastic layer. Compos. B Eng. **151**, 201–214 (2018)
49. Yahaya, R., Sapuan, S.M., Jawaid, M., Leman, Z., Zainudin, E.S.: Effect of fibre orientations on the mechanical properties of kenaf–aramid hybrid composites for spall-liner application. Defence Technology **12**(1), 52–58 (2016)
50. Evans, W.J., Isaac, D.H., Suddell, B.C., Crosky, A.: Natural fibres and their composites: A global perspective. In: Proceedings of the Risø international symposium on materials science, pp. 1–14. (2002)
51. Li, Y., Fan, H., Gao, X.L.: Ballistic helmets: Recent advances in materials, protection mechanisms, performance, and head injury mitigation. Compos. B Eng. **238**, 109890 (2022)
52. Marsyahyo, E., Jamasri, Rochardjo, H.S.B., Soekrisno.: Preliminary investigation on bulletproof panels made from ramie fiber reinforced composites for NIJ Level II, IIA, and IV. J. Ind. Text, **39**(1), 13–26 (2009)
53. Robinson, M.J., Kosmatka, J.B.: Development of a short-span fiber-reinforced composite bridge for emergency response and military applications. J. Bridg. Eng. **13**(4), 388–397 (2008)
54. Umair, M., Khan, R.M.W.U.: Fibers for sports textiles. Fibers Tech. Text., 93–115 (2020)
55. Dicker, M.P., Duckworth, P.F., Baker, A.B., Francois, G., Hazzard, M.K., Weaver, P.M.: Green composites: A review of material attributes and complementary applications. Compos. A Appl. Sci. Manuf. **56**, 280–289 (2014)
56. Yusup, E.M., Mahzan, S., Kamaruddin, M.A.H.: Natural fiber reinforced polymer for the application of sports equipment using mold casting method. In: IOP conference series: materials science and engineering, vol. 494, no. 1, p. 012040. IOP Publishing, (2019)

57. Abiola, O.S.: Natural fibre cement composites. In: Advanced high strength natural fibre composites in construction, pp. 205–214. Woodhead Publishing, (2017)
58. Belarbi, A., Dawood, M., Acun, B.: Sustainability of fiber-reinforced polymers (FRPs) as a construction material. In: Sustainability of construction materials, pp. 521–538. Woodhead Publishing, (2016)
59. Al-Azad, N., Asril, M.F.M., Shah, M.K.M.: A review on development of natural fibre composites for construction applications. J. Mater. Sci. Chem. Eng. **9**(7), 1–9 (2021)
60. Hota, G., Liang, R.: Advanced fiber reinforced polymer composites for sustainable civil infrastructures. In: Proceedings of the international symposium on innovation & sustainability of structures in civil engineering. Xiamen University, (2011)
61. Lila, M., Kumar, F., Sharma, S.: Composites from waste for civil engineering applications. I-Manager's J. Mater. Sci **1**(3), 1–11 (2013)
62. Dweib, M.A., Hu, B., O'donnell, A., Shenton, H.W., Wool, R.P.: All natural composite sandwich beams for structural applications. Compos. Struct., **63**(2), 147–157 (2004)
63. Blok, R., Smits, J., Gkaidatzis, R., Teuffel, P.: Bio-based composite footbridge: design, production and in situ monitoring. Struct. Eng. Int. **29**(3), 453–465 (2019)
64. Poletanovic, B., Dragas, J., Ignjatovic, I., Komljenovic, M., Merta, I.: Physical and mechanical properties of hemp fibre reinforced alkali-activated fly ash and fly ash/slag mortars. Constr. Build. Mater. **259**, 119677 (2020)
65. Castañeda, D., Silva, G., Salirrosas, J., Kim, S., Bertolotti, B., Nakamatsu, J., Aguilar, R.: Production of a lightweight masonry block using alkaline activated natural pozzolana and natural fibers. Constr. Build. Mater. **253**, 119143 (2020)
66. Abdollahiparsa, H., Shahmirzaloo, A., Teuffel, P., Blok, R.: A review of recent developments in structural applications of natural fiber-Reinforced composites (NFRCs). Compos. Adv. Mater. **32**, 26349833221147540 (2023)
67. Ge, S., Ouyang, H., Ye, H., Shi, Y., Sheng, Y., Peng, W.: High-performance and environmentally friendly acrylonitrile butadiene styrene/wood composite for versatile applications in furniture and construction. Adv. Compos. Hybrid Mater. **6**(1), 44 (2023)
68. Aravinthan, T., Manalo, A.: Field applications and case studies of FRP in civil infrastructure: The Australian experience. In: Proceedings of the 6th international conference on frp composites in civil engineering (CICE 2012), pp. 1–8. International Institute for FRP in Construction (IIFC), (2012)
69. Harasymiuk, J.: Marketing system of construction products in Poland after the entry into force regulation (EU) No 305/2011. Technical Sciences/University of Warmia and Mazury in Olsztyn, (2015)
70. Gao, S.L., Mäder, E.: Jute/polypropylene composites I. Effect of matrix modification. Compos. Sci. Technol., **66**(7–8), 952–963(2006)
71. http://www.permali.co.in/wood-based-composite-material.html
72. Mili, M., Singhwane, A., Hada, V., Naik, A., Nair, P., Srivastava, A.K. and Verma, S.: Advances in bamboo composites for structural applications: a review. (2023)
73. Li, Z., Wang, L., Wang, X.: Flexural characteristics of coir fiber reinforced cementitious composites. Fibers Polym. **7**, 286–294 (2006)
74. Md Sadiqu, N., Rahman Sob, H., Shiblee Sa, M., Saiful Isl, M.: The use of coconut fibre in the production of structural lightweight concrete. J. Appl. Sci. **12**(9), 831–839 (2012)
75. Asasutjarit, C., Hirunlabh, J., Khedari, J., Charoenvai, S., Zeghmati, B., Shin, U.C.: Development of coconut coir-based lightweight cement board. Constr. Build. Mater. **21**(2), 277–288 (2007)
76. Corradi, S., Isidori, T., Corradi, M., Soleri, F., Olivari, L.: Composite boat hulls with bamboo natural fibres. Int. J. Mater. Prod. Technol. **36**(1–4), 73–89 (2009)
77. Mohanty, A.K., Misra, M. and Drzal, L.T. eds.: Natural fibers, biopolymers, and biocomposites. CRC press, (2005)
78. Ali, M.: Coconut fibre: A versatile material and its applications in engineering. J. Civ. Eng. Constr. Technol. **2**(9), 189–197 (2011)

79. Das, S.: Jute composite and its applications. In: International workshop, IJSG, Indian Jute Industries Research Association, Kolkata, India (2009). www. jute. org (2009, accessed July 2010).
80. Li, Q., Matuana, L.M.: Surface of cellulosic materials modified with functionalized polyethylene coupling agents. J. Appl. Polym. Sci. **88**(2), 278–286 (2003)
81. Rahem, Z., Mayouf, I., Guessoum, M., Delaite, C., Douibi, A., Lallam, A.: Compatibilization of biocomposites based on sponge-gourd natural fiber reinforced poly (lactic acid). Polym. Compos. **40**(12), 4489–4499 (2019)
82. Lai, D.S., Osman, A.F., Adnan, S.A., Ibrahim, I., Alrashdi, A.A., Ahmad Salimi, M.N., Ul-Hamid, A.: On the use of OPEFB-derived microcrystalline cellulose and nano-bentonite for development of thermoplastic starch hybrid bio-composites with improved performance. Polymers **13**(6), 897 (2021)
83. Varghese, S.A., Pulikkalparambil, H., Rangappa, S.M., Siengchin, S., Parameswaranpillai, J.: Novel biodegradable polymer films based on poly (3-hydroxybutyrate-co-3-hydroxyvalerate) and Ceiba pentandra natural fibers for packaging applications. Food Packag. Shelf Life **25**, 100538 (2020)
84. Carus, M., Partanen, A.: Natural fibre-reinforced plastics: establishment and growth in niche markets. JEC Compos. Mag. **55**(118), 23–24 (2018)
85. Balaji, A., Karthikeyan, B., Raj, C.S.: Bagasse fiber–the future biocomposite material: a review. Int. J. Cemtech Res. **7**(1), 223–233 (2014)
86. Georgios, K., Silva, A., Furtado, S.: Applications of green composite materials. Biodegrad. Green Compos., 312–337 (2016)
87. Dahy, H.: Natural fibre-reinforced polymer composites (NFRP) fabricated from lignocellulosic fibres for future sustainable architectural applications, case studies: segmented-shell construction, acoustic panels, and furniture. Sensors **19**(3), 738 (2019)

Chapter 11
Elastomeric-Based Composite Materials for Engineering Applications

Bibhuti B. Sahu, Srikanta Moharana, and Pravat Kumar Behera

Abstract Due to its unique combination of flexibility, durability, and better mechanical qualities, elastomeric-based composite materials have garnered a great deal of interest. This chapter provides an in-depth analysis of composites based on elastomers, exploring their many uses in many industries. An introduction to elastomeric materials and their numerous technical uses is given at the outset of the chapter. Then the section delves into the characteristics and types of elastomers often used in composites, with a focus on their chemical, mechanical, and thermal capabilities. Improving elastomeric composites' performance attributes relies heavily on using reinforcing materials. Fibres, fillers, and nanoparticles are just a few of the reinforcing elements investigated in this chapter, along with their effects on composite properties as a whole. Calendaring, rubber mixing, and moulding are some of the unique production operations that are examined in this study, which focuses on elastomeric composites. Elastomeric composites' thermal and electrical conductivity qualities are further explored in the chapter, which delves into useful measurement techniques and the many applications that can benefit from these features. Also covered in this chapter are the many technological uses of elastomer composites in fields as diverse as automotive, biomedical, aerospace, and sports. The capacity of these materials to efficiently meet the unique requirements of different industries is their most prominent feature. In its last section, the chapter explores future directions and challenges in the field, providing readers with a glimpse into emerging research areas and potential advancements. This phenomenon encourages further research and development in the field of engineering applications involving elastomeric-based composite materials.

B. B. Sahu (✉)
Department of Basic Science and Humanities, Nalanda Institute of Technology, Bhubaneswar, Odisha, India
e-mail: bbsvssut@gmail.com

S. Moharana (✉)
Department of Chemistry, School of Applied Sciences, Centurion University of Technology and Management, Jatni, Bhubaneswar, Odisha, India
e-mail: srikantanit@gmail.com

P. K. Behera
Atma Ram Sanatan Dharm College, University of Delhi, New Delhi, India

S. Moharana et al. (eds.), *Polymer Composites*, Engineering Materials,
https://doi.org/10.1007/978-981-97-2075-0_11

Keywords Elastomeric-based composites · Reinforcement materials · Mechanical properties · Manufacturing techniques · Structural characterisation · Thermal conductivity · Electrical conductivity · Automotive industry · Aerospace applications · Medical applications · Sports and leisure industry

11.1 Introduction

Elastomeric materials are crucial in many technological applications such as transportation, aeroplanes, industry, healthcare, and more depending on these unique materials [1–3]. Special polymers called elastomers may stretch greatly before returning to form when force is relieved. Its unique elasticity comes from long, coiled polymer chains, which stretch and rebound easily [4, 5]. Elastomers are suited for flexible, durable applications due to their high elongation at break, low modulus of elasticity, and fatigue and impact resistance. Resilience in harsh settings makes elastomeric materials versatile [6, 7]. They resist acids, bases, oils, and solvents. Chemically inert elastomers can operate properly and maintain their structure. Due to their weathering resistance, elastomers can withstand UV radiation, high temperatures, and wetness. Engineering employs elastomers for sealing, dampening, vibration isolation, and shock absorption [8]. Automobile engines, hydraulic systems, and pipe fittings depend on their seals and gaskets [9]. Elastomers, which find application in aircraft fuel system seals, demonstrate remarkable elasticity by undergoing substantial deformation in response to applied force, thereby efficiently mitigating vibration and noise. Medical devices, prosthetics, orthotics, and drug administration systems use elastomers [10]. Biocompatibility, suppleness, and flexibility may enhance human-body applications. Elastomeric coatings, adhesives, and encapsulants are employed in electronics and electrical applications due to their thermal stability and electrical insulation [11]. Engineering applications benefit from elastomers' compounding and formulation capabilities, which offer customisation. Reinforcing fillers, plasticisers, and flame retardants can modify elastomers' mechanical, thermal, electrical, and chemical properties [12, 13]. Elastomeric materials' strength, abrasion resistance, and fire retardancy may be improved by engineers using design flexibility. Due to their unique properties, elastomeric materials are essential in engineering [14]. They are useful in various industries due to their high elasticity, chemical and weather resistance, and sealing, damping, and vibration isolation capabilities. Modifying their properties to suit an application emphasises their worth. Thus, elastomers boost technology, ensuring safer, more efficient, and more reliable engineering systems [15]. The exciting field of elastomeric-based composite materials is at the crossroads of polymer science, materials engineering, and advanced composites, providing an abundant ground for new developments and practical applications [16]. The foundation of these composites is elastomers, which exhibit exceptional elasticity and reversible deformability. A new category of materials with improved mechanical, thermal, and electrical characteristics is created when elastomers are mixed with additional substances like reinforcing fibres or nanoparticles [13, 17].

The synthesis methods used to create elastomeric-based composites, from the more basic mixing procedures to the more complex in-situ polymerisation and 3D printing. Learning these processes is essential for modifying the characteristics of materials for use in electronics, smart fabrics, and other related fields. The appearance, chemical content, and crystalline structure of these materials dig into techniques like scanning electron microscopy, Fourier-transform infrared spectroscopy, and X-ray diffraction [18, 19]. Composites with improved mechanical, thermal, and chemical characteristics are made by combining elastomers which are very pliable and elastic with reinforcing elements. Numerous technical fields, including structural, biomedical, automotive, and aerospace, stand to benefit greatly from this new class of materials [20]. Composites with customised qualities, appropriate for specific applications, are made possible by combining the inherent flexibility of elastomers with the strength and rigidity provided by reinforcing agents. It is important to carefully evaluate the elastomer selection, reinforcing components, and production techniques while synthesising these composites [21]. Recent developments in material processing methods, including vulcanisation, injection moulding, and extrusion, have allowed for the creation of elastomeric composites that exhibit exceptional performance qualities [22].

This introduction is to offer a basic grasp of composite materials based on elastomers, highlighting their adaptability, possible uses, and critical function in solving engineering problems. Our investigation into this area will lead us to examine elastomeric-based composites from every angle, including material choice, processing techniques, reinforcing schemes, and performance assessment. We hope that by delving into these components, we may shed light on the theoretical and practical potential of adding elastomers to composite matrices, which could lead to new developments in materials science and engineering. Providing a comprehensive examination of the possibilities and depth inherent in incorporating elastomers into advanced composite materials is the overriding goal of the chapter on elastomeric-based composite materials for engineering applications. The remarkable elasticity and flexibility of elastomers make them ideal candidates for strategic use in improving the functionality of engineered materials in a wide range of contexts. This chapter mainly aims to introduce elastomers, explain why they are used in composite matrices, and go into detail about how to choose and characterise materials. In addition, the chapter examines the different processing processes employed to synthesise elastomeric composites and also delves into a wide array of engineering applications. The ultimate goal of this chapter is to promote progress in materials science and creative solutions to engineering problems by providing a thorough resource for researchers, academics, and engineers.

11.2 Properties and Categories of Elastomers

An essential area of study in the broader field of materials science is elastomeric-based composite materials, which integrate the advantageous properties of elastomers with the structural support provided by composites [23]. Polymers known as elastomers are characterised by their exceptional elasticity and resilience, which enable them to revert back to their original shape even after being deformed. Elastomeric polymer chains may keep their original shape even after being stretched or deformed significantly because of their amorphous nature [24]. The unique characteristics of elastomers and how they might be used in composites have been the subject of recent studies. There are a number of things that make elastomers unique from other forms of polymers. Their squishy, rubbery feel is caused by their long, flexible, and amorphous macromolecular chains [15]. Although neoprene, polyurethane, and silicone are synthetic elastomers with adjustable properties, natural rubber made from latex is still the most prevalent kind [25]. Weak intermolecular interactions between polymer chains enable elastomers' considerable chain mobility, which is what gives them their elasticity. Real estate with a high molecular weight polymer strand interspersed between the crosslinks, elastomers are crosslinked polymer networks that have a low crosslinking density. The crosslinks in an elastomer inhibit flow, but the long polymer strands change chain conformation under stress, allowing the network to flex significantly. Elastomers often have a tenfold stretchability, exceptionally high elongation at break (e.g., 400–1000%), and the ability to return to their original form upon stress release [26, 27]. Shore durometer measurements represent one type of hardness measurement, and an overview of materials and their Shorehardness are depicted given in Fig. 11.1.

A comprehensive understanding of elastomers is provided by the paper, which incorporates recent research on their molecular structure, mechanical behaviour, and thermal properties. It delves into the role of crosslinking in enhancing the elastic modulus and tensile strength of elastomers, two crucial metrics for assessing their versatility. Tensile testing, compression testing, hardness measures, shear and flexural measurements, tear testing, and friction/abrasion property studies are crucial for characterising elastomers.

Dynamic stress/strain measurements, such as dynamic mechanical analysis (DMA) for solid-like materials or melt rheology for liquid-like samples, can provide valuable viscoelastic information [27]. Fatigue tests remain essential when evaluating an object that may undergo several cycles. Conversely, creep, relaxation, and set tests assess the behaviour of an elastomer in terms of deformation and time [28]. In Fig. 11.2, the stress–strain behaviour of a brittle polymer, thermoplastic elastomer, and plastic is illustrated. The mechanical properties of elastomeric-based composite materials are the result of the combination of the elastomers' inherent elasticity with reinforcing agents. Parathodika et al. have delved into the latest findings about the production and analysis of elastomeric composites. To boost mechanical strength, stiffness, and thermal stability, these composites often incorporate reinforcing elements like nanoparticles, fibres, or fillers. In order to provide insight

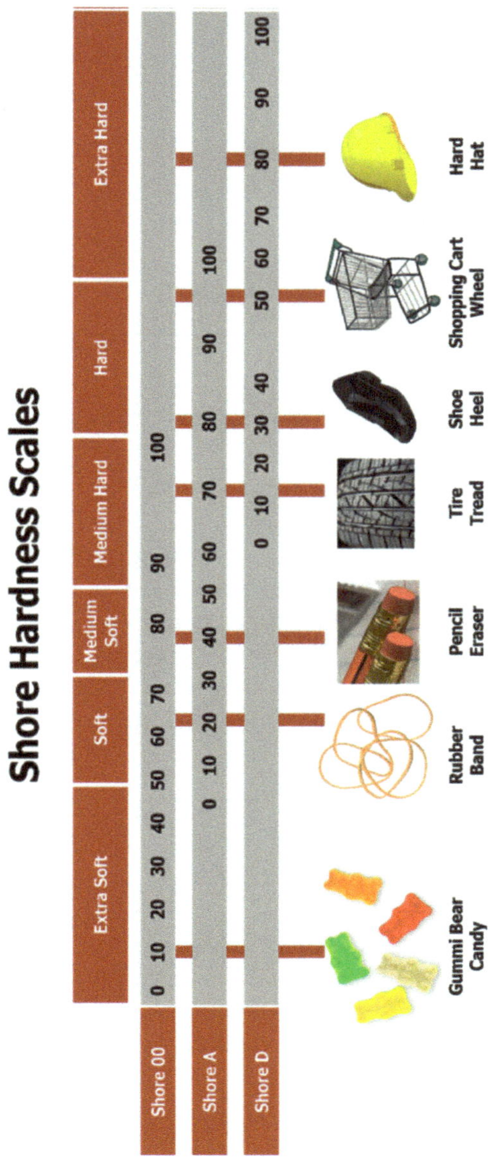

Fig. 11.1 Product surface hardness scale for commonplace items [27]. Adapted from Ref. 28 with permission from [Elsevier], Copyright [2019]

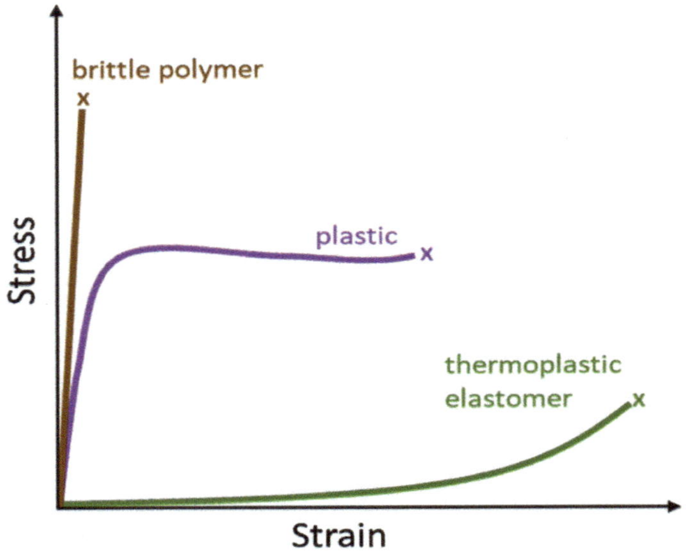

Fig. 11.2 The stress–strain behaviour of a brittle polymer, thermoplastic elastomer, and plastic is illustrated [27]. Adapted from Ref. 28 with permission from [Elsevier], Copyright [2019]

into the optimal combinations for certain uses, the study examines the synergistic effects of reinforcing components and elastomers [29]. In addition, the viscoelastic behaviour of elastomers is affected by their molecular structure. Lipsa and his research group have determined the effect on elastomer viscoelasticity of changing molecular weight, crosslink density, and chemical structure. Elastomeric properties may be tailored to meet the demands of many different applications by gaining a better understanding of these interactions [30]. These applications span from biomedical devices to automotive components.

11.3 Categories of Elastomers

The extraordinary elasticity and durability of elastomer composites are a result of the special properties of these polymers. These materials are vital in a wide variety of industries, from medical devices to automotive parts. Knowing the most prevalent kinds of elastomers used in composites is crucial for optimising material performance and tailoring them to particular applications.

- Natural rubber (NR)

Natural rubber (NR) is derived from the latex of rubber trees and is one of the oldest and most popular kinds of elastomers [31] due to its peculiar chemical structure, which comprises cis-1,4-polyisoprene chains, it possesses remarkable flexibility. Research on composites including natural rubber has recently expanded, such as Mohanty et al. and his research group's research aims to enhance mechanical properties and explore potential environmentally friendly applications of composites [32].

- Styrene-butadiene rubber (SBR)

By combining the monomers butadiene and styrene, a synthetic elastomer called styrene-butadiene rubber (SBR) is created. It possesses both the strength and flexibility of styrene and butadiene, making it an ideal elastomer for a variety of applications [33]. Nanocomposite formulations that include SBR in composite materials with the purpose of enhancing mechanical performance are investigated in publications like "Morphology and mechanical properties of clay/styrene-butadiene rubber nanocomposites" by Zhang et al. [34].

- Polyurethane (PU)

The PU elastomers, made of synthetic polymers, are long-lasting, flexible, and resistant to damage. One of several recent publications that have investigated the production and utilisation of elastomeric composites derived from polyurethane by Akindoyo et al. [35] reported the possible uses of polyurethane composites as coatings, adhesives, and flexible materials.

- Rubber made of silicon

Silicone rubber is an elastomer with exceptional thermal stability and a strong resistance to both high and low temperatures. Its constituent parts are silicon, oxygen, carbon, and hydrogen. Using the title Liu and his research colleagues have investigated the utilisation of silicone rubber in composites designed for usage in high-temperature environments [36]. The study investigates how the unique properties of silicone rubber, such as its resistance to heat and its longevity, are used to the advantage of composites.

- Ethylene Propylene Diene Monomer

Ethylene Propylene Diene Monomer, or EPDM rubber, is a synthetic elastomer that has great weather resistance and electrical insulating properties. Athawale et al. have provided light on how EPDM is used in outdoor applications. The purpose of this research is to find ways to make composite materials with EPDM that are resistant to things like temperature fluctuations and UV radiation [37].

- Fluoroelastomers

Viton and similar fluoroelastomers are renowned for their exceptional resistance to a broad variety of solvents, including fuels and lubricants. Ponnamma et al. have discussed composites that use fluoroelastomers. In particular, the impacts of fluoroelastomers' enhanced chemical resistance and durability in harsh environments are examined in this study [38].

Knowing the strengths and weaknesses of certain elastomers is essential for developing composite materials with tailored properties. Recent research in these areas has deepened our comprehension of the fundamental characteristics of elastomers and shown new approaches to enhancing the performance and practicality of composite materials in many contexts. Engineers and materials scientists continue to be captivated by the synergy that happens in composites when reinforcing components and elastomers collaborate.

11.4 Elastomeric Composite Reinforcement Materials

11.4.1 Reinforcement Materials: An Overview

Reinforcing composites with their strength, stiffness, and durability, fibres are an essential component. Natural fibres, glass, carbon, and aramid are common types of fibres. Glass fibres are inexpensive, lightweight, and offer a good mix of strength and weight, making them popular in industries like construction and automotive. In contrast, carbon fibres are ideal for demanding applications like aerospace due to their low bulk and exceptional strength. Lightweight protective gear and buildings often make use of Aramid fibres due to their famed excellent impact resistance. For sustainable composites to be made, natural fibres sourced from plants are essential. Composites may have their properties improved by adding fillers, which are chemicals that do just that. They make things better in terms of wear resistance, thermal conductivity, and hardness. Common ingredients include clay, silica, talc, and calcium carbonate. Afroze et al. have reported and analysed how talc fillers affect the thermal and mechanical properties of polypropylene composites; to maximise efficiency and tailor composites to individual applications, fillers are needed [39].

Three things must be known in order to obtain a description of the composite material, i.e., the components' qualities and features, the reinforcement's distribution and arrangement, and the nature of the contact between the matrix and reinforcement which is specified in Fig. 11.3. There are a lot of different parts that come together to make the composite material [40]. Thus, the range of fundamental elements has almost infinite possible expressions. What matters most for the finished product's attributes are the matrix's characteristics, the reinforcements' types and quantities, the additives' proportions in the formulation, and the method of implementation.

Nanoparticles are highly engineered reinforcing materials with average sizes below 100 nm. Strength, thermal conductivity, and barrier properties are some of the unique traits they impart. Carbon nanotubes, graphene, and nano-clays are just a few examples of the nanoparticles that have recently attracted attention for their potential to increase the nanoscale strength of composites. Miyagawa et al. have

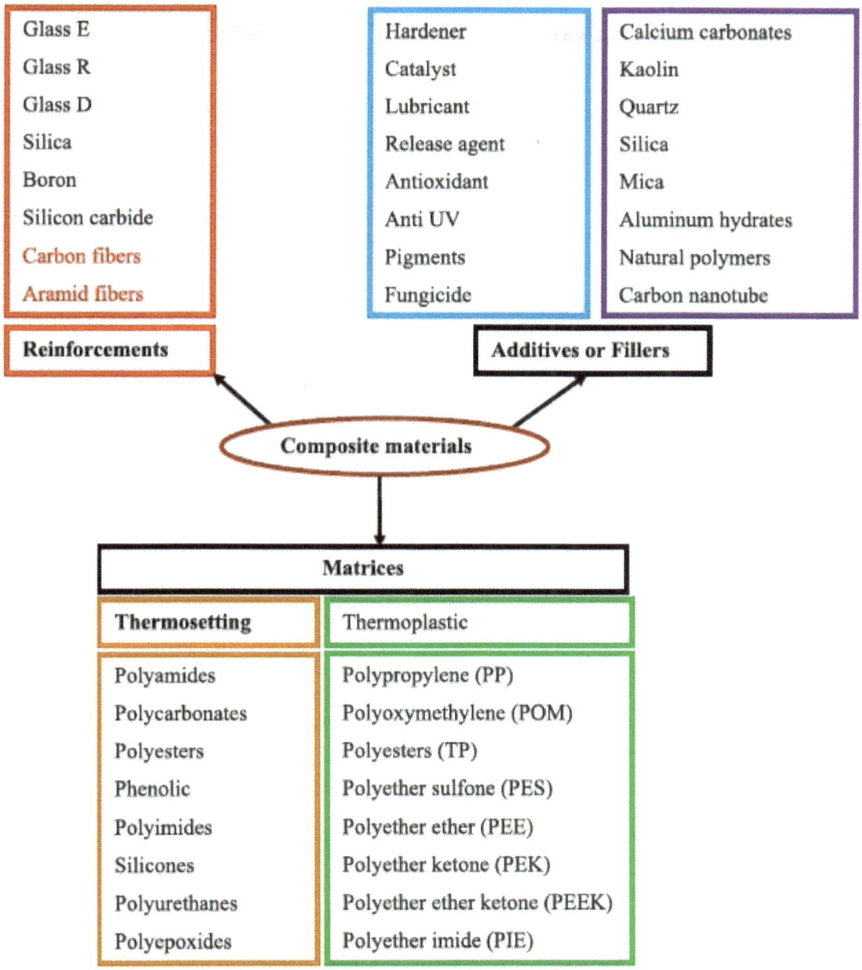

Fig. 11.3 Formulation scheme for composite materials [40]. Adapted from Ref. 41 with permission from [Elsevier], Copyright [2021]

delved into the incorporation of carbon nanotubes into polymer matrices and show the potential for nanoscale strengthening [41].

Particulate Reinforcements: Adding powders or granules as part of a reinforcement system improves a number of composite properties. The increased resistance to wear that composite, like automobile components, got by using aluminium oxide makes them last longer and serves a variety of purposes [42]. The impact of aluminium oxide particle reinforcement on the wear resistance of epoxy composites is examined. The growing concern for environmental impact has led to an uptick in the usage of composites made with natural fibres and bio-based reinforcements [43]. You may replace traditional synthetic fibres with more eco-friendly alternatives like flax, hemp, and sisal. The use of flax fibres as a reinforcing material in biodegradable composites is explored from this study point to sustainable packaging and other fields having bright futures.

Composites may be tailored to meet unique requirements in many industries because of the extensive variety of reinforcing components, including fibres, fillers, nanoparticles, and more [27]. Researchers in this field are now looking into new ways to combine reinforcing components, which is helping to advance the creation and application of composite materials for a wide range of applications.

11.4.2 How Reinforcement Materials Affect Elastomer Composites' Characteristics?

Scientists are making great strides in elastomeric-based composite materials by experimenting with various reinforcing elements to tailor the properties of elastomer composites to specific applications. To get the most out of these composites, you need to know how different reinforcing materials influence their mechanical, thermal, and dynamic properties.

i. Utilisation of Fibres as Substrates

Chin et al. have reported on tensile deformation mechanisms of polypropylene/elastomer blends reinforced with short glass fibre [44]. The effects of various fibre orientations and contents on modulus, tensile strength, and tear resistance are examined in the study. Integrating glass fibres into elastomer composites improved their mechanical properties, which might lead to better structural performance.

Carbon Fibres: Researchers are now investigating the effects of carbon fibre reinforcing on elastomeric composites used in aerospace. This research looks at the effects of carbon fibres on the thermal stability, fatigue resistance, and stiffness of elastomeric matrices [45].

ii. Reinforcement using particulate matter and fillers

The effects of nano-clay reinforcements on elastomeric composites are thoroughly investigated. The research explains how the use of nano-clays improves mechanical strength, tear resistance, and barrier properties. The study also discusses the potential for enhanced thermal stability and flame resistance [46]. Effects of Carbon Black Reinforcement on Dynamic Mechanical Properties of Natural Rubber Composites delves into the topic of carbon black and its effects on the dynamic mechanical properties of conventionally reinforced elastomers. Elastomeric composites' dampening capabilities and dynamic performance are examined in this work in relation to carbon black.

iii. Nanoparticles

Research into using carbon nanotubes (CNTs) as reinforcing elements in elastomeric composites is ongoing. CNTs affect electrical and mechanical conductivity which highlights the potential for tailoring elastomeric composites to applications that need both electrical conductivity and structural integrity [47]. Enhancing the Thermal and Mechanical Properties of Silicone Rubber Composites with Nano-Silica delves into the effects of nano-silica on elastomeric composites, providing information about how these particles enhance mechanical strength and thermal conductivity.

iv. Natural Fibres and Reinforcements Derived from Biology

Recently, research on the effects of natural fibre reinforcing, with an eye towards sustainability, has focused on flax fibre reinforcement. The addition of flax fibre makes elastomeric composites used in automobile parts more environmentally friendly and improves their mechanical properties. Recent studies have shown that elastomeric composites may have their mechanical, thermal, and dynamic properties tailored using a wide range of reinforcements, from simple fillers to complex nanomaterials [48].

11.5 Procedures for Manufacturing Elastomeric Composites

The production processes have a significant effect on the structural integrity, durability, and application-specific properties of elastomeric-based composite materials, which in turn affects their performance and characteristics. In order to maximise the production of elastomeric composites, recent research have concentrated on refining existing methods and developing new techniques [49]. These strategies include layer-by-layer assembly, additive manufacturing (3D printing), casting, extrusion, and moulding.

- Methods for Moulding

Elastomeric composites are often made via moulding operations, and recent studies have focused on ways to make these procedures better. The realm of injection moulding explores novel approaches to improving fibre orientation and distribution inside the elastomeric matrix [50]. This research can help manufacturers make better use of injection moulding by examining the effects of varying these parameters on the mechanical and thermal properties of the finished composite. Compression moulding is also a traditional method for making elastomeric composites. New research demonstrates that increasing the pressure can enhance mechanical properties and filler dispersion.

Moulding processes, such as injection and compression moulding, are ideal for mass production because of their efficiency and scalability. Injection moulding enables precise control over the distribution and orientation of fibres, which improves mechanical properties, and an economical way to generate high-pressure circumstances that enhance the dispersion of fillers is compression moulding. Moulding processes, although highly successful, can provide challenges when it comes to maintaining constant fibre alignment, particularly in complex geometries. Another potential bottleneck in production efficiency is the cycle times linked to traditional moulding methods [51].

- Extrusion Methods

Elastomeric composites rely heavily on extrusion methods because of their efficiency and scalability in their continuous production. The latest developments in continuous extrusion will provide in-depth information on how to manage the alignment and dispersion of fibres when extruding [52]. The results provide light on the process parameters that may be varied to produce elastomeric composites with enhanced strength, modulus, and overall performance. When it comes to making elastomeric composites with consistent cross-sectional profiles, extrusion techniques are ideal since they are unbroken.

- Casting and Calendaring

There has been a lot of interest in studying the possibility of using casting and calendaring techniques to make thin, uniform elastomeric composite sheets recently. Shah et al. investigated the methods of precision casting elastomeric composites with an emphasis on the achievable surface quality and wear resistance [53]. For applications that need precise dimensions and surface characteristics, learning about casting procedures is crucial. It may be challenging to build complex three-dimensional structures, and there may be limits on the ability to scale up casting processes. Reinforcing distribution uniformity may also vary with cast material thickness.

- Additive manufacturing (3D printing)

The use of additive manufacturing, and more specifically 3D printing, has revolutionised the production of elastomeric composites. The potential for intricate, personalised geometries is brought to light in "Additive Manufacturing of Elastomeric Composite Structures". Chen et al. [51] delve into the intricacies of 3D printing elastomeric composites. This study lays forth a plan to solve printability, material compatibility, and post-processing problems so that additive manufacturing may be used to its full potential in the elastomeric composites industry. Uneven features in various orientations might result from 3D printing's sequential nature, and large-scale manufacturing at high speeds remains a challenge. Research on post-processing concerns and material compatibility is ongoing.

- Assembly Method: Layer-by-Layer

These techniques are gaining popularity as they focus on precise material stacking, akin to 3D printing. Zhang et al. [54] have delved into the process of stacking elastomeric and reinforcing materials to create composite films with customised properties. This research presents a high-tech approach to optimising the assembly of elastomeric composites layer-by-layer and examines the effects of different layering patterns on mechanical performance. Elastomeric composites may be uniquely constructed using layer-by-layer construction processes. Achieving uniformity across big constructions may be challenging with layer-by-layer assembly, just like with 3D printing. Certain apps may have limitations when it comes to production speed and scalability.

11.6 Application for Composites Based on Elastomers

11.6.1 Application in the Automobile Sector

The automotive industry is currently experiencing a time of profound change, spurred on by technology advancements that seek to enhance the performance, safety, and environmental friendliness of vehicles. Combining elastomers with reinforcing components and composites based on elastomers is a crucial part of what will allow this dramatic transition to happen. Figure 11.4 is a schematic of typical rubber components used in a passenger vehicle. To begin, the primary role of seals and gaskets is to maintain the integrity of a joint and to stop fluid from leaking out. Various types of rubber may be necessary for their operation, depending on their location and the fluids they come into contact with. Seals and gaskets for the engine and fuel system are typically composed of fluoro-silicones, non-woven rubber, or fluoroelastomers. These materials are known for their high thermal stability and exceptional chemical resistance to fluids derived from petroleum, such as oils, petrol, and more [55].

In order to illuminate the myriad uses of elastomeric composites in the automobile industry, this exhaustive study dives into recent literature and statistics. Tyres, security

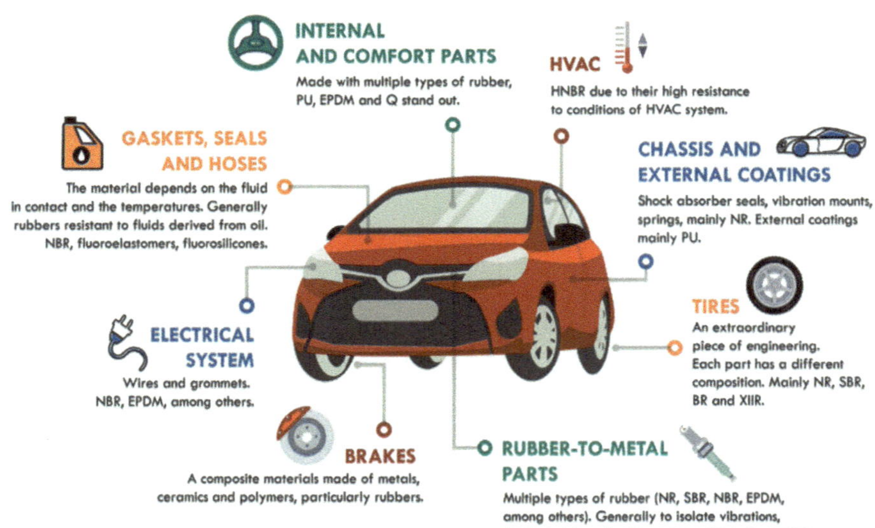

Fig. 11.4 Common rubber components in a passenger automobile [55]. Open access under a CC BY 4.0 license. Adapted from Ref. 57 with permission from [Elsevier], Copyright [2023]

systems, vibration isolation, lightweighting projects, and noise reduction methods are all examples of such applications.

Elastomers used in automobiles often employ the principles summarised in Fig. 11.4. When thinking about the self-healing capacity, it's important to take the polymer's architecture into account, which includes things like its aggregation state, chain structure, and crosslink density [56]. The ultimate healing performance of the material is heavily dependent on the crosslink density, especially in the elastomers group. Soft robotics [57], soft electronics, biomedical, and structural applications [58] are only a few of the advanced areas where the self-healing principles in polymers have been thoroughly examined. Surprisingly, though, the automobile sector—the most important use case for elastomers—has not yet undergone extensive analysis. This is a huge omission because elastomers are most commonly used in the automobile industry, which stands to gain a great deal from using self-healing technology. Resolving this important knowledge gap will pave the way for stronger, more sustainable automotive materials of the future that can mend themselves, bringing about a paradigm shift in the industry.

Elastomers used in automobiles often employ the principles summarised in Fig. 11.5 and things like the polymer's crosslink density, aggregation state, and self-healing capacity should also take chain topology into account [55]. The ultimate healing performance of the material is heavily dependent on the crosslink density, especially in the elastomers group.

- Improved Performance and Decreased Risk: Tyres

Improvements in elastomeric-based composites have contributed significantly to the recent technological boom in tyres [59]. Novel elastomeric materials have recently been discovered to improve tyre performance, according to research studies. Main goals include making it last longer, making it more durable, and making it traction better. The idea that nanocomposites may make tyre constructions last longer with less wear has piqued people's interest in them in recent years. Graphene and other

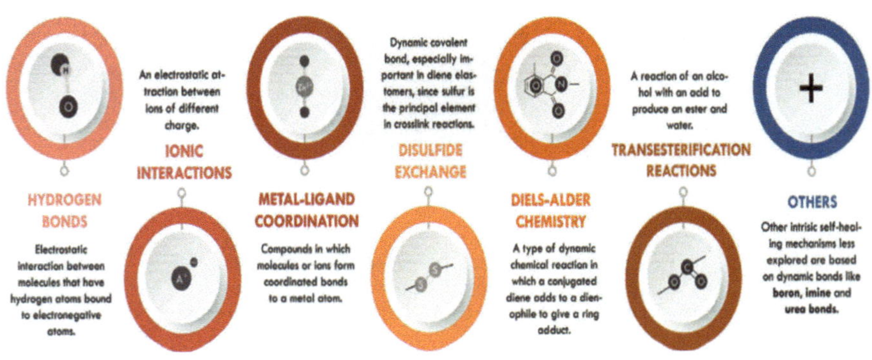

Fig. 11.5 Strategies for intrinsic self-healing in elastomers with potential automotive uses [55]. Open access under a CC BY 4.0 license. Adapted from Ref. 57 with permission from [Elsevier], Copyright [2023]

nanoparticles are present in these materials. Smart tyre technology is a huge step forward as it incorporates elastomeric sensors into the tyre's components. These sensors allow for the continuous monitoring of tyre conditions in real-time. By providing a new perspective on tyre health, this invention enhances vehicle security and maintenance effectiveness.

Gaskets and seals: Improving Reliability and Efficiency

The proper and dependable operation of several vehicle systems is dependent on gaskets and seals. When dealing with issues of fluid and oil seepage, elastomeric composites have shown to be indispensable. Nanofillers have increased the chemical resistance and mechanical strength of gaskets and seals, and elastomeric materials have lately been the centre of attention for their better sealing capabilities that boost engine and gearbox efficiency [60]. Because of this, these parts endure longer in the end. Research on self-healing elastomers is also picking up steam for these specific applications. Elastomeric materials that can self-heal little damage may be a thrilling new innovation in car maintenance. Overall, this has the potential to make cars more dependable and cut down on their downtime.

- Methods for Reducing and Isolating Vibrations

In the quest for a more pleasant and stress-free driving experience, the automobile industry has achieved significant strides in vibration isolation and dampening technologies. For instance, elastomeric materials' vibration-absorbing and vibration-distribution properties make them indispensable in engine mounts and suspension systems [61]. Research is now focused on developing custom-made elastomeric materials that have enhanced damping characteristics.

- The objective of reducing weight is to get improved fuel efficiency

The automobile industry is investing heavily in research on methods to decrease vehicle weight in response to rising concerns about environmental impact and fuel efficiency. Composites based on elastomers are making an ever-greater impact in this field. The addition of carbon fibres and nanoparticles to elastomeric matrices is being considered as a means to reduce vehicle weight and improve energy efficiency without compromising mechanical properties [62]. This would be in line with the industry's dedication to sustainability. Important automobile components rely on elastomeric composites, which may be customised for specific uses, to reduce weight without sacrificing structural integrity.

11.6.2 Applications in the Aerospace and Aviation Sectors

Materials that can maintain their structure in the face of high temperatures while simultaneously improving fuel efficiency are in high demand in the aerospace and aviation sectors. The fact that elastomeric-based composites are used extensively in many aircraft applications shows how versatile and up to these criteria they are

[63]. Examining the impact of elastomeric composites on the aviation and aerospace industries, this comprehensive study examines recent studies and data.

- Improvements in Strength and Weight Efficiency in Structural Components

Composites based on elastomers are extensively used in the aerospace sector for the construction of structural components. New research demonstrates that complicated elastomeric matrices reinforced with carbon and aramid fibres may provide lightweight and strong materials [64]. Integrating these materials into crucial parts, such as wing structures, improves fuel efficiency and the overall performance of the aircraft. A lot of research has gone into finding the sweet spot for the mechanical properties of elastomeric composites so they can fulfil the unique demands of aeroplane applications.

- Improve the Reliability of Aerospace Gaskets and Seals

The extreme heat and stringent performance requirements placed on seals and gaskets make them indispensable in the aircraft industry. More recently, elastomeric composites have emerged as an important factor in extending the life and dependability of these crucial components. Aerospace systems rely heavily on keeping a sealed environment; therefore, elastomeric materials boost the performance of the sealing mechanism, which is critical for their safety and function [65]. Adding nanoparticles to elastomeric compositions has been considered as a potential method to improve the mechanical and thermal properties of gaskets and seals.

- Issues with Vibration Dampening: Longevity of Components and Comfort of Passengers

To ensure the comfort and safety of passengers during flight, vibration mitigation systems rely on elastomer composites. Elastomeric materials with tuned damping capabilities have been the subject of recent research aimed at reducing aircraft vibrations and noise pollution. These composite materials are utilised in crucial structural components and aircraft interiors [66]. Air travel is often more enjoyable since passengers are considerably more comfortable due to reduced vibrations and noise. Aeronautical components have a longer lifespan and less fatigue on structural sections. Elastomeric damping material design and optimisation for a certain stiffness-to-damping-efficiency ratio is the primary goal of this area of study.

- Adhesives for Bonding and Elastomeric Materials: Strengthening Aircraft Components

Aeroplane component assembly and bonding need materials that offer strong but flexible connections. More and more, elastomer-based adhesives are being used for the assembly of aeroplane structures. Recent research has focused on elastomeric adhesives as a means to reduce the total weight of aircraft components while simultaneously improving load transfer efficiency [67]. The objective of this research is to enhance the adhesive's properties, such as its pliability and adhesive strength, so that it can be a dependable function under different operating conditions. The adaptability of elastomeric adhesives makes them ideal for combining various components, including metals and composites, to enhance the overall strength of the aircraft.

- Preventing Extreme Temperatures: The Function of Thermal Protection Systems

Materials utilised in aircraft applications can occasionally undergo significant temperature variations, especially upon re-entry into Earth's atmosphere. Composites based on elastomers allow for thermal protection systems (TPS) [68], which have been vital in protecting spacecraft and aeroplane components from potentially fatally high temperatures. Incorporating TPS materials into elastomeric composites can enhance the heat insulation and absorption characteristics of the composites, according to a recent study. Maintaining structural integrity of aircraft components is of utmost importance since temperatures can reach several thousand degrees Celsius during re-entry or high-speed flight. The purpose of these elastomeric TPS materials is to diffuse heat and shield subsystems from thermal damage [69].

- Advancements in Smart Materials and Long-Term Sustainability

Composites based on elastomers used in aircraft will be defined in the future by innovations in smart materials and a steadfast dedication to sustainability. Researchers have been exploring methods to include sensors and actuators in elastomeric composites in an effort to develop smart materials with the ability to sense and respond to changes in their environment. These materials can provide up-to-the-minute information on the condition of aeroplane parts, which might impede maintenance activities [70]. The use of smart materials in elastomeric composites reflects the growing trend of predictive maintenance in the aviation sector. Constant structural integrity monitoring allows aerospace operators to proactively identify and address potential issues, resulting in reduced idle times and enhanced overall safety.

11.6.3 Applications in Medical and Healthcare

Recent years have seen a surge in interest in elastomer composites due to their extraordinary characteristics, adaptability, and wide range of possible applications. Tissue engineering and medical devices are only two of the numerous medical and healthcare applications that might benefit greatly from these compounds' extraordinary characteristics [71]. What follows is an extensive discussion of the present and future of elastomer-based composite materials in medical and healthcare:

- Biocompatibility and biodegradability

Elastomeric composites, in particular, have demonstrated exceptional biocompatibility when composed of either natural or synthetic polymers, such as elastin or polyurethane [72]. Elastomeric composites for tissue engineering scaffolds have recently become the subject of investigation. The study demonstrates that these chemicals have a beneficial effect on biological systems, facilitating cell adhesion and tissue regeneration.

- Characteristics pertaining to mechanics as they pertain to medical equipment

Many people are familiar with elastomeric composites because of their exceptional mechanical qualities, which encompass both stiffness and flexibility. A number of recent innovations in medical device technology—including wearable sensors, catheters, and prostheses—can be traced back to elastomeric composites [73]. The purpose of this research paper is to provide an optimised composite composition that meets the mechanical requirements of different medical devices.

- A targeted drug delivery system using elastomeric composites

Our final conclusion is that elastomeric composites are a viable material with a lot of potential for controlled medication delivery systems. Using these nanoparticles, new smart drug delivery methods have emerged, which explains elastomeric nanocomposites that respond to stimuli might be utilised to release medications at certain physiological times [74]. This method has the potential to enhance therapeutic efficacy while reducing adverse effects.

- Antimicrobials and Their applications

A key component of infection risk assessment in healthcare facilities is the elimination of opportunities for microbial colonisation of surfaces. The resistance of elastomeric composites to microbial adhesion is significantly enhanced by adding antimicrobial agents to them [75]. The study demonstrates that the growth of bacteria may be efficiently inhibited, which paves the way for the creation of future antimicrobial coatings for surfaces and medical equipment.

- Use as sensors in applications

Elastomeric composites are finding increasing use in the production of sensors that assess a wide range of physiological markers. Elastomeric Composite Sensors for Healthcare Monitoring delve into the use of elastomeric nanocomposites as wearable sensors for real-time health monitoring, as explored by Chen et al. [76]. Continuous monitoring of vital signs and mobility is made possible by these conformal and flexible sensors, which provide crucial information for patient care.

- Environmentally friendly substitutes

There has been a dramatic increase in the quest for biodegradable substitutes, including elastomeric implants because implantable devices are fundamental to several medical procedures. Bat et al. reported on biodegradable elastomers for biomedical applications and regenerative medicine example of a recent academic work that aimed to enhance elastomeric composites with regenerative tissue capabilities after degradation [77]. This puts an end to further surgical procedures often performed to remove implants. When it comes to biodegradable medical implants, this is a huge breakthrough.

- 3D-Printed Biomimetic and other Applications

Exoskeletons of ancient fish, self-shaping cellular structures resembling morphing seedpods, combustion-powered soft robots, biomimetic shark skin with engineered surface roughness, and various structural materials resembling bone and biogenic calcite have all been created using the sophisticated capabilities of multi-material inkjet printing (Fig. 11.6a–c). To replicate the beating of the heart's left ventricle, [78] an actuator with a soft matrix is used (Fig. 11.6d). Figure 11.6e shows the results of an intriguing application of reinforced hydrogels: a highly stretchable and self-healing sensor for strain and temperature sensing. Another intriguing application was developed by Ge et al., who used polyaniline nanofibres and filled them with glycerol. Takahashi et al. devised a simple strategy to fabricate macroscale hydrogel composites by reinforcing soft hydrogels with low melting point alloys (LMA), which can transition from a load-bearing solid state to a free deformable liquid state at relatively low temperature. The authors suggested that these composites could be used for engineering channel structures in hydrogel fluidic devices for filtering and chemical reactor applications.

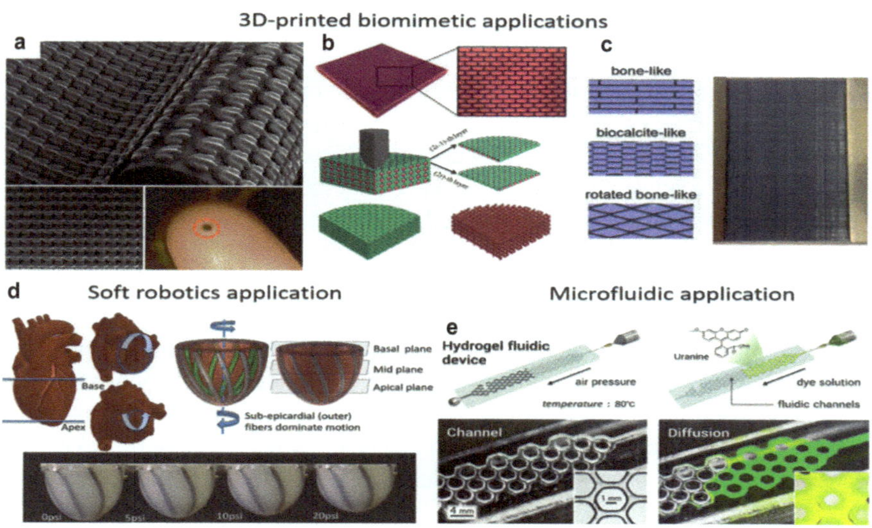

Fig. 11.6 Practical uses of Reinforced elastomers and gels **a** A membrane that resembles shark skin, created using 3D printing technology, consisting of stiff denticles on a flexible substrate. Made with the author's approval. **b** A simulated nacre-like pattern (quarter geometry) and a 3D-printed nacre sample (inspiring). **c** Bio-inspired 3D-printed structural motifs made from a mix of flexible and rigid polymers and an image of a 3D-printed composite that resembles bone. **d** A bio-inspired, soft-actuated material that can mimic the effects of natural movement. **e** Incorporating the LMA skeleton into a PAAm hydrogel fluidic device for channel template demonstration [78]. Adapted from Ref. 80 with permission from [American Chemical Society], Copyright [2020]

11.6.4 Applications in Sports and Leisure

Particularly in the dynamic fields of sports and recreation, elastomer-based composites have seen tremendous growth in both usefulness and attractiveness. Research and development into the potential applications of elastomeric composites in various sports and leisure goods, such as athletic equipment, athletic apparel, and leisure merchandise, could be a massive undertaking because of the remarkable combination of flexibility, durability, and adaptability that these materials exhibit [79].

- Advancements in the Domain of Sports Shoes

In an effort to enhance players' comfort and performance, research has focused on exploring the use of elastomeric composites in sports footwear. Midsole construction is the subject of this research's in-depth examination. When compared to more traditional materials, the study's findings reveal superior stress absorption, energy return, and overall performance [80]. In addition to improving athletic ability, these technological advancements have a major impact on reducing athlete fatigue and injury risk.

- Innovations in Sports Gear

A number of sporting equipment, such as bicycle parts, tennis rackets, and golf clubs, have benefited greatly from the use of elastomeric composites, which have caused a sea change in the design and operation of these items. An in-depth analysis of how elastomeric composites can enhance the stiffness and damping characteristics of tennis racket frames is presented by AL-Oqla et al. [81]. As this study shows, elastomeric composites have the potential to change the game when it comes to athletic gear by improving player comfort, power transfer, and control.

- Impact Mitigation in Safety Gear

Incorporating elastomeric materials into protective gear has been a response to safety concerns in sports aimed at reducing impact forces. Research demonstrates that elastomeric materials are useful for preventing the common head injuries caused by contact sports by efficiently absorbing and dispersing impact energy [82]. This is a huge step forward in making sure that athletes are safe while they play their sports.

- Athletic clothing that is both comfortable and able to stretch

Because of the paramount necessity of breathability and flexibility in athletic apparel, elastomeric composites have emerged as key components. The production of fabrics with the desirable qualities of being breathable, moisture-wicking, and stretchy [83]. Fabrics made with elastomeric composites allow for remarkable movement without sacrificing durability or support, making them ideal for athletes. The groundbreaking potential of elastomeric composites to reimagine performance and comfort standards in sportswear is demonstrated by this research.

- Sensors that Integrate Intelligent Sports Equipment

Elastomeric composites are essential components in the field of intelligent technology integration in sporting clothing, which has seen a large increase in its usage [84]. Instructors and athletes may get priceless insight into training optimisation, injury prevention, and performance measures through the real-time monitoring of biomechanical data provided by these sensors. Intelligent sports equipment that uses elastomeric composites is a step in the right direction towards improving performance and society as a whole [85].

A new era of innovation has begun in the sports and leisure industry thanks to advancements in composite materials based on elastomers. More exciting than ever before is the convergence of state-of-the-art technology with athletic interests, thanks to the ongoing exploration and integration of elastomeric composites into recreational and sporting uses.

11.7 Future Perspectives and Challenges

We discover a dynamic area with tremendous potential for improving engineering applications by investigating the most recent advances in elastomer-based composite materials. Improving material properties by the controlled introduction of nanofillers and nanoparticles into an elastomeric matrix is a significant new development that makes use of nanotechnology. Engineers now have potent new tools at their disposal thanks to nanocomposite technology, which allows them to customise the mechanical, thermal, and electrical properties of materials. Another major development is the self-healing elastomeric composites, which may drastically change the durability of materials. Microcapsules or vascular networks carrying therapeutic compounds allow these composites to self-repair damage, extending their operating lifetime and reducing maintenance expenses in emergency scenarios. This ability to repair itself resolves an old issue in materials research and is crucial in sectors where dependability is key. Sustainable engineering practices have led to an increase in the use of bio-based elastomers in composite material production. This environmentally conscious method not only reduces waste but also maximises the usefulness of bio-based elastomers, which in turn leads to novel uses and helps advance more sustainable technologies. The rise of additive manufacturing processes, such as 3D printing, is having a profound effect on the development and manufacture of elastomeric composites within the field of advanced manufacturing. This additive method enables the production of specific and complicated structures, allowing for the creation of functional components with improved performance characteristics and intricate geometries. Elastomeric composites that aim to integrate numerous functionalities into one material are also gaining popularity due to their multifunctionality. Combining sensing capabilities with mechanical attributes is necessary to make intelligent materials that can adapt to changing environments. Many industries could undergo radical transformation as a result of these breakthroughs, including structural engineering and wearable tech.

Composites based on elastomers have gone a long way, but engineers are still constrained by a number of issues. Finding a happy medium between being overly rigid and being too pliable is a major challenge. Due to their intrinsic elasticity, elastomers could provide a difficulty in applications requiring both great mechanical strength and exceptional flexibility. Because these composites' longevity in the face of extreme environmental stresses like high temperatures, chemicals, and UV radiation is still an important question, more research into protective coatings and resilience-boosting additives is required. One other thing: making elastomeric composites is a complicated procedure. Placing reinforcing materials (such as fibres or nanoparticles) evenly throughout the elastomeric matrix is a major challenge in the pursuit of repeatable and scalable composite production. Exact control of material characteristics, particularly in nanocomposites, requires complex manufacturing procedures, which in turn increases production costs and introduces operational issues. Concerns about the long-term viability of elastomeric composites—particularly those containing synthetic elastomers—have prompted an upsurge in research on greener alternatives. Integrating elastomeric composites with many capacities presents technical and design hurdles in the field of multifunctionality. The compromises needed to synchronise mechanical, thermal, and electrical quality while maintaining component compatibility may limit the variety of potential usage. Despite the increasing demand for self-healing materials, concerns about their lifetime, cost-effectiveness, and scalability continue to hinder their broad adoption in real-world engineering applications.

11.8 Conclusion

Research into engineering composites based on elastomers has produced impressive results, highlighting their critical role in determining the direction of materials science and technology in the future. These composites have been incredibly useful in tackling difficult problems, providing solutions that emphasise resilience, adaptability, and impact resistance in fields like aerospace and biomedicine. A major takeaway from this chapter is the incredible malleability of elastomeric composites, which enables the exact modification of material characteristics to meet unique engineering demands. Wearable tech and soft robotics are only two areas that stand to benefit from the intelligent structures and materials that have been created by making use of this flexibility. Because of its ability to absorb and reduce energy, elastomeric composites are a great choice for applications in automotive engineering that need high levels of safety, such as impact-resistant parts. To help individuals make sophisticated decisions in their own manufacturing processes, this chapter provides the advantages and disadvantages of various approaches. In addition, it details various methods for characterising elastomeric composites, which are useful for evaluating their mechanical properties, morphology, and structure. It also examines how the mechanical properties of these composites are affected by the reinforcing materials. The recyclability and eco-friendly properties of elastomeric materials are also important considerations in light of the continuing push for sustainable engineering techniques. Solutions that are kind to the environment are in high demand right now among engineers all around the world. To that end, composites made of elastomers stand out as an ideal material choice. When used, elastomer composites will undeniably have far-reaching effects and revolutionise engineering and material science. Composites characterised by the use of elastomers and other reinforcing components allow engineers to tackle a wide range of issues. Materials with extraordinary resilience, flexibility, and durability are created by combining the natural elasticity of elastomers with the customised characteristics of other components. Elastomeric composites have found several applications that have shown positive results, including in the aerospace, automotive, civil, and biomedical engineering sectors. Their energy-absorbing and energy-distributing properties make them excellent for use in vibration-reduction systems and crash-resistant structures, enhancing the dependability and security of these critical applications. These composites are extremely resistant to weather and ageing, which means they last longer and require less maintenance during infrastructure projects. Because of its pliability, elastomeric composites are great for creating smart materials and structures that can have their functions customised. Engineers may manipulate mechanical, thermal, and electrical characteristics with additives and fillers to generate opportunities for innovation in responsive and adaptive systems. The extraordinary flexibility and adaptability of elastomeric composites make them indispensable in wearable electronics, soft robotics, and cutting-edge medical equipment. Due to their recyclable nature and environmentally favourable characteristics, elastomeric materials are well-suited to the expanding field of sustainable engineering. In the ever-growing quest for eco-friendly engineering solutions, elastomeric-based composites are playing a pivotal role. Composites built of elastomers are important because of the many benefits they provide, which allow for technological innovation. Composites like this have the potential to revolutionise materials science with their exceptional combination of durability, flexibility, and eco-friendliness. The study's findings highlight the crucial role of elastomeric-based composites in the ongoing advancement of engineering materials research.

References

1. Fazli, A., Rodrigue, D.: Waste rubber recycling: A review on the evolution and properties of thermoplastic elastomers. Materials **13**(3), 782 (2020)
2. Herzberger, J., Sirrine, J.M., Williams, C.B., Long, T.E.: Polymer design for 3D printing elastomers: recent advances in structure, properties, and printing. Prog. Polym. Sci. **97**, 101144 (2019)
3. Drobny, J.G.: Handbook of thermoplastic elastomers. Elsevier (2014)
4. Caruso, M.M., Davis, D.A., Shen, Q., Odom, S.A., Sottos, N.R., White, S.R., Moore, J.S.: Mechanically-induced chemical changes in polymeric materials. Chem. Rev. **109**(11), 5755–5798 (2009)
5. Zhao, Q., Qi, H.J., Xie, T.: Recent progress in shape memory polymer: New behavior, enabling materials, and mechanistic understanding. Prog. Polym. Sci. **49**, 79–120 (2015)

6. Wang, R., Zhang, J., Kang, H., Zhang, L.: Design, preparation and properties of bio-based elastomer composites aiming at engineering applications. Compos. Sci. Technol. **133**, 136–156 (2016)
7. Wang, D., Wang, Z., Ren, S., Xu, J., Wang, C., Hu, P., Fu, J.: Molecular engineering of a colorless, extremely tough, superiorly self-recoverable, and healable poly (urethane–urea) elastomer for impact-resistant applications. Mater. Horiz. **8**(8), 2238–2250 (2021)
8. Li, W., Zhang, X.: Research and applications of MR elastomers. Recent. Pat. Mech. Eng. **1**(3), 161–166 (2008)
9. Skalski, P., Kalita, K.: Role of magnetorheological fluids and elastomers in today's world. Acta Mech. Autom., **11**(4), 267–274 (2017)
10. Hafner, B.J., Cagle, J.C., Allyn, K.J., Sanders, J.E.: Elastomeric liners for people with transtibial amputation: Survey of prosthetists' clinical practices. Prosthet. Orthot. Int. **41**(2), 149–156 (2017)
11. Zhang, T., Liu, N., Xu, J., Liu, Z., Zhou, Y., Yang, Y., Jiang, S.: Flexible electronics for cardiovascular healthcare monitoring. The Innovation, **4**(5), (2023)
12. Bahrami, M., Abenojar, J., Martínez, M.Á.: Recent progress in hybrid biocomposites: Mechanical properties, water absorption, and flame retardancy. Materials **13**(22), 5145 (2020)
13. Peters, E.N.: Thermoplastics, thermosets, and elastomers—Descriptions and properties. Mech. Engineers' Handb., Vol. 1: Mater. Eng. Mech., **1**, 353 (2015)
14. Somarathna, H.M.C.C., Raman, S.N., Mohotti, D., Mutalib, A.A., Badri, K.H.: The use of polyurethane for structural and infrastructural engineering applications: A state-of-the-art review. Constr. Build. Mater. **190**, 995–1014 (2018)
15. Shanks, R.A., Kong, I.: General purpose elastomers: structure, chemistry, physics and performance. Adv. Elastomers I: Blends Interpenetr. Netw., 11–45 (2013)
16. Baghani, M., Baniassadi, M., Rémond, Y.: Computational modeling of intelligent soft matter: shape memory polymers and hydrogels. Elsevier, (2023)
17. Rajak, D.K., Pagar, D.D., Menezes, P.L., Linul, E.: Fiber-reinforced polymer composites: Manufacturing, properties, and applications. Polymers **11**(10), 1667 (2019)
18. Mrlík, M., Ilčíková, M., Mosnáček, J., Pavlínek, V.: Modification of carbon nanotubes: improvement of physical and photo-actuating properties of their elastomeric composites. Carbon Nanotub.-Based Compos, (2015)
19. Rollo, G., Ronca, A., Cerruti, P., Gan, X.P., Fei, G., Xia, H., Ambrosio, L.: On the synergistic effect of multi-walled carbon nanotubes and graphene nanoplatelets to enhance the functional properties of SLS 3D-printed elastomeric structures. Polymers **12**(8), 1841 (2020)
20. Tepylo, N., Huang, X., Patnaik, P.C.: Laser-based additive manufacturing technologies for aerospace applications. Adv. Eng. Mater. **21**(11), 1900617 (2019)
21. Ponnamma, D., Sadasivuni, K.K., Grohens, Y., Guo, Q., Thomas, S.: Carbon nanotube based elastomer composites–an approach towards multifunctional materials. J. Mater. Chem. C **2**(40), 8446–8485 (2014)
22. Ning, N., Li, S., Wu, H., Tian, H., Yao, P., Guo-Hua, H.U., Zhang, L.: Preparation, microstructure, and microstructure-properties relationship of thermoplastic vulcanizates (TPVs): A review. Prog. Polym. Sci. **79**, 61–97 (2018)
23. Gibson, R.F.: A review of recent research on mechanics of multifunctional composite materials and structures. Compos. Struct. **92**(12), 2793–2810 (2010)
24. Zhou, J., Turner, S.A., Brosnan, S.M., Li, Q., Carrillo, J.M.Y., Nykypanchuk, D., Sheiko, S.S.: Shapeshifting: Reversible shape memory in semicrystalline elastomers. Macromolecules **47**(5), 1768–1776 (2014)
25. Odian, G.: Principles of polymerization, pp. 128–130. John Wiley & Sons Inc., New York (2004)
26. Drobny, J.G.: Handbook of thermoplastic elastomers, p. 404. WilliamAndrew Publishing, Norwich NY (2018)
27. Herzberger, Sirrine, J.M., Williams, C.B. et al.: Progress in Polymer Science **97**, 101144 (2019)
28. Persons, A.K., Ball, J.E., Freeman, C., Macias, D.M., Simpson, C.L., Smith, B.K., Burch, V., R.F: Fatigue testing of wearable sensing technologies: Issues and opportunities. Materials **14**(15), 4070 (2021)

29. Parathodika, A.R., Naskar, K.: Ultra-high molecular weight EPDM composites via crossover curing agents: stabilized crosslink network structure and performance properties. Polym. Degrad. Stab., 110603 (2023)
30. Lipsa, L., Rajput, S., Parida, S., Ghosh, S.P., Verma, S.K., Gupta, S.: Influence of hot-press temperature on β-phase formation and electrical properties of solvent-casted PVDF-HFP co-polymer films prepared from two different solvents: a comparison study. Macromol. Chem. Phys. **224**(21), 2300204 (2023)
31. Kadir, A.A.S.: Advances in natural rubber production. Rubber Chem. Technol. **67**(3), 537–548 (1994)
32. Mohanty, A.K., Vivekanandhan, S., Pin, J.M., Misra, M.: Composites from renewable and sustainable resources: Challenges and innovations. Science **362**(6414), 536–542 (2018)
33. Zubov, A., Pokorny, J., Kosek, J.: Styrene–butadiene rubber (SBR) production by emulsion polymerization: Dynamic modeling and intensification of the process. Chem. Eng. J. **207**, 414–420 (2012)
34. Zhang, L., Wang, Y., Wang, Y., Sui, Y., Yu, D.: Morphology and mechanical properties of clay/styrene-butadiene rubber nanocomposites. J. Appl. Polym. Sci. **78**(11), 1873–1878 (2000)
35. Akindoyo, J.O., Beg, M., Ghazali, S., Islam, M.R., Jeyaratnam, N., Yuvaraj, A.R.: Polyurethane types, synthesis and applications–a review. RSC Adv. **6**(115), 114453–114482 (2016)
36. Liu, Y., Ma, C., Li, Y., Yin, Z., Su, J., Gao, J.: Ablation and thermo-mechanical investigation of polyarylacetylene impregnated heat vulcanising silicone rubber for ultrahigh temperature applications. Plast., Rubber Compos. **47**(5), 202–207 (2018)
37. Athawale, A.A., Joshi, A.M.: Electronic applications of ethylene propylene diene monomer rubber and its composites. Flex. Stretchable Electron. Compos., 305–333 (2016)
38. Ponnamma, D., Jose Chirayil, C., Sadasivuni, K. K., Somasekharan, L., Yaragalla, S., Abraham, J., Thomas, S.: x Special purpose elastomers: synthesis, structure-property relationship, compounding, processing and applications. Adv. Elastomers I: Blends Interpenetr. Netw., 47–82 (2013)
39. Afroze, S., Rahman, M.M., Kabir, H., Kabir, M.A., Ahmed, F., Hossain, M.A., Gafur, M.A.: Physical, optical and thermal properties of graphite and talc filler reinforced polypropylene (PP) composites. Int J Adv Sci Technol Res **5**, 40–49 (2012)
40. Hsissou, R., Seghiri, R., Benzekri, Z., Hilali, M., Rafik, M., Elharfi, A.: Polymer composite materials: A comprehensive review. Compos. Struct. **262**, 113640 (2021)
41. Miyagawa, H., Misra, M., Mohanty, A.K.: Mechanical properties of carbon nanotubes and their polymer nanocomposites. J. Nanosci. Nanotechnol. **5**(10), 1593–1615 (2005)
42. Dayanand, S., Boppana, S.B.: Manufacturing of particulate and fiber reinforced composites: a review. Struct. Compos. Mater.: Fabr., Prop., Appl. Chall., 119–142 (2023)
43. Miller, S.A.: Natural fiber textile reinforced bio-based composites: Mechanical properties, creep, and environmental impacts. J. Clean. Prod. **198**, 612–623 (2018)
44. Chin Tjong, S., Xu, S.A., Li, R.K.Y., Mai, Y.W.: Tensile deformation mechanisms of polypropylene/elastomer blends reinforced with short glass fibre. J. Appl. Polym. Sci. **87**(3), 441–451 (2003)
45. Maples, H.A., Wakefield, S., Robinson, P., Bismarck, A.: High performance carbon fibre reinforced epoxy composites with controllable stiffness. Compos. Sci. Technol. **105**, 134–143 (2014)
46. Suresh, G., Vivek, S., Babu, L.G., Bernard, S.S., Akash, R.M., Kanna, S.R., Sethuramalingam, P.: Evaluation of mechanical behaviour of carbon fiber reinforced nanoclay filled IPN matrix composite. Mater. Res. Express **6**(12), 125311 (2019)
47. Mohd Nurazzi, N., Asyraf, M.M., Khalina, A., Abdullah, N., Sabaruddin, F.A., Kamarudin, S.H., Sapuan, S.M.: Fabrication, functionalization, and application of carbon nanotube-reinforced polymer composite: An overview. Polymers **13**(7), 1047 (2021)
48. Bessa, J., Matos, J., Mota, C., Cunha, F., Araújo, I., Silva, L., Fangueiro, R.: Influence of surface treatments on the mechanical properties of fibre reinforced thermoplastic composites. Procedia Eng. **200**, 465–471 (2017)

49. Mohan, N., Senthil, P., Vinodh, S., Jayanth, N.: A review on composite materials and process parameters optimisation for the fused deposition modelling process. Virtual Phys. Prototyp. **12**(1), 47–59 (2017)
50. Koushki, P., Kwok, T.H., Hof, L., Wuthrich, R.: Reinforcing silicone with hemp fiber for additive manufacturing. Compos. Sci. Technol. **194**, 108139 (2020)
51. Safaee, S., Schock, M., Joyee, E.B., Pan, Y., Chen, R.K.: Field-assisted additive manufacturing of polymeric composites. Addit. Manuf. **51**, 102642 (2022)
52. Peeke, L.M.: Material extrusion additive manufacturing of thermoset and thermoplastic composites. (2021)
53. Shah, V., Bhaliya, J., Patel, G.M., Deshmukh, K.: Advances in polymeric nanocomposites for automotive applications: A review. Polym. Adv. Technol. **33**(10), 3023–3048 (2022)
54. Zhang, X., Xu, Y., Zhang, X., Wu, H., Shen, J., Chen, R., Guo, S.: Progress on the layer-by-layer assembly of multilayered polymer composites: Strategy, structural control and applications. Prog. Polym. Sci. **89**, 76–107 (2019)
55. Utrera-Barrios, S., Verdejo, R., López-Manchado, M.Á., Santana, M.H.: Self-healing elastomers: A sustainable solution for automotive applications. Eur. Polym. J., 112023 (2023)
56. Mavila, S., Eivgi, O., Berkovich, I., Lemcoff, N.G.: Intramolecular cross-linking methodologies for the synthesis of polymer nanoparticles. Chem. Rev. **116**(3), 878–961 (2016)
57. Zhao, X., Chen, X., Yuk, H., Lin, S., Liu, X., Parada, G.: Soft materials by design: unconventional polymer networks give extreme properties. Chem. Rev. **121**(8), 4309–4372 (2021)
58. Llerena Zambrano, B., Renz, A.F., Ruff, T., Lienemann, S., Tybrandt, K., Vörös, J., Lee, J.: Soft electronics based on stretchable and conductive nanocomposites for biomedical applications. Adv. Healthcare Mater. **10**(3), 2001397 (2021)
59. Deepalekshmi, P., Visakh, P.M., Mathew, A.P., Chandra, A.K., Thomas, S.: Advances in elastomers: their blends and interpenetrating networks-state of art, new challenges and opportunities. Adv. Elastomers I: Blends Interpenetr. Netw., 1–9 (2013)
60. Singh, H.K.: Lifetime prediction and durability of elastomeric seals for fuel cell applications. Doctoral dissertation, Virginia Tech, (2009)
61. Youssef, G.: Applied mechanics of polymers: properties, processing, and behavior. Elsevier (2021)
62. Araby, S., Meng, Q., Zhang, L., Zaman, I., Majewski, P., Ma, J.: Elastomeric composites based on carbon nanomaterials. Nanotechnology **26**(11), 112001 (2015)
63. Baroutaji, A., Wilberforce, T., Ramadan, M., Olabi, A.G.: Comprehensive investigation on hydrogen and fuel cell technology in the aviation and aerospace sectors. Renew. Sustain. Energy Rev. **106**, 31–40 (2019)
64. Gore, P.M., Kandasubramanian, B.: Functionalized aramid fibers and composites for protective applications: a review. Ind. Eng. Chem. Res. **57**(49), 16537–16563 (2018)
65. Patel, H., Salehi, S., Ahmed, R., Teodoriu, C.: Review of elastomer seal assemblies in oil & gas wells: Performance evaluation, failure mechanisms, and gaps in industry standards. J. Petrol. Sci. Eng. **179**, 1046–1062 (2019)
66. Fan, R., Meng, G., Yang, J., He, C.: Experimental study of the effect of viscoelastic damping materials on noise and vibration reduction within railway vehicles. J. Sound Vib. **319**(1–2), 58–76 (2009)
67. Pethrick, R.A.: Design and ageing of adhesives for structural adhesive bonding–a review. Proceedings of the Institution of Mechanical Engineers, Part L: Journal of Materials: design and applications **229**(5), 349–379 (2015)
68. Uyanna, O., Najafi, H.: Thermal protection systems for space vehicles: A review on technology development, current challenges and future prospects. Acta Astronaut. **176**, 341–356 (2020)
69. Miao, J., Zhong, Q., Zhao, Q., Zhao, X.: Spacecraft thermal control technologies, p. 27. Springer, Singapore (2021)
70. Parveez, B., Kittur, M.I., Badruddin, I.A., Kamangar, S., Hussien, M., Umarfarooq, M.A.: Scientific advancements in composite materials for aircraft applications: a review. Polymers **14**(22), 5007 (2022)

71. Pavan Kalyan, B.G., Kumar, L.: 3D printing: applications in tissue engineering, medical devices, and drug delivery. AAPS PharmSciTech **23**(4), 92 (2022)
72. Wise, S.G., Liu, H., Yeo, G.C., Michael, P.L., Chan, A.H., Ngo, A.K., Weiss, A.S.: Blended polyurethane and tropoelastin as a novel class of biologically interactive elastomer. Tissue Eng. Part A **22**(5–6), 524–533 (2016)
73. Lou, Z., Wang, L., Shen, G.: Recent advances in smart wearable sensing systems. Adv. Mater. Technol. **3**(12), 1800444 (2018)
74. Mertz, D., Harlepp, S., Goetz, J., Bégin, D., Schlatter, G., Bégin-Colin, S., Hébraud, A.: Nanocomposite polymer scaffolds responding under external stimuli for drug delivery and tissue engineering applications. Adv. Ther. **3**(2), 1900143 (2020)
75. Abedin, F.N.J., Yahaya, A.N.A., Khalil, N.A., Zulkifli, M.: Antimicrobial capability of inorganic materials as fillers for thermoplastic Elastomer–A state of the art review. Mater. Today: Proc. **74**, 480–484 (2023)
76. Chen, S., Fan, S., Chan, H., Qiao, Z., Qi, J., Wu, Z., Lim, C.T.: Liquid metal functionalization innovations in wearables and soft robotics for smart healthcare applications. Adv. Funct. Mater., 2309989 (2023)
77. Bat, E., Zhang, Z., Feijen, J., Grijpma, D.W., Poot, A.A.: Biodegradable elastomers for biomedical applications and regenerative medicine. Regen. Med. **9**(3), 385–398 (2014)
78. Zhalmuratova, D., Chung, H.J.: Reinforced gels and elastomers for biomedical and soft robotics applications. ACS Appl. Polym. Mater. **2**(3), 1073–1091 (2020)
79. Duncan, O., Shepherd, T., Moroney, C., Foster, L., Venkatraman, P.D., Winwood, K., Alderson, A.: Review of auxetic materials for sports applications: Expanding options in comfort and protection. Appl. Sci. **8**(6), 941 (2018)
80. Liu, J., Ren, T.: Research on the protection of athletes from injury by flexible conjugated materials in sports events. Front. Chem., **11**, (2023)
81. AL-Oqla, F.M., Hayajneh, M. T., Nawafleh, N.: Advanced synthetic and biobased composite materials in sustainable applications: a comprehensive review. Emergent Materials, 1–18 (2023)
82. Tomin, M., Kmetty, A.: Polymer foams as advanced energy absorbing materials for sports applications—A review. J. Appl. Polym. Sci. **139**(9), 51714 (2022)
83. Kamalha, E., Zeng, Y., Mwasiagi, J.I., Kyatuheire, S.: The comfort dimension; a review of perception in clothing. J. Sens. Stud. **28**(6), 423–444 (2013)
84. Liang, X., Yu, J.: Design method and application of intelligent sports equipment based on stochastic differential equation. Appl. Math. Nonlinear Sciences., (2023)
85. Zang, Y., Zhang, F., Di, C.A., Zhu, D.: Advances of flexible pressure sensors toward artificial intelligence and health care applications. Mater. Horiz. **2**(2), 140–156 (2015)

Chapter 12
State-Of-The-Art in Textile Polymer Composites and Applications

Biswajeet Acharya, Amulyaratna Behera, Srikanta Moharana, and Suchismeeta Behera

Abstract Polymer textile composites represent a cutting-edge field in materials science and engineering, combining the exceptional properties of polymers and textiles to create versatile, high-performance materials. The material used for preparing polymer textile composites can vary depending on the desired application. Commonly used polymers include thermoplastics such as polyethylene and polypropylene, as well as thermosetting resins like epoxy and polyester. Several equipment are used in the fabrication process of polymer textile composites, including melt extruders, weaving machines, and knitting machines. These equipments allow for precise control over the composition and structure of the composites, resulting in materials with tailored properties for specific applications. The potential application of polymer textile composites is vast, ranging from aerospace and automotive industries to sports equipment and medical devices. These composites offer advantages such as high strength-to-weight ratio, corrosion resistance, and design flexibility. Additionally, their ability to be molded into complex shapes makes them suitable for various manufacturing processes. As research and development in the field of polymer textile composites continue to advance, we can expect to see even more innovative applications in the future. The current review aims to highlight the latest advancements in polymer textile composites and their potential impact on various industries. It will also explore the challenges and opportunities that lie ahead in terms of improving manufacturing processes, enhancing material properties, and exploring new applications.

B. Acharya · A. Behera (✉)
School of Pharmacy and Life Sciences, Centurion University of Technology and Management, Jatni, Odisha, India
e-mail: dr.amulyaratnabehera@gmail.com

S. Moharana
Department of Chemistry, School of Applied Sciences, Centurion University of Technology and Management, Jatni, Bhubaneswar, Odisha, India

S. Behera
State Forensic Science Laboratory, Bhubaneswar, Odisha, India

S. Moharana et al. (eds.), *Polymer Composites*, Engineering Materials,
https://doi.org/10.1007/978-981-97-2075-0_12

Keywords Textile Polymer composite · Nanotechnology · Sustainability · Biomedical · Fibers

12.1 Introduction

Textile polymer composites represent a remarkable convergence of two distinct material categories, textiles, and polymers, resulting in advanced materials with a wide array of applications. These composites are engineered by strategically combining textile reinforcements, such as fibers or fabrics, with polymer matrices to create materials that harness the inherent properties of both components [1]. The resulting synergy yields materials that are lightweight, durable, and adaptable to various industries. Researchers and engineers have delved into the utilization of high-performance fibers, including carbon, aramid, and glass. These fibers exhibit exceptional strength-to-weight ratios, making them ideal candidates for enhancing the mechanical properties of composites. The careful selection and arrangement of these fibers within the polymer matrix allow for precise control over properties like tensile strength, stiffness, and impact resistance [2].

Advancements in nanotechnology have also influenced the development of textile polymer composites. The incorporation of nanomaterials, such as carbon nanotubes and graphene, into polymer matrices has opened up new horizons. These nanocomposites exhibit extraordinary electrical conductivity, thermal stability, and mechanical strength, making them suitable for applications demanding cutting-edge performance, including aerospace and electronics [3]. By integrating sensors, actuators, and electronic components directly into the textile structure, these materials become capable of sensing, reacting to environmental changes, and even providing therapeutic functions [4]. Sustainability is an increasingly vital aspect of textile polymer composites. In response to environmental concerns, there's a growing emphasis on utilizing biodegradable and eco-friendly polymer matrices. These eco-conscious choices aim to reduce the environmental footprint associated with composite production, making them more aligned with sustainable practices [5]. In manufacturing, automation and advanced techniques such as 3D weaving and automated fiber placement (AFP) have streamlined production processes, increasing efficiency, precision, and complexity. These techniques allow for the creation of intricate textile structures that were once challenging to produce, further expanding the possibilities of textile polymer composites [6].

The applications of textile polymer composites are diverse and far-reaching. In the aerospace industry, they are used in aircraft components, spacecraft, and drones, where their high strength-to-weight ratios contribute to improved fuel efficiency and overall performance. The automotive sector benefits from these composites as well, as they help reduce vehicle weight, enhance fuel efficiency, and improve safety through the use of composite vehicle components and structural parts [7]. Additionally, they are employed in construction to reinforce concrete structures and create durable, lightweight building materials. In the medical field, textile polymer composites are used in medical implants and orthopedic devices, offering both biocompatibility and structural integrity. Furthermore, protective gear like bulletproof vests and helmets often incorporates these composites to provide enhanced protection without the added bulk and weight [8]. In summary, textile polymer composites exemplify the relentless pursuit of innovation at the crossroads of textiles and polymers. Their unique blend of properties, coupled with ongoing research and advancements, ensures they will continue to play a pivotal role in an ever-expanding range of applications, from aerospace to healthcare and beyond [9].

12.2 History of Textile Polymer Composites

The history of textile polymer composites is a journey of innovation and evolution. It began with the early use of natural fibers in ancient civilizations for basic textile applications. The Industrial Revolution introduced mechanized textile production in the late eighteenth century, setting the stage for more complex fabric structures. The development of synthetic polymers started with Leo Baekeland's invention of Bakelite in 1907 [10]. During World War II, composite materials research intensified, leading to the creation of fiberglass-reinforced composites for military applications. The 1960 and 1970s marked a significant era with the development of high-performance fibers like carbon and Kevlar. These fibers laid the foundation for advanced composites that found applications in various industries, especially aerospace [11]. The 1990s saw the emergence of nanocomposites, incorporating nanoparticles like carbon nanotubes to enhance composite properties. In the 2000s, smart textiles became a significant development, integrating sensors and electronics into textile structures for applications in healthcare, aerospace, and wearable technology. Today, textile polymer composites are widely used in industries such as aerospace, automotive, construction, healthcare, and consumer goods [12]. Table 12.1 depicts a timeline of the evolution of polymer textile composites.

12.3 Materials Used in the Preparation of Polymer Textile Composite

12.3.1 Reinforcement Fibers

Reinforcement fibers play a pivotal role in polymer textile composites, as they are responsible for imparting many of the desired mechanical, thermal, and electrical properties to these advanced materials. Reinforcement fibers are typically much stronger and stiffer than the polymer matrix. When incorporated into the composite, they act as load-bearing components, carrying the majority of the mechanical stresses [15]. This results in composites that are significantly stronger and stiffer than the

Table 12.1 A chronological account of the development of polymer textile composites [7, 13, 14]

Year	Milestone in textile polymer composites development
Ancient	Early use of natural fibers like cotton, wool, and flax for textiles
Late 18th century	The Industrial Revolution introduces mechanized textile production
1907	The invention of Bakelite, the first synthetic plastic, marks the beginning of synthetic polymer development
1940s	Composite materials research intensifies during World War II, leading to the development of fiberglass-reinforced composites
1961	Dr. Roger Bacon invents high-performance carbon fibers, contributing to advanced composites
1965	Stephanie Kwolek invents Kevlar, an aramid fiber with exceptional strength and lightweight properties
1970s	The aerospace industry adopts composite materials for aircraft components, improving fuel efficiency
1980s	Carbon fiber-reinforced composites gain prominence in high-performance applications
1990s	Emergence of nanocomposites incorporating nanoparticles for enhanced properties
2000s	Introduction of smart textiles integrating sensors and electronics for various applications
Present	Ongoing research and development continue to advance textile polymer composites in diverse industries

polymer alone. For example, carbon fibers are renowned for their exceptional tensile strength and stiffness, making them ideal for applications where lightweight yet robust materials are required, such as in aerospace and automotive components. The high strength-to-weight ratio of many reinforcement fibers allows for weight reduction in applications. By using lightweight fibers, the overall weight of a composite component can be significantly reduced compared to traditional materials like metals [16]. This weight reduction is especially critical in aerospace, where every gram saved translates to improved fuel efficiency and performance. Reinforcement fibers, particularly aramid fibers like Kevlar, offer excellent impact resistance. This property is valuable in applications where the composite needs to withstand sudden forces or impacts, such as in protective gear like helmets and bulletproof vests. Certain reinforcement fibers, like carbon fibers, have excellent thermal conductivity [17]. This property can be advantageous in applications where heat dissipation or thermal stability is crucial, such as in high-temperature environments or electronic enclosures. The choice of reinforcement fibers allows engineers to tailor the properties of polymer textile composites to specific applications. By selecting the appropriate fiber type, orientation, and arrangement, composites can be designed to meet a wide range of requirements. For example, hybrid composites, which combine different types of fibers, can strike a balance between strength, cost-effectiveness, and other properties, making them versatile for various applications [18]. Some reinforcement fibers, like carbon fibers and certain conductive nanofibers, possess high electrical conductivity. This property is exploited in applications where electrical conductivity is required, such as in electronic components and conductive textiles. The use of durable reinforcement fibers can significantly enhance the longevity of composite materials. For instance, glass fibers provide excellent resistance to environmental factors like moisture and UV radiation, making them suitable for outdoor and marine applications. Reinforcement fibers can improve the fatigue resistance of composites, allowing them to withstand cyclic loading and extending their operational lifespan [19]. This property is essential in applications exposed to repetitive stress, such as aircraft components. Table 12.2 depicts a brief overview of the reinforcing material, as well as its usual features and uses.

12.3.2 Polymer Matrices

Polymer matrices play a fundamental role in the preparation of polymer textile composites. They serve as the matrix material that encapsulates and binds the reinforcement fibers together, creating a unified composite material. Polymer matrices act as a protective enclosure for the reinforcement fibers. They surround and encapsulate the fibers, providing a stable and durable structure [24]. This encapsulation helps prevent fiber damage, wear, and environmental degradation. Polymer matrices transfer mechanical loads and stresses from one fiber to another within the composite. When external forces are applied, the matrices distribute these forces evenly across the reinforcement fibers, ensuring that the load is shared and absorbed effectively. This load-sharing mechanism helps to enhance the overall strength and stiffness of the composite. The polymer matrix essentially acts as a binder that holds the fibers in place and enables them to work together as a unified structure. Polymer matrices can improve the impact resistance of the composite [25]. When subjected to impact or dynamic loading, the matrix material absorbs and disperses the energy, preventing localized damage to individual fibers. This property is particularly important in applications requiring impact resistance, such as automotive parts. Polymer matrices protect environmental factors such as moisture, chemicals, and UV radiation. This protection helps extend the durability of the composite, especially in outdoor or harsh environments. For example, in marine applications, the polymer matrix can shield the reinforcement fibers (e.g., glass fibers) from the corrosive effects of saltwater. The choice of polymer matrix can influence the thermal properties of the composite. Some matrices are more thermally stable than others, allowing the composite to withstand high temperatures without deformation or degradation [26]. This is crucial in applications exposed to elevated temperatures, such as aircraft engine components. Polymer matrices provide adhesion and bonding between the reinforcement fibers. This ensures that the fibers are firmly held in place and maintain their orientation during manufacturing and use. Different polymer matrices based on the desired properties of the composite. For example, epoxy matrices are chosen for their excellent mechanical properties and adhesion, while thermoplastic matrices like

Table 12.2 A concise description of reinforcement material and its common properties and applications [5, 18, 20–23]

Reinforcement material	Common physical properties	Common chemical composition	Examples of applications
Carbon fiber	• High tensile strength • High modulus of elasticity • Lightweight • Low thermal expansion	• Primarily composed of carbon (C) atoms • Inert and non-reactive • High-temperature stability	• Aerospace components (e.g., aircraft parts) • Automotive components (e.g., body panels) • Sports equipment (e.g., tennis rackets) • Wind turbine blades
Aramid fiber (e.g., Kevlar)	• Exceptional tensile strength • Low weight • Excellent impact resistance	• Organic polymer composed of hydrogen, oxygen, nitrogen, and carbon • Resistant to most chemicals	• Protective gear (e.g., bulletproof vests, helmets) • Aerospace components (e.g., aircraft panels) • Sports equipment (e.g., racing sails) • Industrial applications (e.g., conveyor belts)
Glass fiber	• High tensile strength • Electrical insulating properties • Corrosion resistance	• Mostly composed of silicon dioxide (SiO2) • Inert and non-reactive • Resistant to moisture	• Construction materials (e.g., fiberglass-reinforced concrete) • Marine applications (e.g., boat hulls) • Electrical components (e.g., circuit boards) • Automotive parts (e.g., bumpers)
Natural fibers (e.g., Hemp, Flax)	• Biodegradable and sustainable • Renewable resource • Lower density compared to some synthetics	• Composed of cellulose, lignin, and hemicellulose • Generally biocompatible • Susceptible to moisture and biological degradation	• Eco-friendly composites (e.g., biodegradable packaging) • Textiles (e.g., clothing, upholstery) • Automotive interior components (e.g., door panels) • Building materials (e.g., particleboard)
Hybrid combinations	• Tailored properties based on fiber combinations	• Varied chemical compositions based on fiber types	• Customized applications based on specific requirements

PEEK offer recyclability and ease of processing [27]. The ability to customize the polymer matrix allows engineers to tailor the composite's properties to meet specific application requirements. Polymer matrices play a crucial role in the processing and fabrication of polymer textile composites. They can be melted, cured, or shaped according to the chosen manufacturing technique (e.g., layup and curing, compression molding, resin transfer molding). Table 12.3 demonstrates a detailed, condensed explanation of the Polymer Matrix's widespread properties and applications.

12.4 Additives and Nanomaterials

Additives and nanomaterials play essential roles in the preparation of polymer textile composites. These materials are incorporated into the composites to enhance their performance, modify their properties, and tailor them for specific applications. Nanomaterials like carbon nanotubes, graphene, and nanoclays are added to the polymer matrix to create nanocomposites. These nanoparticles significantly enhance properties such as mechanical strength, stiffness, and thermal and electrical conductivity [33]. Micron-sized fillers, such as silica or alumina, are used to modify the mechanical and thermal properties of the composite. They can improve properties like hardness, wear resistance, and dimensional stability. Plasticizers are added to the polymer matrix to improve its flexibility and processability. They lower the glass transition temperature of the polymer, making it easier to mold and shape during manufacturing. Additives with flame-retardant properties are incorporated to reduce the flammability of the composite [34]. They release flame-inhibiting gases or char when exposed to fire, slowing down combustion. UV stabilizer additives protect the composite from UV radiation and prevent degradation due to sunlight exposure. They are crucial in outdoor applications to maintain the composite's appearance and structural integrity. Antioxidants are used to prevent oxidative degradation of the polymer matrix, extending the composite's service life. They neutralize free radicals that can cause deterioration over time [35]. Colorants and Pigment additives are added to the polymer matrix to provide color and appearance customization. They are commonly used in consumer products and architectural applications. Materials like conductive carbon black or metallic particles are included to impart electrical conductivity to the composite. This is valuable in applications requiring EMI shielding or conductivity. Coupling agents are used to enhance the adhesion between the polymer matrix and reinforcement fibers, ensuring strong bonding and preventing delamination [36]. Lubricants and Slip Agent additives are used to reduce friction between the composite and other surfaces. They are essential in applications involving sliding or wear, such as conveyor belts. Biocides and Antimicrobial Agent additives are incorporated for applications where microbial growth or biofouling needs to be prevented, such as in medical devices or marine structures [37]. Thickeners and Rheology Modifiers: These additives help control the viscosity and flow behavior of the polymer matrix

Table 12.3 A concise description in detail of the polymer matrix with its common properties and applications

Type of polymer matrix	Common physical properties	Common chemical composition	Examples of applications [28–32]
Epoxy resin	• High mechanical strength • Excellent adhesion • Low thermal expansion	• Epoxides, often bisphenol-A and epichlorohydrin	• Aerospace components (e.g., aircraft structures) • Marine applications (e.g., boat hulls) • Composite materials (e.g., carbon fiber composites)
Polyester resin	• Ease of handling and processing • Good corrosion resistance • Low cost	• Variety of esters • Orthophthalic, isophthalic, and unsaturated polyesters are common	• Automotive components (e.g., body panels) • Construction materials (e.g., fiberglass composites) • Consumer goods (e.g., watercraft)
Vinyl ester resin	• Corrosion resistance • Good mechanical properties • Chemical resistance	• Derived from epoxy and unsaturated monocarboxylic acids	• Chemical storage tanks and pipes • Marine applications (e.g., boat parts) • Wastewater treatment components
Thermoplastic polymers (e.g., PEEK)	• Recyclability • High-temperature resistance • Chemical resistance	• Varies based on the specific thermoplastic material	• Aerospace (e.g., aircraft interior components) • Automotive (e.g., fuel system components) • Medical devices (e.g., orthopedic implants)
Phenolic resin	• Excellent heat resistance • Flame resistance • Electrical insulation	• Derived from phenol and formaldehyde	• Automotive brake components • Aircraft interior components • Electrical insulation materials

(continued)

Table 12.3 (continued)

Type of polymer matrix	Common physical properties	Common chemical composition	Examples of applications [28–32]
Polyurethane resin	• Flexibility and elasticity • Good abrasion resistance • Low-temperature properties	• Composed of isocyanate and polyol compounds	• Automotive suspension component • Footwear (e.g., shoe soles) • Coatings and adhesives
Acrylic resin	• Transparency and UV resistance • Weatherability • Good adhesion	• Composed of acrylic acid or acrylate esters	• Optical components (e.g., aircraft windows • Automotive lighting • Signage and displays

during processing, aiding in the manufacturing of complex shapes. Green Additives As sustainability becomes a greater concern, environmentally friendly additives and nanomaterials are being developed to reduce the environmental impact of composite production [38]. Table 12.4 depicts a quick explanation of additives and nanomaterials, as well as their frequent features and uses.

12.5 Types of Equipment Used to Prepare the Polymer Textile Composite

12.5.1 Equipment for Mixing and Preparing Resin

The preparation of polymer textile composites begins with the careful mixing of the polymer resin. High-shear mixers are indispensable for achieving a homogeneous blend of the resin, reinforcement fibers, fillers, and additives. These mixers effectively disperse solid components into viscous resins, ensuring uniformity in the composite [43]. The accuracy of resin mixing is facilitated by resin dispensers, which measure and combine resin and hardener components precisely. Additionally, vacuum degassing equipment is employed to remove entrapped air from the resin mixture. The absence of air bubbles is critical to prevent voids and defects in the final composite.

Table 12.4 A brief description of additives and nanomaterials with their common properties with applications

Type of material	Common physical properties	Common chemical composition	Typical applications [39–42]
Carbon nanotubes	• Exceptional strength and stiffness • High thermal and electrical conductivity	• Carbon (C) atoms arranged in nanotube structures	• Aerospace components • Electrical conductors • Sports equipment
Graphene	• Single layer of carbon atoms arranged in a hexagonal lattice • Exceptional electrical and thermal conductivity	• Carbon (C) atoms arranged in a single-layer structure	• Electronics (e.g., flexible displays) • Energy storage (e.g., batteries) • Composite materials
Nanoclays	• High aspect ratio (large surface area per unit volume) • Improved mechanical properties • Barrier properties against gases	• Natural clay minerals with nanoscale dimensions	• Packaging materials • Automotive components • Coatings
Silica nanoparticles	• High surface area • Excellent reinforcing properties • Transparency	• Silicon dioxide (SiO_2)	• Tire compounds • Transparent coatings • Adhesives
Titanium Dioxide (TiO_2) Nanoparticles	• UV-blocking properties • Improved opacity • Photocatalytic activity	• Titanium dioxide (TiO_2)	• Sunscreen formulations • Paints and coatings • Photocatalysis applications
Plasticizers	• Improved flexibility and low-temperature properties	• Various organic compounds	• PVC products (e.g., flexible PVC pipes) • Plastic films • Textiles
Flame retardants	• Reduced flammability and smoke generation • Enhanced fire resistance	• Various compounds, such as bromine-based or phosphorus-based flame retardants	• Electronics and electrical components • Building materials • Transportation components
Antioxidants	• Protection against oxidation and thermal degradation	• Various organic compounds	• Plastics, rubber, and elastomers • Automotive parts • Wire and cable insulation

(continued)

Table 12.4 (continued)

Type of material	Common physical properties	Common chemical composition	Typical applications [39–42]
Colorants and pigments	• Coloring and visual customization • UV resistance	• Various organic and inorganic compounds	• Consumer products (e.g., colored plastics) • Textiles • Architectural coatings
Biocides and antimicrobial agents	• Inhibition of microbial growth • Preservation of materials	• Various antimicrobial compounds	• Medical devices • Marine structures • Textiles
Rheology modifiers	• Control of viscosity and flow behavior during processing	• Various organic and inorganic compounds	• Coatings and paints • Adhesives and sealants • Food industry

12.5.2 Equipment for Textile Reinforcement

The textile reinforcement phase involves weaving looms and knitting machines. Weaving looms interlace warp and weft fibers to create fabric sheets with specific weave patterns. This process produces strong and flexible textile reinforcements that form the backbone of the composite's mechanical properties. Knitting machines, on the other hand, generate knitted fabrics. These fabrics provide unique properties and are frequently used as reinforcement layers in composites. Both of these machines are vital for creating the textile foundation that gives the composite its strength [43].

12.5.3 Cutting Equipment

Achieving precise shapes and dimensions is paramount in composite manufacturing. Cutting equipment, including cutting tables and computer numerical control (CNC) cutting machines, plays a crucial role in trimming and shaping textile reinforcement. These machines ensure accuracy and consistency in the dimensions of the reinforcement layers, which is essential for achieving the intended composite properties. Properly cut reinforcement materials are then ready for layup [44].

12.5.4 Layup and Molding Equipment

Layup and molding equipment form the core of the composite manufacturing process. Molds and tooling provide the structure around which the composite takes shape.

Molds can be crafted from various materials, including metals, composites, and polymers, depending on the specific requirements of the application. Release agents are applied to molds to prevent the composite material from adhering to the mold surface, ensuring easy demolding [45]. Breather and bleeder materials, porous layers placed atop the composite layup, serve to absorb excess resin and aid in achieving uniform resin distribution. Vacuum bagging and resin infusion systems are utilized to create a vacuum environment around the composite layup. These systems remove air, compact the composite, and infuse resin under controlled pressure, ensuring proper consolidation and distribution of the resin [46].

12.5.5 Curing Equipment

The curing phase is crucial for solidifying the composite. Ovens are commonly employed to cure composites at precise and controlled temperatures. This heating process allows the resin matrix to harden, creating a strong and durable material. For advanced applications, autoclaves are used. Autoclaves provide not only controlled temperature but also controlled pressure. This combination is particularly vital in aerospace applications, ensuring a high-quality composite with exceptional mechanical properties [47].

12.5.6 Quality Control and Testing Instruments

Ensuring the quality and performance of polymer textile composites is paramount. Universal Testing Machines (UTM) are used to evaluate mechanical properties such as tensile strength, compressive strength, and flexural strength. Dynamic Mechanical Analyzers (DMA) assess the viscoelastic properties of composite materials, providing insights into their behavior under various conditions. Thermogravimetric Analyzers (TGA) analyze the thermal stability and decomposition behavior of composites by measuring changes in weight with temperature. Differential Scanning Calorimeters (DSC) help identify thermal transitions like the glass transition temperature [48]. Microscopes allow for microstructural analysis, enabling the inspection of fiber-matrix bonding and the detection of defects. Ultrasonic testing equipment aids in non-destructive evaluation, detecting defects, delaminations, or voids. Moisture content analyzers measure the moisture content in composites, which can affect properties like strength and dimensional stability. All of these instruments contribute to quality assurance and performance validation.

12.5.7 Safety Equipment

Safety is a paramount concern when working with chemicals, resins, and composite materials. Personal protective equipment (PPE) is a critical component of the preparation process. Gloves protect hands, safety glasses shield eyes, and respirators ensure that workers breathe clean air. These precautions are essential for safeguarding personnel throughout the manufacturing process [49].

12.5.8 Environmental Chambers

Environmental chambers are utilized to conduct various tests, evaluating the composite's durability under different conditions. These chambers can simulate temperature extremes, humidity levels, and UV radiation. By subjecting the composite to environmental testing, manufacturers gain insights into its performance and resilience under real-world scenarios. In summary, the preparation of polymer textile composites is a complex and precise process that relies on a wide array of equipment and instruments. These tools ensure the proper combination of materials, the creation of a strong and reliable structure, and the validation of performance and quality [50]. As technology advances, equipment continues to evolve, offering improved precision, efficiency, and control in composite manufacturing processes. Ultimately, the selection of equipment depends on factors such as the composite type, manufacturing process, and specific application requirements. Table 12.5 contains detailed descriptions of the equipment and tools needed to create polymer textile composites, as well as their principal uses.

12.6 Preparation Techniques of Polymer Textile Composite

In the realm of preparing polymer textile composites, a compelling comparison can be drawn between the traditional methods that have stood the test of time and the modern approaches that have emerged as technology advances. Traditional methods, such as hand layup and vacuum bagging, have long been valued for their simplicity and accessibility. Hand layup, for instance, relies on skilled artisans to manually layer reinforcement fabrics and apply resin. It's a technique that's relatively easy to understand and implement, making it suitable for small-scale production and prototyping [56]. Vacuum bagging, often used in conjunction with hand layup, enhances the quality of composite fabrication by leveraging vacuum pressure to consolidate the materials, resulting in better fiber wetting and reduced void content. However, these traditional methods have their limitations. They tend to be labor-intensive, reliant on the skill and experience of the operator, and can lead to variations in quality and fiber orientation control [57]. In contrast, modern methods like Resin Transfer Molding

Table 12.5 Detailed descriptions of the equipment and instruments used to prepare polymer textile composites, along with their primary applications [51–55]

Equipment and instruments	Purpose and function	Typical applications
High-shear mixers	Blend and disperse reinforcement fibers and additives	Mixing of polymer resin and additives
Resin dispensers	Accurately measure and mix resin components	Precise control of resin-to-hardener ratio
Vacuum degassing equipment	Remove air bubbles from resin mixture	Ensuring air-free composite
Weaving looms	Interlace warp and weft fibers to create fabric sheets	Fabrication of textile reinforcements
Knitting machines	Generate knitted textile structures	Production of knitted reinforcement
Cutting tables and CNC machines	Trim and shape reinforcement materials	Precision cutting of prepreg materials
Molds and tooling	Provide structure for composite layup	Shaping composite components
Release agents	Prevent composite from sticking to mold	Easy demolding of cured parts
Breather and bleeder materials	Absorb excess resin and ensure compaction	Uniform resin distribution in layup
Vacuum bagging systems	Create a vacuum environment around the layup	Remove air, compact composite
Resin infusion systems	Infuse resin into layup under vacuum pressure	Controlled resin distribution
Ovens	Cure composites at controlled temperatures	Solidifying the resin matrix
Autoclaves	Control temperature and pressure during curing	Aerospace and critical applications
Universal Testing Machines (UTM)	Evaluate mechanical properties of composites	Tensile, compressive, flexural testing
Dynamic Mechanical Analyzers (DMA)	Assess viscoelastic properties of composite materials	Analysis of material behavior
Thermogravimetric Analyzers (TGA)	Analyze thermal stability and decomposition behavior	Study thermal characteristics
Differential Scanning Calorimeters (DSC)	Identify thermal transitions of composites	Detection of glass transition temperature
Microscopes	Examine microstructure and defects in composites	Quality control and defect detection
Ultrasonic testing equipment	Non-destructive evaluation of composite components	Detect defects and delaminations
Moisture content analyzers	Measure moisture content in composites	Assess material moisture levels

(continued)

Table 12.5 (continued)

Equipment and instruments	Purpose and function	Typical applications
Personal Protective Equipment (PPE)	Ensure safety of personnel while handling materials	Protection against chemical exposure
Environmental chambers	Simulate environmental conditions for testing	Assess durability under real-world scenarios

(RTM) and Compression molding have brought automation and precision to the forefront of composite manufacturing. RTM, for instance, automates the resin injection process, allowing for higher production rates and reduced labor requirements. It offers better control over resin impregnation, resulting in consistently low void content and enhanced composite quality. Compression molding, another modern technique, shares the benefits of efficiency and reduced labor while also offering control over resin distribution, making it particularly well-suited for industries like automotive and construction [58]. However, modern methods typically necessitate a higher initial investment in equipment and tooling, which can pose a challenge for smaller-scale production or prototyping efforts.

The choice between traditional and modern methods hinges on several factors, including production volume, part complexity, budget constraints, and the desired level of automation and quality control. Many manufacturers and researchers today opt for a hybrid approach, leveraging the strengths of both traditional and modern techniques to effectively meet project-specific requirements [59]. As technology continues to advance, composite manufacturing methods evolve, providing an increasingly diverse array of options to create polymer textile composites tailored to a wide range of applications. Table 12.6 shows several methodologies linked with the creation of polymer textile composites, as well as a succinct explanation of their key processes. Figure 12.1 illustrates the different additives and techniques employed in the preparation of polymer textile composites.

12.7 Application of Polymer Textile Composite

Traditional and modern applications of polymer textile composites represent two distinct eras in the evolution of materials technology, each with its unique characteristics and implications for various industries. These societies harnessed natural fibers like cotton, wool, and flax, combined with rudimentary resins and adhesives, to create textiles for clothing and shelter. The utilization of these natural materials was born out of necessity, as ancient humans sought ways to protect themselves from the elements and fashion functional clothing from locally available resources [69]. One of the most enduring traditional applications was the use of wool in textiles. Wool offered remarkable insulation, keeping individuals warm in harsh climates while allowing for breathability, making it a staple material for traditional clothing worldwide. Moreover, indigenous communities harnessed their knowledge of natural fibers to create

Table 12.6 Various techniques associated with the production of polymer textile composites, along with a concise explanation of their essential steps

Preparation technique	Description	Key steps	Typical applications [60–68]
Hand layup	Manual layering of reinforcement and resin	• Prepare mold/tooling • Apply release agent • Place reinforcement fabric • Apply resin • Compact and cure	Prototyping, small-scale production
Vacuum bagging	Consolidation under vacuum pressure	• Layup reinforcement • Seal vacuum bag • Evacuate air • Infuse resin • Cure composite	Aerospace, marine applications
Resin transfer molding (RTM)	Injection of resin under pressure	• Place dry fabric in mold • Seal mold • Inject resin • Impregnate fabric • Cure composite	Automotive, industrial components
Compression molding	Heat and pressure consolidation	• Pre-cut reinforcement fabric • Place fabric in mold • Heat and compact • Cool and solidify	Automotive parts, electrical insulators
Filament winding	Continuous winding of resin-impregnated fibers	• Impregnate fibers with resin • Wind around mandrel • Control tension and winding angle • Continuous curing	Pressure vessels, pipes and tubing
Pultrusion	Continuous pulling of resin-impregnated fibers	• Pull fibers through a resin bath • Mold into shape • Continuous curing	Structural shapes, rod and tube profiles
Prepreg processing	Use of pre-impregnated reinforcement fabrics	• Cut and layer prepreg sheets • Apply heat and pressure • Cure composite	Aerospace components, sports equipment
Infusion techniques	Controlled resin infusion into dry fabric stack	• Place dry fabric in mold • Create vacuum environment • Infuse resin • Cure composite	Various, depending on specific method

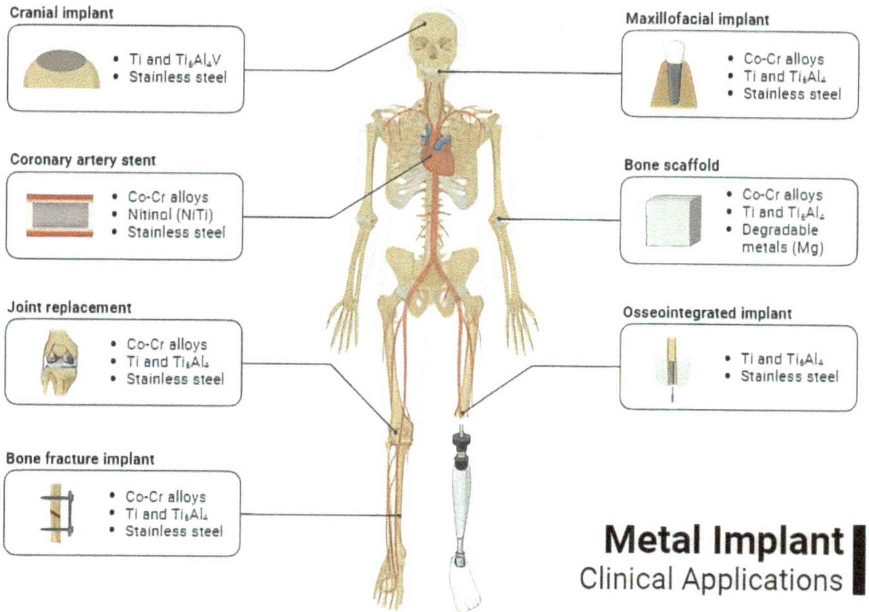

Fig. 12.1 Different Biomedical applications of polymer textile composite in the form of Metal implants [88]

composite materials for constructing shelters and crafting tools. For example, a combination of plant fibers and resins could be used to reinforce structures, enabling them to withstand environmental challenges [70]. In stark contrast, modern applications of polymer textile composites represent the culmination of centuries of technological advancement. The advent of synthetic polymers, coupled with sophisticated manufacturing processes, has catapulted these materials into the forefront of innovation across various industries. The aerospace industry epitomizes the transformative potential of modern applications. Carbon fiber-reinforced polymer composites have become indispensable in aircraft manufacturing [71]. These composites are engineered with precision to achieve exceptional strength-to-weight ratios, enabling the development of lightweight and fuel-efficient aircraft. Advanced composite materials find extensive use in constructing critical components, from aircraft fuselages to wing structures. These materials have revolutionized air travel, making it more sustainable and cost-effective. Similarly, the automotive sector has witnessed a paradigm shift with modern polymer textile composites [72]. Carbon fiber-reinforced body panels not only enhance structural integrity but also reduce overall vehicle weight, leading to improved fuel economy and reduced emissions. In the era of electric vehicles, these composites contribute to the development of lightweight and energy-efficient batteries, further advancing sustainable transportation solutions. The medical field has also been revolutionized by modern polymer textile composites, particularly biocompatible materials [73]. These advanced composites are used in the development of artificial organs, medical implants, and tissue engineering. They offer a harmonious blend of biocompatibility, strength, and the ability to mimic natural tissues, leading to life-altering medical advancements. Table 12.7 depicts numerous kinds of polymer textile composites as well as their diverse uses.

12.7.1 Biomedical Applications

Polymer textile composites have emerged as a game-changing class of materials in the field of biomedicine, ushering in a new era of innovative solutions and enhanced patient care. These advanced materials combine the flexibility and versatility of polymers with the strength and structure of textiles, making them exceptionally well-suited for a wide range of biomedical applications [81]. Biocompatible polymer composites, often reinforced with ceramics or nanoparticles, have become the backbone of artificial joints, bone scaffolds, and dental implants. These materials possess the ideal combination of strength and durability required for long-lasting medical implants, while their biocompatibility ensures they interact harmoniously with the human body. This reduces the risk of adverse reactions, inflammation, and the need for frequent replacements, ultimately improving the quality of life for countless patients. In the field of tissue engineering, polymer textile composites are instrumental in creating scaffolds that mimic the natural extracellular matrix [82]. These scaffolds provide a three-dimensional framework for the growth of living cells and facilitate tissue regeneration. Researchers are harnessing the porous structure and tailored mechanical properties of these composites to engineer replacement tissues and organs, offering hope to patients in need of transplants. Moreover, polymer textile composites have found a crucial role in wound care. Their breathability and flexibility make them ideal candidates for wound dressings and bandages [83]. Additionally, many polymer composites possess antimicrobial properties, which can help prevent infections and expedite the healing process. These materials have become indispensable tools for healthcare professionals in managing various types of wounds, from minor cuts to complex surgical incisions. In the domain of diagnostics and monitoring, polymer textile composites are at the forefront of innovation [84]. They are being used to create biosensors and wearable medical devices that monitor vital signs, biomarkers, and other health-related data in real time. These devices provide valuable insights into a patient's condition, allowing for early intervention and personalized healthcare. They have the potential to revolutionize the way we manage chronic diseases and enhance preventive care. As the field of biomedicine continues to evolve, polymer textile composites are poised to play an even more significant role in driving advancements [85]. Their versatility, biocompatibility, and multifunctionality make them an invaluable resource for researchers, engineers, and medical professionals alike. Whether it's improving the longevity of medical implants, enabling tissue regeneration, or enhancing diagnostics, polymer textile composites are contributing to the development of cutting-edge biomedical solutions that have the power to

Table 12.7 Different classes of polymer textile composite along with their various applications

Serial number	Category of polymer textile composite	Subcategories	Applications [74–80]
1. Fiber-reinforced polymer Composites (FRPC)			
1	Carbon Fiber Reinforced Polymer (CFRP)	Unidirectional CFRP	Aerospace components, automotive parts, sports equipment
	Woven CFRP	Structural components, aircraft interiors	
2	Glass Fiber Reinforced Polymer (GFRP)	E-glass GFRP	Construction, marine, automotive, wind energy
		S-glass GFRP	Aerospace, defense, sports equipment
3	Aramid Fiber Reinforced Polymer (AFRP)	Kevlar AFRP	Ballistic protection, structural reinforcements
		Twaron AFRP	Industrial, automotive, composites
4	Natural Fiber Reinforced Polymer (NFRP)	Hemp NFRP	Textiles, packaging, furniture
		Flax NFRP	Automotive, building materials
2. Polymer-ceramic composites			
1	Polymer-ceramic matrix composites	Alumina matrix	Wear-resistant coatings, dental restorations
		Zirconia matrix	Aerospace components, medical implants
2	Polymer-ceramic nanocomposites	Hydroxyapatite	Biomedical implants, electrical insulators
		Silica	Optical components, sensors
3. Polymer-metal composites			
1	Polymer-metal matrix composites	Aluminum matrix	Lightweight structural components, heat sinks
		Magnesium matrix	Aerospace components, automotive parts
2	Polymer-metal nanocomposites	Copper nanoparticles	Aerospace components, electromagnetic shielding
		Nickel nanoparticles	Electronics, corrosion-resistant coatings
4. Polymer-carbon nanotube composites			
1	Polymer-CNT matrix composites	Single-walled CNTs	Conductive materials, sensors, aerospace components

(continued)

Table 12.7 (continued)

Serial number	Category of polymer textile composite	Subcategories	Applications [74–80]
		Multi-walled CNTs	Energy storage, electronics, nanocomposites
2	Polymer-CNT Nanocomposites	Polymer-CNT nanocomposites	Fuel cells, sensors, advanced materials
		Polymer-CNT composite films	Flexible electronics, coatings
5. Polymer-graphene composites			
1	Polymer-graphene matrix composites	Graphene oxide	Electrically conductive materials, supercapacitors
		Reduced graphene oxide	Sensors, energy storage, coatings
2	Polymer-Graphene Nanocomposites	Polymer-graphene nanocomposites	Flexible electronics, sensors, coatings
		Graphene-based coatings	Corrosion protection, membranes
6. Polymer–clay composites			
1	Polymer–clay matrix composites	Montmorillonite clay	Packaging materials, automotive parts
		Kaolinite clay	Coatings, adhesives, textiles
2	Polymer–clay nanocomposites	Polymer–clay nanocomposites	Water purification membranes, gas barriers
		Polymer–clay nanocomposite films	Food packaging, barrier materials
7. Conductive polymer composites			
1	Polymer-carbon black composites	Carbon black composites	Antistatic materials, EMI shielding
2	Polymer-conductive polymer composites	Polyaniline composites	Flexible circuits, sensors, wearable tech
		Polypyrrole composites	Bioelectronics, actuators
8. Smart polymer composites			
1	Polymer shape memory composites	Shape memory polymers	Biomedical devices, aerospace, robotics
2	Polymer piezoelectric composites	Piezoelectric polymers	Energy harvesting, sensors, actuators
		PVDF-based composites	Medical sensors, touch-sensitive devices
9. Polymer nanocomposites			

(continued)

Table 12.7 (continued)

Serial number	Category of polymer textile composite	Subcategories	Applications [74–80]
1	Polymer-particle nanocomposites	Polymer-ceramic nanocomposites	Coatings, packaging, automotive
		Polymer-metal nanocomposites	Antimicrobial materials, conductive coatings
2	Polymer-filler nanocomposites	Polymer-silica nanocomposites	Flame retardant materials, adhesives
		Polymer-carbon nanotube composites	Structural materials, sensors

transform and save lives [86, 87]. Various biomedical applications of polymer textile composites in the form of metal implants are illustrated in Fig. 12.1.

12.7.2 Medical Implants

The application of polymer textile composites in medical implants represents a groundbreaking advancement in the field of healthcare. These advanced materials, which blend the flexibility and versatility of polymers with the strength and structure of textiles, have transformed the landscape of medical device manufacturing. Medical implants, such as artificial joints (hip and knee replacements), dental implants, and bone scaffolds, have greatly benefited from the unique properties of polymer textile composites [87]. In the realm of orthopedics, where the demand for durable and biocompatible materials is paramount, these composites have become indispensable. They offer the perfect balance of mechanical integrity and biocompatibility, ensuring that implanted devices can withstand the rigors of daily life within the human body while minimizing the risk of adverse reactions or implant failure. This has translated to longer-lasting implants and improved patient outcomes, enhancing the mobility and quality of life for individuals suffering from joint degeneration or bone-related issues. Furthermore, the use of polymer textile composites in medical implants extends beyond orthopedics [89]. These materials have found applications in dental implants, where their strength and biocompatibility make them an ideal choice for replacing missing teeth. Dental implants made from these composites provide patients with aesthetically pleasing, functional, and long-lasting solutions for tooth restoration. In tissue engineering, polymer textile composites serve as scaffolds that facilitate the growth of living cells, enabling the regeneration of damaged or lost tissues and organs [71]. This groundbreaking technology has the potential to revolutionize organ transplantation by providing a viable alternative through tissue engineering and regenerative medicine. Overall, the application of polymer textile composites in medical implants represents a significant leap forward in the quest

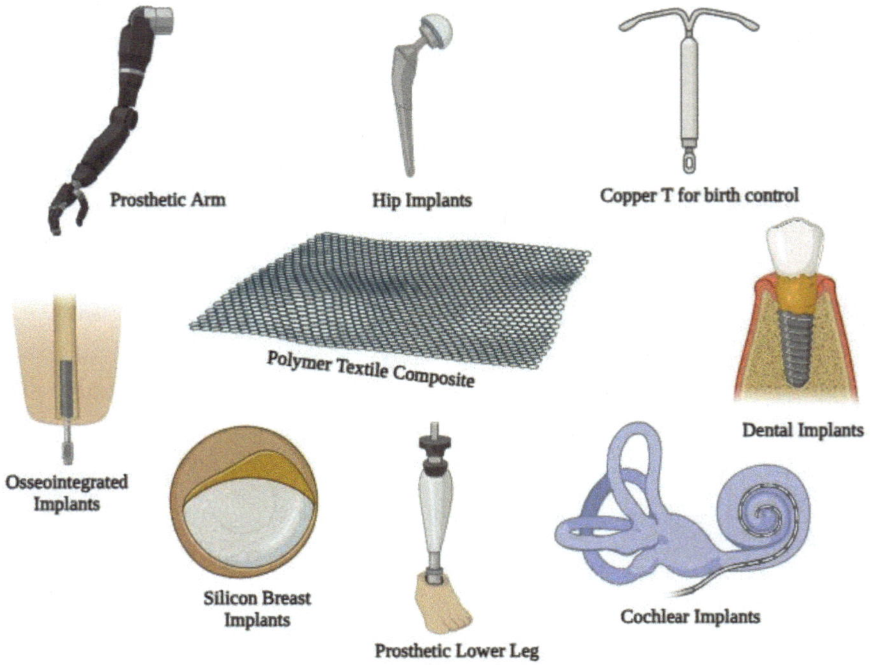

Fig. 12.2 Graphical representation of different medical implants for organ functions (Created with www.biorender.com)

to enhance patient care and well-being. These materials contribute to the development of safer, longer-lasting, and more biocompatible medical devices, ultimately improving the quality of life for countless individuals in need of implantable solutions [90]. Figure 12.2 illustrates the graphical depiction of various medical implants utilized for organ functionality.

12.7.3 Tissue Engineering

The application of polymer textile composites in tissue engineering is at the forefront of biomedical innovation, offering promising solutions to some of the most pressing challenges in regenerative medicine. These advanced materials, which seamlessly blend the versatility of polymers with the structural support of textiles, are instrumental in creating functional biological tissues and organs [91]. In the realm of tissue engineering, polymer textile composites serve as meticulously designed scaffolds that provide a three-dimensional framework for the growth and development of living cells. Researchers have harnessed the unique properties of these composites to mimic the intricate architecture of natural tissues, facilitating the regeneration of damaged or degenerated body parts. One of the most notable applications of polymer textile composites in tissue engineering lies in the development of replacement tissues and organs [92]. Whether its crafting heart valves, skin grafts, or even entire organs like bladders or tracheas, these composites offer the mechanical integrity and biocompatibility required for successful transplantation. By combining the porous structure of textile composites with polymers that encourage cell adhesion and growth, scientists have achieved remarkable success in creating functional biological tissues that integrate seamlessly with the recipient's body. Moreover, the customization potential of polymer textile composites in tissue engineering is a game-changer [93]. These materials can be tailored to meet the specific mechanical, biochemical, and structural requirements of different tissues and organs. For instance, in bone tissue engineering, composites may be designed to mimic the stiffness and mineral content of natural bone, while in cardiac tissue engineering, they can replicate the elasticity and electrical conductivity of heart muscle. This adaptability enables the precise engineering of tissues to suit individual patient needs, potentially reducing rejection rates and post-transplant complications [94]. Additionally, polymer textile composites have also found applications in wound care within the broader scope of tissue engineering. Wound dressings made from these materials provide a conducive environment for tissue regeneration while offering breathability, flexibility, and sometimes even antimicrobial properties to prevent infections. These dressings have significantly improved the wound healing process, particularly for chronic wounds and burns, by facilitating faster and more efficient tissue repair [95]. Figure 12.3 illustrates the potential of augmenting the electrical conductivity of scaffolds in tissue engineering by incorporating diverse conductive materials through in-situ blending synthetic processes. This method enables the seamless integration of conductive polymers or nanoparticles, such as polypyrrole or graphene, into the scaffold matrix during its fabrication. The presence of these conductive elements is anticipated to promote essential cellular processes, including adhesion, proliferation, and differentiation, thereby serving as integral components in the creation of biomimetic and functional tissue engineering constructs.

12.7.4 Wound Dressing

The application of polymer textile composite wound dressings has transformed the landscape of wound care, offering innovative solutions that promote faster and more effective healing while minimizing the risk of complications. These advanced materials, born from the synergy of polymers and textile substrates, have emerged as a vital component in the management of various types of wounds, from minor cuts and abrasions to chronic ulcers and burns. Polymer textile composite wound dressings are designed to create an optimal environment for the wound healing process [97]. They are breathable, allowing for the exchange of oxygen and moisture between the wound site and the external environment, which is essential for tissue repair. Additionally, these dressings are highly flexible, conforming to the contours of the body and ensuring patient comfort. This flexibility is particularly advantageous for

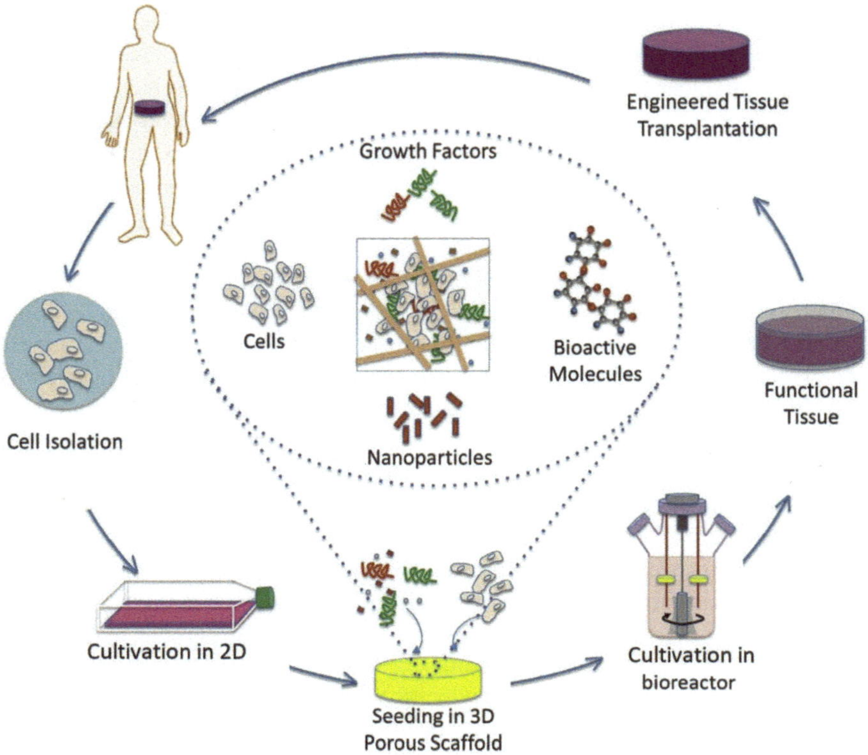

Fig. 12.3 Tissue engineering technology by seeding in 3D porous scaffold for building a functionalized tissue [96]. Open access under a CC BY 4.0 license

complex wound geometries or areas that require constant movement, such as joints. One of the remarkable features of polymer textile composite dressings is their ability to manage wound exudate effectively. Exudate, the fluid that oozes from wounds, can carry harmful bacteria and impede the healing process if not managed properly [71]. These dressings are engineered to absorb and retain exudate, maintaining a moist wound environment that supports tissue regeneration while minimizing the risk of infection. Some composite dressings also have antimicrobial properties, further enhancing their infection-fighting capabilities. Beyond their moisture management and infection control capabilities, polymer textile composite wound dressings can be tailored to address specific wound types. For instance, in the treatment of chronic wounds like diabetic ulcers, these dressings may incorporate bioactive agents or growth factors that stimulate tissue repair and angiogenesis, the formation of new blood vessels [98]. In burn care, they may provide a cooling effect to relieve pain and minimize damage caused by excessive heat. Moreover, these dressings often come with a non-adherent surface that prevents them from sticking to the wound bed, reducing pain and trauma during dressing changes. This feature is particularly advantageous for patients with sensitive or fragile skin. The versatility of polymer

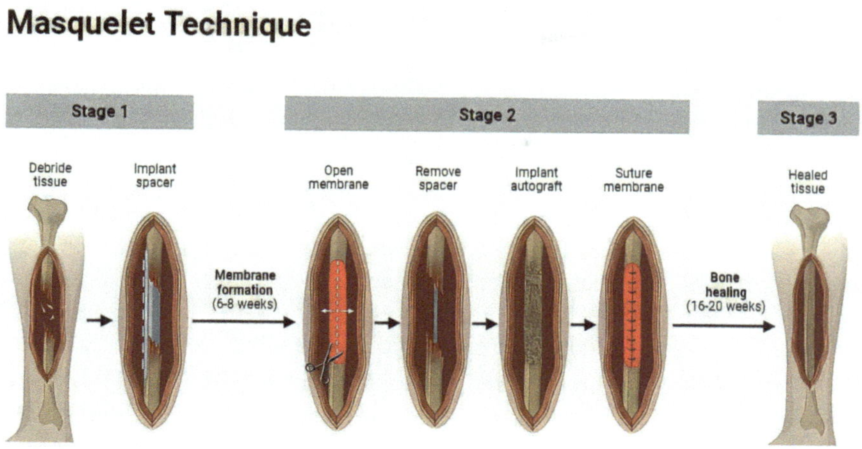

Fig. 12.4 Wound healing process with the implication of polymer textile composite material by Masquelet technique [101]

textile composite wound dressings has led to their adoption in a wide range of clinical settings, including hospitals, clinics, and home care environments [99]. Their ability to facilitate efficient wound healing, reduce the frequency of dressing changes, and improve patient comfort has made them a preferred choice among healthcare professionals. As ongoing research and technological advancements continue to enhance their performance and expand their applications, polymer textile composite wound dressings are poised to play an increasingly pivotal role in modern wound care, ultimately improving the quality of life for patients worldwide [100]. Figure 12.4 illustrates the depiction of the wound healing process involving the utilization of polymer textile composite material through the Masquelet technique.

12.7.5 Drug Delivery System

The application of polymer textile composite drug delivery systems represents a transformative approach in the field of pharmaceuticals and healthcare. These advanced materials, resulting from the ingenious fusion of polymers with textile substrates, have revolutionized the way medications are administered and released within the human body. The primary objective of these drug delivery systems is to improve therapeutic outcomes by precisely controlling the release rate and distribution of pharmaceutical agents, ultimately optimizing treatment efficacy while minimizing potential side effects [102]. One of the notable advantages of polymer textile composite drug delivery systems is their capacity for controlled drug release. By incorporating medications into these composite materials, pharmaceutical companies can engineer precise drug delivery profiles tailored to the specific needs of patients. Whether it's the sustained release of pain medication to manage chronic pain or the targeted delivery of chemotherapeutic agents to cancer cells, these systems allow for the gradual, controlled release of drugs over time. This not only ensures that therapeutic levels of medication are consistently maintained but also reduces the risk of overdosing and minimizes fluctuations in drug concentrations within the body [103]. Furthermore, these drug delivery systems offer exceptional versatility in terms of administration routes. Whether through oral, transdermal, or implantable routes, polymer textile composites can be adapted to suit the most appropriate mode of drug delivery for a particular treatment. Transdermal patches, for example, use these composites to deliver medications through the skin, bypassing the digestive system and minimizing gastrointestinal side effects. Polymer textile composite drug delivery systems also excel in enhancing patient adherence to medication regimens [104]. By reducing the frequency of dosing or the need for multiple administrations, these systems improve patient convenience and compliance. This is particularly crucial for managing chronic conditions, where consistent adherence to prescribed medications is paramount to achieving positive health outcomes. Moreover, these systems are instrumental in minimizing the adverse effects of certain drugs by allowing for targeted drug delivery. In cancer treatment, for instance, composite systems can be designed to release chemotherapy agents directly at tumor sites, sparing healthy tissues and reducing side effects [105]. Similarly, they can be employed in the treatment of localized pain, such as joint injections for arthritis, offering targeted pain relief without affecting the entire body. The representation of the implication of natural polymers in the fabrication of polymer composites for drug delivery systems, specifically for oral, transdermal, and parenteral routes, is depicted in Fig. 12.5.

12.7.6 Diagnostic

The application of polymer textile composites in diagnostic tools represents a remarkable convergence of materials science and healthcare technology, offering innovative solutions for the early detection and monitoring of various medical conditions. These advanced materials, forged by the fusion of polymers and textile substrates, have brought about significant advancements in the development of diagnostic devices and biosensors, revolutionizing the way we approach disease diagnosis and management [107]. One of the key advantages of polymer textile composite-based diagnostic tools lies in their ability to create highly sensitive and specific biosensors. These biosensors are designed to detect specific biomarkers or analytes in bodily fluids, such as blood, urine, or saliva, which can serve as indicators of various medical conditions. By precisely controlling the composition and structure of these composites, researchers can engineer biosensors with tailored properties that facilitate rapid and accurate detection [108]. This is particularly valuable in the early diagnosis of diseases like diabetes, infectious diseases, and certain cancers, where timely detection is critical for effective treatment and management. Polymer textile composites also offer a versatile platform for the development of wearable medical devices that can continuously

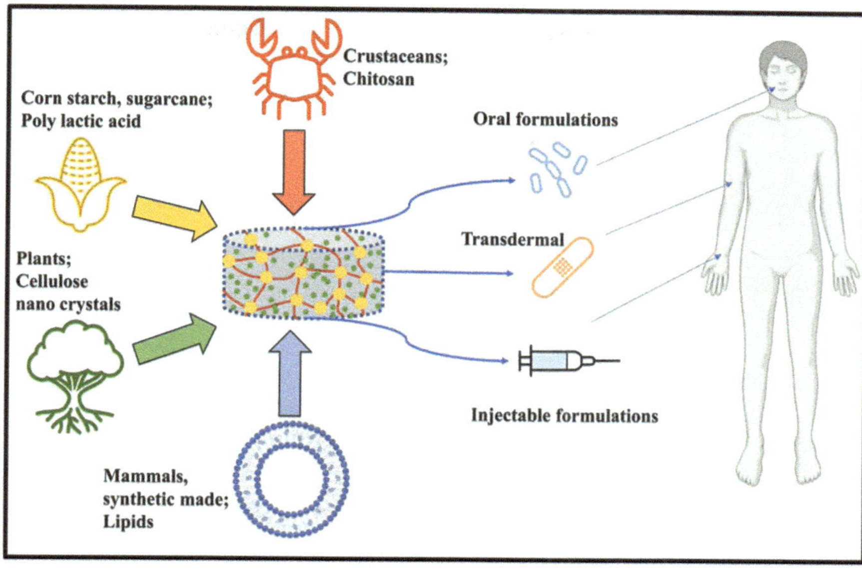

Fig. 12.5 Implication of natural polymer for making polymer composite for drug delivery systems (Oral, transdermal, and parenteral) [106] Adapted from Ref. 108 with permission from [Elsevier], Copyright [2023]

monitor vital signs and health parameters in real time. These devices are designed to seamlessly integrate with the human body, providing comfort and mobility to patients while continuously collecting data. For example, smart textiles incorporating these composites can monitor heart rate, blood pressure, glucose levels, and other vital signs, transmitting this information to healthcare providers for remote monitoring and timely intervention [107]. Such wearable diagnostic tools have the potential to revolutionize the management of chronic diseases and improve the overall quality of patient care. Furthermore, the adaptability of polymer textile composites allows for the creation of point-of-care diagnostic devices that can be easily used by healthcare professionals or even individuals themselves. These handheld devices can provide rapid and accurate results, reducing the need for complex laboratory testing and enabling swift decision-making in clinical settings [108]. Applications range from pregnancy tests to infectious disease diagnostics, providing critical information to guide treatment and care. In addition to their diagnostic capabilities, polymer textile composite-based tools also have the potential to serve as drug delivery systems, further enhancing the field of personalized medicine. By incorporating drug-releasing components into these devices, they can not only detect health parameters but also administer medications in response to specific diagnostic results, offering a truly integrated approach to healthcare [64]. Figure 12.6 illustrates a visual depiction of biosensors utilized for the purpose of diagnosing and detecting various acute and chronic diseases. These biosensors are constructed using a composite material consisting of polymer and textile components.

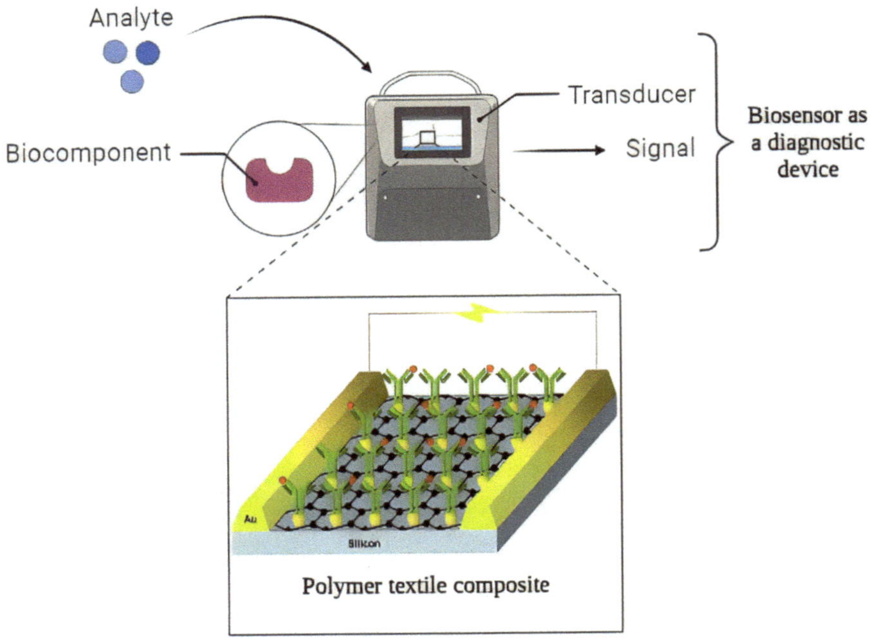

Fig. 12.6 Biosensor for diagnosis and detection of several acute and chronic diseases made from polymer textile composite (Created with www.biorender.com)

12.7.7 Surgical Instruments

The application of polymer textile composites in surgical instruments signifies a significant advancement in modern surgery, offering numerous benefits to both healthcare professionals and patients. These innovative materials, formed through the integration of polymers and textile substrates, have become instrumental in the development of surgical instruments that enhance precision, reduce tissue trauma, and improve overall surgical outcomes [109]. One of the primary advantages of polymer textile composite-based surgical instruments is their exceptional combination of strength and flexibility. These instruments can be designed to possess the necessary rigidity for precise cutting, suturing, or tissue manipulation while maintaining the flexibility required for navigating complex anatomical structures. This attribute is especially valuable in minimally invasive surgery, where surgeons need instruments capable of accessing and working within confined spaces with minimal disruption to surrounding tissues. Additionally, the lightweight nature of polymer textile composite instruments reduces the physical strain on surgeons during lengthy procedures, contributing to reduced fatigue and improved ergonomics [110]. This results in more precise and steady movements, ultimately enhancing the safety and efficiency of surgical interventions. Furthermore, these materials are resistant to corrosion and can withstand the rigors of sterilization processes, ensuring the longevity and reliability of surgical instruments. This durability is particularly important in maintaining the integrity of instruments throughout their lifecycle, contributing to cost-effectiveness in healthcare settings [111]. Moreover, polymer textile composite instruments can be designed with specialized features to address specific surgical needs. For example, laparoscopic instruments used in abdominal surgery often incorporate these materials for their combination of strength and lightweight properties. Likewise, robotic-assisted surgery platforms, which rely on highly flexible and precise instruments, frequently employ polymer textile composites to achieve the desired levels of functionality [112]. In neurosurgery, ophthalmic surgery, and other delicate procedures, the ability to manipulate tissues gently and precisely is critical. Instruments made from these advanced materials provide the necessary dexterity while minimizing tissue damage and post-operative complications. The adaptability and versatility of polymer textile composites in surgical instruments continue to drive innovation in the field [113]. As surgical techniques evolve, these materials enable the development of instruments that cater to the ever-expanding demands of modern medicine, ultimately leading to improved patient care, reduced recovery times, and enhanced surgical outcomes. Figure 12.7 depicts a compilation of images showcasing surgical instruments composed of polymer textile composites, which have been developed to enhance medical diagnostics and treatment procedures.

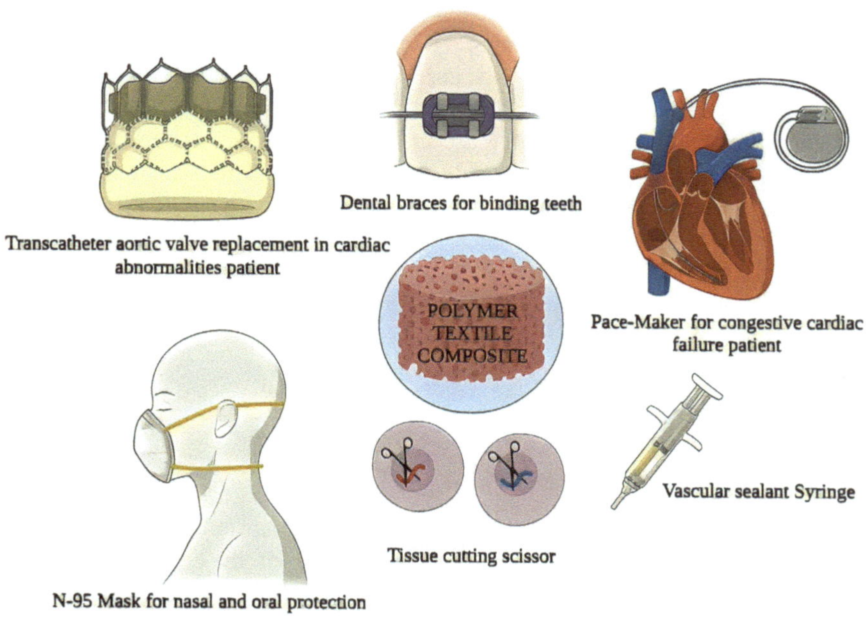

Fig. 12.7 Polymer textile composite-based surgical instruments for better medical diagnosis and treatments (Created with www.biorender.com)

12.7.8 Biodegradable Sutures

The application of polymer textile composites in biodegradable sutures represents a significant advancement in surgical materials, offering a blend of strength, biocompatibility, and sustainability. These innovative sutures are designed to facilitate wound closure while gradually breaking down within the body, eliminating the need for suture removal and reducing tissue irritation. The primary advantage of polymer textile composite-based biodegradable sutures lies in their capacity to provide robust wound closure during the critical early stages of healing when tissues are vulnerable to infection and mechanical stress [114]. These sutures maintain their tensile strength for an essential period and then naturally degrade over time, typically through hydrolysis, into biocompatible byproducts that the body can safely absorb or excrete. The use of polymer textile composites in biodegradable sutures allows for precise control over the suture's degradation rate, ensuring it aligns with the tissue's healing process. This customization is especially valuable in surgeries where extended wound support is required, such as in complex orthopedic procedures, gastrointestinal surgeries, or tissue reconstruction. As the suture degrades gradually, it encourages the natural healing process and minimizes the risk of complications related to suture retention [115]. Biodegradable sutures made from polymer textile composites are particularly well-suited for delicate and sensitive tissues, such as those found in ophthalmic, pediatric, and cardiovascular surgeries. Their soft and pliable nature minimizes tissue trauma and irritation, reducing post-operative discomfort and the potential for scarring. Additionally, the absence of a need for suture removal procedures enhances patient comfort and convenience while simplifying post-operative care. Furthermore, the sustainability aspect of these sutures is noteworthy [116]. The use of biodegradable materials reduces environmental impact by eliminating the need for additional interventions or waste associated with suture removal. This aligns with the growing focus on environmentally responsible medical practices. Figure 12.8 illustrates a visual depiction of a Biodegradable suture composed of a Polymer composite, specifically designed for the purpose of conducting surgical procedures in rats.

12.7.9 Tissue Reinforcement

The application of polymer textile composites in tissue reinforcement represents a pivotal development in medical procedures, particularly in the context of strengthening weakened or damaged biological tissues. These innovative materials, resulting from the amalgamation of polymers and textile substrates, offer a multifaceted approach to enhancing tissue support and durability in various medical scenarios [117]. One prominent application is in the field of hernia repair, where surgical

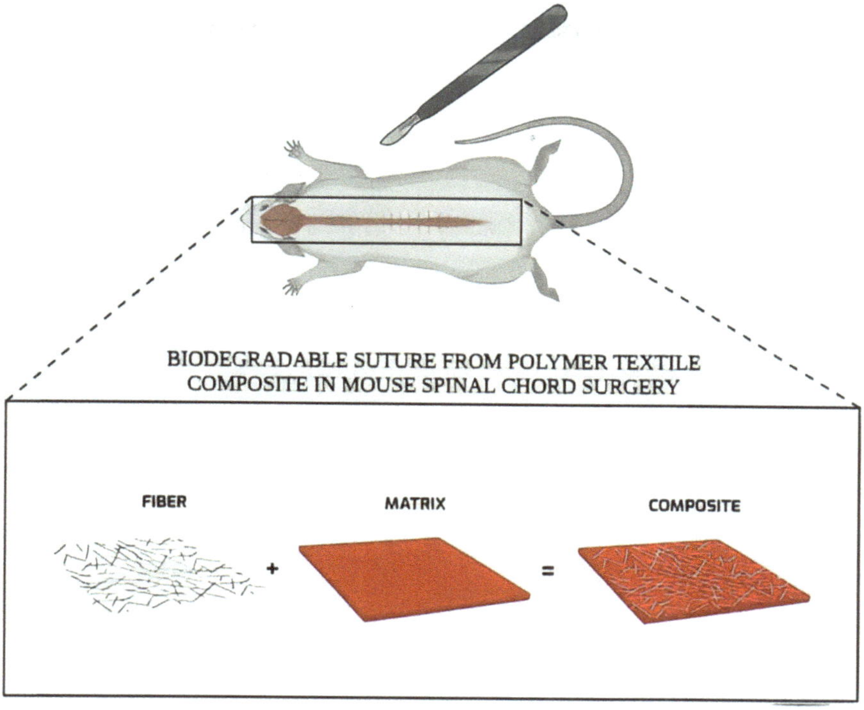

Fig. 12.8 Biodegradable suture made from polymer composite for performing surgical procedures in rats (Created with www.biorender.com)

mesh implants made from polymer textile composites are employed to provide reinforcement to abdominal walls weakened by hernias. These implants offer exceptional strength, biocompatibility, and durability, helping to prevent hernia recurrence and minimizing patient discomfort. Similarly, in the management of pelvic organ prolapse, where weakened pelvic floor muscles and tissues cause organs to descend, polymer textile composite implants serve as a vital solution for tissue reinforcement [118]. These materials offer the necessary structural support to restore anatomical integrity and alleviate the symptoms associated with pelvic organ prolapse. Moreover, in cardiovascular surgery, polymer textile composites are used to reinforce damaged or weakened blood vessels, providing added strength to repair aneurysms or dissections and reduce the risk of further vascular complications. The versatility of these materials extends to applications in plastic and reconstructive surgery, where they are employed to reinforce soft tissues during breast reconstruction or facial surgeries [119]. The integration of polymer textile composites in tissue reinforcement procedures enhances the longevity and effectiveness of surgical interventions, ultimately improving patient outcomes, reducing the risk of complications, and restoring the structural integrity of weakened or damaged tissues. The utilization of a composite material consisting of polyglycolic acid and trimethylene carbonate

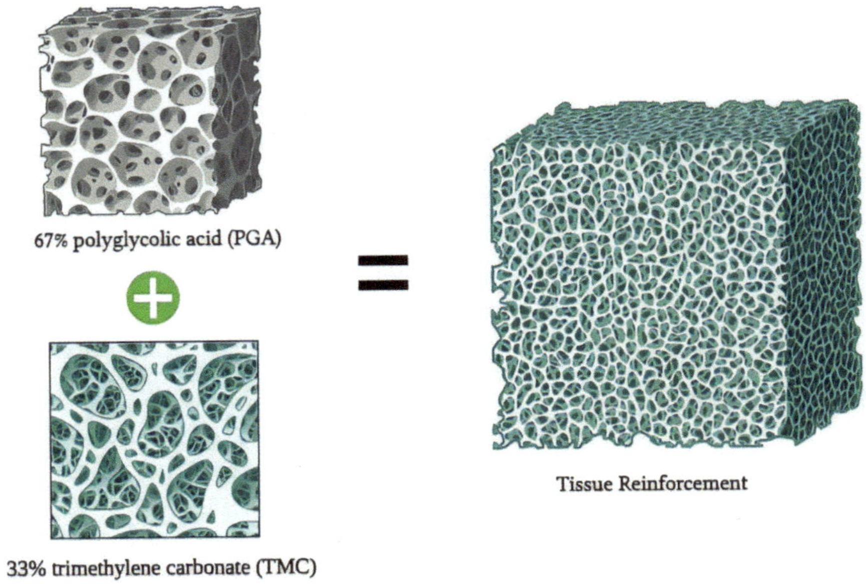

Fig. 12.9 Tissue reinforcement for regeneration of dead tissue by compositing polyglycolic acid and trimethylene carbonate (Created with www.biorender.com)

for tissue reinforcement in the context of tissue regeneration has been referenced in Fig. 12.9.

12.7.9.1 Dental Materials

The application of polymer textile composites in dental materials signifies a remarkable advancement in modern dentistry, offering versatile and aesthetically pleasing solutions for various dental procedures. These innovative materials, born from the combination of polymers and textile substrates, have revolutionized the field by providing superior strength, durability, and aesthetics in dental applications. One of the most notable applications of polymer textile composites in dentistry is in restorative procedures [120]. Composite resin materials reinforced with textile composites have become the material of choice for dental fillings, crowns, and bridges. These materials offer several advantages, including excellent biocompatibility, resistance to wear and staining, and the ability to closely match the natural color and translucency of teeth. This aesthetic resemblance makes them highly appealing for patients seeking restorations that blend seamlessly with their natural dentition. Moreover, polymer textile composite-based dental materials are particularly well-suited for minimally invasive dentistry. These materials can be used to create veneers and inlays/onlays, which require less tooth reduction compared to traditional crowns or fillings [121]. This preservation of healthy tooth structure is advantageous for patients and contributes to improved long-term dental health. Dental implants, another critical advancement in restorative dentistry, also benefit from the integration of polymer textile composites. These materials are used for the fabrication of implant abutments and prosthetic components. The composites' biocompatibility and strength ensure the stability and longevity of dental implants, which have become the gold standard for replacing missing teeth and restoring oral function [122]. In orthodontics, polymer textile composite-based materials are employed in the fabrication of clear aligners and retainers. These materials offer excellent transparency, comfort, and durability, making them a preferred choice for patients seeking discreet orthodontic treatment and long-term retention solutions. Furthermore, in prosthetic dentistry, polymer textile composites play a role in the creation of flexible denture bases. These materials provide improved flexibility and comfort for patients wearing removable partial dentures, reducing the risk of sore spots and discomfort associated with traditional rigid denture bases [123]. Overall, the application of polymer textile composites in dental materials has transformed the field by offering enhanced aesthetics, durability, and patient comfort. These materials continue to drive innovation in restorative, orthodontic, and prosthetic dentistry, contributing to improved oral health and patient satisfaction. As research and development efforts continue to advance, polymer textile composite-based dental materials hold the promise of further elevating the standards of dental care and aesthetics in dentistry [124]. The depiction of the incorporation of polymer composites in the production of maxillofacial metal implants for dental restorations has been effectively illustrated in Fig. 12.10.

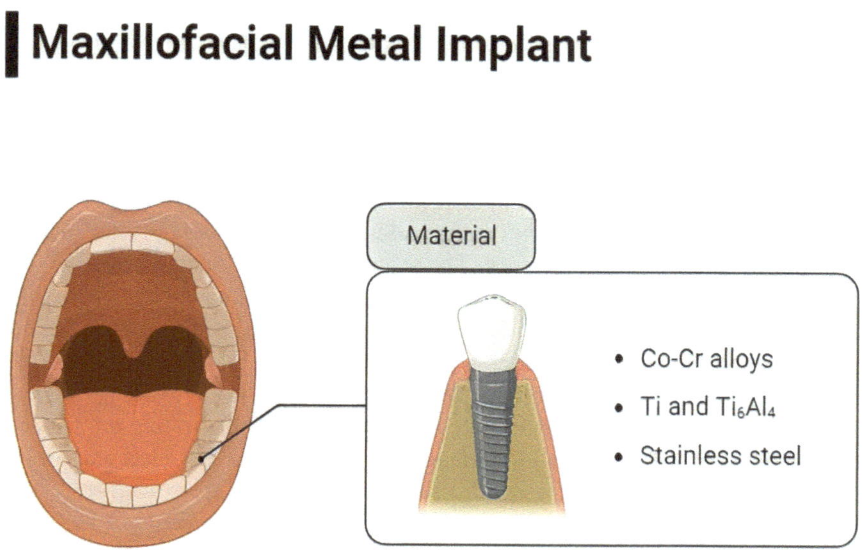

Fig. 12.10 The utilization of polymer composites in the fabrication of maxillofacial metal implants for dental fillings [125]

12.7.9.2 Catheters and Medical Tubing

The application of polymer textile composites in catheters and medical tubing has ushered in a new era of innovation in minimally invasive medical procedures and patient care. These advanced materials, resulting from the fusion of polymers and textile substrates, have become instrumental in the design and manufacturing of catheters and medical tubing used in various clinical applications. One of their primary advantages lies in their exceptional flexibility, durability, and biocompatibility, which make them ideal for navigating through the intricate anatomical pathways of the human body [126]. In the field of interventional cardiology, catheters incorporating polymer textile composites are used for procedures such as angioplasty and stent placement. These composites provide the necessary structural integrity while allowing for precise maneuvrability within the blood vessels, ensuring that the catheter can reach its target site with minimal trauma to the vessel walls. Polymer textile composite-based catheters are also employed in urology, gastroenterology, and vascular surgery [127]. They are used in procedures like urinary catheterization, endoscopy, and venous access. Their flexibility and biocompatibility enable safe and comfortable insertion, reducing the risk of tissue damage or infection. Furthermore, in the field of intravascular ultrasound (IVUS) and electrophysiology, these materials are used in the construction of specialized catheters that provide real-time imaging and mapping capabilities [128]. The composites' durability and resistance to chemical degradation ensure the longevity and reliability of these diagnostic and therapeutic devices. Figure 12.11 depicts polymer composite textile-based medical tube devices and catheters.

12.8 Conclusion

In conclusion, the state of the art in polymer textile composites represents a remarkable convergence of science, engineering, and innovation that has unlocked a multitude of possibilities across industries. The continuous evolution of fabrication techniques and polymer chemistry has led to the development of composites with finely tuned properties, making them indispensable in fields where lightweight, durable, and sustainable materials are imperative. The broad spectrum of applications is a testament to the versatility of polymer textile composites. From revolutionizing the aerospace and automotive sectors by reducing weight and improving efficiency to enhancing sporting equipment for athletes, these composites have left an indelible mark. Furthermore, they play a crucial role in infrastructure reinforcement, extending the lifespan of critical structures, and contributing to advancements in healthcare by providing biocompatible materials for medical devices. As our understanding of materials science deepens and sustainability becomes an even more pressing concern, polymer textile composites are poised to remain at the forefront of innovation. Their potential for reducing environmental impact, while simultaneously enhancing performance, makes them a key player in the pursuit of a greener and more efficient future.

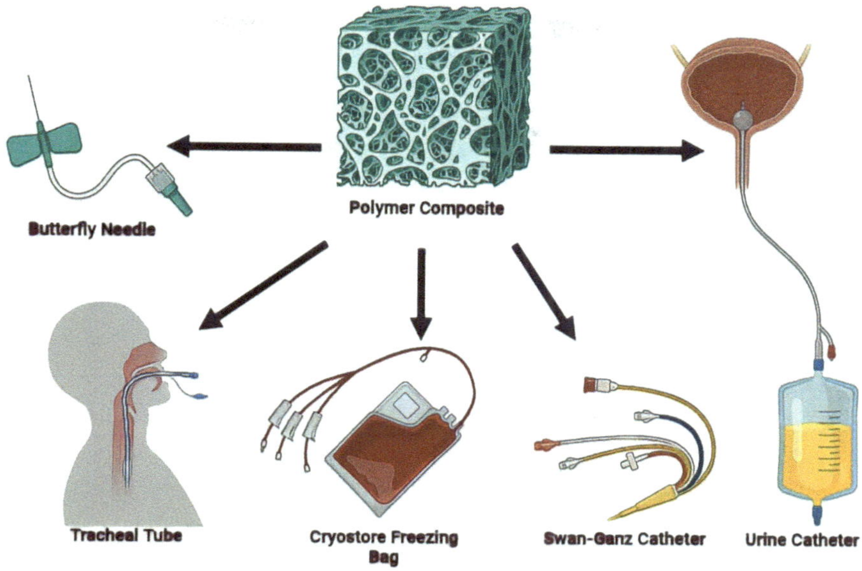

Fig. 12.11 Polymer composite textile-based medical tube devices and catheters (Created with www.biorender.com)

With ongoing research and development, we can anticipate further breakthroughs and novel applications that will continue to redefine the boundaries of what is achievable in the realm of materials engineering.

12.9 Future Prospects

The future prospects for polymer textile composites and their applications hold immense promise, driven by ongoing research, technological advancements, and the growing emphasis on sustainability and performance optimization. As the world increasingly embraces sustainability, the development of biodegradable and recyclable polymer textile composites will gain prominence. These materials will play a crucial role in reducing the environmental footprint of various industries, from automotive to construction. Integrating nanotechnology into polymer textile composites will open doors to the creation of smart materials with unprecedented functionalities. These composites may possess self-healing properties, adaptive responses to environmental changes, and enhanced electrical or thermal conductivity. Advancements in 3D printing and digital design will enable the production of highly customized polymer textile composites, tailored to specific applications and individual preferences. This could revolutionize fields like fashion, sports equipment, and medical devices. Polymer textile composites will become pivotal in space exploration, with their lightweight, durable, and radiation-resistant properties. They will also find applications in extreme terrestrial environments, such as deep-sea exploration and arctic research. Combining polymer textile composites with IoT technologies will enable the creation of smart wearables and textiles with embedded sensors for healthcare monitoring, environmental sensing, and interactive clothing. In summary, the future of polymer textile composites is bright, with innovations driven by sustainability, nanotechnology, customization, and their integration into advanced technologies. These materials will continue to redefine industries and play a vital role in addressing the challenges of the twenty-first century.

References

1. Radulescu, M., Ficai, D., Oprea, O., Ficai, A., Andronescu, E., Holban, M.A.: Antimicrobial chitosan based formulations with impact on different biomedical applications. Curr. Pharm. Biotechnology. **16**(2), 128–36 (2015)
2. Cheung, T.C., Choi, S.Y.: Evaluation of the influence of three-dimensional printing conditions on peel resistance and surface roughness of flexible polymer-textile composites. Text. Res. J. **93**(7–8), 1531–1550 (2023)
3. Idumah, C.I.: Design, fabrication, characterization and properties of metallic and conductive smart polymeric textiles for multifunctional applications. Nano-Struct. & Nano-Objects. **1**(35), 100982 (2023)
4. Wang, Y., Zhang, Z., Chu, F., Liu, M., He, Y., Li, P., Yuan, J., Yang, M.: Designing polydopamine-capped [BMIm] PF6@ halloysite/NaL microcapsule optimize the wear-resistance of polymer composite liner. Tribol. Int. **1**(179), 108104 (2023)
5. Navaratnam, S., Selvaranjan, K., Jayasooriya, D., Rajeev, P., Sanjayan, J.: Applications of natural and synthetic fiber reinforced polymer in infrastructure: a suitability assessment. J. Build. Engineering. **10**, 105835 (2023)
6. Song, Y., Qu, Z., Liao, H., Ai, S.: Material twins generation of woven polymer composites based on ResL-U-Net convolutional neural networks. Compos. Struct. **1**(307), 116672 (2023)
7. Hossain, M.A.: Simultaneous thermoregulated and sensorial effect of smart textiles with artificial composite phase change materials (CPCM) incorporated with Carbon Nano Conductive Materials for special workers and extreme weather conditions. School of Fashions and Textiles, RMIT University, Victoria 3056, Australia (engr. anowar@ yahoo. com). (2023)
8. Priyanka, P., Sharma, P., Mali, H.S., Sharma, P.: High strain rate compression response of kevlar and interyarn hybrid carbon-kevlar polymer composites. J. Mater. Eng. Perform. **27**, 1–4 (2023)
9. Xia, Y., He, Y., Zhang, F., Liu, Y., Leng, J.: A review of shape memory polymers and composites: mechanisms, materials, and applications. Adv. Mater. **33**(6), 2000713 (2021)
10. Hussain, A., Podgursky, V., Viljus, M., Awan, M.R.: The role of paradigms and technical strategies for implementation of the circular economy in the polymer and composite recycling industries. Adv. Ind. Eng. Polym. Research. **6**(1), 1–2 (2023)
11. Pander, M., Gil-San-Millan, R., Delgado, P., Perona-Bermejo, C., Kostrzewa, U., Kaczkowski, K., Kubicki, D.J., Navarro, J.A., Bury, W.: MOF/polymer hybrids through in situ free radical polymerization in metal-organic frameworks. Mater. Horiz. **10**(4), 1301–1308 (2023)
12. Cui, Y., Zheng, G., Jiang, Z., Zhou, M., Wang, P., Yu, Y., Wang, Q.: Preparation of multi-functional composite materials based on PEDOT via enzymatic cascade polymerization. Surf. Interfaces. **1**(39), 102961 (2023)
13. Selvam, S., Yim, J.H.: Effective self-charge boosting sweat electrolyte textile supercapacitors array from bio-compatible polymer metal chelates. J. Power Sources **1**(556), 232511 (2023)

14. Jhang, J.C., Kao, H., Lin, J.H., Lou, C.W.: Composite planks consisting of high resilience buffer nonwoven fabrics and conductive elastic polymer films via hot pressing lamination: manufacturing techniques and mechanical, buffer, and electrical properties. Fibers and Polymers. **24**(5), 1781–1788 (2023)
15. Panneke, N., Ehrmann, A.: Stab-resistant polymers—recent developments in materials and structures. Polymers **15**(4), 983 (2023)
16. Wang, L., Zhang, F., Liu, Y., Leng, J.: Shape memory polymer fibers: materials, structures, and applications. Adv. Fiber Materials. **1**, 1–9 (2022)
17. Mishra, R.K.: Advances in textile structural composites. polymers. **15**(4), 808 (2023)
18. Pl, J., Bose, S.: Evolution of surface engineering in the development of textile-based EMI shields–A review. ACS Appl. Electron. Mater. **5**(4), 1947–1969 (2023)
19. Meena, J.S., Choi, S.B., Khanh, T.D., Shin, H.S., Choi, J.S., Joo, J., Kim, J.W.: Highly stretchable and robust textile-based capacitive mechanical sensor for human motion detection. Appl. Surf. Sci. **15**(613), 155961 (2023)
20. Li, H.Y., Huang, D.N., Ren, K.F., Ji, J.: Inorganic-polymer composite coatings for biomedical devices. Smart Materials in Medicine. **1**(2), 1–4 (2021)
21. Zhan, Y., Santillo, C., Meng, Y., Lavorgna, M.: Recent advances and perspectives on silver-based polymer composites for electromagnetic interference shielding. J. Mater. Chem. C., (2023)
22. Akram, S., Ashraf, M., Javid, A., Abid, H.A., Ahmad, S., Nawab, Y., Rasheed, A., Xue, Z., Nosheen, A.: Recent advances in electromagnetic interference (EMI) shielding textiles: A comprehensive review. Synth. Met. **1**(294), 117305 (2023)
23. Aruchamy, K., Mylsamy, B., Palaniappan, S.K., Subramani, S.P., Velayutham, T., Rangappa, S.M., Siengchin, S.: Influence of weave arrangements on mechanical characteristics of cotton and bamboo woven fabric reinforced composite laminates. J. Reinf. Plast. Compos. **21**, 07316844221140350 (2023)
24. Benny Mattam, L., Bijoy, A., Abraham Thadathil, D., George, L., Varghese, A.: Conducting polymers: a versatile material for biomedical applications. ChemistrySelect **7**(42), e202201765 (2022)
25. Mohit, H., Mavinkere Rangappa, S., Siengchin, S., Gorbatyuk, S., Manimaran, P., Alka Kumari, C., Khan, A., Doddamani, M.: A comprehensive review on performance and machinability of plant fiber polymer composites. Polym. Compos. **43**(1), 608–623 (2022)
26. Ražić, S.E., Ludaš, A., Kaurin, T., Zonjić, T.: Applicability of polymers printed on textiles with a 3D printer for possible use in car interior. In: IOP conference series: earth and environmental science, vol. 1128, No. 1, p. 012027. IOP Publishing, (2023)
27. Erdem, G., Grothe, T., Ehrmann, A.: Adhesion of new thermoplastic materials printed on textile fabrics. Tekstilec. **10**(66), 1–7 (2023)
28. Swain, S.K., Jawaid, M., eds.: Nanostructured polymer composites for biomedical applications. Elsevier, (2019)
29. Das, P., Manna, S., Behera, A.K., Shee, M., Basak, P., Sharma, A.K.: Current synthesis and characterization techniques for clay-based polymer nano-composites and its biomedical applications: A review. Environ. Res. **1**(212), 113534 (2022)
30. Grings, K.J., Carneiro Ribeiro, F.R., Junior, D.V., de Azevedo, A.R., Kulakowski, M.P.: Evaluation of light cementitious matrix with composite textile reinforcement from garment waste. Materials. **16**(2), 733 (2023)
31. Guo, W.T., Tang, X.G., Tang, Z., Sun, Q.J.: Recent advances in polymer composites for flexible pressure sensors. Polymers **15**(9), 2176 (2023)
32. Mitra, D., Kang, E.T., Neoh, K.G.: Polymer-based coatings with integrated antifouling and bactericidal properties for targeted biomedical applications. ACS Appl. Polym. Materials. **3**(5), 2233–2263 (2021)
33. Kozior, T., Ehrmann, A.: First proof-of-principle of polyjet 3D printing on textile fabrics. Polymers **15**(17), 3536 (2023)
34. Idumah, C.I.: Recent advancements in conducting polymer bionanocomposites and hydrogels for biomedical applications. Int. J. Polym. Mater. Polym. Biomater. **71**(7), 513–530 (2022)

35. Jalalah, M., Ahmad, A., Saleem, A., Qadir, M.B., Khaliq, Z., Khan, M.Q., Nazir, A., Faisal, M., Alsaiari, M., Irfan, M., Alsareii, S.A.: Electrospun nanofiber/textile supported composite membranes with improved mechanical performance for biomedical applications. Membranes **12**(11), 1158 (2022)
36. Aisyah, H.A., Paridah, M.T., Sapuan, S.M., Ilyas, R.A., Khalina, A., Nurazzi, N.M., Lee, S.H., Lee, C.H.: A comprehensive review on advanced sustainable woven natural fibre polymer composites. Polymers **13**(3), 471 (2021)
37. Ozimek, J., Pielichowski, K.: Recent advances in polyurethane/POSS hybrids for biomedical applications. Molecules **27**(1), 40 (2021)
38. Joe, M.S., Sudherson, D.P., Suyambulingam, I., Siengchin, S.: Extraction and characterization of novel biomass–based cellulosic plant fiber from Ficus benjamina L. stem for a potential polymeric composite reinforcement. Biomass Convers. Biorefinery., 1–5 (2023)
39. Wankhede, B., Bisaria, H., Ojha, S., Dakre, V.S.: A review on cotton fibre-reinforced polymer composites and their applications. Proceedings of the Institution of Mechanical Engineers, Part L: Journal of Materials: Design and Applications. **237**(6), 1347–1362 (2023)
40. de M de Lima, T.A., de Lima, G.G., Nugent, M.J.: Natural fibre-reinforced polymer composites: manufacturing and biomedical applications. Polym. Nat. Compos.: Mater., Manuf. Biomed. Appl., 25–63 (2022)
41. Tseghai, G.B., Mengistie, D.A., Malengier, B., Fante, K.A., Van Langenhove, L.: PEDOT: PSS-based conductive textiles and their applications. Sensors. **20**(7), 1881 (2020)
42. Oladele, I.O., Omotosho, T.F., Adediran, A.A.: Polymer-based composites: an indispensable material for present and future applications. Int. J. Polym. Science. **19**(2020), 1–2 (2020)
43. Jahangir, H., Soleymani, A., Esfahani, M.R.: Investigating the confining effect of steel-reinforced polymer and grout composites on compressive behaviour of square concrete columns. Iran. J. Sci. Technol., Trans. Civ. Engineering. **47**(2), 775–791 (2023)
44. Akter, M., Uddin, M.H., Anik, H.R.: Plant fibre-reinforced polymer composites: a review on modification, fabrication, properties, and applications. Polym. Bull. **17**, 1–85 (2023)
45. Zhang, D., Liu, L., Xu, P., Zhao, Y., Li, Q., Lan, X., Zou, X., Li, Y., He, Y., Liu, Y., Leng, J.: Ancient papyrus scroll-inspired self-deployable mechanism based on shape memory polymer composites for Mars explorations. Compos. Struct. **15**(304), 116391 (2023)
46. Seidi, F., Yazdi, M.K., Jouyandeh, M., Dominic, M., Naeim, H., Nezhad, M.N., Bagheri, B., Habibzadeh, S., Zarrintaj, P., Saeb, M.R., Mozafari, M.: Chitosan-based blends for biomedical applications. Int. J. Biol. Macromol. **31**(183), 1818–1850 (2021)
47. Ornaghi, H.L., Jr., Bianchi, O.: Temperature-dependent shape-memory textiles: physical principles and applications. Textiles. **3**(2), 257–274 (2023)
48. Nimra, S.S., Rehan, Z.A., Ali, S.H., Atir, S., Fatima, K., Shahzadi, F., Shakir, H.F., Alamir, M.A., EL-Bagory, T.M., Shahid, I.: Electrically conductive fibers fabrication and characterization via in-situ polymerization of aniline for the protection against EMI and thermal imaging signals. J. Mater. Res. Technol. **23**, 2399–409 (2023)
49. Renkler, N.Z., Cruz-Maya, I., Bonadies, I., Guarino, V.: Electro fluid dynamics: a route to design polymers and composites for biomedical and bio-sustainable applications. Polymers **14**(19), 4249 (2022)
50. Kupińska, K., Michalik, M., Krajenta, J., Bielicka, M., Markiewicz, K.H., Kalska-Szostko, B., Wilczewska, A.Z.: An In-Situ Fabrication Method of ZnO and Other Zn (II) Compounds Containing Polypropylene Composites. Int. J. Mol. Sci. **24**(3), 2357 (2023)
51. Kumar, J., Kumar, K., Jaiswal, B., Kumar, K., Verma, R.K.: Investigation on the physio-mechanical properties of carpet waste polymer composites incorporated with multi-wall carbon nanotube (MWCNT). J. Text. Inst. **114**(4), 664–673 (2023)
52. Biswas, B., Mandal, G., Das, A., Majumdar, A., Sinha, A.: Ceramic particle–dispersed polymer composites. In: Advances in biomedical polymers and composites, pp. 399–432. Elsevier, (2023)
53. Mandal, L., Verma, B., Patel, P.K.: Review on polymer nanocomposite for ballistic & aerospace applications. Mater. Today: Proceedings. **1**(26), 3161–3166 (2020)

54. Zhalmuratova, D., Chung, H.J.: Reinforced gels and elastomers for biomedical and soft robotics applications. ACS Appl. Polym. Materials. **2**(3), 1073–1091 (2020)
55. Kiran, S., Adeel, S., Abrar, S., Iqbal, S., Naz, S., Amin, N.: Textile applications of nanofibers and nanocomposites. In: Handbook of nanofibers and nanocomposites, pp. 149–167. Jenny Stanford Publishing, (2024)
56. Parachuru, R.: Selected research briefs on the fabrication of value added composites using recycled textile waste as reinforcement material. Int. J. Mater. Eng. Technology. **22**(1), 43–53 (2023)
57. Agossou, O.G., Homoro, O., Amziane, S.: Application of natural FRCM composites in structural strengthening/rehabilitation: state-of-the-art analysis. In: International conference on bio-based building materials, pp. 613–626. Springer Nature Switzerland, Cham (2023)
58. Maithil, P., Gupta, P., Chandravanshi, M.L.: Study of mechanical properties of the natural-synthetic fiber reinforced polymer matrix composite. Mater. Today: Proc. (2023)
59. Aziz, T., Haq, F., Farid, A., Kiran, M., Faisal, S., Ullah, A., Ullah, N., Bokhari, A., Mubashir, M., Chuah, L.F., Show, P.L.: Challenges associated with cellulose composite material: Facet engineering and prospective. Environ. Res. **4**, 115429 (2023)
60. Cho, C., Kim, D., Lee, C., Oh, J.H.: Ultrasensitive Ionic Liquid Polymer Composites with a Convex and Wrinkled Microstructure and Their Application as Wearable Pressure Sensors. ACS Appl. Mater. Interfaces. **15**(10), 13625–13636 (2023)
61. Singh, S., Ramakrishna, S., Berto, F.: 3D Printing of polymer composites: A short review. Mater. Des. & Process. Communications. **2**(2), e97 (2020)
62. Deng, B., Fang, L., Fang, K., Han, X., Liang, Y.: Scalable preparation of MWCNTs/PAN conductive composite fibers with Tai Chi structure for thermotherapy textiles. Compos. Sci. Technol. **8**(232), 109866 (2023)
63. Kalirajan, C., Dukle, A., Nathanael, A.J., Oh, T.H., Manivasagam, G.: A critical review on polymeric biomaterials for biomedical applications. Polymers **13**(17), 3015 (2021)
64. Saleh Alghamdi, S., John, S., Roy Choudhury, N., Dutta, N.K.: Additive manufacturing of polymer materials: Progress, promise and challenges. Polymers **13**(5), 753 (2021)
65. Biswas, M.C., Chowdhury, A., Hossain, M.M., Hossain, M.K.: Applications, drawbacks, and future scope of nanoparticle-based polymer composites. In: Nanoparticle-based polymer composites, pp. 243–275. Woodhead Publishing, (2022)
66. Park, S., Pal, S.K., Otoufat, T., Kim, G.: Radiative-cooling composites with enhanced infrared emissivity by structural infrared scattering through indium tin oxide nanoparticles in a polymer matrix. ACS Appl. Mater. Interfaces. **15**(12), 16026–16033 (2023)
67. Kumar, S.D., Samvel, R., Aravindh, M., Vibin, R.A., Poovarasu, E., Prasad, M.S.: Ballistic studies on synthetic fibre reinforced polymer composites and it's applications–A brief review. Mater. Today: Proc. (2023)
68. Chakraborty, S., Biswas, M.C.: 3D printing technology of polymer-fiber composites in textile and fashion industry: A potential roadmap of concept to consumer. Compos. Struct. **15**(248), 112562 (2020)
69. Seydibeyoğlu, M.Ö., Dogru, A., Wang, J., Rencheck, M., Han, Y., Wang, L., Seydibeyoğlu, E.A., Zhao, X., Ong, K., Shatkin, J.A., Shams, E.-H.: Review on hybrid reinforced polymer matrix composites with nanocellulose, nanomaterials, and other fibers. Polymers **15**(4), 984 (2023)
70. Rehman, A., Houshyar, S., Wang, X.: Nanodiamond in composite: Biomedical application. J. Biomed. Mater. Res., Part A **108**(4), 906–922 (2020)
71. Tarhini, A., Tehrani-Bagha, A.R.: Advances in preparation methods and conductivity properties of graphene-based polymer composites. Appl. Compos. Mater. **5**, 1–26 (2023)
72. Luo, H.B., Lin, F.R., Liu, Z.Y., Kong, Y.R., Idrees, K.B., Liu, Y., Zou, Y., Farha, O.K., Ren, X.M.: MOF–polymer mixed matrix membranes as chemical protective layers for solid-phase detoxification of toxic organophosphates. ACS Appl. Mater. Interfaces. **15**(2), 2933–2939 (2023)
73. Li, C., Song, J., Xing, W., Wang, L., Cui, Y., Pei, X.: Mechanical properties of interlayer hybrid textile composite materials based on modified aramid and UHMWPE fabrics. Polym. Adv. Technol. **34**(1), 205–216 (2023)

74. Salunkhe, S., Murali, A.P., Mohammed Abdel Moneam, H., Naranje, V., Shanmugam, R.: 3D printing of plant fiber reinforced polymer composites (PFRC's): an insight into methods, challenges and opportunities. Polym.-Plast. Technol. Mater. **62**(6), 816–38 (2023)
75. Zhao, C., Zhang, C., Wang, P., Chen, Z., Wang, Y., Zhu, J., Gao, C., Gao, Q.: Wet-spun PEDOT: PSS/ionic liquid composite fibers for wearable e-textiles. Eur. Polymer J. **25**(190), 112025 (2023)
76. Jagadeshvaran, P.L., Parasuram, S., Bose, S.: Polymer-based nanocomposites for smart and functional textiles. Smart Funct. Textiles. **3**, 183 (2023)
77. Kumar Gurajala, N., Neelapala, R., Kumar, A.V., Sai, Y.K., Manikanta, J.E.: Biodegradable polymer reinforced natural fiber composition for mechanical properties: A review. Mater. Today: Proc., (2023)
78. Nivetha, R., Vennila, A., Dharshini, B.: A review on various properties of textile reinforced concrete. Mater. Today: Proc., (2023)
79. Sonali, S., Farzana, M., Haque, M.M., Saha, A., Khan, R.A., Mollah, M.Z.: Natural fiber reinforced polymer-based composites: importance of jute fiber. GSC Adv. Res. Reviews. **15**(1), 021–029 (2023)
80. Chen, X., Zhao, Y., Li, L., Wang, Y., Wang, J., Xiong, J., Du, S., Zhang, P., Shi, X., Yu, J.: MXene/polymer nanocomposites: preparation, properties, and applications. Polym. Rev. **61**(1), 80–115 (2021)
81. Maity, S., Kunal, S., Pintu, P.: Speciality coatings and laminations on textiles. Smart Funct. Textiles. **3**, 151 (2023)
82. Abubakre, O.K., Medupin, R.O., Akintunde, I.B., Jimoh, O.T., Abdulkareem, A.S., Muriana, R.A., James, J.A., Ukoba, K.O., Jen, T.C., Yoro, K.O.: Carbon nanotube-reinforced polymer nanocomposites for sustainable biomedical applications: A review. J. Sci.: Adv. Mater. Devices. **24**, 100557 (2023)
83. Liu, J., Shi, L., Deng, Y., Zou, M., Cai, B., Song, Y., Wang, Z., Wang, L.: Silk sericin-based materials for biomedical applications. Biomaterials **17**, 121638 (2022)
84. Sikdar, P., Dip, T.M., Dhar, A.K., Bhattacharjee, M., Hoque, M.S., Ali, S.B.: Polyurethane (PU) based multifunctional materials: Emerging paradigm for functional textiles, smart, and biomedical applications. J. Appl. Polym. Sci. **139**(38), e52832 (2022)
85. Mali, H.S., Sharma, P.: Machinability of high-strength fiber-reinforced polymer textile composites: a review. Mech. Compos. Mater. **59**(1), 1–28 (2023)
86. Santos, T.F., Santos, C.M., Hussein, E.K., Zilio, L., Dias, M., Sanjay, M.R., Fonseca, R., Amaral, A., Aquino, M.: Tensile behavior of weft-knitted structure for potential use in composite reinforcement via factorial and 3D Surface. In: Green hybrid composite in engineering and non-engineering applications, pp. 233–259. Springer Nature Singapore, Singapore (2023)
87. Khalid, M.Y., Arif, Z.U., Noroozi, R., Zolfagharian, A., Bodaghi, M.: 4D printing of shape memory polymer composites: A review on fabrication techniques, applications, and future perspectives. J. Manuf. Process. **1**(81), 759–797 (2022)
88. Jiang, W.: Metal implant clinical applications. (2023). https://app.biorender.com/biorender-templates/figures/all/t-6470c9ba874beee9d19864a8-metal-implant-clinical-applications
89. Yu, X., Su, X., Liu, Y., Yu, D., Ren, Y., Liu, X.: Biomass intumescent flame retardant poly-acrylonitrile composite: flame retardancy, smoke suppression and recycling. Compos. A Appl. Sci. Manuf. **7**, 107647 (2023)
90. Lopes, C., Ferreira, A., Correa, M., Machado Rodrigues, A., Mendes, P., Vaz, F.: EMI shielding and conductive textiles functionalized with (Ti, Cu) nanomaterials for biomedical applications. ACS Appl. Mater. & Interfaces., (2023)
91. Mobaraki, M., Liu, M., Masoud, A.R., Mills, D.K.: Biomedical applications of blow-spun coatings, mats, and scaffolds—a mini-review. J. Compos. Science. **7**(2), 86 (2023)
92. Torğut, G., Yazdıç, F.C., Gürler, N.: Synthesis, characterization, pH-sensitive swelling and antimicrobial activities of chitosan–graft-poly (hydroxyethyl methacrylate) hydrogel composites for biomedical applications. Polym. Eng. Sci. **62**(8), 2552–2559 (2022)

93. Radu, E.R., Voicu, S.I., Thakur, V.K.: Polymeric membranes for biomedical applications. Polymers. **15**(3), 619 (2023)
94. Arif, Z.U., Khalid, M.Y., Sheikh, M.F., Zolfagharian, A., Bodaghi, M.: Biopolymeric sustainable materials and their emerging applications. J. Environ. Chem. Eng. **10**(4), 108159 (2022)
95. Min, J.H., Patel, M., Koh, W.-G.: Incorporation of conductive materials into hydrogels for tissue engineering applications. Polymers **10**, 1078 (2018). https://doi.org/10.3390/polym10101078
96. Business Bliss Consultants FZE. Polymers used in tissue engineering [Internet]. (2018). [Accessed 10 October 2023]. Available from https://ukdiss.com/examples/tissue-engineering-polymers.php?vref=1
97. Wang, Y., Zhang, Z., Liu, M., He, Y., Li, P., Yuan, J., Yang, M., Liu, W.: Halloysite–gold core–shell nanosystem synergistically enhances thermal conductivity and mechanical properties to optimize the wear-resistance of a pheonlic-PBO/PTFE textile composite liner. Friction. **29**, 1–5 (2023)
98. Popescu, M., Ungureanu, C.: Green nanomaterials for smart textiles dedicated to environmental and biomedical applications. Materials. **16**(11), 4075 (2023)
99. Athar, A.Z.: Biomedical applications of composite materials: a review
100. Veerabagu, U., Palza, H., Quero, F.: Auxetic polymer-based mechanical metamaterials for biomedical applications. ACS Biomater. Sci. Eng. **8**(7), 2798–2824 (2022)
101. Jones, S.: Masquelet Technique. (2023). https://app.biorender.com/biorender-templates/figures/all/t-63dd838c54ec321a0a15b040-masquelet-technique
102. Merzah, Z.F., Fakhry, S., Allami, T.G., Yuhana, N.Y., Alamiery, A.: Enhancement of the properties of hybridizing epoxy and nanoclay for mechanical, industrial, and biomedical applications. Polymers **14**(3), 526 (2022)
103. Tamulevičius, T.: Nanotechnologies in Textiles. Materials. **15**(4), 1466 (2022)
104. Salifu, S., Ogunbiyi, O., Olubambi, P.A.: Potentials and challenges of additive manufacturing techniques in the fabrication of polymer composites. Int. J. Adv. Manuf. Technology. **122**(2), 577–600 (2022)
105. Dhanasekar, S., Stella, T.J., Thenmozhi, A., Bharathi, N.D., Thiyagarajan, K., Singh, P., Reddy, Y.S., Srinivas, G., Jayakumar, M.: Study of polymer matrix composites for electronics applications. J. Nanomater. **20**, 2022 (2022)
106. Syed, M.H., Zahari, M.A., Khan, M.M., Beg, M.D., Abdullah, N.: An overview on recent biomedical applications of biopolymers: Their role in drug delivery systems and comparison of major systems. J. Drug Deliv. Sci. Technology. **29**, 104121 (2022)
107. Mousavi, S.M., Hashemi, S.A., Kalashgrani, M.Y., Omidifar, N., Bahrani, S., Vijayakameswara Rao, N., Babapoor, A., Gholami, A., Chiang, W.H.: Bioactive graphene quantum dots based polymer composite for biomedical applications. Polymers **14**(3), 617 (2022)
108. Pryadko, A.S., Botvin, V.V., Mukhortova, Y.R., Pariy, I., Wagner, D.V., Laktionov, P.P., Chernonosova, V.S., Chelobanov, B.P., Chernozem, R.V., Surmeneva, M.A., Kholkin, A.L.: Core-shell magnetoactive PHB/gelatin/magnetite composite electrospun scaffolds for biomedical applications. Polymers **14**(3), 529 (2022)
109. Idumah, C.I.: Recently emerging advancements in polymeric cryogel nanostructures and biomedical applications. Int. J. Polym. Mater. Polym. Biomater. **72**(16), 1307–1327 (2023)
110. Afshar, A., Gultekinoglu, M., Edirisinghe, M.: Binary polymer systems for biomedical applications. Int. Mater. Rev. **68**(2), 184–224 (2023)
111. Fatullayeva, S., Tagiyev, D., Zeynalov, N., Mammadova, S., Aliyeva, E.: Recent advances of chitosan-based polymers in biomedical applications and environmental protection. J. Polym. Res. **29**(7), 259 (2022)
112. Chandana, A., Mallick, S.P., Dikshit, P.K., Singh, B.N., Sahi, A.K.: Recent developments in bacterial nanocellulose production and its biomedical applications. J. Polym. Environ. **30**(10), 4040–4067 (2022)

113. Wan, X., Zhao, Y., Li, Z., Li, L.: Emerging polymeric electrospun fibers: From structural diversity to application in flexible bioelectronics and tissue engineering. In: Exploration, vol. 2, no. 1, p. 20210029. (2022)
114. Shahbaz, A., Hussain, N., Mahmood, T., Iqbal, H.M., Emran, T.B., Show, P.L., Bilal. M.: Polymer nanocomposites for biomedical applications. In: Smart polymer nanocomposites, pp. 379–394. Elsevier, (2023)
115. Rai, R., Dhar, P.: Biomedical engineering aspects of nanocellulose: A review. Nanotechnology **33**(36), 362001 (2022)
116. Barman, J., Tirkey, A., Batra, S., Paul, A.A., Panda, K., Deka, R., Babu, P.J.: The role of nanotechnology based wearable electronic textiles in biomedical and healthcare applications. Mater. Today Commun. **16**, 104055 (2022)
117. Kumar, A., Yadav, S., Pramanik, J., Sivamaruthi, B.S., Jayeoye, T.J., Prajapati, B.G., Chaiyasut, C.: Chitosan-based composites: development and perspective in food preservation and biomedical applications. Polymers **15**(15), 3150 (2023)
118. Kumar, S., Manna, A., Dang, R.: A review on applications of natural Fiber-Reinforced composites (NFRCs). Materials Today: Proceedings. **1**(50), 1632–1636 (2022)
119. Mohamed Haneef, I.N., Mohd Shaffiar, N., Buys, Y.F., Syed Shaharuddin, S.I., Abdul Hamid, A.M., Widiyati, K.: Recent advancement in polymer/halloysite nanotube nanocomposites for biomedical applications. J. Biomed. Mater. Res. B Appl. Biomater. **110**(11), 2574–2588 (2022)
120. Naikwadi, A.T., Sharma, B.K., Bhatt, K.D., Mahanwar, P.A.: Gamma radiation processed polymeric materials for high performance applications: A review. Front. Chem. **14**(10), 837111 (2022)
121. Varaprasad, K., Jayaramudu, T., Kanikireddy, V., Toro, C., Sadiku, E.R.: Alginate-based composite materials for wound dressing application: A mini review. Carbohyd. Polym. **15**(236), 116025 (2020)
122. Biswal, T., BadJena, S.K., Pradhan, D.: Synthesis of polymer composite materials and their biomedical applications. Materials Today: Proce. **1**(30), 305–315 (2020)
123. Kim, K., Yoo, H., Lee, E.K.: New Opportunities for organic semiconducting polymers in biomedical applications. Polymers **14**(14), 2960 (2022)
124. Jiang, W.: Maxillofacial metal implant. (2023). https://app.biorender.com/biorender-templates/figures/all/t-64750ea8c4b1599f9c02e392-maxillofacial-metal-implant
125. Singh, S.K., Jameel, Y., Goyal, R.K., Srivastava, A.: Recent trends and future potential for polymeric materials (natural and synthetic). In: Assessment of polymeric materials for biomedical applications, pp. 1–12. CRC Press
126. Sharma, S., Rostamabadi, H., Gupta, S., Nadda, A.K., Kharazmi, M.S., Jafari, S.M.: Nano/micro-formulations of keratin in biocomposites, wound healing and drug delivery systems; recent advances in biomedical applications. Eur. Polymer J. **3**, 111614 (2022)
127. Kanjwal, M.A., Ghaferi, A.A.: Graphene incorporated electrospun nanofiber for electrochemical sensing and biomedical applications: A critical review. Sensors. **22**(22), 8661 (2022)
128. Namathoti, S., Rama Sreekanth, P.S.: A review on progress in magnetic, microwave, ultrasonic responsive Shape-memory polymer composites. Mater. Today: Proc. **56**, 1182–1191 (2022)

Chapter 13
Polymer Composites for Use in Nano-Generators for Energy Harvesting

Debabrata Mishra and Mayank Tiwari

Abstract Polymer composites exhibit unique mechanical, electrical, and thermal properties which makes them a potential candidate in the realm of nanoscale energy harvesting. The incorporation of various materials, such as ceramics, perovskites, and nanomaterials, into polymer matrices enhances the piezoelectric, pyroelectric, and triboelectric capabilities of the composite for utilization in different nano-generators for energy harvesting. Polymer composites-based nano-generators harvest energy from various sources such as vibrations, human movements, and temperature variations. These polymer composite-based nano-generators also hold significant value for the advancements in self-sustaining electronic devices, wearable sensors, and Internet of Things (IoT) applications. This chapter covers the basic principles and workings of different types of nano-generators including piezoelectric, triboelectric, pyroelectric, and hybridized nano-generators, followed by the polymer composites utilized in the development and performance enhancement of those nano-generators.

Keywords Piezoelectric · Triboelectric · Pyroelectric · Hybridized · Nano-generators · Ceramics · Perovskites · And composites

13.1 Introduction

Due to the increasing population, rapid industrialization, advancements in technology, and intractable demand for energy, effective energy harvesting and storage have become increasingly critical [1]. Nano-generators have emerged as a promising solution to energy storage and harvesting challenges. Nano-generators are devices capable of generating electrical energy from various sources such as mechanical, thermal, etc. at nanoscale levels [2]. Nano-generators have drawn significant attention within the arena of nanotechnology and energy harvesting due to their potential

D. Mishra (✉) · M. Tiwari
Department of Physics and Astrophysics, University of Delhi, New Delhi 110007, India
e-mail: dmishra@physics.du.ac.in

S. Moharana et al. (eds.), *Polymer Composites*, Engineering Materials,
https://doi.org/10.1007/978-981-97-2075-0_13

application in powering compact electronic devices, sensors, and low-power electronics, all without reliance on conventional batteries or external power supplies. Nano-generators present a potentially effective resolution for the extraction of energy from diverse environmental sources, including human motion, vibrations, temperature variations, and many more. Consequently, it holds significant value for the development of self-sustaining systems, wearable electronic devices, and Internet of Things (IoT) applications [3]. Polymer composites have been widely utilized in the field of energy storage and harvesting. These composite materials can be synthesized to harvest energy from various sources, including vibrations, human movements, or environmental mechanical stimuli. The unique mechanical, electrical, and thermal properties of polymer composites such as their flexibility, high permittivity, stability at high-temperature ranges, piezoelectricity, triboelectricity, pyroelectricity, etc. make them a potential candidate for the development of nano-generators and energy storage devices.

This pivotal chapter on "Polymer composites for use in nano-generators and energy storage" is divided into four sections. In each section, the basic principles and working a particular type of nano-generator will be discussed followed by the polymer composites utilized in the development of those nano-generators. These sections and subsections also cover the role of polymer composites in enhancing the output performances of the nano-generators. The type of nano-generators incorporating polymer composites discussed in this chapter includes (i) Piezoelectric Nano-generators (ii) Triboelectric Nano-generators (iii) Pyroelectric Nano-generators (iv) Hybridized Nano-generators.

13.2 Polymer Composites for Use in Piezoelectric Nano-Generators (PENGs)

Polymer composite materials can highly affect the growth and advancement of piezoelectric nano-generators. These composite materials including polymer-ceramic-based composites, polymer-nanoparticle composites, etc. provide flexible support for converting mechanical vibrations into electrical energy. These composites offer structural integrity, flexibility, and durability. The integration of nanoscale piezoelectric materials, such as piezoelectric nanoparticles or nanowires, enhances the collective piezoelectric phenomenon.

This first section of the chapter focuses on the basic principle and working of Piezoelectric Nano-generators (PENG) followed by polymers and ceramics utilized in PENGs, and finally discusses the role and application of different polymer composites in enhancing the piezoelectric characteristics of PENGs for energy harvesting applications.

13.2.1 Piezoelectricity and Piezoelectric Nano-Generators (PENGs)

Piezoelectric nano-generators (PENGs) are devices capable of converting mechanical energy to electrical energy using a piezoelectric material. Piezoelectric materials exhibit direct and indirect piezoelectric effects and therefore produce electrical signals on the application of external mechanical stress and can develop strain when subjected to external electric fields [4, 5]. Not all crystal structures possess piezoelectric properties. Twenty out of thirty-two crystals with non-centrosymmetry show piezoelectric abilities. The piezoelectric mechanism can be understood from the schematics shown in Fig. 13.1. Figure 13.1a represents a non-centrosymmetric crystal structure with positive and negative ions. When the crystal is subjected to external pressure, the centers of positive and negative charges separate, resulting in a dipole formation as shown in Fig. 13.1b. Thus, pressure leads to deformations in the crystal structure and the formation of a dipole. This mechanical energy (pressure) conversion to electrical energy (charge separation) can be harvested using PENGs. The two modes in which a piezoelectric material can operate are shown in Fig. 13.1c. Generally, the d_{33} mode is preferred over the d_{31} mode in piezoelectric energy harvesting.

Figures of merit (FOM_S) are important parameters for comparing the characteristics of different piezoelectric materials, PENGs, and other piezoelectricity-based devices. The important figures of merit include piezoelectric charge constant (d), piezoelectric voltage constant (g), and electromagnetic coupling factor (k^2), which are given in Eq. 13.1, 13.2, and 13.3, respectively [6].

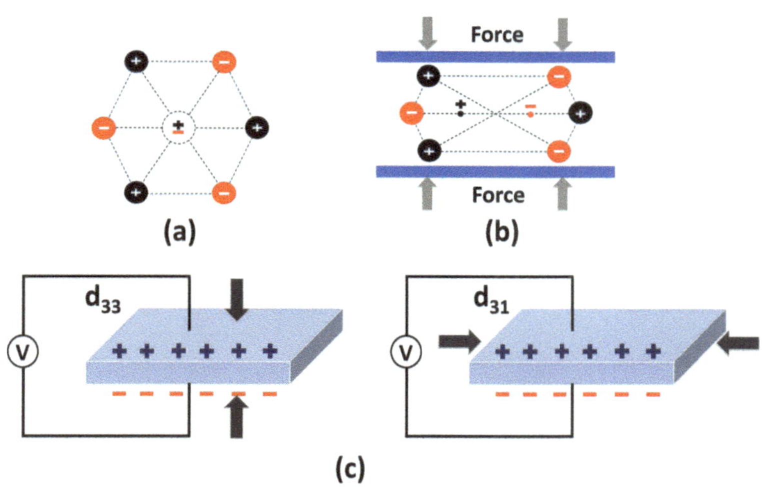

Fig. 13.1 Schematic illustration of the principle of piezoelectric effect and modes utilized in piezoelectric energy harvesting

$$d = \frac{x}{T} = \frac{S}{E} \quad (13.1)$$

$$g = \frac{E}{T} \quad (13.2)$$

$$k^2 = \frac{\text{mechanicalenergystored}}{\text{inputmechanicalenergy}} \quad (13.3)$$

where x represents the charge density, T is the applied stress, S is strain, and E is the external electric field.

13.2.2 Polymers and Ceramics in PENGs

The most widely used piezoelectric materials include ceramics. Piezoelectric ceramics, including lead zirconium titanate (PZT), barium titanate ($BaTiO_3$), calcium titanate ($CaTiO_3$), etc., have a perovskite structure (ABO_3) and demonstrate enhanced piezoelectric properties such as a high piezoelectric constant and a high electrochemical coupling coefficient. However, due to their high stiffness and large resonant frequencies, their usage becomes limited for energy harvesting from vibrations with low frequencies. Another class of piezoelectric materials for practical applications in PENGs includes piezoelectric polymers. Polyvinylidene fluoride (PVDF) and its copolymer polyvinylidene fluoride-co-tetrafluoroethylene (PVDF-TrFE) exhibit good piezoelectric properties and are flexible. PVDF possesses two prominent phases, including the α phase and the β phase. The α phase is most stable but non-polar; however, the β phase exhibits polarity due to the non-zero dipole moment because of the opposite positioning of hydrogen and electronegative fluorine atoms on the carbon chain [7, 8]. Figure 13.2a represents the polar β phase of PVDF and Fig. 13.2b represents the non-polar α phase of PVDF. Piezoelectric polymers are flexible, light, and have good piezoelectric properties; however, they lack mechanical strength and durability, and they do not withstand high-temperature ranges.

13.2.3 Polymer Composites in PENGs

The combination of piezoelectric polymer and ceramic together form piezoelectric polymer composites. These composites can be classified into (a-b) patterns, where a represents the electrochemically active phase and b represents the inactive polymer phase [9]. These piezoelectric composites can enhance the piezoelectric properties and performance of the PENGs. PVDF-PZT, a polymer-ceramic composite is the most commonly utilized piezoelectric composite in PENGs. PVDF-PZT combines

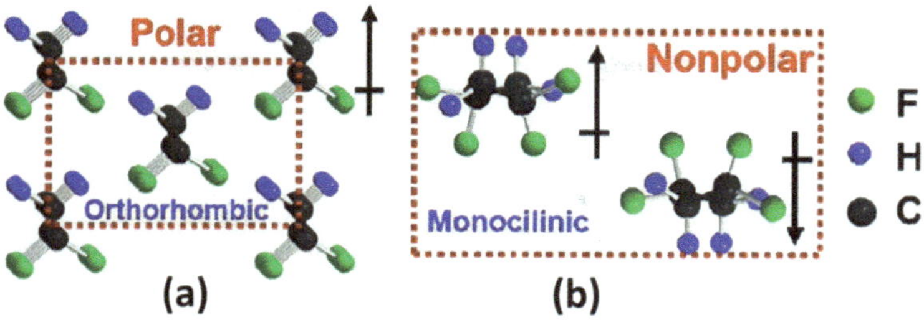

Fig. 13.2 Schematic diagram of **a** polar β phase and **b** non-polar α phase of PVDF. Reprinted from Ref. [8], Copyright (2021), with permission from Springer Nature

the flexibility of PVDF with the high piezoelectric properties of PZT, resulting in superior energy harvesting [9, 10]. In recent work, DU et al. prepared multi-layer PVDF/PZT composites for the construction of PENGs with high performance. PZT particles were embedded in PVDF fibers forming a porous structure and facilitating β phase of PVDF. The PENG utilizing 10 wt% of PVDF/PZT composite demonstrated an output voltage of 62 V with 136.9 μW power. The PENG also showed remarkable flexibility with Young's modulus of 227.2 MPa. Furthermore, the PENG was utilized to charge a capacitor of 1μF and LED lightning. The capacitor was charged to 6.4 V in 65 s and 21 LEDs were illuminated by the PENG utilizing PVDF/PZT composite [11]. Another work on PVD/PZT composite was done by Wankhade et al. The group fabricated PVDF/PZT nanohybrid system-based PENG for energy harvesting. The structured system demonstrated significant durability, affordability, and high efficiency. 30 wt% of PZT in the PVDF matrix was used for the generation of output voltage of 54 V and 36 μW/cm^2 power density. Moreover, the output voltage in the PVDF/PZT hybrid nano-generator can be produced from different motion activities including finger tapping, bending, and tapping of feet [11]. Another ceramic-based polymer composite used in PENGs includes ZnO/PVDF. ZnO is a biocompatible and non-toxic piezoelectric material [12]. ZnO/PVDF composites display enhanced piezoelectric properties and can be used in self-power nano-devices [6]. Joshi et al. in their work, synthesized ZnO flowers and PVDF composite for the fabrication of PENG devices. The device achieved a remarkable open circuit voltage of 25.5 V when subjected to an applied force of 20 N and a frequency of 7 Hz. The exceptional piezoelectric capabilities were achieved due to the enhancement of many interfacial sites between polyvinylidene fluoride (PVDF) and the textured surface of wurtzite zinc oxide (ZnO). The ZnO/PVDF composite demonstrated piezoelectric potentials ranging from 120 mV to 25.5 V. The highest recorded power density was 35.7 μW/cm^2, achieved with a load resistance of 0.7 MΩ. Moreover, the PENG (3 cm × 3 cm), was able to provide illumination to 33 red LEDs, thereby validating its superior energy harvesting abilities [13]. Mahapatra et al. designed a PENG utilizing ZnO/PVDF composite with an open circuit voltage of 14.6 V and a short circuit current of 0.6 μA. The device exhibited a power density of 21.32 μW/cm^2 and was capable of powering LEDs and charging capacitors. Furthermore, 10 wt% of the composite was used to fabricate a shoe insole pedometer which was successfully tested for 5000 steps at various frequencies [14].

Perovskite-based polymer composites have gained much attention from researchers in the arena of piezoelectric nano-energy harvesting. ABO_3/PVDF-based perovskite polymer composites are widely utilized in PENGs. In the ABO_3 perovskite structure, A and B represent cations and O represents an anion. Piezoelectric polymer composites, specifically $BaTiO_3$/PVDF and $CaTiO_3$/PVDF composites, have drawn increasing interest and widespread utilization in the field of energy harvesting in recent years. These polymer perovskite composites exhibit higher energy output without disrupting the flexibility of the nano-generators. $BaTiO_3$/PVDF and $CaTiO_3$/PVDF-based composites can be utilized in several applications including biomechanical energy harvesting, space application, and various other biomedical applications [15, 16]. Panda et al. in their published article, have reported the development of biocompatible $CaTiO_3$/PVDF as PENGs which can be utilized as a power source. The dispersion of $CaTiO_3$ (CTO) particles inside the PVDF matrix was capable of generating an output voltage of 20 V and 250 nA of current. This performance was observed for 8 wt% of the composite film. Moreover, at 10^8 Ω, the device exhibited a high power density of 0.19 μW/cm^2. Figure 13.3a represents the synthesis procedure of the PVDF-CTO composite. Different steps performed in the synthesis process are mentioned in the figure and the detailed process can be found elsewhere [17]. Figure 13.3b demonstrates the working mechanism of the PENG. Initially, the dipoles are oriented randomly, as soon as a force is applied and removed there is a generation of electricity that can be harvested. Practical implementation of the device was utilized in demonstrating the correct way of skipping rope or jumping. With the assistance of an artificial neural network model, the accuracy of the device was evaluated to be 99.63%. Figure 13.3c represents the correct way of skipping demonstrated through the artificial neural network schematic in Fig. 13.3d [17].

$BaTiO_3$ (BTO) is a lead-free perovskite exhibiting a high piezoelectric coefficient. However, its applicability is hindered by its inherent brittleness, which poses limitations on its utilization. Integration of BTO with PVDF can resolve the issue of flexibility and enhance its energy harvesting abilities in PENGs. Sahu et al. designed a PVDF-BTO composite-based PENG which is robust as well as cost-effective. The PENG incorporated 10 wt% of PVDF-BTO composite and demonstrated a significant electrical output voltage of 7.2 V and a current of 38 nA. At a resistance of 100 MΩ, the power density of the nano-generator was measured to be 0.8 μW/cm^2. The PENG was able to harness energy from various vibrations and charge a 10 μF capacitor. The capacitor was in turn used to power up a calculator thus demonstrating its practical application of the fabricated PVDF-BTO composite-based PENG in energy harvesting [18]. Another lead-free perovskite utilized in the PENG is $Cs_3Bi_2I_9$. Currently, extensive research is in progress for $Cs_3Bi_2I_9$-based devices for their energy harvesting usage and functionalization. $Cs_3Bi_2I_9$ belongs to the $A_3Bi_2X_9$ (A: Cs, X: I, Br, Cl) bismuth-based halide perovskites, known for their enhanced efficiency of power conversion [19]. Mondal et al. synthesized $Cs_3Bi_2I_9$ (lead-free perovskite) in PVDF for PENGs. The $Cs_3Bi_2I_9$-PVDF-based PENG exhibited exceptional mechanoelectrical-conversion capabilities, as evidenced by its open

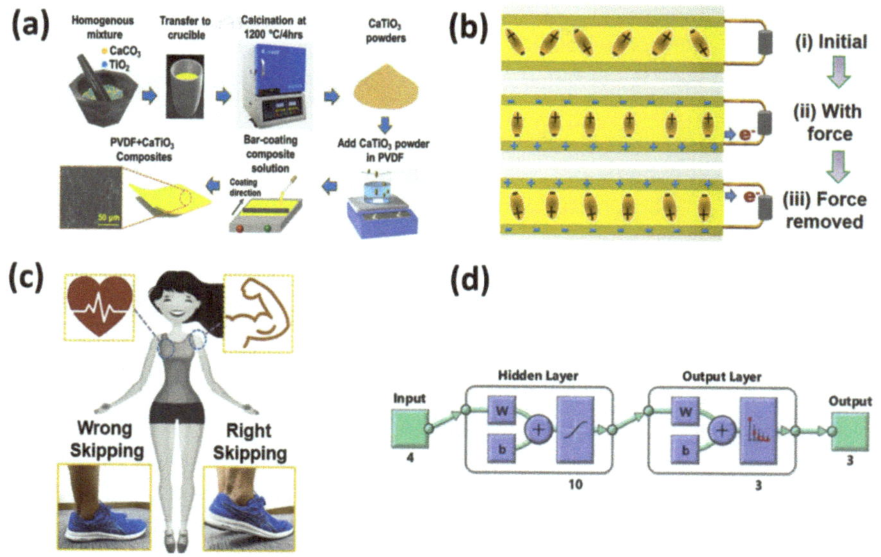

Fig. 13.3 **a** Schematic diagram of the synthesis of $CaTiO_3$/PVDF for utilization in PENG. **b** Working mechanism of PENG. **c** Correct way of jumping demonstrated through **d** artificial neural network. Reprinted from Ref. [17], Copyright (2022), with permission from Elsevier

circuit voltage of approximately 3.9 V and short circuit current of approximately 2 μA when subjected to continuous repeated stress. The device was positioned on different parts of the human anatomy to facilitate the monitoring of physiological signals. These signals incorporate the real-time detection of the movement of joints, speech patterns, and sensing of figure gestures. Moreover, the device also exhibited photodetection properties when subjected to applied strain, thus directing the application of the device toward piezo-photonics [20].

In summary, ceramics exhibit large piezoelectric properties and high-temperature stability, however, they are brittle. On the other hand, polymers exhibit weak piezoelectric properties compared to ceramics, however, they are flexible. Polymer-based composites show various advantages including, strong piezoelectric properties, high flexibility, and significant operating temperature ranges. Table 13.1 illustrates the comparison of some of the important qualitative properties of ceramics, polymers, and composites, as reported in the published article [6].

Table 13.1 Comparison of various properties of ceramics, polymers, and polymer-based composites for utilization in PENGs

Property	Ceramic	Polymers	Polymer composites
Piezoelectric constant	Large	Small	Large
Voltage constant	Small	Large	Large
Flexibility	Small	Large	Moderate
Quality factor (mechanical)	Moderate	Small	Large

13.3 Polymer Composites for Use in Triboelectric Nano-Generators (TENGs)

Polymer composites can boost the triboelectric performance of TENGs. Incorporation of different materials like metal nanoparticles, oxides, perovskites, etc. into the polymer matrix enhances the conductivity and surface area thus enhancing the triboelectric properties. These composites offer many other advantages for triboelectric energy harvesting which will be discussed in the upcoming sections.

This section of the chapter focuses on the basic principle and working of Triboelectric Nano-generators (TENG) followed by polymers and copolymers utilized in TENGs, and finally discusses the advantages and application of different kinds of polymer composites in enhancing the triboelectric characteristics and development of TENGs for energy harvesting applications.

13.3.1 Triboelectricity and Triboelectric Nano-Generators

Triboelectric nano-generators (TENGs) are micro-nano energy harvesting devices that facilitate the conversion of mechanical energy into electrical energy by utilizing the triboelectric effect and electrostatic induction principles. The triboelectric effect is a phenomenon that occurs when two materials make contact (or slide) and subsequently separate, leading to the migration of electrons from one material to the other as a result of their disparate electron affinities. Triboelectric materials are the class of materials representing these triboelectric properties. When these materials are employed in a triboelectric nano-generator (TENG), the reciprocal actions of contact and separation induce the production of an electric charge, which can subsequently be harnessed and stored as electrical energy [21]. Figure 13.4 represents the basic working principle of a TENG based on a double dielectric trilayer. Figure 13.4a demonstrates the mechanism of contact-separation mode in TENG. The two dielectric materials and their conductive support are assembled with an external circuit and load resistance as shown in (i) of Fig. 13.4a. During contact ((ii) of Fig. 13.4a), opposite charges accumulate on the surface of both dielectric materials, and a current (i) starts to flow in the external circuit to compensate for the additional charges. Similarly, a current in the opposite direction flows when the two layers separate, as shown in (iii) of Fig. 13.4a. Repeated cycles of this contact-separation process lead to the generation of an alternating current in the circuit. Similarly, an alternating current can be generated by sliding a dielectric layer over another. The mechanism of sliding is represented in Fig. 13.4b. The other modes of operation of TENGs include single electrode mode and free-standing mode represented in Fig. 13.4c, d, respectively.

The performance of a TENG can be estimated by the concept of the figure of merit for performance (FOM_P) [22]. FOM_P constitutes two parts: structural figure of merit (FOM_S) and material figure of merit (FOM_M). The former is related to the structure of the material and the latter is related to the surface charge density of the triboelectric material. Mathematically:

$$FOM_S = 2\varepsilon_0 E_m/\sigma^2 A x_{max} \quad (13.4)$$

$$FOM_P = FOM_S \times \sigma^2 = 2\varepsilon_0 E_m/A x_{max} \quad (13.5)$$

where E_m is the largest possible output energy, σ is the surface charge density, and x_{max} represents the maximum distance between the two materials.

Maximum surface charge density depends upon the chemical and band structure and is a material property that directly affects the output of TENGs. The other

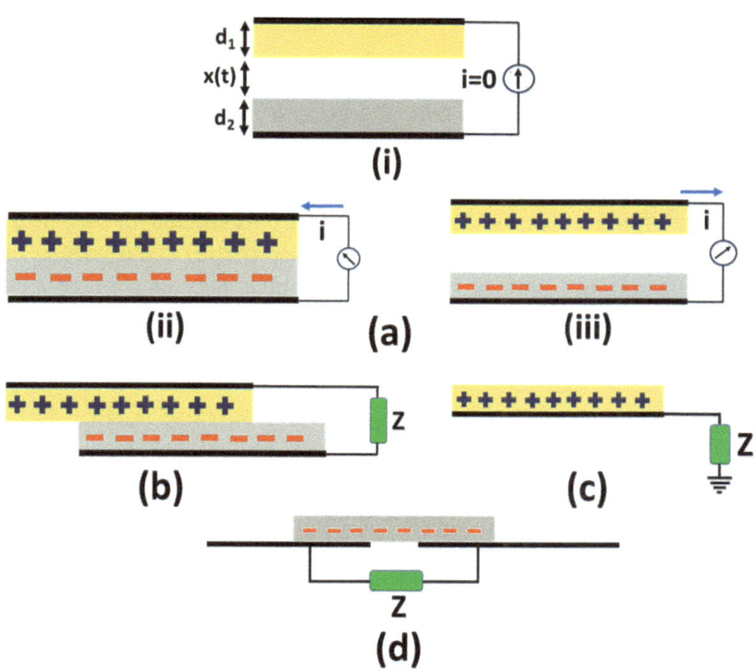

Fig. 13.4 Schematic representation of working of Triboelectric Nano-generators in **a** contact-separation mode **b** linear sliding mode **c** single electrode mode and **d** free-standing mode

parameters for the assessment of the properties of TENGs include open circuit voltage (V_{OC}), short circuit charge transfer (Q_{SC}), and short circuit current (I_{SC}) given by Eqs. 13.6, 13.7, and 13.8, respectively [23].

$$V_{OC} = \frac{\sigma \times x(t)}{\varepsilon_0} \tag{13.6}$$

$$Q_{SC} = \frac{S\sigma x(t)}{d_0 + x(t)} \tag{13.7}$$

$$I_{SC} = \frac{dQ_{SC}}{dt} = \frac{S\sigma d_0}{(d_0 + x(t))^2}\frac{dx}{dt} = \frac{S\sigma d_0 v(t)}{(d_0 + x(t))^2} \tag{13.8}$$

$$d_0 = \sum_{i=1}^{n} \frac{d_i}{\varepsilon_{r,i}} \tag{13.9}$$

where x(t) represents the distance between the triboelectric layers, S is the surface area, σ is the surface charge density, v(t) is the contact speed, ε represents the permittivity, and d_0 is a thickness parameter given by Eq. 13.9. From Eqs. 13.4, 13.5, 13.6, 13.7, and 13.8, it becomes evident that the factors including surface charge density, thickness, distance, contact speed, and surface area are vital for enhancing the current, voltage outputs, and performance of TENGs.

13.3.2 Polymers and Copolymers Utilized in TENGs

Triboelectric polymers incorporate various properties, including high charge density, flexibility, cost-effectiveness, and flexibility, and are thus utilized for the construction of TENGs. One way of enhancing the triboelectric properties of polymers is through the proper selection and synthesis of monomer units. The selection of a positive and a negative polymer in the series of triboelectric layers can enhance the TENG output. Negative materials include high-fluorine-containing polymers with excellent negative polarity and high surface charge densities. Some widely utilized negative triboelectric polymers include PVDF, polytetrafluoroethylene (PTFE), FEP, polydimethylsiloxane (PDMS), etc. Positive materials, on the other hand, have a high tendency to donate electrons. Positive triboelectric polymers offer various advantages, such as flexibility, processability, and low cost. Polyamide, epoxy resin, and amino resin are some of the positive polymers. Another viable approach to enhancing triboelectric properties is copolymerization. Copolymerization is the strategy of combining the properties of two or multiple polymers to form a copolymer. PVDF-based copolymers, including P(VDF-TrFE), P(VDF-HFP), and P(VDF-TFE), can augment dipole orientation and crystallization through the process of copolymerization with different monomers. These copolymers can enhance the triboelectric output of the TENGs.

13.3.3 Polymer Composites in TENGs

The polymers utilized in TENGs possess several disadvantages, such as low mobility of surface charge carriers, a low dielectric constant, and low conductivity. Moreover, the surface charge generated due to the triboelectric effect is lost after a short period. These challenges can be overcome by integrating other materials with the polymers to create polymer composites. For instance, to enhance the dielectric constant of a polymer, a material with a large dielectric constant can be added to the polymer to form a composite. This can also enhance the charge transfer abilities, and the polymer composite can be utilized for high output. Various studies have demonstrated that the incorporation of metal nanoparticles like gold, silver, and copper can enhance the dielectric constant of the composite material. Inside metals, the electric field does not exist, and hence the permittivity is large, which makes the dielectric constant of metals almost infinite [23]. Chen et al. designed a TENG by incorporating gold nanoparticles (Au-NPs) with polytetrafluoroethylene (PTFE) polymer. Integrating PTFE films with Au-NPs enhanced the output current by 70% and the surface charge density rose to 85 $\mu C/m^2$. The AU-NPs work as nano charge traps in the composite, forming a nanocapacitor between Au and PTFE molecules. Moreover, the Au-PTFE composite-based TENG demonstrated excellent stability and durability [24]. A similar work on enhancing the output characteristics of TENG using polymer and metal nanoparticle composites was published by Xia et al. The group utilized PDMS polymer and filled it with Ag nanoparticles, resulting in the formation of an internal cellular structure. The sugar and silver nanoparticles were first stirred in a dish with PDMS. The composite film formed was then shaped in a desired pattern. The film was cured at 330 K and was subsequently washed to remove sugar particles. The composite-based device was then connected to the external copper foil electrodes to form a circuit. Steps (1), (2), (3), (4), (5), and (6) in Fig. 13.5(a) represent this process of embedding Ag-NPs in PDMS-based films for the fabrication of a TENG. The composite-based TENG exhibited four times higher current density than the PDMS-based film. Moreover, the output power density was reported to be 6 W/m^2 at 20 MΩ load resistance. The TENG was also capable of lightning 100 L.E.Ds in series. Figure 13.5b–d represents the output characteristics of Ag/PDMS composite-based TENG [25].

Another work on polymer-metal nanoparticle-based composites in TENG was published by Zhang et al. The group fabricated a TENG with the help of Cu-NPs doped thermoplastic elastomer (TPE) thin film, forming a TPE composite. The TENG exhibited a high open circuit voltage of 600 V and a short circuit current of 36 μA using Cu-NPs-based TPE composite. Integrating Cu-NPs enhanced the output performance of the TENG by 1.5 times. The TENG was capable of generating a power of 12 mW for the external resistor of 3 MΩ when a force of 60 N was applied. Furthermore, this TENG was capable of harvesting energy from different motions generated by the human body, therefore, demonstrating the ability of the composite-based TENG in wearable devices [26]. Another class of materials that can be integrated with polymers to form composites for TENG includes ferroelectric oxide-based materials. Ferroelectric oxide nanoparticles when embedded in polymers enhance the permittivity of the resulting composite due to the spontaneous polarization. Polymer and ferroelectric oxide-based composites can readily enhance the output of TENGs. TiO_2, SiO_2, $BaTiO_3$, and $SrTiO_3$ are the most widely utilized for the formation of polymer-ferroelectric oxide composites for the development of TENGs. Chen et al., in their published work, demonstrated the effect of incorporating oxide with a high dielectric constant, on the output characteristics and performance of TENGs. Integrating the nanoparticles of SiO_2 ($\varepsilon_r = 3$), TiO_2 ($\varepsilon_r = 80$), $BaTiO_3$ ($\varepsilon_r = 150$), and $SrTiO_3$ ($\varepsilon_r = 300$), in PDMS films results in the formation of polymer composite. Table 13.2, as reported in [27], represents the relative permittivity ($\varepsilon_{exp,}$ ε_{theory}) and charge transfer (Q (nC/cm^2)) of different oxide nanoparticles when filled in PDMS films [23, 27]. $SrTiO_3$/PDMS composite exhibited the highest charge transfer and permittivity. The group further utilized the composites for the development of TENG. It was observed that the open circuit voltage and short-current were 305 V and 7.18 μAcm^{-2} when PDMS films were embedded with $SrTiO_3$ NPs (10 vol % in PDMS), which was approximately 1.8 and 2.4 times than that of bare PDMS-based TENG. Furthermore, the integration of SiO_2, TiO_2, $BaTiO_3$, and NPs also enhanced the output characteristics of the bare PDMS-based TENG. Under optimized conditions, SiO_2/PDMS-based TENG exhibited an open circuit voltage of 167.6 V and a short circuit current of 2.8 μAcm^{-2}, when 5 V% of the material was utilized. The open circuit voltage and short circuit current for TiO_2/PDMS (8 V%), were 238.5 V and 5.9 μAcm^{-2}. $BaTiO_3$/PDMS(10 V%), exhibited an open circuit voltage of 270 V and a short circuit current of 6.6 μAcm^{-2}. This study provides a deeper understanding of the triboelectricity generated by the triboelectric nano-generator (TENG) in terms of materials. Additionally, it introduces an innovative approach to boost the efficiency of TENG by tweaking the intrinsic properties of the tribo-material, rather than altering its surface characteristics [23, 27].

13.4 Polymer Composites for Use in Pyroelectric Nano-Generators (PyNGs)

Thermal fluctuation from surroundings can be harvested using PyNGs. Ceramics with perovskite structures and polymers are widely used for the development of PyNGs. Polymer composites for instance 2D nanostructures like graphene nanosheets when embedded in polymer matrix result in the formation of polymer composite. These composites can increase the output performance of the PyNGs.

This third section of the chapter focuses on the basic principle and working of Pyroelectric Nano-generators (PyNGs) followed by different materials such as polymers and ceramics utilized in PyNGs, and finally discusses the advantages and application of different kinds of polymer composites in enhancing the pyroelectric properties and development of PyNGs and pyroelectric energy conversion devices for energy harvesting applications.

Table 13.2 Comparison of experimental (ε_{exp}) and theoretical (,ε_{theory}) relative permittivity and charge transfer (Q (nC/cm^2)) of different oxide nanoparticles when filled in PDMS films

Oxide filler	Volume percentage (%)	ε_{exp}	ε_{theory}	Q (nC/cm^2)
SiO_2	5	2.970	3	7.972
	10	3.035	3	7.586
	15	2.983	3	7.352
	20	2.929	3	7.262
TiO_2	5	3.586	3.535	11.389
	10	4.267	4.166	11.718
	15	4.996	4.909	10.270
	20	5.921	5.785	8.602
$BaTiO_3$	5	3.649	3.648	12.367
	10	4.577	4.436	14.071
	15	5.602	5.395	12.895
	20	6.829	6.56	10.583
$SaTiO_3$	5	4.165	3.919	13.684
	10	4.967	4.755	**16.012**
	15	6.484	5.986	15.229
	20	**7.823**	**7.535**	12.996

13.4.1 Pyroelectricity and Pyroelectric Nano-Generators (PyNGs)

The pyroelectric effect is an inherent characteristic observed in some materials, wherein they exhibit a transient voltage generation in response to alterations in their temperature with time. The fundamental principle that governs the pyroelectric effect is spontaneous polarization (P_s) observed in pyroelectric materials [28]. A temperature change, results in the change in polarization of pyroelectric materials, resulting in redistribution of charges and development of potential across the pyroelectric material. Figure 13.6 provides a clear idea about the pyroelectric effect in pyroelectric materials. Figure 13.6a depicts the dipoles with spontaneous polarization P_s in pyroelectric material. Figure 13.6b represents the connection of two conductive electrodes with the pyroelectric material (formation of a capacitor) at constant temperature. When the temperature of the pyroelectric material is increased ($dT/dt > 0$), the thermal vibrations cause an increase in the dipole oscillation, resulting in a reduction of P_S. This drives the movement of electrons in the external circuit to balance the overall charge, as shown in Fig. 13.6c. Moreover, when the temperature of the pyroelectric material is reduced ($dT/dt < 0$), the P_S increases, which results in the motion of electrons in the opposite direction as represented in Fig. 13.6d. The electrical energy generated due to the pyroelectric effect is known as pyroelectricity. The pyroelectric effect can be utilized to harvest electric energy by utilizing the heat of the environment. Pyroelectric nano-generators (PyNGs) and Pyroelectric energy conversion devices (PEC) are the devices that make use of the pyroelectric effect to harvest heat energy and generate electric output.

Some of the important parameters that characterize the performance of the pyroelectric materials and PEC are pyroelectric current (i_P), output current (i), pyroelectric coefficient (p), and figure of merit of PEC (FoM_{PEC}), mathematically represented by Eq. 13.10, 13.11, 13.12, and 13.13, respectively [28].

$$i_p = \frac{dQ}{dt} = pS\left(\frac{dT}{dt}\right) \tag{13.10}$$

$$i = \frac{da}{dt} = pA\frac{dT}{dt} \tag{13.11}$$

$$p = i_p\left(\frac{dT}{dt}\right) - 1/S \tag{13.12}$$

$$FoM_{PEC} = \frac{\pi^2 T}{C\varepsilon_0\varepsilon_r} \tag{13.13}$$

where T represents the temperature, A is the surface charge of the material, ε_r is the permittivity, a represents pyroelectric charges, and t is the time.

In order to enhance the performance of PEC, it is imperative to have a larger pyroelectric coefficient, low volume-specific heat capacity (C), and relatively low

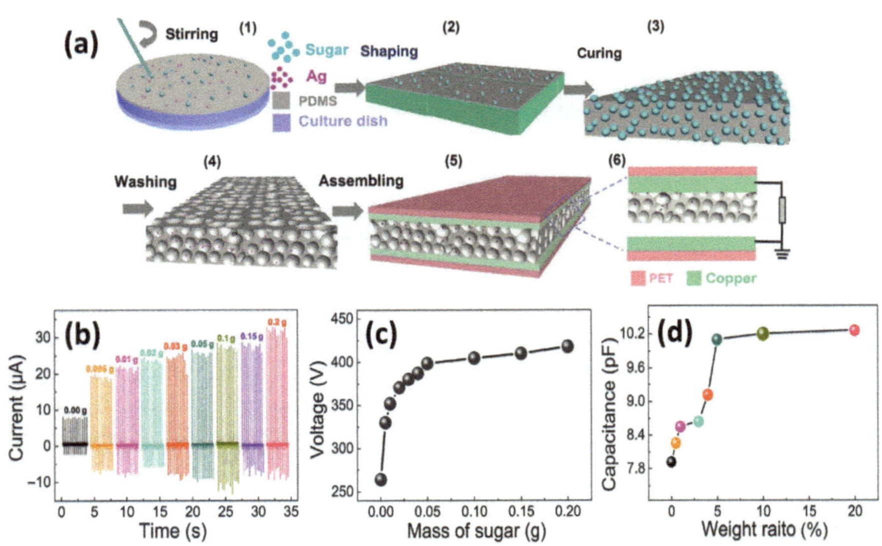

Fig. 13.5 **a** Schematic illustration of fabrication of Ag-NPs-based composite TENG. **b** Short circuit current vs time, **c** open circuit voltage, **d** capacitance, for different Ag/PDMS, composite-based TENG. Reprinted from Ref. [25], Copyright (2016), with permission from Springer Nature.

permittivity. These characteristics enable the generation of a significant amount of energy from temperature variation [28].

13.4.2 Polymers and Ceramics Utilized in PyNGs

Ceramics with perovskite structures are known for their piezoelectric, ferroelectric, and triboelectric properties, as discussed in the previous section. Another application of the perovskite ceramics can be found in PyNGs. Perovskite structured ceramics exhibit ABX_3 structure, where A belongs to the rare earth elements, B to transitional elements and X is an oxide/halide. These structures have a permanent dipole moment, offering an advantage for their utilization in piezoelectric, triboelectric, and pyroelectric energy harvesting applications. Some of the widely utilized perovskite ceramics in pyroelectric energy conversion include $BaNiO_3$, $BaTiO_3$, $BiFeO_3$, etc. Some of these perovskites exhibit low pyroelectric coefficients however, their pyroelectric characteristics can be enhanced by incorporating ions. For instance, the pyroelectric coefficient of $BaTiO_3$ can be uplifted by the substitution of Sr^{2+}, Ca^{2+} on the A site, and Sn^{+}, Zr^{4+} on the B site.

Among the polymers, PVDF is the most widely studied and utilized in pyroelectric energy harvesting. PVDF possesses unique properties such as a high pyroelectric coefficient and low dielectric constant, which can also be enhanced with composite formation and structural variation. Moreover, PVDF is highly flexible and can also withstand high-temperature ranges. These properties make PVDF a potential candidate, among polymers for utilization in pyroelectric energy harvesting applications. Furthermore, PVDF composite films (PVDF-GO, PVDF-ceramics, etc.) can achieve a high pyroelectric coefficient which will be discussed in the next section. Other polymers and copolymers utilized in pyroelectric energy harvesting include P(VDF-TrFE), PTFE, and PDMS.

13.4.3 Polymer Composites in PyNGs and PEC

Among ceramics and polymers, the former exhibit a high pyroelectric coefficient; however, the advantages of polymer-based composites, including high flexibility, ease of synthesis, cheap fabrication, and low permittivity, enable their utilization in pyroelectric energy conversion devices. 2D nanostructures have gathered significant attention in the biomedical and electrochemical fields. 2D nanostructures such as graphene oxide (GO) and reduced graphene oxide (rGO) nanosheets exhibit enhanced electric conductivity, larger surface area, and biocompatibility and hence can be utilized in various energy storage applications. Integration of GO and RGO in polymers can enhance the pyroelectric properties of the composite [29]. Abbasipour et al. in their study demonstrated that incorporating GO nanoparticles with PVDF enhances the pyroelectric properties of the composite [29]. It was observed that 1.6 wt% GO increases the output voltage by four times than the bare PVDF under 360° K. Moreover, the thermal stability of the composite increased by 49% by embedding the GO nanoparticles in PVDF. Furthermore, the thermal stability of the device was raised to 400 °C [29]. Another self-powered pyroelectric breathing sensor was developed by Roy et al. Their study presented a novel self-powered, flexible nano-generator (NG) device that combines piezoelectric and pyroelectric properties [30]. The device was designed to be attached to various areas of the human skin, enabling the detection of both static and dynamic pressure changes. Additionally, the device was capable of monitoring temperature fluctuations during the respiratory process. The group utilized electrospun PVDF/GO nanofiber to design a piezoelectric pressure sensor and pyroelectric breathing sensor. The process of fabrication of the composite and the device is illustrated in Fig. 13.7a, and the detailed process can be found in the published article. This PVDF/GO composite-based PyNG was capable of generating an output power of 1.2 nW/m^2. To test the real-time application of the NG, the device was mounted on an N95 mask which was worn on a human face. During the breathing process, temperature fluctuations occur around the mouth. This temperature fluctuation during the respiration process was monitored by the device. The temperature difference was reported to be around 5 K. The process of respiration can generate the temperature difference and thus nano-energy can be harvested from the respiration process. Figure 13.7b illustrates the temperature changes steered by respiration. Figure 13.7c represents the open circuit voltage of the device embedded in the N95 mask (shown in inset). The device exhibits application in wearable sensors and devices, human health monitoring, and other nano-energy-based systems [30].

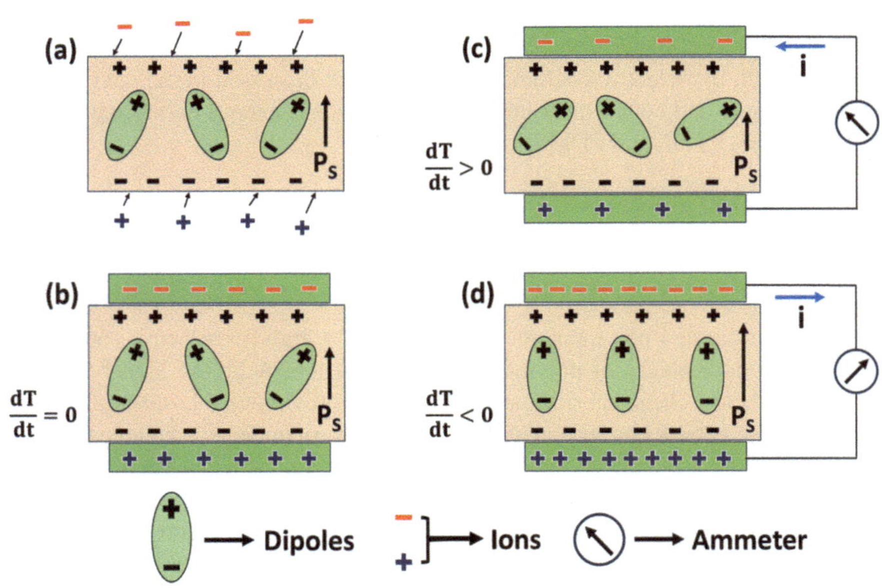

Fig. 13.6 Schematic diagram representing the principle of Pyroelectric effect

Another study on graphene-based PVDF composite was done by Ma et al. The group developed an antibacterial tactile sensor by utilizing a triboelectric-piezoelectric-pyroelectric multiple effect mechanism by utilizing Ag nanowires@PTFE/Graphene/PVDF composite [31]. The sensor exhibited remarkable sensitivity of 0.092 V/KPa and 0.11 V/°C for pressure and temperature tracking, respectively. Furthermore, the sensor was self-powered, easily fabricated, and multifunctional, thus can be utilized in the realm of healthcare and other applications [31].

Another material that can enhance the pyroelectric properties of PVDF polymer is TiO_2. Gan et al. demonstrated that filling TiO_2 with different volume concentrations in PVDF polymer can significantly enhance the pyroelectric characteristics of the resulting composite. The pyroelectric coefficient was enhanced at a significantly smaller poling electric field. The group observed an outstanding increase in the pyroelectric coefficient when 20 wt% of TiO_2 was dispersed into PVDF thin films. The pyroelectric coefficient was increased by 26% to 24.5 $\mu C\ m^{-2}\ K^{-1}$ by the incorporation of TiO_2 [32].

Nanomaterials have gained significant attention in electronic and energy harvesting applications. Nanomaterials can exist in various forms such as quantum dots, nanowires, and nanosheets [33]. Quantum dots (QDs) having particle sizes comparable to the Bohr radius have attracted significant attention due to their unique properties which can be tuned according to their dimensions. QDs are widely utilized in various energy harvesting applications such as solar energy harvesting, supercapacitors, and various nano-generators. Ahmad et al. integrated ZnO QDs in P(VDF-TrFE) to form a P(VDF-TrFE)/ZnO composite which exhibited enhanced pyroelectric properties. The incorporation of 0.15 wt% of ZnO QDs into the polymer enhanced the pyroelectric coefficient of the composite to 49 $\mu C/m^2\ K$ [34]. Yang et al. utilized $KNbO_3$ nanowires in PDMS forming a $KNbO_3$/PDMS composite. The produced composite could modulate the output voltage/current in a controlled manner. The nano-generators exhibited contrasting variations in voltage and current outputs when subjected to heating and cooling settings, with an increase in temperature change leading to an augmentation of these outputs. Pyroelectric nano-generators hold significant potential for widespread implementation in self-sustaining nano-devices and nano-systems [35].

13.5 Polymer Composites for Use in Hybridized Nano-Generators

This section of the chapter includes the hybridization of different kinds of nano-generators for enhanced energy harvesting devices. The basic principle of hybridized nano-generators and polymer composite-based different types of hybridized nano-generators will be discussed.

13.5.1 Hybridized Nano-Generators

Hybridized nano-generators represent advanced energy harvesting systems that integrate diverse technologies to produce electrical energy from a multitude of environmental sources. Hybridized nano-generators can utilize energy from multiple sources available in the environment to enhance the output delivery of the nano-generator [36]. Hybridized nano-generators enhance the effectiveness and versatility of energy harvesting systems by integrating multiple kinds of nano-generators utilizing the approach of hybridization. Hybridized nano-generators have attracted significant attention in the field due to their ability to integrate multiple nano-generators into a unified system. This integration allows for the independent or simultaneous utilization of various energy sources, thereby enabling the harnessing of ambient energy in any available form at any given moment. There has been a notable rise in the overall electrical power generation achieved by the strategic integration of two or more types of harvesters capable of extracting diverse forms of energy from the surrounding environment. Hybridized nano-generator-based devices can be utilized in multiple disciplines, such as sensors with self-power capability, charging batteries supercapacitors, etc. These gadgets have the potential to be utilized in a diverse range of applications. The next section includes the classification of different polymer composite-based hybridized nano-generators [36].

13.5.2 Polymer Composite-Based Hybridized Piezoelectric-Triboelectric Nano-Generators

Hybridized piezoelectric-triboelectric nano-generators (PTNGs) combine both piezoelectric and triboelectric effects to generate electrical energy from mechanical vibrations and movements. In hybrid PTNGs, materials with both piezoelectric and triboelectric properties are integrated into a single device, maximizing energy harvesting efficiency. By combining these two mechanisms, PTNGs can capture mechanical energy from various sources, such as human motion, vibrations, and environmental mechanical disturbances. These devices can significantly power small electronic devices, wearable technologies, and sensors, especially in applications where a constant and renewable energy source is needed. The integration of piezoelectric and triboelectric mechanisms in PTNGs enhances the overall energy conversion efficiency and ensures a more stable power output, making them promising candidates for sustainable energy harvesting solutions in the future. In recent works, Singh et al. developed a MoS_2/PVDF/PDMS-based hybrid piezoelectric-triboelectric nano-generator for harvesting mechanical energy. Integrating MoS_2 as filler into the polymer increased the output of the PENG from 9.5 V to 18 V. Further, the hybridized PTNG exhibited an output voltage of 35.3 V, which was significantly greater than the bare PVDF/PDMS films. Moreover, the hybridized PTNG was capable of lighting 21 LED bulbs by utilizing the mechanical energy produced by tapping fingers. The

PTNG was also able to charge a 10 μF capacitor to 9 V in 400 s. Figure 13.8 exhibits the enhanced performance of the hybrid PTNG for bare and MoS2-incorporated PVDF/PDMS [37].

In another recent work, Rana et al. fabricated a hybrid PTNG using boron/nitrogen-codoped rGO embedded in PVDF (PVDF/BN-rGO) in PDMS film. The PVDF/BN-rGO-based PTNG, demonstrated a significant boost in both output voltage and piezoelectric current. Specifically, the device exhibits approximately a two-fold increase in output voltage and a six-fold increase in piezoelectric current. On the other hand, the hybrid nano-generator, incorporating the composite film, demonstrated superior electrical performance. It generated an output voltage of 57.6 V and a current of 28.8 μA. The enhanced efficiency of the nano-generator was attributed to the combined influence of piezoelectric and triboelectric phenomena, as well as the incorporation of the codoped reduced graphene oxide (rGO) inside the PVDF matrix. This work emphasizes the importance of polymer composites and hybrid PTNG in energy harvesting applications [38]. Yun et al. developed a piezo-triboelectric hybridized nano-generator by incorporating Mxene-based and $BaTiO_3$ fillers in the polymer matrix. Embedding Mxene significantly enhanced the output performance of the hybridized nano-generator. The output voltage of the device was enhanced by 200% after poling. The open circuit voltage, short circuit current, and power density of the device were calculated to be 80 V, 14 μA, and 135 W/m^2, respectively. The group also exhibited the practical application of the device by implementing a

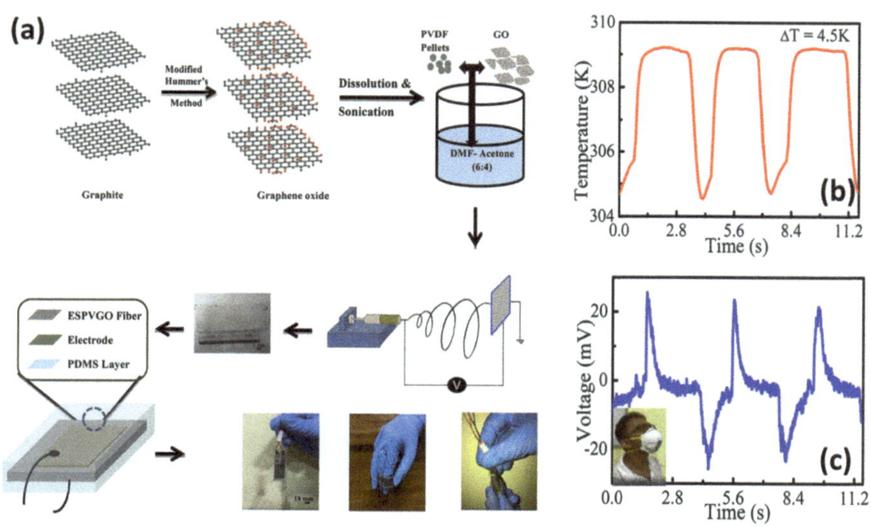

Fig. 13.7 **a** Schematic diagram illustrating the process of fabrication of PVDF/GO composite-based PyNG. **b** Fluctuations in temperature monitored by the PVDF/GO composite-based PyNG. **c** Open circuit voltage vs time plot for the PyNG when embedded in an N95 mask (shown in inset). Reprinted from Ref. [30], Copyright 2019, with permission from American Chemical Society

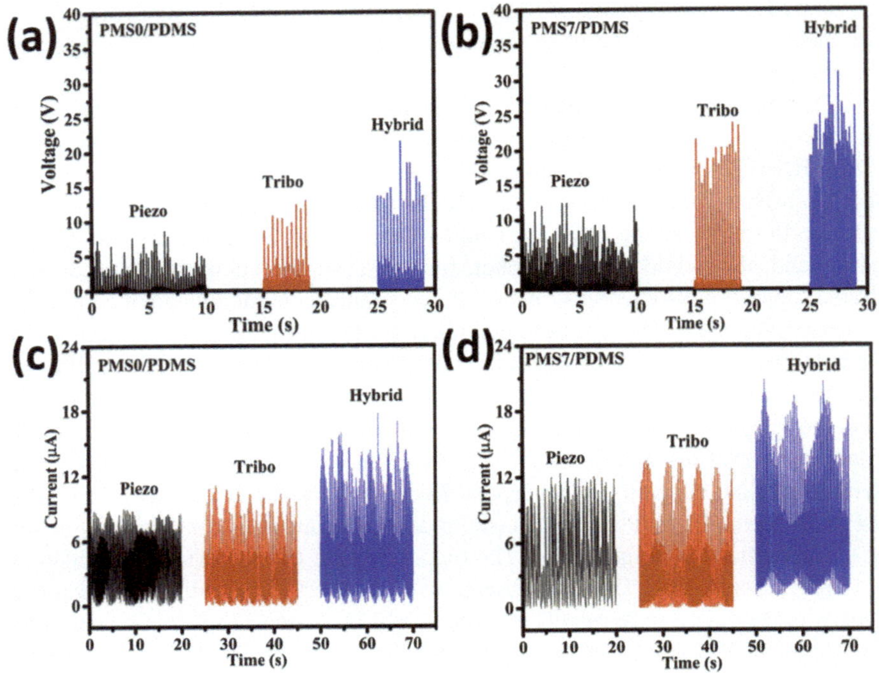

Fig. 13.8 Schematic diagram representing the enhanced output voltage for hybridized nano-generator compared to PENG and TENG, for bare (PMS0/PDMS) and 7 wt% (PMS7/PDMS) MoS_2 nanoflakes in PDMS. Reprinted from Ref. [37], Copyright 2023, with permission from Elsevier

human motion system and material detection system. The combined effects of piezoelectricity and triboelectricity in the hybrid nano-generator were investigated by the utilization of MXene-based fillers in the PDMS films [39].

13.5.3 Polymer Composite-Based Hybridized Piezoelectric-Pyroelectric Nano-Generators

As the name suggests, hybrid piezo-pyro nano-generators represent advanced energy harvesting systems that employ two distinct effects for energy conversion: piezoelectric and pyroelectric. A hybrid piezo-pyro nano-generator integrates both piezoelectric and pyroelectric principles within a single structure. The integration of different components enables the nano-generator to effectively capture and utilize energy derived from mechanical vibrations, including human motion and vibrations from variable sources, as well as temperature fluctuations arising from changes in the surrounding environment. Hybrid piezo-pyro nano-generators have the ability to generate a greater amount of power by harnessing energy from many sources, beyond the power output achievable by utilizing each mechanism in isolation. Gao et al. in their work, constructed a hybrid piezo-pyroelectric nano-generator by utilizing Ag and Cu nanoparticles in a PDMS matrix. The incorporation of magnetic nanoparticles enhances the polarization in the composite. The piezoelectric and the pyroelectric figures of merit were enhanced remarkably by 700%. The device was capable of harvesting mechanical and thermal energy from activity like running and thus was able to charge a Bluetooth headset by 10% in 0.5 h and a smart bracelet to 17% in 1 h. Composite-based materials with versatile energy harvesting abilities can hence increase the output performance of the energy harvesting devices [40]. Sultana et al. in their work, constructed a piezo-pyroelectric nano-generator with the help of methylammonium lead iodide ($CH_3NH_3PbI_3$) embedded with electrospun PVDF nanofibers. The piezo-pyroelectric nano-generator was capable of harvesting thermal as well as mechanical energy. The composite utilized in the nano-generator exhibited a piezoelectric coefficient (d_{33}) of around 19.7 pC/N. The power density of the composite was around 0.8 mW/m^2, which was the maximum output power density, and was adequate for charging various capacitors thus, eliminating the need for an external power supply to operate portable electronics. Furthermore, when subjected to a cyclic thermal process involving alternating heating and cooling at a specific temperature of 38 K, the composite exhibited an output current of 18.2 pA and a voltage of 41.78 mV. The hybrid piezo-pyroelectric nano-generator exhibited a self-powered temperature sensing capacity, as evidenced by its quick response time of 1.14 s, reset time of 1.25 s, and pyroelectric coefficient of around 44 pC/m^2 K. The aforementioned features render the composite well-suited for flexible piezoelectric-pyroelectric energy harvesting, hence enabling its application in self-powered electronic devices [41].

13.5.4 Polymer Composite-Based Other Hybridized Nano-Generators

The integration of solar energy and triboelectricity represents an innovative approach to energy harvesting by combining renewable solar energy with the versatility of mechanical energy. Solar panels are designed to absorb and convert sunlight into usable energy during daylight hours. On the other hand, triboelectric nano-generators (TENGs) are devices that can generate electrical energy by utilizing mechanical sources such as vibrations, wind, and even the impact of raindrops. The integration of these two components results in a mutually beneficial relationship that guarantees a consistent provision of energy, even in situations characterized by little sunlight or unfavorable weather conditions, thereby demonstrating a high level of efficiency and sustainability. Liu et al. integrated a Silicon solar cell with TENG. Water drop TENG was constructed on the solar cell by combining PDMS with poly(3,4- ethylenedioxythiophene):poly(styrenesulfonate) (PEDOT:PSS) film. The output of the TENG was enhanced, with a short circuit current of 33.0 nA and an open circuit voltage of 2.14 V, respectively. The integrated system was capable of converting both the solar and the mechanical energy from water droplets thus enhancing the overall energy harvesting capability of the integrated system. This study also demonstrates the utilization of energy in different weather conditions by combining solar and mechanical harvesting techniques [42].

Harvesting mechanical energy only from vibrations and mechanical stress results in several issues including large energy losses, mechanical fatigue and structural damage to the device, heat dissipation, etc. However, these losses can be minimized and output energy can be enhanced by combining different mechanical energy and thermal energy conversion effects into a single device. Wang et al. developed a highly efficient hybridized electromagnetic-triboelectric-thermoelectric nano-generator. The hybridized nano-generator had an electromagnetic generator (EMG), a triboelectric nano-generator (TENG) for harvesting mechanical energy, and a thermoelectric nano-generator (ThNG) for harvesting thermal energy. The construction of the device included the utilization of different materials including Polyamide, PTFE, copper, magnet, protection layer, acrylic, rotor, and heat transfer plate. The construction of the hybrid device can be found in the published article. The device was capable of producing a constant output voltage of 5 V and a pulsed output peak current of 160 mA. Furthermore, the device was successfully implemented in harvesting biomechanical energy for lighting globe lights and phone charging [43].

13.6 Conclusion

In conclusion, the investigation of polymer composites for use in nano-generators offers an interesting approach to overcome the inherent limits of individual materials. In PENGs, polymer composites offer enhancement in the piezoelectric properties by combining the significant piezoelectric characteristics of ceramics with the flexibility of polymers. Moreover, polymer composites overcome the brittleness of ceramics and lower piezoelectricity of pure polymers, thus resulting in enhanced piezoelectric behavior of the composites for use in piezoelectric nano-generators. Incorporation of nano-fillers including metal nanoparticles and materials with high dielectric constant into polymers improves the charge transfer and the mobility of the polymer composite leading to the potential for high-output in triboelectric TENGs. For PyNGs the integration of 2D nanostructures, such as graphene oxide (GO) and reduced graphene oxide (rGO), along with other nanomaterials, into polymer composites demonstrate remarkable pyroelectric energy harvesting capabilities. Moreover, enhanced pyroelectric properties can be achieved by incorporating materials like GO, TiO_2, and quantum dots into polymers like PVDF, enabling efficient pyroelectric energy harvesting. Furthermore, polymers composite can also be utilized in hybridized nano-generators, wherein they effectively merge the benefits of different types of materials and nano-generators to improve energy harvesting. Incorporating polymers with other materials in hybridized nano-generators optimizes their efficiency by synergistically gathering energy from multiple sources and providing enhanced electrical outputs. Moreover, these polymer composite-based nano-generators can be utilized in the fabrication and development of self-powered electronics, wearable devices, biosensors, etc. Polymer composites used in nano-generators not only surpass the limitations of individual materials but also demonstrate a viable path toward sustainable and efficient energy harvesting solutions.

Acknowledgements D.M. acknowledges the financial support from MHRDDST through Grant No. IMP/2018/001668, SERB through Grant No. CRG/2018/004264, IoE, University of Delhi through Grant No. IoE/2021/12/FRP, and UGC through Grant No. F.4-5(202- FRP)/2015(BSR). M.T. acknowledges UGC and CSIR New Delhi for providing the financial support.

References:

1. Shaqsi, A.Z.A., K. Sopian, and A.J.E.R. Al-Hinai, Review of energy storage services, applications, limitations, and benefits. 2020. 6: p. 288–306.
2. Wang, Z.L., et al.: Progress in nanogenerators for portable electronics. **15**(12), 532–543 (2012)
3. Feng, H., et al.: Nanogenerator for biomedical applications. **7**(10), 1701298 (2018)
4. Ramadan, K.S., et al.: A review of piezoelectric polymers as functional materials for electromechanical transducers. **23**(3), 033001 (2014)
5. Hao, J., et al.: Progress in high-strain perovskite piezoelectric ceramics. **135**, 1–57 (2019)
6. Habib, M., I. Lantgios, and K.J.J.o.P.D.A.P. Hornbostel, A review of ceramic, polymer and composite piezoelectric materials. 2022.
7. Sappati, K.K., Bhadra, S.J.S.: Piezoelectric polymer and paper substrates: A review. **18**(11), 3605 (2018)
8. Zhu, H.J.P.J.: Interfacial preparation of ferroelectric polymer nanostructures for electronic applications. **53**(8), 877–886 (2021)
9. Uchino, K., Advanced piezoelectric materials: Science and technology. 2017: Woodhead Publishing.
10. Pradhan, S.K., et al.: Piezoelectric and mechanical properties of PVDF-PZT composite. **558**(1), 59–66 (2020)
11. Du, X., et al., Porous, multi-layered piezoelectric composites based on highly oriented PZT/PVDF electrospinning fibers for high-performance piezoelectric nanogenerators. 2022. 11: p. 331–344.
12. Bhadwal, N., R. Ben Mrad, and K.J.S. Behdinan, Review of Zinc Oxide Piezoelectric Nanogenerators: Piezoelectric Properties, Composite Structures and Power Output. 2023. 23(8): p. 3859.
13. Joshi, B., et al.: Supersonically sprayed PVDF and ZnO flowers with built-in nanocuboids for wearable piezoelectric nanogenerators. **112**, 108447 (2023)
14. Mahapatra, A., et al., Flexible ZnO: PVDF based Free Standing Piezoelectric Nanogenerator for Vibrational Energy Harvesting and Wearable Shoe Insole Pedometer Sensor. 2023: p. 170898.
15. Deng, W., et al.: Piezoelectric nanogenerators for personalized healthcare. **51**(9), 3380–3435 (2022)
16. Ye, S., et al., High-performance piezoelectric nanogenerator based on microstructured P (VDF-TrFE)/BNNTs composite for energy harvesting and radiation protection in space. 2019. 60: p. 701–714.
17. Panda, S., et al., Biocompatible CaTiO3-PVDF composite-based piezoelectric nanogenerator for exercise evaluation and energy harvesting. 2022. 102: p. 107682.
18. Sahu, M., et al.: Piezoelectric nanogenerator based on lead-free flexible PVDF-barium titanate composite films for driving low power electronics. **11**(2), 85 (2021)

19. Jiang, F., et al., Stretchable, breathable, and stable lead-free perovskite/polymer nanofiber composite for hybrid Triboelectric and piezoelectric energy harvesting. 2022. 34(17): p. 2200042.
20. Mondal, B., et al., Lead-Free Perovskite Cs3Bi2I9-Derived Electroactive PVDF Composite-Based Piezoelectric Nanogenerators for Physiological Signal Monitoring and Piezo-Phototronic-Aided Strain Modulated Photodetectors. 2022. 38(40): p. 12157–12172
21. Tao, X., et al.: Design and synthesis of triboelectric polymers for high performance triboelectric nanogenerators. **16**(9), 3654–3678 (2023)
22. Zi, Y., et al.: Standards and figure-of-merits for quantifying the performance of triboelectric nanogenerators. **6**(1), 8376 (2015)
23. Shanbedi, M., H. Ardebili, and A.J.P.i.P.S. Karim, Polymer-based triboelectric nanogenerators: Materials, characterization, and applications. 2023: p. 101723.
24. Chen, B.D., et al.: Au nanocomposite enhanced electret film for triboelectric nanogenerator. **11**, 3096–3105 (2018)
25. Xia, X., et al.: Embedding variable micro-capacitors in polydimethylsiloxane for enhancing output power of triboelectric nanogenerator. **10**, 320–330 (2017)
26. Zhang, Z., et al.: Facile method and novel dielectric material using a nanoparticle-doped thermoplastic elastomer composite fabric for triboelectric nanogenerator applications. **10**(15), 13082–13091 (2018)
27. Chen, J., et al.: Enhancing performance of triboelectric nanogenerator by filling high dielectric nanoparticles into sponge PDMS film. **8**(1), 736–744 (2016)
28. Mondal, R., et al., Advanced pyroelectric materials for energy harvesting and sensing applications. 2023.
29. Abbasipour, M., et al.: Improving piezoelectric and pyroelectric properties of electrospun PVDF nanofibers using nanofillers for energy harvesting application. **30**(2), 279–291 (2019)
30. Roy, K., et al.: A self-powered wearable pressure sensor and pyroelectric breathing sensor based on GO interfaced PVDF nanofibers. **2**(4), 2013–2025 (2019)
31. Ma, M., et al., Self-powered flexible antibacterial tactile sensor based on triboelectric-piezoelectric-pyroelectric multi-effect coupling mechanism. 2019. 66: p. 104105.
32. Gan, W., W.J.S.M. Abd Majid, and Structures, Effect of TiO2 on enhanced pyroelectric activity of PVDF composite. 2014. 23(4): p. 045026.
33. Tiwari, M., N. Bangruwa, and D.J.T.O. Mishra, 0D, 1D, and 2D Magnetic Nanostructures: Classification and their Applications in Modern Biosensors. 2023: p. 100257.
34. Ahmad, N., et al., Energy harvesting performance of a novel polymer-nanocrystal composite of P (VDF-TrFE)/ZnO QD films. 2023. 289: p. 116256.
35. Yang, Y., et al., Flexible pyroelectric nanogenerators using a composite structure of lead-free KNbO3 nanowires. 2012. 24(39): p. 5357–5362.
36. Zhang, T., et al., Recent progress in hybridized nanogenerators for energy scavenging. 2020. 23(11).
37. Singh, V., B.J.J.o.A. Singh, and Compounds, MoS2-PVDF/PDMS based flexible hybrid piezo-triboelectric nanogenerator for harvesting mechanical energy. 2023. 941: p. 168850.
38. Rana, S., V. Singh, and B.J.A.A.E.M. Singh, Tailoring the Output Performance of PVDF-Based Piezo–Tribo Hybridized Nanogenerators via B, N-Codoped Reduced Graphene Oxide. 2022. 4(12): pp. 5893–5904.
39. Yun, J., et al.: Piezo-triboelectric hybridized nanogenerator embedding MXene based bifunctional conductive filler in polymer matrix for boosting electrical power. **105**, 108018 (2023)
40. Gao, X., et al.: Combination of power generation and flexibility in piezo-pyroelectric hybrid nanogenerators by designing bimetal textured structure. **116**, 108773 (2023)
41. Sultana, A., et al., Methylammonium lead iodide incorporated poly (vinylidene fluoride) nanofibers for flexible piezoelectric–pyroelectric nanogenerator. 2019. 11(30): p. 27279–27287.

42. Liu, Y., et al.: Integrating a silicon solar cell with a triboelectric nanogenerator via a mutual electrode for harvesting energy from sunlight and raindrops. **12**(3), 2893–2899 (2018)
43. Wang, X., Z.L. Wang, Y.J.N.E. Yang, Hybridized nanogenerator for simultaneously scavenging mechanical and thermal energies by electromagnetic-triboelectric-thermoelectric effects. 2016. 26: p. 164–171

Chapter 14
Polymer Nanoparticle Composites for Modern Applications

Ankita Subhrasmita Gadtya, Bibhuti B. Sahu, Suresh Sagadevan, Ram Naresh Mahaling, and Srikanta Moharana

Abstract The integration of nanoparticles into polymeric matrices is a rapidly advancing field of study aimed at creating polymer nanocomposites. These composites are formed by combining polymers with carbon-based nanoparticles, such as quantum dots, nanosheets, and nanotubes, which exhibit outstanding electrical, magnetic, and optical properties. In order to attain a successful composite, it is imperative to ensure that the selected nanoparticles are dispersed uniformly throughout the matrix and exhibit exceptional compatibility with it. Polymer-based composites are extensively utilized in several sectors including electronics, aerospace, cosmetics, and biological fields. They are in great demand for cutting-edge technologies because of their versatility in terms of configuration and structure, resulting in the production of materials with specialized qualities. Nevertheless, it is clear that the incorporation of nanoparticles into a polymer matrix progresses the overall strength, stiffness, and longevity of the resultant composites. Moreover, the precise manipulation of the material's characteristics is facilitated by the controlled dispersion of nanoparticles throughout the polymer matrix. This suggests the possibility of attaining favorable characteristics such as enhanced conductivity or increased biocompatibility. These

A. S. Gadtya
School of Applied Sciences, Centurion University of Technology and Management, Bhubaneswar, Odisha, India
e-mail: srikantanit@gmail.com

B. B. Sahu
Department of Basic Science and Humanities, Nalanda Institute of Technology, Bhubaneswar, Odisha, India

S. Sagadevan
Nanotechnology and Catalysis Research Centre, University of Malaya, 50603 Kuala Lumpur, Malaysia

R. N. Mahaling
Laboratory of Polymeric and Materials Chemistry, School of Chemistry, Sambalpur University, Jyoti Vihar, Burla-768019, Odisha, India

S. Moharana
Department of Chemistry, School of Applied Sciences, Centurion University of Technology and Management, Jatni, Bhubaneswar, Odisha, India

S. Moharana et al. (eds.), *Polymer Composites*, Engineering Materials,
https://doi.org/10.1007/978-981-97-2075-0_14

advancements have created opportunities for revolutionary advancements in materials and technology across several industries. This chapter examines the advanced uses of polymer composites and investigates recent technical advancements that have improved their performance and versatility.

Keywords Polymer · Nanocomposites · Carbon nanoparticles · Radical polymerization · Interfacial polymerization · Bioimaging · Cosmetic application · Biomedical application · Energy storage

14.1 Introduction

In recent times, composites containing polymer nanoparticles gained significant attention among academic and industrial scholars owing to their easy processing, better elasticity, and broad range of potential applications including electronic, aerospace, cosmetic sectors, etc. [1–3]. These composites synergistically integrate the distinct characteristics of polymers and nanoparticles, leading to improved mechanical, electrical, and thermal properties. By manipulating the magnitude, profile, and concentration of nanoparticles embedded in the polymer matrix, these composites provide limitless opportunities for innovation and progress in several disciplines [4]. Polymers are used in the manufacturing of various materials for a wide range of applications [5]. However, polymeric nanoparticles may be manipulated into many forms, such as thin films, monolayers, bilayers, and nanocoated films, utilizing techniques including spin coating, dip coating, film casting, and printing [6]. Unlike metals and ceramics, the majority of pure polymers have inferior thermal, mechanical, and electrical characteristics [7]. The numerous categories of polymers, such as copolymers, homopolymers, blended polymers, and modified polymers, fail to adequately elucidate several challenging properties required [8]. These parts are also important in optoelectronic uses, where they have been used for a variety of purposes. Polymer-based nanocomposites, a type of plastic material, have become a major area of study because they can be used in many ways [9]. The integration of inorganic nanoparticles into polymers has led to a significant improvement in the characteristics of polymer nanocomposites, which has attracted the attention of scientists globally in recent years [3]. It is reported that the particles are thought to be colloidal particles caused by polymers, with a size in the nanometer range. Nanoparticles composed of polymers are frequently associated with distinctive attributes due to their extremely small size. By simply converting bulk polymer to nano-size polymer, several fascinating and innovative properties may be achieved without changing the composition of the materials. The physical and chemical characteristics of polymers undergo changes when they are converted into polymer nanoparticles (PNP), resulting in the emergence of novel advantages in the realms of nanoscience and nanotechnology. The substantial divergence in the properties of polymer nanoparticles in comparison to bulk polymers may largely be ascribed to their reduced molecular dimensions. At the nanoscale, quantum effects are the most important, and the surface-to-volume fraction surges significantly, giving the material several unique features [10–13]. These polymer nanoparticles are seeing tremendous growth in research and are emerging as a prominent area of study in chemistry. This is mostly because of the extensive applications it offers in several sectors like electronics, medicine, and photonics. Moreover, polymeric nanoparticles have played an important protagonist in the production of coatings, paints, adhesives, and several other essential compounds for a noteworthy period. In recent years, its application has been observed in diagnostics, medication delivery, and bio-imaging, as well as other areas of biology. Potential bio-nanomaterials may be derived from polymer nanoparticles, which possess exceptional physical and chemical characteristics [14–17]. A polymer nanocomposite can be formed by incorporating inorganic nanoparticles into organic polymers and utilizing nano-reinforcement. To simplify the resulting nanocomposite, it is essential to comprehend how the features are modified when the molecular size decreases to the nanoscale [18]. Nanoreinforcements have demonstrated superior efficacy as compared to conventional filler materials. On the other hand, polymer nanocomposites may be manufactured using a variety of methods, including in situ polymerization, melt extrusion, and solution processes. Many diverse kinds of polymer nanocomposites occur, each with its inimitable composition [19, 20]. A few examples of nanocomposites include those containing both organic and inorganic components, as well as those including inorganic and organic components. The desirable properties of the polymer nanocomposite are mainly because the components of the added substances are smaller and the surface area-to-mass ratio is higher [21]. The quantity of the supplementary compounds may vary according to the necessity, in contrast to more conventional substitutes. The research of nanocomposites is thus expanding at a rapid pace. Research into the development of nanocomposite materials could find applications in a wide range of fields, such as ceramics, polymers, biomedicine, electronics, metals, and countless more. The research further explores the many methods for synthesizing polymer nanoparticles and their extensive range of applications in scientific and technical fields. Upon concluding the chapter, we will delve into the potential outcomes followed by prospects and challenges for polymer nanoparticles.

14.2 Synthesis Techniques of Polymer Nanoparticles

The production of polymer nanoparticles has gained significant attention due to the growing need for these tiny, adaptable particles and their reliance on several scientific fields. Contemporary techniques for manufacturing polymer nanoparticles encompass various processes such as radical polymerization, salting out, solvent displacement, surfactant-free emulsion polymerization, emulsification diffusion, nanoprecipitation, rapid desertion, ionic gelation, freeze-drying, spray drying, mini-emulsion polymerization, double desolvation, dialysis, and solvent evaporation. Here, we provide a comprehensive overview of some frequently used techniques. Further study is necessary to develop novel preparation techniques that offer many benefits, such as improved efficiency, cost-effectiveness, and time-saving. However, there is still much room for development in this area.

14.2.1 Radical Polymerization

Through radical polymerization courses, new approaches were provided to generate polymeric nanoparticles. These techniques employ a variety of initiators and monomers to accomplish controlled radical polymerization, which yields precisely specified polymer nanoparticles with the required characteristics. Furthermore, the advancement of advanced techniques such as live radical polymerization has enhanced the precision and manipulation of polymer nanoparticle creation to a greater extent. Synthesis of polymer nanoparticles with tightly regulated molecular weights and limited size ranges is now possible because of these cutting-edge methods. However, these polymeric nanoparticles have found new uses in areas like drug delivery and nanotechnology owing to the integration of functional groups into the polymer structure made possible by specialized initiators and monomers. Some of the current methods for synthesizing polymer nanoparticles include atom transfer radical polymerization, nitroxide-mediated polymerization, and reversible addition and fragmentation transfer chain polymerization [22, 23]. For instance, Payami and his co-workers [24] have made up Fe_3O_4@SiO_2-PolyFc nanoparticles (Fig. 14.1) through surface-initiated radical polymerization for supercapacitor applications. It is observed that the surface-modified nanoparticles with superior charge storage capacity of about 170 $mAhg^{-1}$ (612 Cg^{-1}) at 2.5 Ag^{-1}.

Different processes, including mini-emulsions, macro-emulsions, micro-emulsions, suspensions, dispersions, precipitations, etc., display different ways for radical initiation, propagation, and termination when it comes to the polymerization mechanism.

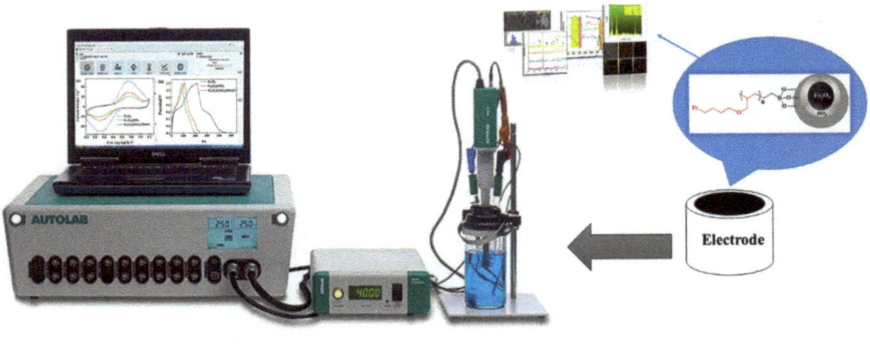

Fig. 14.1 Schematic illustration for preparation of Fe_3O_4@SiO_2-PolyFc nanoparticles [24]. From Ref. [24] by consent from [Elsevier], Copyright [2023]

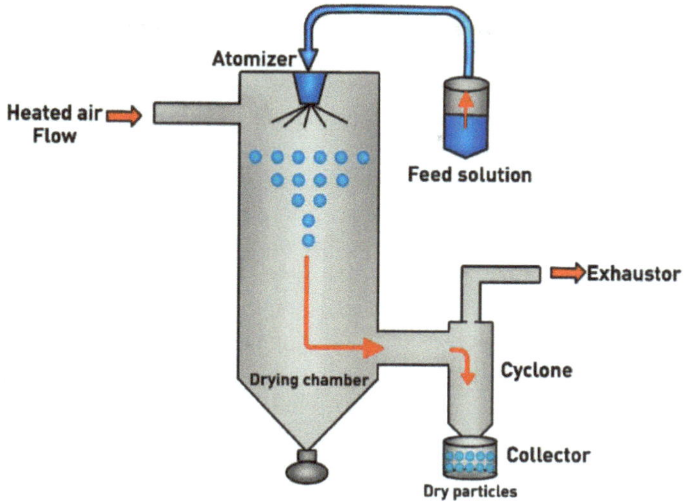

Fig. 14.2 Process flow diagram for spray drying to create dry particles [27]. Open access under a CC BY 4.0 license

14.2.2 Spray Drying Method

Spray drying, a time-honored and impressive technique, is recognized by operative technique for producing nanoparticles. It was pioneered some decades ago. Spray drying is a straightforward and efficient method for converting liquid phases into solid particles while maintaining the physicochemical integrity of nanoparticles. This approach, which is limited to one stage and is exclusive to a particular province, is appropriate for producing a diverse selection of regulated powders designed for specific uses. Nanoparticles, predominantly composed of polymers, are generated by atomizing a solution of solvent and polymer. This process results in the formation of small spherical particles, enabling accurate regulation of their size, shape, and stability [25, 26]. As shown in Fig. 14.2, there are four main stages to the spray-drying process: (i) feeding a warm gas stream into a spray to atomize the liquid; (ii) collecting and sorting the dry particles based on their size; and (iii) forming the particles by evaporating the liquid [27, 28].

14.2.3 Mini-emulsion Polymerization Technique

The mini-emulsion polymerization technique is a versatile method for synthesizing polymer nanoparticles ranging from 50 to 500 nm. The process involves using surfactants and emulsifiers to stabilize the system while dispersing monomers in a continuous phase. This approach has several advantages, including accurate control over particle size, high monomer conversion, and the ability to include a diverse array of functional groups into the polymer chain. Mini-emulsion polymerization is a promising method for producing novel materials with tailored properties. It is easily adaptable for industries involving large-scale production [29]. However, initiators (which create free radicals) and free radical polymerizable monomers are necessary components of mini-emulsion polymerization processes. Emulsion polymerization media, often deionized water, dispersion media, organic solvents, surfactants, and organic oils are also needed. A solvent phase containing an initiator and a dissolvable is mixed continuously with a liquid phase containing water and a surfactant until the two phases are homogeneous [29, 30]. The result is the classic emulsion. A high-energy mechanical source, like an ultrasonicate, is cast-off to create nanodroplets. The mechanical upsetting process causes the monomer beads to rapidly change in size until they settle into a stable equilibrium with little attainable nano-molecule size. For instance, Tovar et al. have reported an active ester comprising surfmer aimed at one-stage polymer nanoparticle superficial functionalization through a mini-emulsion polymerization technique. An enhanced method for producing the active ester containing surfmer p-(11-acrylamido)undecanoyloxyphenyl dimethyl-sulfonium methyl sulfate (AUPDS) is presented in this work. Figure 14.3 shows the whole mini-emulsion polymerization procedure employed in this study. The high concentration of droplets in the mini-emulsion causes various scattering effects, making it impossible to conduct DLS of the monomer droplets before polymerization. It was also not recommended to dilute the mini-emulsion at the stage since doing so might cause the AUPDS to be displaced or the monomer to be lost by diffusion [31].

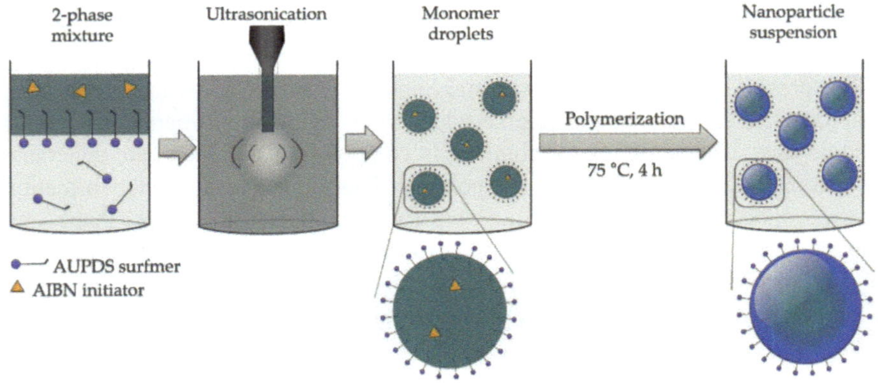

Fig. 14.3 Schematic illustration of the mini-emulsion polymerization procedure with AUPDS surfmer involves a mixture of continuous and dispersed phases, mini-emulsified using probe ultrasonication to produce stable monomer droplets, which are then polymerized [31]. Open access under a CC BY 4.0 license

14.2.4 Micro-emulsion Technique

Method for creating polymer nanoparticles using micro-emulsion. Micro-emulsion is a typical method for producing polymer nanoparticles. To stabilize and regulate the formation of nanoparticles, a stable emulsion of water, oil, and surfactants is required. Among the numerous advantages of this approach are the nanoparticles' tunable properties, narrow size distribution, and high yield. The fact that it can be easily mass-produced in an industrial environment adds to its potential applications in nanomedicine and drug delivery [32, 33]. Figure 14.4 shows a micro-emulsion, which combines a monomer with a water-solvent initiator in a thermodynamically controlled miniature-size emulsion fluid. The rapid initiation of polymerization upon the addition of surfactant-containing water is due to the enormous quantity of surfactant outline that reduces interfacial tension to nearly zero. These emulsions don't need ultrasonication and are made by mixing two immiscible oil/water phases. The application of a high concentration of surfactant also causes the nanoparticles to become completely coated in the solvent. Particles with dimensions between 5 and 50 nm are the end product. It is likely to produce polymer nanoparticles with an extensive variation of shapes using this method. However, micro-emulsions can combine hydrophilic and hydrophobic domains at the nanoscale level due to their surface behavior. Microemulsions are used to synthesize various nanomaterials, including solid lipid nanoparticles, nanostructured lipid carriers, polymeric nanoparticles, mesoporous silica nanoparticles, and metallic nanoparticles [34] (as shown in Fig. 14.4).

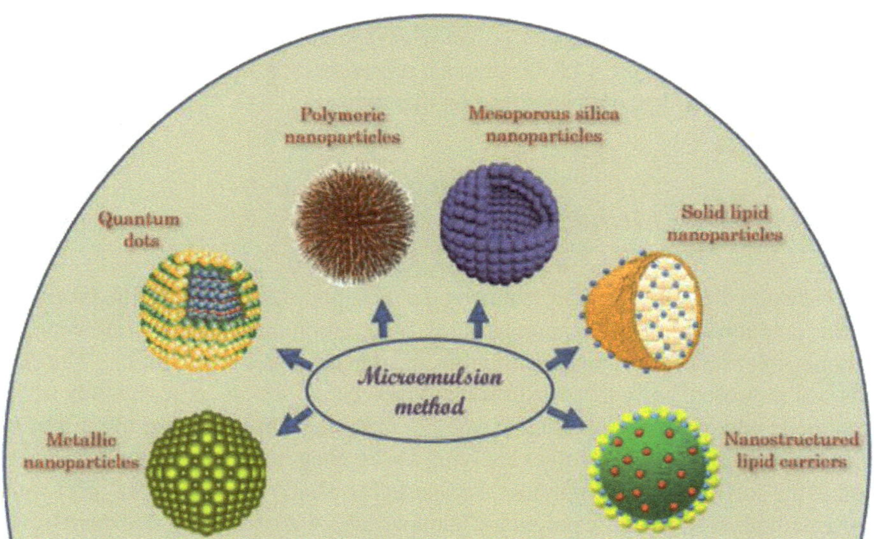

Fig. 14.4 Form of nanoparticles (NPs) made using the micro-emulsion technique [34]. Open access under a CC BY 4.0 license

14.2.5 Interfacial Polymerization

Interfacial polymerization is one of the several approaches for synthesizing polymer nanoparticles. The process creates polymer nanoparticles by combining two incompatible phases at their interface; these phases are usually water and organic. Drug delivery, catalysis, and nanotechnology are a few of the many areas that can benefit from this method's ability to precisely regulate the dimension, profile, and composition of nanoparticles. In addition to being easily adaptable for large-scale synthesis of polymer nanoparticles, interfacial polymerization also offers the benefit of scalability. Step growth polymerization is an alternative term for the interfacial polymerization process. In this case, the two liquid phases that can combine undergo polymerization at their contact. Several types of polymer nanoparticles, including ultra-thin films, nanocapsules, and nanofibers, may be synthesized by means of this method [35, 36]. Emulsions involving liquids, polymerization between different liquids, and polymerization between solids are the most typical types of interfacial polymerization. The immiscible fluid is divided into two separate phases, each containing two reactive monomers. The site of the reaction is the interface between two incompatible phases. For an extended time, the latex nanoparticles float around in liquid form. Moreover, the conducting polymers such as polypyrrole (PPy), poly(3,4-ethylenedioxythiophene), polyaniline (PANI), and polythiophene (PTh) have been produced using interfacial polymerization. For instance, Wu and his co-workers [37] have created nanofiltration layers for dye separation using polyethersulfone ultrafiltration membrane through interfacial polymerization. This work involved the use of PES ultrafiltration membranes and new amine monomers with nitrogen content and less reactive aldehyde monomers to conduct an interaction between tannic acid-assisted polyethylene polyamine (PEPA) and terephthalaldehyde (TPAL). Figure 14.5 illustrates a representation illustrating the synthesis procedure of the composite nanofiltration membrane.

14.2.6 Dialysis Method

The diffusion dialysis technique is among the most popular dialysis methods utilized in the creation of polymer nanoparticles. This technique involves immersing a polymer solution in a solvent bath after placing it in a dialysis bag or tubing. The dialysis bag is subsequently filled with solvent molecules, which cause the polymer to precipitate and create nanoparticles. Another dialysis technique is the ultrafiltration dialysis technique, which involves applying a pressure gradient across a membrane to isolate the solvent from the polymer nanoparticles. Dialysis is one of the most well-liked methods for producing polymeric nanoparticles with a uniform size distribution because of its many benefits, including its ease of use and high efficiency [38–40]. Dialysis provides a straightforward and useful method for producing tiny, constant PNPs, eliminating the need for surfactants. The initial stage involves introducing a

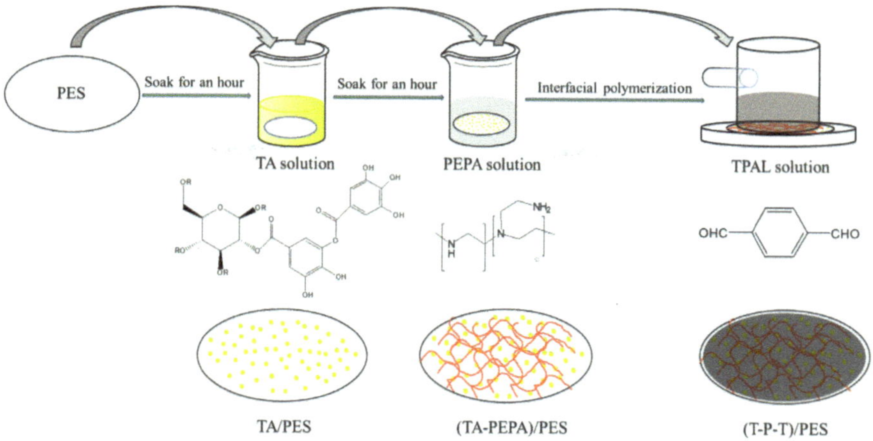

Fig. 14.5 Methods for creating the (TA-PEPA-TPAL)/PES membrane [37]. Open access under a CC BY 4.0 license

polymer solution into a dialysis tube that has a certain molecular weight cutoff. The decrease in solubility and the creation of uniform nanoparticle suspensions result in the buildup of leading-edge polymers as the solvent evaporates from the film. The polymer solvent has a vital impact in influencing the dimensions and morphology of nanoparticles. The standard procedure involves dissolving the polymer in an organic solvent, inserting it into the dialysis membrane, and dialyzing it against a non-solvent (Fig. 14.6). The presence of dilute polymer solutions and solvent miscibility are basic requirements. The capability of mixture to liquefy the polymer reduces when the solvent is expatriate into the membrane. When the interfacial tension rises, the nanoparticles in the suspension form a colloidal network, and the polymers join together. Despite dialysis's prevalence and convenience of use, the lengthy operation and vast amount of counter-dialyzing media pose risks of nanoparticle payload premature release [39].

14.2.7 Nanoprecipitation Technique

One popular way of creating polymer nanoparticles is the nanoprecipitation method. To create nanoparticles, a polymer must precipitate from a solution. This method provides accurate control over the nanoparticles' size, shape, and surface characteristics, which makes it appropriate for several uses including drug administration, sensing, and imaging. In addition, as compared to alternative ways of synthesizing nanoparticles, the nanoprecipitation approach is very easy to use and inexpensive [41]. Moreover, the polymer matrix may be modified to incorporate other medications or imaging agents through the use of the nanoprecipitation process, allowing for controlled and targeted release. Its adaptability makes it a viable strategy for creating

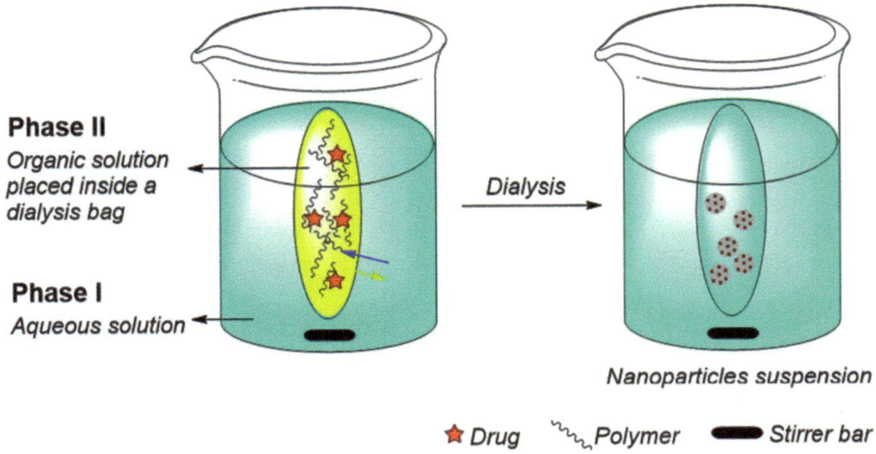

Fig. 14.6 Nanosphere synthesis via dialysis, depicted schematically [39]. Adapted from Ref. [39] with consent from [Elsevier], Copyright [2017]

cutting-edge medications and diagnostic instruments. There are various benefits of this technique including an easy, gentle, and low-energy process. A straightforward, uncomplicated, and dependable method for dealing with yield nanoparticles with a homogenous size distribution and a variety of sizes is the nanoprecipitation approach (Fig. 14.7). It is an inexpensive and easy method. Fessi et al. [42] were the first to invent the nanoprecipitation process, which is also known as solvent displacement. The core idea behind this method is that an organic solvent may be displaced from a lipophilic solution into an aqueous phase, leading to the deposition in the interface of a polymer (Fig. 14.8). One shot, stepwise, dropwise, or precise adding rates are all viable options for dissolving the polymer in an intermediate polarity water-miscible solvent and then adding the resulting solution to a stirred aqueous solution [39].

14.3 Applications of Polymer Nanoparticle Composites

14.3.1 Cosmetic Applications

Cosmetics are chemicals used to enhance the aesthetic look of an individual's skin, face, hair, nails, and eyes. Cosmetics are substances used on the human body to improve beauty, attractiveness, cleanliness, or appearance without causing any negative impact on physiological functions or anatomical characteristics. The majority of individuals nowadays utilize cosmetics to augment their physical appearance, often disregarding the possible adverse consequences on their skin. Using parabens regularly often leads to the development of skin allergies. To avoid these adverse

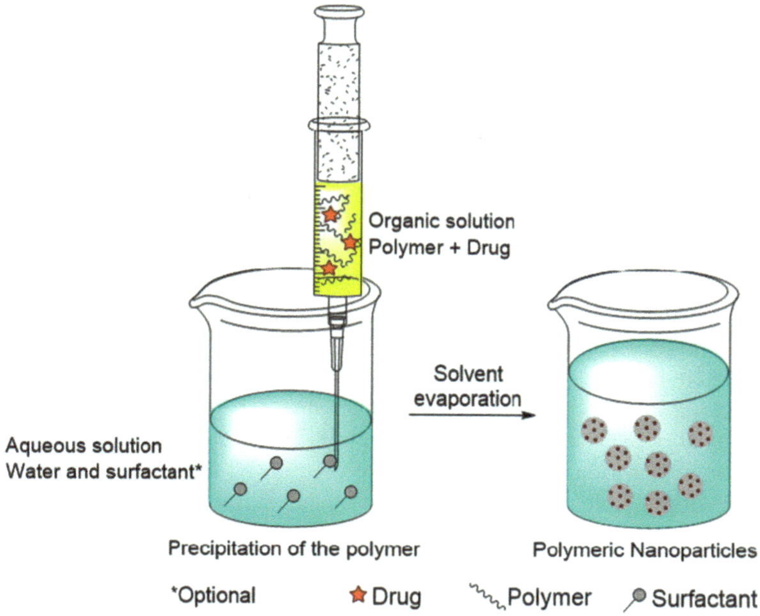

Fig. 14.7 Schematic depiction of the nanoprecipitation method used to create nanospheres [39]. Adapted from Ref. [37] with authorization from [Elsevier], Copyright [2017]

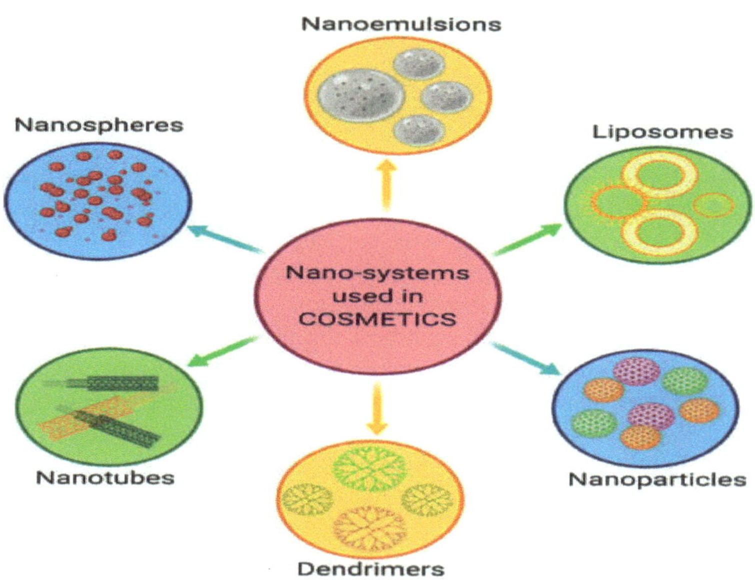

Fig. 14.8 A collection of nanostructured cues found in cosmetics [48]. Open access under a CC BY 4.0 license

effects, customers can go for natural skin care products that are crafted from naturally sourced components or botanicals, such as essential oils, herbs, flowers, and roots. Bacterial cellulose or bio-cellulose (BC) is a type of cellulose that is produced by a few microorganisms like Acetobacter xylinum (A. xylinum), Agrobacterium, Salmonella, Pseudomonas, and Rhizobium [43–45]. BC is used in a wide range of cosmetic products, such as face scrubs, contact lenses, masks, and personal cleaning solutions. BC has superior tensile strength in comparison to vegetal cellulose due to its ultra-fine network architecture. Additionally, BC is more chemically pure, devoid of both hemicellulose and lignin, and possesses enhanced hydrophilicity, water-retention capacity, and polymerization. BC has a higher degree of crystallinity and arranges itself into microfibrils that bear a resemblance to ribbons, in contrast to plant cellulose. BC exhibits significantly smaller and distinctive thin microfibrils in comparison to plant cellulose, resulting in enhanced porosity. BC improves the aesthetic quality of the skin due to its exceptional smoothness, soothing properties, moisturizing effects, and ability to provide comfort to the skin. BC incorporates active substances such as essential oils, herbs, plant extracts, flowers, algae extracts, and panthenol for use in the cosmetics sector. The active components inside the BC matrix are chemically linked, namely, through hydrogen bonds. When chemicals inside the matrix are concentrated and released onto the skin upon application, hydrogen bonds exhibit a high degree of selectivity [43–47]. There are numerous nano-systems or innovative nanocarriers which are frequently employed in cosmetic formulations to encapsulate active chemicals for their effective distribution across skin barriers. These nano-systems can significantly enhance the absorption or penetration of active ingredients in cosmeceuticals inside the layers of the skin, allowing for either a rapid or gradual release of these active ingredients. Figure 14.8 displays a range of nanostructured cues employed in cosmetics. Moreover, this section offers a comprehensive understanding of the distinctive characteristics and functions of various nanocarriers or nanostructured materials used for delivering cosmeceuticals in the field of cosmetics [48]. The nanoparticles have advantages but meet constraints in terms of stability, scalability, toxicity, cost, and safety. Their diminutive dimensions, augmented surface area, and affirmative surface charge intensify physiological contact. Previous studies have shown that the toxicity of drugs can differ based on the amount and manner of administration. The amount of the material has a bigger role in determining how much of it is available for the body to use, rather than its physical and chemical properties. The use of nano-formulations in cosmetics has the potential to improve the transportation of ingredients into the bloodstream and raise toxicity levels [49]. In light of this, challenges arise about the appropriateness of their use in the production of cosmetics and cosmeceuticals (as seen in Fig. 14.9).

Wang and his co-workers had synthesized a BC composite prepared which was incorporated with hyaluronic acid (HA) and silk sericin (SS) by both in situ and ex situ methods. The prepared BC-HA/SS composites had a very porous framework, strong mechanical properties, excellent thermostability, and biocompatibility with fibroblast cells. The composite materials improve oxygen permeability and have interconnected pores that are essential for absorbing wound exudate. By interacting with cells and exerting their mitogenic effect, the composites hasten the healing

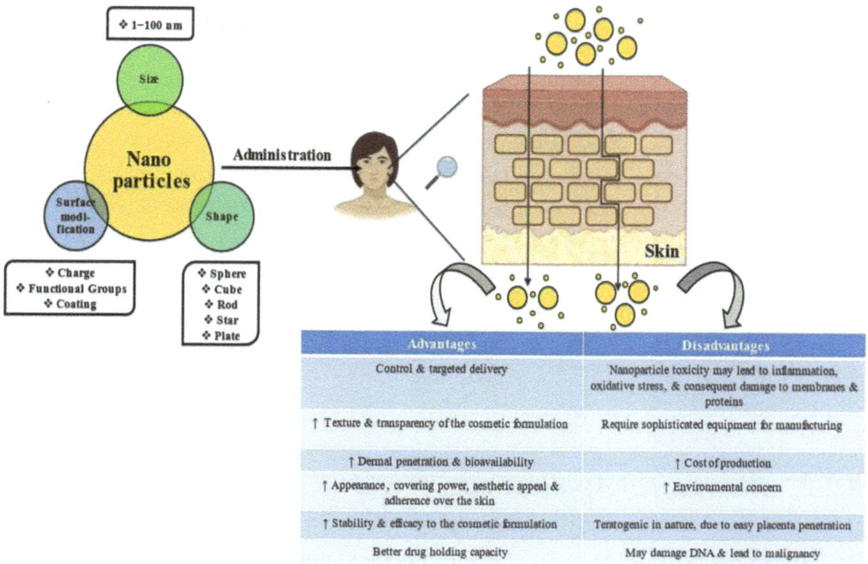

Fig. 14.9 Benefits of nanocosmeceuticals [49]. Open access under a CC BY 4.0 license

process and minimize the formation of scars. All things considered, we can say that the BC-HA/SS films, particularly the ex situ modified composite, would make excellent materials for wound dressings and cosmetics like face masks [50]. A biopolymer called chitosan is prepared by deacetylating chitin. Chitosan and its derivatives have attracted a lot of interest because of their biological activity and possible applications in the food, pharmaceutical, cosmetic, and medical industries. These materials have the potential to be used as agents in the cosmetic industry because of their physicochemical and biological activity, as well as their typical biocompatibility, biodegradability, and non-toxicity. Chitosan and its derivatives could be great options for sunblock, skin, oral, and hair care formulations due to their enhanced biological activities. Because of their antimicrobial and antioxidant activities, regenerative and cell-penetrating qualities, lack of toxicity, and immunogenicity, the natural polymer chitin and its derivative chitosan have potential as ingredients for skin care products [51–53]. Kodolova-Chukhontseva et al. have synthesized chitosan-based new composite materials using chitin nanofibrils to use in the cosmetic field mainly for using it as face masks. The produced composite films with chitin nanofibrils have better mechanical properties with 99 MPa of tensile strength and 14% of tensile strain. The residual particles of loose (non-crosslinked) chitosan may come into contact with skin cells, causing the films to swell easily. Fibroblasts are the primary dermal cells that synthesize extracellular matrix, which gives skin its elasticity and resilience. For this reason, they were chosen to conduct in vitro studies. Degradation of these factors gives rise to hints of skin aging [54]. Chitosan films incorporated with rutin and HA were prepared by Sionkowska and co-workers for cosmetic uses.

With anti-inflammatory and antioxidant qualities, rutin is a polyphenolic flavonoid. Because of its antioxidative qualities, rutin is used in cosmetic dermal products like anti-aging serums and cosmetic preparations for atopic and allergic skin. Glycerin was included as a moisturizing agent to improve film performance and also improve the rate at which active ingredients reach the skin. Following the topical application of the films, skin elasticity remained unchanged, but after 15 min, skin moisturization became slightly better. After 20 min, the optimum release of rutin was discovered at pH 5.5, which is comparable to the pH of typical human skin. Following the topical application of the suggested chitosan-rutin mask, the skin's appearance was enhanced. The maximum discharge of rutin is reduced in the presence of HA. So rutin's bonding in the chitosan-based films is strengthened when HA is added to the formulation [55].

14.3.2 Biomedical Applications

Novel materials possessing prospective biomedical uses have been created to guide, enhance, or replace the functions of living tissues. Chemical, biological, physical, and structural requirements, such as mechanical behavior and integration with the related host tissues, must all be met by biocompatible materials. When it comes to mechanical properties, strength, stiffness, and the best possible load transmission are of particular importance, especially the elastic modulus (also known as Young's modulus). Polymer composites made of ceramics and metals, in addition to polymers, enable customization of the desired physical attributes. These polymer composites are used in many medical fields like dental application, joint replacement, bone plates and fracture repair, prosthetic limbs, etc. [56, 57].

14.3.2.1 GQD-Based Polymer Composite for Biomedical Application

Most drugs can be quickly loaded into bioactive graphene quantum dot (GQD)-based polymer composites because of their ability to bind to a variety of biological materials through π–π and electrostatic interactions. Conversely, bioactive GQD-based polymer composites exhibit strong membrane permeability and biocompatibility, thereby enhancing drug efficiency and increasing efficacy against drug-resistant cells. The prolonged blood circulation time resulting from the high delivery of bioactive GQDs is another factor contributing to their efficiency [58–60]. Also because of the superior mechanical qualities, high differentiation potential, and low toxicity of these particles, bioactive GQD-based polymer composites are valued for their potential to promote tissue regeneration. Furthermore, they can raise calcium levels. Bioactive GQDs serve as reinforcing agents in tissue engineering, improving the mechanical properties of scaffolds suitable for the placement of orthopaedic implants [61, 62].

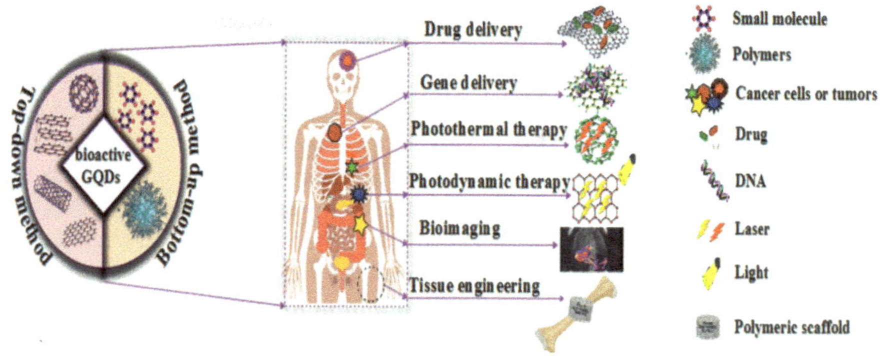

Fig. 14.10 The potential applications of polymer composites based on bioactive GQDs in biomedicine [65]. Open access under a CC BY 4.0 license

Photodynamic therapy (PDT) uses a light source as well as a light-sensitizing agent for the treatment of psoriasis, acne, and cancer. In the context of PDT, light-sensitizing elements like semiconductor GQDs have drawn a lot of attention. Target cell death happens through a PDT mechanism that includes oxidative stress and ROS generation. A bioactive GQD-based polymer composite, or another light-sensitizing agent, mediates the production of reactive oxygen species (ROS) in cancer cells and tumors. Bioactive GQD-based polymer composites induce autophagy and apoptosis in cancer cells via oxidative stress [63, 64]. The potential applications of polymer composites based on bioactive GQDs in biomedicine are shown in Fig. 14.10 [65].

14.3.2.2 MXene-Based Polymer Composite for Biomedical Application

The 2D planar structure and the exceptional properties exhibited by polymer functionalized MXenes are attributable to their applications in the medical sector, including antimicrobial, photothermal therapy (PTT), drug delivery, diagnostic imaging, biosensors, and bone regeneration.

(a) **Antimicrobial activity**: The MXene surface's hydrophilicity and anion properties improve the interaction between bacterial cell membranes. Cell inactivation is caused by the hydrogen bonding between the MXene functional groups and the lipopolysaccharide molecules in the cell membrane, which stops the absorption of nutrients and impedes the growth of bacteria. MXenes have a higher antibacterial activity than GO nanosheets, which can be credited to their higher conductivity [66–69]. Rasool et al. reported that $Ti_3C_2T_x$ was supported by PVDF (polyvinylidene fluoride) and had antimicrobial potency. The PVDF membrane's hydrophilicity (contact angle of 37°) was enhanced and the presence of big pores was lessened upon coating it with the high aspect ratio $Ti_3C_2T_x$. About 73% of the gram-positive B. subtilis and 67% of the gram-negative E. coli bacteria showed growth inhibition and decreased viability. The composition of bacterial cell walls and the surrounding environment are two of the many variables that influence antibacterial efficacy. There was a difference in the rate of resistance to MXene between E. Col and B. subtilis due to the thickness of the peptidoglycan layer, which lies within the outer and inner cell membranes, being 2–3 nm in E. Coli and 20–80 nm in B. subtilis [70]. To create multifunctional, durable, and transparent 3-hydroxybutyrate-co-hydroxyvalerate (PHBV)-graphene oxide (GO)/MXene composite membranes, GO/MXene nanosheets were prepared and grafted onto PHBV with success. Polyvinylpyrrolidone (PVP) served as a pore-forming agent in this process. The inclusion of hydroxyl molecules and terminal oxygen in GO and MXene nanosheets' laminar structure and hydrophilicity contributed to the far better hydrophilicity of PHBV-GO/MXene composite membranes. This led to providing a functional site for the ester bonds in between the PHBV and GO/MXene for free radical polymerization. The PHBV-GO/MXene 0.1% composite membranes demonstrated a favorable bacterial reduction in E. Coli and S. aureus bacteria in contrast to the PHBV composite membranes. The antimicrobial efficacy of the PHBV-GO/MXene 0.5% composite membranes also attained 95%. The platelet adsorption capacity of PHBV-GO/MXene composite membranes was superior to that of pure PHBV. The PHBV-GO/MXene 0.5% composite membranes were one of them, and they adsorbed 1143 $\pm$ 43 $\times$ $10^8/m^2$, which is nearly five times more than the PHBV composite membranes [71].

(b) **Bioimaging**: MXenes are widely recognized for their exceptional ion transport behavior, substantial surface area-to-volume ratio, good biocompatibility, significant electrical conductivity ($Ti_3C_2T_x$ monolayer: 4600 $\pm$ 1100 S/cm), and ease of functionalization. Because of these characteristics, MXenes is acknowledged as a highly sophisticated biosensing instrument that is capable of identifying a wide range of small molecules, big biomolecules, and even cancer cells. Enzymes are typically immobilized on MXene nanosheets to catalyze the chemical reaction of the molecules to detect small molecules [72–74]. Haemoglobin (Hb) has been immobilized using a manufactured TiO_2-nanoparticle modified organ-like Ti_3C_2 MXene (TiO_2-Ti_3C_2) nanocomposite to create a mediator-free biosensor. Owing to the unique hybrid structure of TiO_2-Ti_3C_2, which resembles an organ, direct electron transfer of Hb is made easier. The biosensors that were created performed well in the detection of H_2O_2, exhibiting a broad linear range of 0.1–380 μM (sensitivity of 447.3 μA mM^{-1} cm^{-2}) and an incredibly low detection limit of 14 nM. In particular, a large number of highly biocompatible TiO_2 NPs on the nanocomposite's surface may offer Hb a protective microenvironment, enhancing the biosensor's long-term stability. After 60 days of storage, the TiO_2-Ti_3C_2-based biosensor still possesses 94.6% of its original response to H_2O_2 [75]. Figure 14.11 represents schematic representation of the organ-like structure of TiO_2-Ti_3C_2 nanocomposite encapsulating Hb.

(c) **Cancer theranostics**: MXene-based structures are promising options for cancer therapy and diagnosis with biological safety and clinical applications potential due to their hydrophilicity and low cytotoxicity shown in Fig. 14.12.

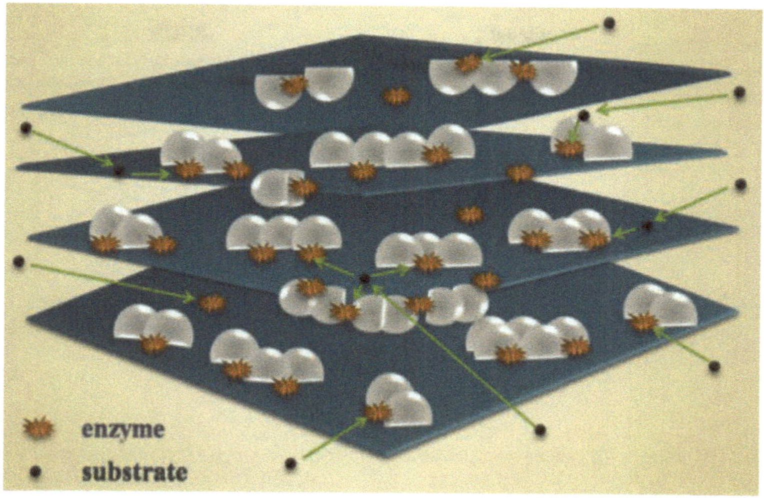

Fig. 14.11 Schematic representation of the organ-like structure of TiO_2-Ti_3C_2 nanocomposite encapsulating Hb [75]. Adapted from Ref. [75] with permission from [Elsevier], Copyright [2015]

Furthermore, these materials should be investigated further for photoacoustic imaging and photothermal therapy due to their vast and potent absorbance in the NIR region and significant light-to-heat conversion efficiency [76–78]. There have been numerous published studies on MXene composites as a contrast agent for bio-imaging combined with photothermal functions. Tantalum carbide (Ta_4C_3) nanosheets containing manganese oxide (MnO_X) were combined with soybean phospholipid (SP) to create a new material known as MnO_X/Ta_4C_3–SP composite in the initial investigation of MXene composite for cancer theranostics. In this composite system, each layer plays a different role. MnO_X is used for magnetic resonance imaging (MRI) that responds to the tumor microenvironment; Ta is preferred for contrast-enhanced CT scans; and SP stabilizes the nanocomposite in physiological settings with minimal cytotoxicity. This work demonstrated the effectiveness of MnO_X/Ta_4C3–SP composite in achieving contrast-enhanced PA imaging and PTT (photothermal therapy) for the inhibition of tumor growth [79]. Zong reported the surface engineering of MXene nanosheets with tumor photothermal therapy and dual MR/CT imaging capabilities (in vivo), using GdW_{10} nanoclusters (as the contrast agents). Further research into these highly biocompatible composites for multifunctional nanotheranostics is necessary, particularly for targeted cancer diagnosis and therapy [80]. Fabrication of MXene-based composite (Au/MXene and Au/Fe_3O_4/MXene) was fabricated by Hussein et al. Their preliminary results on the embryonic toxicity generally showed that the as-prepared (Au/MXene) and (Au/Fe_3O_4/MXene) nanocomposites did not significantly cause developmental toxicity or fatal effects in zebrafish embryos. According to the authors, MXene's low observed effect concentration (LOEC) was 100 μg/mL, and its cumulative

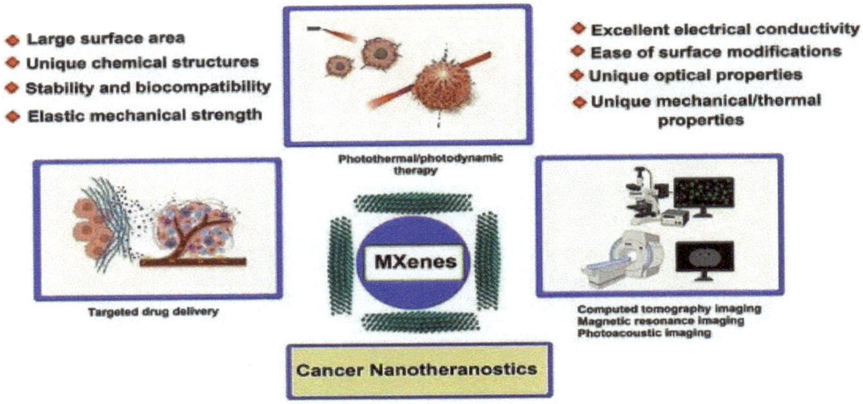

Fig. 14.12 MXenes with cancer nanotheranostic applications [76]. Open access under a CC BY 4.0 license

mortality rate was approximately 21%. It's interesting to note that Au/MXene and Au/Fe_3O_4/MXene showed lower embryonic mortality ($LC_{50} \gg 1000$ μg/mL) than pure MXene ($LC_{50} = 257.46$ μg/mL) in in vivo acute toxicity assays conducted on zebrafish embryos [81].

14.3.2.3 Hydrogel-Based Composite for Biomedical Application

Hydrogels have a propensity to hold onto a significant amount of water in their interstices and continue to bond it while keeping the network structure swollen, because of their unique three-dimensional crosslinked polymer meshwork structure. Hydrogels are essential in biomedical applications because of their unique qualities, which include superabsorber, hydrophilicity, biodegradability, biocompatibility, viscoelasticity, softness, and fluffiness. Applications of hydrogel nanocomposites in the biomedical industry are demonstrated in Fig. 14.13. By developing links with the polymeric chain which may desorb under the influence of stress, NPs are thought to enhance the mechanical properties of hydrogels by reducing a certain amount of the tension within the hydrogel network. Hydrogels' chemical characteristics, such as their bioactivity and sensitivity to stimuli, are essential to their suitability for use in biomedicine. Hydrogel nanocomposites have acquired wide uses throughout biomedical sciences and engineering, involving adhesives, drug delivery agents, antimicrobial dressings, responsive inks for bioprinting, implantable scaffolds, flexible conductive sensors, and cell culture systems. This is due to the nanocomposites' enhanced and expanded physical and biochemical properties, which go beyond their mechanical qualities [82–86].

It has been shown that hydrogels are excellent wound dressings because of their many benefits, which include good air permeability, the capacity to absorb tissue exudates, the ability to maintain a moist environment at the site of the wound, and the

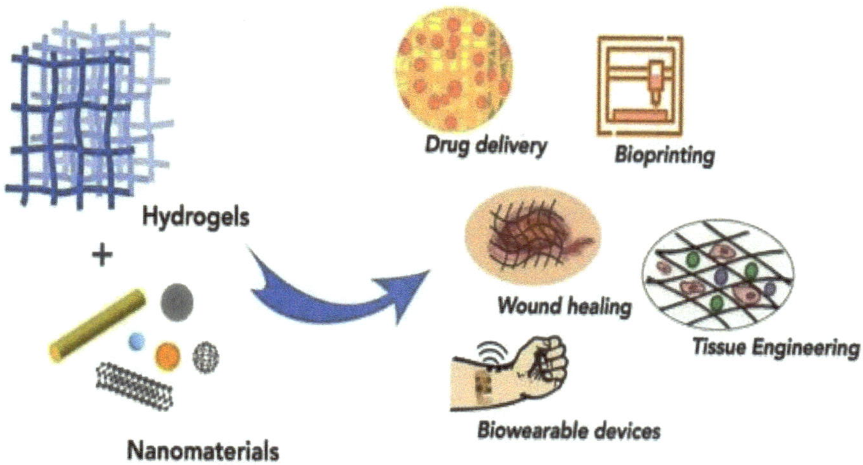

Fig. 14.13 Applications of hydrogel nanocomposites in the biomedical industry [86]. Open access under a CC BY 4.0 license

ability to thwart microbial invasion. In addition to these benefits, poly(N-isopropyl acrylamide) (PNIPAM) hydrogels offer a few more benefits for wound dressings. Because of their thermos-sensitivity, the dressings' characteristics and functions can be controlled by temperature straightforwardly and efficiently. Moreover, the transparent network of PNIPAM hydrogels enables the monitoring of the healing process. Even with these benefits, PNIPAM hydrogels for use as wound dressings still have a lot of space for development. Some of these include improving wound adhesion, achieving the ideal balance between air permeability and water retention, and enhancing the ability to encapsulate and release certain medications over time to prevent infection. The recent efforts to optimize PNIPAM-based hydrogels for use as wound dressings are briefly summarized in this section. The hydrogels based on PNIPAM show promise for use as wound dressings. By optimizing the properties of the wound dressings, such as moisture retention, improved adhesiveness, high comfort, temperature-triggered precise drug delivery, long-term infection inhibition, and the ability to supervise the wound treatment process, composite hydrogel formulation can help hasten the healing process of wounds [87–89]. The composite matrix system made up of sodium alginate (SA) and electrospun polycaprolactone (PCL) nanofiber lattice provided tensile properties that were better than those of the individual scaffold components. Furthermore, suture pull-out strength in the range of 18–20N was demonstrated by composite matrices. The outcomes of the low molecular weight SA were the uniform matrix coating and significantly higher tensile modulus. Due to the composite matrix's similar cross-linking density, the velocity of SA erosion was irrespective of molecular weight. As predicted, composite matrices lost more than 95% of their body weight in 10 days and demonstrated progressive weight loss in vitro. These composite scaffolds outperformed PCL nanofiber matrices in supporting human mesenchymal stem cells (h-MSC) adhesion, viability, proliferation, and neurogenic differentiation under neural induction media. In comparison to high molecular weight (HMW) SA and control PCL with or without laminin, low molecular weight SA has been shown to promote phenotypic marker S-100 protein expression more strongly. These composite biphasic scaffolds demonstrated sufficient mechanical and degradation features as well as cell compatibility, making them appropriate for the regeneration of peripheral nerves [90].

14.3.2.4 Nanogenerators (NGs)

Nanogenerators are typically energy-harvesting devices that utilize surplus mechanical energy from the surrounding environment for producing electricity. NGs are a developing technology that is capable of collecting diverse sorts of mechanical energy, including human motion, vibration, water flowing, raindrops, wind, and even the ability to turn waste heat into electricity (pyroelectric effect) [91, 92]. There are different types of NGs like piezoelectric NGs, triboelectric NGs, and pyroelectric NGs. Piezoelectric nanogenerators (PENGs) have advanced very quickly in the last few years. Because, in contrast to traditional generators, they can transform the environment's erratic and weak mechanical energy into electrical energy. Utilizing the piezoelectric characteristic of piezoelectric materials, PNGs have the capability of collecting mechanical energy with a high energy density and are completely compatible with future microelectronic appliances. Boosting the flexibility and voltage/current output of appliances is the primary objective of the PENGs [93–97]. Many components are impacting the PENGs like the use of piezoelectric materials, altering the typical piezoelectric materials' microscopic morphologies, generating thin-film composite materials, etc. which are shown in Fig. 14.14 [98]. PENG devices based on composites that integrate piezoelectric nanoparticles (NPs) through soft polymer matrices have distinct benefits over alternative methods due to their easy and scalable fabrication process, which uses inexpensive materials to achieve excellent flexibility and enhanced mechanical durability. To generate high output power, an even distribution of piezoelectric particles throughout the polymer matrix is just as important as the characteristics of the piezoelectric material. Metal nanorods and carbon nanomaterials are employed as dispersants in composite systems to produce homogenous dispersions [99].

Jung Lee et al. have proposed a superior-performance PNG device built upon a composite system that is chemically reinforced. Without the use of dispersion enhancers, homogeneous dispersions with uniformly dispersed lead zirconate titanate (PZT) NPs were produced by combining amine-functionalized PZT NPs with a thermoplastic triblock copolymer grafted with maleic anhydride and mixing them. The output voltage and current generated were enough to drive commercial LEDs, reaching 65 V and 1.6 μA, respectively. The flexible PNG developed using PZT-NH_2 NP demonstrated exceptional stability without any degradation even after 5000 cycles of mechanical deformation. The alternating output energy produced by the

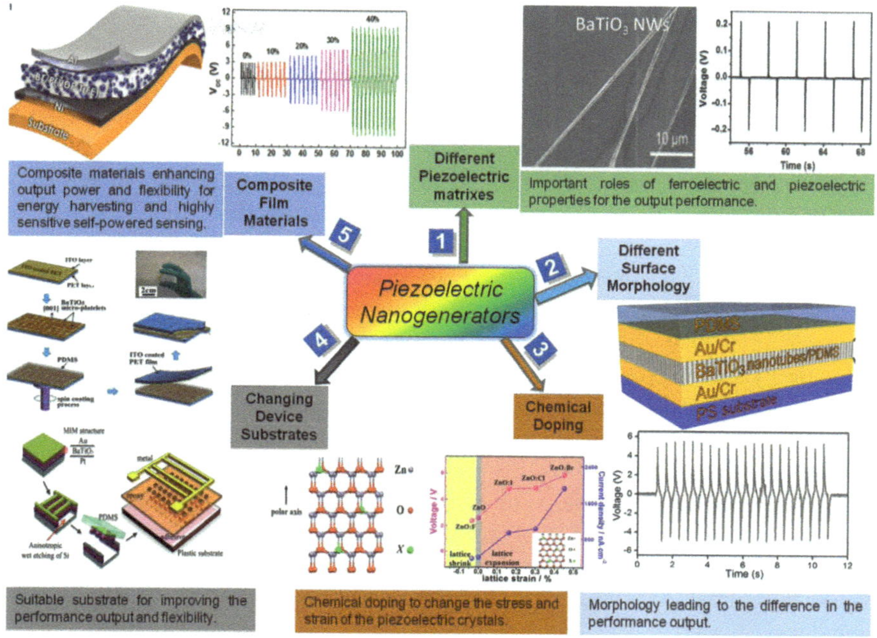

Fig. 14.14 Different factors influencing PENGs' output performance [98]. Adapted from Ref. [98] with consent from [Elsevier], Copyright [2019]

PENG device was preserved in capacitors and then utilized for powering a commercial LED device via the use of a full-wave bridge rectifier circuit. The composite system effectively gets around the limitations imposed by dispersion stability on earlier nanogenerators, making it possible to build an energy-harvesting system that is straightforward, reliable, and large scale [100]. Another nanogenerator grounded on lead-free barium titanate ($BaTiO_3$) and a polydimethylsiloxane (PDMS) polymer composite was prepared by Bouhamed and his co-workers. Multiwalled carbon nanotubes (MWCNTs) were integrated into the $BaTiO_3$/PDMS composite as conductive components to improve its performance. The electrical and mechanical attributes of the materials are enhanced by the inclusion of MWCNTs, particularly at the percolation threshold. Later on, this greatly improved the NG's performance, making it two times better than the NG without MWCNTs. Both the applied frequency and the output voltage are improved by raising the composite thickness. With no further poling process, the improved NG structure facilitates the recovery of biomechanical energy at 5.56 V under light finger tapping and 11.22 V under palm striking. Consequently, there are a lot of chances to develop wearable energy harvesters for self-powered devices with the discovered NG [101]. Different mechanical energies can be efficiently converted into electricity using a triboelectric nanogenerator (TENG). The TENG has numerous benefits, including high conversion efficiency, low cost, big output power, and simple production. Many TENGs, mainly centered on the coupling among the triboelectric effect and electrostatic induction, have been developed. The application of flexible TENGs as power sources has been thoroughly researched due to the rapidly increasing demand for flexible electronics in wearables, bendable displays, and electronic skin [102–104].

In contrast to certain developed polymer generators, Sun et al. suggested a type of eco-friendly triboelectric nanogenerator (TENG) grounded on naturally degradable cellulose. For enhancement of performance, polyamide (PA6)/polyvinylidene fluoride (PVDF) with varying electronegativity and barium titanate ($BaTiO_3$) with an elevated dielectric constant and low dielectric loss are added. Because of its excellent stability and dependability, the nanogenerator device described in this article has broad applications in the energy-harvesting field, significantly advancing the development of natural biomaterials. Since the friction layer's internal charge and permeability are linked with the specific capacitance, adding PA6 increases the friction layer's internal area and capacity and its charge capacity, raising specific capacitance. According to the findings, there is a 2.544-fold improvement in the open-circuit output voltage peak value of 20.155 V and a 5.480-fold improvement in the short-circuit output current peak value of 6.001 μA [105]. The development of porous polydimethylsiloxane (PDMS)/polydopamine (PDA)/Au composites with different pore sizes was done by Biutty et al. to improve TENG devices' electrical performance. In contrast to the TENGs with flat PDMS, the TENGs with PDMS/PDA/Au composites showed elevated electric output. Pen-circuit voltage (V) from TENGs with large, medium, and small pores with PDMS/PDA/Au elastomers was measured as 100, 170, and 200 V, respectively. This was significantly higher than TENGs with flat PDMS which was found to be 32 V (Fig. 14.15a). The electrical efficiency of the TENG can be improved by increasing the triboelectric charge density upon the outer surface of the PDMS/PDA/Au through an expanded contact area. An analogous pattern has been noted regarding the short-circuit current (I_{sc}). In Fig. 14.15b, we can see that as pore size decreases, there is a noticeable increase in Isc values, also related to the effective contact area [106].

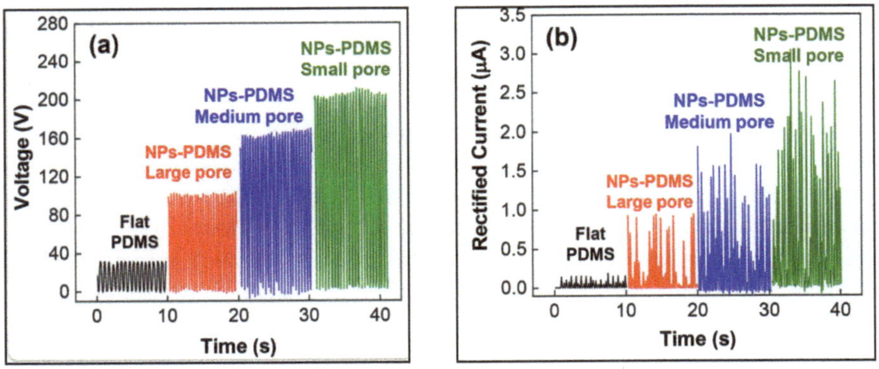

Fig. 14.15 **a** Open-circuit voltage (V_{OC}) and **b** short-circuit voltage (I_{SC}) curves of the TENG having PDMS/PDA/Au composites [106]. Adapted from Ref. [106] with permission from [Springer], Copyright [2021]

The thermal energy may be transformed into electric energy using the pyroelectric nanogenerator (PyNG), which utilizes the Seebeck and pyroelectric effects. No issues with the devices' operational performance were encountered because PyNGs are not subjected to mechanical deformation like PENG and TENG. PyNGs use the temperature variations that occur in the human body, particularly in the winter, and the alterations in temperature that occur around the mouth while breathing to transform thermal energy to electrical energy through the polarization change between electrodes. As they can gather waste heat energy, PyNGs are noteworthy for their prospective uses in temperature monitoring, healthcare diagnosis, and household electronics [107–109]. Mistewicz studied a PyNG based on antimony sulfoiodide (SbSI) nanowires and titanium dioxide (TiO_2) nanoparticles composite. The prepared SbSI–TiO_2 nanocomposite-based PyNG was wrapped between gold (Au)- and indium tin oxide (ITO)-coated polyethylene terephthalate (PET) substrate electrodes. The highest and average surface power densities of the SbSI–TiO_2 nanocomposite-based PyNG were found to be 8.39(2) and 2.57(2) $\mu W/m^2$, respectively. After one thermal cycle with a temperature variation of 25 K, the generated electrical energy was 15.2 $\mu J/cm^3$ per unit volume. It is important to note that when compared with all the ferroelectric composites, the SbSI–TiO_2 nanocomposite possessed the most significant values for the pyroelectric coefficient and surface power density. It is important to note that when compared with all the ferroelectric composites, the SbSI–TiO_2 nanocomposite possessed the most significant values for the pyroelectric coefficient and surface power density. Moreover, the SbSI–TiO_2 nanocomposite exhibited significantly superior pyroelectric performance compared to SbSI thin films and antimony selenoiodide (SbSeI) nanowires, the latter serving as an isostructural material for SbSI. The ferroelectric SbSI–TiO_2 nanocomposite has a lot of exciting possibilities to be used in pyroelectric sensors and thermal energy harvesters, as shown by the findings that are discussed in this paper [110].

14.3.2.5 Electromagnetic Interference (EMI) Shielding

The use of magnetic or conducting materials to stop electric and magnetic waves from traveling from one area to another is known as electromagnetic interference (EMI) shielding. This shielding can be achieved by reducing the amount of signal that enters a system through wave absorption or reflection. EMI has grown to be a major concern for electronic devices as a result of the constant growth in wireless communication. In severe situations, EMI can cause electronic systems to fail as it tends to interfere with the regular operation of electronic devices. Therefore, protecting electronic equipment from EMI has become critical. The material that is most frequently used for EMI shielding is metal, and electromagnetic waves interact with the free electrons that are present in metals. The reflection mechanism will provide shielding against electromagnetic waves (EM) when the composite is vastly conductive. Graphene and CNT with superior specific surface area, relatively small density, high conductivity, and the ability to absorb electromagnetic radiation. Applications for CNT and graphene nanofillers include sensors, electronics, biomedical

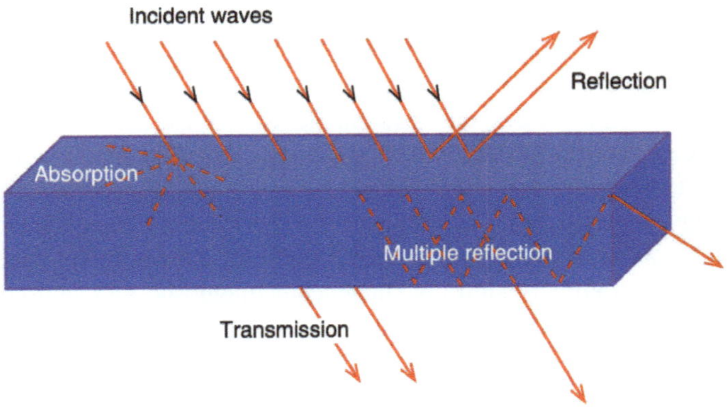

Fig. 14.16 Diagram illustrating the electromagnetic shielding mechanism [113]. Adapted from Ref. [113] with permission from [Wiley], Copyright [2018]

applications, catalysts, and high-performance composites for EMI shielding. Free-moving electrons or holes are necessary for the shielding material to have an effective EMI-shielding effect (EMI-SE). The amount of reflection loss has straight correlation with the shielding material's electrical conductivity (σ). Thus, although a high σ is not required, electrically conductive materials are preferred for use as electromagnetic shielding absorbers [111–115]. Illustration demonstrating the electromagnetic shielding mechanism is shown in Fig. 14.16.

Rao et al. investigated the dielectric and EMI response properties of a novel composite made up of polyvinyl alcohol (PVA) and Fe_3O_4 nanofiller-decorated single-layer graphene-assembled porous carbon (SLGAPC), which has a density and flexibility of 0.75 g/cm^3. In comparison to the PVA-SLGAPC composite, the PVA-SLGAPC-Fe_3O_4 composite exhibits high dielectric permittivity in the 1 Hz–10 MHz range, enhanced by a factor of 4, with a factor of 2 reduction in loss. With a thickness of only 0.3 mm, the composite exhibits a remarkable shielding performance of ~15 dB and a specific shielding effectiveness of 20 dB/(g/cm^3) in the X-band (8.2 GHz–12.4 GHz) for EMI shielding. This material shows promise for use in flexible EMI-shielding applications [116]. Jiang et al. created spherical Fe_3O_4 particles with integrity crystal on the sheet of graphene surface using an easy solvothermal technique, which they then combined with graphene nanoplates to create a magnetic Fe_3O_4. The impedance matching ability of Fe_3O_4@f-GNPs demonstrated a better absorption performance than that of magnetic Fe_3O_4 and dielectric f-GNPs, with a reflection loss of −25 dB at 10 GHz frequency when the sample thickness was 2 mm and the effective absorption bandwidth was 2.4 GHz (below −10 dB). Furthermore, when the Fe_3O_4@f-GNPs demonstrated shielding efficiency properties with 232 nm Fe_3O_4 in the X-band, the absorption and total efficiencies were 32 dB and 25 dB, respectively [117]. Polystyrene (PS)-multiwalled carbon nanotubes (MWCNT)/graphite nanoplate (GNP) nanocomposites with extremely high EMI values were created by polymerizing styrene/MWCNT with PS-GNP beads. The PS/MWCNT/GNP nanocomposites have a commercially applicable EMI-SE value of approximately 20.2 dB, which was achieved at very low filler concentrations (2 weight % loading of MWCNT and 1.5 weight % loading of GNP). As a result, the nanocomposites can be used in a variety of fields, including aerospace, mobile phones, lightweight shielding rooms, and spacecraft. An electrical conductivity of approximately 9.47×10^{-3} S cm^{-1} was obtained at the lowest possible MWCNT loading (0.3 wt%) and 0.29 wt % GNP loading, which has never been reported, by optimum utilization of the ratio of PS – GNP bead and MWCNT in the nanocomposites [118].

Owing to their high specific surface area, high surface atom percentage, high quantum size effect, and high Snoek's limit, magnetic NPs are ideal for use as EMI shielding at high frequency over GHz. These properties give magnetic NPs their strong interface polarization [119]. For example, the nickel-polyetherketone (Ni/PEK) nanocomposites were made by Dravid et al. with the help of hot pressing them using a traditional method after they had been ground in a planetary ball mill. As an exceptionally well semi-crystalline thermoplastic polymer with outstanding mechanical, thermal, wear, and chemical resistance qualities, polyetherketone (PEK) is being researched extensively for a variety of uses. When the concentration of Ni nanoparticles in the nanocomposites increased, the EMI-shielding effectiveness (SE), which was measured in the X-band at 8.2–12.4 GHz, increased gradually up to 12.91 vol.%. However, for the 18.19 vol.% Ni nanocomposite, the EMI-SE increased suddenly to a value of about 25 dB. The results showed that the absorption loss serves as these nanocomposites' main shielding mechanism, not the reflection. The concentration of Ni nanoparticles determines an increase in absorption loss. Because these nanocomposites are lighter, more microhardy, and show more dimensional stability than pure matrix, there is potential to create a high-performance substrate or paint that might be cast-off as superior-performance EMI-shielding materials on electronic equipment [120]. In another experiment, the solvent casting method was used to prepare the polyvinyl chloride/barium titanate/nickel oxide ($PVC/BaTiO_3/NiO$) nanocomposite films whose EMI-SE was examined. The improved performance of the shielding properties of the nanocomposite was made possible by the influence of nanofillers. The PVC matrix now has a larger interaction surface and aspect ratio thanks to the addition of nanofillers, which also improved conductivity. As the concentration of the nanofiller varied, the properties of the nanocomposite improved. In the Ku-band region, the nanocomposite displayed an EMI-SE of −18.7 dB at 30 weight % $BaTiO_3$ + 5 weight % NiO loading. Furthermore, the maximum dielectric constant of 11 was observed in the $PVC/BaTiO_3/NiO$ nanocomposite films at 20 weight % $BaTiO_3$ + 15 weight % NiO loading in the PVC matrix. According to the findings, $PVC/BaTiO_3/NiO$ nanocomposite films have the potential to be an effective EMI-SE material for electromagnetic absorption [121].

14.3.3 Polymer Composites for Energy Storage

The improved dielectric performance achieved by integrating the elevated electrical breakdown strength of the polymer matrix with the superior permittivity of the nanoparticles has made polymer nanocomposite dielectrics a recent trend in the domain of electric energy storage in recent years [122]. Hu et al. have reported on when it comes to designing polymer nanocomposite dielectrics rationally based on models and simulating dielectric energy storage. Polymer nanocomposite dielectrics and their production methods and behavior are also covered. Interfacial design, microstructural engineering, and novel high-dielectric-filler composites are some of the important tactics and analytical models that have been the subject of this review in an effort to significantly increase the energy density of composite dielectrics. New designs have arisen as a result of combining analytical models with machine learning approaches [123] Some examples of electrostatic energy storage capacitors are vacuum capacitors and those using conventional dielectrics as well as dielectrics with high dielectric permittivity, as shown in Fig. 14.17a. The energy needed to polarize the interior dielectrics is responsible for the elevated charge qm seen in Fig. 14.17b. An extra rise in charge qp results from using high-dielectric-permittivity dielectrics in place of typical linear dielectrics. The three energy sources are shown in Fig. 14.17c.

In Fig. 14.18, the process of fabricating compounds by using PVDF and 2D MXene nanosheets as fillers has been illustrated. It was possible to produce a considerable dielectric permittivity of around 80,000 at a frequency of 10 kHz, and dielectric loss of 0.05. It is also possible for the percolation theory procedure to shed light on the large increase in dielectric permittivity [123]. In spite of this, the present challenge is to find solutions to the problems of small breakdown strength, excessive energy indulgence, and inadequate voltage endurance. This is necessary in order to make full advantage of the likely conductive fillers in practical applications.

These are the benefits that the ideal dielectric polymer composites should have: excellent dielectric qualities, high thermal stability, low dissipation factor, high

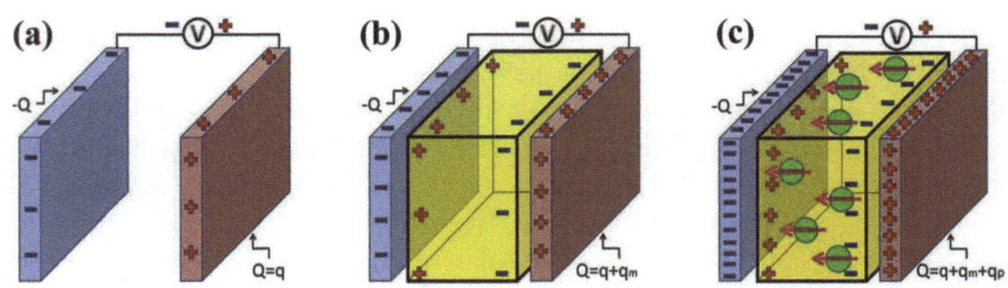

Fig. 14.17 Electrostatic capacitors store energy when an electric field is applied, as seen in schematic designs. **a** A capacitor in a vacuum, **b** a capacitor with dielectrics within a charge of $Q = q + q_m$, and **c** a capacitor entrenched with dielectrics of high dielectric permittivity to a collected charge of $Q = q_m + q_p$ [123]. Adapted from Ref. [123] with consent from [Elsevier], Copyright [2020]

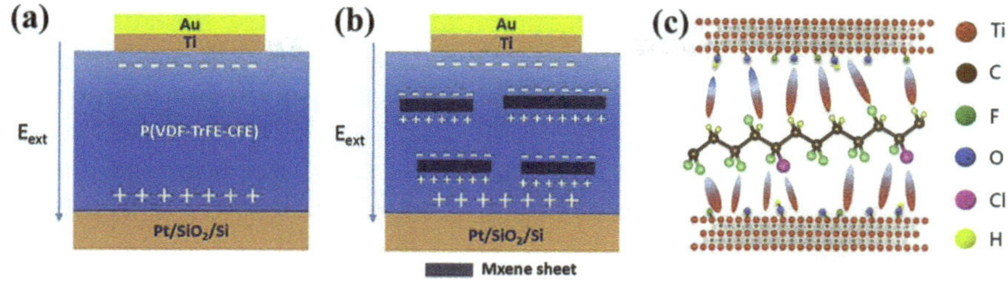

Fig. 14.18 Schematic diagrams that illustrate the presence of polarization charges in **a** P(VDF-TrFE-CFE) capacitors and **b** MXene/P(VDF-TrFE-CFE) metal–insulator–metal (MIM) capacitors with applied electric field. The diagrammatic representation of the dipoles that are produced between the atoms of the MXene surface (F and O) and the atom H of the polymer backbone is shown in **c** [123]. 123 Adapted from Ref. [39] with consent from [Elsevier], Copyright [2020]

dielectric constant, and easy processability [124]. In the future, scientists hope to learn more about the nanoparticle filler's special characteristics, how to modify the filler, and how the filler and polymer matrix are distributed. Polymer composite dielectrics with exceptional thermal durability at elevated temperatures and outstanding energy density are essential for practical applications in aerospace power systems and subsurface oil industries, among others [122].

14.4 Conclusions and Future Prospectives

The integration of nanoparticles into a polymer matrix has facilitated the development of polymer nanocomposites. Polymers infused with carbon nanoparticles, such as quantum dots, nanosheets, and nanotubes, have resulted in materials with exceptional optical, electrical, and magnetic characteristics. Ensuring compatibility between the polymer and carefully chosen nanoparticles and their homogeneous distribution throughout the matrix is crucial for the effective creation of composites. Polymer nanocomposites are extremely versatile in terms of structure and content, making them useful in many different sectors. These include electronics, aerospace, cosmetics, and biomedicine. One notable improvement in the final composites' strength, stiffness, and durability was the use of nanoparticles. The exact spreading of nanoparticles inside the polymer matrix allows for the painstaking regulation of material features, opening up possibilities for modifying particular qualities like conductivity and biocompatibility. These developments have opened up new avenues of possibility, encouraging the creation of innovative materials and enhancing technology in many other fields. Research into cutting-edge applications of polymer composites has significantly expanded our understanding of materials science and facilitated groundbreaking technological breakthroughs.

The future of polymer nanocomposites is promising with great potential for further advancements in technology. Research into ever more advanced composite materials is anticipated to be driven by the continued quest of improved performance and flexibility. New composites with extraordinary characteristics and capabilities will most certainly be created as a result of improvements in manufacturing methods and optimization of nanoparticle–polymer interactions. Smart materials and responsive nanocomposites, when combined, have the potential to usher in a new age of material design characterized by materials' ability to dynamically react to environmental stimuli. This paves the way for potential uses in areas including responsive biological devices, self-healing constructions, and adaptive electronics. The creation of new polymer nanocomposites might be accelerated by interdisciplinary collaboration and the use of AI in material design and characterization. In order to uncover new horizons of potential in this dynamic sector, the synergy between experimental and computational techniques will be crucial. Overall, polymer nanocomposites have promising future possibilities and ongoing research and the discipline of material science is poised to reach new heights owing to continuing research and development, which is projected to generate revolutionary breakthroughs in industry and technology.

Acknowledgements The authors gratefully acknowledge the support provided by Centurion University of Technology and Management, Odisha, India for carrying out the present research work.

References

1. Kumar, S., Nehra, M., Dilbaghi, N., Tankeshwar, K., Kim, K.H.: Recent advances and remaining challenges for polymeric nanocomposites in healthcare applications. Prog. Polym. Sci.Polym. Sci. **80**, 1–38 (2018)
2. Saleh Alghamdi, S., John, S., Roy Choudhury, N., Dutta, N.K.: Additive manufacturing of polymer materials: Progress, promise and challenges. Polymers **13**(5), 753 (2021)
3. Shah, V., Bhaliya, J., Patel, G.M., Deshmukh, K.: Advances in polymeric nanocomposites for automotive applications: A review. Polym. Adv. Technol.. Adv. Technol. **33**(10), 3023–3048 (2022)
4. Chafiq, M., Chaouiki, A., & Ko, Y. G. (2023). Advances in COFs for energy storage devices: harnessing the potential of covalent organic framework materials. *Energy Storage Materials*, 103014.
5. Kangishwar, S., Radhika, N., Sheik, A.A., Chavali, A., Hariharan, S.: A comprehensive review on polymer matrix composites: Material selection, fabrication, and application. Polym. Bull.. Bull. **80**(1), 47–87 (2023)
6. Choi, Y., Phan, B., Tanaka, M., Hong, J., Choi, J.: Methods and applications of biomolecular surface coatings on individual cells. ACS Appl. Bio Mater. **3**(10), 6556–6570 (2020)
7. Siripongpreda, T., Hoven, V.P., Narupai, B., Rodthongkum, N.: Emerging 3D printing based on polymers and nanomaterial additives: Enhancement of properties and potential applications. Eur. Polymer J. **184**, 111806 (2023)
8. Zha, J.W., Zheng, M.S., Fan, B.H., Dang, Z.M.: Polymer-based dielectrics with high permittivity for electric energy storage: A review. Nano Energy **89**, 106438 (2021)
9. Kaur, G., Kaur, A., Kaur, H.: Review on nanomaterials/conducting polymer based nanocomposites for the development of biosensors and electrochemical sensors. Polymer-Plastics Technology and Materials **60**(5), 504–521 (2021)

10. Yi, C., Yang, Y., Liu, B., He, J., Nie, Z.: Polymer-guided assembly of inorganic nanoparticles. Chem. Soc. Rev. **49**(2), 465–508 (2020)
11. Joudeh, N., Linke, D.: Nanoparticle classification, physicochemical properties, characterization, and applications: a comprehensive review for biologists. Journal of Nanobiotechnology **20**(1), 262 (2022)
12. Baig, N., Kammakakam, I., Falath, W.: Nanomaterials: A review of synthesis methods, properties, recent progress, and challenges. Materials Advances **2**(6), 1821–1871 (2021)
13. Jeong, G.H., Sasikala, S.P., Yun, T., Lee, G.Y., Lee, W.J., Kim, S.O.: Nanoscale assembly of 2D materials for energy and environmental applications. Adv. Mater. **32**(35), 1907006 (2020)
14. Romero, M., Macchione, M.A., Mattea, F., Strumia, M.: The role of polymers in analytical medical applications. A review. Microchemical Journal **159**, 105366 (2020)
15. Kumar, N., Chamoli, P., Misra, M., Manoj, M.K., Sharma, A.: Advanced metal and carbon nanostructures for medical, drug delivery and bio-imaging applications. Nanoscale **14**(11), 3987–4017 (2022)
16. Rozbu, M. R., Kabir, A., & Selvakumar, P. M. (2021). Functionalized Bio-carbon Nanomaterials for Environmental Utilizations. *Environmental Applications of Carbon Nanomaterials-Based Devices*, 347–374.
17. Ramya, M., Kumar, P. S., Rangasamy, G., Rajesh, G., Nirmala, K., Saravanan, A., & Krishnapandi, A. (2022). A recent advancement on the applications of nanomaterials in electrochemical sensors and biosensors. *Chemosphere*, 136416.
18. Soni, S.K., Thomas, B., Thomas, S.B., Tile, P.S., Sakharwade, S.G.: Carbon nanotubes as exceptional nanofillers in polymer and polymer/fiber nanocomposites: An extensive review. Materials Today Communications **37**, 107358 (2023)
19. Ramakoti, I.S., Panda, A.K., Gouda, N.: A brief review on polymer nanocomposites: current trends and prospects. J. Polym. Eng.Polym. Eng. **43**(8), 651–679 (2023)
20. Darwish, M.S., Mostafa, M.H., Al-Harbi, L.M.: Polymeric nanocomposites for environmental and industrial applications. Int. J. Mol. Sci. **23**(3), 1023 (2022)
21. Karki, S., Gohain, M.B., Yadav, D., Ingole, P.G.: Nanocomposite and bio-nanocomposite polymeric materials/membranes development in energy and medical sector: A review. Int. J. Biol. Macromol.Macromol. **193**, 2121–2139 (2021)
22. Monteiro, M.J., Cunningham, M.F.: Polymer nanoparticles via living radical polymerization in aqueous dispersions: design and applications. Macromolecules **45**(12), 4939–4957 (2012)
23. Mandal, T.K., Fleming, M.S., Walt, D.R.: Preparation of polymer coated gold nanoparticles by surface-confined living radical polymerization at ambient temperature. Nano Lett. **2**(1), 3–7 (2002)
24. Payami, E., Keynezhad, M.A., Safa, K.D., Teimuri-Mofrad, R.: Development of high-performance supercapacitor based on Fe3O4@ SiO2@ PolyFc nanoparticles via surface-initiated radical polymerization. Electrochim. Acta. Acta **439**, 141663 (2023)
25. Li, X., Anton, N., Arpagaus, C., Belleteix, F., Vandamme, T.F.: Nanoparticles by spray drying using innovative new technology: The Büchi Nano Spray Dryer B-90. J. Control. Release **147**(2), 304–310 (2010)
26. Wang, Y., Kho, K., Cheow, W.S., Hadinoto, K.: A comparison between spray drying and spray freeze drying for dry powder inhaler formulation of drug-loaded lipid–polymer hybrid nanoparticles. Int. J. Pharm. **424**(1–2), 98–106 (2012)
27. Marante, T., Viegas, C., Duarte, I., Macedo, A.S., Fonte, P.: An overview on spray-drying of protein-loaded polymeric nanoparticles for dry powder inhalation. Pharmaceutics **12**(11), 1032 (2020)
28. Sosnik, A., Seremeta, K.P.: Advantages and challenges of the spray-drying technology for the production of pure drug particles and drug-loaded polymeric carriers. Adv. Coll. Interface. Sci. **223**, 40–54 (2015)
29. Musyanovych, A., Schmitz-Wienke, J., Mailänder, V., Walther, P., Landfester, K.: Preparation of biodegradable polymer nanoparticles by miniemulsion technique and their cell interactions. Macromol. Biosci.. Biosci. **8**(2), 127–139 (2008)

30. Baier, M.C., Huber, J., Mecking, S.: Fluorescent conjugated polymer nanoparticles by polymerization in miniemulsion. J. Am. Chem. Soc. **131**(40), 14267–14273 (2009)
31. Albernaz, V.L., Bach, M., Weber, A., Southan, A., Tovar, G.E.: Active ester containing surfmer for one-stage polymer nanoparticle surface functionalization in mini-emulsion polymerization. Polymers **10**(4), 408 (2018)
32. Wu, L.G., Huang, L.L., Yao, Y., Liu, Z.H., Wang, T., Yang, X.Y., Dong, C.Y.: Fabrication of polyvinylidene fluoride blending membrane coupling with microemulsion polymerization and their anti-fouling performance. Polymer **203**, 122767 (2020)
33. Chavda, V. P., Gogoi, N., Vaghela, D. A., Balar, P. C., Daware, S., & Dave, D. J. (2023). Parenteral microemulsion for drug delivery: Advances and update. *Journal of Drug Delivery Science and Technology*, 104991.
34. Kanwar R, Rathee J, Patil MT, Mehta SK. Microemulsions as nanotemplates: a soft and versatile approach. Microemulsion—Chemical Nanoreactor. 2019 Sep 18.
35. Gao, H., Jiang, T., Han, B., Wang, Y., Du, J., Liu, Z., Zhang, J.: Aqueous/ionic liquid interfacial polymerization for preparing polyaniline nanoparticles. Polymer **45**(9), 3017–3019 (2004)
36. Krauel, K., Davies, N.M., Hook, S., Rades, T.: Using different structure types of microemulsions for the preparation of poly (alkylcyanoacrylate) nanoparticles by interfacial polymerization. J. Control. Release **106**(1–2), 76–87 (2005)
37. Liu, L., Wu, W., Jin, X., Luo, X., Wu, L.: Interfacial Polymerization on Polyethersulfone Ultrafiltration Membrane to Prepare Nanofiltration Layers for Dye Separation. Polymers **15**(9), 2018 (2023)
38. Zhang, C., Chung, J.W., Priestley, R.D.: Dialysis nanoprecipitation of polystyrene nanoparticles. Macromol. Rapid Commun.. Rapid Commun. **33**(20), 1798–1803 (2012)
39. Crucho, C.I., Barros, M.T.: Polymeric nanoparticles: A study on the preparation variables and characterization methods. Mater. Sci. Eng. C **80**, 771–784 (2017)
40. Pulingam, T., Foroozandeh, P., Chuah, J.A., Sudesh, K.: Exploring various techniques for the chemical and biological synthesis of polymeric nanoparticles. Nanomaterials **12**(3), 576 (2022)
41. Mainardes, R.M., Evangelista, R.C.: PLGA nanoparticles containing praziquantel: effect of formulation variables on size distribution. Int. J. Pharm. **290**(1–2), 137–144 (2005)
42. Fessi, H.P.F.D., Puisieux, F., Devissaguet, J.P., Ammoury, N., Benita, S.: Nanocapsule formation by interfacial polymer deposition following solvent displacement. Int. J. Pharm. **55**(1), R1–R4 (1989)
43. Hasan, N., Biak, D. R. A., & Kamarudin, S. (2012). Application of bacterial cellulose (BC) in natural facial scrub. *Int. J. Adv. Sci. Eng. Inf. Technol*, *2*(4), 1–4.
44. Ullah, H., Santos, H.A., Khan, T.: Applications of bacterial cellulose in food, cosmetics and drug delivery. Cellulose **23**, 2291–2314 (2016)
45. Bianchet, R.T., Cubas, A.L.V., Machado, M.M., Moecke, E.H.S.: Applicability of bacterial cellulose in cosmetics–bibliometric review. Biotechnology Reports **27**, e00502 (2020)
46. Mbituyimana, B., Liu, L., Ye, W., Boni, B.O.O., Zhang, K., Chen, J., Yang, G.: Bacterial cellulose-based composites for biomedical and cosmetic applications: Research progress and existing products. Carbohyd. Polym.. Polym. **273**, 118565 (2021)
47. Gregory, D.A., Tripathi, L., Fricker, A.T., Asare, E., Orlando, I., Raghavendran, V., Roy, I.: Bacterial cellulose: A smart biomaterial with diverse applications. Mater. Sci. Eng. R. Rep. **145**, 100623 (2021)
48. Bilal, M., Iqbal, H.M.: New insights on unique features and role of nanostructured materials in cosmetics. Cosmetics **7**(2), 24 (2020)
49. Gupta, V., Mohapatra, S., Mishra, H., Farooq, U., Kumar, K., Ansari, M.J., Aldawsari, M.F., Alalaiwe, A.S., Mirza, M.A., Iqbal, Z.: Nanotechnology in cosmetics and cosmeceuticals—A review of latest advancements. Gels **8**(3), 173 (2022)
50. Wang, X., Tang, J., Huang, J., Hui, M.: Production and characterization of bacterial cellulose membranes with hyaluronic acid and silk sericin. Colloids Surf. B **195**, 111273 (2020)
51. Ahn, C.B., Je, J.Y.: Chitosan and Its Derivatives as Potential Cosmetic Agents: A Review. Journal of Chitin and Chitosan **20**(4), 287–296 (2015)

52. Kulka, K., Sionkowska, A.: Chitosan Based Materials in Cosmetic Applications: A Review. Molecules **28**(4), 1817 (2023)
53. Guzmán, E., Ortega, F., Rubio, R.G.: Chitosan: A Promising Multifunctional Cosmetic Ingredient for Skin and Hair Care. Cosmetics **9**(5), 99 (2022)
54. Sionkowska, A., Lewandowska, K., Kurzawa, M.: Chitosan-based films containing rutin for potential cosmetic applications. Polymers **15**(15), 3224 (2023)
55. Hanemann, T., Szabó, D.V.: Polymer-nanoparticle composites: from synthesis to modern applications. Materials **3**(6), 3468–3517 (2010)
56. Guo, Z., Poot, A.A., Grijpma, D.W.: Advanced polymer-based composites and structures for biomedical applications. Eur. Polymer J. **149**, 110388 (2021)
57. Thomas, J., Thomas, S., Kalarikkal, N., & Jose, J. (Eds.). (2018). *Nanoparticles in polymer systems for biomedical applications*. CRC Press.
58. Iannazzo, D., Pistone, A., Celesti, C., Triolo, C., Patané, S., Giofré, S.V., Di Pietro, A.: A smart nanovector for cancer targeted drug delivery based on graphene quantum dots. Nanomaterials **9**(2), 282 (2019)
59. Zhao, C., Song, X., Liu, Y., Fu, Y., Ye, L., Wang, N., Liu, J.: Synthesis of graphene quantum dots and their applications in drug delivery. Journal of Nanobiotechnology **18**, 1–32 (2020)
60. Iannazzo, D., Pistone, A., Salamò, M., Galvagno, S., Romeo, R., Giofré, S.V., Di Pietro, A.: Graphene quantum dots for cancer targeted drug delivery. Int. J. Pharm. **518**(1–2), 185–192 (2017)
61. Hou, T., Ma, H., Gao, X., Sun, H., Wang, L., An, M.: Study on Long-Term Tracing of Fibroblasts on Three-Dimensional Tissue Engineering Scaffolds Based on Graphene Quantum Dots. Int. J. Mol. Sci. **23**(19), 11040 (2022)
62. Liu, C., Xia, Z., Czernuszka, J.T.: Design and development of three-dimensional scaffolds for tissue engineering. Chem. Eng. Res. Des. **85**(7), 1051–1064 (2007)
63. Uprety, B., Abrahamse, H.: Semiconductor quantum dots for photodynamic therapy: Recent advances. Front. Chem. **10**, 946574 (2022)
64. Ge, J., Lan, M., Zhou, B., Liu, W., Guo, L., Wang, H., Han, X.: A graphene quantum dot photodynamic therapy agent with high singlet oxygen generation. Nat. Commun.Commun. **5**(1), 4596 (2014)
65. Mousavi, S.M., Hashemi, S.A., Kalashgrani, M.Y., Omidifar, N., Bahrani, S., Vijayakameswara Rao, N., Chiang, W.H.: Bioactive graphene quantum dots based polymer composite for biomedical applications. Polymers **14**(3), 617 (2022)
66. Lu, Z.H., Tian, Q., Zhou, D.D., Chen, M., Cao, Y.W., Zhuang, L.Y., Senosy, I.A.: Magnetic MXene based metal-organic frameworks composites: Synthesis, characterization and application. J. Environ. Chem. Eng. **10**(3), 108037 (2022)
67. Zamhuri, A., Lim, G.P., Ma, N.L., Tee, K.S., Soon, C.F.: MXene in the lens of biomedical engineering: synthesis, applications and future outlook. Biomed. Eng. Online **20**(1), 1–24 (2021)
68. Xu, X., Wang, S., Wu, H., Liu, Y., Xu, F., Zhao, J.: A multimodal antimicrobial platform based on MXene for treatment of wound infection. Colloids Surf. B **207**, 111979 (2021)
69. Soleymaniha, M., Shahbazi, M.A., Rafieerad, A.R., Maleki, A., Amiri, A.: Promoting role of MXene nanosheets in biomedical sciences: therapeutic and biosensing innovations. Adv. Healthcare Mater. **8**(1), 1801137 (2019)
70. Rasool, K., Mahmoud, K.A., Johnson, D.J., Helal, M., Berdiyorov, G.R., Gogotsi, Y.: Efficient antibacterial membrane based on two-dimensional Ti3C2Tx (MXene) nanosheets. Sci. Rep. **7**(1), 1598 (2017)
71. Wu, Y., Zheng, W., Xiao, Y., Du, B., Zhang, X., Wen, M., Sheng, L.: Multifunctional, Robust, and Porous PHBV—GO/MXene Composite Membranes with Good Hydrophilicity, Antibacterial Activity, and Platelet Adsorption Performance. Polymers **13**(21), 3748 (2021)
72. Szuplewska, A., Kulpińska, D., Dybko, A., Chudy, M., Jastrzębska, A.M., Olszyna, A., Brzózka, Z.: Future applications of MXenes in biotechnology, nanomedicine, and sensors. Trends Biotechnol.Biotechnol. **38**(3), 264–279 (2020)

73. Zhou, L., Wu, F., Yu, J., Deng, Q., Zhang, F., Wang, G.: Titanium carbide (Ti3C2Tx) MXene: a novel precursor to amphiphilic carbide-derived graphene quantum dots for fluorescent ink, light-emitting composite and bioimaging. Carbon **118**, 50–57 (2017)
74. Huang, M., Gu, Z., Zhang, J., Zhang, D., Zhang, H., Yang, Z., Qu, J.: MXene and black phosphorus-based 2D nanomaterials in bioimaging and biosensing: Progress and perspectives. Journal of Materials Chemistry B **9**(26), 5195–5220 (2021)
75. Wang, F., Yang, C., Duan, M., Tang, Y., Zhu, J.: TiO2 nanoparticle-modified organ-like Ti3C2 MXene nanocomposite encapsulating haemoglobin for a mediator-free biosensor with excellent performances. Biosens. Bioelectron.. Bioelectron. **74**, 1022–1028 (2015)
76. Iravani, S., Varma, R.S.: MXenes in cancer nanotheranostics. Nanomaterials **12**(19), 3360 (2022)
77. Iravani, S., Varma, R.S.: MXenes for cancer therapy and diagnosis: recent advances and current challenges. ACS Biomater. Sci. Eng.Biomater. Sci. Eng. **7**(6), 1900–1913 (2021)
78. Maleki, A., Ghomi, M., Nikfarjam, N., Akbari, M., Sharifi, E., Shahbazi, M.A., Chen, Y.: Biomedical applications of MXene-integrated composites: regenerative medicine, infection therapy, cancer treatment, and biosensing. Adv. Func. Mater. **32**(34), 2203430 (2022)
79. Dai, C., Chen, Y., Jing, X., Xiang, L., Yang, D., Lin, H., Wu, R.: Two-dimensional tantalum carbide (MXenes) composite nanosheets for multiple imaging-guided photothermal tumor ablation. ACS Nano **11**(12), 12696–12712 (2017)
80. Zong, L., Wu, H., Lin, H., Chen, Y.: A polyoxometalate-functionalized two-dimensional titanium carbide composite MXene for effective cancer theranostics. Nano Res. **11**, 4149–4168 (2018)
81. Hussein, E. A., Zagho, M. M., Rizeq, B. R., Younes, N. N., Pintus, G., Mahmoud, K. A., ... & Elzatahry, A. A. (2019). Plasmonic MXene-based nanocomposites exhibiting photothermal therapeutic effects with lower acute toxicity than pure MXene. *International journal of nanomedicine*, 4529–4539.
82. Bashir, S., Hina, M., Iqbal, J., Rajpar, A.H., Mujtaba, M.A., Alghamdi, N.A., Ramesh, S.: Fundamental concepts of hydrogels: Synthesis, properties, and their applications. Polymers **12**(11), 2702 (2020)
83. Aswathy, S. H., Narendrakumar, U., & Manjubala, I. (2020). Commercial hydrogels for biomedical applications. *Heliyon*, *6*(4).
84. Lavrador, P., Esteves, M.R., Gaspar, V.M., Mano, J.F.: Stimuli-responsive nanocomposite hydrogels for biomedical applications. Adv. Func. Mater. **31**(8), 2005941 (2021)
85. Zhang, Y., Huang, Y.: Rational design of smart hydrogels for biomedical applications. Front. Chem. **8**, 615665 (2021)
86. Barrett-Catton, E., Ross, M.L., Asuri, P.: Multifunctional hydrogel nanocomposites for biomedical applications. Polymers **13**(6), 856 (2021)
87. Xu, X., Liu, Y., Fu, W., Yao, M., Ding, Z., Xuan, J., Cao, M.: Poly (N-isopropylacrylamide)-based thermoresponsive composite hydrogels for biomedical applications. Polymers **12**(3), 580 (2020)
88. Ansari, M.J., Rajendran, R.R., Mohanto, S., Agarwal, U., Panda, K., Dhotre, K., Pramanik, S.: Poly (N-isopropylacrylamide)-based hydrogels for biomedical applications: A review of the state-of-the-art. Gels **8**(7), 454 (2022)
89. Lin, X., Guan, X., Wu, Y., Zhuang, S., Wu, Y., Du, L., Tu, M.: An alginate/poly (N-isopropylacrylamide)-based composite hydrogel dressing with stepwise delivery of drug and growth factor for wound repair. Mater. Sci. Eng. C **115**, 111123 (2020)
90. Shelke, N.B., Lee, P., Anderson, M., Mistry, N., Nagarale, R.K., Ma, X.M., Kumbar, S.G.: Neural tissue engineering: nanofiber-hydrogel based composite scaffolds. Polym. Adv. Technol.. Adv. Technol. **27**(1), 42–51 (2016)
91. Sripadmanabhan Indira, S., Aravind Vaithilingam, C., Oruganti, K.S.P., Mohd, F., Rahman, S.: Nanogenerators as a sustainable power source: state of art, applications, and challenges. Nanomaterials **9**(5), 773 (2019)
92. Iqbal, M. M., Muhammad, G., Saif, T., Aslam, M. S., Raza, M. A., Hussain, M. A., & Haseeb, M. T. (2022). Nanogenerators as a Sustainable Power Source. In *Nanogenerators* (pp. 301–322). CRC Press.

93. Wang, X.: Piezoelectric nanogenerators-Harvesting ambient mechanical energy at the nanometer scale. Nano Energy **1**(1), 13–24 (2012)
94. Seol, M.L., Choi, J.M., Kim, J.Y., Ahn, J.H., Moon, D.I., Choi, Y.K.: Piezoelectric nanogenerator with a nanoforest structure. Nano Energy **2**(6), 1142–1148 (2013)
95. Deng, W., Zhou, Y., Libanori, A., Chen, G., Yang, W., Chen, J.: Piezoelectric nanogenerators for personalized healthcare. Chem. Soc. Rev. **51**(9), 3380–3435 (2022)
96. Askari, H., Khajepour, A., Khamesee, M.B., Saadatnia, Z., Wang, Z.L.: Piezoelectric and triboelectric nanogenerators: Trends and impacts. Nano Today **22**, 10–13 (2018)
97. Niu, X., Jia, W., Qian, S., Zhu, J., Zhang, J., Hou, X., Chou, X.: High-performance PZT-based stretchable piezoelectric nanogenerator. ACS Sustainable Chemistry & Engineering **7**(1), 979–985 (2018)
98. Hu, D., Yao, M., Fan, Y., Ma, C., Fan, M., Liu, M.: Strategies to achieve high performance piezoelectric nanogenerators. Nano Energy **55**, 288–304 (2019)
99. Zhang, Y., Kim, H., Wang, Q., Jo, W., Kingon, A.I., Kim, S.H., Jeong, C.K.: Progress in lead-free piezoelectric nanofiller materials and related composite nanogenerator devices. Nanoscale Advances **2**(8), 3131–3149 (2020)
100. JungáLee, E., YunáKim, T., YeonáLee, S.: High-performance piezoelectric nanogenerators based on chemically-reinforced composites. Energy Environ. Sci. **11**(6), 1425–1430 (2018)
101. Bouhamed, A., Jöhrmann, N., Naifar, S., Böhm, B., Hellwig, O., Wunderle, B., Kanoun, O.: Collaborative filler network for enhancing the performance of BaTiO3/PDMS flexible piezoelectric polymer composite nanogenerators. Sensors **22**(11), 4181 (2022)
102. Wang, Y., Yang, Y., & Wang, Z. L. (2017). Triboelectric nanogenerators as flexible power sources. *npj Flexible Electronics*, *1*(1), 10.
103. Wu, C., Wang, A.C., Ding, W., Guo, H., Wang, Z.L.: Triboelectric nanogenerator: a foundation of the energy for the new era. Adv. Energy Mater. **9**(1), 1802906 (2019)
104. Wang, Z.L.: Triboelectric nanogenerator (TENG)—sparking an energy and sensor revolution. Adv. Energy Mater. **10**(17), 2000137 (2020)
105. Sun, Z., Yang, L., Liu, S., Zhao, J., Hu, Z., Song, W.: A green triboelectric nano-generator composite of degradable cellulose, piezoelectric polymers of PVDF/PA6, and nanoparticles of BaTiO3. Sensors **20**(2), 506 (2020)
106. Biutty, M.N., Yoo, S.I.: Enhanced performance of triboelectric nanogenerator by controlled pore size in polydimethylsiloxane composites with au nanoparticles. Macromol. Res.. Res. **29**, 98–104 (2021)
107. Korkmaz, S., Kariper, İA.: Pyroelectric nanogenerators (PyNGs) in converting thermal energy into electrical energy: Fundamentals and current status. Nano Energy **84**, 105888 (2021)
108. Yang, Y., Guo, W., Pradel, K.C., Zhu, G., Zhou, Y., Zhang, Y., Wang, Z.L.: Pyroelectric nanogenerators for harvesting thermoelectric energy. Nano Lett. **12**(6), 2833–2838 (2012)
109. Xue, H., Yang, Q., Wang, D., Luo, W., Wang, W., Lin, M., Luo, Q.: A wearable pyroelectric nanogenerator and self-powered breathing sensor. Nano Energy **38**, 147–154 (2017)
110. Mistewicz, K.: Pyroelectric nanogenerator based on an SbSI–TiO2 nanocomposite. Sensors **22**(1), 69 (2021)
111. Kondawar, S. B., & Modak, P. R. (2020). Theory of EMI shielding. In *Materials for Potential EMI Shielding Applications* (pp. 9–25). Elsevier.
112. Geetha, S., Satheesh Kumar, K.K., Rao, C.R., Vijayan, M., Trivedi, D.C.: EMI shielding: Methods and materials—A review. J. Appl. Polym. Sci.Polym. Sci. **112**(4), 2073–2086 (2009)
113. Aswathi, M. K., Rane, A. V., Ajitha, A. R., Thomas, S., & Jaroszewski, M. (2018). EMI shielding fundamentals. *Advanced Materials for Electromagnetic Shielding: Fundamentals, Properties, and Applications*, 1–9.
114. Chung, D.D.L.: Materials for electromagnetic interference shielding. Mater. Chem. Phys. **255**, 123587 (2020)
115. Jiang, D., Murugadoss, V., Wang, Y., Lin, J., Ding, T., Wang, Z., Guo, Z.: Electromagnetic interference shielding polymers and nanocomposites-a review. Polym. Rev.. Rev. **59**(2), 280–337 (2019)

116. Rao, B.B., Yadav, P., Aepuru, R., Panda, H.S., Ogale, S., Kale, S.N.: Single-layer graphene-assembled 3D porous carbon composites with PVA and Fe3O4 nano-fillers: an interface-mediated superior dielectric and EMI shielding performance. Phys. Chem. Chem. Phys. **17**(28), 18353–18363 (2015)
117. Jiang, S., Qian, K., Yu, K., Zhou, H., Weng, Y., Zhang, Z.: Controllable synthesis and microwave absorption properties of Fe3O4@ f-GNPs nanocomposites. Composites Communications **20**, 100363 (2020)
118. Maiti, S., Shrivastava, N.K., Suin, S., Khatua, B.B.: Polystyrene/MWCNT/graphite nanoplate nanocomposites: efficient electromagnetic interference shielding material through graphite nanoplate–MWCNT–graphite nanoplate networking. ACS Appl. Mater. Interfaces **5**(11), 4712–4724 (2013)
119. Lyu, L., Liu, J., Liu, H., Liu, C., Lu, Y., Sun, K., Wujcik, E.K.: An overview of electrically conductive polymer nanocomposites toward electromagnetic interference shielding. Engineered Science **2**(59), 26–42 (2018)
120. Dravid, S.V., Bhosale, S.D., Datar, S., Goyal, R.K.: Nickel nanoparticle-filled high-performance polymeric nanocomposites for EMI shielding applications. J. Electron. Mater. **49**, 1630–1637 (2020)
121. Muzaffar, A., Ahamed, M.B., Deshmukh, K., Faisal, M.: Electromagnetic interference shielding properties of polyvinylchloride (PVC), barium titanate (BaTiO3) and nickel oxide (NiO) based nanocomposites. Polym. Testing. Testing **77**, 105925–105936 (2019)
122. Tan, D.Q.: The search for enhanced dielectric strength of polymer-based dielectrics: a focused review on polymer nanocomposites. J. Appl. Polym. Sci.Polym. Sci. **137**(33), 49379 (2020)
123. Hu, H., Zhang, F., Luo, S., Chang, W., Yue, J., Wang, C.H.: Recent advances in rational design of polymer nanocomposite dielectrics for energy storage. Nano Energy **74**, 104844 (2020)
124. Li, R., Yang, X., Li, J., Shen, Y., Zhang, L., Lu, R., Zhang, T.: Review on polymer composites with high thermal conductivity and low dielectric properties for electronic packaging. Materials Today Physics **22**, 100594 (2022)

Chapter 15
Polymer Composites for Electromagnetic Interference (EMI) Shielding Applications

Ankita Meher, Bibhuti B. Sahu, Ram Naresh Mahaling, Suresh Sagadevan, and Srikanta Moharana

Abstract Polymer composites have recently increased substantial attention due to their exceptional characteristics and a broad range of practical applicability. Especially, they have shown great potential in the domain of electromagnetic interference (EMI) shielding. The term "EMI" stands for electromagnetic interference, which may have a negative influence on the functionality of electronic systems and devices. This chapter seeks to present a comprehensive summary of the most advanced and up-to-date polymer composites used for electromagnetic interference (EMI) shielding applications. Since wireless communication is the norm in modern times, interference and disarray between different electromagnetic waves are the main problems. To reduce cross-talk between electronic equipment, effective shielding against electromagnetic interference (EMI) waves using complex materials becomes an important research topic. The research community places significant importance on systems that provide good shielding against electromagnetic interference (EMI) due to the rapid progress in electronics and telecommunications in contemporary times. To accomplish electromagnetic interference (EMI) shielding effectiveness (SE), one can

A. Meher
Department of Chemistry, School of Applied Sciences, Centurion University of Technology and Management, Bhubaneswar, Odisha, India
e-mail: srikantanit@gmail.com

B. B. Sahu
Department of Basic Science and Humanities, Nalanda Institute of Technology, Bhubaneswar, Odisha, India

R. N. Mahaling
Laboratory of Polymeric and Materials Chemistry, School of Chemistry, Sambalpur University, Jyoti Vihar, Burla, Odisha 768019, India

S. Sagadevan
Nanotechnology & Catalysis Research Centre, University of Malaya, 50603 Kuala Lumpur, Malaysia

S. Moharana
Department of Chemistry, School of Applied Sciences, Centurion University of Technology and Management, Jatni, Bhubaneswar, Odisha, India

S. Moharana et al. (eds.), *Polymer Composites*, Engineering Materials,
https://doi.org/10.1007/978-981-97-2075-0_15

utilize polymer composites that seamlessly integrate metals, ceramics, and carbons. This chapter explores the mechanism of EMI shielding and provides an overview of the current research on polymer-based composite materials specifically designed for EMI shielding. The main goal of this chapter is to present an in-depth review of the potential and developments in polymer-based EMI shielding composites, with a precise emphasis on assessing the materials' shielding efficiency (SE).

Keywords Polymer · Composites · Electromagnetic Interference · Efficiency

15.1 Introduction

A process known as electromagnetic shielding (EMI shielding) occurs when a substance reflects or absorbs electromagnetic waves, acting as a barrier to prevent perilous electromagnetic radiation from penetrating electronic equipment. In other terms, EMI shielding is the process of suppressing EM radiation using a barrier made of either conductive or magnetic materials [1]. In the past, metallic and magnetic constituents were relied upon for their high shielding efficiency (SE) in EMI shielding applications. Materials such as copper, aluminum, steel, iron, etc. have been used in the past as efficient shielding materials due to their superior electrical conductivity and superior permeability [2]. The efficiency of electromagnetic interference shielding (EMI SE) depends on several aspects. Even though metals possess a high EMI SE, their substantial weight and limited flexibility make them unsuitable for use in modern electronic devices. Research focuses on improving polymeric materials for EMI shielding due to their flexibility, ease of processing, scaleability, chemical resistance, and lightweight nature [3]. Moreover, polymer-based materials offer superior protection against electromagnetic waves compared to metal-based materials because of their dominant absorption shielding capabilities. This makes them particularly well-suited for military applications like stealth technology and camouflage. Polymeric materials with inherent electrical conductivity, such as polyaniline and polypyrrole, are well-suited for this persistence because of their elevated electrical conductivity, which may be further augmented via chemical doping [4, 5]. The application of polymer-based electromagnetic interference shielding has expanded over time. To function as an efficient reflection/absorption shield, polymer mixes need to have high electrical conductivity/permeability, which can be achieved by appropriately incorporating magnetic or electrically conducting fillers into the polymer matrix. In the realm of conducting filler materials, carbonaceous elements like carbon nanotubes (CNTs), graphene, carbon black, and graphite fibers have dominated the landscape alongside metallic counterparts like metal nanowires and nanoparticles. These nano or microstructures serve as pivotal components in various applications. Similar to graphene, conducting transition metal 2D carbides (MXenes) are experiencing a growing popularity and are potentially highly efficient fillers for EMI shielding polymer hybrids. These electrically conducting inorganic nanoparticles and their composites demonstrated strong EMI shielding capability as well as excellent efficiency in solar cell electrochemistry, rechargeable lithium-ion batteries, and sensing devices [5–10]. Polymeric nanocomposite study results led to extremely efficient electromagnetic interference (EMI) shielding materials. EMI shields, for example, have been developed using nitrile butadiene rubber in combination with silver flakes [2, 11]. The measured EMI-SE at 1 GHz was ~75 dB. Indeed, the area of electromagnetic interference (EMI) shielding nanocomposites involves the incorporation of inorganic nanoparticles and hybrid nanofillers. It is critical for finding innovative production and utilization procedures for tackling inorganic magnetic nanoparticle accumulation and implementation. The implementation of nanoclay nanofillers importing into polyaniline demonstrated remarkable efficacy in electromagnetic interference (EMI) shielding applications. This improvement resulted in a significant signal-to-noise ratio (SE) range of 40 to 80 dB, notably at 8 GHz. Nanoparticle properties such as size, shape, aspect ratio, and surface area determine the final morphology of nanocomposites. In particular, high aspect ratio nanofillers can form complex interconnected network architectures in polymeric matrices. As these nanofillers exceed a crucial percolation threshold, they form an unified network inside the polymer matrices, which gives rise to increased electrical conductivity. This percolated network shape allows for extraordinary electrical conductivity values, which frequently surpass >104 S/m. Because of this, these nanocomposites have strong EMI shielding properties, usually with a range of 20–50 dB [13–15].

15.2 Mechanism of Electromagnetic Shielding (EMI Shielding)

The shielding mechanisms inside PPR/MWCNT nanocomposites are schematically represented in Fig. 15.1. When it comes to pore-filled substances and three-dimensional composite structures, nonetheless, numerous reflections play a far significant impact. Conversely, 90 ~ samples show flawless morphology, which allows for greater interaction between EM waves and MWCNT. Owing to numerous internal reflections, it shows a larger absorption loss after absorption [16]. The shielding processes that take place within these materials are diagrammatically represented in Fig. 15.2 [17]. Due to certain features including broad phase interfaces, significant unoccupied regions, particularly large areas of surface, and ubiquitous voids, electromagnetic radiation is scattered, reflected, and absorbed in this region. The efficacy of electromagnetic radiation (EMI) shielding is predominantly influenced by three primary mechanisms: absorption (A), reflection (R), and multiple reflections (M) [16–19].

The major approach used in electromagnetic interference (EMI) shielding is predominantly reflection. This fundamental concept relies on the material's capability to reflect electromagnetic (EM) waves from its surface. To effectively reflect electromagnetic interference (EMI), a material has to have available electric charges,

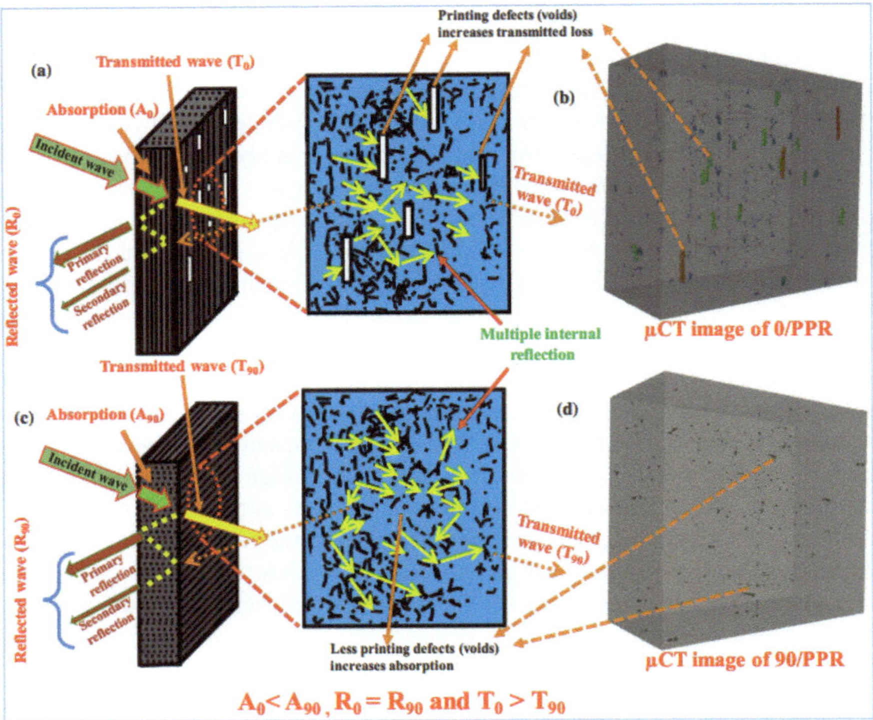

Fig. 15.1 EMI shielding mechanism in PPR/MWCNT nanocomposites, printed in **a**, **b** printed in 0° direction, **c**, **d** 90° directions [16]. Open access under a CC BY 4.0 license

such as electrons or holes that might intermingle with the electromagnetic field contained in the radiation. Though superior conductivity is not necessary (a volume resistivity of around 1 cm is often satisfactory), the shielding material must possess conductivity. Attaining the percolation threshold or establishing connections between conductive fillers inside dielectric matrices is not necessary for effective reflection. However, enhancing these characteristics significantly enhances the effectiveness of shielding. Metals that are abundant in free electrons that may conduct electric current, such as copper, nickel, aluminum, silver, and gold, typically exhibit outstanding efficiency in shielding electromagnetic interference (EMI) by reflection. Composites with conductive fillers like carbon fibers or metal powders, along with materials treated with conductive coatings and layers, provide impressive shielding performance by reflection. In general, the reflectivity improves as the shield's conductivity grows beyond 103 Hz, but it decreases when the shield's permeability and the frequency of electromagnetic radiation increase [18–21].

The secondary process involved in EMI shielding is the absorption of electromagnetic interference (EMI), which is greatly influenced by the width of the shielding substantial. [7, 14]. The presence of electric and magnetic dipoles within the shield

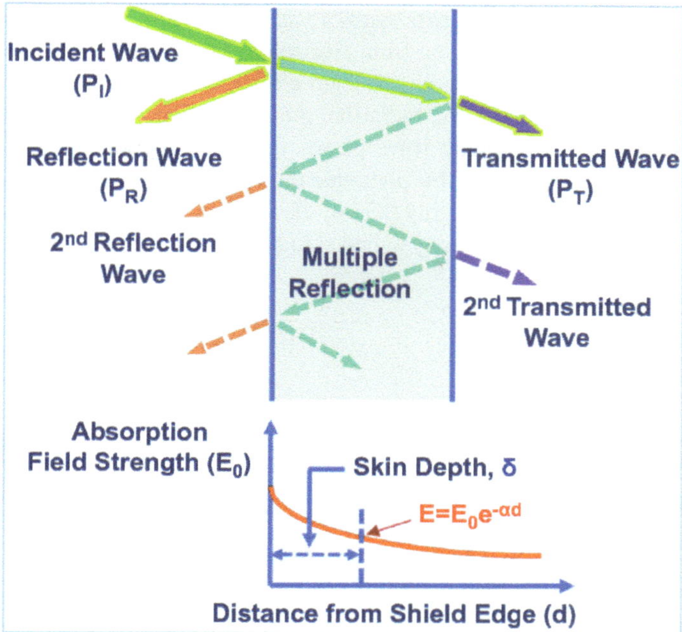

Fig. 15.2 Diagram illustrating the EMI shielding mechanisms [17]. Open access under a CC BY 4.0 license

enhances the shielding efficacy using absorption when interacting with electromagnetic (EM) radiation. Materials with a high dielectric constant (permittivity), like $BaTiO_3$, Fe_2O_3, and ZrO_2, are the foremost bases of electric dipoles that are essential to this process. Conversely, highly permeable materials that include ferrites, metal (an iron and nickel alloy), Fe_3O_4, as well as super permalloy are reported as great absorbers because of their magnetic dipole properties [15–18]. Research findings show that absorption effectiveness increases with rising electromagnetic radiation frequency, increased width, and increased permeability of the shielding material. With the absence of magnetic nature, EMI shielding depends only on the dielectric properties; when dielectric properties are absent, magnetic features are required to provide efficient shielding against electromagnetic interference [22–28].

The third and final mechanism of electromagnetic interference (EMI) shielding is manifold reflection. This denotes the phenomenon of electromagnetic radiation being refracted from several exteriors, phase interfaces, and inhomogeneities inside the shield. Multiple reflection occurs in conjunction with reflection and absorption. Exceptional multiple reflection qualities are often defined by either a substantial specific surface area or a significant phase interface (observed in materials with fillers possessing high specific surface areas). These properties allow for excellent attenuation of the amplitude of incoming radiation via various internal reflections inside the shielding material. Materials that have a higher surface area or bigger interface for reflection are better able to absorb and deflect incoming radiation, which helps to provide effective shielding from electromagnetic interference [29, 30].

Especially the close surface areas beneath electric conductors are susceptible to higher-frequency electromagnetic radiation, this phenomenon is called skin effect. Since incoming electromagnetic wave passes through the conductor, its intensity quickly decreases as a result of this phenomenon. The deepness where electric field declines to 1/e (where 1/e is roughly 0.37 of the original incidence magnitude and e is Euler's number) is known simply called skin depth (δ).

$$\delta = \frac{1}{\sqrt{\pi f \sigma \mu}} \tag{15.1}$$

where, f—frequency of electromagnetic radiation.

σ—shield electric conductivity (Ω^{-1} m^{-1}),

μ is shield magnetic permeability i.e. $\mu = \mu_o \mu_r$,

where $\mu_{r=}$ shield relative magnetic permeability, and μ_o – permeability of air or free space ($\mu_o = 4\ \pi\ 10^{-7}$ H m^{-1}).

The skin depth is influenced by electromagnetic pulse frequency, magnetic permeability, and shield conductivity, with increased factors leading to a decrease in skin depth. This phenomenon emphasizes the importance of composites containing conductive fillers that have tiny sizes as they are more effective at shielding than fillers that have greater unit sizes. Particularly, research results indicate that the unit dimensions of the filler is smaller than or equivalent to the skin depth for best shielding purposes. Fillers that have unit sizes of 1 μm or smaller are therefore frequently recommended. The capacity of a material to attenuate incident electromagnetic radiation (EM radiation) is referred to as its shielding efficiency (SE) when discussing electromagnetic interference (EMI) shielding [31]. The mathematical definition of the ratio between incoming radiation power and transmitted power is often represented in decibels (dB) as.

$$\delta = \frac{1}{\sqrt{\pi f \sigma \mu}} \tag{15.2}$$

The magnitude of the incident and transmitted electromagnetic waves, represented by PI (EI or HI) and PT (ET or HT), is measured in terms of power intensity [32]. Usually, electromagnetic spectrum interference (EMI) takes place in the frequency extent of 10^4–10^{12} Hz. As the degree of EMI shielding efficiency (SE) rises the energy transferred across the shield falls. The minimal threshold for successful shielding in a variety of applications is generally accepted to be between 10 and 30 dB. It is interesting that research has indicated that 99.9% of incoming radiation may be absorbed by a protecting material with 30 dB of EMI SE, which is enough for the widely held of commercial and applications industries. Additionally, shields with a SE of 20 dB may cut 99% of the electromagnetic radiation that enters, proving to be incredibly efficient in mitigating electromagnetic interference in a range of scenarios [33, 34].

The total shielding efficiency (SE) in dB is the sum of the reflection (SE_R), absorption (SE_A), and manifold reflections (SE_{MR}) of EMI:

$$\text{EMI SE} = SE_R + SE_A + SE_{MR} \tag{15.3}$$

The determination of the shielding efficacy of materials may be achieved by many theories, including the plane-wave theory, metal foil, near-field shielding, low-frequency magnetic field source, and scattering characteristics. Among these options, the plane-wave theory is usually used to calculate EMI SE based on the thickness and properties of the composite. The efficacy of reflection, according to the plane-wave theory, is contingent upon the disparity in impedances amongst the incoming electromagnetic wave and the shielding material [35]. This may be determined by:

$$SE_R = 20\log\left|\frac{Z_o}{4Z_s}\right| \tag{15.4}$$

where Z_0 is the impedance of the incident wave and Z_S is the impedance of the shield.

From Eq. (15.4), it may be obtained that when the electromagnetic wave's impedance increases and the shield impedance drops, the efficacy of reflection shielding increases. The impedance of the electromagnetic wave may be found using the radiation frequency, the source's distance from the source, and the impedance source. In general, a wave's impedance is represented as high or low. Because they have moderate shield impedance values, even medium-conductive materials can successfully shield electromagnetic interference (EMI) by reflection. The following equations were developed for absorption and multiple reflection:

$$SE_A = 20\log^{t/\varsigma} \tag{15.5}$$

$$SE_{MR} = 20\log\left(1 - e^{-2t/\varsigma}\right) \tag{15.6}$$

where t is the width of the shielding material and d is the skin depth.

It is evident that when the thickness of the shielding material grows, it also increases the effectiveness of the absorption shielding. Conversely, recurrent reflection has a negative effect and reduces the overall shielding efficacy when $t < d$. Multiple reflections become insignificant under these circumstances and are often disregarded, particularly if the absorption loss (SEA) exceeds 10 dB. Typically, conditions of this nature occur at extremely high frequencies, typically gigahertz (GHz) or greater [36]. The electromagnetic impedance wave (Z) is essentially the ratio of the amplitudes of the electric and magnetic fields:

$$(Z) = \left|\frac{E}{H}\right| = \sqrt{\frac{j\omega\mu}{\sigma + j\omega\mu}} \tag{15.7}$$

where, ω = angular frequency of the radiation,

μ = magnetic permeability,
σ = electric conductivity,
ε = permittivity of the medium.

Considering that the electromagnetic plane wave in the vacuum or blank space (i.e. $\sigma = 0$), Eq. (15.7) may be expressed differently here:

$$Z_o = \sqrt{\frac{j\omega\mu}{\sigma + 0}} = \sqrt{\frac{\mu}{\varepsilon}} \tag{15.8}$$

For the conductive materials $\sigma >> \omega\varepsilon$, therefore, the impedance of the shield Z_S can be calculated from Eq. (15.7), in which f represents the frequency of the incident electromagnetic wave:

As a result, since the conductive materials $\sigma >> \omega\varepsilon$, the impedance of the shield Z_S may be determined by using Eq. (15.7). In the resulting equation, the frequency of an incident electromagnetic wave is represented by f:

$$Z_S = \sqrt{\frac{j\omega\mu}{\sigma + 0}} = \sqrt{\frac{j\omega\mu}{\sigma}} = \sqrt{\frac{2\pi f\mu}{\sigma}} \tag{15.9}$$

Therefore, the effectiveness of shielding by absorption and reflection may be summed up as follows:

$$SER = 39.5 + 10\log\frac{\sigma}{2\pi f\mu} \tag{15.10}$$

$$SE_A = 8.7\frac{t}{1}\sqrt{\pi f\mu\sigma} \tag{15.11}$$

Based on the information provided, the efficacy of shielding by absorption is found by the value of $\sigma\mu$, whereas shielding through reflection depends on the ratio of σ to μ. The aforementioned formulae lead to the inference that the effectiveness of shielding by reflection is contingent upon the magnetic permeability, electrical conductivity, and their ratio, whereas shielding through absorption depends on $\sigma\mu$ [29, 33, 37].

T represents the measurement of thickness. The symbol μ represents the permeability of the shield material, whereas σ represents its electrical conductivity. The variable "f" denotes the frequency at which shielding occurs. Therefore, strong electrical conductivity is required allowing the material that shields to reflect electromagnetic waves, and a sufficient thickness of material having excellent electrical conductivity as well as high permeability may capture more electromagnetic radiations.

The depth to which electromagnetic waves penetrate a conductor diminishes with increasing depth within the material. This phenomenon is quantified by the skin depth, denoting the extent at which the intensity of the electromagnetic wave decreases to 1/e or 37% of its initial value. Mathematically expressed as, where f represents the frequency in Hertz, μ denotes the magnetic permeability ($\mu = \mu_0\,\mu_r$, where μ_0 is the absolute permeability of free space, equating to 4π), and σ stands for the electrical conductivity of the material in Siemens per meter (S/m).

Research has shown that the conductivity mismatch between the conducting filling agent and insulating matrix can lead to enhanced dielectric permittivity in composite materials, which subsequently in turn can magnify dielectric loss. This difference in electrical conductivity frequently causes polarization and charge buildup at composite material surfaces. Polymer materials with elevated dielectric values have potential as effective absorbers of electromagnetic (EM) waves. Furthermore, studies have shown that the polarization capacity of these composites may be increased by carefully selecting fillers to use inside the matrix and by orienting them strategically. Furthermore, permeability of material remains a crucial part of efficient shielding. High permeability or magnetic loss materials show promise as good options for absorbing electromagnetic waves; permeability frequently varies in response to temperature, magnetic field strength, and others. The efficiency of electromagnetic interference (EMI) shielding has an unbreakable connection with the frequency of radiation [38, 39]. Research gives the theoretical investigations that show absorption shielding capacities to rise together with an increase in radiation frequency. Furthermore, crucial criteria impacting shielding performance include shielding material thickness. Greater EMI shielding capacity is often correlated with thicker material. Furthermore, the size of fillers and aspect ratio have a big influence on effective shielding. These characteristics are most prominently displayed by nanomaterials, which are known for having exceptional electrical, magnetic, and mechanical capabilities. Considerably more electrical conductivity was shown by larger area graphene sheets with a significant aspect ratio than by smaller graphene sheets due to the lower inter-sheet contact resistance. Furthermore, materials with larger aspect ratios typically develop more precise and tightly packed films. More electrical conductivity in graphene thin films with a larger aspect ratio has therefore been associated with better EMI shielding effectiveness when compared to traditional small-size graphene sheets. Numerous factors, particularly electrical conductivity, permittivity, permeability, filler size, and shielding material thickness, all impact EMI shielding performance, especially electrical conductivity emerging as a significant criterion for successful shielding materials [40–43].

15.3 Polymer Composite for EMI Shielding

Polymer composites are becoming an attention of growing research interest due to their excellent characteristics such as amazing stability, lightweight composition, vast absorption capabilities, flexibility, and simplicity of processing. The essential attributes necessary for an EMI shielding material to function well are its lightweight composition and pliability, both of which are crucial technological variables that ensure successful and advantageous applications. This section explores the gradual growth of polymer composites specifically developed for electromagnetic interference (EMI) shielding [44]. A wide range of conducting polymers, such as polyaniline (PANI), polypyrrole (PPy), polystyrene sulfonate (PSS), and polythiophene (PTP), together with associated polymer composites, have remarkable electrical conductivity. Significant research efforts have been spent on investigating EMI shielding techniques based on these materials. When compared to their traditional metallic equivalents, these materials provide several advantages, such as reduced density and improved corrosion resistance. In addition, insulating polymers such as epoxy, polystyrene (PS), poly(vinylidene fluoride) (PVDF), polypropylene (PP), poly(methyl methacrylate) (PMMA), poly(vinyl alcohol) (PVA), polyethylene (PE), and polystyrene (PS) have garnered significant interest in EMI shielding applications due to their notable characteristics, including easy processing, strong mechanical properties, and resistance to chemical degradation [45, 46].

The investigation demonstrates a significant impact on the electromagnetic interference (EMI) shielding of flexible electronic sheets made from polymer PVA thin films with varying filler content percentages of rGO (reduced graphene oxide) and MXene. The study reveals that both the increase and decrease matrix in a polymer composite have substantial effects on the EMI shielding properties of the resulting materials. This discovery highlights the crucial significance of filler content in influencing the electromagnetic shielding effectiveness of the created flexible electronic sheets, underlining the importance of careful compositional balance to obtain optimal EMI shielding performance [47].

15.4 Polyaniline-Based Polymer Composite

Numerous research in EMI shielding have thoroughly investigated conducting polymers due to their advantageous properties, particularly their lightweight nature and excellent corrosion resistance. Polyaniline (PANI) stands out as a highly sought-after material in this category due to its outstanding stability and capacity to modify its electrical and optical characteristics. The electromagnetic interference (EMI) shielding efficiency (SE) of polyaniline (PANI) typically ranges from 16 to 50 dB across a frequency range of 0.1–1 GHz. The EMI SE values range from 31.9 to 81 dB at higher frequencies (1–18 GHz) [33, 48, 49].

Taj et al. [50] synthesized a PANI-molybdate nanocomposite by employing two different solvents, specifically dodecyl benzene sulphonic acid (DBSA) and camphor sulphonic acid (CSA), for assessment purposes. The materials obtained exhibited a notable enhancement in morphological characteristics, as the naturally conducting three-dimensional interconnected networks of PANI acted as an efficient matrix. This matrix enclosed stable MoO_3 particles, causing a dramatic change in shape and allowing them to operate as effective attenuation for wave points. Their comparison research revealed interesting findings: samples treated with CSA had an outstanding shielding response of −14 dB, while those prepared with DBSA had a shielding response of −12 dB. The enhanced shielding effectiveness seen in these composites can be attributable to the significant interfacial area and porous network facilitated by the presence of MoO_3 inside the PANI/DBSA/CSA molecular chains, as depicted in

Figs. 15.3 and 15.4. The conflict between magnetic and electric dipoles inside these composites appears to have led to the reduction of electromagnetic wave intensity. In summary, the overall shielding performance of these composites in the X-band, mostly by absorption, demonstrates their cost-effective potential for PANI/DBSA/MoO_3 (30 wt%) and PANI/CSA/MoO_3 (40 wt%) composites. This makes them well-suited for applications involving electromagnetic interference shielding [50].

Kumar et al. conducted a study on the electromagnetic interference (EMI) shielding properties of in-situ polymerized ferrite-polyaniline nano-composites applied to textiles. A similar improvement in shielding effectiveness (SE) with increased thickness was seen across all samples. In particular, for the $Ni0._7Zn0.3Fe_2O_4$-PANI sample, SE values increased dramatically from 40 dB at 1 mm thickness to a staggering 60 dB at 3 mm thickness as shown in (Fig. 15.5a). Two different kinds of materials army cloth 1 and white cotton cloth 2, were carefully coated with the synthesized nanocomposites (as shown in Fig. 15.5b). The

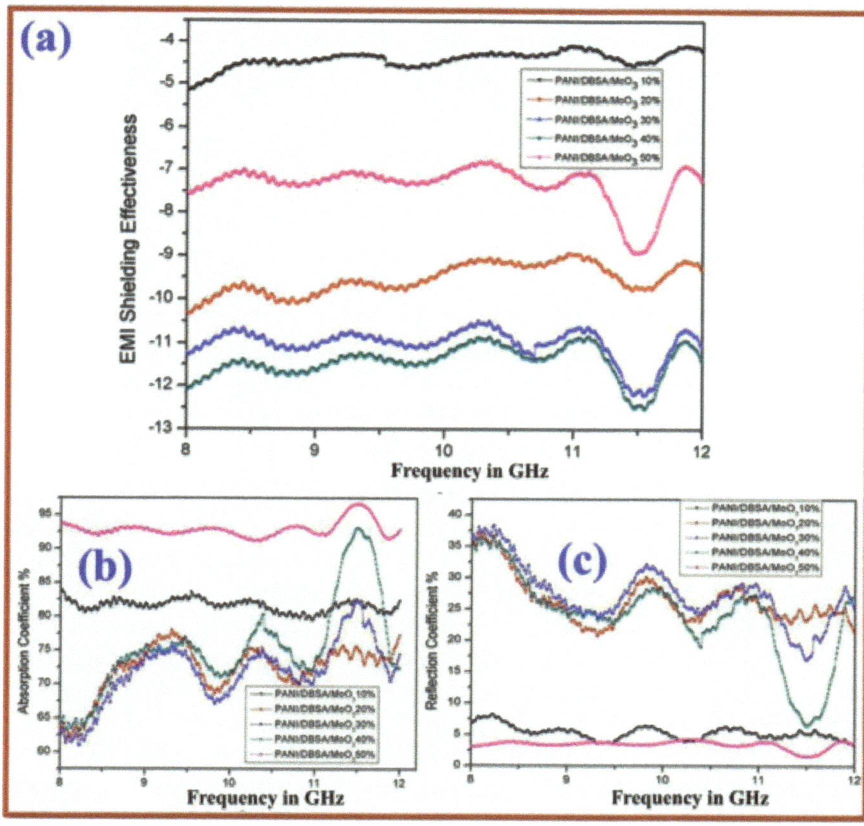

Fig. 15.3 Effectiveness of EMI shielding in PANI-DBSA composite—with varied wt% of MoO_3 **a** The overall shielding efficacy, **b** Absorption coefficient percentage, and **c** Reflection coefficient (%) with different frequency [50]. Open access under a CC BY 4.0 license

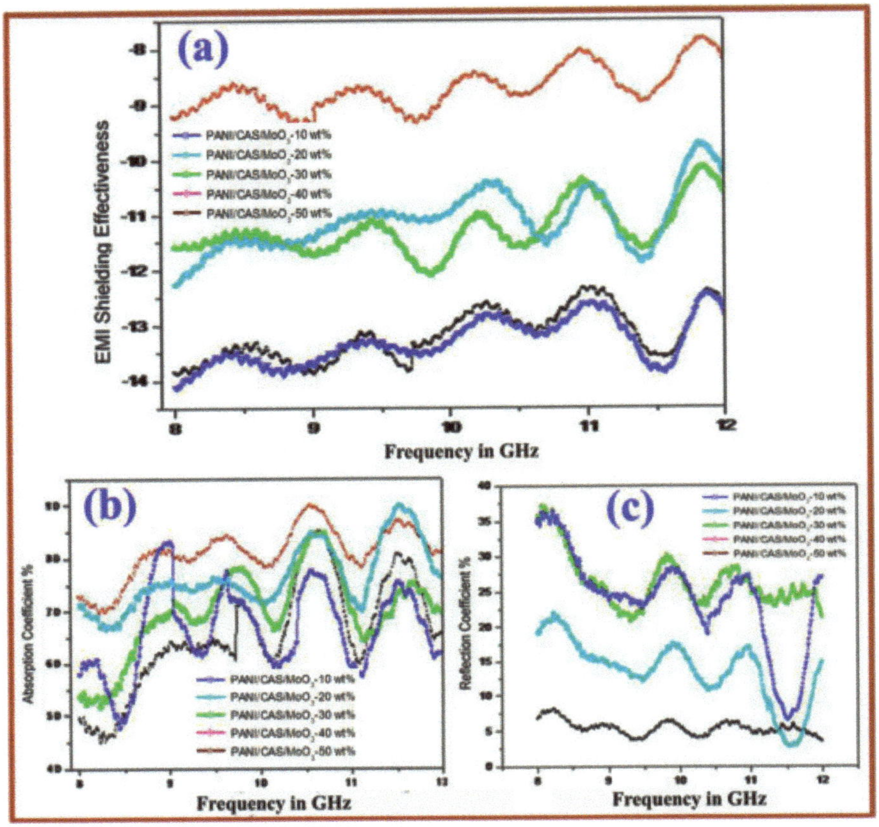

Fig. 15.4 EMI shielding performance of PANI-CSA composite at varying MoO_3 weight percentages **a** Total shielding efficacy **b** Absorption coefficient as a percentage. **c** Reflection coefficient (%) alterations with respect to frequency [50]. Open access under a CC BY 4.0 license

frequency-dependent EMI shielding tests, as shown in Fig. 15.5c, were carried out at room temperature in the X-band frequency range. Notably, the findings revealed that the textiles coated with $Ni_{0.7}Zn_{0.3}Fe_2O_4$-PANI composite material had convincing SE values. Army fabric (cloth 1) had the greatest SE of 48 dB, whereas white cotton cloth (cloth 2) had a significant SE of 45 dB. This reflected the SE patterns identified in the composite materials, which corresponded to their magnetic properties. These findings highlight the potential usefulness of Ni0.7Zn0.3Fe_2O_4-PANI composites as effective EMI shielding materials when applied to a variety of fabric substrates, opening up new possibilities for customized electromagnetic protection [50]

Akram et al. developed a textile made of cotton coated with silver electroless plated PANI/AGO, which effectively shields against electromagnetic interference (EMI) to protect both human life and electrical equipment from harmful electromagnetic radiation. Figure 15.6a illustrates the EMI shielding mechanism of the redesigned fabric, showcasing three distinct mechanisms: reflection, absorption, and multiple

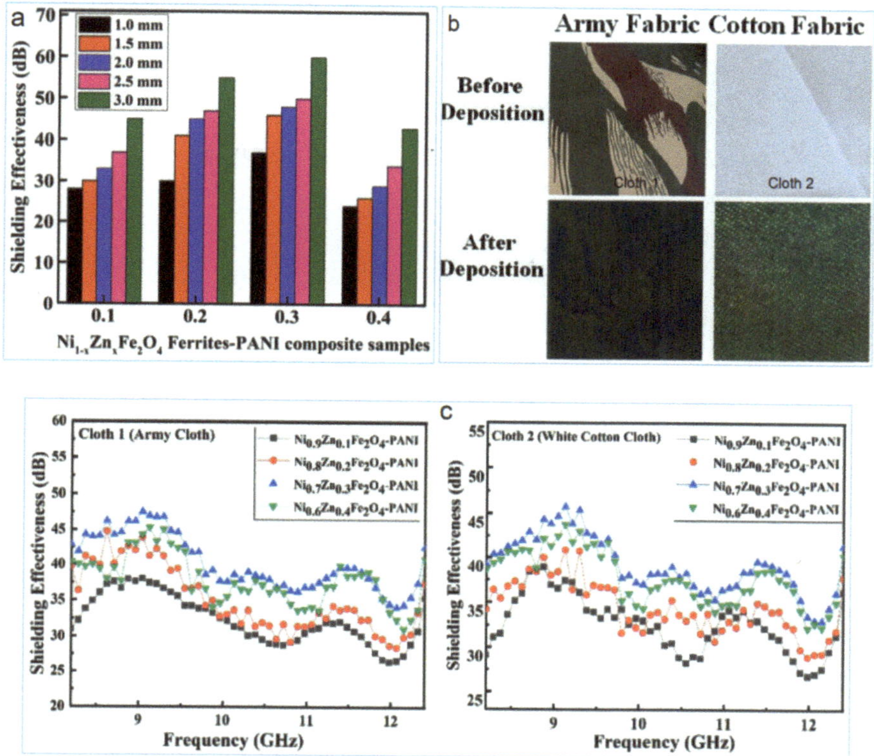

Fig. 15.5 **a** Differences in the SE value across various thicknesses of the composite material, **b** Fabrics before and after deposition of Ferrite-PANI composite. **c** The efficacy of electromagnetic shielding for fabrics coated with nanocomposites with varying frequency ranges [50]. Open access under a CC BY 4.0 license

internal reflections when exposed to incident waves. The silver on the top layer of the cloth is responsible for the reflection mechanism's supremacy. PANI and carbon-based materials can absorb radiation because of the conducting mechanism through conjugation and delocalized pi-pi electrons. Every single sample exhibits a progressive decline in the EMI shielding effectiveness ratio (SE_R) with increasing frequency, although the EMI shielding effectiveness attenuation (SE_A) improves at lower frequencies. Sample A-6 showed the greatest SE_A at around 83 dB and a SE_R of about 10 dB, highlighting the importance of absorption loss over reflection in its shielding behavior. Furthermore, the comparison of total shielding effectiveness (SE_T), SE_A, and SE_R in Fig. 15.6b confirms the high SE_A values across all samples. As a result, these modified materials perform well in EMI shielding applications, indicating potential in minimizing electromagnetic radiation pollution [51, 52].

Zhang et al. created a composite material consisting of Ti3C2Tx/c-PANI for electromagnetic interference (EMI) shielding [52, 53]. The initial stage of his study was co-doping hydrochloric acid and dodecyl benzene sulfonic acid with modified

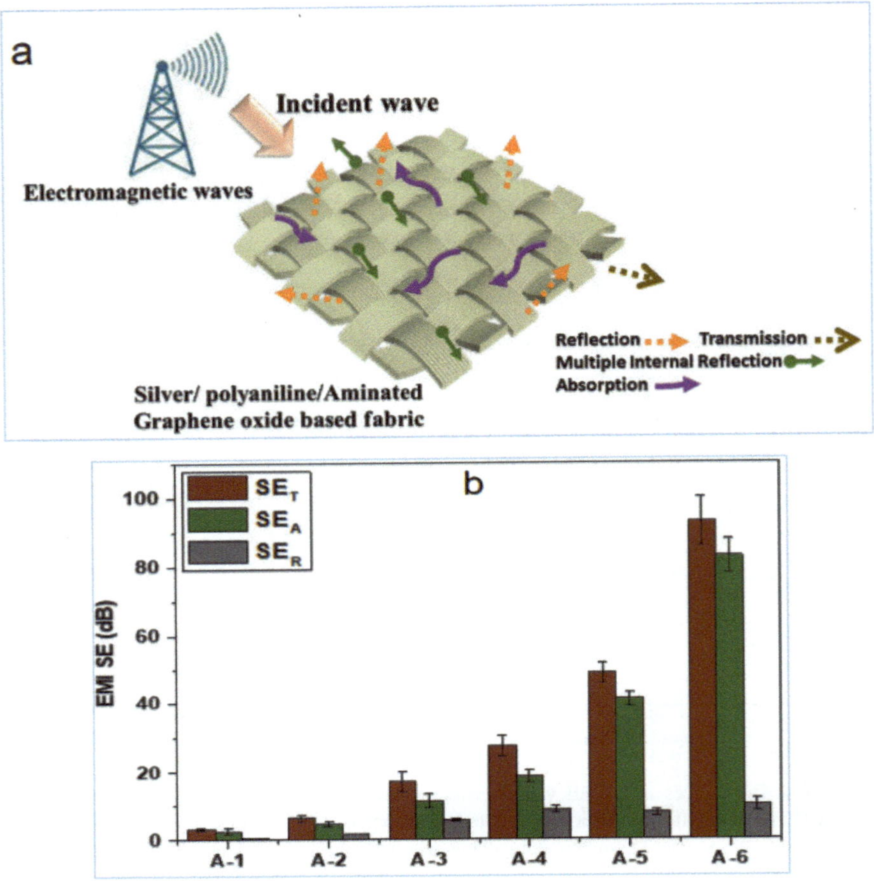

Fig. 15.6 Investigation of the electromagnetic interference (EMI) shielding mechanism of the modified fabric samples SET, SEA, and SER [52]. Open access under a CC BY 4.0 license

polyaniline (c-PANI). They then used ionic intercalation and ultrasonic methods to generate fewer layers of MXene. The last stage was to create MXene/c-PANI EMI shielding films using a vacuum filtering technique, as shown in Fig. 15.7a. The pure c-PANI film had a certain amount of flexibility and was dark green in color (Fig. 15.7b). But when Ti_3C_2Tx was added, the Ti_3C_2Tx/c-PANI EMI shielding composite film retained its high flexibility and took on a black-purple color (Fig. 15.7c). SEM images (Fig. 15.7d, e)) showed that the c-PANI in the pure film was tightly packed together [53]. These films demonstrated remarkable electrical conductivity (σ) of 24.4 S cm^{-1} and remarkable electromagnetic interference (EMI) shielding efficiency (SE) of 36 dB at a thickness of 0.040 mm. The results showed significant improvements, with conductivity being 81-fold higher and EMI shielding effectiveness being 2.3 times greater compared to pure c-PANI sheets. The use of vacuum filtering offers advantages in terms of controlling film thickness and ensuring efficient utilization of raw materials. However, limitations were observed due to the dimensions of the equipment, which restricted the size of the film. Furthermore, the process of vacuum filtration for producing films from two-dimensional sheet materials was labor-intensive [53–55].

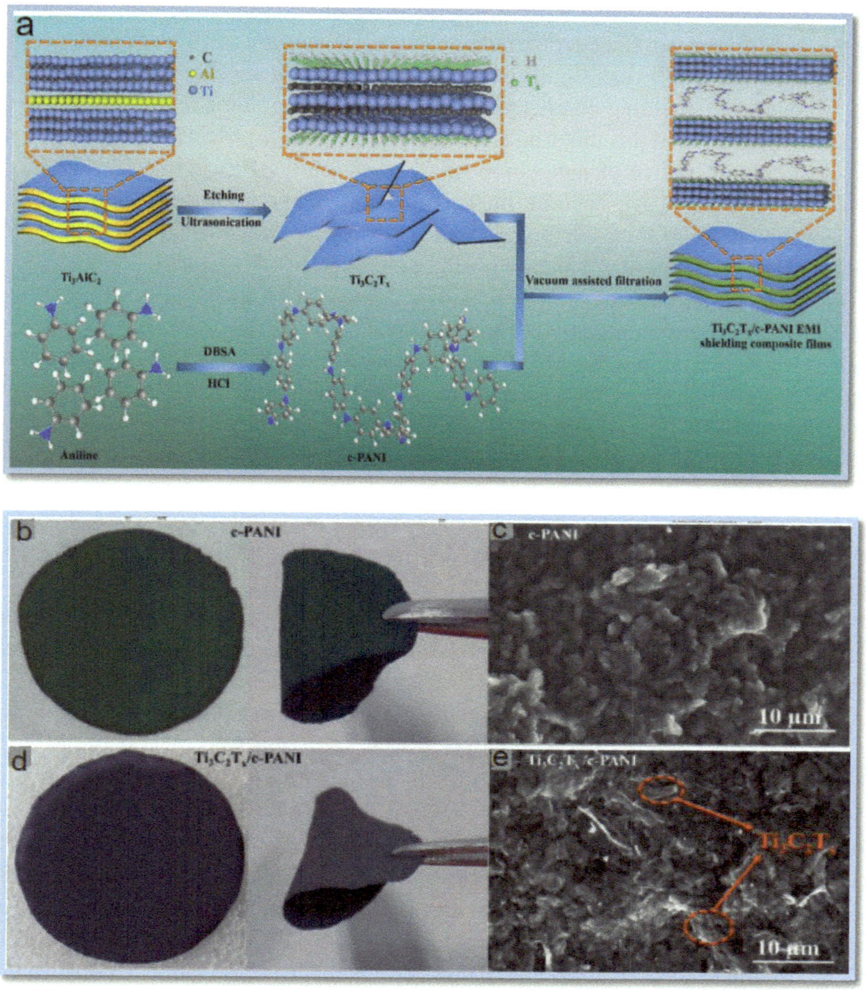

Fig. 15.7 **a** Diagrammatically representation method of preparation of Ti_3C_2Tx/c-PANI EMI shielding composite **b** Picture of pure c-PANI composite, **c** SEM image of pure c-PANI composite, **d** Picture of Ti_3C_2Tx/c-PANI composite and **e** SEM image of Ti_3C_2Tx/c-PANI composite [53]. Adapted from Ref.54 with permission from [Elsevier], Copyright [2019]

15.5 Polyvinyl Alcohol-Based Polymer Composite

PVA polymer is unique among other polymers because of its exceptional and adaptable qualities, which make it very desirable for a wide range of applications [56]. Its simplicity of formulation, environmental stability, cost-effectiveness, and capacity to reversibly alter mechanical, thermal, and electromagnetic shielding qualities make it particularly appealing. Recent research indicates that the addition of various nanoparticles to the PVA matrix can significantly alter the electrical, mechanical, thermal, and electromagnetic shielding characteristics of the resulting nanocomposites. Furthermore, the inclusion of this material enhances the mechanical, electrical, thermal, and physical characteristics of nanocomposites, particularly at their interface regions. In addition, CQDs have become important components among zero-dimensional materials because of their intriguing physical characteristics, which result from the quantum confinement effect and edge effect phenomena [47, 57–60].

Lai et al. created flexible composites by coating carbon with a layer of polyvinyl alcohol/reduced graphene oxide. These composites were designed for electromagnetic shielding, as seen in Fig. 15.8a. The presence of ultrathin individual sheets or a few-layer morphology of decorated reduced graphene oxide (RGO) is demonstrated by the obvious AC pores on the surface of RGO@AC, as shown in Fig. 15.8b. When incorporated into PVA polymer, these complexes form flexible films that exhibit exceptional flexibility and torsional capability, as seen in Fig. 15.8a. However, the PVA/RGO@AC composites are designed to create a large interface area between the PVA domains and conductive RGO@AC, which helps allow the many reflections, scattering, and absorption of incoming microwaves within the composites. Wave attenuation is further facilitated by the pore-rich nature of AC. Thus, the incident electromagnetic microwaves have a tough time escaping before they are absorbed and converted to heat. This total mechanism is diagrammatically represented in Fig. 15.8 c. Figure 15.8d demonstrates that when the sample thickness increases from 0.8 mm to 1.5 mm and 2.0 mm, the average electromagnetic interference shielding effectiveness (EMI SE) improves from 14.0 dB to 20.3 dB and 25.6 dB, respectively. This indicates that only 0.93% and 0.28% of the transmission passes through the shielding material. Therefore, it is suggested that the PVA/RGO@AC composites start to demonstrate shielding capabilities (EMI SE more than 20 dB) when the thickness reaches a crucial range of 1.5–2.0 mm [62]. Shamy et al. successfully produced nano-composite films made of polyvinyl alcohol (PVA) and colloidal quantum dots (CQD) to utilize them as a defense against electromagnetic radiation. Figure 15.9A illustrates the spectral efficiency (SE) of PVA/CQDs nano-composite films as a function of frequency. The PVA/CQD films exhibited a range of SE values, from 13 to 36.8 dB, at a frequency of 1 GHz. These values were observed for CQD concentrations ranging from 2 to 8 wt%. As electromagnetic waves strike a film, some of them are reflected as they approach the surface, while the remaining portion of the waves enter the film itself (Fig. 15.9b). A portion of those electromagnetic waves were captured via CQD nanoparticles and PVA chains. The CQDs in the films will frequently reflect and absorb the internal reflected waves until they ultimately dissipate. Figure 15.10a, b depicts the frequency dependence of the absorption SEA and reflection SER of a PVA/CQDs nanocomposite. Higher frequency and concentration of CQD NPs throughout the nanocomposite bring about a drop in the SEA value. PVA/CQDs NCs exhibit a feature that makes it a promising material for electromagnetic wave absorption at microwave frequencies. This characteristic also makes it applicable to electromagnetic shielding technologies [61].

An excellent platform for the fabrication of electromagnetic shielding (EMI) devices was created by Atta et al. using their PVA-NiNdxFe$_2$-X O$_4$ film. The FE-SEM picture of the PND2 sample is shown in Fig. 15.11. Figure 15.11a shows the square-shaped flexible composite. In ImageJ software, the width of the composite

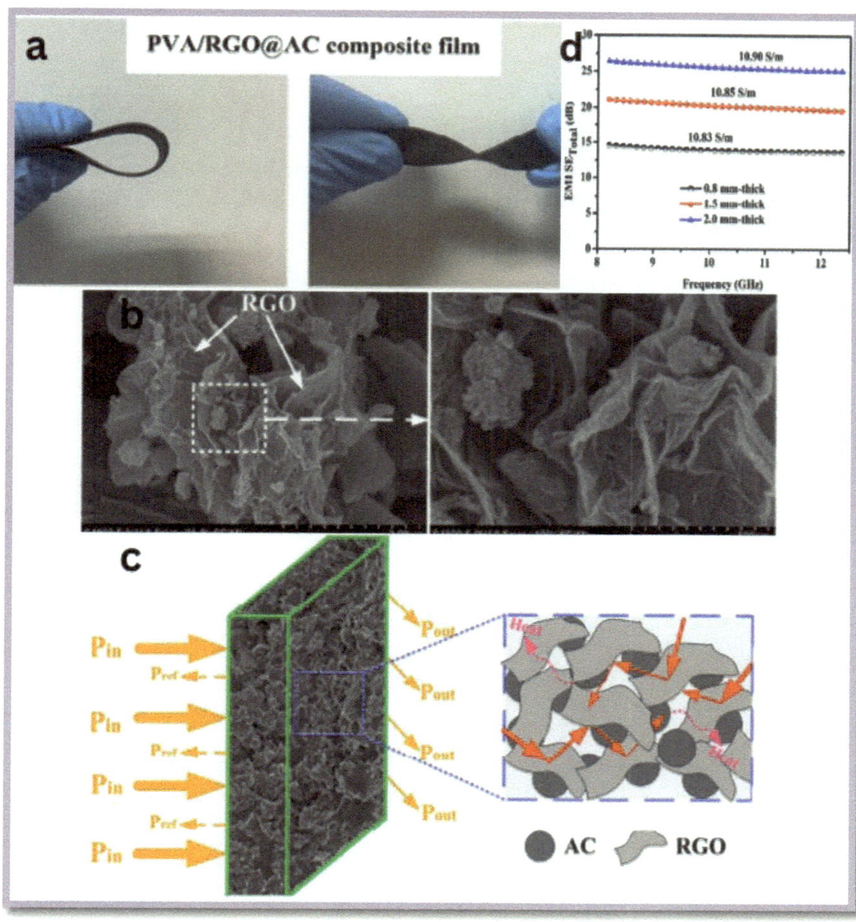

Fig. 15.8 **a** 10wt% filler contains PVA/RGO@AC composite, **b** Magnifying image of RGO@AC-10 complex, **c** Mechanism of EMI shielding, **d** total EMI SE (SE$_{Total}$) [47]. Adapted from Ref. [62] with permission from [American Chemical Society], Copyright [2018]

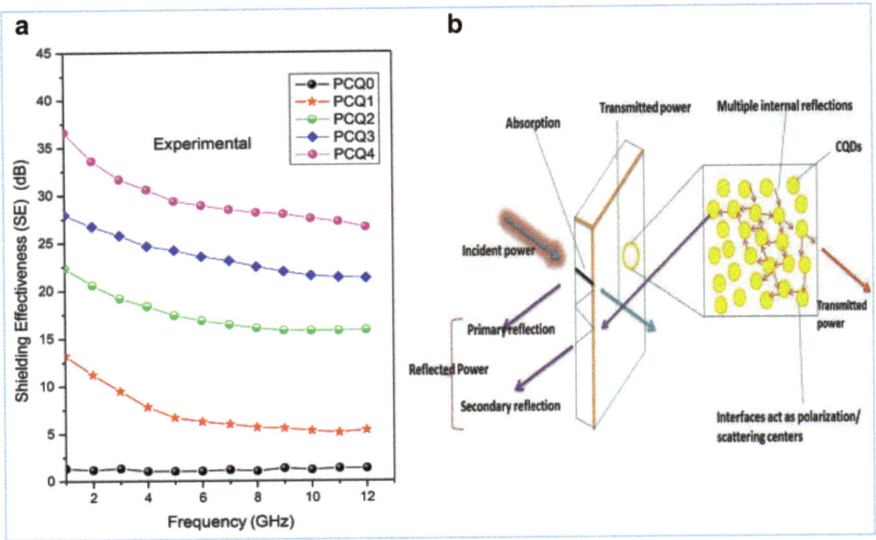

Fig. 15.9 **a** The PVA/CQD films exhibit shielding efficacy values. **b** The study examines the way electromagnetic waves are transmitted, reflected internally, and absorbed in films [61]. Adapted from Ref. [63] with permission from [Elsevier], Copyright [2019]

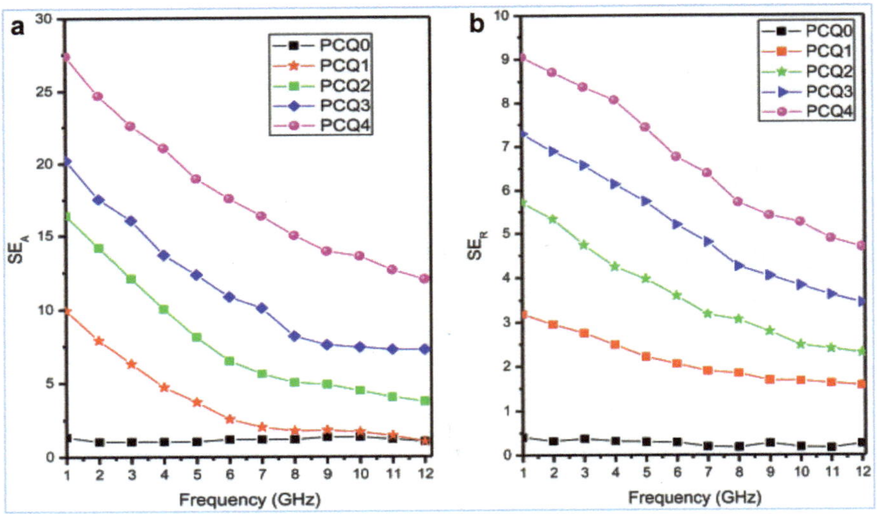

Fig. 15.10 PVA/CQDs nanocomposite frequency ranges **a** SEA and **b** SER of the PVA/CQDs [61]. [63] Adapted from Ref. with permission from [Elsevier], Copyright [2019]

was analyzed, which is 182 μm. The EDS analysis, as seen in Fig. 15.11b, c, it provides a comprehensive identification of the chemical composition and purity of the PVA-Nd-NFO samples. The EDS analysis of the PND1 sample confirms the presence of elements including Ni, Fe, Nd, O, and C. The electron energy loss spectroscopy (EDS) of the carbon atom is fully saturated in this spectrum since the polyvinyl alcohol (PVA) polymer serves as the matrix material in the nanocomposite material. As the level of Nd doping grows in the PND2 samples, the intensities of the Nd peaks in the EDS spectra become significantly stronger. The low skin depth of this nanocomposite (as illustrated in Fig. 15.11d, e) indicates that it may be useful for the development of effective electromagnetic shielding devices, implying beneficial properties [62].

Yang et al. effectively utilized a thermally generated PVA/GNP nanocomposite to create objects with significantly improved electromagnetic interference (EMI) shielding capabilities using Fused Deposition Modeling 3D Printing (FDM). The GNP (graphene nanoplatelets) content in the PVA matrix raised correspondingly enhanced the electromagnetic interference shielding effectiveness (EMI SE) of the printed W-mPVA/GNP components. Notably, the printed W-mPVA/GNP component with 8 wt% GNP at a thickness of 2.43 mm had an EMI SE range of 26–32 dB over a frequency range of 8–12.4 GHz (as shown in Fig. 15.12). In comparison to pure PVA samples, the EMI performance of these printed W-mPVA/GNP nanocomposite components was significantly improved. The shielding efficacy attained not only

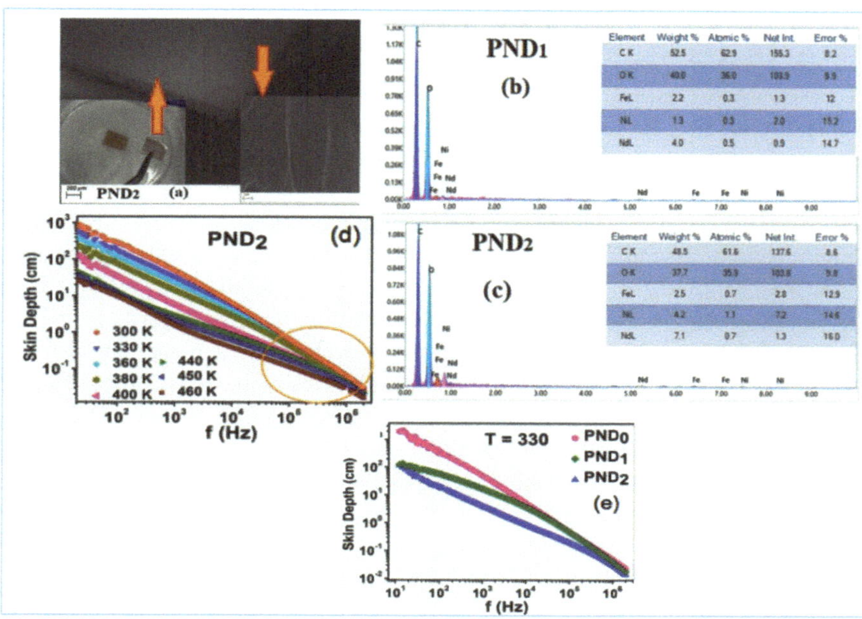

Fig. 15.11 **a** The SEM image of the PND2 sample, **b**, **c** EDX analysis of PND1 and PND2 samples, and **d**, **e** variations of skin depth over different frequencies [62]. Adapted from Ref. [64] with permission from [Elsevier], Copyright [2023]

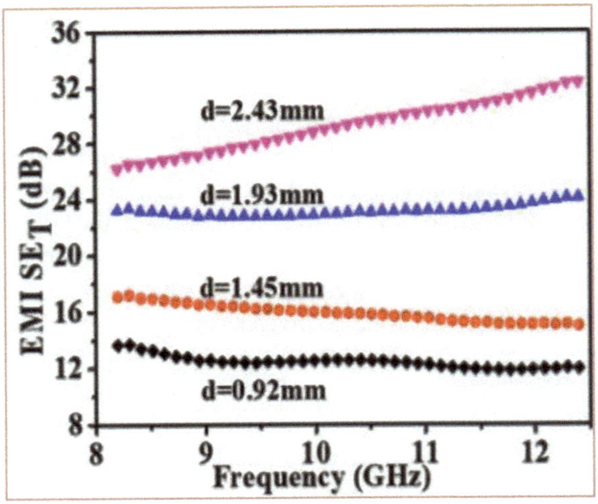

Fig. 15.12 GNP parts' thicknesses at 12 GHz frequency SER, SEA, and SET of 8 wt% [63]. [63] Adapted from Ref. with permission from [American Chemical Society], Copyright [2020]

showed significant improvement but also met the parameters required for practical applications, indicating the practical feasibility and usefulness of these composite materials in real-world circumstances [63].

15.6 Polyethersulfone (PES)-Based Polymer Composite

Polyethersulfone (PES) is a specialized engineering plastic known for its outstanding chemical resistance, formidable heat endurance, and unwavering stability in fluctuating temperatures. It is a standout polymer, particularly sought after for the creation of superhydrophobic coatings. However, direct amalgamation of standard inorganic fillers with PES frequently results in impaired mechanical characteristics due to a lack of significant interfacial effects. This difficulty has been recognized and addressed within the discipline, driving novel techniques to fully realize potential of PES. PES has been widely used in the creation of asymmetrical membranes as well as hollow-fiber materials by utilizing its unmatched chemical and thermal stability. These innovative efforts leverage the phase inversion technique to create new porous structures that take advantage of the intrinsic features of PES [64–66]. Polyethersulfone (PES) has emerged as an excellent composite matrix, selected intentionally to improve load-stress transmission while simultaneously decreasing composite weight and related raw material and processing costs [67]. By combining a planetary ball mill with a hot compaction technique, Verma et al. carried out groundbreaking research on the creation of PES/MWCNT (Multi-Walled Carbon Nanotube) nanocomposites. They found a strong link between the dielectric characteristics of the composite and the milling duration. This enhancement indirectly led to an increase in the electromagnetic interference shielding effectiveness (EMI-SE) for a specific volume % of MWCNT. Remarkably, their research uncovered a distinct correlation between the dielectric constant and the thickness of the composite. The dielectric constant increased as the thickness increased, increasing electromagnetic interference shielding effectiveness (EMI-SE), maybe due to enhanced absorption qualities inside the composite structure [68]. In their previous research, the focus shifted to crafting poly (ether sulfone)/multiwalled carbon nanotube (PES/MWCNT) nanocomposites, capitalizing on PES as the matrix and employing untreated MWCNT as reinforcement, achieved through an environmentally friendly ball milling and hot-pressing process. Interestingly, the pure PES showed just 2 dB of total SE, which saw a spectacular increase, rising around 15 times to 29 dB for the 6.76 vol% MWCNT-filled nanocomposite (NC10). The formation of electrically conductive networks scattered across the insulating matrix is responsible for this significant increase in EMI-SE.). An average total EMI-SE of around 30 dB was noted for the 6.76 vol% nanocomposite, which reflects a consistency in the EMI-SE performance in both the X-band and the Ku-band levels. This uniformity, which shows very little fluctuation with varying frequencies, emphasizes how resilient and dependable these nanocomposites are at preserving their capacity to protect at various frequencies [69].

Zhang et al. created an alternating multi-layer film of RGOx and PESy using a rigorous thermal pressure technique. The purposeful design of this multilayer film conferred onto the material an array of superior electromagnetic shielding efficiency mechanisms. The EMI SE values of the produced RGOx, PESy composites were investigated with different numbers of RGO layers. Interestingly, differences in SER values were not as large as they may have been when there were more layers, but the SEA value showed a steady, gradual increase [70]. Surprisingly, the shielding performance of the RGO_7 and PES_6 multilayer film remained robust at approximately 45 dB across a broad range of temperatures spanning from 298 to 473 K, placing it as a promising contender for electromagnetic shielding applications in elevated temperature environments. However, the comparison study revealed that the RGO_7 film with the stacked structure in place of alternation had a lower SEA value. Furthermore, the constant maintenance of effective shielding performance throughout a wide temperature range of 298 K to 473 K highlights the immense potential of RGO7 and PES6 composites for high-temperature EMI shielding applications [70, 71].

15.7 Polypyrrole (PPy)-Based Polymer Composite

Among conductive polymers, polypyrrole (PPy) is a prominent player because of its remarkable electrical conductivity, ease of synthesis, porous structure, and strong resistance to environmental stress. Because of its exceptional qualities, PPy is now widely used as a material for electromagnetic interference (EMI) shielding in modern applications [71–73]. Improving PPy's conductivity performance entails strategically incorporating nanofillers inside its matrix, such as metals or carbon-based compounds, to promote a conducive structure with increased surface area. Notably, there is growing interest in iron-decorated PPy composites, which exhibit higher conductivity to virgin PPy, spurring research into synthesizing iron-decorated PPy for examining its interactions with graphite. Graphite, a carbon-based substance, considerably contributes to these composite materials by providing conductive pathways and a large surface area. Extensive literature searches highlight remarkable electrical, thermal, and mechanical.

Madhusudhan et al. have conducted a study on PPy/graphite nanocomposites for dielectric and electromagnetic interference (EMI) shielding applications. In their analysis, it encompassing frequencies from 12 to 18 GHz, a similar trend appeared demonstrating a decline in shielding efficiency (SE) across all samples with increasing frequencies, particularly in the MSNG 3 specimens. The analysis found improved SE in all composite samples when compared to pure PPy, attributing the improvement to increased graphite concentrations, which boosted the SE from roughly 10 dB to over 18 dB. [78].

A significant breakthrough in environmentally benign and incredibly effective electromagnetic interference (EMI) shielding materials was made possible by Nazir et al.'s invention of ferrocene-based polypyrrole/carbon (PPyFc) composites. Multiwalled carbon nanotubes (MWCNTs), reduced graphene oxide (RGO), and carbon black (CB) are successfully incorporated in these composites. Significantly, the EMI shielding efficiency (SE) was obtained for several composite formulations: PPyFc/CB had an estimated SE of −11.08 dB, while PPyFc/RGO had an approximate SE of −11.44 dB. However, the best result was obtained with the PPyFc/MWCNT composite, which has a striking SE value of about −23.74 dB (Fig. 15.13). This highlights the promising potential of these composites in the field of EMI shielding materials [79].

Using thermoplastic polyurethane (TPU) as the matrix and polypyrrole/multiwalled carbon nanotubes (PCNT) as the filler, Gahlout et al. created a

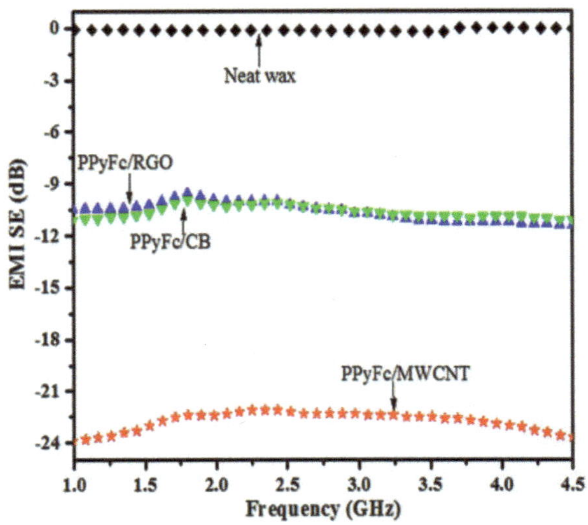

Fig. 15.13 EMI SE of pure wax, PPyFc/MWCNT, PPyFc/RGO, and PPyFc/CB mixtures between 1 and 4.5 GHz [79]. [81] Adapted from Ref. with permission from [Springer], Copyright [2020]

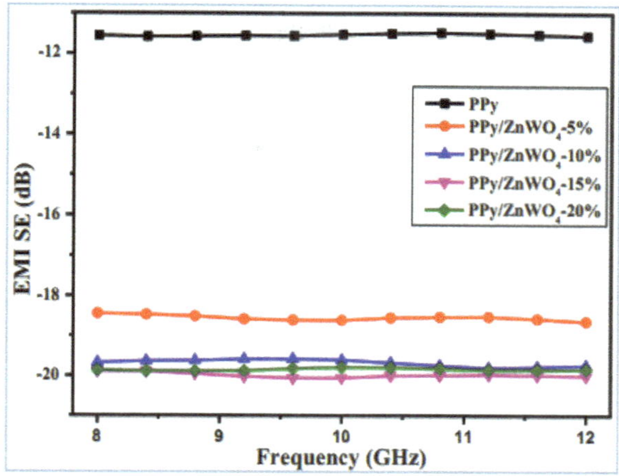

Fig. 15.14 EMI shielding efficiency of PPy/$ZnWO_4$ composites [82]. [84] Adapted from Ref. with permission from [Springer], Copyright [2022]

polypyrrole-multiwalled carbon nanotube/polyurethane composite. In this innovative approach, he studied conducting, dielectric, and magnetic properties in the polypyrrole/MWCNT composite. The X-band microwave absorber usefulness of these composites was proven across a frequency range of 8.2–12.4 GHz. A thorough evaluation and comparison of polyurethane composites including neat polypyrrole (PPy) as the filler was conducted in order to assess the EMI shielding effectiveness. Specifically, PCNT-filled polyurethane composites demonstrated noticeably increased shielding effectiveness, peaking at -46 dB at 40 weight percent PCNT filler concentration. Conversely, polyurethane composites with PPy as the only filler had a worse shielding effectiveness of −18 dB [80].

The innovative nanocomposite formulation was developed by Gairola et al. and comprises polypyrrole, barium ferrite, and tantalum pentoxide. Their remarkable microwave absorption qualities result in an exceptional overall shielding effectiveness of 50.0 dB, which is divided into shielding efficiency due to absorption (SE_A) of 42.4 dB and shielding effectiveness owing to reflection (SE_R) of 8.2 dB. The SE values of PBT-125, PB-11, PBT-152, PBT-122, and PT-11 achieved 50 dB, 37 dB, 23 dB, 22 dB, and 19 dB, respectively. In order to better understand the individual roles of SE_R and SE_A, notably, PBT-125 had an excellent SE_A of 42.4 dB [81].

Kadar et al created nanocomposites employing polypyrrole/zinc tungstate (PPy/$ZnWO_4$), revealing a multifunctional material with remarkable anticorrosive characteristics and interesting implications for broadband EMI shielding applications. The addition of $ZnWO_4$ to the PPy matrix had a transformational effect, resulting in customized electrical conductivity and dynamic dielectric behavior that favored absorption-centric EMI shielding efficacy greatly. The thorough analysis demonstrated that the PPy/$ZnWO_4$ nanocomposite had remarkable EMI shielding efficiency, reaching a peak efficacy of −20 dB, as shown in Fig. 15.14. This technique offers a high level of electromagnetic noise reduction, up to 99%. It is specifically designed for usage in the X-band frequency region, which spans from 8 to 12 GHz [82].

Kadar et al. have created the first composites of PPy/$BaFe_{12}O_{19}$ that possess changeable electrical properties and robust resistance against corrosion. As a result, these composites emerge as promising contenders for applications involving broadband electromagnetic interference (EMI) shielding. The SEM images are presented in Fig. 15.15. Here clearly demonstrate the morphological characteristics of the PPy/$BaFe_{12}O_{19}$ composites, namely samples BR1 (10 wt% of $BaFe_{12}O_{19}$), BR2 (20 wt% of $BaFe_{12}O_{19}$), and BR3 (25 wt% of $BaFe_{12}O_{19}$). The composites were extensively researched to determine their effectiveness in shielding against electromagnetic interference (EMI) and reducing microwave transmission within the 8–12 GHz frequency range (X-band). The study findings confirmed the practicality of these composites. The observed shielding efficiency of these anticorrosive composites was very high, ranging from −23 dB to −27 dB shown in Fig. 15.15d, surpassing attenuation values of over 99%. This demonstrated their exceptional ability to alleviate the increasing challenge. This showcased their remarkable proficiency in mitigating the escalating electromagnetic interference (EMI) problem, suggesting their capacity to function as effective remedies in real-world EMI shielding scenarios [83].

Fig. 15.15 **a–c** PPy /$BaFe_{12}O_{19}$ composites. **d** shielding effectiveness varies composites with X-band (8–12 GHz) frequency in PPy and PPy/$BaFe_{12}O_{19}$ [83]. Adapted from Ref. [83] with permission from [Springer], Copyright [2023]

Lin et al. developed a nanocomposite consisting of graphene nanoplatelets (GNP), multi-walled carbon nanotubes (MWCNT), and polypyrrole (PPy). This nanocomposite exhibits significantly enhanced electrical conductivity and electromagnetic interference (EMI) shielding ability compared to both GNP/PPy and MWCNT/PPy composites. Their formulation, consisting of a combination of GNP and MWCNT with PPy, was prepared with a mass ratio of 25:75 for GNP to MWCNT and a ratio of 4:4 for PPy. The resulting nanostructure exhibited a distinctive arrangement, with GNP/PPy nanolayers forming a two-dimensional structure that was extensively coated with MWCNT/PPy nanowires, resulting in a three-dimensional hybrid filler with remarkable properties. When this hybrid filler was incorporated into a PU matrix, $\pi - \pi$ interactions between GNPs, MWCNTs, and PPy enabled the development of a nanohybrid. FE-SEM analysis revealed that this interaction led to the creation of a cohesive network consisting of GNP – PPy – CNT – PPy – GNP routes within the PU insulator. This intricate network significantly increased the reflectance and attenuation of electromagnetic radiation. The nanohybrid structure exhibited outstanding electromagnetic interference (EMI) shielding capabilities, achieving a shielding effectiveness of 35–40 dB over frequencies spanning from 30 to 1800 MHz. This indicates its ability to block over 99.9% of electromagnetic waves within this frequency range [83]. This groundbreaking advancement showcases its capacity for various applications that require dependable electromagnetic interference (EMI) shielding solutions.

15.8 Conclusion

In summary, polymer composites have demonstrated significant efficacy in delivering electromagnetic interference (EMI) shielding across several applications. Their distinct amalgamation of characteristics, including as elevated electrical conductivity and robustness, renders them well-suited for electromagnetic interference (EMI) shielding. Given the ongoing technological advancements and the growing need for electromagnetic interference (EMI) protection, polymer composites will surely have a significant impact in assuring dependable and effective electromagnetic compatibility. This section explores the present status of research on materials for electromagnetic interference (EMI) shielding, specifically focusing on polymer nanocomposites. Modifying the filler content in polymer nanocomposites, which utilize various polymers such as polyaniline, polyvinyl alcohol, polyethersulfone, and polypyrrole, yields remarkable results in terms of electromagnetic interference (EMI) shielding. Furthermore, this research encompasses a comprehensive examination of several strategies aimed at optimizing morphological, structural, and processing elements to enhance the overall effectiveness of EMI shielding. The nuances of various methods are covered in detail, illuminating the complex facets of material design. This study conducts a comprehensive analysis of existing literature to determine the necessary future actions required to progress polymer-nanocomposite materials as advanced electromagnetic interference (EMI) shielding materials for impending applications.

Acknowledgements The authors gratefully acknowledge the support provided by Centurion University of Technology and Management, Odisha, India for carrying out the present research work.

References

1. Sankaran, S., Deshmukh, K., Ahamed, M.B., Pasha, S.K.: Recent advances in electromagnetic interference shielding properties of metal and carbon filler reinforced flexible polymer composites: A review. Compos. A Appl. Sci. Manuf. **114**, 49–71 (2018)
2. Safdar, F., Ashraf, M., Javid, A., & Iqbal, K. (2022). Polymeric textile-based electromagnetic interference shielding materials, their synthesis, mechanism and applications–A review. *Journal of Industrial Textiles*, *51*(5_suppl), 7293S-7358S.
3. Zhang, M., Zhang, P., Wang, Q., Li, L., Dong, S., Liu, J., Rao, W.: Stretchable liquid metal electromagnetic interference shielding coating materials with superior effectiveness. Journal of Materials Chemistry C **7**(33), 10331–10337 (2019)
4. Chen, L., Zhao, J., Wang, L., Peng, F., Liu, H., Zhang, J., Guo, Z.: In-situ pyrolyzed polymethylsilsesquioxane multi-walled carbon nanotubes derived ceramic nanocomposites for electromagnetic wave absorption. Ceram. Int. **45**(9), 11756–11764 (2019)
5. Soares, B.G., Barra, G.M., Indrusiak, T.: Conducting polymeric composites based on intrinsically conducting polymers as electromagnetic interference shielding/microwave absorbing materials—A review. Journal of Composites Science **5**(7), 173 (2021)
6. Song, P., Liu, B., Qiu, H., Shi, X., Cao, D., Gu, J.: MXenes for polymer matrix electromagnetic interference shielding composites: A review. Composites Communications **24**, 100653 (2021)
7. Vlad, A., Singh, N., Galande, C., Ajayan, P.M.: Design considerations for unconventional electrochemical energy storage architectures. Adv. Energy Mater. **5**(19), 1402115 (2015)
8. Kuila, C., Maji, A., Murmu, N.C., Kuila, T., Srivastava, S.K.: Recent advancements in carbonaceous nanomaterials for multifunctional broadband electromagnetic interference shielding and wearable devices. Carbon **210**, 118075 (2023)
9. Kumar, D., Moharana, A., Kumar, A.: Current trends in spinel based modified polymer composite materials for electromagnetic shielding. Materials Today Chemistry **17**, 100346 (2020)
10. Wei, H., Wang, H., Li, A., Cui, D., Zhao, Z., Chu, L., Guo, Z.: Multifunctions of polymer nanocomposites: environmental remediation, electromagnetic interference shielding, and sensing applications. ChemNanoMat **6**(2), 174–184 (2020)
11. Lin, T.A., Shiu, B.C., Lin, J.H., Lou, C.W.: Tensile, electromagnetic, and far-infrared properties of stainless steel/far-infrared polyester composite materials. Polym. Compos.. Compos. **40**(6), 2219–2230 (2019)
12. Kwon, S., Ma, R., Kim, U., Choi, H.R., Baik, S.: Flexible electromagnetic interference shields made of silver flakes, carbon nanotubes and nitrile butadiene rubber. Carbon **68**, 118–124 (2014)
13. Gass, J., Poddar, P., Almand, J., Srinath, S., Srikanth, H.: Superparamagnetic polymer nanocomposites with uniform Fe3O4 nanoparticle dispersions. Adv. Func. Mater. **16**(1), 71–75 (2006)
14. Gelves, G.A., Al-Saleh, M.H., Sundararaj, U.: Highly electrically conductive and high performance EMI shielding nanowire/polymer nanocomposites by miscible mixing and precipitation. J. Mater. Chem. **21**(3), 829–836 (2011)
15. Xu, L., Lu, H., Dong, Y., Fu, Y., Ni, Q.: Graphene fibre film/polydimethylsiloxane nanocomposites for high-performance electromagnetic interference shielding. Nanoscale Advances **4**(18), 3804–3815 (2022)

16. Verma, P., Bansala, T., Chauhan, S.S., Kumar, D., Deveci, S., Kumar, S.: Electromagnetic interference shielding performance of carbon nanostructure reinforced, 3D printed polymer composites. J. Mater. Sci. **56**, 11769–11788 (2021)
17. Cheng, J., Li, C., Xiong, Y., Zhang, H., Raza, H., Ullah, S., Che, R.: Recent advances in design strategies and multifunctionality of flexible electromagnetic interference shielding materials. Nano-micro letters **14**(1), 80 (2022)
18. Pradhan, S.S., Unnikrishnan, L., Mohanty, S., Nayak, S.K.: Thermally conducting polymer composites with EMI shielding: a review. J. Electron. Mater. **49**, 1749–1764 (2020)
19. Kondawar, S. B., & Modak, P. R. (2020). Theory of EMI shielding. In *Materials for Potential EMI Shielding Applications* (pp. 9–25). Elsevier.
20. do Amaral Junior, M. A., Marcuzzo, J. S., da Silva Pinheiro, B., Lopes, B. H. K., de Oliveira, A. P. S., Matsushima, J. T., & Baldan, M. R. (2019). Study of reflection process for nickel coated activated carbon fiber felt applied with electromagnetic interference shielding. *Journal of Materials Research and Technology*, *8*(5), 4040–4047.
21. Pandey, R., Tekumalla, S., & Gupta, M. (2020). EMI shielding of metals, alloys, and composites. In *Materials for potential EMI shielding applications* (pp. 341–355). Elsevier.
22. Ma, Z., Mang, C., Weng, X., Zhang, Q., Si, L., Zhao, H.: The Influence of different metal ions on the absorption properties of nano-nickel zinc ferrite. Materials **11**(4), 590 (2018)
23. Los, P., Lukomska, A., Jeziorska, R.: Metal-polymer composites for electromagnetic interference shielding applications. Polimery **61**(10), 663–669 (2016)
24. Li, Y., Sun, N., Liu, J., Hao, X., Du, J., Yang, H., Cao, M.: Multifunctional BiFeO3 composites: absorption attenuation dominated effective electromagnetic interference shielding and electromagnetic absorption induced by multiple dielectric and magnetic relaxations. Compos. Sci. Technol. **159**, 240–250 (2018)
25. Yang, Y., Li, M., Wu, Y., Zong, B., Ding, J.: Size-dependent microwave absorption properties of Fe 3 O 4 nanodiscs. RSC Adv. **6**(30), 25444–25448 (2016)
26. Dong, C., Wang, X., Zhou, P., Liu, T., Xie, J., Deng, L.: Microwave magnetic and absorption properties of M-type ferrite BaCoxTixFe12− 2xO19 in the Ka band. J. Magn. Magn. Mater. **354**, 340–344 (2014)
27. Xia, Y., Gao, W., Gao, C.: A review on graphene-based electromagnetic functional materials: electromagnetic wave shielding and absorption. Adv. Func. Mater. **32**(42), 2204591 (2022)
28. Seng, L.Y., Wee, F.H., Rahim, H.A., Malek, F., You, K.Y., Liyana, Z., Ezanuddin, A.A.M.: EMI shielding based on MWCNTs/polyester composites. Appl. Phys. A **124**, 1–7 (2018)
29. Tao, Y., Li, P., Shi, S.Q.: Effects of carbonization temperature and component ratio on electromagnetic interference shielding effectiveness of woodceramics. Materials **9**(7), 540 (2016)
30. Zhao, B., Hamidinejad, M., Wang, S., Bai, P., Che, R., Zhang, R., Park, C.B.: Advances in electromagnetic shielding properties of composite foams. Journal of Materials Chemistry A **9**(14), 8896–8949 (2021)
31. Kunkel, G. M., & Kunkel, G. M. (2020). Penetration of electromagnetic wave through shielding barrier. *Shielding of Electromagnetic Waves: Theory and Practice*, 13–18.
32. Rathi, V., Panwar, V.: Electromagnetic interference shielding analysis of conducting composites in near-and far-field region. IEEE Trans. Electromagn. Compat.Electromagn. Compat. **60**(6), 1795–1801 (2018)
33. Kumar, P., Narayan Maiti, U., Sikdar, A., Kumar Das, T., Kumar, A., Sudarsan, V.: Recent advances in polymer and polymer composites for electromagnetic interference shielding: review and future prospects. Polym. Rev.. Rev. **59**(4), 687–738 (2019)
34. Devi, N., Ray, S.S.: Electromagnetic interference cognizance and potential of advanced polymer composites toward electromagnetic interference shielding: A review. Polym. Eng. Sci.. Eng. Sci. **62**(3), 591–621 (2022)
35. Wanasinghe, D., Aslani, F., Ma, G., Habibi, D.: Review of polymer composites with diverse nanofillers for electromagnetic interference shielding. Nanomaterials **10**(3), 541 (2020)
36. Jiang, D., Murugadoss, V., Wang, Y., Lin, J., Ding, T., Wang, Z., Guo, Z.: Electromagnetic interference shielding polymers and nanocomposites-a review. Polym. Rev.. Rev. **59**(2), 280–337 (2019)

37. Portes, R.C., Lopes, B.H.K., do Amaral Junior, M. A., Florez-Vergara, D. E., Quirino, S. F., & Baldan, M. R.: Effect of granulometric distribution on electromagnetic shielding effectiveness for polymeric composite based on natural graphite. Sci. Eng. Compos. Mater. **26**(1), 531–539 (2019)
38. Xie, Z., Liu, D., Xiao, Y., Wang, K., Zhang, Q., Wu, K., Fu, Q.: The effect of filler permittivity on the dielectric properties of polymer-based composites. Compos. Sci. Technol. **222**, 109342 (2022)
39. Lan, D., Li, H., Wang, M., Ren, Y., Zhang, J., Zhang, M., ... & Wang, Y. (2023). Recent Advances in Construction Strategies and Multifunctional Properties of Flexible Electromagnetic Wave Absorbing Materials. *Materials Research Bulletin*, 112630.
40. Lu, L., Xing, D., Teh, K.S., Liu, H., Xie, Y., Liu, X., Tang, Y.: Structural effects in a composite nonwoven fabric on EMI shielding. Mater. Des. **120**, 354–362 (2017)
41. Sambhudevan, S.: Ferrite-based polymer nanocomposites as shielding materials: A review. Chem. Pap. **75**(8), 3697–3710 (2021)
42. Joseph, K., Wilson, R., & Gejo, G. (Eds.). (2019). *Materials for potential EMI shielding applications: processing, properties and current trends.* Elsevier.
43. Kumar, P., Shahzad, F., Yu, S., Hong, S.M., Kim, Y.H., Koo, C.M.: Large-area reduced graphene oxide thin film with excellent thermal conductivity and electromagnetic interference shielding effectiveness. Carbon **94**, 494–500 (2015)
44. Wang, Y., Zhao, W., Tan, L., Li, Y., Qin, L., Li, S.: Review of Polymer-Based Composites for Electromagnetic Shielding Application. Molecules **28**(15), 5628 (2023)
45. Wu, F., Xu, Z., Wang, Y., Wang, M.: Electromagnetic Interference Shielding Properties of Solid-State Polymerization Conducting Polymer. RSC Adv. **4**, 38797–38803 (2014). https://doi.org/10.1039/C4RA05340K
46. Dong, M., Li, Q., Liu, H., Liu, C., Wujcik, E.K., Shao, Q., Ding, T., Mai, X., Shen, C., Guo, Z.: Thermoplastic Polyurethane-Carbon Black Nanocomposite Coating: Fabrication and Solid Particle Erosion Resistance. Polymer **158**, 381–390 (2018). https://doi.org/10.1016/j.polymer.2018.11.003
47. Lai, D., Chen, X., Liu, X., Wang, Y.: Flexible Poly (vinyl alcohol)/reduced graphene oxide coated carbon composites for electromagnetic interference shielding. ACS Applied Nano Materials **1**(10), 5854–5864 (2018)
48. Wang, M., Tang, X.H., Cai, J.H., Wu, H., Shen, J.B., Guo, S.Y.: Construction, mechanism and prospective of conductive polymer composites with multiple interfaces for electromagnetic interference shielding: A review. Carbon **177**, 377–402 (2021)
49. Shukla, V.: Advances in hybrid conducting polymer technology for EMI shielding materials. In: Advances in Hybrid Conducting Polymer Technology, pp. 201–247. Springer International Publishing, Cham (2021)
50. Taj, M.N., Prasad, B.D., Narapareddy, R., Nagabhushana, H., Ramakrishna, G., Mahesh, B., Dadami, S.T.: PANI-molybdate nanocomposites: Structural, morphological and dielectric properties for the effective electromagnetic interference (EMI) shielding applications in X-band. Applied Surface Science Advances **7**, 100203 (2022)
51. Kumar, S., Kumar, P., Gupta, R., Verma, V.: Electromagnetic interference shielding behaviors of in-situ polymerized ferrite-polyaniline nano-composites and ferrite-polyaniline deposited fabrics in X-band frequency range. J. Alloy. Compd. **862**, 158331 (2021)
52. Akram, S., Aziz, H., Imran, A., Javid, A., Nosheen, A., Ashraf, M., Raza, M.: Fabrication of silver/polyaniline/aminated graphene oxide coated textiles for electromagnetic interference shielding application within the different bands of frequency. Synth. Met. **298**, 117440 (2023)
53. Zhang, Y., Wang, L., Zhang, J., Song, P., Xiao, Z., Liang, C., Gu, J.: Fabrication and investigation on the ultra-thin and flexible Ti3C2Tx/co-doped polyaniline electromagnetic interference shielding composite films. Compos. Sci. Technol. **183**, 107833 (2019)
54. Zhan, Y., Lago, E., Santillo, C., Castillo, A.E.D.R., Hao, S., Buonocore, G.G., Bonaccorso, F.: An anisotropic layer-by-layer carbon nanotube/boron nitride/rubber composite and its application in electromagnetic shielding. Nanoscale **12**(14), 7782–7791 (2020)

55. Yang, W., Liu, J.J., Wang, L.L., Wang, W., Yuen, A.C.Y., Peng, S., Wang, C.H.: Multifunctional MXene/natural rubber composite films with exceptional flexibility and durability. Compos. B Eng. **188**, 107875 (2020)
56. Aslam, M., Kalyar, M.A., Raza, Z.A.: Polyvinyl alcohol: A review of research status and use of polyvinyl alcohol based nanocomposites. Polym. Eng. Sci.. Eng. Sci. **58**(12), 2119–2132 (2018)
57. El-Shamy, A.G.: Composite (PVA/Cu nano) films: Two yield points, embedding mechanism and thermal properties. Prog. Org. Coat. **127**, 252–259 (2019)
58. Beygisangchin, M., Abdul Rashid, S., Shafie, S., Sadrolhosseini, A.R., Lim, H.N.: Preparations, properties, and applications of polyaniline and polyaniline thin films—A review. Polymers **13**(12), 2003 (2021)
59. Rani, P., Ahamed, M.B., Deshmukh, K.: Dielectric and electromagnetic interference shielding properties of carbon black nanoparticles reinforced PVA/PEG blend nanocomposite films. Materials Research Express **7**(6), 064008 (2020)
60. El-Shamy, A.G., Maati, A.A., Attia, W., Abd El-Kader, K.M.: Promising method for preparation the PVA/Ag nanocomposite and Ag nano-rods. J. Alloy. Compd. **744**, 701–711 (2018)
61. El-Shamy, A.G.: Novel conducting PVA/Carbon quantum dots (CQDs) nanocomposite for high anti-electromagnetic wave performance. J. Alloy. Compd. **810**, 151940 (2019)
62. Atta, S., Halder, M., Dey, S., Raul, C.K., Meikap, A.K.: Study of dielectric and magnetodielectric behavior of flexible PVA-NiNdxFe2-X O4 film for EMI shielding application. Curr. Appl. Phys. **47**, 15–23 (2023)
63. Yang, L., Chen, Y., Wang, M., Shi, S., Jing, J.: Fused deposition modeling 3D printing of novel poly (vinyl alcohol)/graphene nanocomposite with enhanced mechanical and electromagnetic interference shielding properties. Ind. Eng. Chem. Res. **59**(16), 8066–8077 (2020)
64. Zhao, Z., Wang, H., Liu, Z., Zhang, X., Zhang, W., Chen, X., Zhu, Y.: Durable fluorine-free superhydrophobic polyethersulfone (PES) composite coating with uniquely weathering stability, anti-corrosion and wear-resistance. Prog. Org. Coat. **127**, 16–26 (2019)
65. Kumar, R., Isloor, A.M., Ismail, A.F., Rashid, S.A., Matsuura, T.J.R.A.: Polysulfone-Chitosan blend ultrafiltration membranes: preparation, characterization, permeation and antifouling properties. RSC Adv. **3**(21), 7855–7861 (2013)
66. Neelakandan, S., Kanagaraj, P., Sabarathinam, R.M., Muthumeenal, A., Nagendran, A.: Effect of sulfonated graphene oxide on the performance enhancement of acid–base composite membranes for direct methanol fuel cells. RSC Adv. **6**(57), 51599–51608 (2016)
67. Belgin, E.E., Aycik, G.A.: Preparation, characterization and comparative ionizing electromagnetic radiation performances: part I—metal oxide reinforced polymeric composites. J. Radioanal. Nucl. Chem.Radioanal. Nucl. Chem. **325**, 503–513 (2020)
68. Verma, R., Rathod, M.J., Goyal, R.K.: Effect of milling parameters on EMI shielding of the PES/MWCNT nanocomposites. Materials Today: Proceedings **43**, 3169–3172 (2021)
69. Verma, R., Rathod, M.J., Goyal, R.K.: High electromagnetic interference shielding of poly (ether-sulfone)/multi-walled carbon nanotube nanocomposites fabricated by an eco-friendly route. Nanotechnology **31**(38), 385702 (2020)
70. Zhang, Y., Han, J., Chang, J., Lu, X., Zhang, Y., Gong, C.: Multilayer RGOx&PESy films for efficient electromagnetic shielding performance in a wide temperature range. Synth. Met. **293**, 117287 (2023)
71. Tat'yana, V.V., Efimov, O.N.: Polypyrrole: a conducting polymer; its synthesis, properties and applications. Russ. Chem. Rev. **66**(5), 443 (1997)
72. Sambyal, P., Dhawan, S.K., Gairola, P., Chauhan, S.S., Gairola, S.P.: Synergistic effect of polypyrrole/BST/RGO/Fe3O4 composite for enhanced microwave absorption and EMI shielding in X-Band. Curr. Appl. Phys. **18**(5), 611–618 (2018)
73. Wang, Y., Gu, F.Q., Ni, L.J., Liang, K., Marcus, K., Liu, S.L., Feng, Z.S.: Easily fabricated and lightweight PPy/PDA/AgNW composites for excellent electromagnetic interference shielding. Nanoscale **9**(46), 18318–18325 (2017)
74. Xiong, C., Zou, Y., Peng, Z., Zhong, W.: Synthesis of morphology-tunable electroactive biomass/graphene composites using metal ions for supercapacitors. Nanoscale **11**(15), 7304–7316 (2019)

75. Krupa, I., Chodak, I.: Physical properties of thermoplastic/graphite composites. Eur. Polymer J. **37**(11), 2159–2168 (2001)
76. Ezquerra, T.A., Kulescza, M., Balta-Calleja, F.J.: Electrical transport in polyethylene-graphite composite materials. Synth. Met. **41**(3), 915–920 (1991)
77. Blaszkiewicz, M., McLachlan, D.S., Newnham, R.E.: The volume fraction and temperature dependence of the resistivity in carbon black and graphite polymer composites: an effective media—percolation approach. Polym. Eng. Sci.. Eng. Sci. **32**, 421–425 (1992)
78. Madhusudhan, C.K., Mahendra, K., Madhukar, B.S., Somesh, T.E., Faisal, M.: Incorporation of graphite into iron decorated polypyrrole for dielectric and EMI shielding applications. Synth. Met. **267**, 116450 (2020)
79. Nazir, A., Yu, H., Wang, L., He, Y., Chen, Q., Amin, B.U., Shen, D.: Electromagnetic interference shielding properties of ferrocene-based polypyrrole/carbon material composites. Appl. Phys. A **126**, 1–15 (2020)
80. Gahlout, P., Choudhary, V.: EMI shielding response of polypyrrole-MWCNT/polyurethane composites. Synth. Met. **266**, 116414 (2020)
81. Gairola, P., Purohit, L.P., Gairola, S.P., Bhardwaj, P., Kaushik, S.: Enhanced electromagnetic absorption in ferrite and tantalum pentoxide based polypyrrole nanocomposite. Progress in Natural Science: Materials International **29**(2), 170–176 (2019)
82. Kadar, C.A., Faisal, M., Raghavendra, N., Maruthi, N., Prasanna, B.P., Nandan, K.R.: Enhancing electromagnetic interference shielding effectiveness (EMI SE) of anticorrosive polypyrrole/zinc tungstate composites: multifunctional approach. J. Mater. Sci. Mater. Electron. **33**(17), 14188–14201 (2022)
83. Kadar, C.A., Faisal, M., Maruthi, N., Raghavendra, N., Prasanna, B.P., Nandan, K.R., Madhusudhan, C.K.: Anticorrosive Polypyrrole/Barium Ferrite (PPy/BaFe12O19) Composites with Tunable Electrical Response for Electromagnetic Wave Absorption and Shielding Performance. J. Electron. Mater. **52**(3), 2080–2093 (2023)
84. Lin, C.L., Li, J.W., Chen, Y.F., Chen, J.X., Cheng, C.C., Chiu, C.W.: Graphene Nanoplatelet/ Multiwalled Carbon Nanotube/Polypyrrole Hybrid Fillers in Polyurethane Nanohybrids with 3D Conductive Networks for EMI Shielding. ACS Omega **7**(49), 45697–45707 (2022)

Chapter 16
Polymer Composites for Biomedical Applications

Rinmayee Praharaj and Tapash R. Rautray

Abstract In order to replace or restore the function of damaged or degenerating organs or tissues, promote healing, enhance function, rectify abnormalities, and ultimately improve patient quality of life, medical practice today makes extensive use of biomaterials in the form of implants and devices. Clinical experience, however, makes it abundantly evident that not all readily available materials—commonly referred to as engineering biomaterials—are appropriate for use in biological contexts. Their processability, biocompatibility, and mechanical qualities are frequently subpar. Due to this, polymeric composites have recently received more interest as researchers look into their potential for use in biological applications, including tissue engineering, drug delivery systems, biosensors, bioimaging, antiviral and antibacterial agents, and more. The characteristics of polymeric composites, such as their elasticity, tensile strength, biocompatibility, conductivity, antibacterial activity, and non-cytotoxicity, are enhanced when they contain a combination of biostable and biocompatible polymers and biologically active nanofillers. In this chapter, we have emphasized the non-cytotoxicity and biocompatibility of various polymeric composite materials. Additionally, numerous polymer composite types being researched or currently in use for various biological purposes are discussed. This chapter also highlighted the critical issues and scientific challenges that still need to be overcome for polymer composite materials to be widely recognized in the biomedical industry.

R. Praharaj (✉)
Department of Physics, Bengaluru City University, Bengaluru, Karnataka 560001, India
e-mail: rinmayee@gmail.com

T. R. Rautray
Biomaterials and Tissue Regeneration Laboratory, Centre of Excellence in Theoretical and Mathematical Sciences, Siksha 'O' Anusandhan University, Khandagiri Square, Bhubaneswar, Odisha 751030, India

S. Moharana et al. (eds.), *Polymer Composites*, Engineering Materials,
https://doi.org/10.1007/978-981-97-2075-0_16

16.1 Demand of Polymer Composites in Biomedicine

Implants and devices are used in medical practice nowadays in significant quantities. Implants, such as bone plates, sutures, ligaments, joint replacements, dental implants, heart valves, vascular grafts, intraocular lenses, etc., and medical devices, such as biosensors, pacemakers, artificial hearts, anticancer therapies, bioimaging, blood tubes, etc., that are restoring or replace damaged or degenerated organs or tissues, promote healing, enhance function, rectify abnormalities, and ultimately enhance patients' quality of life are all examples of biomaterials in use [1–3].

Understanding the interactions between tissues and materials has advanced significantly during the past three decades. At first, some researchers used metals and alloys as implant materials [3–5]. Due to their superior mechanical qualities, substantial load capacity, resilience, and bioinert character. However, several metals have drawbacks, such as (1) being overly stiff, which can lead to tissue degeneration, (2) corroding aggressively in the body's internal environment after a few years, (3) releasing toxic ions that can induce allergic reactions in the skin or tissue, (4) excessive density, and (5) limited biocompatibility [4–7]. Subsequently, hydroxyapatite (HAp), carbon nanotubes (CNTs), titania (TiO_2), zirconia (ZrO_2), alumina (Al_2O_3), and bioactive glasses (BG) are widely used as biocompatible materials due to their strong corrosion resistance, high compression resistance, osteoinductive, and osteoconductive qualities [6–10]. However, the primary drawbacks of these materials are their excessive density and poor mechanical dependability: fracture hardness, brittleness, fragmentation, lack of robustness, and fabrication complexity [11, 12]. Simultaneously, a multitude of polymers is employed in diverse biomedical applications, including chitosan (CS) [45, 46], gelatin (Gel), collagen (Col), silk fibroin (SF), silk sericin (SC), nanocellulose, alginate (Alg), polycaprolactone (PCL), polyhydroxybutyrate-cohydroxyvaletare (PHBV), poly-L-lactic acid (PLLA), polymethylmethacrylate (PMMA), polyurethanes (PU), polyether ether ketone (PEEK), Ppy, PEDOT, PANI, PTh, PLGA, poly(lactide-co-glycolide) (PLAGA), PDLA, polyesters, silicone rubber (SR), polysulfone (PS), ultra-high molecular weight polyethylene (UHMWPE), polytetrafluoroethylene (PTFE), polyacetal (PA), and polyethylene terephthalate (PET) [13–18]. This is due mainly to their vast range of compositions, characteristics, and forms (solids, fibers, textiles, films, and gels) and their ease of fabrication into intricate shapes and structures. Nevertheless, they frequently lack the strength and flexibility needed to meet the mechanical requirements of some applications, such as orthopedic and dental implants.

Tissues are typically classified as hard or soft. Soft tissues like skin, blood vessels, cartilage, and ligaments accompany hard tissues like bone and teeth. As their names imply, hard tissues often have higher tensile strength and are stiffer (elastic modulus) than soft tissues [21, 22]. Metals or ceramics are selected for hard tissue applications based on their structural or mechanical compatibility with tissues, while polymers are used for soft-tissue applications [21–23]. In this regard, using low-modulus materials—like polymers—appears intriguing, yet their practical application is typically limited by the poor strength that goes along with low modulus [23, 24]. Numerous

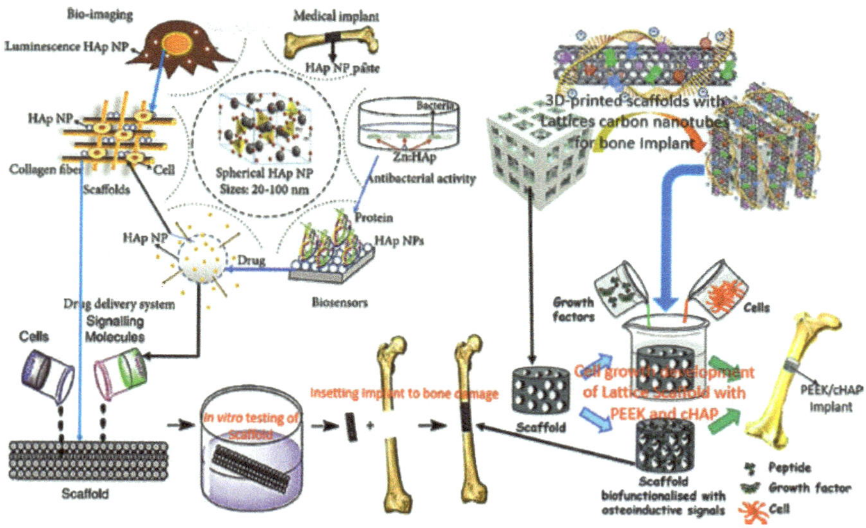

Fig. 16.1 Cell growth development of lattice scaffold with PEEK and cHAp nanoparticles, coating on polymer and 3D-printed scaffolds with carbon nanotube lattices to create effective bone-implant bioprinting [2]. Copyright@2021, European Polymer Journal

orthopedic uses have been suggested for fiber-reinforced polymers, also known as polymer nanocomposite materials, due to their simultaneous low elastic modulus and high strength [84, 175]. Several examples of polymer composite biomaterials are HA/PE, silica/SR, CF/UHMWPE, CF/epoxy, and cHAp/PEEK (shown in Fig. 16.1). Every material type has unique positive qualities that make it especially well-suited for a given application [13–18].

Another benefit of composite materials is their ability to modify and customize an implant's design and qualities to the mechanical and physiological characteristics of the host tissues by adjusting the volume fractions and local and global reinforcement phase arrangement [22, 23]. Therefore, it is proposed that, in comparison to homogenous, monolithic materials, composite materials offer a more significant possibility for structural biocompatibility [22–25]. Human tissues are an excellent illustration of a composite material since they are anisotropic and rely on the structural configurations and functions of different tissue constituents, such as hydroxyapatite, collagen, and elastin [22–25]. Additional factors that led to the development of polymer composite biomaterials include the low fracture toughness of ceramic materials, which makes them a difficult choice for load-bearing applications; lack of metal alloy fatigue failure and corrosion, as well as the release of ions from metals like chromium or nickel, which can lead to implant loosening, patient pain, and allergic skin responses [13–18]. In addition to these noteworthy benefits, composite materials outperform metal alloys and ceramics in addressing perceived or previously mentioned shortcomings. Because they are radio-opaque, metals, alloys, and ceramics can occasionally produce undesired artifacts in X-ray radiography. Mixing

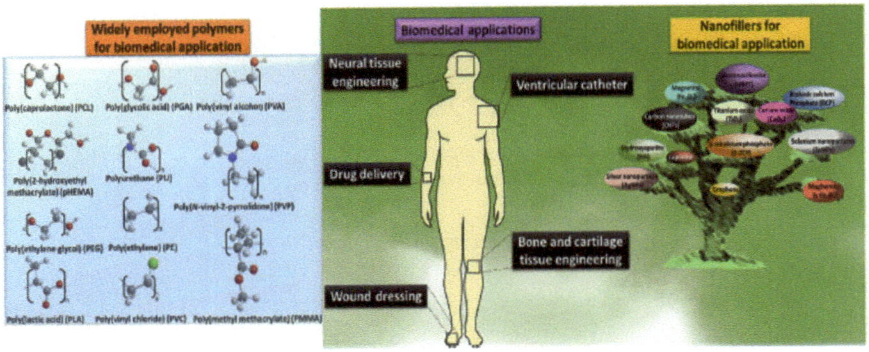

Fig. 16.2 Common polymers and nanofillers for biomedical applications [297]. Copyright@2019, Elsevier Books

contrast medium into the polymer can change the radio transparency of polymer composite materials. Moreover, due to their non-magnetic nature, polymer-composed materials are completely compatible with modern diagnostic methods such as CT (computed tomography) and MRI (Magnetic resonance imaging [13–18, 22–25]. The structural elements of these imaging devices are also composed of polymer composites due to their superior mechanical properties and lightweight nature [22–25]. Numerous polymer composite materials have been created over time for various biomedical uses (shown in Fig. 16.2)[13–18, 22–25]. Details on polymer composite materials for biocompatibility, non-cytotoxicity, tissue engineering, drug delivery, biosensing, antimicrobial, antiviral, and bioimaging applications are provided in the following sections.

16.2 Biocompatibility and Non-cytotoxicity

The leading causes of polymeric nanobiocomposites' failure are degradation, inadequate osseointegration, decreased cell proliferation, and infection. The biomaterials industries widely employ carbon nanotubes, nanoscale metal oxides, and stem cells to solve these issues [26, 27]. Some researchers have gone one step further by employing customized nanoadditives/fillers or another specific agent(s) to create biomaterials that mimic those found in the human body [26, 27] (shown in Fig. 16.3). Numerous studies have been carried out over the past few decades to enhance human mesenchymal stem cell (hMSC) adhesion and differentiation on polymeric nanobiocomposites. For example, it was discovered that the hMSC-coated polystyrene nanocomposites may undergo osteogenic differentiation even without growth hormones. They have elongated-shaped cells that may also undergo osteoblast development formed on these microinjected polystyrene nanocomposites. Additionally, it was demonstrated by osteocalcin mRNA expression and calcium deposition that the increase in nanocomposites' diameter and interpillar distance had a favorable

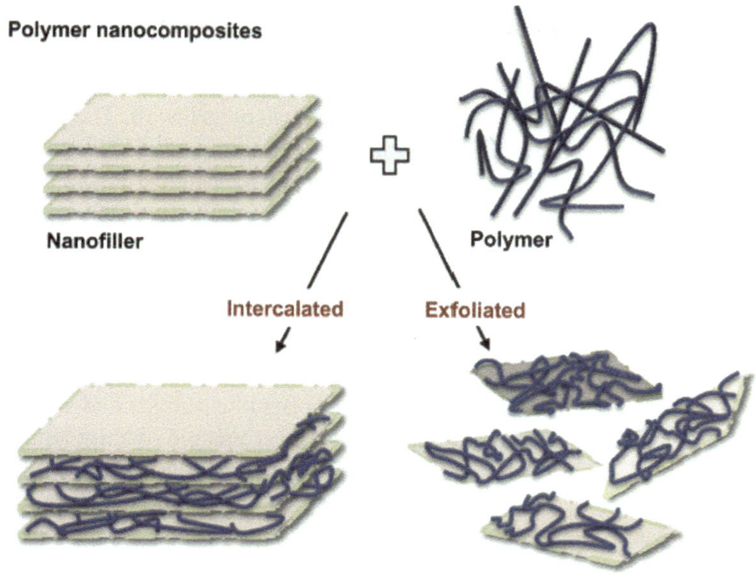

Fig. 16.3 Representation of intercalated or exfoliated polymeric nanocomposites [297]. Copyright@2019, Elsevier Books

impact on the expression of osteocalcin genes and mineralization [26–28]. Similarly, by using nanofillers in the composites, cell adhesion can also be enhanced [28]. Cell proliferation, differentiation, and adhesion are often critical for successfully developing scaffolds resembling the extracellular matrix. It is well established that the polymeric matrix's diameter and pore size significantly impact stem cell adhesion and incorporation [28]. Consequently, researchers worldwide are still searching for the appropriate new technique or techniques to regulate the polymer matrix's pore size and diameter. Yang et al. proposed nanocomposites made from polymers (PLLA) and nanofibers employing the electrospun process under ideal circumstances; however, despite having effective control of the electrospun fibers' diameter, no substantial change in cell reactions has been observed [29].

Surgical infection is a significant factor contributing to the failure of polymeric nanobiocomposites following implantation into the human body, along with inadequate cell adhesion and proliferation. The leading causes of these infections include Escherichia coli, Staphylococcus aureus, Pseudomonas aeruginosa, and Candida albicans [30–33]. Using nanofillers that prevent surgical injection is one strategy to lessen the early rejection of the composites. Adding antimicrobial or antifungal compounds to polymeric nanocomposites is a second way to eliminate infection associated with them [30–33]. For instance, the most effective nanofillers for lowering infection have been discovered to be carbon nanotubes, silver, gold, and titanium dioxide. For example, Goodwin et al. investigated the antibacterial activity of polymeric composites enhanced by CNTs [34]. These nanocomposites demonstrated their antimicrobial potential against P. aeruginosa for an extended period [34]. Since the antibacterial properties of these nanocomposites only become active when they come into direct contact with microbial cells, another intriguing feature of these nanocomposites is that their cytotoxicity toward living cells was at its lowest. These dead cells shield the surviving cells once the microbial cells are eliminated [33, 34]. Similarly, Khan et al. created nanocomposites using titanium dioxide nanoparticles and regenerated bacterial cellulose (RBC) [35]. These nanocomposites demonstrated outstanding cell adhesion and proliferation capabilities and antibacterial impact against E. coli [35]. Additionally, in an attempt to eradicate two groups of bacteria simultaneously, some researchers have combined two different types of nanofillers to increase the antibacterial activity of the nanobiocomposites. In order to demonstrate this point, Shah et al. simultaneously used titanium dioxide and silver to create nanocomposites that could simultaneously inhibit the functions of Bacillus subtilis and E. coli [36].

Using antibiotics or antimicrobial agents is another method of eliminating infection; however, the oral dosage of these drugs is occasionally not target-specific and may metabolize before reaching the intended spot. Antibiotics of various kinds are used to prevent infections that arise after surgery; amoxicillin, penicillin, and cyclosporine are among the often-used varieties [37–40]. These antibiotics are administered by encasing the medications in polymeric nanocomposites or coating the polymeric composites. Ghatti-graft-poly(methacrylic acid-aniline) interpenetrating network (IPN) hydrogel nanocomposites were used by Sharma et al. to study the controlled delivery of amoxicillin trihydrate. With a pH of 9.2, these nanocomposites proved to be efficient drug transporters, primarily to the colon [41]. Similarly, Chao et al. coated a catheter with electrospun polymeric nanofibers and penicillin G sodium; their effectiveness in penetrating the pleural cavity was assessed and contrasted, and it was determined that the antibiotic-eluting catheter could treat empyema [41]. In addition, several attempts have been made to use metal oxides and antibiotics together to eradicate surgical infections. For instance, Lee et al. coated an amoxicillin-specific antibody onto TNTs/PGLA composites to prevent the early rejection of bone plates and teeth following surgery, and they found that this effectively inhibited S. aureus growth [41].

16.3 Medical Applications of Polymer Composites

16.3.1 Hard Tissue Engineering

Tissue engineering is a concept that seeks to repair damaged tissues by using scaffold biomaterials as a template to promote the development of new cells [42]. Tissue- and cell-specific materials known as biocompatible scaffolds, derived from the natural extracellular matrix (ECM), facilitate the quick integration of tissues [43]. Human cells can adhere and organize efficiently around polymer composites that are smaller in size than the cells themselves [44, 45]. Polymer composites have an important morphology that promotes cell migration, growth, and seeding. Polymer composites are commonly employed in both soft-tissue engineering and hard/bone tissue engineering [41]. Therefore, enhancing polymer composites for cell adhesion and proliferation is crucial to restore organs and tissues.

Researchers have focused mainly on tissue engineering to create biodegradable alternatives to bone grafts. For bone tissue engineering, various natural and synthetic biodegradable polymers have been used as scaffold systems. These include chitosan (CS), gelatin (Gel), collagen (Col), silk fibroin (SF), silk sericin (SC), cellulose, alginate (Alg), polycaprolactone (PCL), polyhydroxybutyrate-cohydroxyvaletare (PHBV), poly-L-lactic acid (PLLA), hyaluronic acid (HA), polymethylmethacrylate (PMMA), polyurethanes (PU), polyether ether ketone (PEEK), Ppy, PEDOT, PANI, PTh, PLGA, poly(lactide-co-glycolide) (PLAGA), PDLA, PDLLA, polyesters, and poly(propylene fumarate) [1, 41, 44–49]. The choice of polymer matrix for orthopedic and dental implants is typically determined by the type of sickness or defect that needs to be treated and the filler materials utilized in polymer composites. For instance, while improving the mechanical characteristics of dental and orthopedic implants, the key factors to examine are the degree of swelling in water, the effect on strain, the hardness, Young's modulus, tensile strength, and loading capacity [50–52]. Several polymer and nanofiller combinations (shown in Table 16.1) have been studied in order to demonstrate this. Carbon nanotube, silver, graphene dioxide, titanium dioxide, nano-hydroxyapatite, bioactive glass, silicon dioxide, and so on are a few of the most effective fillers [53–59]. Since these fillers are easily absorbed into a polymeric matrix using normal calendaring techniques, on-site precipitation, or simply depositing the nanofillers onto the polymeric matrix, they can be successfully employed in orthopedic and dental applications. Among these fillers, the capacity to improve mechanical characteristics and function as superior antibacterial agents makes silver bromide, titanium dioxide, and nanographene regarded as dual-action tunable nanomaterials [54–56]. Similarly, carbon nanotubes may significantly improve mechanical qualities, including youthful modulus, elastic modulus, and tensile strength when added to composite materials in as little as 0.1%, so they are widely utilized in orthopedic and dental applications. Another illustration would be hydroxyapatite in nanoscale form, one of the best materials for fixing bone deformities (shown in Fig. 16.4) [53]. One osteoconductive substance that may facilitate osteoprogenitor cell proliferation is hydroxyapatite [57]. Its enhanced biocompatibility, bioactivity, mechanical strength, and high aspect ratio are only a few of its advantageous qualities [57].

As an illustration, Jeon et al. and Leeuwenburgh have developed bone substitutes using polymers and hydroxyapatite [60–62]. The mechanical characteristics of poly-lactic-coglycolic acid supplemented with various ceramic fillers were investigated by Liu et al. [61]. Similar to this, Olkhov et al. examined the results of poly-(3-hydroxybutyrate) nanobiocomposites employing a variety of nanofillers, such as silicon and titanium dioxide particles, using the electrospinning process [41]. The titanium and silicon dioxide nanoparticles significantly increased the polymeric matrix's tensile strength and modulus. However, the hydroxyapatite at the nanoscale reduces the water adsorption capacity of the composites and increases their ductility [41,

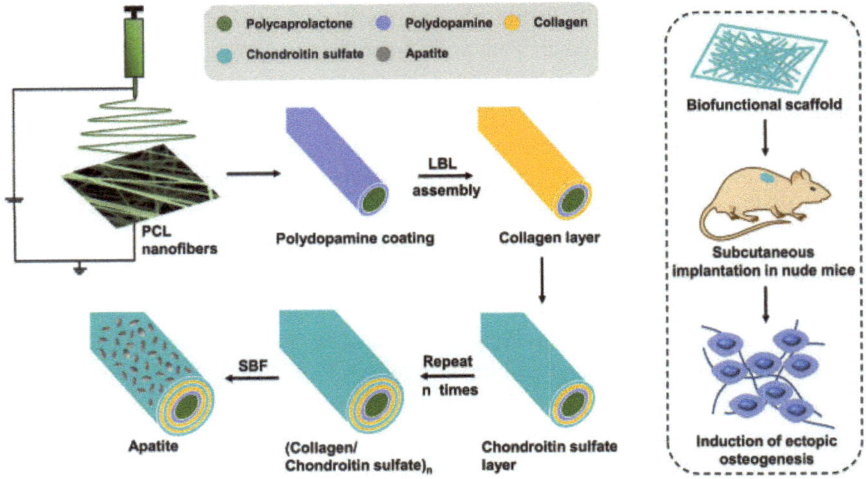

Fig. 16.4 Schematic of the biofunctional bone scaffold fabrication with LBL self-assembly on electrospun fiber membranes followed by apatite deposition [164]. Adopted with permission from Ref. [226] copyright@2021, Materials Science and Engineering: C

56–59] Consequently, research has concentrated on combining polymer matrices or two or more nanofillers to achieve the best mechanical and biological properties that resemble genuine bone. Hydrogels are widely utilized due to their chemical and biological characteristics that align with the cellular matrix. However, the hydrogels do not exhibit the intended mechanical qualities of bone substitutes [63, 64]. Chitin/HAp and PEG-silica have been successfully incorporated into hydrogels by researchers like Chang et al. [65] and Gaharwar et al. [41], respectively. Their presence improves the mechanical characteristics and toughness and even increases the cell adhesion of hydrogel. Combining bone morphogenetic protein with nanoparticles is an additional strategy to enhance bone regeneration. Chitosan-based biocomposite scaffolds for bone tissue engineering are depicted in Fig. 16.5.

Advances in nanotechnology have allowed researchers to create polymer nanocomposites containing drugs and growth factors, allowing for the sustained release of growth factors. They have also employed surface modification techniques to achieve this goal. Using selective laser sintering and other surface changes, Duan et al. [41] created 3D-calcium phosphate/poly(hydroxybutyrate-co-hydroxyvalerate) nanocomposites, which is one example of the development of multifunctional scaffolds for tissue engineering. Due to their potential to enhance osteoconductivity and provide regulated delivery mechanisms, these nanocomposites may represent a promising direction in bone tissue regeneration. Table 16.1 lists possible combinations of nanofillers and polymer composites for bone tissue engineering.

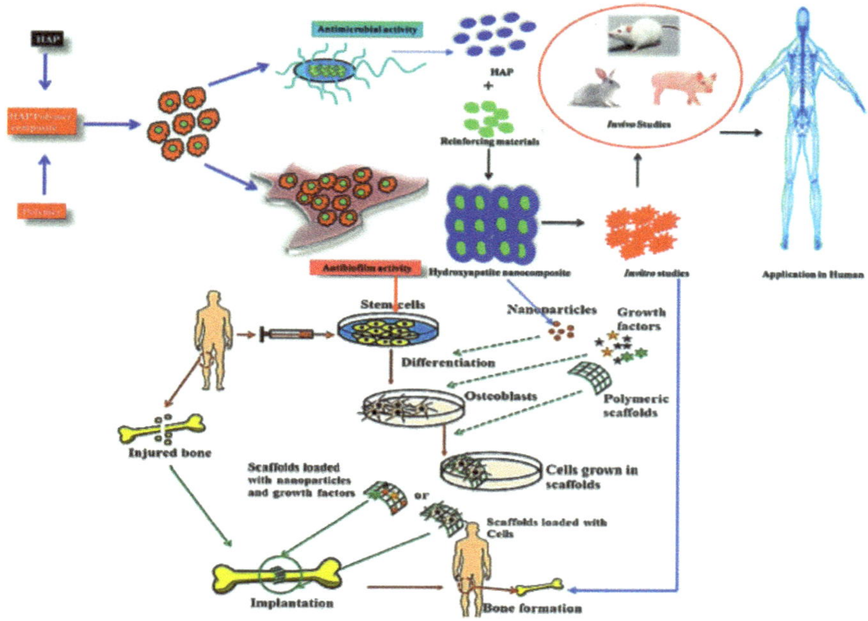

Fig. 16.5 Chitosan-based biocomposite scaffolds for bone tissue engineering, using cells and biomaterials or a mix of cells stacked onto biodegradable platforms used to treat essential measured bone imperfections [10]. Adopted with permission from Ref. [34–37] copyright@2021, European Polymer Journal

16.3.1.1 Soft-Tissue Engineering

Developing soft-tissue engineering is the most demanding and challenging tissue engineering application. However, soft-tissue applications did not include additional tissue engineering uses, such as skin, cartilage, or bone. For esthetic purposes, where substantial quantities and a high degree of vascularization are typically required, the liver, kidney, adipose tissue, or muscle are the primary targets for developing soft-tissue engineering. It has been discovered in earlier studies that the lack of oxygen transport in cartilage and cortical bone limits their ability to form functional tissue in vitro to thicknesses of less than a few hundred micrometers [128–130]. Many studies that aim to create scaffolding materials that resemble extracellular matrix (ECM) for soft-tissue engineering have been published in the past ten years. Soft-tissue engineering creates synthetic natural tissues to alter or replace missing, damaged, or diseased tissues. Tissue replacements are made in tissue culture bioreactors or as regulated tissue-mimetic microenvironments, used as scaffolding materials in vitro or in vivo for in situ tissue development. This process's investigation aims to comprehend how pharmacological and biomechanical factors affect cell behavior (shown in Fig. 16.6) [128–130].

Table 16.1 Lists possible combinations of nanofillers and polymer composites for bone tissue engineering

Polymer Ref. matrix	Nanoadditives/Fillers	Composite form	Potential biomedical application	Refs.
Chitosan (CS)	45S5 Bioglasss®	Film	Bone tissue engineering	[67]
	Hydroxyapatite(HAp)	Scaffold	Bone tissue engineering	[68]
	BBG	Pellet	Osteomyelitis treatment	[69]
	Graphene oxide/r-GO	Scaffold	Cartilage tissue engineering	[70]
	CNT + HAp + Col	Scaffold	Bone tissue engineering	[71]
	Biphasic CaP	Scaffold	Bone tissue engineering	[72]
	HAp + silica	Scaffold	Bone regeneration	[73]
	TEOS + silica	Film	Coating in bone tissue	[74]
Gelatin	Nano-HAp	Scaffold	Bone tissue engineering	[75]
(Gel)	PCL/PEG/PCL	Scaffold	Cartilage tissue engineering	[76]
	PCL/MWCNT	Scaffold	Cartilage tissue engineering	[77]
Gel + CS	Decellularized meniscus	ECM Scaffold	Meniscus tissue engineering	[78]
Collagen	3D-nano-HAp	Scaffold	Bone tissue engineering	[79]
(Col)	β-TCP/LGA/ECM	Scaffold	Osteochondral tissue	[80]
	CaPs from shark teeth	Scaffold	Bone tissue	[81]
	Denatured Col	Scaffold	Cartilage tissue engineering	[82]
	α-TCP	Scaffold	Bone tissue engineering	[83]
	dECM/silk fibroin (SF)	Scaffold	Bone tissue engineering	[84]
Silk Fibroin	HAp + Alginate	Scaffold	Bone tissue engineering	[85]
(SF)	Na-alginate	Scaffold	Bone tissue engineering	[86]
	Wollastonite	Scaffold	Osteochondral tissue	[87]

(continued)

Table 16.1 (continued)

Polymer Ref. matrix	Nanoadditives/Fillers	Composite form	Potential biomedical application	Refs.
	β-TCP	Scaffold	Bone tissue engineering	[88]
Silk Sericin	CS + HAp	Scaffold	Bone tissue engineering	[89]
(SS)	Gelatin	Scaffold/Film	Tissue engineering	[90]
Cellulose	PANI	Scaffold	Tissue engineering	[91]
	PLA + HAp	Scaffold	Bone tissue engineering	[92]
	Alginate/HNTs/PVDF	Scaffold	Cartilage applications	[93]
Alginate	Alginate-sulfate	Scaffold	Bone tissue engineering	[94]
Na-alginate	Gel	Scaffold	Cartilage repair	[95]
Polycaprolactone	BBG	Film	Bone tissue engineering	[96]
(PCL)	HAp	Scaffold	Osteochondral tissue	[97]
	β-TCP + dECM	Scaffold	Bone tissue engineering	[98]
PCL	MgO NP	Scaffold	Bone—soft-tissue engineering	[99]
	Sr-SBG	Scaffold	Bone graft substitutes	[100]
	PBG	Film	Tissue regeneration	[101]
PHBV	Bredigite	Scaffold	Bone tissue engineering	[102]
Poly-L-lactic acid	β-TCP	Scaffold	Bone tissue engineering	[103]
(PLLA)	Ca-deficient HAp	Scaffold	Bone substitute	[104]
	Nanodiamond	Scaffold	Bone substitute	[105]
	Mg rod	Scaffold	Bone tissue engineering	[106]
	Bioglass fibers (BGF)	Scaffold	Bone tissue engineering	[107]
Hyaluronic acid (HA)	MWCNT + MSCs	Scaffold	Skeletal tissue engineering	[108]
PMMA	MWCNTs-COOH	Scaffold	Bone substitute	[109]
	Polyurethane microshells	Scaffold	Bone substitute	[110]
	Ag	Scaffold	Bone cement	[111]

(continued)

Table 16.1 (continued)

Polymer Ref. matrix	Nanoadditives/Fillers	Composite form	Potential biomedical application	Refs.
PEEK	Nanocalcium silicate	Scaffold	Bone tissue	[112]
	HAp doped with fluorine	Scaffold	Bone tissue	[113]
	Nano-BG	Scaffold	Bone tissue	[114]
	β-tricalcium phosphate	Scaffold	Bone tissue	[115]
	Sr-HAp	Scaffold	Bone tissue	[116]
	nano-TiO_2	Scaffold	Bone tissue	[117]
PPy + PCL	Ferric chloride	Scaffold	Bone tissue engineering	[118]
PANI	Gelatin nanofibers	Scaffold	Bone tissue engineering	[119]
PTh	Oligosiloxane	Scaffold	Biomedical applications	[120]
PLAGA	Calcium phosphate	Scaffold	Bone tissue engineering	[121]
PLGA	Collagen gel	Scaffold	Bone regeneration	[122]
	Nano-HAp	Scaffold	Bone tissue engineering	[123]
PDLLA	45S5 Bioglasss®	Scaffold	Regeneration of soft/ hard	[124]
Tissue Defects	45S5 Bioglasss®	Scaffold	Tissue engineering	[125]
PDLA + PPC	TCP	Scaffold	Bone tissue engineering	[126]
MeSDCC	BG + HAp nanofibers	Scaffold	Bone regeneration	[127]
PHB	45S5 Bioglasss®	Scaffold	Bone tissue engineering	[128]

Natural or artificial polymers and a few increasingly popular nanoadditives have recently been used in soft-tissue engineering. It is commonly known that using nanostructured polymers as a scaffold material in soft-tissue engineering provides better qualities than using microfibers. For example, CNTs have the same viscoelastic properties as soft-tissue membranes [131]. Therefore, they can increase the hybrid material's tensile strength and Young's modulus [131]. Additionally, CNTs can help in the growth of neurons. Studies have demonstrated, for instance, that conjugating nanotubes to diverse substrates can promote growth, attachment, and differentiation, affecting neuronal behavior and long-term survival [131]. In their study, Naseri et al. [132] used a genipin crosslink matrix that was loaded with gelatin and chitosan to examine a 3D-porous scaffold that was built on freeze-dried cellulose nanofiber and had varying fiber loading starting from 70 to 90% by weight [132]. Previous investigations indicate that the scaffolds have linked pores with an average pore diameter of 75–200 μm and nanoscaled pore wall roughness, making them appropriate for cell interactions in cartilage repair [132, 133]. The 3D-consistency and compression modulus of the scaffold (13 MPa) at room temperature were also demonstrated by X-ray examination to be greater than those of natural cartilage (1 MPa) [132, 133]. Chondrogenesis benefited from the phosphate-buffered saline (PBS) solution's decreased compression modulus of 1060 kPa at 37C. High PBS absorption, high porosity (95%), and good chondrocyte cytocompatibility are favorable characteristics for ECM and production cell attachment to scaffolds. For instance, the structural characteristics of native tissues can be imitated by combining different phases inside nanocomposite hydrogel networks [133]. Kawaguchi et al.'s [134] study examined the mechanical and biocompatibility characteristics of a nanocomposite gel made of sodium alginate and CNTs. Their results demonstrated that nanocomposite gels are noncytotoxic and exhibit a modest inflammatory response. Additionally, they proposed that nanocomposite gels could be a helpful scaffolding material in tissue engineering [134]. Furthermore, the addition of CNTs to the nanocomposite and gelatine gel was discovered to be photo-patternable, enabling simple microscale production without needing a problematic process [131–134]. This is because, when used in several complicated 3D-biomimetic tissue engineering applications, nanocomposite gels demonstrated tenable rigidity and high compatibility, as well as the ability to modulate osteogenic, neurogenic, and myogenic changes in human mesenchymal stem cells [134, 135]. Table 16.2 lists the number of polymer composites with nanostructures for soft-tissue engineering.

16.3.2 Drug Delivery System

The deadly illness known as cancer has received a lot of attention lately and is becoming more common as people get older and more urbanized. Despite advancements in medical science and biomedical engineering, administration issues with medications frequently hinder the therapeutic development of anticancer therapies [181, 182]. The development of nanostructured films with enhanced drug loading has taken the front stage in recent years. The drug loading capacity and release per stimulation were found to be ten times higher in nanoporous films fabricated with colloidal templates and nanowire-based films fabricated with anodic aluminum oxide templates than in conventional flat films [183]. In contrast to films, nanoparticles make drug loading, scaling, and processing simpler. Samanta et al. have reported a varied drug release profile based on the oxidizing agents (hydrogen peroxide, ferric chloride, and chloroauric acid) used in the manufacture of films derived from polypyrrole nanoparticles [184]. Metal-based oxidizing agents FeC13 and AuC14 increase the amount of medicine released per stimulation. Furthermore, it was demonstrated that medicines could be released at voltages as low as – 0.05 V, almost an order of magnitude lower than typically used voltages when FeC13 is used as the oxidizing molecule

Table 16.2 Lists possible combinations of nanofillers and polymer composites for Soft-tissue engineering

Polymer matrix	Nanoadditives/Fillers	Composite form	Potential biomedical application	Refs.
Chitosan	Titanium dioxide (TiO2)	Scaffold	Wound dressing	[137]
(CS)	Ag NPs + Agarose	Scaffold	Soft-tissue engineering	[138]
	Carbon dot	Scaffold	Wound dressing	[139]
	Polypyrrole (PPy)	Scaffold	Biomedical Application	[140]
	Polylactic acid (PLA)	Scaffold	Cardiac tissue engineering	[141]
	Banana peel powder (BPP)	Film	Wound dressing	[142]
	Caffeic acid (CCA) + PCL	Scaffold	Wound dressing	[143]
	GO + SWC nanohorns	Scaffold	Biomedical devices	[144]
	Montmorillonite GO + PLA	Scaffold	Tissue engineering	[145]
Gelatin	PCL	Scaffold	Blood vessel tissue	[146]
(Gel)	Gr-doped electrospun PCL	Scaffold	Nerve tissue engineering	[147]
Gel + CS	- -	Scaffold	Blood vessel tissue engineering	[148]
Collagen	Cobalt BG	Scaffold	Skin tissue engineering	[149]
(Col)	PANI Nanofibers	Film	Biomedical applications	[150]
	PCL	Scaffold	Soft-tissue engineering	[151]
	PLA + PCL	Scaffold	Soft-tissue engineering	[152]
Silk Fibroin	GO	Scaffold	Soft-tissue engineering	[153]
(SF)	GO	Scaffold	Wound dressing	[154]
Silk Sericin	PNIPAAm/PLCL/CIF	Scaffold	Wound dressing	[155]
Cellulose	PVA	Scaffold	Blood vessel tissue	[156]
	PVP + CS	Scaffold	Wound dressing	[157]
	Ultrapure alginate	Scaffold	Pancreatic islets transplantation	[158]
Alginate (Alg)	Ibuprofen	Film	Wound dressing	[159]
PCL	BG NPs	Scaffold	Wound dressing	[160]
	PBG	Film	Tissue regeneration and wound- Healing	[161]
	lignin	Scaffold	Soft-tissue engineering	[162]

(continued)

Table 16.2 (continued)

Polymer matrix	Nanoadditives/Fillers	Composite form	Potential biomedical application	Refs.
	GO	Scaffold	Neural tissue engineering	[163]
Hyaluronic acid (HA)	Silica nanohybrid	Scaffold	Soft-tissue engineering	[164]
polyurethanes	Bisphosphonate	Scaffold	Cardiac tissue-engineering	[165]
	Kaolin	Scaffold	Wound dressing	[166]
	Silver	Scaffold	Ventricular catheter	[167]
Polymeric filamentous	Endothelial cell	Scaffold	Cardiac tissue-engineering	[168]
POSS–PCU	Poly(carbonate-urea)urethan	Scaffold	Cardiac tissue	[169]
	Silver	Scaffold	Cardiac tissue	[170]
	Under the influence of nitric acid	Scaffold	Cardiac tissue	[171]
	Anti-CD 34	Scaffold	Cardiac tissue	[172]
PEDOT	Silicon Microelectrode Arrays	Film	Nerve tissue engineering	[173]
Ppy	PSS/NaDBS	Scaffold	Nerve tissue engineering	[174]
PLGA + PGA	Porcine tooth buds	Scaffold	Oral tissues	[175]
PLGA	N-hydroxysuccinimide(NHS)	Scaffold	Soft-tissue engineering	[176]
	Mg(OH)2 + renal dECM	Scaffold	Kidney tissue	[177]
Glycerol (Gly) + Gellan gum (Gg)	HNTs	Scaffold	Soft-tissue engineering	[178]
PEG	NaCl	Scaffold	Pancreatic islets transplantation	[179]
PVA	Silica	Scaffold	Soft-tissue engineering	[180]
PCE	Graphene	Scaffold	Soft-tissue engineering	[181]

[185]. Similarly, cyclodextrin, polylactide, and chitosan are included in a graphene-based composite of polymers that improve the biocompatibility of the matrix for the controlled release of therapeutic molecules, and folic acid is a ligand that targets cell membranes and is used as an efficient means of delivering anticancer medications [186]. The anticancer medication doxorubicin (Dox) has been produced using transferrin (Tf)-conjugated polyethylene glycol (PEG)-based nanoscaled graphene oxide. Tf-PEG-GO-Dox investigations in vivo and in vitro demonstrated a notable increase in therapeutic efficiency in transporting Dox to brain tumors (Fig. 16.6) [187]. With

GO nanoparticles, Zhao et al. have created a 3D-biocompatible, reduction-responsive cytamine (Cy)-modified PEGylated alginate (ALG) that is loaded with Dox (GON-Cy-ALG-PEG) [188]. The in vitro release demonstrated that the platform could, in response to glutathione (GSH) (Fig. 16.7), both detach the Cy-modified PEGylated alginate (Cy-ALG-PEG) moieties and stop the loaded DOX from leaking under physiological conditions [188]. Depan et al. created a folic acid-conjugated chitosan–GO composite, while Qin et al. created a folic acid-GO-poly (vinyl pyrrolidone) composite to deliver doxorubicin, an anticancer medication. It is implied that GO and doxorubicin interact through hydrogen bonds in the in vitro drug delivery of both composites, which exhibit pH dependence. Owing to the decreased interaction between doxorubicin and composites, the composite exhibits enhanced drug delivery at low pH levels of roughly 5.3–5.5 [189, 190]. The anticancer drug doxorubicin has been delivered for cancer treatment using several biocompatible GNS-based composites, such as GO–Fe3O4, chitosan–GO, PEG–grafted GO, and GP–thiol grafted pluronic [189, 190].

Conductive polymers can be added to nanoadditive-laden polymer composites to make them permeable. Their delocalized charge carriers help linked molecules disperse, which makes them appropriate for drug release [41, 191]. Numerous therapeutic drugs, including dexamethasone, dopamine, 2-ethylhexyl phosphate, nerve growth factor (NGF), heparin, and naproxen, have been effectively released from polymers that have been synthesized by electrochemical synthesis, graft co-polymerization, and electro-polymerization in studies [41, 191]. These polymers are subjected to electrical stimulations, such as CV-triggered release [41, 191]. PPy effectively released neurotrophin-3, NGF, and brain-derived neurotrophic factor (BDNF), leading to the growth and differentiation of neurons [192]. In one experiment, heparin was released from a poly (vinyl alcohol) hydrogel formed on a PPy film; in another, heparin was exposed on the surface following 90 s of exposure to an electrical potential between 0.4 and 0.7 V [193, 194]. For the specific voltammetric analysis of dopamine mixtures with uric and ascorbic acids, as well as samples of human urine with real interferents, Fabregat G et al. developed electrodes by combining three-layered films made of different conductive polymers, such as PNCPy, PNMPy, or PEDOT, and by using electrochemical synthesis techniques [195]. Voltammetric investigations of mixtures of solutions revealed that electrodes consisting of a poly (N-methyl pyrrole) intermediate layer and alternating internal and external PEDOT

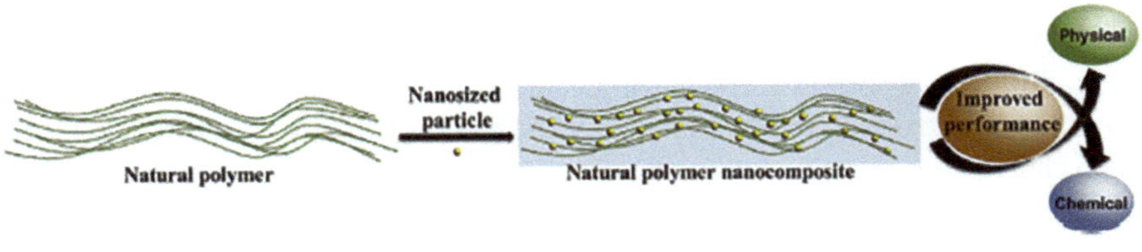

Fig. 16.6 The combination of natural polymer with nanosized particles providing natural polymer-based nanocomposites with improved physical and chemical properties, promoting enhanced biological performance [297]. Copyright@2019, Elsevier Books

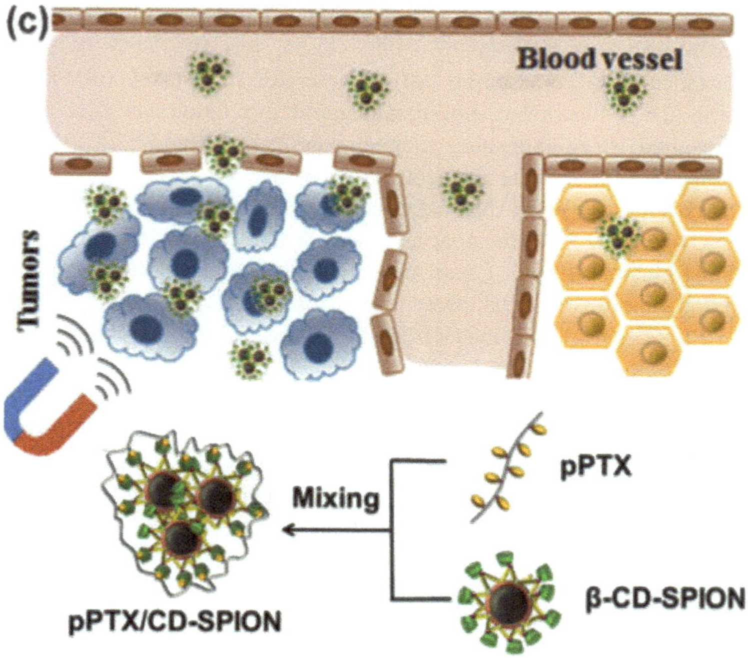

Fig. 16.7 Schematic illustration of the fabrication of pPTX/CD-SPION nano-assemblies and their application in magnetically guided drug delivery [298]. Copyright@2016, Journal of Controlled Release

layers demonstrated excellent sensitivity and resolution [196]. Furthermore, the three-layered electrodes' sensitivity somewhat increased when gold nanoparticles (AuNPs) were applied to their surface [195, 196].

Winter et al. employed an oxidative mini-emulsion polymerization technique to synthesize PEDOT nanoparticles while stabilizing the resulting HA effectively [197]. This method improved the hydrophilicity and bioactivity of the nanoparticles [197]. With a higher amount of oxidant, it was demonstrated that the conductivities of the nanoparticles were up to ten times larger than those of pure PEDOT: PSS [191, 196]. This easy-to-manufacture approach with better features increased the promise of conductive polymers for flexible biomedical applications, such as cardiac patches, nerve conduits, and tissue scaffolds [191, 196]. By employing conductive polymer nanoparticulate backbones, Nassab et al. could controllably release insulin, a high molecular weight polypeptide [198]. A simple Langmuir-type adsorption model was used to simulate the interactions between insulin and the polymer scaffold. They found that the proportion of medication loading might reach 51 weight percent by varying the NPs to insulin ratio. Using a model of in vivo mice, the electrically triggered insulin release's activity was verified [198]. Superparamagnetic polymer nanocomposites are especially appealing because they can be targeted and confined using an external magnetic field to deliver magnetically targeted medication (shown in Fig. 16.7) [199].

For instance, the co-assembly of polymerized paclitaxel (pPTX) and β-cyclodextrin (β-CD) decorated superparamagnetic iron oxide nanoparticles (β-CD-SPION) resulted in an intriguing magnetic delivery system [200]. The potential for drug delivery using a nanocomposite hydrogel (NCH) based on poly(N-isopropyl acrylamide) (PNIPAM), alginate interpenetrating polymer, and graphene oxide–Fe3O4 (GO–Fe3O4) nanoparticles was described by Cao et al. [201]. This hydrogel can transport drugs at near-infrared (NIR) light, magnetic, and pH levels. Increasing temperature over the lower critical solution temperature under NIR light and alternating magnetic field, the GO-Fe3O4 nanoparticles incorporated in the interpenetrating polymer stimulate doxorubicin (DOX) release and facilitate PNIPAM hydrogel deswelling [201]. Notably, NIR light is favored because it better penetrates deep tissues and harms normal tissues less [201]. Compared to pure PNIPAM hydrogels, the incorporation of alginate into PNIPAM hydrogels demonstrated improved mechanical properties, and the presence of alginate hydrogels caused the NCH system to release more quickly in an acidic environment [201]. Table 16.3 presents a concise overview of the polymer composites used in DDS.

16.3.3 In Biosensors

Analytical tools for analyte detection are biosensors. A biosensor typically consists of a biological sensing device and a transducer. The sensing material chemically reacts with the target analyte to send a signal to the transducer, ultimately transforming the input signal into an electrical signal. The biological sensing element may consist of cell receptors, enzymes, antibodies, and DNA probes interacting with a specific analyte [1, 223, 224]. Conducting polymers are frequently used as transducers to integrate signals from various biological sensing components. For example, conducting polymer-based biosensors can also be modified using CNTs [224]. Luo et al. used CNTs to modify a PANI-based biosensor. The CNTs may enhance the mechanical robustness and conductivity of the PANI biosensor [225]. It was demonstrated in this study that CNTs may improve the efficiency of enzyme absorption and immobilization [225]. Horseradish peroxidase (HRP) was added to the PANI layer during the electrochemical polymerization process utilizing glassy carbon electrodes [226]. Subsequently, carbon nanotubes were added to one of the electrode's surfaces by dropping uniformly distributed dimethyl formamide-containing carbon nanotubes. Researchers found that the amount of HRP incorporated into the biosensors increased when carbon nanotubes were added to the PANI electrodes. The PANI/CNTs and PANI biosensors were compared for performance, and the biosensor with carbon nanotubes produced a stronger signal [223–226]. This finding may have been caused by the stable bonding of HRP with PANI and CNTs [223–226]. According to Kuila et al. [227], Nguyen and Berry [228–230], GNS-based polymer composite is employed in electronic and sensor applications due to its electroactivity, transparency,

Table 16.3 Presents a concise overview of the polymer composites used in DDS

Polymer matrix	Additives	Drug	Refs.
Gelatin	GNS	Drug-delivery	[203]
CS	Graphene nanogel	Drug-delivery	[204]
Ppy	Sodium dodecyl sulfate (0.1 M)	Insulin (5 mg/mL)	[205]
PEDOT	Electrospun nanoscale fibers	Anti-inflammatory drug	[206]
Neutral PEDOT	Sodium dodecyl sulfate (0.009 M)	Curcumin (0.027 M)	[207]
Ppy	Sodium dodecyl sulfate (0.1 M)	Fluorescein sodium salt and Methotrexate (1–2.2 mM)	[208]
Ppy	PVA	Heparin release	[209]
Anionic and	DDAB (0.162 M)	Fluorescein and daunorubicin	[210]
Cationic Ppy	and decanol (0.236 M)	(1–2 mM)	
Ppy	Sodium dodecyl sulfate (0.1 M)	Fluorescein, piroxicam and insulin (3 mg/mL)	[211]
CMCS + OAlg	CMs + HAp NPs	Drug delivery and bone tissue Engineering	[212]
PLGA	Sparfloxacin	Ocular drug delivery	[213]
PLA	Retinal pigment epithelium cells	Ocular drug delivery	[214]
Polyurethane	Bevacizumab (1.5 mg)	Ocular drug delivery	[215]
CS	Aluminosilicate	Drug delivery	[216]
CS	ZnO quantum dots	Drug delivery	[217]
PMMA	MMT	Verapamil release from MMT-Verapamil hybrid	218]
U-PEO	γ-Fe_2O_3	Cancer therapy	[219]
U-PEO	CeO_2	Biomedical and optical devices fields	[220]
PEG-PLA	Laponite	Anticancer Drug-delivery	[221]
PEG	Ce6/Dox/GO	Photodynamic anticancer therapy	[222]
Polyoxypropylene	MMT	Colorectal therapy	[223]
Polystyrene	Fe_3O_4 + GO	Targeted drug delivery applications	[224]
PVP	NGO + FA	Chemo-photothermal therapy	[225]
ALG-PEG	DOX loaded NGO + Cy	Tumor therapy	[226]

and flexibility. Because of this, GNS-based polymer composite has been applied to a wide range of biosensing schemes, such as biomolecular labels, electrochemical biosensors, impedance biosensors, electrochemiluminescence, electrochemical immunosensors (Fig. 16.5), and transducer in bio-field-effect transistors [228–232]. The list of sensing polymer composites as biosensors is shown in Table 16.4.

Biosensors can detect various biological compounds by encapsulating various biological sensing chemicals inside the CPs. PPy has been extensively studied for biosensor applications because of its strong electrical conductivity and resistance to environmental changes [253]. Furthermore, it is easily synthesized using chemical or electrochemical methods [253]. Gooding et al. reported a label-free immunosensor that uses an anti-rabbit IgG antibody on the polypyrrole's surface to detect the IgG

Table 16.4 List of sensing polymer composites as biosensors

Bio sensing Polymer Composites	Types of sensing material	Refs.
Chitosan (CS-ferrocene/glucose oxidase (GOX)/GO Glucose		[237]
GNS-AuNPs-GOX-chitosan	Glucose	[238]
GOX-GNS-CS	Glucose	[239]
GNS-AuNPs-CS	H_2O_2 and O_2	[238]
Hemoglobin/CS-ionic liquid-ferrocene/	H_2O_2	[240]
GNS/ GCE/ GNS-PANI	Dopamine	[241]
GO-AuNPs-CS/silica gel-GCE	Dopamine and uric acid	[242]
GNS-CS-GCE	Acetaminophen	[243]
GNS-ionic liquid-CS	Adenine and guanine	[244]
GO-PANI-GCE	DNA	[245]
Polypyrrole-polyvinyl sulfonate films (PPY-PVS)	Urea (urease)	[246]
Polypyrrole and Diaminonaphthalene	Cholesterol (cholesterol oxidase/esterase)	[247]
polyaniline/cholesterol (oxidase (ChOx)/ esterase (ChEt)) films	Cholesterol (cholesterol oxidase/esterase)	[248]
PPy/ChOx/ChEt films	Cholesterol (cholesterol oxidase/esterase)	[249]
GOX/poly(*o*-amino benzoic acid) (PAB)	Glucose	[250]
PPy/GOX	Glucose	[251]
PPy film	DNA hybridization	[252]
Au–Ag- PPy film	DNA	[253]
Polythiophene (PTh) film	DNA hybridization	[254]
PANI/Lactate oxidase (LOD)–lactate dehydrogenase (LDH)	L-lactate (LOD/LDH)	[255]
Polypyrrole–Polyvinylsulphonate film	L-lactate (LOD/LDH)	[256]
PEDOT:PSS/PU (aqueous dispersion)	High pressure sensitivity	[257]
PAni nanobre/bacterial	Electrical signals on cell activities	[258]

antigen of rabbits [254]. When the ion flux within and outside of the CP membrane changed in response to the antigen's interaction with the polymer that the antibody had changed, the observed signals became visible [254]. Gao et al. created biosensors that combine Ppy and CNTs to detect glucose [255]. They used CNT-aligned substrate materials to provide a large surface area. Electropolymerization of PPy was used to cover the CNTs [255]. The electrolyte solution, monomers, and an enzyme known as glucose oxidase—a biological sensor for glucose—were mixed during the polymerization process. Thus, the polypyrrole contained the glucose oxidase. Signal transduction was made possible by the tight bond between the glucose oxidase and the polypyrrole film. As a result of the electrochemical reaction, hydrogen peroxide (H_2O_2) was oxidized, and the electrical current generated during this process was measured using an amperometric technique [255, 256].

Abidian et al. reported a novel type of glucose biosensor using different conducting polymers [257]. PEDOT was utilized in this innovative glucose biosensor due to its exceptional conductivity and strong chemical stability. Electrochemical synthesis of PEDOT nanofiber and film biosensors was done using platinum (Pt) brain microelectrodes [257]. A PEDOT film was directly applied to the Pt microelectrode array surface. PEDOT nanofiber biosensors were created by electrolyzing PEDOT on Pt microelectrodes and near electrospun nanofibers. During the electropolymerization processes, the glucose oxidase enzyme was immobilized on PEDOT films and nanofibers [257, 258]. The two biosensors utilized an electrochemical quartz crystal microbalance (EQCM)' and variations in glucose oxidase incorporation were compared [259]. Two biosensors were used to measure the amperometric change in current to determine their sensitivity when additional glucose was added to the electrolyte solution. They found that compared to PEDOT film biosensors, PEDOT nanofiber biosensors demonstrated greater sensitivity and more glucose oxidase integrated into the nanofibers [256–259]. Gu et al. investigated boron-doped diamond (BDD) DNA biosensors with polyaniline/polyacrylate modifications [260]. Diamond electrodes were coated thinly with polyaniline/polyacrylate using an electrochemical polymerization technique. Carboxylic groups in polyaniline/polyacrylate were used to immobilize the DNA-sensing probes [260]. DNA probes may identify DNA whose nucleotide sequences are complementary to their own [260].

16.3.4 Bioimaging

Bioimaging is an indispensable tool for tracking drugs and monitoring tumors in biomedicine. It is a potent technique that allows one to see biological processes in situ at the cellular or even subatomic level [261]. The supramolecular polymer nanocomposites-based bioimaging probes frequently simultaneously possess numerous significant qualities, including stimulus responsiveness, target selectivity, biocompatibility, and variety, compared to tiny molecular and standard polymer probes. Differentiable bioimaging probes can be prepared by adding specific contrast agents or functional components [261, 262]. For example, fluorescent components can be used for fluorescence imaging [263], infrared absorbing agents for photoacoustic (PA) imaging [264], paramagnetic components for MRI [98], and X-ray absorbing agents for computed tomography (CT) imaging [265]. Fluorescence imaging is currently a hot research topic for QD-reinforced supramolecular polymer nanocomposites. Wang et al. [266], for instance, created a matrix metalloproteinase (MMP)-sensitive fluorescent imaging probe for cancer cells by encasing QDs within supramolecular gelatin nanoparticles (SGNs). For the preparation process in this work, a microfluidic platform was used. Size-controlled QDs@SGNs were produced through modifications to constituents' physiochemical properties and flow rates. Due to their pre-crosslinking with glutaraldehyde, these QDs@SGNs nanocomposites showed adequate stability. Due to SGNs' sensitivity to MMP, this probe has a high tumor cell specificity. Cells treated with QDs@SGNs have fluorescence intensity 1.5 times higher in the absence of MMP inhibitors than in the presence of MMP inhibitors [263, 266]. Furthermore, QDs can be released and internalized to light up cells if MMPs break down the gelatin matrix. This work presents a typical example of proteinase-responsive fluorescence imaging, wherein the QDs@SGNs probes concentrate at specific target sites and then self-activate their optical signals [263, 266]. An everyday use for UCNPs is fluorescence imaging due to their distinct upconversion luminous characteristics. Lin et al. [267] coated UCNPs with the self-assembled poly(acrylic acid) (PAA) shells to create a hybrid luminous probe (Fig. 16.6b) and demonstrated a highly efficient upconversion luminescent imaging capacity.

Magnetic supramolecular polymer nanocomposites can be employed as contrast agents to improve MRIs. Ai et al. [268] synthesized an ultrasensitive MRI contrast agent for liver imaging by using Mn-doped superparamagnetic iron oxide (Mn-SPIO) nanoparticles as a filler (Fig. 16.8). This work used amphiphilic methoxy poly(ethylene glycol)-b-poly(3-caprolactone) (mPEG-b-PCL) to assist hydrophobic Mn-SPIO nanocrystals self-assemble into tiny clusters, which ultimately generated micellar clustering nanocomposites [268]. An in vivo magnetic resonance imaging investigation revealed that the mice received intravenous delivery of Mn-SPIO micelles and that the temporal window for enhanced MRI extended up to 36 h. T2-weighted signal intensity in the liver dropped by approximately 80% at 5 min. These benefits enable the application of these nanocomposites to detect minor liver abnormalities [268].

The potential for improved anticancer efficaciousness has drawn much interest to the simultaneous delivery of drugs and bioimaging functions for both therapeutic and diagnostic applications. Li et al. [269] provided a typical example of such a system when they created a unique multifunctional supramolecular hybrid nanocarrier with simultaneous cellular imaging capabilities and synergistic gene and medication co-delivery (Fig. 16.7) [269]. This design allows for the fabrication of a core–shell fluorescent probe (β-CD-OEI@QD) by electrostatically coating a red QD with a cationic β-CD polymer (β-CD-OEI) in a star shape. The resulting β-CD-OEI@QD complex can act as a carrier for the co-delivery of paclitaxel (PTX) and gene by utilizing the strong guest binding ability of the β-CD cavity and their positive surfaces. Effective cancer therapy is made possible by a significant increase in gene expression brought

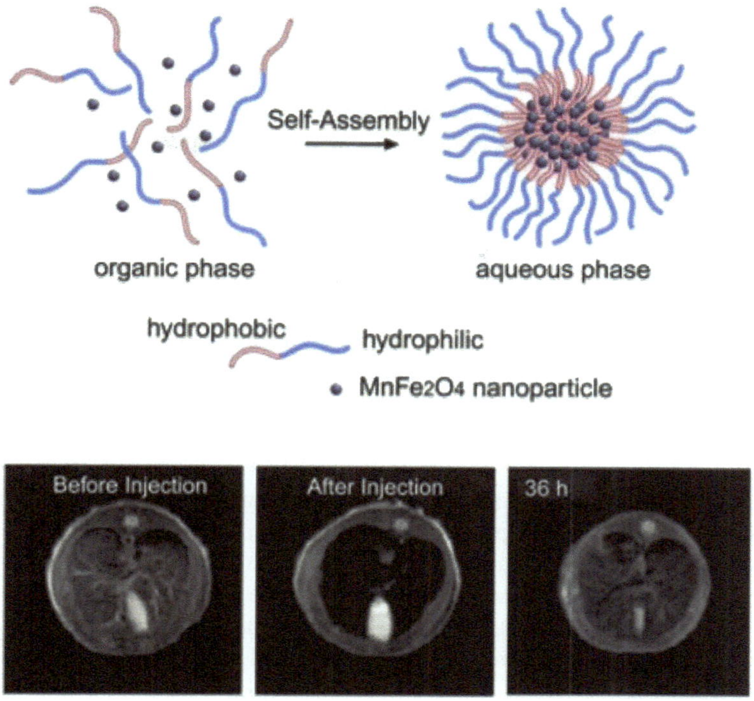

Fig. 16.8 The formation of hybrid micelles via self-assembly of Mn-dope superparamagnetic iron oxide (Mn-SPIO) nanoparticles and amphiphilic polymer (mPEG-*b*-PCL). MRI imaging results with the Mn-SPIO micelles [299]. Copyright@2009, Biomaterials

about by the cooperative behavior of PTX and the gene [269]. Chu and colleagues [270] have documented the preparation of multifunctional theranostic microcapsules using layer-by-layer self-assembly of magnetite (Fe3O4) and supramolecular polymers. These microcapsules exhibit dual functionality, including pH-responsive magnetic resonance imaging and UV light-responsive drug administration [270].

16.3.5 Antiviral and Antibacterial Applications

16.3.5.1 Antiviral

Recently, antiviral, antibacterial, and antifungal agents known as polymer nanocomposites (PNCs) have been created for a wide range of biomedical applications and have found use in medical devices such as personal protective equipment (PPE) [271, 272]. Since the discovery of COVID-19 in December 2019, polymer composite's antiviral activity has risen to the top of the list of research topics in polymer, nanotechnology, and materials science [272]. This significantly affects all spheres of life, including social, educational, and economic. polymer composite's scope activity, centered on nanoscale measurements, gives it a great chance to combat coronavirus [272]. Furthermore, the components and structure of the coronavirus are nanoscale, supporting polymer composite's strong recommendation. polymer composite has been applied in a variety of methods to COVID-19 therapy [272, 273]. One possible method to avoid virus infection of the respiratory system is to incorporate nanoadditives into personal protective equipment (PPE). A further approach would be to use inhalators to introduce nanoparticles into the drug delivery system for the pulmonary system. Additionally, polymer composite targets viral S protein in the virus membrane or angiotensin-converting enzyme 2 (ACE2) receptors to inhibit viruses from binding to cells. For example, blood plasma from recovered COVID-19-infected patients is collected using polyethersulfone (PES)-, polymethylmethacrylate (PMMA)-, and polypropylene (PP)-based polymer composites, which are then given to newly infected patients [209, 272, 273]. In order to develop, produce, and investigate antiviral polymer composite for use in coating applications for life goods and human cells, several filled nanoparticles, including silver, copper oxide, titanium dioxide, and zinc oxide nanoparticles, have been used [1, 41, 209, 272, 273]. The basic mechanisms by which antiviral activity functions are the formation of a barrier between the virus-host cells and other body cells or the application of repulsive solid forces to the product and surface surfaces to repel virus droplets. These factors may be applied today to reduce coronavirus infections and better manage economic and social life activities since they have been shown to play a substantial role in many viruses' activities in vivo and on product surfaces [209, 273].

16.3.5.2 Antibacterial Activities

Nanotechnology is now being used in biomedical fields to combat and prevent infectious diseases by employing functional materials at the atomic scale. These days, a wide range of functional materials have been created by materials scientists and engineers to serve as scaffolds in tissue engineering and prevent infections of the sinuses, chronic otitis, lungs, biliary tracts, endocarditis, urinary tracts, osteomyelitis, endocarditis, or chronic wounds [274–276]. However, compared to nanoadditive polymer composite material, none exhibit bacterial inhibitory activity. Using various techniques, antimicrobial and antibacterial additives are employed to change the polymer matrix to create polymer antibacterial materials. The most popular methods are those that use biodegradable and biocompatible polymers that are loaded with antibacterial particles (Fig. 16.9) [275, 276].

For example, at various Ag concentrations, PLA/Ag ultrafine fibers demonstrated efficient antibacterial activity (microorganism decrease) against S. aureus and E. coli [277]. This nanocomposite has an antibacterial efficacy of 94–58% in 12 h and lasts more than 20 days. Because of this, these fibers have useful uses in anti-adhesion membranes and wound dressings [277]. Argon surface plasma treatment (APT) of PP/Ag and PP/Cu nanocomposites enhances their antibacterial properties against pathogenic bacteria, according to Espana-Sanchez et al. [278]. This is because

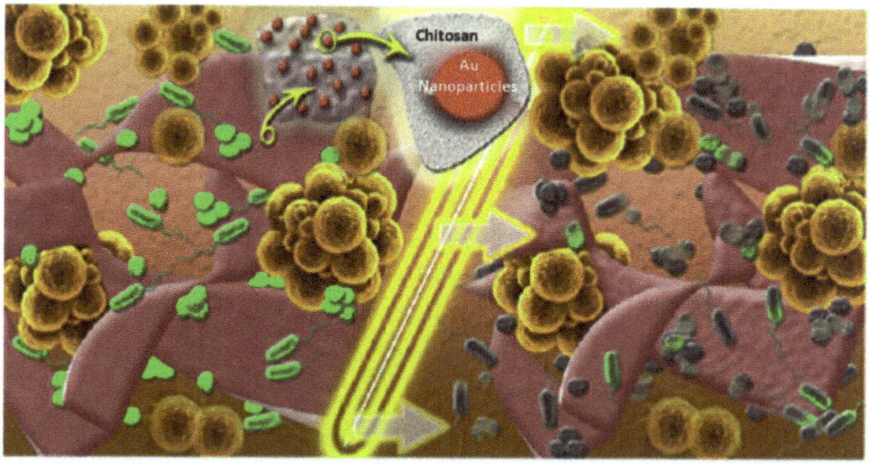

Fig. 16.9 Mechanism of antibacterial activity of gold nanoparticles released from chitosan scaffold [300]. Adopted with permission from Ref. [133] copyright@2020, Elsevier Books

APT increases the surface increment of exposed metal nanoparticles and increases the roughness and hydrophilicity of the film surface. In situ polymerization was also used to create PE nanocomposites with antibacterial silver nanoparticles [278]. According to reports, compared to clean PE, the sample with a higher concentration of silver exhibits a greater release of silver ions. Other investigations have reported the antibacterial characteristics of PE/Ag and PE/Cu nanocomposites [277, 278]. In a different study, zinc oxide (ZnO) nanoparticle-filled, flexible, osteoconductive, and antibacterial poly(caprolactone) (PCL) films were created via electrospinning [278]. Porous films with regulated mechanical characteristics were created to use guided tissue regeneration (GTR) to treat periodontitis. In a defect made on a rat model, the PLC/ZnO membrane is positioned between the gum flap and the alveolar bone. The addition of ZnO enhanced the periodontal membrane's osteoconductivity and antibacterial activity [279]. Similarly, it has been demonstrated that chitosan materials containing selenium nanoparticles are efficacious against burn wound infections [280]. Furthermore, scaffolds for skin tissue regeneration based on silk fibroin-collagen and loaded with titania nanoparticles were reported [281]. For skin regeneration applications, ciprofloxacin-modified hybrid polyurethane-polylactide porous scaffolds were developed as an antibacterial scaffold [282]. Due to its exceptional antibacterial properties, TiO2 was added to the scaffolds to act as a strengthening agent to enhance cell development [283]. For example, Beheraa et al. synthesized an economical and effective chitosan (CS)/TiO2 composite membrane to develop the survival and proliferation of mouse L929 fibroblast cells and wound healing [283]. The membranes exhibited superior mechanical strength, porosity, flexibility, and crystallinity upon formation. Regarding Staphylococcus aureus, the composite materials demonstrated outstanding antibacterial activity [282, 283]. Comparably, substances like banana peel show promise due to their high mineral content, phenolic structure, antioxidant properties, and ability to combat bacteria, fungi, and yeast [284]. Kamel et al. recently used Banana peel powder (BPP) and chitosan to create novel nanocomposite membranes [284]. BPP acts as a strengthening agent and ionic crosslinker for chitosan. After adding BPP, chitosan's swelling characteristics decreased. Moreover, glycerol shortened the revolving units' relaxation time and molar volume, decreasing the frequency. By raising the BPP loading, the frequency reduction effect was strengthened. The polymer/filler interaction, which lengthens the relaxation period and the relaxed units, could cause this decrease. The membrane CS-BPP at 10 weight percent BPP showed the highest sensitivity and activity against Candida albicans. Chitosan is an exception to the rule that most polymers lack intrinsic antibacterial qualities; its antibacterial activities have been demonstrated [284]. Caffeic acid also exhibits a variety of characteristics, including antiviral, antioxidant, antibacterial, anticoagulatory, and anti-inflammatory actions, and two vicinal hydroxyl groups on an aromatic ring, which is known as a catechol group [285]. Using chitosan (CS) and chitosan–caffeic acid (CCA) electrospun PCL, Oh et al. [285] created fibers devoid of bead formation for a wound healing strategy. PCL/CS, PCL/CCA, and 1.20 ± 1.22 μm were the average diameters, while 0.94 ± 0.68 μm was the average diameter for PCL fibers. It was highlighted by in vitro measurements of morphology, cell viability, and antimicrobial behavior that PCL/CCA mats can encourage normal human dermal fibroblast-neonatal (NHDF-neo) cell spreading and proliferation with remarkable microbial behavior against S. aureus concerning PCL and PCL/CS fibers [285]. The high mechanical strength, water-holding capacity, conformability, and nanocellulose (NC) elasticity have led to its usage in wound-care dressings [286]. Through the use of their high surface-to-volume ratio, NC particles were able to promote cell migration and proliferation. In contrast to pathogenic bacteria, the NC particles exhibited weak antimicrobial activity [286]. Poonguzhali et al. created a promising bio-nanocomposite for in vitro wound dressings using NC, poly(vinyl pyrrolidone) (PVP), and chitosan [283]. The hydroxyl groups of nanocellulose formed hydrogen bonds with the carbonyl groups of PVP and the amino and hydroxyl groups of chitosan to form a biocompatible composite with nanocellulose particles [283]. PVP and chitosan were morphologically miscible and compatible with each other. The outcome showed a notable improvement in the mechanical and barrier properties of PVP/CS/NC films [287]. The swelling behavior was better at 5 weight percent NC. At 3, 5, and 10% weight percentage NC, the tensile strength rose by 35.6 ± 5.8, 38.4 ± 2.7, and 39.7 ± 6.9, respectively. Blends containing 3 and 5 weight percent NC also demonstrated outstanding antibacterial activity against gram-positive and gram-negative bacteria, good cytotoxicity, and blood compatibility [287]. Silk sericin, a naturally occurring protein generated from silkworms with important qualities like antibacterial, UV, and oxidation resistance, as well as moisture absorption and release, is another example of a natural material for wound dressing [288]. Yang et al.'s [289] use of silk sericin composites allowed for an expedited wound healing procedure. The potential of a nanoclay hydrogel nanocomposite with cross-linked lithium magnesium silicate hydrate (LMSH) and poly((N-isopropyl acrylamide)(NIPAm)/LMSH) (HSP) as a wound dressing was examined by the authors [290]. The cross-sectional structures changed from a honeycomb to layered morphology because of the enhanced hydrophilicity. In contrast, the hydrogels, balanced at 37 ∘C, had a pyknotic morphology with pore diameters ranging from 0.8 to 1.5 μm. When the concentration of sericin rose, the water vapor transfer rate (WVTR) values increased by 100 g/(m2day), and the hydrophilic qualities enhanced the reduction of the compression modulus by 3.859 kPa. Furthermore, the addition of sericin led to the development of the antibacterial activity [290]. Furthermore, fiber composites exhibit hemostatic, absorption, and semipermeability properties [291]. This idea states that a powerful wound dressing was created by electrospinning thermosensitive poly(N-isopropylacrylamide) (PNIPAAm) with poly(L-lactic acid-o-ε-caprolactone) (PLCL) and the antibiotic ciprofloxacin (CIF) [291]. After adding CIF, the diameter of the cylindrical nanofibers shrank. The wettability of the PNIPAAm/PLCL nanofibers declined at a temperature of about 32 °C, and the water angle of the produced mats varied with temperature. In vitro drug release measurements showed that CIF was released from the mats for 200 h [291]. The drug-loaded mats demonstrated outstanding antibacterial activity against S. aureus and E. coli, with high biocompatibility and proliferation of L929 fibroblasts. PCL still lacks bioactivity despite having distinct qualities, including a prolonged breakdown rate, excellent hydrophobicity, and crystallinity [292, 293]. Applying bioactive inorganic additives, including bioactive glasses (BGs), is one of the most popular methods to lessen this disadvantage [292, 293]. A prime example is the electrospun PCL membrane processing membranes for soft-tissue wound healing by Moura et al. [294] using BG nanoparticles and silver–cobalt-doped bioactive glass nanoparticles (DB-NPs). Subtly adding the nanoparticles caused the fibers to lose weight more quickly. Silver and cobalt ions produced the antibacterial and angiogenic qualities of these fibrous membranes, making the processed DB-NPs/PCL fibers attractive candidates for soft-tissue repair [294]. It was discovered that adding the nanoparticles improved the final tensile strength and decreased the elongation at fracture. These fiber composites could be an alternative for wound healing in tissue as specialized as skin, given the tensile properties of the skin. For applications in wound healing, Wang et al. [295] recently synthesized SF/graphene oxide (GO) nanofibers with a bioinspired nanostructure. Graphene oxide enhanced SF nanofibers' antibacterial and biocompatibility properties [295]. In addition to being used as scaffolds, 3D-printed materials can be used as external prostheses or to create inexpensive medical devices, such as surgical instruments, with antibacterial or antimicrobial qualities [149]. J.M. Zuniga provided a concise study that included a real-world example of creating a prosthetic material using PLACTIVETM, a brand-name antibacterial PLA filament [149]. Little amounts of copper particles (1–3%) were used to achieve the antibacterial characteristics and demonstrated that PLACTIVETM containing 1% antibacterial nanoparticle additions was 99.99% efficient against Staphylococcus [149, 296].

16.3.6 Challenges and Future Prospective

Recent research in synthetic and natural materials has concentrated more on composites than on using individual materials in biomedical implants and devices. Differentiating between materials with distinct functionalities is challenging. Researchers still need to be able to compare these materials in vivo or perform in vivo characterization trials. In parallel, many people are steadily creating new composites that would satisfy the distinct mechanical and biocompatibility needs of every area of the body. Pure synthetic polymeric materials have already peaked in their application to medical implants and devices. Combining several materials to create composites that offer novel functionality or better mechanical strength and flexibility is the current trend in this subject. Furthermore, non-biodegradable implants face problems such as stress shielding, wear debris, and the possibility of surgical removal following use, raising concerns about their biodegradability. A more comfortable and practical healing process can be accomplished by using polymers that can break down in the body and lessen the need for the implant itself while promoting the growth and self-supportability of the surrounding muscle or bone. For this reason, biodegradable bone implants, for instance, are being researched [144]. These worries about biodegradable materials include their mechanical strength, the length of time needed for degradation, and the waste products created when they come into touch with human tissue. For example, silk fibroin films have been shown to bind well to mammalian cells, and biodegradable silk has been used as a suture material for millennia.146 As a result, they have been applied to enhance cell adhesion and create composites that aid in creating bones [147, 148]. Bi-layer silk fibroid scaffolds were used by Algarrahi et al. 149 during onlay esophagoplasty in rats, and it was noted that these scaffolds encouraged the development of innervated, vascularized epithelial and muscle tissues within the implantation sites. This indicates that silk can expedite the healing process.

The field of shape memory, which deals with materials that may change their physical properties in response to external input, presents a novel facet of biomedical devices. These intelligent materials are helpful for minimally invasive operations and can be tailored to specific uses. The material's mechanical qualities would be crucial in this regard. Many synthetic polymers, such as polyurethane shape memory polymer, polytetrafluoroethylene, polyacrylonitrile, etc., are currently employed as materials for these "smart devices" [151, 152]. Utilizing shape memory polymers in medical implants and devices still presents several difficulties. However, increased research on this area of shape memory polymers is anticipated to lead to further advancements, given the growing need for minimally invasive operations. With all of the attention that polymeric materials are receiving in the medical business due to the emphasis placed on health care globally, the future of these materials seems bright. Aside from that, research is still being done to develop implants and devices with increased functionality, such as biodegradability and MRI safety. Because so many different materials may be combined and yet be compatible, the possibilities for material combinations as composites are practically limitless. Worldwide, efforts are continuously underway to develop new composite materials, including blends of synthetic and natural polymers that may one day perform mechanical tasks akin to human bodies. An increasing number of applications are becoming possible thanks to advancements in fabrication technologies that use these polymeric materials to fabricate unique parts quickly and affordably. In retrospect, with the development of numerous medical gadgets and implants, what was once known to cause permanent malfunction has been reduced to limited incapacity with greater comfort. A significant turning point has been reached for the biomedical sector (including antiviral, antibacterial properties, bioimaging, medication delivery, tissue engineering, and biological sensors). It is possible to ultimately realize the goal of improving people's quality of life if polymeric material research and development continue.

References

1. Ramakrishna, S., Mayer, J., Wintermantel, E., Leong, K.W.: Biomedical applications of polymer-composite materials: a review. Compos. Sci. Technol. **61**(9), 1189–1224 (2001)
2. Oladapo, B.I., Zahedi, S.A., Omigbodun, F.T.: A systematic review of polymer composite in biomedical engineering. Eur. Polymer J. **154**, 110534 (2021)
3. Zamel, D., Khan, A. U., Khan, A. N., Waris, A., Ilyas, M., Ali, A., & Baset, A. (2023). Regenerated Cellulose and Composites for Biomedical Applications. In *Regenerated Cellulose and Composites: Morphology-Property Relationship* (pp. 265–311). Singapore: Springer Nature Singapore.
4. Praharaj, R., Mishra, S., Misra, R.D.K., Rautray, T.R.: Biocompatibility and adhesion response of magnesium-hydroxyapatite/strontium-titania (Mg-HAp/Sr-TiO2) bilayer coating on titanium. Mater. Technol. **37**(4), 230–239 (2022)
5. Dziadek, M., Stodolak-Zych, E., & Cholewa-Kowalska, K. (2017). Biodegradable
6. ceramic-polymer composites for biomedical applications: A review. *Materials Science and Engineering: C*, *71*, 1175–1191.
7. Christman, K.L.: Biomaterials for tissue repair. Science **363**(6425), 340–341 (2019)
8. Senra, M.R., Marques, M.D.F.V., Monteiro, S.N.: Poly (Ether-Ether-Ketone) for Biomedical Applications: From Enhancing Bioactivity to Reinforced-Bioactive Composites—An Overview. Polymers **15**(2), 373 (2023)
9. Woźniak-Budych, M. J. (2021). Polymeric membranes for biomedical applications. Physical Sciences Reviews, (0), 20210052.
10. Zagho, M.M., Hussein, E.A., Elzatahry, A.A.: Recent overviews in functional polymer composites for biomedical applications. Polymers **10**(7), 739 (2018)
11. Guo, Z., Poot, A.A., Grijpma, D.W.: Advanced polymer-based composites and structures for biomedical applications. Eur. Polymer J. **149**, 110388 (2021)
12. Zhang, X., Geven, M.A., Grijpma, D.W., Gautrot, J.E., Peijs, T.: Polymer-polymer composites for the design of strong and tough degradable biomaterials. Materials today communications **8**, 53–63 (2016)
13. Choudhary, R., Saraswat, M., & Venkatraman, S. K. (2019). A fundamental approach toward polymers and polymer composites: Current trends for biomedical applications. *Polymer nanocomposites in biomedical engineering*, 1–28.
14. Saini, R. K., Bajpai, A. K., & Jain, E. (2017). " 13 Advances in bionanocomposites for biomedical applications. *Biodegradable and biocompatible polymer composites: Processing, properties and applications*, *379*.
15. Jagur-Grodzinski, J. (2006). Polymers for tissue engineering, medical devices, and regenerative medicine. Concise general review of recent studies. *Polymers for advanced technologies*, *17*(6), 395–418.

16. Praharaj, R., Mishra, S., Rautray, T.R.: The structural and bioactive behaviour of strontium-doped titanium dioxide nanorods. J. Korean Ceram. Soc. **57**, 271–280 (2020)
17. Biswal, T., BadJena, S.K., Pradhan, D.: Synthesis of polymer composite materials and their biomedical applications. Materials Today: Proceedings **30**, 305–315 (2020)
18. Zadehnazari, A.: Metal oxide/polymer nanocomposites: A review on recent advances in fabrication and applications. Polymer-Plastics Technology and Materials **62**(5), 655–700 (2023)
19. Hsissou, R., Seghiri, R., Benzekri, Z., Hilali, M., Rafik, M., Elharfi, A.: Polymer composite materials: A comprehensive review. Compos. Struct. **262**, 113640 (2021)
20. Vijay Kumar, V., Balaganesan, G., Lee, J.K.Y., Neisiany, R.E., Surendran, S., Ramakrishna, S.: A review of recent advances in nanoengineered polymer composites. Polymers **11**(4), 644 (2019)
21. Praharaj, R., Mishra, S., Rautray, T.R.: Growth mechanism of aligned porous oxide layers on titanium by anodization in electrolyte containing Cl. Materials Today: Proceedings **62**, 6216–6220 (2022)
22. Salernitano, E., Migliaresi, C.: Composite materials for biomedical applications: a review. J. Appl. Biomater. Biomech.Biomater. Biomech. **1**(1), 3–18 (2003)
23. Swain, S., Mishra, S., Patra, A., Praharaj, R., Rautray, T.: Dual action of polarised zinc hydroxyapatite-guar gum composite as a next generation bone filler material. Materials Today: Proceedings **62**, 6125–6130 (2022)
24. Atiqah, A., & Ansari, M. N. M. (2019). Nanostructure–polymer composites for soft-tissue engineering. In *Nanostructured Polymer Composites for Biomedical Applications* (pp. 105–115). Elsevier.
25. Pina, S., Ribeiro, V.P., Marques, C.F., Maia, F.R., Silva, T.H., Reis, R.L., Oliveira, J.M.: Scaffolding strategies for tissue engineering and regenerative medicine applications. Materials **12**(11), 1824 (2019)
26. Mozumder, M.S., Mairpady, A., Mourad, A.H.I.: Polymeric nanobiocomposites for biomedical applications. J. Biomed. Mater. Res. B Appl. Biomater.Biomater. **105**(5), 1241–1259 (2017)
27. Shameem, M.M., Sasikanth, S.M., Annamalai, R., Raman, R.G.: A brief review on polymer nanocomposites and its applications. Materials Today: Proceedings **45**, 2536–2539 (2021)
28. Silva AD, Stocco TD, Granato AE, Harb SV, Afewerki S, Bassous NJ, Hammer P, Webster TJ, Marciano FR, Lobo AO. Recent advances in nanostructured polymer composites for biomedical applications. Nanostructured polymer composites for biomedical applications. 2019 Jan 1:21–52
29. Kordas, G. (2019). Nanotechnology in cancer treatment as a Trojan Horse: from the bench to preclinical studies. In *Nanostructured Polymer Composites for Biomedical Applications* (pp. 323–365). Elsevier.
30. Yang, Z., Li, X., Si, J., Cui, Z., & Peng, K. (2019). Morphological, mechanical and
31. thermal properties of poly (lactic acid)(PLA)/cellulose nanofibrils (CNF) composites
32. nanofiber for tissue engineering. *Journal of Wuhan University of Technology-Mater. Sci. Ed.*, *34*, 207–215.
33. Ahmed, T., Saleem, A., Ramyakrishna, P., Rajender, B., Gulzar, T., Khan, A., & Asiri, A. M. (2019). Nanostructured polymer composites for bio-applications. In *Nanostructured Polymer Composites for Biomedical Applications* (pp. 167–188). Elsevier.
34. Williams, D.F.: Challenges with the development of biomaterials for sustainable tissue engineering. Frontiers in bioengineering and biotechnology **7**, 127 (2019)
35. Idumah, C. I., Zurina, M., Hassan, A., Norhayani, O., & Shuhadah, I. N. (2019). Recently emerging trends in bone replacement polymer nanocomposites. *Nanostructured Polymer Composites for Biomedical Applications*, 139–166.
36. Swain, S., Patra, A., Kumari, S., Praharaj, R., Mishra, S., Rautray, T.: Corona poled gelatin-Magnesium hydroxyapatite composite demonstrates osteogenicity. Materials Today: Proceedings **62**, 6131–6135 (2022)
37. Goodwin Jr, D. G., Marsh, K. M., Sosa, I. B., Payne, J. B., Gorham, J. M., Bouwer, E. J.,

38. & Fairbrother, D. H. (2015). Interactions of microorganisms with polymer nanocomposite
39. surfaces containing oxidized carbon nanotubes. *Environmental science & technology, 49*(9), 5484–5492.
40. Khan, S., Ul-Islam, M., Khattak, W.A., Ullah, M.W., Park, J.K.: Bacterial cellulose-titanium dioxide nanocomposites: nanostructural characteristics, antibacterial mechanism, and biocompatibility. Cellulose **22**, 565–579 (2015)
41. Shah, K.W., Huseien, G.F.: Inorganic nanomaterials for fighting surface and airborne pathogens and viruses. Nano Express **1**(3), 032003 (2020)
42. Liu, H.H.: Antibiotics and infectious diseases. Primary Care: Clinics in Office Practice **17**(4), 745–774 (1990)
43. Calderón, C.B., Sabundayo, B.P.: Antimicrobial classifications. Antimicrobial susceptibility testing protocols **7**, 60–88 (2007)
44. Fahim, M., & Namias, N. (2006). Complications of Antibiotic Therapy. In *Complications in Surgery and Trauma* (pp. 55–66). CRC Press.
45. Mozumder, M. S., Mairpady, A., & Mourad, A. H. I. (2017). Polymeric
46. nanobiocomposites for biomedical applications. *Journal of Biomedical Materials Research Part B: Applied Biomaterials, 105*(5), 1241–1259.
47. Lee, J. H., Moon, S. K., Kim, K. M., & Kim, K. N. (2013). Modification of TiO2
48. nanotube surfaces by electro-spray deposition of amoxicillin combined with PLGA for bactericidal effects at surgical implantation sites. *Acta Odontologica Scandinavica, 71*(1), 168–174.
49. Kretlow, J.D., Mikos, A.G.: From material to tissue: Biomaterial development, scaffold fabrication, and tissue engineering. AIChE J. **54**(12), 3048–3067 (2008)
50. Alaribe, F.N., Manoto, S.L., Motaung, S.C.: Scaffolds from biomaterials: advantages and limitations in bone and tissue engineering. Biologia **71**(4), 353–366 (2016)
51. Bakhshandeh, B., Zarrintaj, P., Oftadeh, M.O., Keramati, F., Fouladiha, H., Sohrabi-Jahromi, S., Ziraksaz, Z.: Tissue engineering; strategies, tissues, and biomaterials. Biotechnol. Genet. Eng. Rev.. Genet. Eng. Rev. **33**(2), 144–172 (2017)
52. Chen, F.M., Liu, X.: Advancing biomaterials of human origin for tissue engineering. Prog. Polym. Sci.Polym. Sci. **53**, 86–168 (2016)
53. Guo, B., Ma, P.X.: Synthetic biodegradable functional polymers for tissue engineering: a brief review. SCIENCE CHINA Chem. **57**, 490–500 (2014)
54. Geevarghese, R., Sajjadi, S.S., Hudecki, A., Sajjadi, S., Jalal, N.R., Madrakian, T., Łos, M.J.: Biodegradable and Non-Biodegradable Biomaterials and Their Effect on Cell Differentiation. Int. J. Mol. Sci. **23**(24), 16185 (2022)
55. Filippi, M., Born, G., Chaaban, M., Scherberich, A.: Natural polymeric scaffolds in bone regeneration. Frontiers in bioengineering and biotechnology **8**, 474 (2020)
56. Bajpai, A.K., Bajpai, J., Saini, R.K., Agrawal, P., Tiwari, A.: Smart biomaterial devices: polymers in biomedical sciences. CRC Press (2016)
57. Mantripragada, V.P., Lecka-Czernik, B., Ebraheim, N.A., Jayasuriya, A.C.: An overview of recent advances in designing orthopedic and craniofacial implants. J. Biomed. Mater. Res., Part A **101**(11), 3349–3364 (2013)
58. Mahanty, A., Shikha, D.: Changes in the morphology, mechanical strength and biocompatibility of polymer and metal/polymer fabricated hydroxyapatite for orthopaedic implants: a review. J. Polym. Eng.Polym. Eng. **42**(4), 298–322 (2022)
59. Aggarwal, D., Kumar, V., Sharma, S.: Drug-loaded biomaterials for orthopedic applications: A review. J. Control. Release **344**, 113–133 (2022)
60. Szeluga, U., Kumanek, B., Trzebicka, B.: Synergy in hybrid polymer/nanocarbon composites. A review. Compos. A Appl. Sci. Manuf. **73**, 204–231 (2015)
61. Homan, K. A., Chen, J., Schiano, A., Mohamed, M., Willets, K. A., Murugesan, S., ... & Emelianov, S. (2011). Silver–polymer composite stars: synthesis and applications. *Advanced functional materials,21*(9), 1673–1680.
62. Gong, L.X., Pei, Y.B., Han, Q.Y., Zhao, L., Wu, L.B., Jiang, J.X., Tang, L.C.: Polymer grafted reduced graphene oxide sheets for improving stress transfer in polymer composites. Compos. Sci. Technol. **134**, 144–152 (2016)

63. Tekin, D., Birhan, D., Kiziltas, H.: Thermal, photocatalytic, and antibacterial properties of calcinated nano-TiO2/polymer composites. Mater. Chem. Phys. **251**, 123067 (2020)
64. Xiao, X., Liu, R., & Huang, Q. (2008). Preparation and characterization of
65. nano-hydroxyapatite/polymer composite scaffolds. *Journal of Materials Science: Materials in Medicine*, *19*, 3429–3435.
66. Sergi, R., Bellucci, D., Cannillo, V.: A review of bioactive glass/natural polymer composites: State of the art. Materials **13**(23), 5560 (2020)
67. Zengin, H., Erkan, B.: Synthesis and characterization of polyaniline/silicon dioxide composites and preparation of conductive films. Polym. Adv. Technol.. Adv. Technol. **21**(3), 216–223 (2010)
68. Jeon, I.S., Lee, M.H., Choi, H.H., Lee, S., Chon, J.W., Chung, D.J., Jho, J.Y.: Mechanical properties and bioactivity of Polyetheretherketone/Hydroxyapatite/Carbon fiber composite prepared by the mechanofusion process. Polymers **13**(12), 1978 (2021)
69. Liu, Y., Wang, M.: Fabrication and characteristics of hydroxyapatite reinforced polypropylene as a bone analogue biomaterial. J. Appl. Polym. Sci.Polym. Sci. **106**(4), 2780–2790 (2007)
70. Bongio, M., Van Den Beucken, J. J., Leeuwenburgh, S. C., & Jansen, J. A. (2010). Development of bone substitute materials: from 'biocompatible'to 'instructive'. Journal of Materials Chemistry, 20(40), 8747-8759.
71. Cao, H., Duan, L., Zhang, Y., Cao, J., Zhang, K.: Current hydrogel advances in physicochemical and biological response-driven biomedical application diversity. Signal Transduct. Target. Ther.Transduct. Target. Ther. **6**(1), 426 (2021)
72. González-Díaz, E.C., Varghese, S.: Hydrogels as extracellular matrix analogs. Gels **2**(3), 20 (2016)
73. Chang, C., Peng, N., He, M., Teramoto, Y., Nishio, Y., Zhang, L.: Fabrication and properties of chitin/hydroxyapatite hybrid hydrogels as scaffold nano-materials. Carbohyd. Polym.. Polym. **91**(1), 7–13 (2013)
74. Caridade, S. G., Merino, E. G., Alves, N. M., de Zea Bermudez, V., Boccaccini, A. R., & Mano, J. F. (2013). Chitosan membranes containing micro or nano-size bioactive glass particles: evolution of biomineralization followed by in situ dynamic mechanical analysis. *journal of the mechanical behavior of biomedical materials*, *20*, 173–183.
75. Thein-Han, W.W., Misra, R.D.K.: Biomimetic chitosan–nanohydroxyapatite composite scaffolds for bone tissue engineering. Acta Biomater.Biomater. **5**(4), 1182–1197 (2009)
76. Zhang, X., Jia, W., Gu, Y., Xiao, W., Liu, X., Wang, D., Zhou, N.: Teicoplanin-loaded borate bioactive glass implants for treating chronic bone infection in a rabbit tibia osteomyelitis model. Biomaterials **31**(22), 5865–5874 (2010)
77. Kosowska, K., Domalik-Pyzik, P., Nocuń, M., Chłopek, J.: Chitosan and graphene oxide/ reduced graphene oxide hybrid nanocomposites–Evaluation of physicochemical properties. Mater. Chem. Phys. **216**, 28–36 (2018)
78. Türk, S., Altınsoy, I., Efe, G.Ç., İpek, M., Özacar, M., Bindal, C.: 3D porous collagen/ functionalized multiwalled carbon nanotube/chitosan/hydroxyapatite composite scaffolds for bone tissue engineering. Mater. Sci. Eng. C **92**, 757–768 (2018)
79. Depan D, Surya PK, Girase B, Misra R.D.K.(2011). Organic/inorganic hybrid network structure nanocomposite scaffolds based on grafted chitosan for tissue engineering. Acta biomater 7(5):2163–2175.
80. Dong, Y., Liang, J., Cui, Y., Xu, S., Zhao, N.: Fabrication of novel bioactive hydroxyapatite-chitosan-silica hybrid scaffolds: Combined the sol-gel method with 3D plotting technique. Carbohyd. Polym.. Polym. **197**, 183–193 (2018)
81. Trujillo, S., Pérez-Román, E., Kyritsis, A., Gomez Ribelles, J.L., Pandis, C.: Organic–inorganic bonding in chitosan–silica hybrid networks: Physical properties. J. Polym. Sci., Part B: Polym. Phys. **53**(19), 1391–1400 (2015)
82. Wenz, A., Borchers, K., Tovar, G.E., Kluger, P.J.: Bone matrix production in hydroxyapatite-modified hydrogels suitable for bone bioprinting. Biofabrication **9**(4), 044103 (2017)
83. Asadi, N., Alizadeh, E., Rahmani Del Bakhshayesh, A., Mostafavi, E., Akbarzadeh, A., Davaran, S.: Fabrication and in vitro evaluation of nanocomposite hydrogel scaffolds based on gelatin/PCL–PEG–PCL for cartilage tissue engineering. ACS Omega **4**(1), 449–457 (2019)

84. Zadehnajar, P., Akbari, B., Karbasi, S., & Mirmusavi, M. H. (2019). Preparation and characterization of poly ε-caprolactone-gelatin/multi-walled carbon nanotubes electrospun scaffolds for cartilage tissue engineering applications. *International Journal of Polymeric Materials and Polymeric Biomaterials.*
85. Yu, Z., Lili, J., Tiezheng, Z., Li, S., Jianzhuang, W., Haichao, D., Tianqing, L.: Development of decellularized meniscus extracellular matrix and gelatin/chitosan scaffolds for meniscus tissue engineering. Bio-Med. Mater. Eng. **30**(2), 125–132 (2019)
86. Du, C., Cui, F.Z., Zhu, X.D., De Groot, K.: Three-dimensional nano-HAp/collagen matrix loading with osteogenic cells in organ culture. Journal of Biomedical Materials Research: An Official Journal of The Society for Biomaterials, The Japanese Society for Biomaterials, and The Australian Society for Biomaterials **44**(4), 407–415 (1999)
87. Da, H., Jia, S.J., Meng, G.L., Cheng, J.H., Zhou, W., Xiong, Z., Liu, J.: The impact of compact layer in biphasic scaffold on osteochondral tissue engineering. PLoS ONE **8**(1), e54838 (2013)
88. Tanase, C.E., Sartoris, A., Popa, M.I., Verestiuc, L., Unger, R.E., Kirkpatrick, C.J.: In vitro evaluation of biomimetic chitosan–calcium phosphate scaffolds with potential application in bone tissue engineering. Biomed. Mater. **8**(2), 025002 (2013)
89. Jiang, L.B., Su, D.H., Liu, P., Ma, Y.Q., Shao, Z.Z., Dong, J.: Shape-memory collagen scaffold for enhanced cartilage regeneration: native collagen versus denatured collagen. Osteoarthritis Cartilage **26**(10), 1389–1399 (2018)
90. Kim, W.J., Yun, H.S., Kim, G.H.: An innovative cell-laden α-TCP/collagen scaffold fabricated using a two-step printing process for potential application in regenerating hard tissues. Sci. Rep. **7**(1), 3181 (2017)
91. Lee, H., Yang, G.H., Kim, M., Lee, J., Huh, J., Kim, G.: Fabrication of micro/nanoporous collagen/dECM/silk-fibroin biocomposite scaffolds using a low temperature 3D printing process for bone tissue regeneration. Mater. Sci. Eng. C **84**, 140–147 (2018)
92. Jo, Y.Y., Kim, S.G., Kwon, K.J., Kweon, H., Chae, W.S., Yang, W.G., Seok, H.: Silk fibroin-alginate-hydroxyapatite composite particles in bone tissue engineering applications in vivo. Int. J. Mol. Sci. **18**(4), 858 (2017)
93. Zheng, A., Cao, L., Liu, Y., Wu, J., Zeng, D., Hu, L., Jiang, X.: Biocompatible silk/calcium silicate/sodium alginate composite scaffolds for bone tissue engineering. Carbohyd. Polym.. Polym. **199**, 244–255 (2018)
94. Shen, T., Dai, Y., Li, X., Xu, S., Gou, Z., Gao, C.: Regeneration of the osteochondral defect by a wollastonite and macroporous fibrin biphasic scaffold. ACS Biomater. Sci. Eng.Biomater. Sci. Eng. **4**(6), 1942–1953 (2017)
95. Pina, S., Canadas, R.F., Jiménez, G., Perán, M., Marchal, J.A., Reis, R.L., Oliveira, J.M.: Biofunctional ionic-doped calcium phosphates: silk fibroin composites for bone tissue engineering scaffolding. Cells Tissues Organs **204**(3–4), 150–163 (2017)
96. Chen, L., Hu, J., Ran, J., Shen, X., Tong, H.: A novel nanocomposite for bone tissue engineering based on chitosan–silk sericin/hydroxyapatite: biomimetic synthesis and its cytocompatibility. RSC Adv. **5**(69), 56410–56422 (2015)
97. Mandal, B.B., Priya, A.S., Kundu, S.C.: Novel silk sericin/gelatin 3-D scaffolds and 2-D films: fabrication and characterization for potential tissue engineering applications. Acta Biomater.Biomater. **5**(8), 3007–3020 (2009)
98. Shi, Z., Zang, S., Jiang, F., Huang, L., Lu, D., Ma, Y., Yang, G.: In situ nano-assembly of bacterial cellulose–polyaniline composites. RSC Adv. **2**(3), 1040–1046 (2012)
99. Mao, D., Li, Q., Bai, N., Dong, H., Li, D.: Porous stable poly (lactic acid)/ethyl cellulose/ hydroxyapatite composite scaffolds prepared by a combined method for bone regeneration. Carbohyd. Polym.. Polym. **180**, 104–111 (2018)
100. Zineh, B.R., Shabgard, M.R., Roshangar, L.: Mechanical and biological performance of printed alginate/methylcellulose/halloysite nanotube/polyvinylidene fluoride bio-scaffolds. Mater. Sci. Eng. C **92**, 779–789 (2018)
101. Park, J., Lee, S.J., Lee, H., Park, S.A., Lee, J.Y.: Three dimensional cell printing with sulfated alginate for improved bone morphogenetic protein-2 delivery and osteogenesis in bone tissue engineering. Carbohyd. Polym.. Polym. **196**, 217–224 (2018)

102. Kankala, R.K., Lu, F.J., Liu, C.G., Zhang, S.S., Chen, A.Z., Wang, S.B.: Effect of icariin on engineered 3d-printed porous scaffolds for cartilage repair. Materials **11**(8), 1390 (2018)
103. Murphy, C., Kolan, K., Li, W., Semon, J., Day, D., & Leu, M. (2017). 3D bioprinting of stem cells and polymer/bioactive glass composite scaffolds for bone tissue engineering. *International Journal of Bioprinting*, *3*(1).
104. Du, Y., Liu, H., Yang, Q., Wang, S., Wang, J., Ma, J., Zhang, S.: Selective laser sintering scaffold with hierarchical architecture and gradient composition for osteochondral repair in rabbits. Biomaterials **137**, 37–48 (2017)
105. Kim, J.Y., Ahn, G., Kim, C., Lee, J.S., Lee, I.G., An, S.H., Shim, J.H.: Synergistic effects of beta tri-calcium phosphate and porcine-derived decellularized bone extracellular matrix in 3D-printed polycaprolactone scaffold on bone regeneration. Macromol. Biosci.. Biosci. **18**(6), 1800025 (2018)
106. Suryavanshi, A., Khanna, K., Sindhu, K.R., Bellare, J., Srivastava, R.: Magnesium oxide nanoparticle-loaded polycaprolactone composite electrospun fiber scaffolds for bone–soft tissue engineering applications: in-vitro and in-vivo evaluation. Biomed. Mater. **12**(5), 055011 (2017)
107. Ren, J., Blackwood, K.A., Doustgani, A., Poh, P.P., Steck, R., Stevens, M.M., Woodruff, M.A.: Melt-electrospun polycaprolactone strontium-substituted bioactive glass scaffolds for bone regeneration. J. Biomed. Mater. Res., Part A **102**(9), 3140–3153 (2014)
108. Kim, H.W., Lee, E.J., Jun, I.K., Kim, H.E., Knowles, J.C.: Degradation and drug release of phosphate glass/polycaprolactone biological composites for hard-tissue regeneration. Journal of Biomedical Materials Research Part B: Applied Biomaterials: An Official Journal of The Society for Biomaterials, The Japanese Society for Biomaterials, and The Australian Society for Biomaterials and the Korean Society for Biomaterials **75**(1), 34–41 (2005)
109. Kouhi, M., Fathi, M., Reddy, V.J., Ramakrishna, S.: Bredigite reinforced electrospun nanofibers for bone tissue engineering. Materials Today: Proceedings **7**, 449–454 (2019)
110. Lee, S., Joshi, M.K., Tiwari, A.P., Maharjan, B., Kim, K.S., Yun, Y.H., Kim, C.S.: Lactic acid assisted fabrication of bioactive three-dimensional PLLA/β-TCP fibrous scaffold for biomedical application. Chem. Eng. J. **347**, 771–781 (2018)
111. Deng, X., Hao, J., Wang, C.: Preparation and mechanical properties of nanocomposites of poly (D, L-lactide) with Ca-deficient hydroxyapatite nanocrystals. Biomaterials **22**(21), 2867–2873 (2001)
112. Zhang, Q., Mochalin, V.N., Neitzel, I., Hazeli, K., Niu, J., Kontsos, A., Gogotsi, Y.: Mechanical properties and biomineralization of multifunctional nanodiamond-PLLA composites for bone tissue engineering. Biomaterials **33**(20), 5067–5075 (2012)
113. Butt, M.S., Bai, J., Wan, X., Chu, C., Xue, F., Ding, H., Zhou, G.: Mg alloy rod reinforced biodegradable poly-lactic acid composite for load bearing bone replacement. Surf. Coat. Technol. **309**, 471–479 (2017)
114. Mehboob, H., Bae, J.H., Han, M.G., Chang, S.H.: Effect of air plasma treatment on mechanical properties of bioactive composites for medical application: Composite preparation and characterization. Compos. Struct. **143**, 23–32 (2016)
115. Solchaga, L.A., Dennis, J.E., Goldberg, V.M., Caplan, A.I.: Hyaluronic acid-based polymers as cell carriers for tissue-engineered repair of bone and cartilage. J. Orthop. Res.Orthop. Res. **17**(2), 205–213 (1999)
116. Ormsby, R., McNally, T., O'Hare, P., Burke, G., Mitchell, C., Dunne, N.: Fatigue and biocompatibility properties of a poly (methyl methacrylate) bone cement with multi-walled carbon nanotubes. Acta Biomater.Biomater. **8**(3), 1201–1212 (2012)
117. Kannan, R.Y., Salacinski, H.J., Edirisinghe, M.J., Hamilton, G., Seifalian, A.M.: Polyhedral oligomeric silsequioxane–polyurethane nanocomposite microvessels for an artificial capillary bed. Biomaterials **27**(26), 4618–4626 (2006)
118. Alt, V., Bechert, T., Steinrücke, P., Wagener, M., Seidel, P., Dingeldein, E., Schnettler, R.: An in vitro assessment of the antibacterial properties and cytotoxicity of nanoparticulate silver bone cement. Biomaterials **25**(18), 4383–4391 (2004)

119. Ma, R., Yu, Z., Tang, S., Pan, Y., Wei, J., & Tang, T. (2016). Osseointegration of nanohydroxyapatite-or nano-calcium silicate-incorporated polyetheretherketone bioactive composites in vivo. *International journal of nanomedicine*, 6023–6033.
120. Wang, L., He, S., Wu, X., Liang, S., Mu, Z., Wei, J., Wei, S.: Polyetheretherketone/nano-fluorohydroxyapatite composite with antimicrobial activity and osseointegration properties. Biomaterials **35**(25), 6758–6775 (2014)
121. Zhang, J., Wei, W., Yang, L., Pan, Y., Wang, X., Wang, T., Wei, J.: Stimulation of cell responses and bone ingrowth into macro-microporous implants of nano-bioglass/polyetheretherketone composite and enhanced antibacterial activity by release of hinokitiol. Colloids Surf. B **164**, 347–357 (2018)
122. Von Wilmonsky, C., Lutz, R., Meisel, U., Srour, S., Rupprecht, S., Toyoshima, T., Schmidt, M.: In vivo evaluation of ß-TCP containing 3D laser sintered poly (ether ether ketone) composites in pigs. J. Bioact. Compat. Polym.Bioact. Compat. Polym. **24**(2), 169–184 (2009)
123. Wong, K.L., Wong, C.T., Liu, W.C., Pan, H.B., Fong, M.K., Lam, W.M., Lu, W.W.: Mechanical properties and in vitro response of strontium-containing hydroxyapatite/polyetheretherketone composites. Biomaterials **30**(23–24), 3810–3817 (2009)
124. Wu, X., Liu, X., Wei, J., Ma, J., Deng, F., & Wei, S. (2012). Nano-TiO2/PEEK bioactive composite as a bone substitute material: in vitro and in vivo studies. *International journal of nanomedicine*, 1215–1225.
125. Maharjan, B., Kaliannagounder, V.K., Jang, S.R., Awasthi, G.P., Bhattarai, D.P., Choukrani, G., Kim, C.S.: In-situ polymerized polypyrrole nanoparticles immobilized poly (ε-caprolactone) electrospun conductive scaffolds for bone tissue engineering. Mater. Sci. Eng. C **114**, 111056 (2020)
126. Li, M., Guo, Y., Wei, Y., MacDiarmid, A.G., Lelkes, P.I.: Electrospinning polyaniline-contained gelatin nanofibers for tissue engineering applications. Biomaterials **27**(13), 2705–2715 (2006)
127. Waugaman, M., Sannigrahi, B., McGeady, P., Khan, I.M.: Synthesis, characterization and biocompatibility studies of oligosiloxane modified polythiophenes. Eur. Polymer J. **39**(7), 1405–1412 (2003)
128. Khan, Y.M., Cushnie, E.K., Kelleher, J.K., Laurencin, C.T.: In situ synthesized ceramic–polymer composites for bone tissue engineering: bioactivity and degradation studies. J. Mater. Sci. **42**, 4183–4190 (2007)
129. Ochi, K., Chen, G., Ushida, T., Gojo, S., Segawa, K., Tai, H., Umezawa, A.: Use of isolated mature osteoblasts in abundance acts as desired-shaped bone regeneration in combination with a modified poly-DL-lactic-co-glycolic acid (PLGA)-collagen sponge. J. Cell. Physiol. **194**(1), 45–53 (2003)
130. Li, H., Chang, J.: PH-compensation effect of bioactive inorganic fillers on the degradation of PLGA. Compos. Sci. Technol. **65**(14), 2226–2232 (2005)
131. Gerhardt, L.C., Widdows, K.L., Erol, M.M., Burch, C.W., Sanz-Herrera, J.A., Ochoa, I., Boccaccini, A.R.: The pro-angiogenic properties of multi-functional bioactive glass composite scaffolds. Biomaterials **32**(17), 4096–4108 (2011)
132. Maquet, V., Boccaccini, A.R., Pravata, L., Notingher, I., Jérôme, R.: Preparation, characterization, and in vitro degradation of bioresorbable and bioactive composites based on Bioglass®-filled polylactide foams. Journal of Biomedical Materials Research Part A: An Official Journal of The Society for Biomaterials, The Japanese Society for Biomaterials, and The Australian Society for Biomaterials and the Korean Society for Biomaterials **66**(2), 335–346 (2003)
133. Chang, G.W., Tseng, C.L., Tzeng, Y.S., Chen, T.M., Fang, H.W.: An in vivo evaluation of a novel malleable composite scaffold (polypropylene carbonate/poly (D-lactic acid)/tricalcium phosphate elastic composites) for bone defect repair. J. Taiwan Inst. Chem. Eng. **80**, 813–819 (2017)
134. Townsend, J.M., Zabel, T.A., Feng, Y., Wang, J., Andrews, B.T., Nudo, R.J., Detamore, M.S.: Effects of tissue processing on bioactivity of cartilage matrix-based hydrogels encapsulating osteoconductive particles. Biomed. Mater. **13**(3), 034108 (2018)

135. Misra, S.K., Mohn, D., Brunner, T.J., Stark, W.J., Philip, S.E., Roy, I., Boccaccini, A.R.: Comparison of nanoscale and microscale bioactive glass on the properties of P (3HB)/ Bioglass® composites. Biomaterials **29**(12), 1750–1761 (2008)
136. Zarrintaj, P., Seidi, F., Azarfam, M.Y., Yazdi, M.K., Erfani, A., Barani, M., Mozafari, M.: Biopolymer-based composites for tissue engineering applications: A basis for future opportunities. Compos. B Eng. **258**, 110701 (2023)
137. Novosel, E.C., Kleinhans, C., Kluger, P.J.: Vascularization is the key challenge in tissue engineering. Adv. Drug Deliv. Rev. **63**(4–5), 300–311 (2011)
138. Place, E.S., Evans, N.D., Stevens, M.M.: Complexity in biomaterials for tissue engineering. Nat. Mater. **8**(6), 457–470 (2009)
139. Huang, B.: Carbon nanotubes and their polymeric composites: The applications in tissue engineering. Biomanufacturing Reviews **5**(1), 3 (2020)
140. Naseri, N. (2014). *Porous structures based on nanopolysaccharides for medical applications* (Doctoral dissertation, Luleå tekniska universitet).
141. Balčiūnas, E. (2020). *Development of bioactive scaffolds for tissue engineering applications* (Doctoral dissertation, Vilniaus universitetas).
142. Bambo, M. F., Mokhena, T. C., & Matabola, K. P. (2023). Synthesis of Natural Polymer–Carbonaceous Filler Composites. In *Polymer-Carbonaceous Filler Based Composites for Wastewater Treatment* (pp. 29–67). CRC Press.
143. Ashammakhi, N., GhavamiNejad, A., Tutar, R., Fricker, A., Roy, I., Chatzistavrou, X., Caterson, E.J.: Highlights on advancing frontiers in tissue engineering. Tissue Eng. Part B Rev. **28**(3), 633–664 (2022)
144. Behera, S.S., Das, U., Kumar, A., Bissoyi, A., Singh, A.K.: Chitosan/TiO2 composite membrane improves proliferation and survival of L929 fibroblast cells: Application in wound dressing and skin regeneration. Int. J. Biol. Macromol.Macromol. **98**, 329–340 (2017)
145. Kumar, N., Desagani, D., Chandran, G., Ghosh, N.N., Karthikeyan, G., Waigaonkar, S., Ganguly, A.: Biocompatible agarose-chitosan coated silver nanoparticle composite for soft tissue engineering applications. Artificial Cells, Nanomedicine, and Biotechnology **46**(3), 637–649 (2018)
146. Omidi, M., Yadegari, A., Tayebi, L.: Wound dressing application of pH-sensitive carbon dots/ chitosan hydrogel. RSC Adv. **7**(18), 10638–10649 (2017)
147. Lee, R.J., Temmer, R., Tamm, T., Aabloo, A., Kiefer, R.: Renewable antioxidant properties of suspensible chitosan–polypyrrole composites. React. Funct. Polym.Funct. Polym. **73**(8), 1072–1077 (2013)
148. Liu, Y., Wang, S., Zhang, R.: Composite poly (lactic acid)/chitosan nanofibrous scaffolds for cardiac tissue engineering. Int. J. Biol. Macromol.Macromol. **103**, 1130–1137 (2017)
149. Kamel, N.A., Abd El-messieh, S.L., Saleh, N.M.: Chitosan/banana peel powder nanocomposites for wound dressing application: Preparation and characterization. Mater. Sci. Eng. C **72**, 543–550 (2017)
150. Oh, G.W., Ko, S.C., Je, J.Y., Kim, Y.M., Oh, J., Jung, W.K.: Fabrication, characterization and determination of biological activities of poly (ε-caprolactone)/chitosan-caffeic acid composite fibrous mat for wound dressing application. Int. J. Biol. Macromol.Macromol. **93**, 1549–1558 (2016)
151. Depan, D., Misra, R.D.K.: The interplay between nanostructured carbon-grafted chitosan scaffolds and protein adsorption on the cellular response of osteoblasts: structure–function property relationship. Acta Biomater.Biomater. **9**(4), 6084–6094 (2013)
152. Depan, D., Pratheep Kumar, A., Singh, R.P., Misra, R.D.K.: Stability of chitosan/ montmorillonite nanohybrid towards enzymatic degradation on grafting with poly (lactic acid). Mater. Sci. Technol. **30**(5), 587–592 (2014)
153. Jiang, Y.C., Jiang, L., Huang, A., Wang, X.F., Li, Q., Turng, L.S.: Electrospun polycaprolactone/gelatin composites with enhanced cell–matrix interactions as blood vessel endothelial layer scaffolds. Mater. Sci. Eng. C **71**, 901–908 (2017)
154. Heidari, M., Bahrami, S.H., Ranjbar-Mohammadi, M., Milan, P.B.: Smart electrospun nanofibers containing PCL/gelatin/graphene oxide for application in nerve tissue engineering. Mater. Sci. Eng. C **103**, 109768 (2019)

155. Badhe, R.V., Bijukumar, D., Chejara, D.R., Mabrouk, M., Choonara, Y.E., Kumar, P., Pillay, V.: A composite chitosan-gelatin bi-layered, biomimetic macroporous scaffold for blood vessel tissue engineering. Carbohyd. Polym.. Polym. **157**, 1215–1225 (2017)
156. Quinlan, E., Partap, S., Azevedo, M.M., Jell, G., Stevens, M.M., O'Brien, F.J.: Hypoxia-mimicking bioactive glass/collagen glycosaminoglycan composite scaffolds to enhance angiogenesis and bone repair. Biomaterials **52**, 358–366 (2015)
157. Kim, H.S., Hobbs, H.L., Wang, L., Rutten, M.J., Wamser, C.C.: Biocompatible composites of polyaniline nanofibers and collagen. Synth. Met. **159**(13), 1313–1318 (2009)
158. Sharif, S., Ai, J., Azami, M., Verdi, J., Atlasi, M.A., Shirian, S., Samadikuchaksaraei, A.: Collagen-coated nano-electrospun PCL seeded with human endometrial stem cells for skin tissue engineering applications. J. Biomed. Mater. Res. B Appl. Biomater.Biomater. **106**(4), 1578–1586 (2018)
159. Rahmani Del Bakhshayesh, A., Mostafavi, E., Alizadeh, E., Asadi, N., Akbarzadeh, A., Davaran, S.: Fabrication of three-dimensional scaffolds based on nano-biomimetic collagen hybrid constructs for skin tissue engineering. ACS Omega **3**(8), 8605–8611 (2018)
160. Wang, S.D., Ma, Q., Wang, K., Chen, H.W.: Improving antibacterial activity and biocompatibility of bioinspired electrospinning silk fibroin nanofibers modified by graphene oxide. ACS Omega **3**(1), 406–413 (2018)
161. Balu, R., Reeder, S., Knott, R., Mata, J., de Campo, L., Dutta, N.K., Choudhury, N.R.: Tough photocrosslinked silk fibroin/graphene oxide nanocomposite hydrogels. Langmuir **34**(31), 9238–9251 (2018)
162. Li, H., Williams, G.R., Wu, J., Wang, H., Sun, X., Zhu, L.M.: Poly (N-isopropylacrylamide)/poly (l-lactic acid-co-ε-caprolactone) fibers loaded with ciprofloxacin as wound dressing materials. Mater. Sci. Eng. C **79**, 245–254 (2017)
163. Tang, J., Bao, L., Li, X., Chen, L., Hong, F.F.: Potential of PVA-doped bacterial nano-cellulose tubular composites for artificial blood vessels. Journal of Materials Chemistry B **3**(43), 8537–8547 (2015)
164. Poonguzhali, R., Basha, S.K., Kumari, V.S.: Synthesis and characterization of chitosan-PVP-nanocellulose composites for in-vitro wound dressing application. Int. J. Biol. Macromol.Macromol. **105**, 111–120 (2017)
165. Duin, S., Schütz, K., Ahlfeld, T., Lehmann, S., Lode, A., Ludwig, B., Gelinsky, M.: 3D bioprinting of functional islets of langerhans in an alginate/methylcellulose hydrogel blend. Adv. Healthcare Mater. **8**(7), 1801631 (2019)
166. Thu, H.E., Zulfakar, M.H., Ng, S.F.: Alginate based bilayer hydrocolloid films as potential slow-release modern wound dressing. Int. J. Pharm. **434**(1–2), 375–383 (2012)
167. Moura, D., Souza, M.T., Liverani, L., Rella, G., Luz, G.M., Mano, J.F., Boccaccini, A.R.: Development of a bioactive glass-polymer composite for wound healing applications. Mater. Sci. Eng. C **76**, 224–232 (2017)
168. Mohammadkhah, A., Marquardt, L.M., Sakiyama-Elbert, S.E., Day, D.E., Harkins, A.B.: Fabrication and characterization of poly-(ε)-caprolactone and bioactive glass composites for tissue engineering applications. Mater. Sci. Eng. C **49**, 632–639 (2015)
169. Salami, M.A., Kaveian, F., Rafienia, M., Saber-Samandari, S., Khandan, A., Naeimi, M.: Electrospun polycaprolactone/lignin-based nanocomposite as a novel tissue scaffold for biomedical applications. Journal of medical signals and sensors **7**(4), 228 (2017)
170. Shah, S., Yin, P. T., Uehara, T. M., Chueng, S. T. D., Yang, L., & Lee, K. B. (2014). Guiding Stem Cell Differentiation into Oligodendrocytes Using Graphene-Nanofiber Scaffolds. *Advanced materials (Deerfield Beach, Fla.)*, *26*(22), 3673.
171. Lee, H.Y., Kim, J., Hwang, C.H., Kim, H.E., Jeong, S.H.: Strategy for Preparing Mechanically Strong Hyaluronic Acid-Silica Nanohybrid Hydrogels via In Situ Sol-Gel Process. Macromol. Mater. Eng.. Mater. Eng. **303**(9), 1800213 (2018)
172. Joshi, R.R., Frautschi, J.R., Phillips, R.E., Jr., Levy, R.J.: Phosphonated polyurethanes that resist calcification. J. Appl. Biomater.Biomater. **5**(1), 65–77 (1994)
173. Lundin, J.G., McGann, C.L., Daniels, G.C., Streifel, B.C., Wynne, J.H.: Hemostatic kaolin-polyurethane foam composites for multifunctional wound dressing applications. Mater. Sci. Eng. C **79**, 702–709 (2017)

174. Domènech, B., Muñoz, M., Muraviev, D. N., & Macanás, J. (2013). Polymer-silver nanocomposites as antibacterial materials. *Microbial pathogens and strategies for combating them: science, technology and education. Badajoz: Formatex*, 630–640.
175. Mohammadi, H.: Nanocomposite biomaterial mimicking aortic heart valve leaflet mechanical behaviour. Proc. Inst. Mech. Eng. H **225**, 718–722 (2011)
176. Hsu, S.H., Lin, Z.C.: Biocompatibility and biostability of a series of poly (carbonate) urethanes. Colloids Surf. B **36**(1), 1–12 (2004)
177. Ghanbari, H., Kidane, A.G., Burriesci, G., Ramesh, B., Darbyshire, A., Seifalian, A.M.: The anti-calcification potential of a silsesquioxane nanocomposite polymer under in vitro conditions: potential material for synthetic leaflet heart valve. Acta Biomater.Biomater. **6**(11), 4249–4260 (2010)
178. de Mel, A., Naghavi, N., Cousins, B.G., Clatworthy, I., Hamilton, G., Darbyshire, A., Seifalian, A.M.: Nitric oxide-eluting nanocomposite for cardiovascular implants. J. Mater. Sci. - Mater. Med. **25**, 917–929 (2014)
179. Tan, A., Goh, D., Farhatnia, Y., Lim, J., Teoh, S.H., Rajadas, J., Seifalian, A.M.: An anti-CD34 antibody-functionalized clinical-grade POSS-PCU nanocomposite polymer for cardiovascular stent coating applications: a preliminary assessment of endothelial progenitor cell capture and hemocompatibility. PLoS ONE **8**(10), e77112 (2013)
180. Ludwig, K.A., Uram, J.D., Yang, J., Martin, D.C., Kipke, D.R.: Chronic neural recordings using silicon microelectrode arrays electrochemically deposited with a poly (3, 4-ethylenedioxythiophene)(PEDOT) film. J. Neural Eng. **3**(1), 59 (2006)
181. George, P.M., Lyckman, A.W., LaVan, D.A., Hegde, A., Leung, Y., Avasare, R., Sur, M.: Fabrication and biocompatibility of polypyrrole implants suitable for neural prosthetics. Biomaterials **26**(17), 3511–3519 (2005)
182. Duailibi, M.T., Duailibi, S.E., Young, C.S., Bartlett, J.D., Vacanti, J.P., Yelick, P.C.: Bioengineered teeth from cultured rat tooth bud cells. J. Dent. Res. **83**(7), 523–528 (2004)
183. Pandey, N., Hakamivala, A., Xu, C., Hariharan, P., Radionov, B., Huang, Z., Hong, Y.: Biodegradable nanoparticles enhanced adhesiveness of mussel-like hydrogels at tissue interface. Adv. Healthcare Mater. **7**(7), 1701069 (2018)
184. Lih, E., Park, W., Park, K.W., Chun, S.Y., Kim, H., Joung, Y.K., Han, D.K.: A bioinspired scaffold with anti-inflammatory magnesium hydroxide and decellularized extracellular matrix for renal tissue regeneration. ACS Cent. Sci. **5**(3), 458–467 (2019)
185. Bonifacio, M.A., Gentile, P., Ferreira, A.M., Cometa, S., De Giglio, E.: Insight into halloysite nanotubes-loaded gellan gum hydrogels for soft tissue engineering applications. Carbohyd. Polym.. Polym. **163**, 280–291 (2017)
186. Rios, P.D., Skoumal, M., Liu, J., Youngblood, R., Kniazeva, E., Garcia, A.J., Shea, L.D.: Evaluation of encapsulating and microporous nondegradable hydrogel scaffold designs on islet engraftment in rodent models of diabetes. Biotechnol. Bioeng.. Bioeng. **115**(9), 2356–2364 (2018)
187. Zhang, C., Liang, K., Zhou, D., Yang, H., Liu, X., Yin, X., Xiao, P.: High-performance photopolymerized poly (vinyl alcohol)/silica nanocomposite hydrogels with enhanced cell adhesion. ACS Appl. Mater. Interfaces **10**(33), 27692–27700 (2018)
188. Du, Y., Ge, J., Li, Y., Ma, P.X., Lei, B.: Biomimetic elastomeric, conductive and biodegradable polycitrate-based nanocomposites for guiding myogenic differentiation and skeletal muscle regeneration. Biomaterials **157**, 40–50 (2018)
189. Dristant, U., Mukherjee, K., Saha, S., Maity, D.: An Overview of Polymeric Nanoparticles-Based Drug Delivery System in Cancer Treatment. Technol. Cancer Res. Treat. **22**, 15330338231152084 (2023)
190. Taghipour-Sabzevar, V., Sharifi, T., Moghaddam, M.M.: Polymeric nanoparticles as carrier for targeted and controlled delivery of anticancer agents. Ther. Deliv. **10**(8), 527–550 (2019)
191. Samanta, D., Hosseini-Nassab, N., McCarty, A.D., Zare, R.N.: Ultra-low voltage triggered release of an anti-cancer drug from polypyrrole nanoparticles. Nanoscale **10**(20), 9773–9779 (2018)

192. Paramshetti, S., Angolkar, M., Al Fatease, A., Alshahrani, S.M., Hani, U., Garg, A., Osmani, R.A.M.: Revolutionizing Drug Delivery and Therapeutics: The Biomedical Applications of Conductive Polymers and Composites-Based Systems. Pharmaceutics **15**(4), 1204 (2023)
193. Beg, S., Rahman, M., Jain, A., Saini, S., Midoux, P., Pichon, C., Akhter, S.: Nanoporous metal organic frameworks as hybrid polymer–metal composites for drug delivery and biomedical applications. Drug Discovery Today **22**(4), 625–637 (2017)
194. Gooneh-Farahani, S., Naimi-Jamal, M.R., Naghib, S.M.: Stimuli-responsive graphene-incorporated multifunctional chitosan for drug delivery applications: a review. Expert Opin. Drug Deliv.Opin. Drug Deliv. **16**(1), 79–99 (2019)
195. Liu, G., Shen, H., Mao, J., Zhang, L., Jiang, Z., Sun, T., Zhang, Z.: Transferrin modified graphene oxide for glioma-targeted drug delivery: in vitro and in vivo evaluations. ACS Appl. Mater. Interfaces **5**(15), 6909–6914 (2013)
196. Zhao, X., Liu, L., Li, X., Zeng, J., Jia, X., Liu, P.: Biocompatible graphene oxide nanoparticle-based drug delivery platform for tumor microenvironment-responsive triggered release of doxorubicin. Langmuir **30**(34), 10419–10429 (2014)
197. Goenka, S., Sant, V., Sant, S.: Graphene-based nanomaterials for drug delivery and tissue engineering. J. Control. Release **173**, 75–88 (2014)
198. Tiwari, H., Karki, N., Pal, M., Basak, S., Verma, R.K., Bal, R., Sahoo, N.G.: Functionalized graphene oxide as a nanocarrier for dual drug delivery applications: The synergistic effect of quercetin and gefitinib against ovarian cancer cells. Colloids Surf. B **178**, 452–459 (2019)
199. Raj, G.K., Singh, E., Hani, U., Ramesh, K.V.R.N.S., Talath, S., Garg, A., Osmani, R.A.M.: Conductive polymers and composites-based systems: An incipient stride in drug delivery and therapeutics realm. J. Control. Release **355**, 709–729 (2023)
200. Willerth, S.M., Sakiyama-Elbert, S.E.: Approaches to neural tissue engineering using scaffolds for drug delivery. Adv. Drug Deliv. Rev. **59**(4–5), 325–338 (2007)
201. Li, Y., Neoh, K. G., & Kang, E. T. (2005). Controlled release of heparin from polypyrrole-poly (vinyl alcohol) assembly by electrical stimulation. *Journal of Biomedical Materials Research Part A: An Official Journal of The Society for Biomaterials, The Japanese Society for Biomaterials, and The*
202. *Australian Society for Biomaterials and the Korean Society for Biomaterials*, *73*(2), 171–181.
203. Garner, B., Georgevich, A., Hodgson, A.J., Liu, L., Wallace, G.G.: Polypyrrole–heparin composites as stimulus-responsive substrates for endothelial cell growth. J. Biomed. Mater. Res. **44**(2), 121–129 (1999)
204. Fabregat, G., Armelin, E., Aleman, C.: Selective detection of dopamine combining multilayers of conducting polymers with gold nanoparticles. J. Phys. Chem. B **118**(17), 4669–4682 (2014)
205. Basim, P., Ajjarapu, S., & Kurakula, M. (2023). Conductive polymers for drug and bioactives delivery. In *Smart Polymer Nanocomposites* (pp. 263–278). Elsevier.
206. Samanta, D., Hosseini-Nassab, N., Zare, R.N.: Electroresponsive nanoparticles for drug delivery on demand. Nanoscale **8**(17), 9310–9317 (2016)
207. Diaconu, A., Chiriac, A. P., Neamtu, I., & Nita, L. E. (2019). Magnetic Polymeric
208. Nanocomposites. *Polymeric Nanomaterials in Nanotherapeutics*, 359–386.
209. Li, X., Xu, W., Xin, Y., Yuan, J., Ji, Y., Chu, S., Luo, Q.: Supramolecular polymer nanocomposites for biomedical applications. Polymers **13**(4), 513 (2021)
210. . Xu, X., Liu, Y., Fu, W., Yao, M., Ding, Z., Xuan, J., ... & Cao, M. (2020). Poly
211. (N-isopropylacrylamide)-based thermoresponsive composite hydrogels for biomedical applications. *Polymers*, *12*(3), 580.
212. An, J., Gou, Y., Yang, C., Hu, F., Wang, C.: Synthesis of a biocompatible gelatin functionalized graphene nanosheets and its application for drug delivery. Mater. Sci. Engg. C **33**, 2827–2837 (2013)
213. Wang, C., Mallela, J., Garapati, U.S.: A chitosan-modified graphene nanogel for noninvasive controlled drug release. Nanomed. Nanotechnol. Biol. Med.. Nanotechnol. Biol. Med. **9**, 903–911 (2013)
214. Hosseini-Nassab, N., Samanta, D., Abdolazimi, Y., Annes, J.P., Zare, R.N.: Electrically controlled release of insulin using polypyrrole nanoparticles. Nanoscale **9**(1), 143–149 (2017)

215. Abidian, M.R., Kim, D.H., Martin, D.C.: Conducting-polymer nanotubes for controlled drug release. Adv. Mater. **18**(4), 405–409 (2006)
216. Syritski, V., Öpik, A., Forsén, O.: Ion transport investigations of polypyrroles doped with different anions by EQCM and CER techniques. Electrochim. Acta. Acta **48**(10), 1409–1417 (2003)
217. Li, Y., Neoh, K.G., Kang, E.T.: Controlled release of heparin from polypyrrole-poly (vinyl alcohol) assembly by electrical stimulation. Journal of Biomedical Materials Research Part A: An Official Journal of The Society for Biomaterials, The Japanese Society for Biomaterials, and The Australian Society for Biomaterials and the Korean Society for Biomaterials **73**(2), 171–181 (2005)
218. Ge, J., Neofytou, E., Cahill, T.J., III., Beygui, R.E., Zare, R.N.: Drug release from electric-field-responsive nanoparticles. ACS Nano **6**(1), 227–233 (2012)
219. Ren, B., Chen, X., Du, S., Ma, Y., Chen, H., Yuan, G., Niu, X.: Injectable polysaccharide hydrogel embedded with hydroxyapatite and calcium carbonate for drug delivery and bone tissue engineering. Int. J. Biol. Macromol.Macromol. **118**, 1257–1266 (2018)
220. Gupta, H., Aqil, M., Khar, R. K., Ali, A., Bhatnagar, A., & Mittal, G. (2010). Sparfloxacin-loaded PLGA nanoparticles for sustained ocular drug delivery. *Nanomedicine: nanotechnology, biology and medicine*, *6*(2), 324–333.
221. Bourges, J.L., Gautier, S.E., Delie, F., Bejjani, R.A., Jeanny, J.C., Gurny, R., Behar-Cohen, F.F.: Ocular drug delivery targeting the retina and retinal pigment epithelium using polylactide nanoparticles. Invest. Ophthalmol. Vis. Sci.Ophthalmol. Vis. Sci. **44**(8), 3562–3569 (2003)
222. Paula, J.S., Ribeiro, V.R.C., Chahud, F., Cannellini, R., Monteiro, T.C., de Lima Gomes, E.C., Silva-Cunha, A.: Bevacizumab-loaded polyurethane subconjunctival implants: effects on experimental glaucoma filtration surgery. J. Ocul. Pharmacol. Ther.Ocul. Pharmacol. Ther. **29**(6), 566–573 (2013)
223. Yuan, Q., Shah, J., Hein, S.R.D.K., Misra, R.D.K.: Controlled and extended drug release behavior of chitosan-based nanoparticle carrier. Acta Biomater.Biomater. **6**(3), 1140–1148 (2010)
224. Yuan, Q., Hein, S., Misra, R.D.K.: New generation of chitosan-encapsulated ZnO quantum dots loaded with drug: synthesis, characterization and in vitro drug delivery response. Acta Biomater.Biomater. **6**(7), 2732–2739 (2010)
225. Mohamed, W.S., Mostafa, A.B., Nasr, H.E.: Characterization and application of intercalated montmorillonite with verapamil and its polymethyl methacrylate nanocomposite in drug delivery. Polym.-Plast. Technol. Eng..-Plast. Technol. Eng. **53**(14), 1425–1433 (2014)
226. Caetano, B.L., Guibert, C., Fini, R., Fresnais, J., Pulcinelli, S.H., Ménager, C., Santilli, C.V.: Magnetic hyperthermia-induced drug release from ureasil-PEO-γ-Fe2O3 nanocomposites. RSC Adv. **6**(68), 63291–63295 (2016)
227. Truffault, L., Magnani, M., Hammer, P., Santilli, C.V., Pulcinelli, S.H.: Structural and optical features of ureasiloxane–polyethylene oxide hybrids containing CeO2 nanoparticles. Colloids Surf. A **471**, 73–80 (2015)
228. Wang, G., Maciel, D., Wu, Y., Rodrigues, J., Shi, X., Yuan, Y., Li, Y.: Amphiphilic polymer-mediated formation of laponite-based nanohybrids with robust stability and pH sensitivity for anticancer drug delivery. ACS Appl. Mater. Interfaces **6**(19), 16687–16695 (2014)
229. Miao, W., Shim, G., Lee, S., Lee, S., Choe, Y.S., Oh, Y.K.: Safety and tumor tissue accumulation of pegylated graphene oxide nanosheets for co-delivery of anticancer drug and photosensitizer. Biomaterials **34**(13), 3402–3410 (2013)
230. Salahuddin, N., Kenawy, E.R., Abdeen, R.: Polyoxypropylene–montmorillonite nanocomposites for drug-delivery vehicles: Preparation and characterization. J. Appl. Polym. Sci.Polym. Sci. **125**(S1), E157–E166 (2012)
231. Wang, J., Tang, B., Tsuzuki, T., Liu, Q., Hou, X., Sun, L.: Synthesis, characterization and adsorption properties of superparamagnetic polystyrene/Fe3O4/graphene oxide. Chem. Eng. J. **204**, 258–263 (2012)
232. Qin, X.C., Guo, Z.Y., Liu, Z.M., Zhang, W., Wan, M.M., Yang, B.W.: Folic acid-conjugated graphene oxide for cancer targeted chemo-photothermal therapy. J. Photochem. Photobiol. B.Photochem. Photobiol. B. **120**, 156–162 (2013)

233. Prajapati, D. G., & Kandasubramanian, B. (2019). Progress in the development of intrinsically
234. conducting polymer composites as biosensors. *Macromolecular chemistry and physics, 220*(10), 1800561
235. Lei, W., Si, W., Xu, Y., Gu, Z., Hao, Q.: Conducting polymer composites with grapheme for use in chemical sensors and biosensors. Microchim. Acta. Acta **181**, 707–722 (2014)
236. Luo, X., Killard, A.J., Morrin, A., Smyth, M.R.: Enhancement of a conducting polymer-based biosensor using carbon nanotube-doped polyaniline. Anal. Chim. Acta **575**(1), 39–44 (2006)
237. Dahiya, T., Yadav, S., Yadav, N., Mann, A., Sharma, M., Rana, J.S.: Monitoring of BNP cardiac biomarker with major emphasis on biosensing methods: a review. Sensors International **2**, 100103 (2021)
238. Kuila, T., Bose, S., Khanra, P., Mishra, A.K., Kim, N.H., Lee, J.H.: Recent advances in graphene-based biosensors. Biosens. Bioelectron.. Bioelectron. **26**(12), 4637–4648 (2011)
239. Nguyen, A.H., Sim, S.J.: Nanoplasmonic biosensor: Detection and amplification of dual bio-signatures of circulating tumor DNA. Biosens. Bioelectron.. Bioelectron. **67**, 443–449 (2015)
240. Nguyen, P., Berry, V.: Graphene interfaced with biological cells: opportunities and challenges. The Journal of Physical Chemistry Letters **3**(8), 1024–1029 (2012)
241. Kim, J., Park, S.J., Min, D.H.: Emerging approaches for graphene oxide biosensor. Anal. Chem. **89**(1), 232–248 (2017)
242. Liu, F., Choi, K.S., Park, T.J., Lee, S.Y., Seo, T.S.: Graphene-based electrochemical biosensor for pathogenic virus detection. BioChip J. J. **5**, 123–128 (2011)
243. Maharana, P.K., Jha, R.: Chalcogenide prism and graphene multilayer based surface plasmon resonance affinity biosensor for high performance. Sens. Actuators, B Chem. **169**, 161–166 (2012)
244. Qiu, J.-D., Huang, J., Liang, R.-P.: Nanocomposite film based on graphene oxide for high performance flexible glucose biosensor. Sens. Actuat. B **160**, 287–294 (2011)
245. Chan, C., Yang, H., Han, D., Zhang, Q., Ivaska, A., Niu, L.,(2010). Graphene/AuNPs/chitosan nano□composites film for glucose biosensing. Biosens. Bioelectr., 25:1070–1074. (2 times ref.)
246. Kang, X., Wang, J., Wu, H., Aksay, I.A., Liu, J., Lin, Y.: Glucose oxidase-graphene-chitosan modified electrode for direct electrochemistry and glucose sensing. Biosens. Bioelect. **25**, 901–905 (2009)
247. Huang, K.-J., Miao, Y.-X., Wang, L., Gan, T., Yu, M., Wang, L.-L.: Direct electrochemistry of hemoglobin based on chitosan-ionic liquid-ferrocene/graphene composite film. Process Biochem.Biochem. **47**, 1171–1177 (2012)
248. Liu, S., Xing, X., Yu, J., Lian, W., Li, J., Cui, M., Huang, J.: A novel label-free electrochemical aptasensor based on graphene–polyaniline composite film for dopamine determination. Biosens. Bioelectron.. Bioelectron. **36**(1), 186–191 (2012)
249. Liu, X., Xie, L., Li, H.: Electrochemical biosensor based on reduced graphene oxide and Au nanoparticles entrapped in chitosan/silica sol–gel hybrid membranes for determination of dopamine and uric acid. J. Electroanal. Chem.Electroanal. Chem. **682**, 158–163 (2012)
250. Zheng, M., Gao, F., Wang, Q., Cai, X., Jiang, S., Huang, L.: Electrocatalytical oxidation and sensitive determination of acetaminophen on glassy carbon electrode modified with graphene–chitosan composite. Mater. Sci. Eng. C **33**, 1514–1520 (2013)
251. Niu, X., Yang, W., Renet, J., et al.: Electrochemical behaviors and simultaneous determination of guanine and adenine based on graphene-ionic liquid-chitosan composite film modified glassy carbon electrode. Electrochim. Acta. Acta **80**, 346–353 (2012)
252. Bao, Y., Yuan, H., Hu, Y., Yao, T., Huang, H.: A novel electrochemical DNA biosensor based on graphene and polyaniline nanowires. Electrochim. Acta. Acta **56**, 2676–2681 (2011)
253. Gambhir, A., Gerard, M., Mulchandani, A.K., Malhotra, B.D.: Coimmobilization of urease and glutamate dehydrogenase in electrochemically prepared polypyrrole-polyvinyl sulfonate films. Appl. Biochem. Biotechnol.Biochem. Biotechnol. **96**, 249–257 (2001)
254. Vidal, J.C., Garcia-Ruiz, E., Espuelas, J., Aramendia, T., Castillo, J.R.: Comparison of biosensors based on entrapment of cholesterol oxidase and cholesterol esterase in electropolymerized

films of polypyrrole and diaminonaphthalene derivatives for amperometric determination of cholesterol. Anal. Bioanal. Chem.Bioanal. Chem. **377**, 273–280 (2003)
255. Singh, S., Solanki, P.R., Pandey, M.K., Malhotra, B.D.: Cholesterol biosensor based on cholesterol esterase, cholesterol oxidase and peroxidase immobilized onto conducting polyaniline films. Sens. Actuators, B Chem. **115**(1), 534–541 (2006)
256. Singh, S., Chaubey, A., Malhotra, B.D.: Preparation and characterization of an enzyme electrode based on cholesterol esterase and cholesterol oxidase immobilized onto conducting polypyrrole films. J. Appl. Polym. Sci.Polym. Sci. **91**(6), 3769–3773 (2004)
257. Ramanathan, K., Pandey, S.S., Kumar, R., Gulati, A., Murthy, A.S.N., Malhotra, B.D.: Covalent immobilization of glucose oxidase to poly (O-amino benzoic acid) for application to glucose biosensor. J. Appl. Polym. Sci.Polym. Sci. **78**(3), 662–667 (2000)
258. Zhu, M., Jiang, Z., Jing, W.: Fabrication of polypyrrole–glucose oxidase biosensor based on multilayered interdigitated ultramicroelectrode array with containing trenches. Sens. Actuators, B Chem. **110**(2), 382–389 (2005)
259. dos Santos Riccardi, C., Yamanaka, H., Josowicz, M., Kowalik, J., Mizaikoff, B., Kranz, C.: Label-free DNA detection based on modified conducting polypyrrole films at microelectrodes. Anal. Chem. **78**(4), 1139–1145 (2006)
260. Fu, Y., Yuan, R., Chai, Y., Zhou, L., Zhang, Y.: Coupling of a reagentless electrochemical DNA biosensor with conducting polymer film and nanocomposite as matrices for the detection of the HIV DNA sequences. Anal. Lett. **39**(3), 467–482 (2006)
261. Gautier, C., Cougnon, C., Pilard, J.F., Casse, N.: Label-free detection of DNA hybridization based on EIS investigation of conducting properties of functionalized polythiophene matrix. J. Electroanal. Chem.Electroanal. Chem. **587**(2), 276–283 (2006)
262. Chaubey, A., Pande, K.K., Singh, V.S., Malhotra, B.D.: Co-immobilization of lactate oxidase and lactate dehydrogenase on conducting polyaniline films. Anal. Chim. Acta **407**(1–2), 97–103 (2000)
263. Chaubey, A., Gerard, M., Singhal, R., Singh, V.S., Malhotra, B.D.: Immobilization of lactate dehydrogenase on electrochemically prepared polypyrrole–polyvinylsulphonate composite films for application to lactate biosensors. Electrochim. Acta. Acta **46**(5), 723–729 (2001)
264. Choong, C.L., Shim, M.B., Lee, B.S., Jeon, S., Ko, D.S., Kang, T.H., Chung, U.I.: Highly stretchable resistive pressure sensors using a conductive elastomeric composite on a micropyramid array. Adv. Mater. **26**(21), 3451–3458 (2014)
265. Fang, Y., Ni, Y., Zhang, G., Mao, C., Huang, X., Shen, J.: Biocompatibility of CS–PPy nanocomposites and their application to glucose biosensor. Bioelectrochemistry **88**, 1–7 (2012)
266. Liu, G., Gooding, J.J.: Towards the fabrication of label-free amperometric immunosensors using SWNTs. Electrochem. Commun.. Commun. **11**(10), 1982–1985 (2009)
267. Gao, M., Dai, L., & Wallace, G. G. (2003). Biosensors based on aligned carbon nanotubes coated with inherently conducting polymers. *Electroanalysis: An International Journal Devoted to Fundamental and Practical Aspects of Electroanalysis*, *15*(13), 1089–1094.
268. Xu, B., Ye, M.L., Yu, Y.X., Zhang, W.D.: A highly sensitive hydrogen peroxide amperometric sensor based on MnO2-modified vertically aligned multiwalled carbon nanotubes. Anal. Chim. Acta **674**(1), 20–26 (2010)
269. Yi, N., & Abidian, M. R. (2016). Conducting polymers and their biomedical applications. In *Biosynthetic polymers for medical applications* (pp. 243–276). Woodhead Publishing.
270. Layton, K. N., & Abidian, M. R. (2011, April). Conducting polymer nanofiber-based biosensor for detection of neurochemicals. In *2011 5th International IEEE/EMBS Conference on Neural Engineering* (pp. 298–301). IEEE.
271. Deng, C., Li, M., Xie, Q., Liu, M., Yang, Q., Xiang, C., Yao, S.: Construction as well as EQCM and SECM characterizations of a novel Nafion/glucose oxidase-glutaraldehyde/poly (thionine)/Au enzyme electrode for glucose sensing. Sens. Actuators, B Chem. **122**(1), 148–157 (2007)
272. Gu, H., Su, X., Loh, K.P.: Conductive polymer-modified boron-doped diamond for DNA hybridization analysis. Chem. Phys. Lett. **388**(4–6), 483–487 (2004)

273. Yang, G., Liang, J., Hu, X., Liu, M., Zhang, X., Wei, Y.: Recent Advances on Fabrication of Polymeric Composites Based on Multicomponent Reactions for Bioimaging and Environmental Pollutant Removal. Macromol. Rapid Commun.. Rapid Commun. **42**(6), 2000563 (2021)
274. Wang, W., Wang, L., Liu, S., Xie, Z.: Metal–organic frameworks@ polymer composites containing cyanines for near-infrared fluorescence imaging and photothermal tumor therapy. Bioconjug. Chem.. Chem. **28**(11), 2784–2793 (2017)
275. Fu, Q., Zhu, R., Song, J., Yang, H., Chen, X.: Photoacoustic imaging: contrast agents and their biomedical applications. Adv. Mater. **31**(6), 1805875 (2019)
276. Liu, S., Wei, X., Chu, M., Peng, J., Xu, Y.: Synthesis and characterization of iron oxide/ polymer composite nanoparticles with pendent functional groups. Colloids Surf. B **51**(2), 101–106 (2006)
277. Garcea, S.C., Wang, Y., Withers, P.J.: X-ray computed tomography of polymer composites. Compos. Sci. Technol. **156**, 305–319 (2018)
278. Wang, K.H., Wang, Y.M., Chiu, L.H., Chen, T.C., Tsai, Y.H., Zuo, C.S., Lai, W.F.T.: Optical imaging of ovarian cancer using a matrix metalloproteinase-3-sensitive near-infrared fluorescent probe. PLoS ONE **13**(2), e0192047 (2018)
279. Cheng, Z., Lin, J.: Synthesis and application of nanohybrids based on upconverting nanoparticles and polymers. Macromol. Rapid Commun.. Rapid Commun. **36**(9), 790–827 (2015)
280. Ai, H., Flask, C., Weinberg, B., Shuai, X.T., Pagel, M.D., Farrell, D., Gao, J.: Magnetite-loaded polymeric micelles as ultrasensitive magnetic-resonance probes. Adv. Mater. **17**(16), 1949–1952 (2005)
281. Li, Yi, Thavasyappan Thambi, and Doo Sung Lee. "Co-delivery of drugs and genes using polymeric nanoparticles for synergistic cancer therapeutic effects." *Advanced healthcare materials* 7, no. 1 (2018): 1700886.
282. Zheng, D.W., Lei, Q., Chen, S., Qiu, W.X., Liu, M.Y., Chen, X., Chu, P.K.: Supermolecular theranostic capsules for pH-sensitive magnetic resonance imaging and multi-responsive drug delivery. Journal of Materials Chemistry B **3**(43), 8499–8507 (2015)
283. Liu, M., Bauman, L., Nogueira, C.L., Aucoin, M.G., Anderson, W.A., Zhao, B.: Antimicrobial polymeric composites for high-touch surfaces in healthcare applications. Current Opinion in Biomedical Engineering **22**, 100395 (2022)
284. Mouritz, A.P., Galos, J., Linklater, D.P., Ladani, R.B., Kandare, E., Crawford, R.J., Ivanova, E.P.: Towards antiviral polymer composites to combat COVID-19 transmission. Nano Select **2**(11), 2061–2071 (2021)
285. Mallakpour, S., Azadi, E., Hussain, C.M.: Recent breakthroughs of antibacterial and antiviral protective polymeric materials during COVID-19 pandemic and after pandemic: Coating, packaging, and textile applications. Curr. Opin. Colloid Interface Sci.Opin. Colloid Interface Sci. **55**, 101480 (2021)
286. Balasubramaniam, B., Prateek, Ranjan, S., Saraf, M., Kar, P., Singh, S. P., ... & Gupta, R. K. (2020). Antibacterial and antiviral functional materials: chemistry and biological activity toward tackling COVID-19-like pandemics. *ACS Pharmacology & Translational Science, 4*(1), 8-54
287. Bhattacharyya, S. K., Maiti, S., Das, N. C., & Banerjee, S. (2023). Antibacterial and Antiviral Functional Materials Based on Polymer Nanocomposites. In *Antibacterial and Antiviral Functional Materials, Volume 1* (pp. 171–202). American Chemical Society.
288. Ippili, S., Jung, J.S., Thomas, A.M., Vuong, V.H., Lee, J.M., Sha, M.S., Yoon, S.G.: An Overview of Polymer Composite Films for Antibacterial Display Coatings and Sensor Applications. Polymers **15**(18), 3791 (2023)
289. Kim, E.S., Kim, S.H., Lee, C.H.: Electrospinning of polylactide fibers containing silver nanoparticles. Macromol. Res.. Res. **18**, 215–221 (2010)
290. Zare, Y., Shabani, I.: Polymer/metal nanocomposites for biomedical applications. Mater. Sci. Eng. C **60**, 195–203 (2016)

291. Olmos, D., González-Benito, J.: Polymeric materials with antibacterial activity: A Review. Polymers **13**(4), 613 (2021)
292. Dorazilová, J., Muchová, J., Šmerková, K., Kočiová, S., Diviš, P., Kopel, P., Vojtová, L.: Synergistic effect of chitosan and selenium nanoparticles on biodegradation and antibacterial properties of collagenous scaffolds designed for infected burn wounds. Nanomaterials **10**(10), 1971 (2020)
293. Khalid, H., Iqbal, H., Zeeshan, R., Nasir, M., Sharif, F., Akram, M., Khan, A.F.: Silk fibroin/collagen 3D scaffolds loaded with TiO2 nanoparticles for skin tissue regeneration. Polym. Bull.. Bull. **78**, 7199–7218 (2021)
294. Carayon, I., Szarlej, P., Gnatowski, P., Piłat, E., Sienkiewicz, M., Glinka, M., Kucińska-Lipka, J.: Polyurethane based hybrid ciprofloxacin-releasing wound dressings designed for skin engineering purpose. Adv. Med. Sci. **67**(2), 269–282 (2022)
295. Khan, Z. A., Jamil, S., Akhtar, A., Bashir, M. M., & Yar, M. (2019). Chitosan based hybrid materials used for wound healing applications-A short review. *International Journal of Polymeric Materials and Polymeric Biomaterials.*
296. Lee, S.J., Kang, M.S., Oh, J.S., Na, H.S., Lim, Y.J., Jeong, Y.I., Lee, H.C.: Caffeic acid-conjugated chitosan derivatives and their anti-tumor activity. Arch. Pharmacal Res. **36**, 1437–1446 (2013)
297. Shoseyov, O., Kam, D., Ben Shalom, T., Shtein, Z., Vinkler, S., & Posen, Y. (2019). Nanocellulose composite biomaterials in industry and medicine. *Extracellular sugar-based biopolymers matrices*, 693–784.
298. Boonpavanitchakul, K., Pimpha, N., Kangwansupamonkon, W., Magaraphan, R.: Processing and antibacterial application of biodegradable sponge nano-composite materials of silver nanoparticles and silk sericin. Eur. Polymer J. **130**, 109649 (2020)
299. Kamalathevan, P., Ooi, P.S., Loo, Y.L.: Silk-based biomaterials in cutaneous wound healing: a systematic review. Adv. Skin Wound Care **31**(12), 565–573 (2018)
300. Yang, C., Xue, R., Zhang, Q., Yang, S., Liu, P., Chen, L., Wei, Y.: Nanoclay cross-linked semi-IPN silk sericin/poly (NIPAm/LMSH) nanocomposite hydrogel: An outstanding antibacterial wound dressing. Mater. Sci. Eng. C **81**, 303–313 (2017)
301. Li, H., Williams, G. R., Wu, J., Wang, H., Sun, X., & Zhu, L. M. (2017). Poly
302. (N-isopropylacrylamide)/poly (l-lactic acid-co-ε-caprolactone) fibers loaded with ciprofloxacin as wound dressing materials. *Materials Science and Engineering: C, 79*, 245–254.
303. Wang, X., Li, W.: Biodegradable mesoporous bioactive glass nanospheres for drug delivery and bone tissue regeneration. Nanotechnology **27**(22), 225102 (2016)
304. Fereshteh, Z., Nooeaid, P., Fathi, M., Bagri, A., Boccaccini, A.R.: Mechanical properties and drug release behavior of PCL/zein coated 45S5 bioactive glass scaffolds for bone tissue engineering application. Data Brief **4**, 524–528 (2015)
305. Zuniga, J.M., Cortes, A.: The role of additive manufacturing and antimicrobial polymers in the COVID-19 pandemic. Expert Rev. Med. Devices **17**(6), 477–481 (2020)
306. Praharaj, R., Swain, S., & Raûtray, T. R. (2023). Bioceramics for antibacterial and antiviral applications. *Advanced Ceramic Coatings for Biomedical Applications*, 347.
307. Swain, S. K., & Jawaid, M. (Eds.). (2019). *Nanostructured Polymer Composites for Biomedical Applications*. Elsevier.
308. Zheng, J., Rahman, N., Li, L., Zhang, J., Tan, H., Xue, Y., Zhao, Y., Zhai, J., Zhao, N., Xu, F., et al.: Biofunctionalization of electrospun fiber membranes by LbL-collagen/chondroitin sulfate nanocoating followed by mineralization for bone regeneration. Mater. Sci. Eng. C **128**, 112295 (2021)
309. Lu, J., Ma, S., Sun, J., Xia, C., Liu, C., Wang, Z., Gu, Z.: Manganese ferrite nanoparticle micellar nanocomposites as MRI contrast agent for liver imaging. Biomaterials **30**(15), 2919–2928 (2009)
310. Akbar, M. U., Athar, M. M., Bhatti, I. A., Bhatti, H. N., Khosa, M. K., Zia, K. M., ... & Ahmad, M. N. (2020). Biomedical applications of bionanocomposites. In *Bionanocomposites* (pp. 457–483). Elsevier.

Chapter 17
Piezoelectric Polymer Composites for Energy Harvesting

Siddharth Parthasarathy, Saubhagyalaxmi Behera, and Srikanta Moharana

Abstract Polymer composites are becoming increasingly attractive for energy harvesting applications because they possess a distinctive blend of mechanical flexibility, low weight, and adjustable electrical characteristics. The composites can effectively transform different types of energy, such as mechanical vibrations or thermal gradients, into electrical energy by utilizing piezoelectric or thermoelectric devices that are included in the polymer matrix. Moreover, the capacity to customize these composites' composition and structure enables enhancing their performance in particular energy harvesting situations. This chapter seeks to offer a thorough examination of the latest progressions in polymer composites utilized for energy harvesting. It explores several categories of polymers utilized, the methods employed in their processing, and the crucial elements that impact their performance. Moreover, it emphasizes the difficulties and potential opportunities in this area, such as the advancement of innovative substances and the incorporation of these combinations into wearable or portable devices for autonomous functions. In summary, this analysis explains the potential of piezoelectric polymer-based composites as very promising energy harvesters. This greatly enhances the advancement of sustainable and efficient energy technology.

Keywords Polymer · Composites · Energy harvesting · Energy efficiency · Piezoelectric · Thermoelectric devices

---

S. Parthasarathy · S. Behera (✉)
School of Applied Sciences, Centurion University of Technology and Management, Bhubaneswar, Odisha, India
e-mail: saubhagyalaxmi.behera@cutm.ac.in

S. Moharana
Department of Chemistry, School of Applied Sciences, Centurion University of Technology and Management, Jatni, Bhubaneswar, Odisha, India

S. Moharana et al. (eds.), *Polymer Composites*, Engineering Materials,
https://doi.org/10.1007/978-981-97-2075-0_17

17.1 Introduction

In recent few times, piezoelectric polymer composites have been very promising materials among academic and industrial researchers owing to their distinctive blend of mechanical flexibility, lightweight characteristics, adjustable electrical properties, and wide range of energy harvesting applications [1–3]. These composites consist of a polymer matrix that is strengthened by including several types of fillers, like carbon nanotubes, graphene, or piezoelectric materials. Polymer composites possess a high degree of mechanical flexibility, which facilitates their seamless integration into wearable devices or flexible electronics [4, 5]. This integration enables the extraction of energy from human body motions or environmental vibrations. Moreover, their inherent lightness renders them well-suited for portable energy harvesting systems that may be conveniently transported or deployed in distant regions. Moreover, by the manipulation of filler type and concentration, it is possible to tailor the electrical characteristics of polymer composites to optimize their performance in energy conversion and storage applications [6, 7]. In general, polymer composites have significant potential for the advancement of efficient and adaptable energy harvesting systems. The capacity to extract energy from ambient vibrations or motions presents novel opportunities for energizing small-scale electronic devices and sensors, therefore diminishing the dependence on conventional batteries and prolonging their operational lifespan [8]. Nevertheless, to effectively utilize the capabilities of polymer composites for energy harvesting, it is imperative to comprehend their fundamental principles and enhance their efficiency by employing customized design methodologies. This study is to offer a comprehensive summary of the latest progress in energy harvesting systems based on polymer composites. It will emphasize the main obstacles and prospects in this swiftly developing area [8–10]. Polymer composites are increasingly becoming recognized as highly advantageous materials for energy harvesting, primarily because of their distinctive characteristics, including flexibility and low weight. These materials may be incorporated into a range of applications, such as wearable gadgets and Internet of Things (IoT) sensors, facilitating the creation of autonomous systems [11, 12]. Nevertheless, there are other obstacles that require attention, such as enhancing the efficacy and durability of energy conversion mechanisms in piezoelectric polymer composites. Furthermore, the design techniques should prioritize improving the compatibility across various components to guarantee optimal performance [13, 14]. In summary, comprehending the principles and enhancing the efficiency of energy harvesting systems based on polymer composites would facilitate their extensive use across diverse sectors. These matters can be addressed by conducting research and development activities focused on enhancing the material characteristics of polymer composites and optimizing their energy conversion efficiency [15, 16]. Moreover, it is imperative to foster cooperation among researchers, engineers, and industry experts to foster the creation of cutting-edge solutions and propel the development of self-powered systems using IoT sensors [17]. In the present chapter, we will investigate several composites based on PVDF and P(VDF-TrFE) that are used for energy harvesting applications.

17.2 Background and State of an Art

17.2.1 Overview of Energy Harvesting in Polymer Composites

The technique of extracting a range of energy (nW-μW-mW range) from outside sources including vibration, thermal, and optical that might otherwise disappear into the surroundings is known as energy harvesting. Because energy harvesting only captures small amounts of energy, its applications are mostly limited to wearable electronics and wearable sensors. Energy harvesting from mechanical motion is considered the most dependable method for offering an alternative to traditional rechargeable batteries. This is due to the fact that the majority of human activities depend on mechanical movement that is not influenced by the surrounding environment. Recently, piezoelectric-based technologies have garnered significant attention among various energy harvesting procedures because of their effective energy conversion, affordability, and ease of use. Because of their non-centrosymmetric crystal structure, piezoelectric can create electrical dipoles, which are localized separations of charge. The commonly known piezoelectric materials are:

- Naturally existing biological piezoelectric materials include deoxyribonucleic acid, cellulose, collagen, human bone, and tendon.
- Quartz, Rochelle's salt, topaz, tourmaline group minerals, and other piezoelectric crystals.
- Synthetic piezoelectric ceramics, including Zinc Oxide (ZnO), Potassium Niobate ($KNbO_3$), Bismuth Ferrite ($BiFeO_3$), Lead Zirconium Titanate (PZT), and Barium Titanate ($BaTiO_3$).
- Synthetic piezoelectric polymers cover a wide range of materials, including polyimide, cellular polypropylene, poly(vinylidene fluoride) and its copolymer like poly (vinylidene fluoride-co-trifluoroethylene) P(VDF-TrFE), and odd-numbered polyamides.

Ceramic perovskite oxide piezoelectric energy harvesters can generate more power, but they aren't as mechanically flexible as piezoelectric polymers, which are extremely flexible [18–20]. At now, polymer-based materials are being examined due to their enormous potential in a wide range of advanced technological applications. These materials possess a remarkable quality known as high dielectric permittivity. Electromechanical transducers, condensers, piezoelectric generators, and integrated parts are appealing because they can adapt to new technologies. Their outstanding mechanical qualities, amazing breakdown strength, low density, high permittivity, ease of processing, and lightweight nature all contribute to their extensive application. Furthermore, polymer composites may be easily and quickly fabricated, which adds to their appeal in a variety of applications. It is important to remember that, in comparison to dielectric oxides, polymers have a comparatively lower dielectric permittivity. As a result, obtaining a high dielectric permittivity in polymer composites persists despite the addition of high permittivity components. Since polymers have lower piezoelectricity than current piezoceramics, this represents a substantial obstacle to their application in energy harvesting. Adding conductive or piezoceramic particles to the polymer matrix is one method for increasing the polymer's piezoelectricity (two-phase composite). Interfacial polarization increases the piezoelectricity of the composite because of the addition of piezoceramics. The overall electrical characteristics improve with the growing concentration of ceramics material in the matrix. Several attempts have been undertaken thus far to improve the piezoelectric efficiency of nanogenerators composed of different piezoelectric polymer composites. So continuously researchers have focused on improving the properties of polymers or fillers by the addition of dopant ions to the nanofiller or by modifying the polymer chain, to achieve even higher performance.

17.2.2 *Piezoelectric Composite Materials Used in Energy Harvesting*

The primary factor used to characterize piezoelectric materials is the piezoelectric charge constant, denoted as d. This constant represents the amount of electric charge generated in response to applied mechanical stress. It is a crucial parameter in designing and optimizing piezoelectric devices for various applications, such as sensors, actuators, and energy harvesting systems. The following formulas define the electromechanical coupling of piezoelectric materials.

$$S_i = S_{ij}^E T_j + d_{ij} E_k \tag{17.1}$$

$$D_{ij} = \varepsilon_{ik}^T E_k + d_{ij} T_j \tag{17.2}$$

where: s^E is the elastic compliance matrix evaluated at a constant electric field; d is a tensor of piezoelectric strain coefficients; ε^T is a tensor of permittivity values evaluated at constant stress; and the subscripts i, j, and k represent the three spatial dimensions. The mechanical and electrical effects are included in the strain tensor, denoted as T, and the stress tensor, denoted as S [21]. Throughout the material operation, a single piezoelectric strain coefficient predominates in the output. Piezoelectric materials often operate in one of two modes: the 33-mode, also known as the stock configuration, and the 31-mode, also known as the bender configuration. It is believed that the poling direction for both is always in the direction of '3'. The material is stretched in the poling or "3" direction and the electric voltage is recovered in the "3" direction when the voltage and stress are acting in the "3" direction in the 33 modes. The material is polarized in the "3" direction and the mechanical tension operates in the "1" direction in the 31 modes. The 33 arrangements typically exhibit greater values of the piezoelectric charge constant (d_{33}) than the other configuration. Materials with substantial electromechanical coupling coefficients and easy deformation to create greater strains are very attractive for energy harvesting applications. Furthermore, for low-frequency (~1 Hz) human mechanical movement-enabled energy harvesting applications, such as breathing or walking, the material's maximum stress or strain can serve as a representation of the elastic energy W contained in it.

$$w = \frac{1}{2}S \times T \times V = \frac{1}{2}\frac{T^2}{Y}V \tag{17.3}$$

where Y represents Young's modulus and V stands for volume. The stress of the piezoelectric element produces electrical energy, which is supplied by:

$$E_C = \frac{1}{2}D \times E \times V = \frac{1}{2}\frac{\mathrm{d}_{33}^2 T^2}{\varepsilon}V = \frac{1}{2}\frac{\mathrm{d}_{33}^2 S^2}{\varepsilon}Y^2 V \tag{17.4}$$

The electromechanical coupling factor, which represents the efficiency of energy conversion, is derived by using Eqs. 13.3 and 13.4.

$$k^2 = \frac{E_c}{w} = y\frac{\mathrm{d}_{33}^2}{\varepsilon} \tag{17.5}$$

The d_{33} constant signifies the piezoelectric coefficient, which governs the material's capacity to transform mechanical energy into electrical energy. Hence, it is imperative to choose a material with a substantial d33 constant in order to optimize the energy harvesting potential of the system. Furthermore, it is crucial to take into account other variables, such as the material's mechanical robustness and its compatibility with the intended use, in order to guarantee sustained effectiveness and dependability. For the majority of wearable applications; strain-driven operation mode is preferred. The choice of matrix and filler, as well as the filler content, can change a composite's property. This allows modifying different features for a variety of applications. So, to get maximum conversion efficiency both the filler and polymer matrix should have large d_{33} values. When compared to other bulk polymers, PVDF exhibits the greatest piezoelectric coefficient of around 33pCN^{-1}, out of all the piezoelectric polymers previously discussed [22]. Among polymeric systems, PVDF offers excellent mechanical and thermal stability along with remarkable ferroelectric, pyroelectric, and piezoelectric capabilities [23, 24]. As a semi-crystalline polymer, PVDF can take one of four distinct forms determined by the relative positions of its carbon-hydrogen and carbon-F bond orientations [4]. These are named α-phase (TGTG', form II), β-phase (TTTT, from I), γ-phase (T3GT3G', form III), and δ-phase [25]. Out of all these polymorph phases, the β-phase possesses a planar zigzag structure that is all-trans (TTTT) (as shown in Fig. 17.1). The effective dipole moment in the direction perpendicular to the carbon backbone is the sum of the dipole moments of the C-F and C-H bonds [26]. The resultant dipole moment of a γ-phase is significantly lower than that of a β-phase, even though both phases are polar. Hence, the β-phase of PVDF is very desirable for ferroelectric and piezoelectric uses.

Because of the large electro-negativity difference between its hydrogen and fluorine atoms, PVDF exhibits good dielectric characteristics. The PVDF unit's

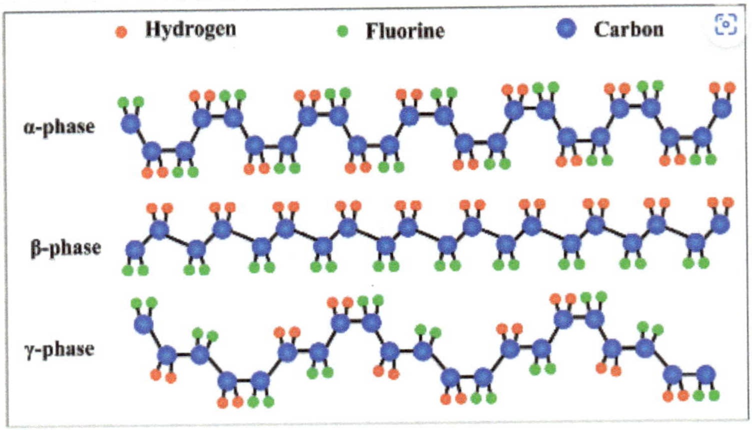

Fig. 17.1 Schematic depiction of the chain configuration for the α, β, and γ-phases of PVDF [26]. Open access under a CC BY 4.0 license

monomer's electric dipole moment exhibits significant variation between phases, namely for α-phase as 5×10^{-30}C.m. and β-phase as 8×10^{-30}C m. respectively [27]. Researchers have focused on polyvinylidene fluoride (PVDF) as the most researched piezoelectric polymer because of its low acoustic impedance, rapid electromechanical response, and flexibility, which permits its usage without a substrate layer [28, 29]. The majority of researchers are interested in this, particularly those who study sensor and actuator devices [30]. The intrinsic piezoelectric effect in PVDF is caused by the dipolar orientation of the molecule segments in the crystal phase for the molecular chain axis. The β-phase has the greatest piezoelectric effect since it is the most electroactive phase. PVDF's piezoelectric response is influenced by several variables linked to fluctuations in the dipolar moment in addition to dimensional changes in the sample [31]. Research conducted by Broadhurst et al. indicates that dimensional changes account for only two-thirds of piezoelectricity. The other one-third is derived from various sources that change dipole moments at constant dimensional parameters, including reaction fields and reversible crystallization at the lamellar interface. Furthermore, the copolymerization process is used to generate a variety of PVDF copolymers with comparable piezoelectric qualities. Poly (vinylidene fluoride trifluoroethylene) (PVDF-TrFE) is a useful PVDF copolymer for the all-trans conformation because of its $(-CH_2-CF_2-)n-(CF_2-CFH-)m$ repeating unit. This is because the polymer chains' potential energies are altered and steric blockage is caused by the substitution of an additional fluorine atom for a hydrogen atom [32, 33]. Compared to PVDF, it exhibits a greater piezoelectric action and has a more polar β -phase in its crystal structure [34]. The P(VDF-TrFE) polymer rapidly solidifies into the β-phase upon cooling from its molten state. This polymer has remarkable piezoelectric, pyroelectric, and ferroelectric capabilities due to the inhibition of functional group rotation along the carbon chain backbone when the TrFE unit is introduced into

PVDF [24, 35–37]. So, it is concluded that the PVDF and its copolymer P(VDF-TrFE) are the most preferable polymer matrix for energy harvesting applications. In view of the above this chapter mainly reviews PVDF and P(VDF-TrFE) based polymer composites for energy harvesting purposes. The non-ferroelectric oxides like ZnO or the ferroelectric ceramic oxides are found most suitable fillers in the above matrixes. However, the flexibility could be hampered by the polymer matrix's excessive ceramic component [38]. Adding a conductive filler could be one way to solve this problem [39]. The addition of a conducting filler to a polymer matrix enhances the dielectric, ferroelectric, and piezoelectric properties of micro-capacitor devices. A potential option to consider instead of using two-phase composites made of PVDF and either ceramic or conducting filler is to use three-phase composites made of PVDF, ceramic, and conducting filler. These three-phase composites have the advantage of enhancing piezoelectric capabilities while maintaining a limited concentration of ceramic material [40]. Choosing the best conductive filler and piezoelectric ceramic is crucial in this regard. In the subsequent sections, our objective is to elucidate the characteristics of PVDF or P(VDF-TrFE) based polymer composites with two-phase or ternary-phase compositions, specifically designed for energy harvesting purposes.

17.3 PZT/PVDF and CNTs Composites

Hou et al. and their coworkers recently utilized the phase-inversion method to fabricate a PZT skeleton with a compact microscopic structure and a vertically aligned 3D microchannel architecture [41]. A composite material consisting of conductive carbon nanotubes (CNTs) and low-dielectric polyvinylidene fluoride (PVDF) was formed by injecting a combination of CNTs and PVDF into the pores using a vacuum impregnation process [41]. The process of creating PZT/PVDF&CNT composites involves the combination of carbon nanotubes (CNTs) and polyvinylidene fluoride (PVDF) as fillers. The superior energy harvesting performance of a material is generally correlated with its high transduction coefficient ($d_{33} \times g_{33}$), which depends on the material's poor dielectric constant with strong piezocoefficient. Nevertheless, it is challenging to obtain both low ε_r and high d_{33} at the same time in piezoceramics. Here, the lead zirconate titanate (PZT) with a microchannel structure oriented vertically prepared using the phase-inversion approach is chosen as a matrix. PZT/PVDF&CNTs composites are made by combining carbon nanotubes (CNTs) and polyvinylidene fluoride (PVDF) as fillers. The distinct structure and evenly spaced CNT network improve the d_{33} by enhancing polarization. The ε_r is effectively reduced by the PVDF filler. This led to the exceptional dielectric constant ($\varepsilon r = 1{,}603$) and good piezoelectric coefficient ($d_{33} = 595$ pC/N) in PZT/ PVDF&CNT composites, which produced a large value of $d_{33} \times g_{33}$ as 24,942 × 10–15 m^2 /N. Consequently, the above composites achieve outstanding energy harvesting capability (66 V of output voltage and 39.22 μA of short current). With the addition of CNTs, the ferroelectric and piezoelectric characteristics of the composite dramatically enhance, showing that the poling electric field is imposed on the PZT particles successfully. In order to lower resistance, the highly conductive CNTs can provide a lot of conduction percolation channels in the polymer which subsequently significantly improves the partial potential placed on the composites. The PZT/PVDF&CNTs piezoelectric composites also achieve a high piezoelectric coefficient (d_{33} = 595 pC/N), which is similar to the parent ceramics (d_{33} = 679 pC/N), despite the small increment of ε_r slightly upon the rise of P_{max}. Cantilever beam type PEHs detect the high d33 × g33 of the prepared composite, which should result in outstanding power generating capacity. The waveform signals of Voc and Isc produced by three different piezoelectric materials at mechanical resonance frequency and 1 g (~9.8 m/s^2) of acceleration are shown in Fig. 17.2a, b.

However, as demonstrated with PZT ceramics, piezoelectric materials that include PVDF and CNTs enhance Voc and Isc. The PZT/PVDF&CNTs material exhibits a voltage (Voc) of 66 V and a current (Isc) of 39.22 mA at the resonant frequency of 68.1 Hz. On the other hand, at 67.6 Hz, the Voc value of the PZT/PVDF material is 50 V and the Isc value is 24.55 mA. In contrast to PZT ceramics, which had lower dielectric constants (εr) and greater open-circuit voltages (Voc = 37 V, Isc = 17.47 μA, @ 82.5 Hz), PVDF greatly decreased the dielectric constants of the piezoelectric materials.

17.4 MWCNTs/BSTO/PVDF Nanocomposites

A three-phase MWCNTs/BSTO/PVDF nanocomposites made with a straightforward drop-casting technique was described by Das et al. [42]. The PVDF is the most suitable piezoelectric energy harvester because the fillers encourage > 80% electroactive phase nucleation in the material. The addition of conductive fillers, such as carbon nanotubes (CNTs), and dielectric fillers, such as barium strontium titanate (BSTO), to the polyvinylidene fluoride (PVDF) matrix, enhances the dielectric, ferroelectric, and output properties of the piezoelectric nanogenerator. The use of a highly efficient PVDF composite with filler concentration of 20% BSTO and 0.20% of CNTs results in piezoelectric nanogenerator PENG generating an output power of 31.5 21 μW, a large open-circuit voltage and a short-circuit current of 42 V and 9 μA respectively. These standards are enough to provide electricity for compact portable electronic equipment. The piezoelectric energy harvesting properties of nanocomposites are mainly dominated by ferroelectric properties. The inclusion of BSTO fillers in the polymer matrix increases both the maximum polarization and remnant polarization values of PVDF. The utilization of MWCNTs as conducting fillers greatly enhances the amplification of these polarization values. It is reported that the particular composition has the greatest polarization value at 160 MV/m electric field, which is 0.043 μC/cm^2. This can be ascribed to the additional MWCNTs' greater conductivity in comparison to the polymer, which alters particle mobilization and raises the composite's hetero polarization. Piezoelectric nanogenerators were built using 1 × 1 cm^2 films to measure the applied periodic force and the piezo output voltage at various filler weight percentages. Several nanogenerators are created by filling the material with parent BSTO, which acts as a dielectric, and MWCNTs, which act as a conductive filler. The output voltage (V_{out}), the greatest peak-to-peak output voltage of the nanocomposites was found to increase with the rise of concentration of the MWCNT as presented in Fig. 17.3. According to Fig. 17.3, the average short-circuit current (Isc) is approximately 0.5, 1.1, 2.3, 4.5, and 9 μA, in that order. When compared to other optimized devices, the 0.20% added MWCNTs based composite show excellent output characteristics.

17.5 PVDF—BF33BT—GO Three-Phase Composite

Muduli et al. used a solvent-casting method followed by hot-pressing to create three-phase PVDF—BF33BT—GO composite films. All of the composite films were found to have a β-phase content of more than 60%, and BF33BT particles and GO sheets were distributed uniformly [43]. Both the dielectric parameters rise with the addition of GO and BF33BT loading; however, a sudden escalation is observed at the threshold GO concentration of 0.08 weight percent. High conductivity is linked to three-phase composites containing more than 0.08 weight percent of GO. G-0.08 has a tan δ of 0.68 and an ε′ of about 303. Likewise, with G-0.08 from pure PVDF, the ferroelectric Pr improves by approximately 166 times and the Ur by 7 times. The corresponding values for d_{33}, Voc, and Isc of G-0.08 are 9 pC·N − 1, 3.9 V, and 1.3 μA. The G-0.08 composite film's application was investigated through rectification, capacitor charging, and calculator powering. When BF33BT concentration rises, the d33 value also rises because BF33BT has a greater d33 value (about 180 $pC{\cdot}N^{-1}$) than pure PVDF (3 $pC{\cdot}N^{-1}$). The composite's d33 is improved by interfacial polarization brought about by the addition of BF33BT to the PVDF matrix. More dipoles are drawn to the interface where polarization develops as a result of BF33BT's increased piezoelectricity. Figure 17.4 explains the effect of both the filler (ceramic and GO) on the energy harvesting characteristics of PVDF polymer. When GO is added to the PVDF − BF33BT matrix, d33 rises to 9 $pC{\cdot}N^{-1}$ for 0.08 weight percent of the GO concentration, and d_{33} decreases with additional concentration increases. As demonstrated in Fig. 17.4d, the addition of GO increases Voc once more to ~3.9 V at a GO concentration of 0.08 weight percent. Because of the creation of electric conductive percolative pathways through GO sheets, increasing the GO loading over 0.08 weight percent results in a decrease in d_{33} and a reduction in Voc. Similar to Voc, the short-circuit current (Isc) exhibits the similar pattern.

In order to verify the device's reliability as a piezoelectric energy harvester, the Voc of the harvester was measured up to approximately 3000 s. The Voc was found to stay consistent (~3.9 V) during the measurement. So this composite is mostly fitted for piezoelectric energy harvesting.

17.6 CaZ/P(VDF-TrFE) Composite Film

A further investigation was carried out on a PVDF-TrFE/Ca-doped ZnO composite material, which improved the dielectric, ferroelectric, and piezoelectric characteristics [44]. Under a coercive field of about 41 MV/m, the composite film shows a maximum saturated polarization of 1.32 $\mu C/cm^2$ and a residual polarization of 0.604 μC/cm2. By subjecting the films to an external mechanical force, their piezoelectric performance may be empirically assessed. Because CaZ nanofillers are included in the polymer matrix, the composite film-based device displays a greater potential voltage in comparison to the pure PVDF-TrFE counterpart. The CaZ/PVDF-TrFE composite film energy harvesting device demonstrated an outstanding rectified output voltage of 3 V when subjected to continuous human hand tapping. At 0.5 N force, it produced an output voltage of 0.289 V peak-to-peak, demonstrating its good performance under oscillating force as well. We used a hand tool to repeatedly probe the voltage responses of both of the manufactured devices. Thanks to the enhanced electroactive phase and electrical characteristics of the composite structure, energy harvesting devices made of composite film are more voltage sensitive than those made of pure film.

17.7 P(VDF-TrFE)/BTO Composite Films

A novel study was conducted by a different research team to investigate the impact of BaTiO3 nanoparticle size on the piezoelectric properties of P(VDF-TrFE)/BTO films after being treated with PDA [45]. The researchers conducted a systematic investigation to examine how the size of functionalized BTO nanoparticles affects the β-phase composition, as well as the electrical characteristics (dielectric, piezoelectric, and ferroelectric). BTO nanoparticles are specially chosen as filler because of their large d33 values (191 pC/N) and non-toxic nature. The addition of BTO nanoparticles to the composites improves their piezoelectric properties. Nevertheless, the aggregation of BTO nanoparticles (NPs) and the presence of interface hole defects between the BTO NPs and the PVDF matrix occasionally impede the achievement of the highest level of piezoelectric performance. This can be fixed by adding polydopamine (PDA) to the surface of BTO NPs.

A pushing test (Fig. 17.5a) was done to see how the PDA layer and the size of the BTO NPs affected the piezoelectric performance of the composites. The composites' piezoelectric performance was assessed by monitoring the current circulating in the external circuit, which was generated by the piezoelectric potential. A possible explanation for the significantly higher *i*sc of the P(VDF-TrFE)/BTO@PDA composites (Fig. 17.5f) is the reduced leakage current and trapped charges that result from including the BTO@PDA NPs. At last, piezoelectric pressure sensors built from P(VDF-TrFE)/BTO@PDA composites were developed to track the radial artery pulse waveform. The sensor was able to capture the unique waveform of the radial pulse.

Furthermore, the sensitivity of the current signals was significantly improved by choosing the tiniest NPs and functionalizing their surface. A self-powered, real-time health monitoring gadget might use the best piezoelectric composite film, as suggested by this research.

17.8 $BiFeO_3$—$BaTiO_3$/P(VDF-TrFE) Multifunctional Polymer Nanocomposites

In order to build composites with multiple functionalities, another group investigated the optical response and ferromagnetism of the nanofibers, together with the ferroelectric characteristics of the polymer P(VDF-TrFE) and $BiFeO_3$ –$BaTiO_3$ nanofibers [46]. The addition of nanofibers not only improves the pyroelectric, piezoelectric, and dielectric properties of the polymer but also results in a range of other properties, including ferromagnetism, magnetoelectric effect, photothermal effect, photostriction, and photo-electric response. The addition of nanofibers improves the nanocomposites' piezoelectric response due to their better dielectric characteristics. According to the plot, adding less than five volume percent of nanofibers caused d_{33} to rise from 14 to 21 pC/N. The composites' increased conductivity accounts for the lower d_{33} value at high concentrations. As seen in the inset of Fig. 17.6, a flexible pressure sensor was created utilizing the composite containing 2.5 vol% nanofibers thanks to the nanocomposites' improved piezoelectric response. Figure 17.6 illustrates how the sensor affixed to the skin above the wrist artery produces a significant electrical response (>400 mV) in response to a human heartbeat. As a result, sensors and energy harvesting devices made from a single material may detect and collect mechanical, thermal, magnetic, and optical energies.

Ramadoss and his coworkers [47] observed that PNG based on PVDF-TrFE/BFO (6 wt%) had a greater maximum output voltage of 18.5 V and a maximum current of 300 nA compared to prior efforts. The experimental results of the current work demonstrate that including the optimal quantity of BFO nanofillers into the copolymers P(VDF-TrFE) and PVDF-HFP enhances the ferroelectric, dielectric, and piezoelectric characteristics of the nanocomposite films. This study indicates that the suggested flexible device has the ability to operate low-power electronic devices.

Many implantable energy harvesting and biomedical applications have been designed thanks to self-powered technologies and smart materials. Piezoelectric materials are those that, when subjected to strain or stress, can produce an electrical output. In contrast to other technologies like triboelectric and electromagnetic, piezoelectric energy harvesters (PEHs) can convert a wide range of mechanical energies into electricity. The primary goal of this work is to summarize the manufacturing of PEHs and to display different piezoelectric materials. Our attention has then turned to the many biological and clinical applications that have made use of PEHs. Finally, thoughts and plans for the future of biological sensors based on PEH are also covered. Figure 17.8 summarizes the primary biomedical uses of PEHs and the roadmap [48].

Figure 17.9a depicts the processes and substances employed in the production of PVDF-based Mxene-Sm-PMN-PT nonwoven piezoelectric composites. These composites were electrospuned from a suspension of MXene powder, Sm-PMN-PT particles, and PVDF. In contrast to traditional piezoelectric ceramics like BTO (~190 pC/N), PZT (560 pC/N), and PMN-PT (~620 pC/N) (bottom inset of Fig. 17.2a), Sm-PMN-PT was chosen as the ceramic fillers due to its high piezoelectric coefficient (~1500 pC/N). Electrospinning is made easier with the addition of MXene lamellae to the precursor, which increases its mechanical ductility and conductivity. This, in turn, assists with in-situ mechanical stretching of polymer composites and local electric poling. Figure 17.9a shows the upper inset of the polymer chain, which aids in achieving the zigzag alignment of − CH2 and − CF2, as well as the all-trans conformation and the intermolecular bond with the PVDF polymer chain. Figure 17.9b displays the structural and geometric details of electrospun piezoelectric nonwoven fabrics constructed from randomly stacked molten PVDF-based MXene nanofibers. The shape of the fibrous structure is little affected by the integration of MXene nanosheets. The high-resolution transmission electron microscopy photos (Fig. 17.9c) indicate that the PVDF polymeric chains align along the inorganic–organic interface, which validates the zigzag alignment of CH_2 and CF_2. Furthermore, the X-ray photoelectron spectrometer (XPS) spectra shown in Fig. 17.9d clearly show that the as-synthesized nanofibers had oxygen functionalities and hydrogen bonds attached to their surface. The creation of strong interactions between the $-CF_2$ groups in the PVDF and the hydrogen bonds (−OH and − Ox functional groups) is shown by the existence of $C-O$ and $C=O$ groups of Mxene. Figure 17.9 e shows the X-ray diffraction (XRD) pattern of the virgin MXene and Sm-PMN-PT powders, as well as the as-synthesized nanofibers. The 2θ peak of the Ti3C2Tx lamellae, which measures 6.48°, is associated with the (002) lattice plane with a d-spacing of 13.66 Å. This indicates that the chemical etching process was effective in removing the Al atom layer from the multilayer Ti_3AlC_2 MAX phase. Moreover, relative dielectric permittivity and dielectric loss of synthesized MPC fabrics and the undoped fabrics exhibit frequency dependency (as shown in Figs. 17.9f and 17.9g). The relaxation effect causes the dielectric permittivity to drop somewhat with increasing frequency (Fig. 17.9f). In the frequency range of 1 kHz to 1 MHz, dielectric permittivity peaks at 2.5 wt%, and dielectric loss reaches its maximum at 5 wt%, demonstrating a clear link with the MXene doping level. Adding the right quantity of conductive MXene nanosheets to polymer composites improves matrix-filler interfacial polarization, whereas increasing leakage current and dielectric loss is the result of using too much Mxene [49].

A technique for producing PVDF/$BaTiO_3$/CNT ternary composites (as shown in Fig. 17.10) using selective laser sintering (SLS) and ultrasonic dispersion has been revealed by Chen and coworkers. A wider sintering window was seen in the PVDF/BaTiO3 composite powders that were coated with CNTs, as compared to the PVDF/$BaTiO_3$/CNT ternary powders that were milled conventionally. The 3D isolated conductive channel in the devices was enhanced by the coupling approach of coating and SLS processing, which in turn increased the conductivity, dielectric, and

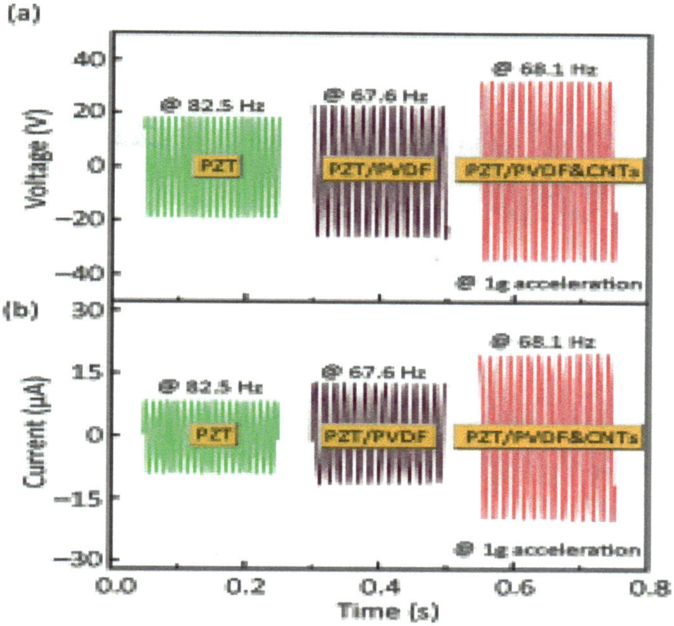

Fig. 17.2 The energy harvester produces open-circuit voltages (**a**) and short-circuit currents (**b**) at varying frequencies [41]. Open access under a CC BY 4.0 license

ferroelectric characteristics of SLS composites. This study introduces a novel material for SLS processing and points the way for improved comprehensive performance devices with high voltages [50].

Khalid and his coworkers [51] have reported flexible nanocomposite-based piezoelectric nanogenerators (PENGs) are created by evenly distributing several types of piezoelectric nanoparticles ($BaTiO_3$, ZnO, and PZT) and graphene nanopowder within a silicone matrix (As shown in Fig. 17.11). The findings demonstrated that the PZT-based composites had a higher level of performance when compared to other ceramics. Following that, the practical use of PENGs was showcased by the creation of a completely operational shoe-insole nanogenerator (SING). The SING produced a significant open-circuit voltage of around 27 V, a short-circuit current of 429.23 microamperes, and a power density of 402 milliwatts per square meter during real-time human walking. The findings indicate that the created PENGs have the potential to serve as self-sustaining biomechanical energy harvesters/sensors in wearable electronics, haptic sensing, or applications connected to the internet of humans.

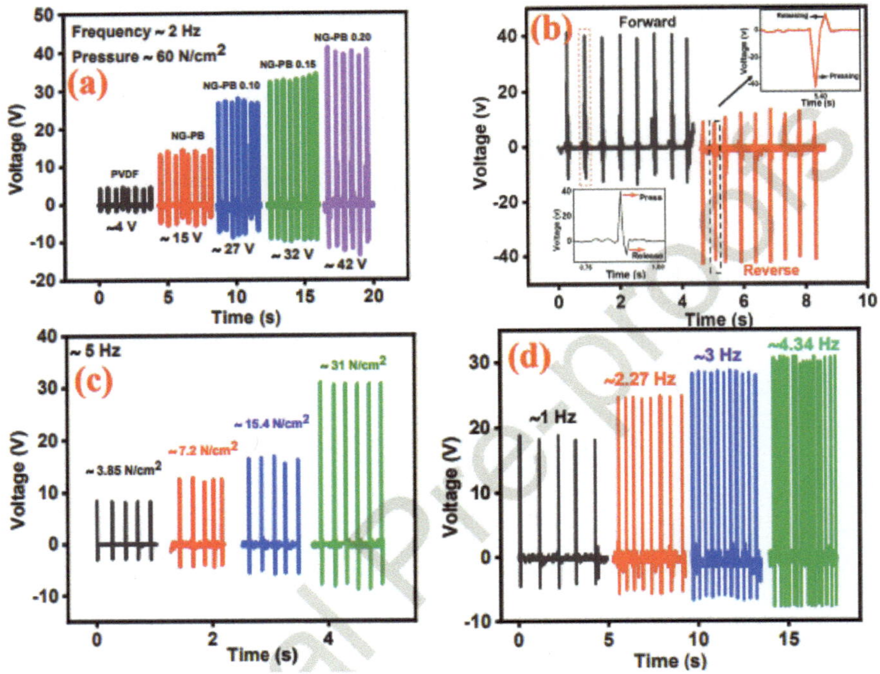

Fig. 17.3 Shows the following: **a** the Voc of several nanogenerators; **b** both the output characteristics of the device formed from the optimized composite **c** & **d** and the output characteristic of the optimized device at various pressures and frequencies [42]. Adapted from Ref. [43] with permission from [Elsevier], Copyright [2024].

17.9 Future Prospective and Challenges

Polymer nanocomposites with piezoelectric properties have significant promise for many future applications. These materials have distinctive characteristics that render them appropriate for energy harvesting, sensing, and actuation technologies. Nevertheless, there are still several obstacles that must be resolved before their extensive adoption. Increasing the durability and efficiency of piezoelectric polymer nanocomposites is an ongoing problem. Material deterioration, fatigue, and piezoelectric property loss are all problems that need fixing. To mass-produce these nanocomposites affordably, scalable manufacturing procedures are also required. Another challenge is making sure these materials won't deteriorate over time. Piezoelectric polymer nanocomposites may degrade over time and in response to specific environmental stressors. Coming up with strategies to make them more durable and resistant to degradation is essential if they are to be successfully implemented in real-world applications. In the end, although piezoelectric polymer nanocomposites show significant potential for the future, it is crucial to overcome these issues to fully realize their capabilities and facilitate their broad use in many industries.

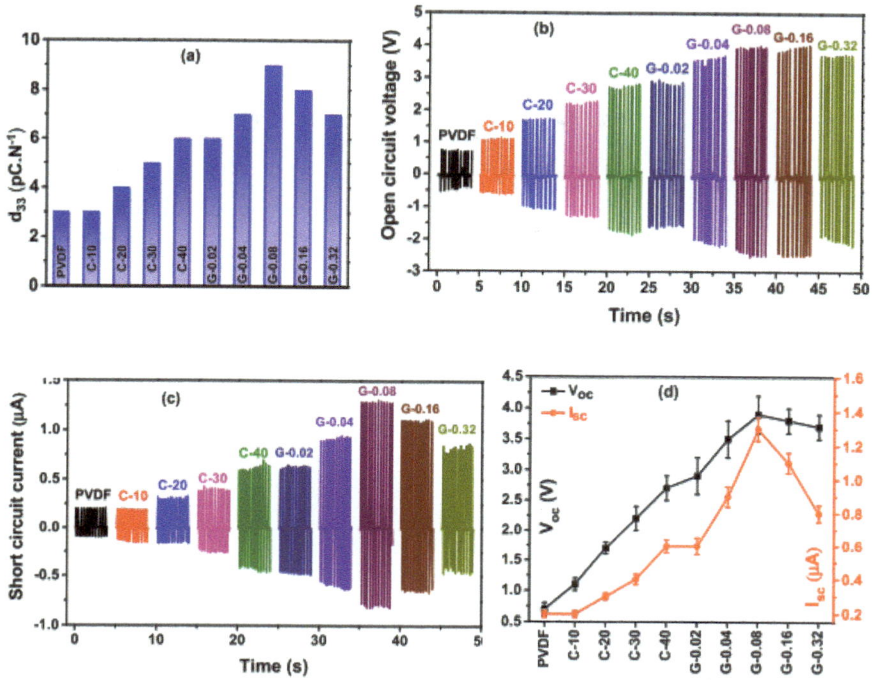

Fig. 17.4 Effect of BF33BT and GO percentage on composites **a** d_{33} **b** Voc **c** Isc, **d** the reverse and forward polarity measurement of Voc [43]. Adapted from Ref. [44] with permission from [American Chemical Society], Copyright [2023]

17.10 Conclusions

This chapter examines the use of polymer composites in energy harvesting applications. Energy harvesting is a burgeoning discipline that seeks to gather and transform diverse energy sources from the surroundings into practical electrical energy. Due to their distinctive features and adaptability, polymer composites have garnered considerable attention in this field. Several piezoelectric PVDF and P(VDF-TrFE) based polymer composites are discussed for energy harvesting applications. This text discusses the manufacturing procedures, methods for evaluating performance, and possible uses of these composites. It is also reviewed that the addition of conductive filler in optimized proportion is beneficial for higher conversion efficiency. In summary, this chapter offers a thorough examination of the progress and obstacles associated with polymer composites for energy harvesting.

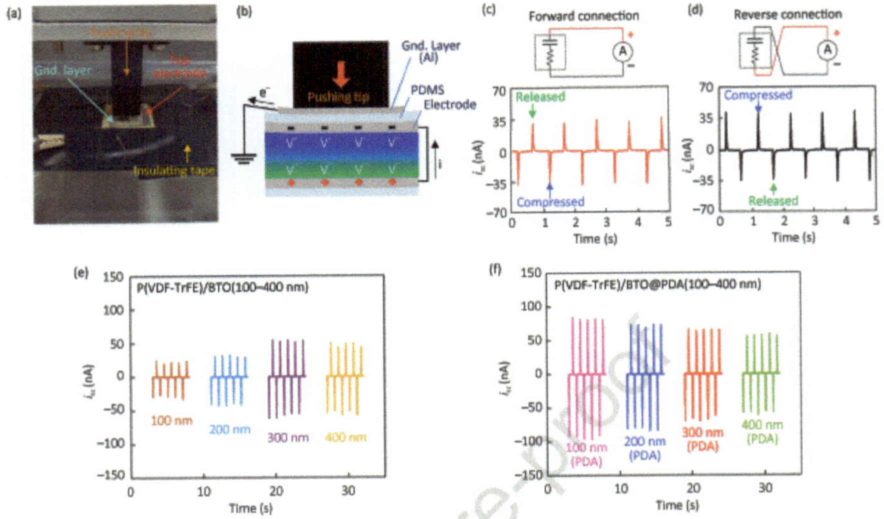

Fig. 17.5 **a** Experimental arrangement for piezoelectric performance, and **b** Composite under the pushing tip. **c** forward connection and **d** reverse connection modes. **e** and **f** Isc of the composite under different pushing conditions [45]. Open access under a CC BY 4.0 license.

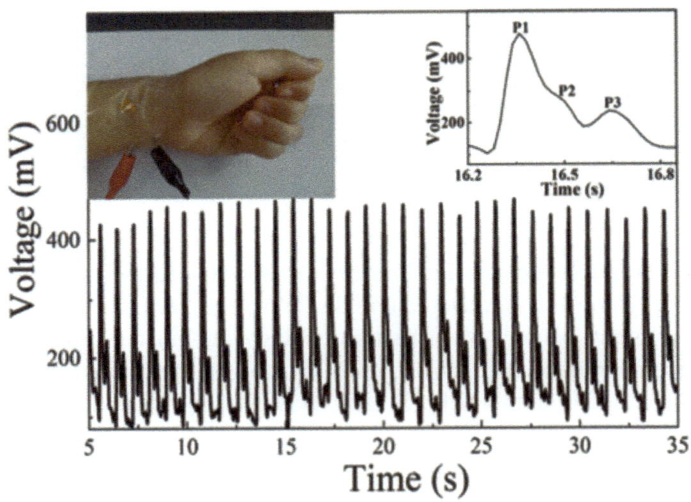

Fig. 17.6 Evaluation of the electrical properties of P(VDF-TrFE)/BFO – BTO [46]. Adapted from Ref.47 with permission from [American Chemical Society], Copyright [2021]

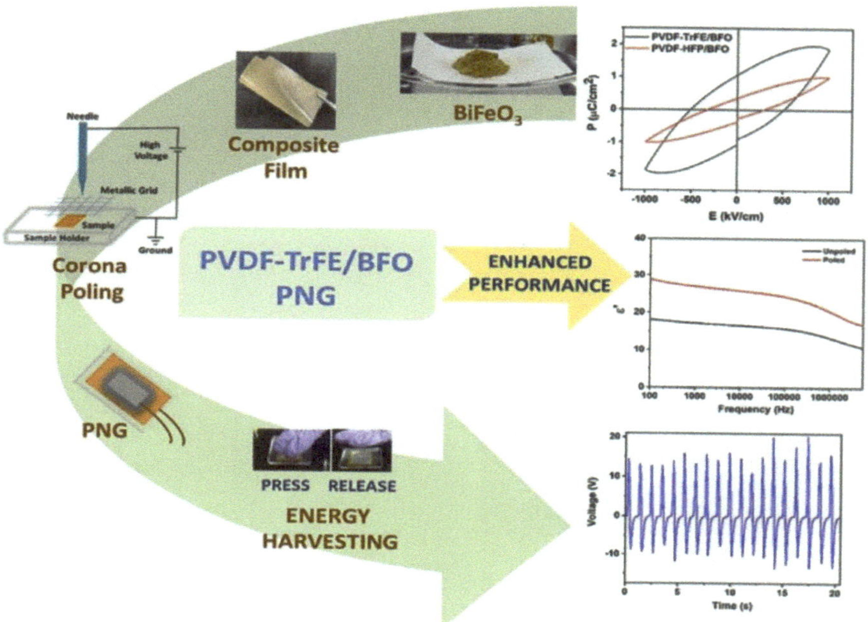

Fig. 17.7 Schematic illustration for preparation of PVDF-TrFE-BFO-PNG composites [47]. Adapted from Ref.48 with permission from [Elsevier], Copyright [2023]

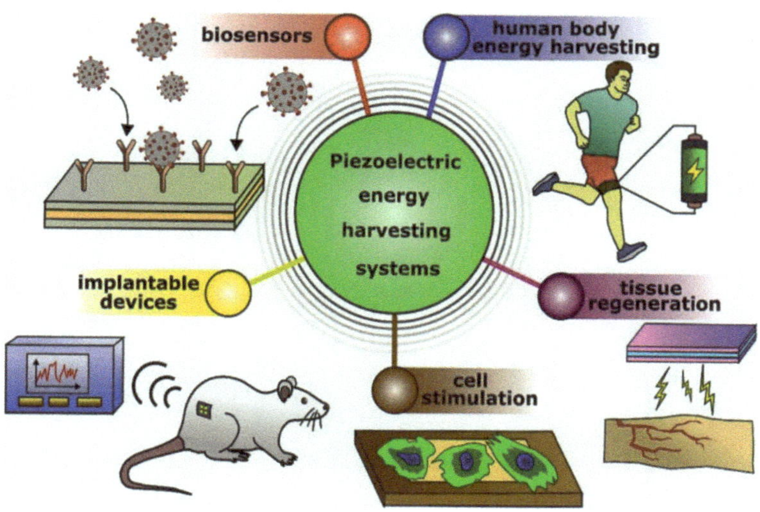

Fig. 17.8 Schematic illustration of piezoelectric energy harvesting systems [48]. Adapted from Ref. 49 with permission from [Elsevier], Copyright [2022]

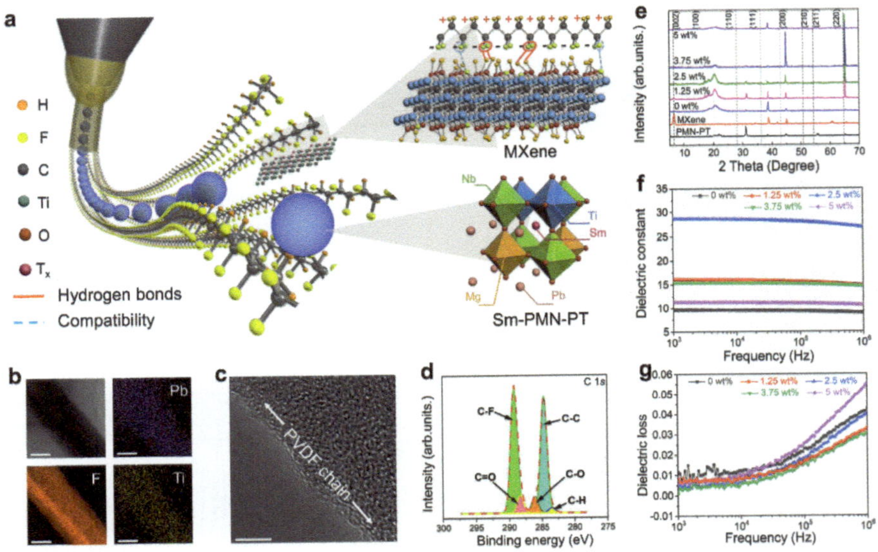

Fig. 17.9 The soft piezoelectric textile (PT) based on MXene-enabled piezoelectric composite (MPC) and its structural design and characterization [49]. Open access under a CC BY 4.0 license

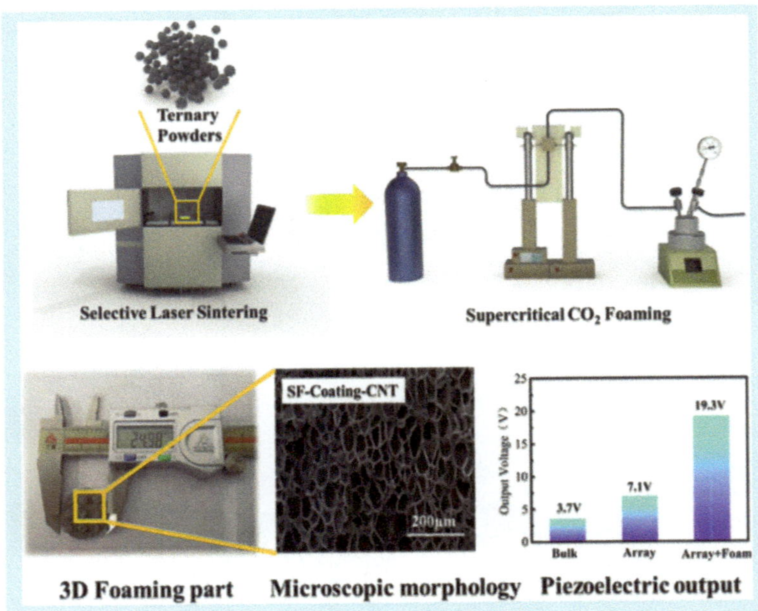

Fig. 17.10 Schematic diagram of PVDF/BaTiO_3/CNT ternary composites for piezoelectric energy harvesting [50]. Adapted from Ref.51 with permission from [American Chemical Society], Copyright [2021]

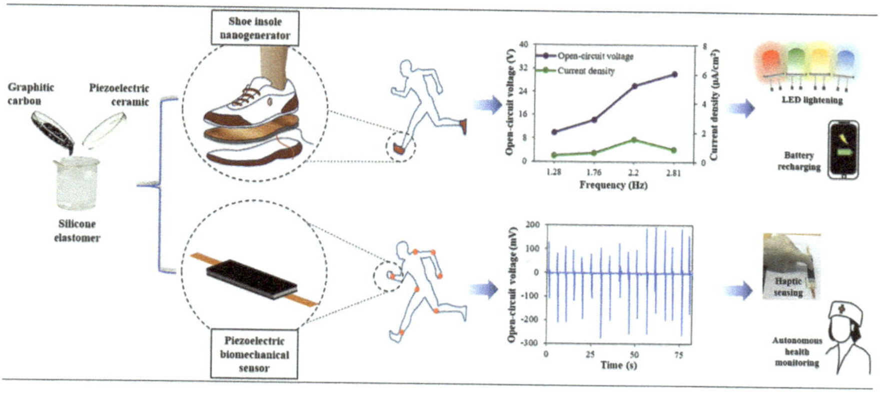

Fig. 17.11 Schematic diagram for piezoelectric transducers for energy harvesting [51]. Adapted from Ref. [52] with permission from [American Chemical Society], Copyright [2020]

References

1. Mishra, S., Unnikrishnan, L., Nayak, S.K., Mohanty, S.: Advances in piezoelectric polymer composites for energy harvesting applications: a systematic review. Macromol. Mater. Eng.. Mater. Eng. **304**(1), 1800463 (2019)
2. Sukumaran, S., Chatbouri, S., Rouxel, D., Tisserand, E., Thiebaud, F., Ben Zineb, T.: Recent advances in flexible PVDF based piezoelectric polymer devices for energy harvesting applications. J. Intell. Mater. Syst. Struct.Intell. Mater. Syst. Struct. **32**(7), 746–780 (2021)
3. Nunes-Pereira, J., Sencadas, V., Correia, V., Rocha, J.G., Lanceros-Mendez, S.: Energy harvesting performance of piezoelectric electrospun polymer fibers and polymer/ceramic composites. Sens. Actuators, A **196**, 55–62 (2013)
4. Wang, C., Xia, K., Wang, H., Liang, X., Yin, Z., Zhang, Y.: Advanced carbon for flexible and wearable electronics. Adv. Mater. **31**(9), 1801072 (2019)
5. Mokhtari, F., Cheng, Z., Raad, R., Xi, J., Foroughi, J.: Piezofibers to smart textiles: A review on recent advances and future outlook for wearable technology. Journal of Materials Chemistry A **8**(19), 9496–9522 (2020)
6. Bocchetta, P., Frattini, D., Ghosh, S., Mohan, A.M., Kumar, Y., Kwon, Y.: Soft materials for wearable/flexible electrochemical energy conversion, storage, and biosensor devices. Materials **13**(12), 2733 (2020)
7. Rani, M., Sehrawat, M., Sharma, S., Singh, B.P.: Recent advancement and challenges in multifunctional carbon nanotube buckypaper and its composites for energy storage and conversion applications. Journal of Energy Storage **73**, 109063 (2023)
8. Khan, A., Joshi, R., Sharma, M.K., Ganguly, A., Parashar, P., Wang, T.W., Lee, S., Kao, F.C., Lin, Z.H.: Piezoelectric and triboelectric nanogenerators: promising technologies for self-powered implantable biomedical devices. Nano Energy 109051 (2023)
9. Ibrahim, I.D., Sadiku, E.R., Jamiru, T., Hamam, Y., Alayli, Y., Eze, A.A.: Prospects of nanostructured composite materials for energy harvesting and storage. Journal of King Saud University-Science **32**(1), 758–764 (2020)
10. Song, Y., Shi, Z., Hu, G.H., Xiong, C., Isogai, A., Yang, Q.: Recent advances in cellulose-based piezoelectric and triboelectric nanogenerators for energy harvesting: a review. Journal of Materials Chemistry A **9**(4), 1910–1937 (2021)
11. Liu, L., Guo, X., Liu, W., Lee, C.: Recent progress in the energy harvesting technology—from self-powered sensors to self-sustained IoT, and new applications. Nanomaterials **11**(11), 2975 (2021)

12. Hasan, M.N., Nafea, M., Nayan, N., Mohamed Ali, M.S.: Thermoelectric generator: materials and applications in wearable health monitoring sensors and internet of things devices. Advanced Materials Technologies **7**(5), 2101203 (2022)
13. Costa, C.M., Cardoso, V.F., Martins, P., Correia, D.M., Gonçalves, R., Costa, P., Correia, V., Ribeiro, C., Fernandes, M.M., Martins, P.M., Lanceros-Méndez, S.: Smart and multifunctional materials based on electroactive poly (vinylidene fluoride): recent advances and opportunities in sensors, actuators, energy, environmental, and biomedical applications. Chem. Rev. **123**(19), 11392–11487 (2023)
14. Donate, R., Paz, R., Moriche, R., Sayagués, M.J., Alemán-Domínguez, M.E., Monzón, M.: An overview of polymeric composite scaffolds with piezoelectric properties for improved bone regeneration. Mater. Des. **11**, 112085 (2023)
15. Yang, J., Shen, X., Yang, W., Kim, J.K.: Templating strategies for 3D-structured thermally conductive composites: Recent advances and thermal energy applications. Prog. Mater. Sci. **133**, 101054 (2023)
16. Tripathi, B.M., Shukla, S.K., Rathore, P.K.: A comprehensive review on solar to thermal energy conversion and storage using phase change materials. Journal of Energy Storage **72**, 108280 (2023)
17. Tao, Z., Zou, H., Li, M., Ren, S., Xu, J., Lin, J., Yang, M., Feng, Y., Wang, G.: Polypyrrole coated carbon nanotube aerogel composite phase change materials with enhanced thermal conductivity, high solar-/electro-thermal energy conversion and storage. J. Colloid Interface Sci. **629**, 632–643 (2023)
18. Park, K.I., Son, J.H., Hwang, G.T., Jeong, C.K., Ryu, J., Koo, M., Choi, I., Lee, S.H., Byun, M., Wang, Z.L., Lee, K.J.: Highly-efficient, flexible piezoelectric PZT thin film nanogenerator on plastic substrates. Adv. Mater. **26**(16), 2514–2520 (2014)
19. Hwang, G.T., Park, H., Lee, J.H., Oh, S., Park, K.I., Byun, M., Park, H., Ahn, G., Jeong, C.K., No, K., Kwon, H.: Self-powered cardiac pacemaker enabled by flexible single crystalline PMN-PT piezoelectric energy harvester. Adv. Mater. **26**(28), 4880–4887 (2014)
20. Hwang, G.T., Yang, J., Yang, S.H., Lee, H.Y., Lee, M., Park, D.Y., Han, J.H., Lee, S.J., Jeong, C.K., Kim, J., Park, K.I.: A reconfigurable rectified flexible energy harvester via solid-state single crystal grown PMN–PZT. Adv. Energy Mater. **5**(10), 1500051 (2015)
21. Qi, Y., McAlpine, M.C.: Nanotechnology-enabled flexible and biocompatible energy harvesting. Energy Environ. Sci. **3**(9), 1275–1285 (2010)
22. Kawai, H.: The piezoelectricity of poly (vinylidene fluoride). Jpn. J. Appl. Phys.. J. Appl. Phys. **8**(7), 975 (1969)
23. Chen, X., Han, X., Shen, Q.D.: PVDF-based ferroelectric polymers in modern flexible electronics. Advanced Electronic Materials **3**(5), 1600460 (2017)
24. Ruan, L., Yao, X., Chang, Y., Zhou, L., Qin, G., Zhang, X.: Properties and applications of the β phase poly (vinylidene fluoride). Polymers **10**(3), 228 (2018)
25. Cai, X., Lei, T., Sun, D., Lin, L.: A critical analysis of the α, β and γ phases in poly (vinylidene fluoride) using FTIR. RSC Adv. **7**(25), 15382–15389 (2017)
26. Wang, X., Sun, F., Yin, G., Wang, Y., Liu, B., Dong, M.: Tactile-sensing based on flexible PVDF nanofibers via electrospinning: A review. Sensors **18**(2), 330 (2018)
27. Abdalla, S., Obaid, A., Al-Marzouki, F.M.: Preparation and characterization of poly (vinylidene fluoride): A high dielectric performance nano-composite for electrical storage. Results in physics **6**, 617–626 (2016)
28. Song J, Zhao G, Li B, Wang J (2017) Design optimization of PVDF-based piezoelectric energy harvesters. Heliyon 3(9).
29. Sappati, K.K., Bhadra, S.: Piezoelectric polymer and paper substrates: A review. Sensors **18**(11), 3605 (2018)
30. Xin, Y., Sun, H., Tian, H., Guo, C., Li, X., Wang, S., Wang, C.: The use of polyvinylidene fluoride (PVDF) films as sensors for vibration measurement: A brief review. Ferroelectrics **502**(1), 28–42 (2016)
31. Broadhurst, M.G., Davis, G.T.: Physical basis for piezoelectricity in PVDF. Ferroelectrics **60**(1), 3–13 (1984)

32. Wang, A., Hu, M., Zhou, L., Qiang, X.: Self-powered well-aligned P (VDF-TrFE) piezoelectric nanofiber nanogenerator for modulating an exact electrical stimulation and enhancing the proliferation of preosteoblasts. Nanomaterials **9**(3), 349 (2019)
33. Chen, S., Yao, K., Tay, F.E., Chew, L.L.: Comparative investigation of the structure and properties of ferroelectric poly (vinylidene fluoride) and poly (vinylidene fluoride–trifluoroethylene) thin films crystallized on substrates. J. Appl. Polym. Sci.Polym. Sci. **116**(6), 3331–3337 (2010)
34. Wan, C., Bowen, C.R.: Multiscale-structuring of polyvinylidene fluoride for energy harvesting: the impact of molecular-, micro-and macro-structure. Journal of Materials Chemistry A **5**(7), 3091–3128 (2017)
35. Hao, X.: A review on the dielectric materials for high energy-storage application. J. Adv. Dielectr.Dielectr. **3**(01), 1330001 (2013)
36. Satapathy, S., Pawar, S., Gupta, P.K., Varma, K.B.: Effect of annealing on phase transition in poly (vinylidene fluoride) films prepared using polar solvent. Bull. Mater. Sci. **34**, 727–733 (2011)
37. Arshad AN, Wahid MH, Rusop M, Majid WH, Subban RH, Rozana MD (2019) Dielectric and structural properties of poly (vinylidene fluoride)(PVDF) and poly (vinylidene fluoride-trifluoroethylene)(PVDF-TrFE) filled with magnesium oxide nanofillers. Journal of Nanomaterials 2019.
38. Singh P, Borkar H, Singh BP, Singh VN, Kumar A (2014) Ferroelectric polymer-ceramic composite thick films for energy storage applications. AIP advances 4(8).
39. Yaqoob, U., Uddin, A.I., Chung, G.S.: The effect of reduced graphene oxide on the dielectric and ferroelectric properties of PVDF–BaTiO 3 nanocomposites. RSC Adv. **6**(36), 30747–30754 (2016)
40. Ghosh B, Calderón RM, Espinoza-González R, Hevia SA (2017) Enhanced dielectric properties of PVDF/CaCu3Ti4O12: Ag composite films. Materials Chemistry and Physics 302–9.
41. Hou, L., Huan, Y., Zheng, M., Liu, Y., Wang, C., Wang, X., Li, C., Wang, Z., Wei, T.: 3D vertically aligned microchannel structure to enhance piezoelectric energy harvesting performance of PZT/PVDF&CNTs piezoelectric composites. Journal of Materiomics (2023). https://doi.org/10.1016/j.jmat.2023.11.006
42. Das, T., Biswas, P., Dev, A., Mallick, J., Kar, M.: Multi-functional piezoelectric nanogenerator based on relaxor ferroelectric materials (BSTO) and conductive fillers (MWCNTs) for self-powered memristor and optoelectronic devices. Chem. Eng. J. **479**, 147900 (2024)
43. Muduli SP, Lipsa L, Choudhary A, Rajput S, Parida S (2023) Modulation of Electrical Characteristics of Polymer–Ceramic–Graphene Hybrid Composite for Piezoelectric Energy Harvesting. ACS Applied Electronic Materials.
44. Sahoo, R., Mishra, S., Ramadoss, A., Mohanty, S., Mahapatra, S., Nayak, S.K.: An approach towards the fabrication of energy harvesting device using Ca-doped ZnO/PVDF-TrFE composite film. Polymer **205**, 122869 (2020)
45. Si H, Lee D, Park D, Oh J, Gu J, Lee I, Kim K (2023) Size effects of polydopamine-coated BaTiO3 nanoparticles on the piezoelectric performance of P (VDF-TrFE)/BaTiO3 composite. Journal of Materiomics.
46. Hu, X., Che, Y., Zhang, Z., Shen, Q.D., Chu, B.: BiFeO3–BaTiO3/P (VDF-TrFE) multifunctional polymer nanocomposites. ACS Appl. Electron. Mater. **3**(2), 743–751 (2021)
47. Tripathy, A., Raj, N.P., Saravanakumar, B., Kim, S.J., Ramadoss, A.: Tuning of highly piezoelectric bismuth ferrite/PVDF-copolymer flexible films for efficient energy harvesting performance. J. Alloy. Compd. **932**, 167569 (2023)
48. Panda, S., Hajra, S., Mistewicz, K., In-na, P., Sahu, M., Rajaitha, P.M., Kim, H.J.: Piezoelectric energy harvesting systems for biomedical applications. Nano Energy **100**, 107514 (2022)
49. Su, Y., Li, W., Cheng, X., Zhou, Y., Yang, S., Zhang, X., Chen, C., Yang, T., Pan, H., Xie, G., Chen, G.: High-performance piezoelectric composites via β phase programming. Nat. Commun.Commun. **13**(1), 4867 (2022)

50. Yang, C., Song, S., Chen, F., Chen, N.: Fabrication of PVDF/BaTiO3/CNT piezoelectric energy harvesters with bionic balsa wood structures through 3D printing and supercritical carbon dioxide foaming. ACS Appl. Mater. Interfaces **13**(35), 41723–41734 (2021)
51. Choudhry, I., Khalid, H.R., Lee, H.K.: Flexible piezoelectric transducers for energy harvesting and sensing from human kinematics. ACS Appl. Electron. Mater. **2**(10), 3346–3357 (2020)

FSC
www.fsc.org
MIX
Papier | Fördert
gute Waldnutzung
FSC® C083411

Kolibri 360 GmbH
Ferdinand-Jühlke-Straße 7
99095 Erfurt, Deutschland
produktsicherheit@kolibri360.de